Birkhäuser

Operator Theory: Advances and Applications

Volume 228

Founded in 1979 by Israel Gohberg

Yuri I. Karlovich
Luigi Rodino
Bernd Silbermann
Ilya M. Spitkovsky
Editors

Operator Theory, Pseudo-Differential Equations, and Mathematical Physics

The Vladimir Rabinovich Anniversary Volume

Birkhäuser

Editors
Yuri I. Karlovich
Facultad de Ciencias
Universidad Autónoma del Estado de Morelos
Cuernavaca, Morelos
Mexico

Luigi Rodino
Dipartimento di Matematica
Università di Torino
Torino
Italy

Bernd Silbermann
Fakultät für Mathematik
Technische Universität Chemnitz
Chemnitz
Germany

Ilya M. Spitkovsky
Department of Mathematics
College of William and Mary
Williamsburg, VA
USA

ISBN 978-3-0348-0536-0 ISBN 978-3-0348-0537-7 (eBook)
DOI 10.1007/978-3-0348-0537-7
Springer Basel Heidelberg New York Dordrecht London

Library of Congress Control Number: 2012951622

Printed on acid-free paper

Springer Basel is part of Springer Science+Business Media (www.springer.com)

Contents

Preface vii

S. Samko
Vladimir Rabinovich: a Mathematician, Colleague and Friend ix

A. Antoniouk and N. Tarkhanov
The Dirichlet Problem for the Heat Equation in Domains with Cuspidal Points on the Boundary 1

U. Battisti, T. Gramchev, L. Rodino and S. Pilipović
Globally Bisingular Elliptic Operators 21

A. Böttcher and A. Perälä
The Index Formula of Douglas for Block Toeplitz Operators on the Bergman Space of the Ball 39

G. Burlak and V. Rabinovich
The Green Function and Optical Field Enhancement in a Multilayered Microsphere with Metamaterial 57

O. Chkadua, R. Duduchava and D. Kapanadze
Potential Methods for Anisotropic Pseudo-Maxwell Equations in Screen Type Problems 73

S. Dobrokhotov, D. Minenkov, V. Nazaikinskii and B. Tirozzi
Functions of Noncommuting Operators in an Asymptotic Problem for a 2D Wave Equation with Variable Velocity and Localized Right-hand Side 95

S. Grudsky and A. Rybkin
On Toeplitz and Hankel Operators with Oscillatory Symbols Containing Blaschke Products and Applications to the KdV Equation 127

N. Habal and B.-W. Schulze
Mellin Quantisation in Corner Operators 151

A.Yu. Karlovich and I.M. Spitkovsky
Pseudodifferential Operators on Variable Lebesgue Spaces 173

Yu.I. Karlovich and I. Loreto Hernández
On Convolution Type Operators with Piecewise Slowly Oscillating Data 185

V.V. Kravchenko and S.M. Torba
Transmutations and Spectral Parameter Power Series in Eigenvalue Problems 209

V.V. Kucherenko
Estimate from Below for the Growth of Solution to the Navier-Stokes Equation When the Solution Blows Up 239

A. Kufner, L.-E. Persson and N. Samko
Some New Scales of Weight Characterizations of Hardy-type Inequalities 261

G.M. N'Guérékata and A. Pankov
Almost Periodic Elliptic Equations: Sub- and Super-solutions 275

H. Rafeiro, N. Samko and S. Samko
Morrey-Campanato Spaces: an Overview 293

S. Roch
Arveson Dichotomy and Essential Fractality 325

Y. Safarov
Lower Bounds for the Counting Function of Integral Operators 343

A. Sánchez-Nungaray and N. Vasilevski
Toeplitz Operators on the Bergman Spaces with Pseudodifferential Defining Symbols 355

M. Seidel and B. Silbermann
Finite Sections of Band-dominated Operators – Norms, Condition Numbers and Pseudospectra 375

F.-O. Speck
On the Reduction of Linear Systems Related to Boundary Value Problems 391

Preface

This volume is a collection of papers devoted to the 70th birthday of Professor Vladimir Rabinovich. The opening article (by Stefan Samko) contains a short biography of Vladimir Rabinovich, along with some personal recollections and bibliography of his work. It is followed by twenty research and survey papers in various branches of Analysis (pseudodifferential operators and partial differential equations, Toeplitz, Hankel, and convolution type operators, variable Lebesgue spaces, etc.) close to Professor Rabinovich's research interests. Many of them are written by participants of the International workshop "Analysis, Operator Theory, and Mathematical Physics" (Ixtapa, Mexico, January 23–27, 2012) having a long history of scientific collaboration with Vladimir Rabinovich, and are partially based on the talks presented there.

The volume will be of great interest to researchers and graduate students in Differential Equations, Operator Theory, Functional and Harmonic Analysis.

The Editors

Operator Theory:
Advances and Applications, Vol. 228, ix–xxv

Vladimir Rabinovich: a Mathematician, Colleague and Friend

Stefan Samko

Dedicated to the 70th anniversary of Professor Vladimir Rabinovich

It was my pleasure to accept the invitation to write this introductory paper to the volume dedicated to the 70th anniversary of Vladimir Samuilovich Rabinovich, my university colleague in the "previous life" in the Soviet Union, collaborator and friend for more than 40 years.

Vladimir Rabinovich, known to his friends and most of the colleagues as Volodya Rabinovich, was born in Kiev on September 2, 1940, where his childhood passed. In the beginning of the Nazi invasion of USSR in 1941, when many people were evacuated from the Western regions to the interior parts of the country, his

V. Rabinovich at his desk, México City, May 2012.

family went from Kiev to the city of Kuibyshev on Volga river (the city of Samara before 1935 and after 1991). Most likely I could not write this article otherwise. They returned to Kiev in 1947.

In 1961 he became a student of the Department of Mechanics and Mathematics of the *Rostov State University* in the Soviet Union. He graduated from this department with Diploma of Honour in 1966. He started his mathematical career at the Chair "Differential and Integral Equations" of the same department as a Ph.D. student in 1966–1969 years, under the guidance of Professor V.A. Kakichev, who noticed Volodya Rabinovich as a capable student and drew him into the world of mathematics. His scientific interests, already during the Ph.D. studies were heavily influenced by the professor of the same Department, well known mathematician I.B. Simonenko. In 1969 Volodya defended Ph.D. Theses and took a position of assistant professor at the same chair, but later moved to the chair "Algebra and discrete mathematics", guided by I.B. Simonenko.

In 1972 he became Associate Professor of the same Department of Mechanics and Mathematics, and Full Professor in 1994.

In 1998 Volodya leaves Russia and moves to Mexico where he took position of the full professor at the *National Polytechnic Institute of Mexico* in Mexico-City, where he continuous to work till present.

Under supervisorship of V. Rabinovich there were defended 8 Ph.D. thesis in Russia, and 3 in Mexico. He is a member of Editorial Boards of various international journals, in particular, "*Complex Variables and Elliptic Equations*", "*Communications in Mathematical Analysis*", "*Journal of Pseudodifferential Operators*", "*Mathematics in Engineering, Science and Aerospace*".

The first studies of V. Rabinovich were related to the investigation of the Fredholm properties of the multidimensional Wiener-Hopf equations in unbounded domains in $\mathbb{R}^n$ with the cone type structure at infinity. These results were published in the papers [37] (1967), [38] (1968) and [40] (1969). But the most important results of V.Rabinovich of that time were obtained during the last years of his Ph.D. studies. They were about the Fredholm properties of the general boundary value problems for pseudodifferential operators in such unbounded domains, published in the leading Soviet journals *Mathem. Sbornik* and *Doklady Akademii Nauk*, see [41] (1969) and [44] (1971). These papers were the first ones where the general boundary value problems for pseudodifferential operators were considered in unbounded domains. These results were included in his Ph.D.

Among the most important results obtained by V. Rabinovich in 1972–1977 there were the solvability of the Cauchy and Goursat problems for parabolic pseudodifferential operators, Fredholm properties of pseudodifferential operators and boundary value problems for them on non compact manifolds [46] (1972), [48] (1973), [52] (1975) and [56] (1979).

His paper [51] (1974) devoted to the multi-dimensional convolution operators in the space with exponential weights is worth of special mentioning. In this paper he extended the well-known results of I. Gohberg and M. Krein to a class of multidimensional convolution operators. It gave a start to his further studies of partial differential operators and pseudodifferential operators in spaces with exponential weights. Thus in the paper [28] (1978) he introduced a class of pseudodifferential operators with analytical symbols in a tube domain in $\mathbb{C}^n$ and obtained effective results on the boundedness of pseudodifferential operators together with the study of their Fredholm properties in spaces with exponential weights and exponential decreasing of solutions of pseudodifferential equations at infinity.

In the series of his papers [60] (1982), [61] (1983), continued with a Ph.D. student R. Babadjanian [1] (1985), [2] (1986) and [3] (1987), he studied Fredholm properties of pseudodifferential-difference operators, integral-difference and differential-difference operators. In particular in [2] there was proved the important theorem on the Wiener-Hopf factorization of the operator-valued functions in the Wiener algebra.

The next important scientific results of the V. Rabinovich are connected with the so-called method of limit operators. The idea of the limit operators historically goes back to a paper of J. Favard of 1927 on the existence of solutions to ordinary differential equations with almost periodic coefficients. These results of Favard were extended to the case of elliptic partial differential equations by E. Muhamadiev in a paper of 1981.

V. Rabinovich in fact turned this approach into a powerful general method, nowadays known as the "*method of limit operators*" by extending it and giving its wide applications to the investigation of the Fredholm properties of pseudodifferential operators, convolution type operators on $\mathbb{Z}^n$ and $\mathbb{R}^n$, general boundary values problems of the Boutet de Monvel type on manifolds with conical structure at infinity and pseudodifferential operators with shifts, etc, in his paper [61] (1985) and in the series of his papers [22, 24] (1985) and [25] (1986) with the Ph.D. student B. Lange and later in his papers [73] (1992), [76] (1993), [77] (1994), [84] (1998), [87] (1999) and [89, 90] (2001).

These investigations were also elaborated and continued in collaboration with S. Roch (Darmstadt) and B. Silbermann (Chemnitz) in the papers [136] (1998), [137, 138] (2001), [113] (2002), [114–116] (2003), [117] (2004), [124, 125] (2007), [141] (2008) and in his papers [95–97] (2003), [98] (2004).

In 2004 the book [139] by V. Rabinovich, S. Roch and B. Silbermann was published, in which there were presented both the techniques of the method of limit operators and the main results on its applications to various problems of operator theory including convolution type operators, discrete and continuous pseudodifferential operators, singular integral operators on Carleson curves and finite sections method.

V. Rabinovich and S. Roch discovered that the method of limit operators is a powerful tool for the investigation of the essential spectra of the electromagnetic Schrödinger operators on $\mathbb{R}^n, \mathbb{Z}^n$, and on periodic graphs, as was realized and developed in the papers [98] (2004), [101, 102] (2005), [103, 120] (2006), [123]

(2007) [127–129, 141] (2008) [130–132] (2009), [134] (2010). In particular, by means of the method of limit operators there was obtained a simple and short proof of the well-known Hunziker-van Winter-Zhislin Theorem on the essential spectra of the multiparticle Schrödinger operators, along with some new results on the essential spectra of the Schrödinger and Dirac operators.

The area of mathematical interests of V. Rabinovich, as can be already seen from the above, includes various topics from both Analysis and Mathematical Physics. In reality it is even wider than has been described in the previous lines. We could mention a lot more. For instance, it is worthwhile mentioning his studies of the exponential decrease of solutions of differential and pseudodifferential equations. In the papers [12] (1997), [99] (2004), [134] (2006), [127, 128] (2008) and [106, 130, 131, 133] (2009) there were obtained strong and exact results on the behavior of solutions of the differential and pseudodifferential equations with increasing discontinuous coefficients, which in particular included a far going generalizations of the well-known S. Agmon's results on the exponential decrease of solutions of elliptic second-order partial differential equations. Note that there were also given applications to the study of exponential decrease of eigenfunctions corresponding to the discrete spectra of the electromagnetic Schrödinger and Dirac operators for wide classes of potentials.

London, Conference dedicated to the 80th birthday of M.Z. Solomyak, September 2011.

Special words should be said about his studies of algebras of singular integral operators on a class of composed Carleson curves with coefficients having oscillating discontinuity, which go back to his interests of his research at the Department of Mechanics and Mathematics of the Rostov State University and continue through his life up to the present time. In the papers [71] (1991), [80] (1995) and [81, 82] (1996). there was shown that in the case where curves, coefficients and weights oscillate, the usual Mellin transform, which is the effective tool in the case of Lyapunov curves and piecewise continuous coefficients and non-oscillating weights, should be replaced by the Mellin pseudodifferential operators with variable and non stabilized symbols.

These investigations were continued with A. Böttcher (Chemnitz) and Yu. Karlovich (Cuernavaca, Mexico) in the papers [4] (1996), [5] (1998), [6] (2000), [7] (2001), where in a crystal clear form there was explained the appearance of the logarithmic spirals and logarithmic horns in the local spectra of singular integral operators on a class of composed Carleson curves.

Recently he turned to the studies in a new and rapidly developing area known as "*Variable Exponent Analysis*". In the papers [143] (1997), [144] (2008) and [145] (2011) joint with the author of this article there were studied singular integral operators and also pseudodifferential in variable exponents Lebesgue spaces, including the case of composite Carleson curves. In particular, in [145] (2011) the Simonenko local principle was extended to the case of variable exponent Lebesgue spaces where the main challenge was the localization of the space itself.

The task to overview all the studies of Volodya Rabinovich is too enormous for this introductory article, but we still mention a few. In the papers [147, 154] (2000), [148, 149] (2001), [150] (2002) and [151] (2004) with B.-W. Schulze and N. Tarkhanov (Potsdam) there were studied Fredholm properties of boundary value problems in domains with cuspidal points and cuspidal edges and also was described the behavior of solutions near singular manifolds of the boundary.

In another cycle of papers [29, 30] (2008) and [31] (2009) with Ya. Lutsky (Karmiel, Israel) he investigated the invertibility of the homogeneous Cauchy problem for parabolic pseudodifferential operators with discontinuous and increasing symbols, along with the study of the behavior of solutions at infinity and near the sets of discontinuities of the symbols.

His interests vary from rather pure mathematical topics in Operator Theory and Mathematical Physics to very applied fields, such as acoustic problems, wave propagation etc. In the papers [69] (1990), [14] (1996) and [33] (1998) with his Ph.D. student O. Obrezanova and the colleague S.M. Grudskii there were solved some theoretical and applied problems of the underwater sound long distance propagation in the ocean. In particular, there were obtained effective asymptotic formulas for the acoustic fields in the ocean generated by non uniformly moving sources. These investigations were continued later after he moved to Mexico, with his Mexican Master and Ph.D. students in the papers [146] (2003), [34] (2005), [35] (2007), [36] (2009), [109] (2010).

Caucasus, Mount Elbrus region, July 2010.

The reader can also find other topics of Vladimir Rabinovich's interests in the titles of his publications in the end of the article.

Volodya is a happy person having a nice family. He and his wife Nelli have two beautiful daughters Katya and Masha, now living in Israel. He has two granddaughters and a grandson and enjoys being their grandfather.

He is very sociable and liked by friends as the life and soul of the party. He has an an outward-looking personality which enables him to get along with people from all walks of life and easily interact with colleagues and all the people around. He is also active and outdoorsy. I remember him playing football when a student at the Rostov State University. He was a member of the student football team of the department and also of a combined team of the university and till his move to Mexico played football on the professors' teams at the Rostov State University. From his student's studies till these days, every year he spends some time in mountains, his hiking there being at a serious alpinist's level. These days, his friends wish him to keep in a top physical shape, and keep in general a keen interest to mathematics, mountains, social life, for many and many years ahead.

References

[1] R.G. Babadzhanyan and V.S. Rabinovich, *A system of integral-difference equations on the half line* (in Russian). Akad. Nauk Armyan. SSR Dokl. **81** (1985), no. 3, 107–111.

[2] R.G. Babadzhanyan and V.S. Rabinovich, *Factorization of almost periodic operator functions* (in Russian). In: Differential and Integral Equations and Complex Analysis (Russian), pp. 13–22, Kalmytsk. Gos. Univ., Elista, 1986.

[3] R.G. Babadzhanyan and V.S. Rabinovich, *Systems of integro-difference equations in a half-space* (in Russian). Izv. Vyssh. Ucheb. Zaved. Mat. (1987), no. 1(296), 63–66, 88; Engl. transl.: Sov. Math. **31** (1987), 84–88.

[4] A. Böttcher, Yu.I. Karlovich, and V.S. Rabinovich, *Emergence, persistence, and disappearance of logarithmic spirals in the spectra of singular integral operators.* Integral Equations Operator Theory **25** (1996), no. 4, 406–444.

[5] A. Böttcher, Yu.I. Karlovich, and V.S. Rabinovich, *Mellin pseudodifferential operators with slowly varying symbols and singular integrals on Carleson curves with Muckenhoupt weights.* Manuscripta Math. **95** (1998), no. 3, 363–376.

[6] A. Böttcher, Yu.I. Karlovich, and V.S. Rabinovich, *The method of limit operators for one-dimensional singular integrals with slowly oscillating data.* J. Operator Theory **43** (2000), no. 1, 171–198.

[7] A. Böttcher, Yu.I. Karlovich, and V.S. Rabinovich, *Singular integral operators with complex conjugation from the viewpoint of pseudodifferential operators.* In: Problems and Methods in Mathematical Physics, Oper. Theory Adv. Appl., vol. 121, pp. 36–59. Birkhäuser, Basel, 2001.

[8] G. Burlak, Yu. Karlovich, and V. Rabinovich, *Fredholm property and essential spectrum of pseudodifferential operators with operator-valued symbols.* Commun. Math. Anal. (electronic only) (2011), 50–65.

[9] R. Duduchava, I. Gohberg, S. Grudsky, and V. Rabinovich, editors. *Recent Trends in Toeplitz and Pseudodifferential Operators, The Nikolai Vasilevski Anniversary Volume.* Oper. Theory Adv. Appl., vol. 210. Birkhäuser Verlag, Basel, 2010.

[10] Ya. Erusalimsky, I. Gohberg, S. Grudsky, V. Rabinovich, and N. Vasilevski, editors. *Modern Operator Theory and Applications, dedicated to* 70*th anniversary of I.B. Simonenko.* Oper. Theory Adv. Appl., vol. 170. Birkhäuser Verlag, Basel, 2006.

[11] Ya. Erusalimskii, A. Kilbas, V. Kokilashvili, V. Rabinovich, and S. Samko, *In memoriam: Nikolai Karapetyants.* Fract. Calc. Appl. Anal. **7** (2004), no. 4, 475–478.

[12] I.M. Gontmakher and V.S. Rabinovich, *Exponential estimates of solutions of partial differential equations near singularities of coefficients.* Linear Topol. Spaces Complex Anal. Dedicated to Professor Vyacheslav Pavlovich Zahariuta. **3** (1997), 65–79.

[13] A.N. Gorbanev, S.M. Grudskiĭ, V.S. Rabinovich, E.A. Rivelis, Yu.V.S. Khokha, and S.L. Èdel′shteĭn, *Asymptotics of an acoustic field generated by a moving aerial sound source in an oceanic waveguide.* In: Oceanic Acoustics (Russian), pp. 9–20. "Nauka", Moscow, 1993.

[14] S.M. Grudskii, O.A. Obrezanova, and V.S. Rabinovich, *Propagation of sound from a moving air-borne source in the ice-covered ocean.* Acoustic J. **27** (1996), no. 2, 191–196; Engl. transl.: Physical Acoustics (1996), no. 2, 166–171.

[15] V.I. Il′ichev, V.S. Rabinovich, E.A. Rivelis, and Yu.V. Hoha, *The far field of narrow-band sound source that moves in a waveguide* (in Russian). Dokl. Akad. Nauk SSSR, **304** (1989), no. 5, 1123–1127.

[16] A.N. Karapetyants, V.S. Rabinovich, and N.L. Vasilevski, *On algebras of two dimensional singular integral operators with homogeneous discontinuities in symbols.* Integr. Equat. Oper. Theory **40** (2001), no. 3, 278–308.

[17] Yu.I. Karlovich, V.S. Rabinovich, and N.L. Vasilevski, *Algebras of pseudodifferential operators with discontinuous oscillating symbols.* Commun. Math. Anal. (electronic only) (2011), 108–130.

[18] K.V. Khmelnytskaya, V.V. Kravchenko, and V.S. Rabinovich, *Quaternionic fundamental solutions for electromagnetic scattering problems and application.* Z. Anal. Anwendungen **22** (2003), no. 1, 147–166.

[19] M.N. Kostjukovich and V.S. Rabinovich, *Discrete analogues of boundary value problems.* Izv. Vysš. Učebn. Zaved., Matematika (1972), no. 11(126), 32–38.

[20] V.D. Kryakvin and V.S. Rabinovich, *Noethericity of the operator of a boundary value problem for a domain with an unbounded boundary in weighted spaces of Hölder continuous functions* (in Russian). In: Differential and Integral Equations and their Applications, pp. 79–86. Kalmytsk. Gos. Univ., Elista, 1983.

[21] B.V. Lange and V.S. Rabinovich, *Discrete operators of convolution type with oscillating coefficients* (in Russian). In: Differential and Integral Equations and their Applications, pp. 86–92. Kalmytsk. Gos. Univ., Elista, 1983.

[22] B.V. Lange and V.S. Rabinovich, *Criterion for Noethericity of pseudodifferential operators on* $\mathbf{R}^n$ *with symbols of class* $C_b^\infty(\mathbf{R}^n \times \mathbf{R}^n)$. Dokl. Akad. Nauk SSSR **285** (1985), no. 6, 1317–1320.

[23] B.V. Lange and V.S. Rabinovich, *Noethericity of multidimensional discrete convolution operators* (in Russian). Mat. Zametki **37** (1985), no. 3, 407–421.

[24] B.V. Lange and V.S. Rabinovich, *Noethericity of multidimensional operators of convolution type with measurable bounded coefficients.* Izv. Vyssh. Uchebn. Zaved., Mat. (1985), no. 6(277), 22–30, 86.

[25] B.V. Lange and V.S. Rabinovich, *Pseudodifferential operators in* $\mathbf{R}^n$ *and limit operators.* Mat. Sb. (N.S.) **129(171)** (1986), no. 2, 175–185, 303.

[26] M. Lindner, V.S. Rabinovich, and S. Roch, *Finite sections of band operators with slowly oscillating coefficients.* Linear Algebra Appl. **390** (2004), 19–26.

[27] Ja.A. Lutsky and V.S. Rabinovich, *Pseudodifferential operators in classes of exponentially growing functions* (in Russian). In: Theory of Functions. Differential Equations and Their Applications, no. 1, pp. 171–181, 220. Kalmytsk. Gos. Univ., Elista, 1976.

[28] Ja.A. Lutsky and V.S. Rabinovich, *Pseudodifferential operators in spaces of generalized functions with exponential behavior at infinity* (in Russian). Funkcional. Anal. i Priložen. **12** (1978), no. 1, 79–80.

[29] Ya. Lutsky and V.S. Rabinovich, *Invertibility of parabolic pseudodifferential operators with rapidly increasing symbols.* Russ. J. Math. Phys. **15** (2008), no. 2, 267–279.

[30] Ya. Lutsky and V.S. Rabinovich, *Parabolic pseudodifferential operators in exponential weighted spaces.* In: Complex Analysis and Dynamical Systems III, Contemp. Math., vol. 455, pp. 279–295. Amer. Math. Soc., Providence, RI, 2008.

[31] Ya. Lutsky and V.S. Rabinovich, *Exponential estimates of solutions of parabolic pseudodifferential equations with discontinuous and growing symbols.* Complex Var. Elliptic Equ., **54** (2009), no. 8, 757–778.

[32] O.A. Obrezanova and V.S. Rabinovich, *Wave propagation from moving source in the ocean.* In ECCOMAS Comput. Fluid Dynamics Conf., pp. 754–759. Paris, 1996.

[33] O.A. Obrezanova and V.S. Rabinovich, *Acoustic field generated by moving sources in stratified waveguides.* Wave Motion **27** (1998), no. 2, 155–167.

[34] R.M. Perez and V.S. Rabinovich, *Asymptotics of the green functions for Helmholtz operators in almost stratified medias.* In: Proceeding of the International Seminar "Day of Diffraction" 2005, June 28-July 1, St. Petersburg, Russia, pp. 209–216. IEEE; Catalog, No. 05EX1151, 2005.

[35] R.M. Perez, V.S. Rabinovich, and I.M. Sanchez, *Adiabatic approximation of the Green function of the Helmholtz operator in almost stratified medias.* Russ. J. Math. Phys. **14** (2007), no. 2, 201–212.

[36] M. Quino-Cerdán and V.S. Rabinovich, *Análisis de propagación de onda acústica en océanos estratificados mediante el método de modos normales.* Lat. Am. J. Phys. Educ. **3** (2009), no. 2, 452–459.

[37] V.S. Rabinovich, *The multidimensional Wiener-Hopf equation for cones* (in Russian). Teor. Funkciĭ, Funkcional. Anal. i Priložen. (1967), no. 5, 59–67.

[38] V.S. Rabinovich, *Certain estimates for translation-invariant operators in* L_p*-spaces with weight* (in Russian). Izv. Vysš. Učebn. Zaved., Matematika (1968), no. 10(77), 72–80.

[39] V.S. Rabinovich, *Convolution type equations of the first kind in conic domains.* In: Mathematical Analysis and its Applications (in Russian), pp. 152–158. Izdat. Rostov. Univ., Rostov-on-Don, 1969.

[40] V.S. Rabinovich, *A multidimensional equation of convolution type whose symbol has singularities of the form of a complex power function of a linearly homogeneous cone* (in Russian). Izv. Vysš. Učebn. Zaved., Matematika (1969), no. 8(87), 64–74.

[41] V.S. Rabinovich, *Pseudodifferential equations in unbounded domains with conical structure at infinity* (in Russian). Mat. Sb. (N.S.), **80(122)** (1969), 77–96.

[42] V.S. Rabinovich, *A multidimensional equation of convolution type with a symbol that has singularities* (in Russian). Teor. Funkciĭ, Funkcional. Anal. i Priložen. (1970), no. 10, 48–56.

[43] V.S. Rabinovich, *A certain class of pseudodifferential equations for unbounded domains in the spaces $H^{s,p}$ of Bessel potentials* (in Russian). Teor. Funkciĭ, Funkcional. Anal. i Priložen. (1971), no. 13, 88–101.

[44] V.S. Rabinovich, *Pseudodifferential equations in unbounded domains* (in Russian). Dokl. Akad. Nauk SSSR **197** (1971), 284–287.

[45] V.S. Rabinovich, *Quasielliptic pseudodifferential operators, and the Cauchy problem for parabolic equations* (in Russian). Dokl. Akad. Nauk SSSR **201** (1971), 1055–1058.

[46] V.S. Rabinovich, *The Cauchy problem for parabolic pseudodifferential equations with nonstabilizing symbol* (in Russian). Izv. Vysš. Učebn. Zaved., Matematika (1972), no. 7(122), 85–94.

[47] V.S. Rabinovich, *Pseudodifferential operators on a certain class of noncompact manifolds.* Mat. Sb. (N.S.), **89(131)** (1972), 46–60; Engl. transl.: Math. USSR, Sbornik **18** (1972), no. 1, 45–59.

[48] V.S. Rabinovich, *A priori estimates and the Fredholm property of a certain class of pseudodifferential operators.* Mat. Sb. (N.S.) **92(134)** (1973), 195–208; Engl. transl.: Math. USSR, Sbornik **21** (1973), no. 2, 191–206.

[49] V.S. Rabinovich, *Elliptic boundary value problems for unbounded domains in spaces of functions with exponential behavior at infinity.* Izv. Severo-Kavkaz. Naučn. Centra Vysš. Školy Ser. Estestv. Nauk. (1974), no. 4, 121–123, 140.

[50] V.S. Rabinovich, *Multidimensional equations of convolution type with operator symbol.* Teor. Funkciĭ, Funkcional Anal. i Priložen. (1974), no. 19, 44–53, 159.

[51] V.S. Rabinovich, *Multidimensional operators of convolution type in spaces of functions that are integrable with a weight.* Mat. Zametki **16** (1974), 267–276.

[52] V.S. Rabinovich, *Quasielliptic and parabolic boundary value problems for unbounded cylindrical domains* (in Russian). Teor. Funkciĭ, Funkcional. Anal. i Priložen. (1975), no. 24, 138–150.

[53] V.S. Rabinovich, *Elliptic problems with a parameter and parabolic problems for operator pseudodifferential equations* (in Russian). Izv. Severo-Kavkaz. Naučn. Centra Vysš. Školy Ser. Estestv. Nauk. (1976), no. 3, 8–11, 116.

[54] V.S. Rabinovich, *Boundary value problems for a class of differential operators in domains with conical points* (in Russian). Differencial′nye Uravnenija, **13** (1977), no. 9, 1727–1729, 1736.

[55] V.S. Rabinovich, *The solvability of differential-difference equations on* $\mathbf{R}^n$ *and in a half-space.* Dokl. Akad. Nauk SSSR **243** (1978), no. 5, 1134–1137; Engl. transl.: Soviet Math. Dokl. **19** (1978), no. 6, 1498–1502.

[56] V.S. Rabinovich, *The Cauchy and Goursat problems for a class of non-hypoelliptic pseudodifferential operators* (in Russian). Izv. Severo-Kavkaz. Nauchn. Tsentra Vyssh. Shkoly Ser. Estestv. Nauk. (1979), no. 1, 4–7, 104.

[57] V.S. Rabinovich, *Differential-difference equations in a half space* (in Russian). Differentsial′nye Uravneniya, **16** (1980), no. 11, 2030–2038, 2110.

[58] V.S. Rabinovich, *Noether property of pseudodifferential operators with symbols of the class* $S^m_{\rho,\delta}$ $(0 \leq \delta = \rho < 1)$ (in Russian). Mat. Zametki, **27** (1980), no. 3, 457–467.

[59] V.S. Rabinovich, *A* C^**-algebra of pseudodifferential operators* (in Russian). In: Differential and Integral Equations and their Applications, pp. 147–150. Kalmytsk. Gos. Univ., Elista, 1982.

[60] V.S. Rabinovich, *On the algebra generated by the pseudodifferential operators on* $\mathbf{R}^n$*, the operators of multiplication by almost periodic functions and the shift operators.* Dokl. Akad. Nauk SSSR **263** (1982), no. 5, 1066–1070; Engl. transl.: Soviet Math. Dokl. **25** (1982), no. 2, 498–502.

[61] V.S. Rabinovich, *Algebras of pseudodifferential operators connected with differential-difference operators. I.* Teor. Funktsiĭ, Funktsional. Anal. i Prilozhen. (1983), no. 40, 119–125.

[62] V.S. Rabinovich, *The Cauchy problem for parabolic differential-difference operators with variable coefficients* (in Russian). Differentsial′nye Uravneniya, **19** (1983), no. 6, 1032–1039.

[63] V.S. Rabinovich, *Integral-difference operators in a quarter plane* (in Russian). In: Integral and Differential Equations and Approximate Solutions, pp. 144–149. Kalmytsk. Gos. Univ., Elista, 1985.

[64] V.S. Rabinovich, *On existence and uniqness of a solution of a problem of propagation of acoustic waves in a two-layered non-homogeneous fluid* (in Russian). In: Mathematical Methods in Applied Acoustics, pp. 107–112. Rostov State University, Rostov-na-Donu, 1986.

[65] V.S. Rabinovich, *The Fredholm property of pseudodifferential operators on* $\mathbf{R}^n$ *in the scale of spaces* $L_{2,p}$ (in Russian). Sibirsk. Mat. Zh. **29** (1988), no. 4, 149–161; Engl. transl.: Siberian Math. J. **29** (1988), no. 4, 635–646.

[66] V.S. Rabinovich, *Elliptic boundary value problems in domains with logarithmic whirl points on the boundary* (in Russian). Dokl. Akad. Nauk **361** (1988), no. 1, 17–20; Engl. transl.: Dokl. Math. **58** (1988), no. 1, 11–14.

[67] V.S. Rabinovich, *The limiting absorption principle for a class of second order differential operators* (in Russian). In: Complex Analysis, Differential and Integral Equations, pp. 113–119. Kalmytsk. Gos. Univ., Elista, 1990.

[68] V.S. Rabinovich, *On spectral properties of a difference model the acoustic operator in a non homogeneous half-space* (in Russian). In: Mathematical Methods in Applied Acoustics, pp. 163–166. Rostov State University, Rostov-na-Donu, 1990.

[69] V.S. Rabinovich, *On the solvability of problems of acoustics of open waveguides.* Differentsial′nye Uravneniya **26** (1990), no. 12, 2178–2180, 2207–2208.

[70] V.S. Rabinovich, *Scattering theory for the acoustic operator in an inhomogeneous half-space.* Izv. Severo-Kavkaz. Nauchn. Tsentra Vyssh. Shkoly Estestv. Nauk (1991), no. 1, 50–56.

[71] V.S. Rabinovich, *Singular integral operators on a composite contour with an oscillating tangent, and pseudodifferential Mellin operators.* Dokl. Akad. Nauk SSSR, **321** (1991), no. 4, 692–696; Engl. transl.: Soviet Math. Dokl. **44** (1992), no. 3, 791–796.

[72] V.S. Rabinovich, *Spectral properties of a class of difference operators* (in Russian). Dokl. Akad. Nauk SSSR **320** (1991), no. 1, 45–48; Engl. transl.: Soviet Math. Dokl. **44** (1992), no. 2, 402–406.

[73] V.S. Rabinovich, *The Fredholm property of general boundary value problems on noncompact manifolds, and limit operators.* Dokl. Akad. Nauk **325** (1992), no. 2, 237–241.

[74] V.S. Rabinovich, *Operator-valued discrete convolutions and some of their applications* (in Russian). Mat. Zametki **51** (1992), no. 5, 90–101; Engl. transl.: Math. Notes, **51** (1992), no. 5, 484–492.

[75] V.S. Rabinovich, *Spectral and scattering theory for acoustic operators in nonhomogeneous fluids. Continuous and discrete models.* In: Symposium "Analysis on Manifolds with Singularities" (Breitenbrunn, 1990), Teubner-Texte Math., vol. 131, pp. 158–167. Teubner, Stuttgart, 1992.

[76] V.S. Rabinovich, *A criterion for the local invertibility of Mellin pseudodifferential operators with operator symbols, and some of its applications* (in Russian). Dokl. Akad. Nauk **333** (1993), no. 2, 147–150; Engl. transl.: Dokl. Math. **48** (1994), no. 3, 465–469.

[77] V.S. Rabinovich, *Pseudodifferential operators with operator symbols: local invertibility and limit operators.* Linear Topol. Spaces Complex Anal. **1** (1994), 58–73.

[78] V.S. Rabinovich, *Mellin pseudodifferential operators with operator symbols and its applications.* In: Partial Differential Operators and Mathematical Physics (Holzhau, 1994), Oper. Theory Adv. Appl., vol. 78, pp. 271–279. Birkhäuser, Basel, 1995.

[79] V.S. Rabinovich, *Pseudodifferential operators with analytical symbols and some of its applications.* Linear Topol. Spaces Complex Anal. **2** (1995), 79–98.

[80] V.S. Rabinovich, *Singular integral operators on composite contours, and pseudodifferential operators.* Mat. Zametki, **58** (1995), no. 1, 67–85; Engl. transl.: Math. Notes **58** (1995), No. 1, 722–734.

[81] V.S. Rabinovich, *Algebras of singular integral operators on composed contours with nodes that are logarithmic whirl points.* Izv. Ross. Akad. Nauk Ser. Mat. **60** (1996), no. 6, 169–200; Engl. transl.: Izvestia: Mathematics **60** (1996), no. 6, 1261–1292.

[82] V.S. Rabinovich, *Algebras of singular integral operators on composite contours with whirl points.* Funktsional. Anal. i Prilozhen. **30** (1996), no. 3, 85–87; Engl. transl.: Funct. Anal. Appl. **30** (1996), no. 3, 213–215.

[83] V.S. Rabinovich, *Elliptic boundary value problems in domains with whirl points on the boundary.* In: Differential Equations, Asymptotic Analysis, and Mathematical Physics (Potsdam, 1996), Math. Res., vol. 100, pp. 276–286. Akademie Verlag, Berlin, 1997.

[84] V.S. Rabinovich, *C^*-algebras of pseudodifferential operators and limit operators.* In: Operator Theory for Complex and Hypercomplex Analysis (Mexico City, 1994), Contemp. Math., vol. 212, pp. 169–183. Amer. Math. Soc., Providence, RI, 1998.

[85] V.S. Rabinovich, *An Introductionary Course on Pseudodifferential Operators.* Instituto Superior Técnico. Textos de Matemática. Centro de Matemática Aplicada, Lisbon, 1998.

[86] V.S. Rabinovich, *Mellin pseudodifferential operators techniques in the theory of singular integral operators on some Carleson curves.* In: Differential and Integral Operators (Regensburg, 1995), Oper. Theory Adv. Appl., vol. 102, pp. 201–218. Birkhäuser, Basel, 1998.

[87] V.S. Rabinovich, *Criterion for local invertibility of pseudodifferential operators with operator symbols and some applications.* In: Proceedings of the St. Petersburg Mathematical Society, Vol. V, Amer. Math. Soc. Transl. Ser. 2, vol. 193, pp. 239–259, Providence, RI, Amer. Math. Soc. 1999.

[88] V. Rabinovich, *Potential type operators on curves with vorticity points.* Z. Anal. Anwendungen, **18** (1999), no. 4, 1065–1081.

[89] V.S. Rabinovich, *An abstract scheme of the method of limit operators and some of its applications.* Dokl. Akad. Nauk, **381** (2001), no. 2, 167–170.

[90] V.S. Rabinovich, *Stability of inverse operators of boundary value problems in expanding domains.* Russ. J. Math. Phys. **8** (2001), no. 3, 336–345.

[91] V.S. Rabinovich, *Stability of inverse operators of boundary value problems in nonsmooth expanding domains.* In: Partial Differential Equations and Spectral Theory (Clausthal, 2000), Oper. Theory Adv. Appl., vol. 126, pp. 261–274. Birkhäuser, Basel, 2001.

[92] V.S. Rabinovich, *Stability of inverse operators of boundary value problems in smooth expanding domains* (in Russian). Funktsional. Anal. i Prilozhen. **35** (2001), no. 4, 85–87.

[93] V.S. Rabinovich, *An abstract scheme for the method of limit operators on a group of unitary operators.* In: Semigroups of Operators: Theory and Applications (Rio de Janeiro, 2001), pp. 220–227. Optimization Software, New York, 2002.

[94] V. Rabinovich, *Pseudodifferential operators with analytic symbols and estimates for eigenfunctions of Schrödinger operators.* Z. Anal. Anwendungen **21** (2002), no. 2, 351–370.

[95] V.S. Rabinovich, *The Fredholm property for pseudodifferential operators with shifts on $\mathbb{R}^n$, and limit operators.* Dokl. Akad. Nauk, **388**(2003), no. 6, 734–737; Engl. transl.: Dokl. Math. **67** (2003), no. 1, 81–84.

[96] V.S. Rabinovich, *Multidimensional pseudodifference operators: Fredholm property, essential spectrum, and exponential estimates of solutions.* Russ. J. Math. Phys. **10** (2003), no. 3, 296–318.

[97] V.S. Rabinovich, *Pseudodifferential operators on $\mathbf{R}^n$ with variable shifts.* Z. Anal. Anwendungen **22** (2003), no. 2, 315–338.

[98] V.S. Rabinovich, *The essential spectrum of differential operators, and limit operators. Applications to the Schrödinger operator.* Dokl. Akad. Nauk, **399** (2004), no. 1, 18–22; Engl. transl.: Dokl. Math. **204** (2004), no. 3, 852–856.

[99] V.S. Rabinovich, *Exponential estimates for eigenfunctions of Schrödinger operators with rapidly increasing and discontinuous potentials.* In: Complex Analysis and Dynamical Systems, Contemp. Math., vol. 364, pp. 225–236. Amer. Math. Soc., Providence, RI, 2004.

[100] V. Rabinovich, *Limiting absorption principle for a class of difference equations.* In: Operator Theoretical Methods and Applications to Mathematical Physics, Oper. Theory Adv. Appl., vol. 147, pp. 403–422. Birkhäuser, Basel, 2004.

[101] V.S. Rabinovich, *Essential spectrum of perturbed pseudodifferential operators. Applications to the Schrödinger, Klein-Gordon, and Dirac operators.* Russ. J. Math. Phys. **12** (2005), no. 1, 62–80.

[102] V.S. Rabinovich, *On the essential spectrum of electromagnetic Schrödinger operators.* In: Complex Analysis and Dynamical Systems II, Contemp. Math., vol. 382, pp. 331–342. Amer. Math. Soc., Providence, RI, 2005.

[103] V.S. Rabinovich, *Essential Spectrum and Exponential Estimates of Perturbed Pseudodifferential Operators.* Izvestia Vuzov, Severo-Kavkazskiy Region. Estestvennie Nauki, Issue dedicated to 70-birthday of Professor I.B. Simonenko, pp. 159–164, 2006.

[104] V.S. Rabinovich, *Fredholm property of pseudo-differential operators on weighted Hölder-Zygmund spaces.* In: Pseudo-differential Operators and Related Topics, Oper. Theory Adv. Appl., vol. 164, pp. 95–114. Birkhäuser, Basel, 2006.

[105] V. Rabinovich, *Invertibility of parabolic pseudodifferential operators.* In: Pseudo-Differential Operators: Partial Differential Equations and Time-Frequency Analysis, Fields Inst. Commun., vol. 52, pp. 201–212. Amer. Math. Soc., Providence, RI, 2007.

[106] V.S. Rabinovich, *Local exponential estimates for h-pseudodifferential operators and tunneling for Schrödinger, Dirac, and square root Klein-Gordon operators.* Russ. J. Math. Phys. **16** (2009), no. 2, 300–308.

[107] V. Rabinovich, *Exponential estimates of solutions of pseudodifferential equations on the lattice* $(h\mathbb{Z})^n$*: applications to the lattice Schrödinger and Dirac operators.* J. Pseudo-Differ. Oper. Appl. **1** (2010), no. 2, 233–253.

[108] V.S. Rabinovich, *Local exponential estimates for h-pseudo-differential operators with operator-valued symbols.* In: Pseudo-Differential Operators: Analysis, Applications and Computations, Oper. Theory Adv. Appl., vol. 213, pp. 173–190. Birkhäuser Verlag, Basel, 2011.

[109] V.S. Rabinovich and M.Q. Cerdan, *Invertibility of Helmholtz operators for nonhomogeneous medias.* Math. Methods Appl. Sci. **33** (2010), no. 4, 527–538.

[110] V.S. Rabinovich and B. Lange, *Fredholmness of pseudodifferential operators with symbols from the class* $C_b^\infty(\mathbf{R}^{2n})$. In Seminar Analysis, 1984/85 (Berlin, 1984/85), pp. 147–161. Akad. Wiss. DDR, Berlin, 1985.

[111] V.S. Rabinovich and I. Miranda, *Non-uniformly moving source in electromagnetic waveguides.* WSEAS Transactions on Systems **1** (2002), no. 2, 109–114.

[112] V.S. Rabinovich and I.S. Miranda, *Radiation from nonuniformly moving sources in the plasma.* In: Proceeding of the International Seminar "Day of Diffraction", 2004, St. Petersburg, Russia, pp. 163–174. IEEE; Catalog, No. 04EX837, 2005.

[113] V.S. Rabinovich and S. Roch, *Local theory of the Fredholmness of band-dominated operators with slowly oscillating coefficients.* In: Toeplitz Matrices and Singular

Integral Equations (Pobershau, 2001), Oper. Theory Adv. Appl., vol. 135, pp. 267–291. Birkhäuser, Basel, 2002.

[114] V.S. Rabinovich and S. Roch, *Algebras of approximation sequences: spectral and pseudospectral approximation of band-dominated operators.* In: Linear Algebra, Numerical Functional Analysis and Wavelet Analysis, pp. 167–188. Allied Publ., New Delhi, 2003.

[115] V.S. Rabinovich and S. Roch, *An axiomatic approach to the limit operators method.* In: Singular Integral Operators, Factorization and Applications, Oper. Theory Adv. Appl., vol. 142, pp. 263–285. Birkhäuser, Basel, 2003.

[116] V.S. Rabinovich and S. Roch, *Integral operators with shifts on homogeneous groups.* In: Factorization, Singular Operators and Related Problems (Funchal, 2002), pp. 255–271. Kluwer Acad. Publ., Dordrecht, 2003.

[117] V.S. Rabinovich and S. Roch, *Fredholmness of convolution type operators.* In: Operator Theoretical Methods and Applications to Mathematical Physics, Oper. Theory Adv. Appl., vol. 147, pp. 423–455. Birkhäuser, Basel, 2004.

[118] V.S. Rabinovich and S. Roch, *Pseudodifference operators on weighted spaces, and applications to discrete Schrödinger operators.* Acta Appl. Math. **84** (2004), no. 1, 55–96.

[119] V.S. Rabinovich and S. Roch, *Wiener algebras of operators, and applications to pseudodifferential operators.* Z. Anal. Anwendungen **23** (2004), no. 3, 437–482.

[120] V.S. Rabinovich and S. Roch. *The essential spectrum of Schrödinger operators on lattices.* J. Phys. A, **39** (2006), no. 26, 8377–8394.

[121] V.S. Rabinovich and S. Roch, *The Fredholm property of pseudo-difference operators in weighted spaces.* Funktsional. Anal. i Prilozhen **40** (2006), no. 1, 83–86.

[122] V.S. Rabinovich and S. Roch, *Reconstruction of input signals in time-varying filters.* Numer. Funct. Anal. Optim. **27** (2006), no. 5–6, 697–720.

[123] V.S. Rabinovich and S. Roch, *Essential spectra of difference operators on $\mathbb{Z}^n$-periodic graphs.* J. Phys. A **40** (2007), no. 33, 10109–10128.

[124] V.S. Rabinovich and S. Roch, *The Fredholm property of pseudodifferential operators with non-smooth symbols on modulation spaces.* In: The Extended Field of Operator Theory, Oper. Theory Adv. Appl., vol. 171, pp. 259–280. Birkhäuser, Basel, 2007.

[125] V.S. Rabinovich and S. Roch, *Exact and numerical inversion of pseudo-differential operators and applications to signal processing.* In: Modern Trends in Pseudo-Differential Operators, Oper. Theory Adv. Appl., vol. 172, pp. 259–277. Birkhäuser, Basel, 2007.

[126] V.S. Rabinovich and S. Roch, *The Fredholm index of locally compact band-dominated operators on $L^p(\mathbb{R})$.* Integral Equations Operator Theory **57** (2007), no. 2, 263–281.

[127] V.S. Rabinovich and S. Roch, *Essential spectra of pseudodifferential operators and exponential decay of their solutions. Applications to Schrödinger operators.* In: Operator Algebras, Operator Theory and Applications, Oper. Theory Adv. Appl., vol. 181, pp. 355–384. Birkhäuser Verlag, Basel, 2008.

[128] V.S. Rabinovich and S. Roch, *Essential spectrum and exponential decay estimates of solutions of elliptic systems of partial differential equations. Applications to Schrödinger and Dirac operators.* Georgian Math. J.**15** (2008), no. 2, 333–351.

[129] V.S. Rabinovich and S. Roch, *Fredholm properties of band-dominated operators on periodic discrete structures.* Complex Anal. Oper. Theory **2** (2008), no. 4, 637–668.

[130] V.S. Rabinovich and S. Roch, *Essential spectra and exponential estimates of eigenfunctions of lattice operators of quantum mechanics.* J. Phys. A **42** (2009), no. 38, 385–207.

[131] V.S. Rabinovich and S. Roch, *The essential spectrum and exponential estimates for eigenfunctions of lattice Schrödinger and Dirac operators.* Dokl. Akad. Nauk **428** (2009), no. 1, 25–29.

[132] V.S. Rabinovich and S. Roch, *The essential spectrum of difference operators on periodic metric spaces.* Funktsional. Anal. i Prilozhen. **43** (2009), no. 2, 83–87.

[133] V.S. Rabinovich and S. Roch, *Exponential estimates for the eigenfunctions of matrix elliptic differential operators, and limit operators.* Dokl. Akad. Nauk **424** (2009), no. 3, 318–321; Engl. transl.: Dokl. Math. **79** (2009), no. 1, 66–69.

[134] V.S. Rabinovich and S. Roch, *Exponential estimates of eigenfunctions of matrix Schrödinger and Dirac operators.* In: Recent Trends in Toeplitz and Pseudodifferential Operators, The Nikolai Vasilevskii Anniversary Volume, Oper. Theory Adv. Appl., vol. 210, pp. 203–216. Birkhäuser Verlag, Basel, 2010.

[135] V.S. Rabinovich, S. Roch, and John Roe, *Fredholm indices of band-dominated operators.* Integral Equations and Operator Theory **49** (2004), no. 2, 221–238.

[136] V.S. Rabinovich, S. Roch, and B. Silbermann, *Fredholm theory and finite section method for band-dominated operators.* Integral Equations Operator Theory **30** (1998), no. 4, 452–495. Dedicated to the memory of Mark Grigorievich Krein (1907–1989).

[137] V.S. Rabinovich, S. Roch, and B. Silbermann, *Algebras of approximation sequences: finite sections of band-dominated operators.* Acta Appl. Math. **65** (2001), no. 1–3, 315–332. Special issue dedicated to Antonio Avantaggiati on the occasion of his 70th birthday.

[138] V.S. Rabinovich, S. Roch, and B. Silbermann, *Band-dominated operators with operator-valued coefficients, their Fredholm properties and finite sections.* Integral Equations Operator Theory **40** (2001), no. 3, 342–381.

[139] V.S. Rabinovich, S. Roch, and B. Silbermann, *Limit Operators and Their Applications in Operator Theory.* Oper. Theory Adv. Appl., vol. 150. Birkhäuser Verlag, Basel, 2004.

[140] V.S. Rabinovich, S. Roch, and B. Silbermann, *Finite sections of band-dominated operators with almost periodic coefficients.* In: Modern Operator Theory and Applications, Oper. Theory Adv. Appl., vol. 170, pp. 205–228. Birkhäuser, Basel, 2007.

[141] V.S. Rabinovich, S. Roch, and B. Silbermann, *On finite sections of band-dominated operators.* In: Operator Algebras, Operator Theory and Applications, Oper. Theory Adv. Appl., vol. 181, pp. 385–391. Birkhäuser Verlag, Basel, 2008.

[142] V.S. Rabinovich, N.G. Samko, and S.G. Samko, *Local Fredholm spectra and Fredholm properties of singular integral operators on Carleson curves acting on weighted Hölder spaces.* Integral Equations Operator Theory **56** (2006), no. 2, 257–283.

[143] V.S. Rabinovich and S.G. Samko, *The essential spectrum of pseudodifferential operators in Sobolev spaces with variable smoothness and variable Lebesgue exponents.* Dokl. Akad. Nauk **417** (2007), no. 2, 167–170.

[144] V.S. Rabinovich and S.G. Samko, *Boundedness and Fredholmness of pseudodifferential operators in variable exponent spaces.* Integral Equations Operator Theory **60** (2008), no. 4, 507–537.

[145] V.S. Rabinovich and S.G. Samko, *Pseudodifferential operators approach to singular integral operators in weighted variable exponent Lebesgue spaces on Carleson curves.* Integral Equations Operator Theory **69** (2011), no. 3, 405–444.

[146] V.S. Rabinovich and I.M. Sanches, *Non-uniformly moving source in electromagnetic waveguides.* In: Proceeding of the International Seminar "Day of Diffraction" 2003, June 24–27, St. Petersburg, Russia, pp. 193–202. IEEE; Catalog, No. 03EX661, 2003.

[147] V.S. Rabinovich, B.-W. Schulze, and N. Tarkhanov, *A calculus of boundary value problems in domains with non-Lipschitz singular points.* Math. Nachr. **215** (2000), 115–160.

[148] V.S. Rabinovich, B.-W. Schulze, and N. Tarkhanov, *Boundary value problems in domains with non Lipschitz boundary.* In: Proceedings of the Third World Congress of Nonlinear Analysts, Part 3 (Catania, 2000), vol. 47, pp. 1881–1891, 2001.

[149] V.S. Rabinovich, B.-W. Schulze, and N. Tarkhanov, *Local algebra of a nonsymmetric corner.* In: Partial Differential Equations and Spectral Theory (Clausthal, 2000), Oper. Theory Adv. Appl., vol. 126, pp. 275–280. Birkhäuser, Basel, 2001.

[150] V.S. Rabinovich, B.-W. Schulze, and N. Tarkhanov, *C^*-algebras of singular integral operators in domains with oscillating conical singularities.* Manuscripta Math. **108** (2002), no. 1, 69–90.

[151] V.S. Rabinovich, B.-W. Schulze, and N. Tarkhanov, *Boundary value problems in oscillating cuspidal wedges.* Rocky Mountain J. Math. **34** (2004), no. 4, 1399–1471.

[152] V.S. Rabinovich and O.N. Shamaylo, *On the solvability of some operator equations* (in Russian). In: Differential and Integral Equations and Complex Analysis, pp. 98–107. Kalmytsk. Gos. Univ., Elista, 1988.

[153] V.S. Rabinovich and O.N. Shamaylo, *On solvability of the Helmholtz equation in a non-homogeneous layer* (in Russian). In: Mathematical Methods in Applied Acoustics **2**, pp. 120–126. Rostov State University, Rostov-na-Donu, 1990.

[154] V.S. Rabinovich, N.N. Tarkhanov, and B.-V. Schulze, *Elliptic boundary value problems in domains with intersecting cuspidal edges.* Dokl. Akad. Nauk **375** (2000), no. 6, 737–741.

[155] V.S. Rabinovich and N. Vasilevski, *Bergman-Toeplitz and pseudodifferential operators.* In: Complex Analysis and Related Topics (Cuernavaca, 1996), Oper. Theory Adv. Appl., vol. 114, pp. 207–234. Birkhäuser, Basel, 2000.

Stefan Samko
Universidade do Algarve
Campus de Gambelas
8005–139 Faro, Portugal
e-mail: ssamko@ualg.pt

Operator Theory:
Advances and Applications, Vol. 228, 1–20

The Dirichlet Problem for the Heat Equation in Domains with Cuspidal Points on the Boundary

Alexandra Antoniouk and Nikolai Tarkhanov

Dedicated to V. Rabinovich on the occasion of his 70 th birthday

Abstract. We treat the Dirichlet problem for the **1D** heat equation in a bounded domain $\mathcal{G} \subset \mathbb{R}^2$. The boundary of $\mathcal{G}$ is assumed to be smooth and noncharacteristic except for two points where it has contact of degree less than 2 with lines orthogonal to the t-axis. At these points the boundary has cuspidal singularities which have to be treated with particular care. We prove that this problem fits into the framework of analysis on manifolds with singular points elaborated by V. Rabinovich et al. (2000). The results extend to general parabolic equations.

Mathematics Subject Classification (2010). Primary 35K35; Secondary 35G15, 58J35.

Keywords. Heat equation, the first boundary value problem, characteristic boundary points, cusps.

1. Preliminaries

Boundary value problems for parabolic equations in a bounded domain with smooth boundary fail to be regular in general, for there are characteristic points on the boundary. Petrovskii in his paper [22] found conditions on the behavior of the boundary nearby points of inflection with horizontal tangents which are necessary and sufficient for the first boundary value problem for the heat equation to be well posed. Most criteria of regularity beginning with that of [22] appeal to the contact degree at which outer characteristics meet the boundary. If the contact degree coincides with the order of equation, the analysis reveals many common features with analysis on manifolds with conical points. This situation was well understood in the 1960 s, see [13, 14, 20, 21, 27], etc. If the contact degree is different from the order of operator, the problem can be handled within the framework of degenerate elliptic equations.

In [13] one studies the first boundary value problem for a single second-order equation

$$Lu := - \sum_{\substack{i=1,\dots,n \\ j=1,\dots,n}} a^{i,j}(x) u''_{x^i,x^j} + \sum_{i=1}^{n} a^i(x) u'_{x^i} + a^0(x) u = f \tag{1.1}$$

with

$$\sigma^2(L)(x,\xi) := \sum_{\substack{i=1,\dots,n \\ j=1,\dots,n}} a^{i,j}(x) \xi_i \xi_j \geq 0$$

for all $\xi = (\xi_1, \dots, \xi_n)$ in $\mathbb{R}^n$. Here $u(x)$ is a real function defined in a bounded domain $\mathcal{G}$ in $\mathbb{R}^n$ with C^∞ boundary, and $x = (x^1, \dots, x^n)$ represents the coordinates. The coefficients are real and of class C^∞ in the closure of $\mathcal{G}$. The first boundary value problem consists in prescribing the values of u on a certain portion of the boundary $\partial\mathcal{G}$. One wishes to obtain unique solutions of the problem which are smooth up to and including the boundary. If the leading part is elliptic, i.e., $\sigma^2(L)(x,\xi) > 0$ for $\xi \neq 0$, we have the usual Dirichlet problem. Another well-known example of (1.1) is the heat equation $-u''_{x,x} + u'_t = 0$. For this classical equation, however, certain aspects of the first boundary value problem have never been adequately studied. It is customary to call operators L, with $\sigma^2(L)(x,\xi) \geq 0$, degenerate elliptic. The systematic study of the general class of such equations was initiated by Fichera [8] who established estimates in L^p norms, and proved the existence of generalized solutions. Oleynik [21] proved under certain conditions that "weak solutions are strong" and that solutions are actually smooth up to the boundary.

Following [8], the boundary is divided into three portions, on two of which the boundary values of u will be given. Let Σ_3 be the set of noncharacteristic boundary points, i.e., those where $\sigma^2(L)(x,\nu) > 0$, Σ_2 the set of characteristic boundary points where

$$\sum_{i=1}^{n} (a^i(x) + \operatorname{div} a^{i,\cdot}(x)) \, \nu_i > 0,$$

and $\Sigma_1 = \partial\mathcal{G} \setminus (\Sigma_2 \cup \Sigma_3)$. As usual, we use $\nu = (\nu_1, \dots, \nu_2)$ to denote the unit exterior normal at $\partial\mathcal{G}$. The first boundary problem is that of finding a solution of (1.1) which has given values on $\Sigma_2 \cup \Sigma_3$. After subtraction of a function with the same values, one may assume that the given boundary values on $\Sigma_2 \cup \Sigma_3$ are zero. Under certain conditions [13] establishes that this problem has a smooth solution in $\overline{\mathcal{G}}$. The proof of regularity in [13] is based on a global argument, which can not prove local regularity.

There are simple examples showing that, if Σ_1 touches Σ_2 or Σ_3, then the solution need not be smooth. On the other hand, there are interesting cases, such as the heat equation, in which they do touch and where, nevertheless, the solutions are smooth.

2. On the heat equation

Consider the heat equation for one space variable $u''_{x,x} - u'_t = 0$ in the plane domain (with C^∞ boundary except at the corners shown) of the type represented in Figure 1.

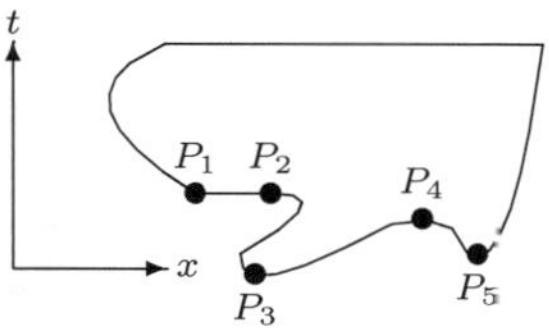

FIGURE 1. A general domain $\mathcal{G}$.

In the first boundary problem the value of the solution is prescribed on the whole boundary except for the top segment. There is an extensive literature devoted to this problem, however, most of the research treats only domains of the kind of horizontal strip limited from the left and from the right by disjoint smooth curves whose angular coefficients never vanish. At the lower corners the boundary values have to satisfy certain compatibility conditions. Concerning domains of the type of Figure 1, Levi in his paper [17] pointed out that the problem of the behavior of the solution at the characteristic points P_3, P_4, P_5, and the characteristic segment $[P_1, P_2]$ (all of which belong to Σ_2) is a very difficult one, and there has been little further study of this problem. Kohn and Nirenberg proved in [13] that, if the solution is smooth in the closure of the domain $\mathcal{G}$ of Figure 1, then the boundary values of u may have to satisfy compatibility conditions at the point P_4 depending on the value of the curvature of the boundary there. Furthermore, if the curvature is not zero, the solution need not be C^∞ there, but the smaller the curvature the smoother is the solution at that point. It is C^∞ if the curvature vanishes.

It is expected that a solution there might be non-smooth at $P_4 = (t_0, x_0)$, since for $t < t_0$ on the two sides of P_4 the solution is determined by different data, and there may not be matching of smoothness at P_4. At all other points, in particular the points P_3 and P_5, where the boundary curve is convex, the solution is C^∞. Kondrat'ev [14] studied boundary value problems for general parabolic equations in domains like Figure 1. In the case of second-order parabolic equations he had noted that at convex boundary points like P_3 and P_5, the smaller the curvature the smoother is the solution. Kohn and Nirenberg said in their paper [13], "He informed us in October 1966 that he could prove that for the heat equation the solution is C^∞ at convex boundary points $P_3 = (x_0, t_0)$ provided that at these points the boundary curve has the form $t - t_0 = c(x - x_0)^p$, where $c > 0$ and $p \geq 2$ is an integer." To our best knowledge, this result has not been published except for the case $p = 2$. In [13] the solution u is proved to be C^∞ at points like P_3 and P_5, where the boundary has positive curvature. This is proved for general second-order parabolic equations in n dimensions. The proof applies to a singular transformation of variables not unlike one used by Kondrat'ev in [14] which "blows up" the points P_3 and P_5.

Both [14] and [13] assume that the boundary is C^∞ in a neighborhood of the characteristic points under study. For the heat equation, the existence of a classical solution to the first boundary values problem in non-cylindrical domains was first obtained by Gevrey [10]. This result applies in particular to the plane domains $\mathcal{G}$ consisting of all $(x,t) \in \mathbb{R}^2$, such that $|x| < 1$ and $\mathfrak{f}(|x|) < t < \mathfrak{f}(1)$, where $\mathfrak{f}(r)$ is a C^1 function on $(0,1]$ with $\mathfrak{f}(r) > 0$, $\mathfrak{f}'(r) \neq 0$ for all $r \in (0,1]$ and $\mathfrak{f}(0+) = 0$. The boundary point $(0,0)$ is regular if $\mathfrak{f}^{-1}(t)$ satisfies the Hölder condition of exponent larger than $1/2$. When applied to the function $\mathfrak{f}(r) = r^p$, this implies $0 < p < 2$. In [3] a more intricate situation is treated when the domain $\mathcal{G}$ nearby the origin consists of those $(x,t) \in \mathbb{R}^2$ which satisfy $x > 0$ and $-ax^2 < t < bx^2$, where a and b are fixed positive numbers. The boundary of $\mathcal{G}$ is therefore not smooth and it has a cuspidal singularity at the origin which can actually be thought of as characteristic point.

For a recent account of the theory along more classical lines using the concept of (ir) regularity of a boundary point for a partial differential equation we refer the reader to [9].

The present paper is aimed to study the first boundary value problem for second-order parabolic equations in the case when the contact degree of outer characteristics with the boundary is less than the order of equation. The problem is shown to fit into analysis on compact manifolds with cuspidal points elaborated by V. Rabinovich et al. [24]. We restrict our discussion to the **1D** heat equation. We hope that the methods employed here may prove useful in treating more general systems.

3. Blow-up techniques

Consider the first boundary value problem for the heat equation in a domain $\mathcal{G} \subset \mathbb{R}^2$ of the type of Figure 2. The boundary of $\mathcal{G}$ is assumed to be C^∞ except for a finite number of characteristic points. At points like P_1 and P_2 the boundary curve possesses a tangent which is horizontal, hence $\partial\mathcal{G}$ is characteristic for the heat equation at such points. The characteristic touches the boundary with the degree ≥ 2, which is included in the treatise [14]. At points like P_2 the boundary curve is not smooth but it touches smoothly a characteristic from below and above. Such points are therefore cuspidal singularities of the boundary, implicit treatable cases have been studied in [3].

We restrict our discussion to the boundary points like P_3 and P_5. These are cuspidal singularities of the boundary curve which touches smoothly a vertical line at P_3 and P_5. Thus, the boundary meets a characteristic at P_3 and P_5 at contact

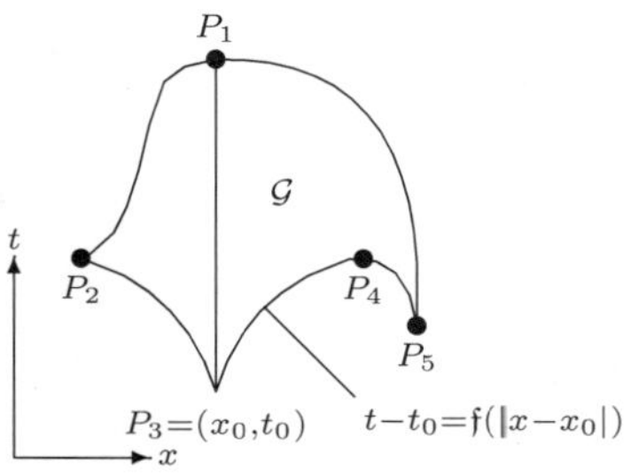

FIGURE 2. Typical domain.

degree < 2. As mentioned, the study of regularity of such points for solutions of the first boundary value problem for the heat equation goes back at least as far as [10]. While the approach of [10] is based on potential theory, we apply the so-called blow-up techniques developed in [24]. This allows one not only to get a regularity theorem in a sharper form including asymptotics of solutions but also to prove the Fredholm property in suitable weighted Sobolev spaces for more general cusps.

The first boundary value problem for the heat equation in $\mathcal{G}$ is formulated as follows: Write Σ_1 for the set of all characteristic points $P_1, P_2, \ldots$ on the boundary of $\mathcal{G}$. Given functions f in $\mathcal{G}$ and u_0 on $\partial\mathcal{G} \setminus \Sigma_1$, find a function u on $\overline{\mathcal{G}} \setminus \Sigma_1$ which satisfies

$$\begin{aligned} -u''_{x,x} + u'_t &= f \quad \text{in } \mathcal{G}, \\ u &= u_0 \quad \text{on } \partial\mathcal{G} \setminus \Sigma_1, \end{aligned} \tag{3.1}$$

cf. Section 1. By the local principle of Simonenko [25], the Fredholm property of problem (3.1) in suitable function spaces is equivalent to the local invertibility of this problem at each point of the closure of $\mathcal{G}$. Here we focus upon the points like P_3.

Assume that the domain $\mathcal{G}$ is described in a neighborhood of the point $P_3 = (x_0, t_0)$ by the inequality

$$t - t_0 > \mathfrak{f}(|x - x_0|), \tag{3.2}$$

where $\mathfrak{f}$ is a monotone increasing C^∞ function of $r \in (0, 1]$ with $\mathfrak{f}(0+) = 0$. We take P_3 to be the origin.

We now blow up the domain $\mathcal{G}$ at the point P_3 by introducing new coordinates (ω, r) with the aid of

$$\begin{aligned} x &= \mathfrak{f}^{-1}(r)\, \omega, \\ t &= r, \end{aligned} \tag{3.3}$$

where $|\omega| < 1$ and $r \in (0, \mathfrak{f}(1)]$. By definition, the new coordinates are singular at $r = 0$, for the entire segment $[-1, 1]$ on the ω-axis is blown down into the origin by (3.3). The rectangle $(-1, 1) \times (0, \mathfrak{f}(1)]$ transforms under the change of coordinates (3.3) into the part of the domain $\mathcal{G}$ nearby P_3 lying below the line $t = \mathfrak{f}(1)$.

In the domain of coordinates (ω, r) problem (3.1) reduces to an ordinary differential equation with respect to the variable r with operator-valued coefficients.

More precisely, under transformation (3.3) the derivatives in t and x change by the formulas

$$\begin{aligned}\frac{\partial u}{\partial t} &= \frac{\partial u}{\partial r} - \frac{1}{\mathfrak{f}^{-1}(r)\mathfrak{f}'(\mathfrak{f}^{-1}(r))}\,\omega\frac{\partial u}{\partial \omega},\\ \frac{\partial u}{\partial x} &= \frac{1}{\mathfrak{f}^{-1}(r)}\frac{\partial u}{\partial \omega},\end{aligned}$$

and so (3.1) transforms into

$$\begin{array}{rcll}(\mathfrak{f}^{-1}(r))^2 U'_r - U''_{\omega,\omega} - \dfrac{\mathfrak{f}^{-1}(r)}{\mathfrak{f}'(\mathfrak{f}^{-1}(r))}\,\omega U'_\omega &=& (\mathfrak{f}^{-1}(r))^2 F & \text{in } (-1,1)\times(0,\mathfrak{f}(1)),\\ U &=& U_0 & \text{on } \{\pm 1\}\times(0,\mathfrak{f}(1)],\end{array} \qquad (3.4)$$

where $U(\omega, r)$ and $F(\omega, r)$ are pullbacks of $u(x, t)$ and $f(x, t)$ under transformation (3.3), respectively.

We are interested in the local solvability of problem (3.4) near the edge $r = 0$ in the rectangle $(-1, 1) \times (0, \mathfrak{f}(1))$. Note that the ordinary differential equation degenerates at $r = 0$, since the coefficient $(\mathfrak{f}^{-1}(r))^2$ of the higher-order derivative in r vanishes at $r = 0$. In order to handle this degeneration in an orderly fashion, we find a change of coordinate $s = \delta(r)$ in an interval $(0, r_0]$ with some $r_0 < \mathfrak{f}(1)$, such that

$$(\mathfrak{f}^{-1}(r))^2\frac{d}{dr} = \frac{d}{ds}.$$

Such a function δ is determined uniquely up to a constant from the equation $\delta'(r) = (\mathfrak{f}^{-1}(r))^{-2}$ and is given by

$$\delta(r) = \delta(r_0) + \int_{r_0}^{r} \frac{d\vartheta}{(\mathfrak{f}^{-1}(\vartheta))^2} \qquad (3.5)$$

for $r \in (0, \mathfrak{f}(1)]$. The constant $\delta(r_0)$ is not essential, one can choose $\delta(r_0) = 0$. Problem (3.4) becomes

$$\begin{array}{rcll}U'_s - U''_{\omega,\omega} + \dfrac{d}{ds}\log\sqrt{\delta'(\delta^{-1}(s))}\,\omega U'_\omega &=& \dfrac{F}{\delta'(\delta^{-1}(s))} & \text{in } (-1,1)\times\delta(0,\mathfrak{f}(1)),\\ U &=& U_0 & \text{on } \{\pm 1\}\times\delta(0,\mathfrak{f}(1)],\end{array} \qquad (3.6)$$

where we use the same letter to designate U and the push-forward of U under the transformation $s = \delta(r)$, and $\delta(0, \mathfrak{f}(1)) = (\delta(0), \delta(\mathfrak{f}(1)))$.

Example. After [10], consider $\mathfrak{f}(r) = r^p$ with $p > 0$. Then

$$\delta(r) = \begin{cases} \delta(r_0) + \dfrac{p}{p-2}\left(r^{\frac{p-2}{p}} - r_0^{\frac{p-2}{p}}\right), & \text{if } p \neq 2,\\ \delta(r_0) + \log\dfrac{r}{r_0}, & \text{if } p = 2.\end{cases}$$

For $p > 2$, the value $\delta(0)$ is obviously finite. For $p \in (0, 2]$, we get $\delta(0+) = -\infty$. Choosing

$$\delta(r_0) = \begin{cases} \dfrac{p}{p-2} r_0^{\frac{p-2}{p}}, & \text{if} \quad p \neq 2, \\ \log r_0, & \text{if} \quad p = 2, \end{cases}$$

we arrive at the local boundary value problem

$$\begin{array}{rcll} U'_s - U''_{\omega,\omega} + \dfrac{1}{2-p}\dfrac{1}{s}\,\omega U'_\omega &=& \left(\dfrac{p-2}{p}\, s\right)^{\frac{2}{p-2}} F & \text{in } (-1,1) \times (\delta(0), p/(p-2)), \\ U &=& U_0 & \text{on } \{\pm 1\} \times (\delta(0), p/(p-2)], \end{array} \tag{3.7}$$

for $p \neq 2$, and

$$\begin{array}{rcll} U'_s - U''_{\omega,\omega} - \dfrac{1}{2}\,\omega U'_\omega &=& e^s F & \text{in } (-1,1) \times (-\infty, 0), \\ U &=& U_0 & \text{on } \{\pm 1\} \times (-\infty, 0], \end{array} \tag{3.8}$$

for $p = 2$.

This example demonstrates rather strikingly that the value $\delta(0)$ actually characterizes the threshold of weak singularities in the first boundary value problem for the heat equation. If $\delta(0+)$ is finite, one can certainly assume that $\delta(0+) = 0$, for if not, we take $r_0 = 0$ and choose $\delta(r_0) = 0$. Then $\delta^{-1}(0+) = 0$ and so $\delta'(\delta^{-1}(0+)) = +\infty$, i.e., the coefficient of U'_ω in (3.6) blows up at $\delta(0+)$. This manifests singularities of solutions at $s = 0$. On the other hand, if $\delta(0+) = -\infty$, then the coefficient of U'_ω in (3.6) need not blow up at $\delta(0+)$, as is seen from (3.7) and (3.8). In the case $\delta(0) = -\infty$ the boundary value problem (3.4) can be specified within the calculus of pseudodifferential operators with operator-valued symbols on the real axis $s \in \mathbb{R}$ developed by Rabinovich et al. in [24]. The symbols under considerations take their values in the space of boundary value problems on the interval $\omega \in [-1, 1]$.

For those pseudodifferential operators whose symbols are slowly varying at the point $s = -\infty$, the paper [24] gives a criterion of local solvability at $-\infty$. Note that this criterion does not apply directly to problem (3.6), for [24] deals with classical polyhomogeneous symbols while our problem requires quasihomogeneous symbols. However, the approach of [24] still works in the anisotropic case while the derivatives in s are counted with weight factor 2.

The symbol of problem (3.6) is slowly varying at $s = -\infty$ if and only if, for each $j = 1, 2, \ldots$, an inequality

$$\sup_{s \in (-\infty, \delta(r_0))} \left| \left(\frac{d}{ds}\right)^j \log \sqrt{\delta'(\delta^{-1}(s))} \right| < \infty \tag{3.9}$$

is valid. Inequalities (3.9) can be easily reformulated in terms of the original function $\mathfrak{f}$, namely,

$$\sup_{r \in (0, r_0)} \left| \left((\mathfrak{f}^{-1}(r))^2 \frac{d}{dr} \right)^j \log \mathfrak{f}^{-1}(r) \right| < \infty$$

for all $j = 1, 2, \ldots$. From the Hardy-Littlewood inequality it follows that, under these conditions, if moreover $j > 1$, then the left-hand side is arbitrarily small if r_0 is small enough.

4. Further reduction

From now on we will tacitly assume that the coefficients of (3.6) are slowly varying at $s = -\infty$, i.e., (3.9) is fulfilled.

Using transformations rather standard in Sturm-Liouville's theory we reduce problem (3.6) to a simpler form. Set

$$\begin{aligned} b(s) &= \frac{d}{ds} \log \sqrt{\delta'(\delta^{-1}(s))}, \\ \tilde{F} &= \frac{F}{\delta'(\delta^{-1}(s))}, \end{aligned}$$

then (3.6) rewrites as

$$\begin{aligned} U'_s - U''_{\omega,\omega} + b(s)\,\omega U'_\omega = \tilde{F} \quad &\text{in} \ \ (-1,1) \times \delta(0, \mathfrak{f}(1)), \\ U = U_0 \quad &\text{on} \ \ \{\pm 1\} \times \delta(0, \mathfrak{f}(1)]. \end{aligned}$$

Introduce

$$a(\omega, s) = \exp\Big(-\frac{1}{2}\omega^2 b(s)\Big)$$

which is a bounded C^∞ function with positive values on the closed cylinder $[-1, 1] \times \delta(0, \mathfrak{f}(1)]$. An easy computation shows that problem (3.6) transforms to

$$\begin{aligned} U'_s - \frac{1}{a}\,(aU'_\omega)_\omega = \tilde{F} \quad &\text{in} \ \ (-1,1) \times \delta(0, \mathfrak{f}(1)), \\ U = U_0 \quad &\text{on} \ \ \{\pm 1\} \times \delta(0, \mathfrak{f}(1)]. \end{aligned}$$

On replacing the unknown function by $U = \frac{1}{\sqrt{a}} v$ we finally arrive at the boundary value problem

$$\begin{aligned} v'_s - v''_{\omega,\omega} + cv = \sqrt{a}\,\tilde{F} \quad &\text{in} \ \ (-1,1) \times \delta(0, \mathfrak{f}(1)), \\ v = \sqrt{a} U_0 \quad &\text{on} \ \ \{\pm 1\} \times \delta(0, \mathfrak{f}(1)], \end{aligned} \tag{4.1}$$

where

$$c(\omega, s) = \frac{(\sqrt{a})''_{\omega,\omega} - (\sqrt{a})'_s}{\sqrt{a}},$$

cf. [6, v. I, p. 250].

Example. If $\mathfrak{f}(r) = r^p$, then

$$c(\omega, s) = \begin{cases} -\frac{1}{2}b(s) + \frac{1}{4}\frac{p-1}{(p-2)^2}\frac{\omega^2}{s^2}, & \text{if} \quad p \neq 2, \\ -\frac{1}{2}b(s) + \Big(\frac{1}{4}\omega\Big)^2, & \text{if} \quad p = 2. \end{cases}$$

In the general case we get

$$c(\omega, s) = -\frac{1}{2}b(s) + \frac{1}{4}\omega^2 \left((b(s))^2 + b'(s)\right) \tag{4.2}$$

where b is a C^∞ function slowly varying at $s = -\infty$. It follows that c inherits this behavior at $s = -\infty$ uniformly in $\omega \in [-1, 1]$.

Our approach to solving problem (4.1) is fairly standard in the theory of linear equations. On choosing a proper scale of weighted Sobolev spaces in the strip $\mathcal{C} = [-1, 1] \times \mathbb{R}$ and taking the data $v_0 = \sqrt{a}U_0$ in the corresponding trace spaces on the boundary $\omega = \pm 1$ of $\mathcal{C}$ we can assume without loss of generality that $v_0 \equiv 0$. We think of (4.1) as a perturbation of the problem with homogeneous boundary conditions

$$\begin{aligned} v'_s - v''_{\omega,\omega} &= \sqrt{a}\,\tilde{F} \quad \text{in } \mathcal{C}, \\ v &= 0 \quad\quad\;\; \text{on } \partial\mathcal{C}. \end{aligned} \tag{4.3}$$

This is exactly the first boundary value problem for the heat equation in the cylinder $\mathcal{C}$ which is nowadays well understood, cf. for instance Chapter 3 in [28]. If $g = \sqrt{a}\,\tilde{F}$ vanishes, problem (4.3) possesses infinitely many linearly independent solutions of the form

$$v_n(\omega, s) = c_n \exp\left(-\left(\frac{\pi}{2}n\right)^2 s\right) \sin\frac{\pi}{2}n(\omega + 1) \tag{4.4}$$

with n a natural number. In order to eliminate the solutions with n large enough it is necessary to pose growth restrictions on $v(\omega, s)$ for $s \to -\infty$. As but one possibility to do that we mention Sobolev spaces with exponential and powerlike weight factors, see [24]. Since the coefficients of the operator are stationary, the Fourier transform in s applies to reduce the problem to a Sturm-Liouville eigenvalue problem on the interval $(-1, 1)$, see Chapter 5 in [6, v. 1]. Instead of the Fourier transform one can use orthogonal decompositions over the eigenfunctions, which leads immediately to asymptotics of solutions of the unperturbed problem at $s = -\infty$.

On returning to problem (4.1) we observe that it differs from the unperturbed problem by the multiplication operator $v \mapsto cv$. If the unperturbed problem is Fredholm and the perturbation $v \mapsto cv$ is compact, then the perturbed problem is Fredholm as well. The local version of this assertion states that if the unperturbed problem is invertible and the perturbation $v \mapsto cv$ is small, then the perturbed problem is also invertible. Since, under our assumptions, $c(\omega, s) \to 0$ uniformly in $\omega \in [-1, 1]$, as $s \to -\infty$, the operator $v \mapsto cv$ is compact in natural scales of weighted Sobolev spaces.

5. The unperturbed problem

In this section we treat problem (4.3) in the infinite strip $\mathcal{C} = (-1, 1) \times \mathbb{R}$. We are interested in a solution of this problem in a half-strip $s \in (-\infty, S)$, where $S = \delta(\mathfrak{f}(1))$.

A solution can be found by the Fourier method of separation of variables, see for instance § 2 of Chapter 3 in [28]. We first look for a solution of the corresponding homogeneous problem of the form $v(\omega, s) = v_1(\omega)v_2(s)$, obtaining two eigenvalue problems for determining the functions $v_1(\omega)$ and $v_2(s)$. The first of the two looks like

$$\begin{aligned} v_1'' &= \lambda v_1 \quad \text{for } \omega \in (-1,1), \\ v_1(\pm 1) &= 0. \end{aligned} \tag{5.1}$$

It has a nonzero solution only for the values $\lambda_n = -\left(\frac{\pi}{2}n\right)^2$, where $n \in \mathbb{N}$. The solution is

$$v_{1,n}(\omega) = \sin\sqrt{-\lambda_n}(\omega+1) \tag{5.2}$$

up to a constant factor. Substituting $\lambda = \lambda_n$ into the equation for $v_2(s)$, we readily find $v_{2,n}(s) = \exp(\lambda_n s)$ up to a constant factor. We have thus constructed a sequence of solutions

$$v_n(\omega, s) = c_n \exp(\lambda_n s) \sin\sqrt{-\lambda_n}(\omega+1)$$

to the homogeneous problem (4.3), cf. (4.4). Note that each solution v_n is unbounded at $s = -\infty$.

This is a general property of Sturm-Liouville eigenvalue problems that system (5.2) is orthogonal and complete in $L^2(-1,1)$. Moreover, this system is orthonormal, as is easy to check.

Let now g be an arbitrary function on $\mathcal{C}$, such that $g(\cdot, s) \in L^2(-1,1)$ for each $s < S$. For any fixed $s < S$, we represent g as Fourier series over the orthonormal basis (5.2)

$$g(\omega, s) = \sum_{n=1}^{\infty} g_n(s) \sin\sqrt{-\lambda_n}(\omega+1),$$

where

$$g_n(s) = \int_{-1}^{1} g(\omega, s) \sin\sqrt{-\lambda_n}(\omega+1)\, d\omega.$$

We seek for a solution v of problem (4.3) in the form of Fourier series over the eigenfunctions of problem (5.1), i.e.,

$$v(\omega, s) = \sum_{n=1}^{\infty} v_n(s) \sin\sqrt{-\lambda_n}(\omega+1),$$

s being thought of as parameter. The function $v(\omega, s)$ satisfies the boundary conditions of (4.3), since all summands of the series satisfy them. Substituting the series into (4.3) yields

$$\sum_{n=1}^{\infty} \left(v_n'(s) - \lambda_n v_n(s) - g_n(s)\right) \sin\sqrt{-\lambda_n}(\omega+1) = 0$$

for all $\omega \in (-1,1)$. This equation is satisfied if and only if all the coefficients vanish, i.e.,

$$v_n'(s) - \lambda_n v_n(s) = g_n(s) \tag{5.3}$$

for $s < S$.

It is worth pointing out that no initial conditions for $v_n(s)$ are available, and so $v_n(s)$ is not determined uniquely. On solving this ordinary differential equation we get

$$v_n(s) = \int_{s_n}^{s} e^{\lambda_n (s-s')} g_n(s') ds', \tag{5.4}$$

where $s_n < S$ is an arbitrary constant. The change of s_n results in an additional multiple of $e^{\lambda_n s}$, for

$$\int_{s_n}^{s} e^{\lambda_n (s-s')} g_n(s') ds' - \int_{s_n+\Delta s_n}^{s} e^{\lambda_n (s-s')} g_n(s') ds' = c_n \, e^{\lambda_n s}$$

with $c_n = \displaystyle\int_{s_n}^{s_n+\Delta s_n} e^{-\lambda_n s'} g_n(s') ds'$.

We have thus proved

Lemma 5.1. *Suppose that g is an arbitrary function on the cylinder $\mathcal{C}$ satisfying $g(\cdot, s) \in L^2(-1,1)$ for all $s < S$. Then problem (4.3) has formal solution of the form*

$$v(\omega, s) = \sum_{n=1}^{\infty} \Big(\int_{s_n}^{s} e^{\lambda_n (s-s')} g_n(s') ds' \Big) \sin \sqrt{-\lambda_n} (\omega + 1).$$

If we pose the additional condition $v(\omega, s_0) = 0$ for some $s_0 < S$, then the functions $v_n(s)$ should fulfill the initial condition $v_n(s_0) = 0$. In this case v_n are uniquely determined by formulas (5.4) with $s_n = s_0$ for all $n \in \mathbb{N}$, which leads to the uniqueness of the formal solution. In our setting the elimination of all nontrivial solutions of the homogeneous problem except for a finite number is achieved by requiring the solution to belong to a scale of Sobolev spaces with exponential weight functions.

6. Asymptotic solutions

We can now return to the study of perturbed problem (4.1). The corresponding equation we write in the form

$$v_s' + C(s) v = g \tag{6.1}$$

where

$$C(s) = -\Big(\frac{d}{d\omega}\Big)^2 + c(\omega, s)$$

is a continuous function on $(-\infty, S)$ with values in second-order ordinary differential operators on $(-1,1)$. We think of $C(s)$ as unbounded operator in $L^2(-1,1)$ whose domain consists of all $v \in H^2(-1,1)$ satisfying $v(-1) = v(1) = 0$. As but one result of the theory of Sturm-Liouville boundary value problems we mention that $C(s)$ is closed.

As usual in the theory of ordinary differential equations with operator-valued coefficients, we associate the operator pencil $\mathfrak{s}(s,\sigma) = (\imath\sigma) - C(s)$ with (6.1). It depends on parameters $s \in (-\infty, S)$ and $\sigma \in \mathbb{C}$. Our basic assumption is that $\mathfrak{s}(s,\sigma)$ stabilizes to an operator pencil $\mathfrak{s}(-\infty,\sigma)$ independent of s, as $s \to -\infty$. This just amounts to saying that the coefficient $c(\omega, s)$ extends continuously to $s = -\infty$. We tacitly assume that $c(-\infty,\omega) \equiv 0$, for we are interested in true cusps, see Figure 2.

Lemma 6.1. *Let $k \geq 1$ be integer. When acting from $H^{2k}(-1,1) \cap \overset{\circ}{H}{}^1(-1,1)$ to $H^{2(k-1)}(-1,1)$, the operator $\mathfrak{s}(-\infty,0) = C(-\infty)$ is invertible.*

Proof. See Section 5. □

Moreover, $\mathfrak{s}(-\infty,\sigma)$ acting from $H^{2k}(-1,1) \cap \overset{\circ}{H}(-1,1)$ to $H^{2(k-1)}(-1,1)$ has a bounded inverse everywhere in the entire complex plane $\mathbb{C}$ except for the discrete set

$$\sigma_n = -\imath\lambda_n = \imath\Big(\frac{\pi}{2}n\Big)^2$$

with $n \in \mathbb{N}$. It is worth pointing out that $\mathfrak{s}(-\infty,\sigma)^{-1} = \mathcal{R}_{C(-\infty)}(-\imath\sigma)$, the resolvent of $C(-\infty)$ at $-\imath\sigma$.

Lemma 6.2. *There exists a constant c with the property that, for all complex σ lying away from any angular sector containing the positive imaginary axis, the inequality*

$$\begin{aligned} &\|v\|^2_{H^{2k}(-1,1)} + |\sigma|^{2k}\|v\|^2_{L^2(-1,1)} \\ &\leq c\Big(\|\mathfrak{s}(-\infty,\sigma)v\|^2_{H^{2(k-1)}(-1,1)} + |\sigma|^{2(k-1)}\|\mathfrak{s}(-\infty,\sigma)v\|^2_{L^2(-1,1)}\Big) \end{aligned}$$

is fulfilled whenever $v \in H^{2k}(-1,1) \cap \overset{\circ}{H}{}^1(-1,1)$ with $k \geq 1$.

Proof. The operator pencils $\mathfrak{s}(-\infty,\sigma)$ with this property are said to be anisotropic elliptic. See [2] for a more general estimate. □

If $\mathfrak{s}(s,\lambda)$ stabilizes at $s = -\infty$ then the singularity at $s = -\infty$ gives rise to a finite number of singular solutions. However, an irregular singular point is a complicated conglomeration of singularities, which does not allow one to construct explicit asymptotic formulas.

By a solution of (6.1) is meant any function $v(s)$ with values in $H^2(-1,1)$ satisfying $v(-1) = v(1) = 0$, which has a strong derivative in $L^2(-1,1)$ for almost all $s < S$, and which fulfills (6.1).

Lemma 5.1 suggests readily a scale of Hilbert spaces to control the solutions. For any $k = 0,1,\dots$ and $\gamma \in \mathbb{R}$, we introduce $H^{k,\gamma}(-\infty,S)$ to be the space of all functions on $(-\infty, S)$ with values in $H^{2k}(-1,1)$, such that the norm

$$\|v\|_{H^{k,\gamma}(-\infty,S)} := \Big(\int_{-\infty}^{S} e^{-2\gamma s} \sum_{j=0}^{k} \|v^{(j)}(s)\|^2_{H^{2(k-j)}(-1,1)} ds\Big)^{1/2}$$

is finite, cf. Slobodetskii [26]. In particular, $H^{0,\gamma}(-\infty, S)$ consists of all square integrable functions on $(-\infty, S)$ with values in $L^2(-1,1)$ with respect to the measure $e^{-2\gamma s}ds$.

Recall that the numbers σ_n are called eigenvalues of the operator pencil $\mathfrak{s}(-\infty,\sigma)$, for there are nonzero functions $\varphi_n = v_{1,n}$ in $H^2(-1,1)$ vanishing at ± 1 and satisfying $\mathfrak{s}(-\infty,\sigma_n)\varphi_n = 0$. The functions φ_n are called eigenfunctions of $\mathfrak{s}(-\infty,\sigma)$ at σ_n.

We now bring three theorems on asymptotic behavior of solutions of homogeneous problem (6.1) as $s \to -\infty$. They fit well the abstract theory of [19]. However, [19] is a straightforward generalisation of the asymptotic formula of Evgrafov [7] for solutions of first-order equations to equations of an arbitrary order. Our results go thus back at least as far as [7] while we refer to the more available paper [19].

Theorem 6.3. *Let $c(\omega, s) \to c(\omega, -\infty)$ in the $L^2(-1,1)$-norm when $s \to -\infty$. Suppose that in the strip $-\mu < \Im\sigma < -\gamma$ there lie exactly N of the eigenvalues σ_n, and that there are no eigenvalues σ_n on the lines $\Im\sigma = -\mu$ and $\Im\sigma = -\gamma$. Then the solution $v \in H^{1,\gamma}(-\infty, S)$ of problem (6.1) with $g \in H^{0,\mu}(-\infty, S)$ has the form*

$$v(s) = c_1\, s_1(s) + \dots + c_N\, s_N(s) + R(s)$$

where $s_1, \dots, s_N$ are solutions of the homogeneous problem in $H^{1,\gamma}(-\infty, S)$ which do not depend on v, $c_1, \dots, c_N$ constants, and $R \in H^{1,\mu}(-\infty, S)$.

Proof. An easy computation using the continuous embedding

$$H^1(-1,1) \hookrightarrow C[-1,1]$$

shows that from the convergence of $c(s,\cdot)$ to $c(-\infty,\cdot)$ in the $L^2(-1,1)$-norm it follows that $C(s) \to C(-\infty)$ in the operator norm of $\mathcal{L}(H^2(-1,1), L^2(-1,1))$, as $s \to -\infty$. Hence the desired conclusion is a direct consequence of Theorem 3 in [19] with

$$H_0 = L^2(-1,1),$$
$$H_1 = H^2(-1,1) \cap \overset{\circ}{H}{}^1(-1,1).$$ □

Thus, any solution $v \in H^{1,\gamma}(-\infty, S)$ of (6.1) with a "good" right-hand side g can be written as the sum of several singular functions and a "remainder" which behaves better at infinity. The singular functions $s_1, \dots, s_N$ are linearly independent and do not depend on the particular solution v. What is still lacking is that they are not explicit.

The concept of stabilization we have so far used falls outside the framework of "small perturbations." To meet this heuristic concept, we need some further restrictions on the speed at which $C(s)$ tends to $C(-\infty)$ when $s \to -\infty$. Let

σ_n be a fixed eigenvalue of the limit pencil $\mathfrak{s}(-\infty,\sigma)$. Assume moreover that the pencil $\mathfrak{s}(s,\sigma)$ stabilizes to $\mathfrak{s}(-\infty,\sigma)$ as $s \to -\infty$. Since σ_n is a simple eigenvalue of $\mathfrak{s}(-\infty,\sigma)$, for s sufficiently large there exists a simple eigenvalue $\sigma_n(s)$ of the pencil $\mathfrak{s}(s,\sigma)$ which tends to σ_n as $s \to -\infty$. We write $\varphi_n(s)$ for the corresponding eigenfunction with $\|\varphi_n(s)\|_{L^2(-1,1)} = 1$.

Theorem 6.4. *Suppose*

$$\int_{-\infty}^{s_0} s^2 \, \|c'(\cdot,s)\|^2_{L^2(-1,1)} ds < \infty$$

for some (and so for all) $s_0 \le S$. Let $v(s)$ be a solution of homogeneous equation (6.1) for $s < S$, such that $v \in H^{1,\gamma}(-\infty,S)$ with $\lambda_{n+1} < \gamma < \lambda_n$. Then,

$$v(s) = e^{-\imath \int_s^S \sigma_n(s')ds'} \big(c\,\varphi_n(s) + R(s)\big), \tag{6.2}$$

where c is a constant and $R \in H^{1,0}(-\infty,S)$.

Proof. By Theorem 1 of [19], it suffices to verify if, under the assumption of Theorem 6.4, the integral

$$\int_{-\infty}^{s_0} s^2 \, \|C'(s)\|^2_{\mathcal{L}(H_1,H_0)} \, ds$$

is finite, where H_0 and H_1 are the same spaces as in the proof of Theorem 6.3. To this end we pick any $v \in H^2(-1,1)$. The Sobolev embedding theorem implies that v is actually continuous on the interval $[-1,1]$ and the $C[-1,1]$-norm of v is dominated by $C\,\|v\|_{H^1(-1,1)}$ with C a constant independent of v. By Hölder's inequality,

$$\begin{aligned} \|C'(s)v\|_{H_0} &= \Big(\int_{-1}^{1} |c'(\omega,s)v(\omega)|^2 \, d\omega\Big)^{1/2} \\ &\le \|c'(\cdot,s)\|_{L^2(-1,1)} \, \|v\|_{C[-1,1]}, \end{aligned}$$

and so

$$\|C'(s)v\|_{H_0} \le C \, \|c'(\cdot,s)\|_{L^2(-1,1)} \, \|v\|_{H_1}.$$

Hence it follows that $\|C'(s)\|_{\mathcal{L}(H_1,H_0)} \le C\,\|c'(\cdot,s)\|_{L^2(-1,1)}$, establishing the desired estimate. □

Were $\sigma_n(s)$ independent of s, we would deduce under the assumptions of Theorem 6.4 that

$$\begin{aligned} e^{-\imath \int_s^S \sigma_n(s')ds'} R(s) &= e^{-\lambda_n(S-s)} R(s) \\ &\in H^{1,\lambda_n}(-\infty,S) \end{aligned}$$

which belongs to $H^{1,\gamma}(-\infty,S)$. Hence, the remainder in formula (6.2) behaves better than $v(s)$ itself, as $s \to -\infty$, showing the asymptotic character of this formula.

If the coefficient $c(\omega, s)$ bears a transparent structure close to the point at (minus) infinity, then the asymptotic behavior of solutions can be described more precisely. Suppose

$$c(\omega, s) = \sum_{j=0}^{J} c_j(\omega)\, \frac{1}{s^j} + c_{J+1}(\omega, s)\, \frac{1}{s^{J+1}} \tag{6.3}$$

on the interval $(-\infty, S)$, where c_j are smooth functions on $[-1, 1]$ for $j \leq J$, and c_{J+1} a smooth function on $[-1, 1] \times (-\infty, S)$ satisfying

$$\begin{aligned} \|c_{J+1}(\cdot, s)\|_{L^2(-1,1)} &\leq C, \\ \|c'_{J+1}(\cdot, s)\|_{L^2(-1,1)} &\leq \frac{C}{s} \end{aligned}$$

for $s \to -\infty$.

Theorem 6.5. *Under the above assumptions, any solution $v(s)$ of homogeneous problem (6.1) which belongs to the space $H^{1,\gamma}(-\infty, S)$ with $\lambda_{n+1} < \gamma < \lambda_n$, has the form*

$$v(s) = s^{\imath\sigma_0}\, e^{\lambda_n s} \Big(c\, \sin \sqrt{-\lambda_n} (\omega + 1) + c \sum_{j=1}^{J-1} \psi_j(\omega)\, \frac{1}{s^j} + R(s)\, \frac{1}{s^J} \Big)$$

where c is a constant depending on the solution $v(s)$, the constant σ_0 and the functions $\psi_j \in H^2(-1, 1)$ vanishing at ± 1 do not depend on the solution, and $R \in H^{1,0}(-\infty, S)$.

Proof. This follows from Theorem 2 of [19] if one takes into account the computations of Section 5. □

The constant σ_0 and the functions ψ_j are computed by means of a finite number of algebraic operations.

7. Local solvability at a cusp

Changing the coordinates by

$$\begin{aligned} \omega &= \frac{x}{\mathfrak{f}^{-1}(t)}, \\ s &= \delta(t), \end{aligned}$$

we return to the coordinates (x, t) in the domain $\mathcal{G}$ close to the boundary point $P_3 = (0, 0)$, see Figure 2. Then Theorems 6.3 and 6.4 are traced back to solutions of the heat equation $u'_t - u''_{x,x} = f$ with zero Dirichlet data near the cuspidal point in $\mathcal{G}$. We get

$$\begin{aligned} v(\omega, s) &= \sqrt{a(\omega, s)}\, u(x, t), \\ g(\omega, s) &= \sqrt{a(\omega, s)}\, (\mathfrak{f}^{-1}(t))^2 f(x, t), \end{aligned}$$

where $a(\omega, s) = \exp \dfrac{1}{4} \dfrac{x^2}{\mathfrak{f}^{-1}(t) \mathfrak{f}'(\mathfrak{f}^{-1}(t))}$.

Let $H^{k,\gamma}(0,T)$ consist of all functions $u(t,x)$ defined for $0 < t < T = \mathfrak{f}(1)$ and $|x| < \mathfrak{f}^{-1}(t)$, such that $\sqrt{a}\, u((\delta \circ \mathfrak{f})^{-1}(s)\omega, \delta^{-1}(s))$ belongs to $H^{k,\gamma}(-\infty, S)$. We endow $H^{k,\gamma}(0,T)$ with a norm in an obvious way. This scale of Hilbert spaces fits well to control the solutions of the heat equation near the singular point P_3 in the domain $\mathcal{G}$.

Since

$$\begin{aligned} \frac{\partial}{\partial s} &= (\mathfrak{f}^{-1}(t))^2 \frac{\partial}{\partial t} + \frac{\mathfrak{f}^{-1}(t)}{\mathfrak{f}'(\mathfrak{f}^{-1}(t))}\, x \frac{\partial}{\partial x}, \\ \frac{\partial}{\partial \omega} &= \mathfrak{f}^{-1}(t) \frac{\partial}{\partial x}, \end{aligned}$$

the norm in $H^{k,\gamma}(0,T)$ under natural assumptions on $\mathfrak{f}$ proves to be equivalent to the norm

$$\begin{aligned} &\|u\|_{H^{k,\gamma}(0,T)} \\ &:= \Big(\iint_{\mathcal{G}_0} e^{-2\gamma\delta(t)} \sum_{2j+|\alpha|\le 2k} |((\mathfrak{f}^{-1}(t))^2 \partial_t)^j (\mathfrak{f}^{-1}(t)\partial_x)^\alpha \left(\sqrt{a}u\right)|^2 \frac{dxdt}{(\mathfrak{f}^{-1}(t))^3} \Big)^{1/2}, \end{aligned}$$

where $\mathcal{G}_0$ is the part of $\mathcal{G}$ nearby P_3 lying below the line $t = \mathfrak{f}(1)$.

Theorem 7.1. *Let $c(\omega, s) \to c(\omega, -\infty)$ in the $L^2(-1,1)$-norm when $s \to -\infty$. Suppose in the strip $-\mu < \Im\sigma < -\gamma$ there lie exactly N of the eigenvalues σ_n and there are no eigenvalues σ_n on the lines $\Im\sigma = -\mu$ and $\Im\sigma = -\gamma$. Then the solution $u \in H^{1,\gamma}(0,T)$ of problem (3.1) with $(\mathfrak{f}^{-1}(t))^2 f \in H^{0,\mu}(0,T)$ has the form*

$$u(t) = c_1\, u_1(t) + \ldots + c_N\, u_N(t) + R(t)$$

where $u_1, \ldots, u_N$ are linearly independent solutions of the homogeneous problem in $H^{1,\gamma}(0,T)$ which do not depend on u, $c_1, \ldots, c_N$ constants, and $R \in H^{1,\mu}(0,T)$.

Proof. This follows from Theorem 6.3 with

$$u_j(x,t) = s_j \Big(\frac{x}{\mathfrak{f}^{-1}(t)}, \delta(t) \Big)$$

for $j = 1, \ldots, N$. □

Theorem 7.1 shows that any solution $u \in H^{1,\gamma}(0,T)$ of (3.1) with a "good" right-hand side f can be written as the sum of several singular functions and a "remainder" which behaves better at the cuspidal point P_3. The singular functions $u_1, \ldots, u_N$ prove to be independent of the particular solution u. Unfortunately, they are not explicit.

Let $\sigma_n = -\imath\lambda_n$ be a fixed eigenvalue of the limit pencil $\mathfrak{s}(-\infty, \sigma)$. Suppose the pencil $\mathfrak{s}(s,\sigma)$ stabilizes to $\mathfrak{s}(-\infty,\sigma)$ as $s \to -\infty$. For s sufficiently large there exists a simple eigenvalue $\sigma_n(s)$ of the pencil $\mathfrak{s}(s,\sigma)$ which tends to σ_n as $s \to -\infty$. We write $\varphi_n(s)$ for the corresponding eigenfunction normalized by $\|\varphi_n(s)\|_{L^2(-1,1)} = 1$.

Theorem 7.2. *Suppose*

$$\int_{-\infty}^{s_0} s^2 \, \|c'(\cdot, s)\|^2_{L^2(-1,1)} ds < \infty$$

for some (and so for all) $s_0 \le S$. Let $u(t)$ be a solution of homogeneous equation (3.1) on the interval $(0,T)$, such that $u \in H^{1,\gamma}(0,T)$ with $\lambda_{n+1} < \gamma < \lambda_n$. Then,

$$u(x,t) = e^{-\imath \int_{\delta(t)}^{\delta(T)} \sigma_n(s')ds'} \Big(c \, \frac{1}{\sqrt{a}} \, \varphi_n \Big(\frac{x}{\mathfrak{f}^{-1}(t)}, \delta(t) \Big) + R(t) \Big) \qquad (7.1)$$

where c is a constant and $R \in H^{1,0}(0,T)$.

Proof. For the proof it suffices to apply formula (6.2) and pass to the coordinates (x,t). □

We now look for restrictions on the geometry of the singular point P_3 under which Theorem 7.2 is applicable. To this end, let $\mathfrak{f}(r) = r^p$ close to $r=0$, where $p > 0$. Then,

$$c(\omega, s) = \frac{1}{2} \frac{1}{p-2} \frac{1}{s} + \frac{1}{4} \frac{p-1}{(p-2)^2} \frac{\omega^2}{s^2},$$

$$c(\omega, s) = \frac{1}{4} + \Big(\frac{1}{4} \omega \Big)^2$$

for $p \neq 2$ and $p = 2$, respectively. Hence, the stabilization condition of Theorem 7.2 is fulfilled for all p.

We finish the paper with local solvability of the Dirichlet problem for the heat equation nearby the boundary point P_3 in $\mathcal{G}$. By the local solvability at P_3 is meant that there is a disk B of small radius around P_3, such that for each f in $\mathcal{G}$ with $(\mathfrak{f}^{-1}(t))^2 f \in H^{0,\gamma}(0,T)$ there is a function $u \in H^{1,\gamma}(0,T)$ satisfying $u'_t - u''_{x,x} = f$ in $\mathcal{G} \cap B$ and $u = 0$ on $\partial\mathcal{G} \cap B$. Yet another designation for the local solvability is the local invertibility from the right at P_3. For a deeper discussion of local invertibility we refer the reader to [24]. Recall that local solvability at each point of $\overline{\mathcal{G}}$ is equivalent to the Fredholm property, which is due to the local principle of [25].

Theorem 7.3. *Suppose that $\gamma \in \mathbb{R}$ is different from λ_n for all $n = 1, 2, \ldots$. Then the Dirichlet problem for the heat equation is locally solvable at the cuspidal point P_3.*

Proof. As mentioned above, condition (3.9) just amounts to saying that our problems fits into the framework of analysis of pseudodifferential operators with slowly varying symbols. Hence, the desired result follows in much the same way as Corollary 23.2 of [24]. □

If $u', u'' \in H^{1,\gamma}(0,T)$ are two solutions to the Dirichlet problem in $\mathcal{G} \cap B$, then their difference $u = u' - u''$ belongs to the space $H^{1,\gamma}(0,T)$ and satisfies the Dirichlet problem with right-hand side f being zero. By Theorem 7.2, u has the form $u = c_1 u_1 + \cdots + c_N u_N + R$, where N is the greatest number with

$\lambda_N > \gamma$, $u_1, \dots, u_N$ are linearly independent solutions of the homogeneous Dirichlet problem in $H^{1,\gamma}(0,T)$, and R a solution of the homogeneous Dirichlet problem in $H^{1,\infty}(0,T)$. The regularity theory of [24] gives even more, namely that $R \in H^{k,\infty}(0,T)$ for all $k \in \mathbb{N}$.

In particular, if $\gamma > \lambda_1 := -(\pi/2)^2$, then the solution u of the Dirichlet problem nearby P_3 is determined uniquely up to a solution of the homogeneous Dirichlet problem which belongs to $H^{k,\infty}(0,T)$ for each $k = 1, 2, \dots$. For $f = 0$, the solution u itself belongs to $H^{k,\infty}(0,T)$ for all $k = 1, 2, \dots$. Hence it follows that $u(0, 0+) = 0$, i.e., the boundary point P_3 is regular in Wiener's sense, see [29]. This viewpoint sheds very surprisingly some new light on the connection between regularity criteria of boundary points in Dirichlet problems and the concept of differential operators with slowly varying coefficients. For a thorough treatment we refer the reader to [9].

Acknowledgment

The first author acknowledges financial support of the DFG (German Research Society), grant TA 289/4-1.

References

[1] S. Agmon and L. Nirenberg, *Properties of solutions of ordinary differential equations in Banach spaces*, Communs. Pure Appl. Math. **16** (1963), 121–239.

[2] M.S. Agranovich and M.I. Vishik, *Elliptic problems with a parameter and parabolic problems of general form*, Russ. Math. Surv. **19** (1964), No. 3, 53–157.

[3] V.N. Aref'ev and L.A. Bagirov, *Asymptotic behavior of solutions of the Dirichlet problem for a parabolic equation in domains with singularities*, Mat. Zam. **59** (1996), No. 1, 12–23.

[4] V.N. Aref'ev and L.A. Bagirov, *On solutions of the heat equation in domains with singularities*, Mat. Zam. **64** (1998), No. 2, 163–179.

[5] Th. Buchholz and B.-W. Schulze, *Anisotropic edge pseudodifferential operators with discrete asymptotics*, Math. Nachr. **184** (1997), 73–125.

[6] R. Courant and D. Hilbert, *Methoden der Mathematischen Physik*, B. I u. II, 3. Auflage, Springer-Verlag, Berlin et al., 1968.

[7] M.A. Evgrafov, *Structure of solutions of exponential growth for some operator equations*, Trudy Mat. Inst. im. V.A. Steklova **60** (1961), 145–180.

[8] G. Fichera, *Sulle equazioni differenziali lineari ellittico-paraboliche del secondo ordine*, Atti Acc. Naz. Lincei Mem., Ser. 8 **5** (1956), 1–30.

[9] V.A. Galaktionov, *On regularity of a boundary point for higher-order parabolic equations: Towards Petrovskii-type criterion by blow-up approach*, arXiv: 0901.3986vI [math.AP] 26 Jan. 2009, 50 pp.

[10] M. Gevrey, *Sur les équations partielles du type parabolique*, J. math. pure appl. Paris **9** (1913), 305–471; **10** (1914), 105–148.

[11] I. Gokhberg and E. Sigal, *An operator generalisation of the logarithmic residue theorem and the theorem of Rouché*, Math. USSR Sbornik **13** (1971), No. 4, 603–625.

[12] S.D. Ivasishen, *Estimates of Green function of the first boundary value problem for a second order parabolic equation in a non-cylindrical domain*, Ukr. Mat. Zh. **21** (1969), No. 1, 15–27.

[13] J.J. Kohn and L. Nirenberg, *Degenerate elliptic-parabolic equations of second order*, Comm. Pure and Appl. Math. **20** (1967), No. 4, 797–872.

[14] V.A. Kondrat'ev, *Boundary problems for parabolic equations in closed domains*, Trans. Moscow Math. Soc. **15** (1966), 400–451.

[15] Th. Krainer, *On the inverse of parabolic boundary value problems for large times*, preprint 12, Institut für Mathematik, Universität Potsdam, 2002, 67 S.

[16] O.A. Ladyzhenskaya, V.A. Solonnikov and N.N. Uraltseva, *Linear and Quasilinear Equations of Parabolic Type*, Translations of Mathematical Monographs, V. 23, Amer. Math. Soc., Providence, Rhode Island, 1968.

[17] E.E. Levi, *Sull'equazione del calore*, Annali di Mat. **14** (1907), 187–264.

[18] J.L. Lions and E. Magenes, *Problèmes aux limites non homogènes et applications*, V. 1–3, Dunod, Paris, 1968–1970.

[19] V.G. Maz'ya and V.A. Plamenevskii, *On the asymptotic behavior of solutions of differential equations in Hilbert space*, Math. USSR Izvestiya **6** (1972), no. 5, 1067–1116.

[20] V.P. Mikhaylov, *On the Dirichlet problem for a parabolic equation. I, II*, Mat. Sb. **61 (103)** (1963), 40–64; **62 (104)** (1963), 140–159.

[21] O.A. Oleynik, *On linear second order equations with non-negative characteristic form*, Mat. Sb. **69** (1966), 111–140.

[22] I.G. Petrovskii, *Zur ersten Randwertaufgabe der Wärmeleitungsgleichung*, Compositio Math. **1** (1935), 389–419.

[23] A. Piriou, *Problèmes aux limites généraux pour des opérateurs différentiels paraboliques dans un domaine borné*, Ann. Inst. Fourier (Grenoble) **21** (1971), No. 1, 59–78.

[24] V. Rabinovich, B.-W. Schulze and N. Tarkhanov, *A calculus of boundary value problems in domains with non-Lipschitz singular points*. Math. Nachr. **215** (2000), 115–160.

[25] I.B. Simonenko, *A new general method for investigating linear operator equations of the singular integral operator type. I, II*, Izv. Akad. Nauk SSSR, Ser. Mat. **29** (1965), 567–586; 757–782.

[26] L.N. Slobodetskii, *The generalized spaces of S.L. Sobolev and their application to boundary value problems for partial differential equations*, Uch. Zapiski Leningr. Ped. Inst. im. A.I. Gertsena **197** (1958), 54–112.

[27] V.A. Solonnikov, *Boundary value problems for linear parabolic systems of differential equations in general form*, Proc. Steklov Inst. Math. **83** (1965), 3–162.

[28] A.N. Tikhonov and A.A. Samarskii, *Equations of Mathematical Physics*, 4th Edition, Nauka, Moscow, 1972.

[29] N. Wiener, *The Dirichlet problem*, J. Math. and Phys. Mass. Inst. Tech. **3** (1924), 127–146.

Alexandra Antoniouk
Department Nonlinear Analysis
Institute of Mathematics NAS
Tereschchenkivska str. 3
01 601 Kiev, Ukraine
e-mail: antoniouk.a@gmail.com

Nikolai Tarkhanov
Institut für Mathematik
Universität Potsdam
Am Neuen Palais 10
D-14469 Potsdam, Germany
e-mail: tarkhanov@math.uni-potsdam.de

Operator Theory:
Advances and Applications, Vol. 228, 21–38

Globally Bisingular Elliptic Operators

U. Battisti, T. Gramchev, L. Rodino and S. Pilipović

Abstract. The main goal of this work is to extend the notion of bisingular pseudo-differential operators, already introduced on compact manifolds, to Shubin type operators on $\mathbb{R}^n = \mathbb{R}^{n_1} \oplus \mathbb{R}^{n_2}$, $n_1 + n_2 = n$. First, we prove global calculus (an analogue of the Γ calculus in the work of Shubin) for such operators, we introduce the notion of bisingular globally elliptic operators and we derive estimates for the action in anisotropic weighted Sobolev spaces, recently introduced by Gramchev, Pilipović, Rodino. Next, we investigate the complex powers of such operators and we demonstrate a Weyl type theorem for the spectral counting function of positive self-adjoint unbounded bisingular globally elliptic operators. The crucial ingredient for the proof is the use of the spectral zeta function. For particular classes of operators, defined as polynomials of $P_1 \times P_2$, $P_1 \times I_{\mathbb{R}^{n_2}}$, $I_{\mathbb{R}^{n_1}} \times P_2$, P_j being globally elliptic in $\mathbb{R}^{n_j}$, $j = 1, 2$, we are able to estimate and, in some cases, calculate explicitly the lower-order term in the asymptotic expansion of the spectral function.

Mathematics Subject Classification (2010). Primary 35S05.

Keywords. Pseudo-differential operators, counting function, ζ-function, bisingular operators.

1. Introduction

Let us recall the expression of a Shubin type differential operators with polynomial coefficients in $\mathbb{R}^n$ ([36], see also [4, 18, 28]):

$$P = \sum_{|\alpha|+|\beta|\leq m} c_{\alpha\beta} x^\beta D_x^\alpha, \quad D^\alpha = (-i)^{|\alpha|} \partial_x^\alpha. \tag{1.1}$$

We assume that P is an L^2- self-adjoint operator and satisfies the global ellipticity condition

$$p_m(x,\xi) = \sum_{|\alpha|+|\beta|=m} c_{\alpha\beta} x^\beta \xi^\alpha \neq 0 \quad \text{for} \quad (x,\xi) \neq (0,0). \tag{1.2}$$

This guarantees the existence of a basis of orthonormal eigenfunctions u_j, $j \in \mathbb{N}$, with eigenvalues λ_j, $\lim_{j\to\infty} |\lambda_j| = +\infty$, see Shubin [36] for the asymptotics of the counting function. If $u \in L^2(\mathbb{R}^n)$, or $u \in \mathcal{S}'(\mathbb{R}^n)$, then

$$u = \sum_{j=1}^{\infty} a_j u_j, \quad a_j = (u, u_j)_{L^2(\mathbb{R}^n)}, \quad j = 1, 2, \dots, \tag{1.3}$$

with convergence in $L^2(\mathbb{R}^n)$ or $\mathcal{S}'(\mathbb{R}^n)$.

Let us address to [5–7, 10, 13, 22] for further information on the regularity of the eigenfunctions. Basic examples of operators in the class considered in this paper are tensor products of Shubin operators. Namely, let $P(x, D)$ be a linear partial differential operator with polynomial coefficients of the form

$$\begin{aligned} P(x, D) &= P_1(x_1, D_{x_1}) P_2(x_2, D_{x_2}) \\ &= \left(\sum_{|\alpha|+|\beta|\le m_1} c^1_{\alpha\beta} x_1^\beta D^\alpha_{x_1} \right) \left(\sum_{|\alpha|+|\beta|\le m_2} c^2_{\alpha\beta} x_2^\beta D^\alpha_{x_2} \right), \end{aligned} \tag{1.4}$$

$x_1 \in \mathbb{R}^{n_1}, x_2 \in \mathbb{R}^{n_2}$, so that P_1 and P_2 are self-adjoint, invertible and globally elliptic on $\mathbb{R}^{n_1}$ and $\mathbb{R}^{n_2}$, that is (1.2) holds for both operators. Spectrum and eigenfunctions of P are easily detected from those of P_1, P_2, if we note that

$$u_{(j_1,j_2)}(x_1, x_2) = u^1_{j_1}(x_1) u^2_{j_2}(x_2), \quad j_1, j_2 \in \mathbb{N},$$

is an orthonormal basis of $L^2(\mathbb{R}^{n_1+n_2})$, and

$$P u_{j_1,j_2} = \lambda^1_{j_1} \lambda^2_{j_2} u_{j_1,j_2}, \quad j_1, j_2 \in \mathbb{N}.$$

The study of the counting function is interesting, and challenging. In Section 2 we shall embed example (1.4) into a general pseudo-differential calculus, including also the case when $p_i(x, D) \in G^{m_i}(\mathbb{R}^{n_i})$ with symbol $p_i(x_i, \xi_i)$ in the classes of Shubin

$$|\partial^{\beta_i}_{x_i} \partial^{\alpha_i}_{\xi_i} p(x_i, \xi_i)| \le C \langle x_i, \xi_i \rangle^{m_i - |\alpha_i| - |\beta_i|}, \quad \langle x_i, \xi_i \rangle = (1 + |x_i|^2 + |\xi_i|^2)^{\frac{1}{2}}.$$

In Section 3 we shall introduce a general notion of ellipticity, inspired by (1.4), see Definition 3.1. As a consequence of Theorem 3.11, using a generalization of Tauberian Theorem due to J. Aramaki [1], we will be able to study the counting function of operators of the form (1.4), see Theorem 3.12. In Section 4 we focus on the tensor product of Hermite-type operators, and we evaluate directly the first term of the asymptotic expansion of the counting function.

Motivations of the present paper, and connection with existing literature, are twofold. On one hand, the case when in (1.4) we have the tensor product of two Hermite operators, or more generally tensor product of real powers of Hermite operators in several distinct variables, is relevant in Probability, see for example [25], and other applications. Tensorized Hermite operators were treated in [12] from a sequential point of view, i.e., basing on eigenfunction expansions. In [12] the authors observed also a connection with the twisted Laplacian of Wong [37], which was proved to be unitarily equivalent to the tensor product of the one-dimensional Hermite operator and the identity operator. Similar ideas are present in the subsequent papers [9, 13, 15, 16, 26, 27]. On the other hand, the structure of our pseudo-differential class is strictly connected with the pioneering work [31], and [29] where similar operators were dubbed as bisingular operators. Recently, bisingular operators on compact manifolds were studied in [2] and [28], see also [3] for analogue results in the SG-setting. In particular, our results of Section 3 can be seen as a version of [2] for global operators on $\mathbb{R}^{n_1} \times \mathbb{R}^{n_2}$. In conclusion, we may also observe that our class of symbols, in the case of zero orders, is included in the Hörmader class $S^0_{0,0}(\mathbb{R}^{n_1+n_2})$. Hence our Theorem 3.3 enters the very general results of [23], see also [24] and [32], where necessary and sufficient condition for the Fredholm property were expressed in terms of invertibility of limit operators. Let us address in particular to Theorem 1.1 in the recent paper [33].

2. Γ calculus for bisingular operators

Definition 2.1. We define $\Gamma^{m_1,m_2}(\mathbb{R}^{n_1+n_2})$, $m_1 \in \mathbb{R}, m_2 \in \mathbb{R}$, as the subset of $C^\infty(\mathbb{R}^{2n_1+2n_2})$ functions such that for all multiindex α_i, β_i $(i = 1, 2)$ there exists a constant C so that

$$|\partial_{x_1}^{\beta_1}\partial_{x_2}^{\beta_2}\partial_{\xi_1}^{\alpha_1}\partial_{\xi_2}^{\alpha_2} a(x_1, x_2, \xi_1, \xi_2)| \le C\langle x_1, \xi_1\rangle^{m_1-|\alpha_1|-|\beta_1|}\langle x_2, \xi_2\rangle^{m_2-|\alpha_2|-|\beta_2|}, \quad (2.1)$$

for all x_1, ξ_1, x_2, ξ_2.

We define

$$\Gamma^{-\infty,-\infty}(\mathbb{R}^{n_1,n_2}) = \bigcap_{m_1,m_2\in\mathbb{R}^2} \Gamma^{m_1,m_2}(\mathbb{R}^{n_1+n_2})$$

as the set of smoothing symbols.

Definition 2.2. A linear operator $A : C_c^\infty(\mathbb{R}^{n_1+n_2}) \to C^\infty(\mathbb{R}^{n_1+n_2})$ is a globally bisingular operator if it can be written in this way[1]

$$\begin{aligned} A(u)(x_1, x_2) &= \mathrm{Op}(a)(u)(x_1, x_2) \\ &= \iint e^{ix_1\cdot\xi_1+ix_2\cdot\xi_2} a(x_1, x_2, \xi_1, \xi_2)\hat{u}(\xi_1, \xi_2) đ\xi_1 đ\xi_2 \end{aligned} \quad (2.2)$$

where $a \in \Gamma^{m_1,m_2}(\mathbb{R}^{n_1+n_2})$. We define $G^{m_1,m_2}(\mathbb{R}^{n_1+n_2})$ as the set of operators as in (2.2) with symbol in $\Gamma^{m_1,m_2}(\mathbb{R}^{n_1+n_2})$.

The $\mathcal{S}$-continuity of globally bisingular operators is immediate, we just have to check all seminorms. More interesting is the Sobolev continuity.

Theorem 2.3. *An operator $A \in G^{m_1,m_2}(\mathbb{R}^{n_1+n_2})$ can be extended for every $s_1 \in \mathbb{R}, s_2 \in \mathbb{R}$ continuously as an operator*

$$A : Q^{s_1,s_2}(\mathbb{R}^{n_1+n_2}) \to Q^{s_1-m_1,s_2-m_2}(\mathbb{R}^{n_1+n_2}).$$

[1] $đ\xi_i = (2\pi)^{-n_i} d\xi_i$.

Where, for positive integers s_1, s_2, we define $Q^{s_1,s_2}(\mathbb{R}^{n_1+n_2})$ as the space of all $u \in L^2(\mathbb{R}^{n_1+n_2})$ such that

$$\|u\|_{Q^{s_1,s_2}} = \sum_{\substack{|\alpha_1|+|\beta_1|\le s_1,\\ |\alpha_2|+|\beta_2|\le s_2}} \|x_1^{\beta_1} x_2^{\beta_2} D_{x_1}^{\alpha_1} D_{x_2}^{\alpha_2} u\|_{L^2}.$$

For general s_1, s_2, we set

$$Q^{s_1,s_2}(\mathbb{R}^{n_1+n_2}) = \{u \in \mathcal{S}'(\mathbb{R}^{n_1+n_2}) \mid \\ u = \mathrm{Op}(\langle x_1, \xi_1\rangle^{-s_1}\langle x_2, \xi_2\rangle^{-s_2})(v),\ v \in L^2(\mathbb{R}^{n_1+n_2})\}.$$

The proof of Theorem 2.3 follows by the remark that $\Gamma^{0,0}(\mathbb{R}^{n_1} \times \mathbb{R}^{n_2}) \subseteq \Gamma_0^0(\mathbb{R}^{n_1+n_2})$. Then we use the well-known results of L^2-continuity and the definition of $Q^{s_1,s_2}(\mathbb{R}^{n_1+n_2})$. We prove now that globally bisingular operators form an algebra.

Theorem 2.4. *Let $A \in G^{m_1,m_2}(\mathbb{R}^{n_1+n_2})$ and $B \in G^{l_1,l_2}(\mathbb{R}^{n_1+n_2})$ then $A \circ B \in G^{m_1+l_1,m_2+l_2}(\mathbb{R}^{n_1+n_2})$.*

Proof. With a simple evaluation we obtain

$$(A \circ B)u(x_1, x_2) = \iint e^{ix_1\xi_1+ix_2\xi_2} c(x_1, x_2, \xi_1, \xi_2)\hat{u}(\xi_1, \xi_2) đ\xi_1 đ\xi_2,$$

where

$$c(x_1, x_2, \xi_1, \xi_2) = \int e^{-i\mu_1-i\mu_2} a(x_1, x_2, \eta_1, \eta_2) b(y_1, y_2, \xi_1, \xi_2) dy_1 dy_2 đ\eta_1 đ\eta_2 \\ \mu_1 = \langle y_1 - x_1, \eta_1 - \xi_1\rangle, \quad \mu_2 = \langle y_2 - x_2, \eta_2 - \xi_2\rangle. \tag{2.3}$$

We divide ab into four parts, for a fixed integer $N > 0$:

$$a(x_1, x_2, \eta_1, \eta_2) b(y_1, y_2, \xi_1, \xi_2) = (ab)_1^N + (ab)_2^N + (ab)_3^N + r_N,$$

where

$$\begin{aligned}
(ab)_1^N &= \sum_{|\beta_1|+|\alpha_1|<2N} \frac{1}{\beta_1!\alpha_1!}(y_1 - x_1)^{\beta_1}(\eta_1 - \xi_1)^{\alpha_1} \partial_{\eta_1}^{\alpha_1} a(x_1, x_2, \xi_1, \eta_2) \\
&\quad \partial_{y_1}^{\beta_1} b(x_1, y_2, \xi_1, \xi_2), \\
(ab)_2^N &= \sum_{|\beta_2|+|\alpha_2|<2N} \frac{1}{\beta_2!\alpha_2!}(y_2 - x_2)^{\beta_2}(\eta_2 - \xi_2)^{\alpha_2} \partial_{\eta_2}^{\alpha_2} a(x_1, x_2, \eta_1, \xi_2) \\
&\quad \partial_{y_2}^{\beta_2} b(y_1, x_2, \xi_1, \xi_2), \\
(ab)_3^N &= -\sum_{\substack{|\alpha_1|+|\beta_1|<2N \\ |\alpha_2|+|\beta_2|<2N}} \frac{1}{\beta_1!\beta_2!\alpha_1!\alpha_2!}(y_1 - x_1)^{\beta_1}(y_2 - x_2)^{\beta_2}(\eta_1 - \xi_1)^{\alpha_1} \\
&\quad (\eta_2 - \xi_2)^{\beta_2} \partial_{\eta_1}^{\alpha_1}\partial_{\eta_2}^{\alpha_2} a(x_1, x_2, \xi_1, \xi_2) \partial_{y_1}^{\beta_1}\partial_{y_2}^{\beta_2} b(x_1, x_2, \xi_1, \xi_2),
\end{aligned}$$

$$r_N = \sum_{\substack{|\alpha_1|+|\beta_1|<2N \\ |\alpha_2|+|\beta_2|<2N}} \frac{1}{\beta_1!\beta_2!\alpha_1!\alpha_2!}(y_1-x_1)^{\beta_1}(y_2-x_2)^{\beta_2}$$

$$(\eta_1-\xi_1)^{\alpha_1}(\eta_2-\xi_2)^{\alpha_2}\int_0^1\int_0^1(1-t_1)^{N-1}(1-t_2)^{N-1}$$

$$\partial_{\eta_1}^{\alpha_1}\partial_{\eta_2}^{\alpha_2}a(x_1,x_2,\xi_1+t_1(\eta_1-\xi_1),\xi_2+t_2(\eta_2-\xi_2))$$

$$\partial_{y_1}^{\beta_1}\partial_{y_2}^{\beta_2}b(x_1+t_1(y_1-x_1),x_2+t_2(y_2-x_2),\xi_1,\xi_2)dt_1dt_2.$$

Dividing the integral (2.3) in four parts, one defines

$$c_i^N = \int e^{-i\mu_1-i\mu_2}(ab)_i^N dy_1dy_2đ\eta_1đ\eta_2,$$

$$R_N = \int e^{-i\mu_1-i\mu_2}r_N dy_1dy_2đ\eta_1đ\eta_2.$$

Now, we only focus on c_1^N. Notice that

$$(y_1-x_1)^{\beta_1}e^{-i\langle y_1-x_1,\eta_1-\xi_1\rangle} = (-i)^{\beta_1}D_{\eta_1}^{\beta_1}e^{-i\langle y_1-x_1,\eta_1-\xi_1\rangle}, \tag{2.4}$$

$$(\eta_1-\xi_1)^{\alpha_1}e^{-i\langle y_1-x_1,\eta_1-\xi_1\rangle} = (-i)^{\alpha_1}D_{y_1}^{\alpha_1}e^{-i\langle y_1-x_1,\eta_1-\xi_1\rangle}. \tag{2.5}$$

If $\alpha_1 \neq \beta_1$, there exists an index i such that, for example, $(\alpha_1)_i > (\beta_1)_i$. So, using relation (2.5) and integrating by parts, we derive $(\alpha_1)_i$ times w.r.t. y_1 the expression $(y_1-x_1)^{\beta_1}$, and, since $(\alpha_1)_i > (\beta_1)_i$, the derivative is zero. Clearly the same scheme can be used if $(\alpha_1)_i < (\beta_1)_i$ using (2.4). This implies that we can restrict ourself to consider the case $\alpha_1 = \beta_1$, so we will just write α_1. Now, integrating by parts and using relation (2.5), we get

$$c_1^N = \frac{1}{\alpha!}\int\int e^{-i\langle y_2-x_2,\eta_2-\xi_2\rangle}\sum_{|\alpha_1|<N}\partial_{\xi_1}^{\alpha_1}a(x_1,x_2,\xi_1,\eta_2) \\ D_{x_1}^{\alpha_1}b(x_1,y_2,\xi_1,\xi_2)dy_2d\eta_2. \tag{2.6}$$

The expression (2.6) can be written in the form

$$c_1^N = \sum_{|\alpha_1|<N}\frac{1}{\alpha_1!}\partial_{\xi_1}^{\alpha_1}a \circ_2 D_{x_1}^{\alpha_1}b,$$

where the symbol $\circ_2$ means the composition of the operators acting on $\mathbb{R}^{n_2}$. With the same scheme we can prove that

$$c_2^N = \sum_{|\alpha_2|<N}\frac{1}{\alpha_2!}\partial_{\xi_2}^{\alpha_2}a \circ_1 D_{x_2}^{\alpha_2}b.$$

Integrating by parts two times, we get

$$c_3^N = -\sum_{\substack{|\alpha_1|<N \\ |\alpha_2|<N}}\frac{1}{\alpha_1!\alpha_2!}\partial_{\xi_1}^{\alpha_1}\partial_{\xi_2}^{\alpha_2}aD_{x_1}^{\alpha_1}D_{x_2}^{\alpha_2}b.$$

We have now to analyze the remainder. Consider this identity

$$\langle y_1,\eta_1\rangle^{2M}\langle y_2,\eta_2\rangle^{2M}(1-\Delta_{y_1}-\Delta_{\eta_1})^M(1-\Delta_{y_2}-\Delta_{\eta_2})^M e^{-i\mu_1-i\mu_2}=e^{-i\mu_1-i\mu_2}. \tag{2.7}$$

By Peetre inequality, we have

$$\begin{aligned}|r_N| \le \langle x_1,\xi_1\rangle^{m_1+l_1-2N}\langle x_2,\xi_2\rangle^{m_2+l_2-2N}\langle y_1-x_1\rangle^{|l_1|+2N}\langle y_2-x_2\rangle^{|l_2|+2N}\\ \langle \eta_1-\xi_1\rangle^{|m_1|+2N}\langle \eta_2-\xi_2\rangle^{|m_2|+2N}.\end{aligned}$$

Using (2.7) with M big enough and integrating by parts, we prove that $R_N \in \Gamma^{m_1+l_2-2N,m_2+l_2-2N}(\mathbb{R}^{n_1+n_2})$. □

Remark 2.5. It is useful to write c in this way

$$c \sim \sum_{j=0}^{\infty} c_{m_1+l_1-2j,m_2+l_2-2j},$$

where

$$\begin{aligned}c_{m_1+l_1-2j,m_2+l_2-2j} = c^1_{m_1+l_1-2j,m_2+l_2-2j} + c^2_{m_1+l_1-2j,m_2+l_2-2j}\\ + c^3_{m_1+l_1-2j,m_2+l_2-2j},\end{aligned}$$

and

$$\begin{aligned}c^1_{m_1+l_1-2j,m_2+l_2-2j} &= \sum_{|\alpha_1|=j}\frac{1}{\alpha_1!}\Big(\partial^{\alpha_1}_{\xi_1}a\circ_2 D^{\alpha_1}_{x_1}b-\sum_{|\alpha_2|\le j}\frac{1}{\alpha_2!}\partial^{\alpha_1}_{x_1}\partial^{\alpha_2}_{x_2}aD^{\alpha_1}_{x_1}D^{\alpha_2}_{x_2}b\Big),\\ c^2_{m_1+l_1-2j,m_2+l_2-2j} &= \sum_{|\alpha_2|=j}\frac{1}{\alpha_2!}\Big(\partial^{\alpha_2}_{\xi_2}a\circ_1 D^{\alpha_2}_{x_2}b-\sum_{|\alpha_1|\le j}\frac{1}{\alpha_1!}\partial^{\alpha_1}_{x_1}\partial^{\alpha_2}_{x_2}aD^{\alpha_1}_{x_1}D^{\alpha_2}_{x_2}b\Big),\\ c^3_{m_1+l_1-2j,m_2+l_2-2j} &= \sum_{|\alpha_1|=|\alpha_2|=j}\frac{1}{\alpha_1!\alpha_2!}\partial^{\alpha_1}_{x_1}\partial^{\alpha_2}_{x_2}aD^{\alpha_1}_{x_1}D^{\alpha_2}_{x_2}.\end{aligned}$$

In the following, we will study a subclass of globally bisingular operators, namely operators with homogeneous principal part.

Definition 2.6. A symbol $a \in \Gamma^{m_1,m_2}(\mathbb{R}^{n_1+n_2})$ has homogeneous principal part if

i) there exists a function $a_{m_1,\cdot}(x_1,x_2,\xi_1,\xi_2)$ homogeneous w.r.t. (x_1,ξ_1) of order m_1 such that

$$a-\psi_1(x_1,\xi_1)a_{m_1,\cdot}\in\Gamma^{m_1-1,m_2}(\mathbb{R}^{n_1+n_2}),$$

ψ_1 cut-off function at the origin, and the operator $a(x_1,x_2,\xi_1,D_2)$, with (x_1,ξ_1) *frozen*, is a classical global operator in $\mathbb{R}^{n_2}$;

ii) there exists $a_{\cdot,m_2}$ homogeneous w.r.t. (x_2,ξ_2) of order m_2 such that

$$a-\psi_2(x_2,\xi_2)a_{\cdot,m_2}\in\Gamma^{m_1,m_2-1}(\mathbb{R}^{n_1,n_2}),$$

ψ_2 cut-off function at the origin, and the operator $a(x_1,x_2,D_1,\xi_2)$, with (x_2,ξ_2) *frozen*, is a classical global operator in $\mathbb{R}^{n_1}$;

iii) there exists a function $a_{m_1,m_2}(x_1,x_2,\xi_1,\xi_2)$ bihomogeneous w.r.t. (x_1,ξ_1) of order m_1 and w.r.t. (x_2,ξ_2) of order m_2, such that a_{m_1,m_2} is equal to the principal symbol of $a_{m_1,\cdot}(x_1,x_2,\xi_1,D_2)$ and of $a_{\cdot,m_2}(x_1,x_2,D_1,\xi_2)$ and

$$a - \psi_1(x_1,\xi_1)a_{m_1,\cdot} - \psi_2(x_2,\xi_2)(a_{\cdot,m_2}) + \psi_1(x_1,\xi_1)\psi_2(x_2,\xi_2)a_{m_1,m_2}$$

belongs to $\Gamma^{m_1-1,m_2-1}(\mathbb{R}^{n_1+n_2})$.

In the following, the class of symbols with homogeneous principal part is written as $\Gamma^{m_1,m_2}_{pr}(\mathbb{R}^{n_1+n_2})$, and the operators with homogeneous principal symbol as $G^{m_1,m_2}_{pr}(\mathbb{R}^{n_1+n_2})$. We introduce three functions associated to an operator $A \in G^{m_1,m_2}_{pr}(\mathbb{R}^{n_1+n_2})$:[2]

$$\begin{aligned}
\sigma_1^{m_1}(A) &: T^*(\mathbb{R}^{n_1}) \setminus \{0\} \to G^{m_2}_{cl}(\mathbb{R}^{n_2}) \\
&\quad (x_1,\xi_1) \mapsto a_{m_1,\cdot}(x_1,x_2,\xi_1,D_2), \\
\sigma_2^{m_2}(A) &: T^*(\mathbb{R}^{n_2}) \setminus \{0\} \to G^{m_1}_{cl}(\mathbb{R}^{n_1}) \\
&\quad (x_2,\xi_2) \mapsto a_{\cdot,m_2}(x_1,x_2,D_1,\xi_2), \\
\sigma^{m_1,m_2}(A) &: T^*(\mathbb{R}^{n_1}) \setminus \{0\} \times T^*(\mathbb{R}^{n_2}) \setminus \{0\} \to \mathcal{H}^{m_1,m_2}_{\xi_1,\xi_2}(\mathbb{R}^{n_1+n_2}) \\
&\quad (x_1,x_2,\xi_1,\xi_2) \mapsto a_{m_1,m_2}(x_1,x_2,\xi,\xi_2).
\end{aligned}$$

3. Globally elliptic bisingular operators and the Weyl formula

Definition 3.1. Let $A \in G^{m_1,m_2}_{pr}(\mathbb{R}^{n_1+n_2})$, A is globally elliptic bisingular operator if there exist constants R_1, R_2 such that

i) the operator

$$a_{m_1,\cdot}(x_1,x_2,\xi_1,D_2) : \mathcal{S}(\mathbb{R}^{n_2}) \to \mathcal{S}(\mathbb{R}^{n_2})$$

is invertible for every $(x_1,\xi_1) \in T^*\mathbb{R}^{n_1} \setminus \{0\}$;

ii) the operator

$$a_{\cdot,m_2}(x_1,x_2,D_1,\xi_2) : \mathcal{S}(\mathbb{R}^{n_1}) \to \mathcal{S}(\mathbb{R}^{n_1})$$

is invertible for every $(x_2,\xi_2) \in T^*\mathbb{R}^{n_2} \setminus \{0\}$;

iii) there exists a positive constant C such that

$$\begin{aligned}|a_{m_1,m_2}(x_1,x_2,\xi_1,\xi_2)| \geq C\langle x_1,\xi_1\rangle^{m_1}\langle x_2,\xi_2\rangle^{m_2}, \\ \forall |x_i|^2 + |\xi_i|^2 \geq R_i, i = 1,2.\end{aligned} \tag{3.1}$$

Since $a_{m_1,m_2}(x_1,x_2,\xi_1,\xi_2)$ is bihomogeneous it is enough to require that (3.1) is fulfilled for $(x_1,\xi_1) \in T^*\mathbb{R}^{n_1} \setminus \{0\}$, $(x_2,\xi_2) \in T^*\mathbb{R}^{n_2} \setminus \{0\}$.

Remark 3.2. If an operator $A \in G^{m_1,m_2}_{pr}(\mathbb{R}^{n_1+n_2})$ satisfies item *iii)* of Definition 3.1 then both the operators $a_{m_1,\cdot}(x_1,x_2,\xi_1,D_2)(x_2,\xi_2) \in G^{m_2}(\mathbb{R}^{n_2})$ and $a_{\cdot,m_2}(x_1,x_2,D_1,\xi_2)(x_1,\xi_1) \in G^{m_1}(\mathbb{R}^{n_1})$ are elliptic Shubin type operators. If moreover A satisfies items *i)* and *ii)* one can prove that both $a_{m_1,\cdot}(x_1,x_2,\xi_1,D_2)(x_2,\xi_2)$

[2] $\mathcal{H}^{m_1,m_2}_{\xi_1,\xi_2}(\mathbb{R}^{n_1+n_2})$ is the set of homogeneous function of order m_i w.r.t. ξ_i.

and $a_{\cdot,m_2}(x_1, x_2, D_1, \xi_2)(x_1, \xi_1)$ are injective Fredholm operator with zero index, therefore invertible operators also in the scale of Q^s spaces. Thus, in Definition 3.1, it is equivalent to require the invertibility of the operators on the Schwartz spaces or on the Sobolev spaces Q^s. For this reason, in the following we will not specify the space in which the operators are assumed to be invertible.

Theorem 3.3. *If an operator A is globally elliptic bisingular then it is a Fredholm operator.*

Proof. It is a consequence of Theorem 2.3. From Remark 2.5, if A is elliptic one can define an operator B as the operator with symbol

$$b = \psi_1(x_1, \xi_1) a^{-1}_{m_1,\cdot} + \psi_2(x_2, \xi_2) a^{-1}_{\cdot,m_2} - \psi_1(x_1, \xi_1)\psi_2(x_2, \xi_2) a^{-1}_{m_1,m_2}.$$

Applying the calculus, one can check that B is an inverse of A modulo compact operator. □

Using a Neumann series procedure, by Theorem 3.3, one can prove that, if an operator is globally elliptic bisingular, then there exists an inverse modulo smoothing operators. So we have this immediate corollary:

Corollary 3.4. *Let $A \in G^{m_1,m_2}_{pr}(\mathbb{R}^{n_1+n_2})$ be globally elliptic then*

i) *if $Au \in Q^{s_1,s_2}(\mathbb{R}^{n_1+n_2})$ then $u \in Q^{s_1+m_1,s_2+m_2}(\mathbb{R}^{n_1+n_2})$;*
ii) *if $Au \in \mathcal{S}(\mathbb{R}^{n_1+n_2})$ then $u \in \mathcal{S}(\mathbb{R}^{n_1+n_2})$.*

Our aim is now to study the counting function of positive self-adjoint globally bisingular operators. We will use Tauberian techniques, so we need to define complex powers of globally bisingular operators.

First we define parameter ellipticity:

Definition 3.5. Let Λ be a sector of the complex plane and a be a symbol belonging to $\Gamma^{m_1,m_2}_{pr}(\mathbb{R}^{n_1+n_2})$; a is called Λ-elliptic w.r.t. Λ if there exists a constant R such that

i)
$$\sigma_1^{m_1}(A)(x_1, \xi_1) - \lambda I_{\mathbb{R}^{n_2}} \in G^{m_2}_{cl}(\mathbb{R}^{n_2})$$
is invertible for all $|x_1| + |\xi_1| > R$, for all $\lambda \in \Lambda$.

ii)
$$\sigma_2^{m_2}(A)(x_2, \xi_2) - \lambda I_{\mathbb{R}^{n_1}} \in G^{m_1}_{cl}(\mathbb{R}^{n_1})$$
is invertible for all $|x_2| + |\xi_2| > R$, for all $\lambda \in \Lambda$.

iii)
$$\left(\sigma^{m_1,m_2}(A)(x_1, x_2, \xi_1, \xi_2) - \lambda\right)^{-1} \in \Gamma^{-m_1,-m_2}(\mathbb{R}^{n_1+n_2})$$
for all $|x_i| + |\xi_i| > R$, for all $\lambda \in \Lambda$.

In the following, we consider sector of the complex plane Λ with vertex at the origin as in the figure below.

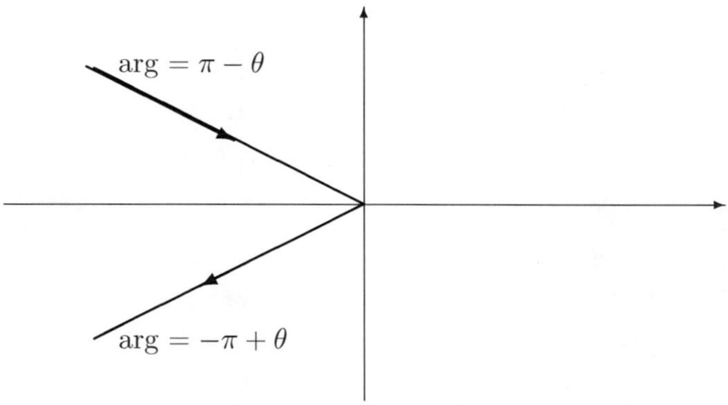

It is an exercise to prove that, if $A \in G^{m_1,m_2}_{pr}(\mathbb{R}^{n_1+n_2})$ is Λ-elliptic, then the operator is sectorial. Follow for example the scheme of Theorem 2 in [2].

We make now some natural assumptions in order to perform the functional calculus.

Assumptions 3.6.

i) $A \in G^{m_1,m_2}_{pr}(\mathbb{R}^{n_1+n_2})$ is Λ-elliptic,
ii) $\sigma(A) \cap \Lambda = \emptyset$, in particular A is invertible.

Remark 3.7. In item ii) of Assumptions 3.6, we assume that the operator is invertible. We have made these assumptions in order to get a simpler theory. It is nevertheless possible to handle functional calculus of operators with non trivial kernel, even with infinite-dimensional kernel, the crucial requirement is that the origin must be an isolated point of the spectrum. Roughly speaking, instead of considering the operator A, one studies the operator $A \circ (I - P_{\ker A})$; $P_{\ker A}$ being the projection into the kernel of A. Clearly this operator is invertible, cf. [8].

Definition 3.8. Let A be a globally bisingular operator that satisfies Assumptions 3.6, we can define

$$A_z := \frac{i}{2\pi} \int_{\partial\Lambda_\epsilon^+} \lambda^z (A - \lambda \mathrm{Id})^{-1} d\lambda, \quad \mathrm{Re}(z) < 0, \tag{3.2}$$

where $\Lambda_\epsilon = \Lambda \cup \{z \in \mathbb{C} \mid |z| < \epsilon\}$. The complex power of A is defined in this way

$$A^z = \begin{cases} A_z & \mathrm{Re}(z) < 0, \\ A_{z-k} \circ A^k & k \in \mathbb{N}, \mathrm{Re}(z-k) < 0. \end{cases}$$

Since the operator A is sectorial, the Dunfort integral in (3.2) converges. As usual, one can prove that the Definition 3.8 does not depend on k.

Theorem 3.9. *If $A \in G^{m_1,m_2}_{pr}(\mathbb{R}^{n_1,n_2})$ fulfils Assumptions 3.6, then $A^z \in G^{m_1 z, m_2 z}(\mathbb{R}^{n_1+n_2})$. Moreover,*[3]

$$\sigma_1^{m_1 z}(A^z)(x_1,\xi_1) = \left(\sigma_1^{m_1}(A)(x_1,\xi_1)\right)^z, \tag{3.3}$$

$$\sigma_2^{m_2 z}(A^z)(x_2,\xi_2) = \left(\sigma_2^{m_2}(A)(x_2,\xi_2)\right)^z, \tag{3.4}$$

$$\sigma^{m_1 z, m_2 z}(A^z)(x_1,x_2,\xi_1,\xi_2) = \left(\sigma^{m_1,m_2}(A)(x_1,x_2,\xi_1,\xi_2)\right)^z, \tag{3.5}$$

where the complex power in (3.3), (3.4) is the complex power of operators, while in (3.5) is the standard complex power of a function.

We now introduce the ζ-function of suitable bisingular operators, then we will study the meromorphic extension of the ζ-function and we will analyze its first *left* pole. We do not write the proofs of the following statements, they are similar to Theorem 4 and Corollary 1 of [2].

Definition 3.10. Let $A \in G^{m_1,m_2}(\mathbb{R}^{n_1+n_2})$ be a bisingular operator that satisfies Assumptions 3.6, then

$$\zeta(A,z) = \iint_{\mathbb{R}^{n_1+n_2}} K_{A^z}(x_1,x_2,x_1,x_2)dx_1dx_2, \quad \operatorname{Re}(z) < 2\min\left\{-\frac{n_1}{m_1}, -\frac{n_2}{m_2}\right\},$$

where K_{A^z} is the kernel of A^z.

Theorem 3.11. *Let $A \in G^{m_1,m_2}(\mathbb{R}^{n_1+n_2})$ be an operator that satisfies Assumptions 3.6. Then $\zeta(A,z)$ can be extended as a meromorphic function on $\{z \in \mathbb{C} \mid \operatorname{Re}(z) < 2\min\{-\frac{n_1}{m_1}, -\frac{n_2}{m_2}\} + \epsilon\}$. Moreover, the Laurent coefficients at pole $z_{pole} = 2\min\{-\frac{n_1}{m_1}, -\frac{n_2}{m_2}\}$ depend on $\frac{n_1}{m_1}$ and $\frac{n_2}{m_2}$.*

In the case $\frac{n_1}{m_1} > \frac{n_2}{m_2}$:

$$\lim_{z\to -\frac{2n_1}{m_1}} \left(z + \frac{2n_1}{m_1}\right)\zeta(A,z) = \frac{(2\pi)^{-n_1-n_2}}{m_1}\int_{\mathbb{R}^{2n_2}}\int_{\mathbb{S}^{2n_1-1}} (a_{m_1,\cdot})^{-\frac{2n_1}{m_1}} d\theta_1 dx_2 d\xi_2. \tag{3.6}$$

In the case $\frac{n_2}{m_2} > \frac{n_1}{m_1}$:

$$\lim_{z\to -\frac{2n_2}{m_2}} \left(z + \frac{2n_2}{m_2}\right)\zeta(A,z) = \frac{(2\pi)^{-n_1-n_2}}{m_2}\int_{\mathbb{R}^{2n_1}}\int_{\mathbb{S}^{2n_2-1}} (a_{\cdot,m_2})^{-\frac{2n_2}{m_2}} d\theta_2 dx_1 d\xi_1. \tag{3.7}$$

In the case $\frac{n_1}{m_1} = \frac{n_2}{m_2} = l$:

$$res^2(A) = \lim_{z\to -l}(z+l)^2\zeta(A,z) = \frac{(2\pi)^{-n_1-n_2}}{m_1m_2}\int_{\mathbb{S}^{2n_2-1}}\int_{\mathbb{S}^{2n_1-1}} (a_{m_1,m_2})^{-l}d\theta_1 d\theta_2, \tag{3.8}$$

$$\lim_{z\to -l}(z+l)\left(\zeta(A,z) - \frac{res^2(A)}{(z+l)^2}\right), = -TR_{1,2}(A) + TR_\theta(A), \tag{3.9}$$

[3] We have just defined symbols $\Gamma^{m_1,m_2}(\mathbb{R}^{n_1+n_2})$ with $m_1, m_2 \in \mathbb{R}^2$. It is nevertheless possible to define the same class with complex numbers z_1, z_2, in the inequality (2.1) instead of m_i we use $\operatorname{Re}(z_i)$.

where

$$TR_{1,2}(A) = (2\pi)^{-n_1-n_2}$$
$$\left(\lim_{\tau\to\infty}\left(\frac{1}{m_1}\int_{|x_2|+|\xi_2|<\tau}\int_{\mathbb{S}^{2n_1-1}}((a_{m_1,\cdot})^{-l}d\theta_1 dx_2 d\xi_2 - res^2(A)\log\tau)\right)\right.$$
$$\left.+\lim_{\tau\to\infty}\left(\frac{1}{m_2}\int_{|x_1|+|\xi_1|<\tau}\int_{\mathbb{S}^{2n_2-1}}((a_{m_2,\cdot})^{-l}d\theta_2 dx_1 d\xi_1 - res^2(A)\log\tau)\right)\right),$$

and

$$TR_\theta(A) = \frac{(2\pi)^{-n_1-n_2}}{m_1 m_2}\int_{\mathbb{S}^{2n_2-1}}\int_{\mathbb{S}^{2n_1-1}}(a_{m_1,m_2})^{-l}\log(a_{m_1,m_2})d\theta_1 d\theta_2.$$

Now, applying a generalization of Tauberian Theorem due to J. Aramaki [1], we easily obtain the following:

Theorem 3.12. *Let $A \in G^{m_1,m_2}(\mathbb{R}^{n_1+n_2})$ be self-adjoint and positive, suppose moreover that A satisfies Assumptions 3.6. Then*

$$N_A(\lambda) = \begin{cases} C_1\lambda^l\log\lambda + C_1'\lambda^l + O(\lambda^{l-\delta_1}) & \frac{2n_1}{m_2} = \frac{2n_2}{m_2} = l, \\ C_2\lambda^{2\frac{n_2}{m_2}} + O(\lambda^{2\frac{n_2}{m_2}-\delta_2}) & \frac{2n_2}{m_2} > \frac{2n_1}{m_1}, \\ C_3\lambda^{2\frac{n_1}{m_2}} + O(\lambda^{2\frac{n_1}{m_1}-\delta_3}) & \frac{2n_1}{m_1} > \frac{2n_2}{m_2}, \end{cases}$$

for certain $\delta_i > 0$. It is moreover possible to find the exact value of the constants in terms of $\{a_{m_1,\cdot}, a_{\cdot,m_2}, a_{m_1,m_2}\}$, the principal symbol of A.

$$C_1 = \frac{1}{(2\pi)^{n_1+n_2}2n_1 m_2}\int_{\mathbb{S}^{2n_2-1}}\int_{\mathbb{S}^{2n_1-1}}(a_{m_1,m_2})^{-l}d\theta_1 d\theta_2,$$
$$C_1' = \frac{TR_{1,2}(A) - TR_\theta(A)}{l} - \frac{1}{4n_1 n_2}\int_{\mathbb{S}^{2n_2-1}}\int_{\mathbb{S}^{2n_1-1}}(a_{m_1,m_2})^{-l}d\theta_1 d\theta_2,$$
$$C_2 = \frac{1}{(2\pi)^{n_1+n_2}2n_2}\int_{\mathbb{R}^{2n_1}}\int_{\mathbb{S}^{2n_2-1}}(a_{\cdot,m_2})^{-\frac{2n_2}{m_2}}d\theta_2 dx_1 d\xi_1,$$
$$C_3 = \frac{1}{(2\pi)^{n_1+n_2}2n_1}\int_{\mathbb{R}^{2n_2}}\int_{\mathbb{S}^{2n_1-1}}(a_{m_1,\cdot})^{-\frac{2n_1}{m_1}}d\theta_1 dx_2 d\xi_2.$$

4. Tensor products of Hermite-type operators

We use the notation in the Introduction. We consider globally elliptic self-adjoint bisingular differential operators of the special form

$$P(x, D_x) = P_1(x_1, D_{x_1})P_2(x_2, D_{x_2}),$$

so that $Pu_{\mathbf{j}} = \lambda^1_{j_1}\lambda^2_{j_2}u_{\mathbf{j}}$, $\mathbf{j} = (j_1, j_2) \in \mathbb{Z}^2_+$, cf. (1.4). Hence we have:

Proposition 4.1. *Let u be a tempered distribution in $\mathbb{R}^{n_1+n_2}$. If*

$$u = \sum_{\mathbf{j}\in\mathbb{Z}^2_+} a_{\mathbf{j}}u_{\mathbf{j}} \quad \textit{in } \mathcal{S}'(\mathbb{R}^{n_1-n_2}),$$

then

$$P(x, D_x)u = \sum_{\mathbf{j}=(j_1,j_2)\in\mathbb{Z}_+^2} \lambda_{j_1}^1 \lambda_{j_2}^2 a_{\mathbf{j}} u_{\mathbf{j}}, \tag{4.1}$$

where $P(x, D_x)u \in L^2(\mathbb{R}^{n_1+n_2})$ *is equivalent to*

$$\sum_{\mathbf{j}\in\mathbb{Z}_+^2} (\lambda_{j_1}^1 \lambda_{j_2}^2)^2 |a_{\mathbf{j}}|^2 \quad < \quad +\infty. \tag{4.2}$$

With the notation at the end of Section 2, we obtain

$$\begin{aligned} \sigma_1^{m_1}(P)(x_1, \xi_1) &= p_{m_1}(x_1, \xi_1) P_2(x_2 D_2), \\ \sigma_2^{m_2}(P)(x_2, \xi_2) &= p_{m_2}(x_2, \xi_2) P_1(x_1, D_1), \\ \sigma^{m_1,m_2}(x_1, x_2, \xi_1, \xi_2) &= p_{m_1}(x_1, \xi_1) p_{m_2}(x_2, \xi_2). \end{aligned}$$

Thus, Definition 3.1 before amounts to assume global ellipticity of p_1, p_2, cf. (1.2), and invertibility of $P_1(x_1, D_1)$ and $P_2(x_1, D_2)$ as required in the Introduction. From Corollary 3.4, we have that $Pu \in Q^{s_1-m_1, s_2-m_2}(\mathbb{R}^{n_1+n_2})$ implies $u \in Q^{s_1,s_2}(\mathbb{R}^{n_1+n_2})$, for every $s_1 \in \mathbb{R}, s_2 \in \mathbb{R}$. In particular, if $Pu \in \mathcal{S}(\mathbb{R}^{n_1+n_2})$ then $u \in \mathcal{S}(\mathbb{R}^{n_1+n_2})$.

Assuming further that P_1, P_2 are strictly positive, we may apply Theorem 3.12 to estimate the counting function $N(\lambda)$ of P. By direct calculation, we shall give now more precise results in the case when P_1, P_2 are Hermite-type operators.

We first recall a classical result of L. Dirichlet for the first summatory function of $\tau(n)$, see [20, 21] for an overview on the subject:

$$D(\lambda) = \sum_{\substack{n\le\lambda,\\ n\in\mathbb{N}}} \tau(n) = \sum_{n=1}^{[\lambda]} \tau(n), \qquad \lambda \ge 1, \tag{4.3}$$

where $\tau(n)$ denotes the number of divisors of n and $[\lambda]$ stands for the integer part of λ. In 1849, Dirichlet proved that

$$D(\lambda) = \lambda \ln \lambda + (2\tilde{\gamma} - 1)\lambda + E(\lambda), \qquad \lambda \ge 1, \tag{4.4}$$

where $\tilde{\gamma}$ is the Euler-Mascheroni constant and

$$E(\lambda) = O(\lambda^{1/2}), \quad \lambda \to +\infty. \tag{4.5}$$

It is still an open problem to evaluate the optimal order of the reminder $E(\lambda)$ in the asymptotic expansion (4.4). In 1916 [17] Hardy discovered that $O(\lambda^{\frac{1}{4}})$ is a lower bound. Then a lot of upper bound have been proved, the better one has been given by Huxley in [19]. He proved that $E(\lambda)$ is $O(\lambda^c(\log\lambda)^d)$, where

$$c := \frac{131}{416} \sim 0,3149038462 \quad d := \frac{18627}{8320} + 1 \sim 3,238822115.$$

The conjecture is that the $E(\lambda)$ is $O(\lambda^{\frac{1}{4}})$. One can recast the issue of computing $D(\lambda)$ as a lattice point problem. More precisely, since

$$D(\lambda) = \sum_{n\leq\lambda, n\in\mathbb{N}} \sum_{d|n, d\in\mathbb{N}} 1 = \sum_{\substack{n_1 n_2 \leq \lambda \\ n_1, n_2 \in \mathbb{N}}} 1, \tag{4.6}$$

we readily obtain that $D(\lambda)$ is the number of positive integers lattice points in the first quadrant between the axes and hyperbola $x_1 x_2 \leq \lambda$. We will use this result for the proof of the second part of the next proposition.

Proposition 4.2. *Assume that A_i are self-adjoint operators with spectrum $n^{m_i}, n \in \mathbb{N}$, and eigenfunctions u_i^n being an orthonormal basis of $\mathbb{R}^{n_i}$, $i = 1, 2$. Denote by $N(\lambda)$, $\lambda > 0$ the counting function,*

$$N(\lambda) = \operatorname{card}\{(n_1, n_2) : n_1^{m_1} n_2^{m_2} \leq \lambda\}, \quad \lambda > 0,$$

where card *means cardinal number. Then we have the following assertions.*

a) *Let $m_1 > m_2 > 0$. Then*

$$N(\lambda) \sim \zeta\left(\frac{m_1}{m_2}\right) \lambda^{1/m_2} + C(m_1, m_2, \lambda)\lambda^{1/m_1} + O(1), \tag{4.7}$$

where

$$-1 - \frac{m_2}{(m_1 - m_2)} \leq C(m_1, m_2, \lambda) \leq -\frac{m_2}{(m_1 - m_2)}.$$

b) *Let $m = m_1 = m_2$. Then*

$$N(\lambda) = \frac{1}{m}\lambda^{\frac{1}{m}} \ln \lambda + \frac{2\tilde{\gamma} - 1}{m}\lambda + O(\lambda^{\frac{1}{2m}}). \tag{4.8}$$

Proof. a) First we recall from [11] the next identity, for $\alpha > 1, q > 0$:

$$\begin{aligned} \sum_{n=0}^{N} \frac{1}{(n+q)^\alpha} \\ = \sum_{n=0}^{\infty} \frac{1}{(n+q)^\alpha} + \frac{1}{(1-\alpha)(N+q)^{\alpha-1}} + \alpha \sum_{n=N}^{\infty} \int_n^{n+1} \frac{t-n}{(t+q)^{\alpha+1}} dt. \end{aligned} \tag{4.9}$$

One can find that

$$\alpha \sum_{n=N}^{\infty} \int_n^{n+1} \frac{t-n}{(t+q)^{\alpha+1}} = O(1/N^\alpha).$$

Note that with $q = 1$ one obtains the formula for the partial sum of the Riemann zeta function $\zeta(\alpha)$. Set

$$R(\lambda, m_1, m_2) = \sum_{n=1}^{\lambda^{1/m_1}} \left(\frac{\lambda^{1/m_2}}{n^{m_1/m_2}} - \left[\frac{\lambda^{1/m_2}}{n^{m_1/m_2}} \right] \right).$$

Clearly $0 \leq R(\lambda, m_1, m_2) \leq \lambda^{1/m_1}$ but we are not able to find the exact behavior of $R(\lambda, m_1, m_2)$ as $\lambda \to \infty$.

Now we calculate

$$N(\lambda) = \sum_{n_1^{m_1} n_2^{m_2} \leq \lambda} 1 = \sum_{n_1=1}^{[\lambda^{1/m_1}]} \sum_{n_2=1}^{\left[\frac{\lambda^{1/m_2}}{n_1^{m_1/m_2}}\right]} 1.$$

We have

$$\sum_{n=1}^{[\lambda^{1/m_1}]} \left[\frac{\lambda^{1/m_2}}{n^{m_1/m_2}}\right] = \sum_{n=1}^{[\lambda^{1/m_1}]} \frac{\lambda^{1/m_2}}{n^{m_1/m_2}} - R(\lambda, m_1, m_2)$$

and, using (4.9) with $\alpha = \frac{m_1}{m_2}$, $N = [\lambda^{1/m_1}]$, we obtain

$$\lambda^{1/m_2} \sum_{n=1}^{\lambda^{1/m_1}} \frac{1}{n^{m_1/m_2}} \sim \zeta\left(\frac{m_1}{m_2}\right) \lambda^{1/m_2} - \frac{m_2}{(m_1 - m_2)} \lambda^{1/m_1} + O(1).$$

This implies

$$\sum_{n_1^{m_1} n_2^{m_2} \leq \lambda} 1 \sim \zeta\left(\frac{m_1}{m_2}\right) \lambda^{1/m_2} - \frac{m_2}{(m_1 - m_2)} \lambda^{1/m_1} - R(\lambda, m_1, m_2) + O(1).$$

This and the estimate of R proves (4.7).

b) Since

$$n_1^m n_2^m \leq \lambda \text{ is equivalent to } n_1 n_2 \leq \lambda^{1/m},$$

we directly obtain $N(\lambda) = D(\lambda^{1/m})$ which gives (4.8). □

Example. An example of on operator that satisfies the hypothesis of Proposition 4.2 is the following. Let $m_1, m_2 \in \mathbb{N}, m_1 > m_2$, $k_1, k_2 > 0$ and

$$A_1 = k_1 \left(-\frac{\partial^2}{\partial_x^2} + x^2\right)^{m_1}, \quad A_2 = k_2 \left(-\frac{\partial^2}{\partial_y^2} + y^2\right)^{m_2}, \quad x, y \in \mathbb{R}.$$

Then, we know that the Hermite basis of $L^2(\mathbb{R}^2)$,

$$h_{j_1,j_2}(x,y) = h_{j_1}(x) h_{j_2}(y), \quad j_1, j_2 = 0, 1, \ldots,$$

is the set of eigenfunctions and that $k_i(2n+1)^{m_i}, n = 0, 1, \ldots$, are the eigenvalues for A_i, $i = 1, 2$.

We use the result of Proposition 4.2 to calculate the counting function for $A_1 A_2$. With the transformation $\lambda/(k_1 k_2) \to \lambda$, we can, and we will, assume that $k_1 = k_2 = 1$.

Put

$$I_\lambda = \{n \in \mathbb{N} \cup 0; 1 \leq 2n+1 \leq [\lambda^{1/m_1}]\},$$

$$I_{\lambda,n_1} = \left\{n_2 \in \mathbb{N} \cup 0; 1 \leq 2n_2 + 1 \leq \left[\frac{\lambda^{1/m_2}}{(2n_1+1)^{m_1/m_2}}\right]\right\}.$$

Hence, the cardinal numbers of these sets are not greater than

$$[\lambda^{1/m_1}]/2+1 \text{ and } \frac{1}{2}\left[\frac{\lambda^{1/m_2}}{(2n_1+1)^{m_1/m_2}}\right]+1,$$

respectively. With this notation we have

$$\begin{aligned} N(\lambda) &= \sum_{(2n_1+1)^{m_1}(2n_2+1)^{m_2}\le\lambda} 1 = \sum_{n_1\in I_\lambda}\sum_{n_2\in I_{\lambda,n_1}} 1 \\ &= \frac{1}{2}\sum_{n\in I_\lambda}\left[\frac{\lambda^{1/m_2}}{(2n+1)^{m_1/m_2}}\right]+r_n, \end{aligned}$$

where r_n takes values 0 and 1. We obtain

$$0\le S(\lambda,m_1,m_2)=\sum_{n\in I_\lambda} r_n \le \frac{1}{2}[\lambda^{1/m_1}]+1.$$

Next

$$\frac{1}{2}\sum_{n\in I_\lambda}\left[\frac{\lambda^{1/m_2}}{(2n+1)^{m_1/m_2}}\right]=\frac{1}{2}\sum_{n\in I_\lambda}\frac{\lambda^{1/m_2}}{(2n+1)^{m_1/m_2}}-R(\lambda,m_1,m_2),$$

where

$$0\le R(\lambda,m_1,m_2)\le\frac{1}{2}[\lambda^{1/m_1}]+1.$$

By the proof of the previous proposition, with $r=1$ or $r=0$,

$$\begin{aligned} \sum_{n\in I_\lambda}\frac{1}{(2n+1)^{m_1/m_2}} &= \sum_{t=1}^{\frac{1}{2}[\lambda^{1/m_1}]+r}\frac{1}{t^{m_1/m_2}} \\ &= \zeta\left(\frac{m_1}{m_2}\right)-\frac{m_2}{m_1-m_2}\left(\frac{1}{2}([\lambda^{1/m_1}]+r)\right)^{\frac{m_2-m_1}{m_2}}+O(1/\lambda^{1/m_2}). \end{aligned}$$

This implies

$$\begin{aligned} N(\lambda) &= \frac{1}{2}\lambda^{1/m_2}\left(\zeta\left(\frac{m_1}{m_2}\right)-\frac{m_2}{m_1-m_2}\left(\frac{1}{2}([\lambda^{1/m_1}]+r)\right)^{\frac{m_2-m_1}{m_2}}\right. \\ &\quad \left.+O(1/\lambda^{1/m_2})\right)-R(\lambda,m_1,m_2)+S(\lambda,m_1,m_2) \\ &= \frac{\lambda^{1/m_2}}{2}\zeta\left(\frac{m_1}{m_2}\right)-\frac{m_2}{m_1-m_2}2^{\frac{m_1-2m_2}{m_2}}\lambda^{1/m_1} \\ &\quad -R(\lambda,m_1,m_2)+S(\lambda,m_1,m_2)+O(1) \\ &= \frac{\lambda^{1/m_2}}{2}\zeta\left(\frac{m_1}{m_2}\right)+C(\lambda,m_1,m_2)\lambda^{1/m_1}+O(1), \end{aligned}$$

where

$$-\frac{2^{\frac{m_1-2m_2}{m_2}} m_2}{m_1-m_2} - \frac{[\lambda^{1/m_1}]}{2\lambda^{1/m_1}} - \frac{1}{\lambda^{1/m_1}} \leq C(\lambda, m_1, m_2)$$
$$\leq -\frac{2^{\frac{m_1-2m_2}{m_2}} m_2}{m_1-m_2} + \frac{[\lambda^{1/m_1}]}{2\lambda^{1/m_1}} + \frac{1}{\lambda^{1/m_1}}.$$

One can consider in a similar way

$$A_1 = k_1(-\Delta_{x_1} + \|x_1\|^2)^{m_1} + r_1, A_2 = k_2(-\Delta_{x_2} + \|x_2\|^2)^{m_2} + r_2, x_i \in \mathbb{R}^{n_1},$$

$k_1 > k_2 > 0, r_i > 0, i = 1, 2$; but the computation is much more complicate.

Acknowledgment

The authors thank the unknown referee for the critical remarks and suggestions which led to the improvement of the paper.

References

[1] J. Aramaki, *On an extension of the Ikehara Tauberian theorem.* Pacific J. Math. **133** (1988), no. 1, 13–30.

[2] U. Battisti, *Weyl asymptotics of Bisingular Operators and Dirichlet Divisor Problem.* preprint, http://arxiv.org/abs/1012.1518.

[3] U. Battisti and S. Coriasco, *Wodzicki residue for operators on manifolds with cylindrical ends.* Ann. Global Anal. Geom. **40** (2021), no. 2, 223–249.

[4] P. Boggiatto, E. Buzano, and L. Rodino, *Global Hypoellipticity and Spectral Theory.* Mathematical Research, vol. 92, Akademie Verlag, Berlin, 1996.

[5] D. Calvo and L. Rodino, *Iterates of operators and Gelfand-Shilov functions.* Int. Transf. Spec. Funct. **22** (2011), no. 5-6, 269–276.

[6] M. Cappiello, T. Gramchev, and L. Rodino, *Super-exponential decay and holomorphic extensions for semilinear equations with polynomial coefficients.* J. Funct. Anal. **237** (2006), no. 2, 634–654.

[7] M. Cappiello, T. Gramchev, and L. Rodino, *Entire extensions and exponential decay for semilinear elliptic equations.* J. Anal. Math. **111** (2010), 339–367.

[8] S. Coriasco, E. Schrohe, and J. Seiler, *Bounded imaginary powers of differential operators on manifolds with conical singularities.* Math. Z. **244** (2003), no. 2, 235–269.

[9] M. de Gosson and F. Luef, *Spectral and regularity properties of a pseudo-differential calculus related to Landau quantization.* J. Pseudo-Differ. Oper. Appl. **1** (2010), no. 1, 3–34.

[10] I.M. Gelfand and G.E. Shilov, *Generalized Functions. Vol. 2: Spaces of Fundamental and Generalized Functions.* Academic Press [Harcourt Brace Jovanovich Publishers], New York, 1968 [1977].

[11] I.S. Gradshteyn and I.M. Ryzhik, *Table of Integrals, Series, and Products.* 7th ed. Elsevier/Academic Press, Amsterdam, 2007.

[12] T. Gramchev, S. Pilipović, and L. Rodino, *Classes of degenerate elliptic operators in Gelfand-Shilov spaces.* New developments in pseudo-differential operators. Operator Theory: Advances and Applications **189** (2009), 15–31.

[13] T. Gramchev, S. Pilipović, and L. Rodino, *Global regularity and stability in S-spaces for classes of degenerate Shubin operators.* Pseudo-Differential Operators: Complex Analysis and Partial Differential Equations. Operator Theory: Advances and Applications **205** (2010), 81–90.

[14] T. Gramchev, S. Pilipović, and L. Rodino, *Eigenfunction expansion in* $\mathbb{R}^n$, Proc. AMS. Soc. **139** (2011), 4361–4368.

[15] Todor Gramchev, Stevan Pilipović, Luigi Rodino, and M.W. Wong, *Spectral properties of the twisted bi-Laplacian.* Arch. Math. (Basel), **93** (2009), no. 6, 565–575.

[16] T. Gramchev, S. Pilipović, L. Rodino and M.W. Wong, *Spectra of Polynomials of the Twisted Laplacian*, to appear in Accad. Sci. Torino.

[17] G.H. Hardy, *On Dirichlet's divisor problem.* Proc. London Math. Soc. (2) **15** (1916), 1–25.

[18] B. Helffer, *Théorie spectrale pour des opérateurs globalement elliptiques.* Astérisque, vol. 112, Société Mathématique de France, Paris, 1984.

[19] M.N. Huxley, *Exponential sums and lattice points III*, Proc. London Math. Soc. (3), **87** (2003), no. 3, 591–609.

[20] Aleksandar Ivić, *The Riemann zeta-function.* Dover Publications Inc., Mineola, NY, 2003, Theory and Applications, Reprint of the 1985 original.

[21] H. Iwaniec and C.J. Mozzochi, *On the divisor and circle problems.* J. Number Theory **29** (1988), no. 1, 60–93.

[22] M. Langenbruch, *Hermite functions and weighted spaces of generalized functions.* Manuscripta Math. **119** (2006), 269–285.

[23] B.V. Lange and V.S. Rabinovich, *Criterion for Noethericity of pseudodifferential operators on* $\mathbf{R}^n$ *with symbols of class* $C_b^\infty(\mathbf{R}^n \times \mathbf{R}^n)$. Dokl. Akad. Nauk SSSR **285** (1985), no. 6, 1317–1320.

[24] B.V. Lange and V.S. Rabinovich, *Pseudodifferential operators in* $\mathbf{R}^n$ *and limit operators.* Mat. Sb. (N.S.) **129(171)** (1986), no. 2, 175–185, 303.

[25] T. Levajkovic, S. Pilipović, and D. Selesi, *Chaos expansion transforms appliedons to equations with Malliavins derivative.* Int. Transf. Spec. Funct. **22** (2011), 97–105.

[26] S. Molahajloo and M.W. Wong, *Hierarchical Weyl transforms and the heat semigroup of the hierarchical twisted Laplacian.* J. Pseudo-Differ. Oper. Appl. **1** (2010), 35–53.

[27] S. Molahajloo, L. Rodino, and M.W. Wong, *Global hypoellipticity of the hierarchical twisted Laplacian*, preprint, (2011).

[28] F. Nicola and L. Rodino, *Global Pseudo-Differential Calculus on Euclidean Spaces.* Pseudo-Differential Operators. Theory and Applications, vol. 4, Birkhäuser Verlag, Basel, 2010.

[29] V.S. Pilidi, *Multidimensional bisingular operators.* Dokl. Akad. Nauk SSSR **201** (1971), 787–789.

[30] S. Pilipović, *Tempered ultradistributions*, Boll. Un. Mat. Ital., B (7) **2** (1988), no. 2, 235–251.

[31] V.S. Rabinovich, *A priori estimates and the Fredholm property of a certain class of pseudodifferential operators*, Mat. Sb., 92 **134** (1973), 195–208, 343.

[32] V.S. Rabinovich, *Pseudodifferential operators with operator symbols: local invertibility and limit operators*, Linear Topol. Spaces Complex Anal., **1** (1994), 58–73.

[33] V.S. Rabinovich and Steffen Roch, *Wiener algebras of operators, and applications to pseudodifferential operators*, Z. Anal. Anwendungen, **23** (2004), no. 3, 437–482.

[34] L. Rodino, *A class of pseudo differential operators on the product of two manifolds and applications*, Ann. Scuola Norm. Sup. Pisa Cl. Sci., (4) **2** (1975), no. 2, 287–302.

[35] R.T. Seeley, *Complex powers of an elliptic operator*. Singular Integrals (Proc. Sympos. Pure Math., Chicago, Ill., 1966), Amer. Math. Soc., Providence, R.I., 1967, pp. 288–307.

[36] M.A. Shubin, *Pseudodifferential Operators and Spectral Theory*. Springer Series in Soviet Mathematics, Springer-Verlag, Berlin, 1987, Translated from the Russian by Stig I. Andersson.

[37] M.W. Wong, *The heat equation for the Hermite operator on the Heisenberg group*. Hokkaido Math. J. **34** (2005), no. 2, 393–404.

U. Battisti and L. Rodino
Università degli Studi di Torino
Dipartimento di Matematica,
Via Carlo Alberto 10
I-10123, Torino, Italia
e-mail: `ubertino.battisti@unito.it`
`luigi.rodino@unito.it`

T. Gramchev
Università di Cagliari
Dipartimento di Matematica
Via Ospedale 72
I-09124 Cagliari, Italia
e-mail: `todor@unica.it`

S. Pilipović
Faculty of Sciences and Mathematics
University of Novi Sad
Trg Dositeja Obradovica 4
21000, Novi Sad, Serbia
e-mail: `stevan.pilipovic@uns.dmi.ac.rs`

Operator Theory:
Advances and Applications, Vol. 228, 39–55

The Index Formula of Douglas for Block Toeplitz Operators on the Bergman Space of the Ball

Albrecht Böttcher and Antti Perälä

For Vladimir Rabinovich on his 70th birthday

Abstract. The index formula of Douglas is a formula which expresses the index of a Fredholm Toeplitz operator with a discontinuous symbol as the limit of the indices of a family of Fredholm Toeplitz operators with continuous symbols. This paper is concerned with Toeplitz operators on the Bergman space A^p of the unit ball of $\mathbb{C}^m$. The symbols are supposed to be matrix functions with entries in $C+H^\infty$ or to be certain discontinuous matrix functions which are locally elliptic in a sense. The main result reduces the index computation for the Toeplitz operators under consideration to the case of continuous matrix symbols.

Mathematics Subject Classification (2010). Primary 47B35. Secondary 30H20, 47B53.

Keywords. Bergman space, Toeplitz operator, index formula, discontinuous symbol.

1. Introduction and main results

Let $B = \mathbb{B}_m$ the unit ball in $\mathbb{C}^m$ and denote by $S = \mathbb{S}^{2m-1}$ the boundary of B. The $\mathbb{C}^N$-valued Bergman space $[A^p(B)]^N$ $(1 < p < \infty)$ is the Banach space of all holomorphic $\mathbb{C}^N$-valued functions in B which belong to $[L^p(B)]^N$ with normalized volume measure. Given a $\mathbb{C}^{N\times N}$-valued function $a \in [L^\infty(B)]^{N\times N}$, the Toeplitz operator $T(a)$ is the bounded linear operator on $[A^p(B)]^N$ which sends f to $P(af)$, where $P : [L^p(B)]^N \to [A^p(B)]^N$ is the Bergman projection. The matrix function a is called the symbol of the operator $T(a)$.

Toeplitz operators with continuous symbols are fairly well understood. If $a \in [C(\overline{B})]^{N\times N}$, then $T(a)$ is Fredholm if and only if $a|S$ is invertible, and one has

The second author acknowledges support by the Academy of Finland project no. 75166001.

nice formulas for the Fredholm index $\operatorname{Ind} T(a)$; see [3], [6], [11], [22], [25]. A first reasonable Banach algebra beyond $C(\overline{B})$ is $C(\overline{B}) + H^\infty(B)$, where $H^\infty(B)$ is the algebra of all bounded holomorphic functions in B. Douglas [8] was the first to consider Toeplitz operators $T(a)$ with $a \in [C+H^\infty]^{N\times N}$ on the $\mathbb{C}^N$-valued Hardy space of $\mathbb{S}^1$, and he proved that $T(a)$ is Fredholm if and only if the determinant of the harmonic extension $\mathcal{H}a$ of a into $\mathbb{B}_1$ is bounded away from zero near $\mathbb{S}^1$ and that in this case

$$\operatorname{Ind} T(a) = \lim_{r\to 1} \operatorname{Ind} T(a_r) = -\lim_{r\to 1} \operatorname{wind} \det a_r, \tag{1}$$

where $a_r(t) = (\mathcal{H}a)(rt)$. See also [2], [10].

Toeplitz operators on $[A^2(B)]^N$ with symbols in $[C(\overline{B}) + H^\infty(B)]^{N\times N}$ were studied by McDonald [13], and the passage to $[A^p(B)]^N$ was performed in [4], [15], [16]. A matrix function $a \in [C(\overline{B}) + H^\infty(B)]^{N\times N}$ has radial limits almost everywhere on S and hence defines a function a_S on S. We may therefore consider the harmonic extension $\mathcal{H}a$ of a_S into B. Arbitrary functions in $L^\infty(B)$ need not to have radial limits almost everywhere on S and so it is not clear what the harmonic extension should be. Consequently, when working in the Bergman space, one prefers using the Berezin transform, which is defined for all matrix functions in $[L^\infty(B)]^{N\times N}$. We let $\widetilde{a}$ denote the Berezin transform of a. Given a matrix function a on B and a number $r \in (0,1)$, we define a_r on B by $a_r(z) = a(rz)$. Note that if $a \in [C(\overline{B}) + H^\infty(B)]^{N\times N}$, then $a_r \in [C(\overline{B})]^{N\times N}$ for all $r \in (0,1)$. Let $\|\cdot\|$ be any matrix norm on $\mathbb{C}^{N\times N}$. It turns out that for symbols a in $[C(\overline{B}) + H^\infty(B)]^{N\times N}$ the following are equivalent:

(i) $T(a)$ is Fredholm on $[A^p(B)]^N$,
(ii) a_S is invertible in $[C(S) + H^\infty(S)]^{N\times N}$;
(iii) there are numbers $r_0 \in (0,1)$ and $M \in (0,\infty)$ such that $a_r|S$ is invertible and $\|a_r^{-1}|S\| \le M$ for all $r_0 < r < 1$;
(iv) there are numbers $r_0 \in (0,1)$ and $M \in (0,\infty)$ such that $(\mathcal{H}a)_r|S$ is invertible and $\|(\mathcal{H}a)_r^{-1}|S\| \le M$ for all $r_0 < r < 1$;
(v) there are numbers $r_0 \in (0,1)$ and $M \in (0,\infty)$ such that $\widetilde{a}_r|S$ is invertible and $\|\widetilde{a}_r^{-1}|S\| \le M$ for all $r_0 < r < 1$.

McDonald [13] proved that the index is zero if $N = 1$ and $m \ge 2$, but did curiously say nothing about the index for $N = m = 1$. The latter case was disposed of in [16]. Except for these two results, we have not seen any extension of the index formula (1) to general N and m in the literature. This problem was mentioned in [15], [16], and there it was also pointed out that the formula $\operatorname{Ind} A = \operatorname{Ind} \det A$, which holds for operator matrices A whose entries commute modulo trace class operators, does not suffice to reach all of $C+H^\infty$ even for the Bergman space on the disk $\mathbb{B}_1$. Moreover, as Fredholm Toeplitz operators with continuous scalar-valued symbols always have index zero if $m \ge 2$, it seems to be principally impossible to reduce the index computation of $T(a)$ to that of $\det T(a)$.

However, there is another approach to index formulas. This approach is due to Silbermann [21], and it is based on combining harmonic approximation with so-called stable convergence; see also [2]. The purpose of this paper is to demonstrate that this approach yields the first equality of Douglas' formula (1) for all N and m. Of course, the second equality must be replaced by any of the index formulas known for continuous matrix symbols.

We note that the method of [21] gives the desired index formula for $C + H^\infty$ symbols almost immediately in the Hardy space setting. The same is not true in the context of Bergman spaces. The reason is that if we are in the Hardy space and one of the symbols a and b is continuous, then $T(a_r b_r) = T((ab)_r) + K + C_r$ where K is compact and $\|C_r\| \to 0$ as $r \to 1$. This nice circumstance is the foundation of the Banach algebraic approach of [21]. In the Bergman space, we can merely show that $T(a_r b_r) = T((ab)_r) + K_r + C_r$ where K_r is compact for each r and $\|C_r\| \to 0$. This makes things a little difficult even for $C + H^\infty$ symbols.

Herewith our first main result.

Theorem 1.1. *Let $1 < p < \infty$ and $a \in [C(\overline{B}) + H^\infty(B)]^{N\times N}$. Suppose the Toeplitz operator $T(a)$ is Fredholm on the Bergman space $[A^p(B)]^N$. Then there is a number $r_0 \in (0,1)$ such that the Toeplitz operators $T(a_r)$ are Fredholm on $[A^p(B)]^N$ for all $r_0 < r < 1$ and*

$$\operatorname{Ind} T(a) = \lim_{r\to 1} \operatorname{Ind} T(a_r). \tag{2}$$

In particular, $\operatorname{Ind} T(a) = 0$ *whenever* $N < m$.

We remark that the theorem remains literally true with a_r replaced by $(\mathcal{H}a)_r$ or $\widetilde{a}_r$.

We secondly consider another class of discontinuous matrix symbols in the algebra $[L^\infty(B)]^{N\times N}$. This class is motivated by paper [1], which was devoted to the case $N = m = 1$ and $p = 2$, and this class contains symbols with jumps. Note that paper [1] had in turn one source of motivation in McDonald's paper [14]. While in the case of Hardy spaces jumps and locally sectorial symbols come along with lenses bounded by two circular arcs when passing to general $p \in (1,\infty)$, we will encounter certain ellipses with their foci in the endpoints of the jumps when considering the Bergman spaces with $p \in (1,\infty)$. However, note that we will establish only sufficient conditions for Fredholmness, so that we cannot exclude that eventually the ellipses may be replaced by smaller lentiform domains, perhaps even circular lenses. Our main result in this context is as follows. The precise theorem requires a series of definitions and explanations and will be deferred to Section 4.

Theorem 1.2. *Let $a \in [L^\infty(B)]^{N\times N}$ be locally μ-elliptic, where μ is a number determined by the norm of the Bergman projection. Then $T(a)$ is Fredholm on $[A^p(B)]^{N\times N}$ and*

$$\operatorname{Ind} T(a) = \lim_{r\to 1} \operatorname{Ind} T(\widetilde{a}_r).$$

In particular, $\operatorname{Ind} T(a) = 0$ *if* $N < m$.

Section 2 contains the proof of Theorem 1.1. Section 3 is devoted to the notion of μ-ellipticity. In Section 4 we state and prove the precise version of Theorem 1.2.

2. $C + H^\infty$ symbols

Each of the four Fredholm criteria quoted in Section 1 shows that if $T(a)$ is Fredholm on $[A^p(B)]^N$ for some $p \in (1,\infty)$, then $T(a)$ is Fredholm on $[A^p(B)]^N$ for all exponents $p \in (1,\infty)$. Let $M(a)$ denote the operator of multiplication by a. The Fredholm properties of $T(a)$ on $[A^p(B)]^N$ are clearly the same as those of $PM(a)P + I - P$ on $[L^p(B)]^N$. A classical result by Shneiberg [20] implies that if a bounded linear operator A is Fredholm on $[L^p(B)]^N$ for all p in some open subset U of $(1,\infty)$, then the index of A is constant on the connected components of U. Therefore it suffices to prove Theorem 1.1 in the case $p = 2$.

Our proof is based on the following result. All operators occurring therein are bounded linear operators on a Banach space. We denote strong convergence (= pointwise convergence) of operators by $\to$ and uniform convergence (= convergence in the norm) by $\rightrightarrows$.

Lemma 2.1. (Silbermann [21]) *Let A be Fredholm and $\operatorname{Ind} A = 0$. Suppose F is another operator and for each $r \in (0,1)$ we are given operators A_r and F_r. Suppose also that the operators A_r are Fredholm for r close enough to 1. Then, with convergence understood as convergence for $r \to 1$, we have the following.*

(a) *If $A_r^* \to A^*$, $F_r \to F$, and $F_rA_r = I + K + E_r$ with a compact operator K and with $E_r \rightrightarrows 0$, then there is an $r_0 \in (0,1)$ such that $\operatorname{Ind} A_r \le 0$ for $r_0 < r < 1$.*

(b) *If $A_r \to A$, $F_r^* \to F^*$, and $A_rF_r = I + K + E_r$ with a compact operator K and with $E_r \rightrightarrows 0$, then there is an $r_0 \in (0,1)$ such that $\operatorname{Ind} A_r \ge 0$ for $r_0 < r < 1$.*

Proof. (a) There is a compact operator L such that $A + L$ is invertible. It follows that

$$\begin{aligned} F_r(A_r + L) &= I + K + E_r + F_rL \\ &= I + K + FL + E_r' \end{aligned}$$

with $E_r' \rightrightarrows 0$. Put

$$D_r = F_r - (K + FL)(A + L)^{-1}.$$

Then

$$\begin{aligned} D_r(A_r + L) &= I + K + FL + E_r' - (K + FL)(A + L)^{-1}(A_r + L) \\ &= I + K + FL - (K + FL) + E_r'' \\ &= I + E_r'' \end{aligned}$$

with $E_r'' \rightrightarrows 0$. Therefore $A_r + L$ is left invertible for all r close enough to 1, which implies that $\operatorname{Ind}(A_r + L) \le 0$ and thus $\operatorname{Ind} A_r \le 0$. Part (b) is analogous. □

Suppose now that $a \in [C(\overline{B}) + H^\infty(B)]^{N\times N}$ and that $T(a)$ is Fredholm. Then there are numbers $r_1 \in (0,1)$ and $M \in (0,\infty)$ such that $\|a^{-1}(z)\| \le M$ for $r_1 < |z| < 1$. Letting $r_2 = \sqrt{r_1}$, we obtain that $\|a_r^{-1}(z)\| \le M$ for $r_2 < r < 1$ and $r_2 < |z| < 1$. We define χ on B by $\chi(z) = 1$ if $|z| > r_2$ and $\chi(z) = 0$ if $|z| < r_2$. Then χa_r^{-1} and χa^{-1} are well defined matrix functions in $[L^\infty(B)]^{N\times N}$.

Lemma 2.2. *We have* $M(\chi a_r^{-1}) \to M(\chi a^{-1})$ *and* $M(a_r) \to M(a)$ *as* $r \to 1$.

Proof. We begin with the first assertion. The norms

$$\|M(\chi a_r^{-1})\| = \|\chi a_r^{-1}\|_\infty$$

remain bounded as $r \to 1$. Since $L^\infty(B)$ is dense in $L^2(B)$, it therefore suffices to show that $M(\chi a_r^{-1})v - M(\chi a^{-1})v \to 0$ for every $v \in [L^\infty(B)]^N$. As

$$M(\chi a_r^{-1}) - M(\chi a^{-1}) = M(\chi a_r^{-1})M(a - a_r)M(\chi a^{-1}),$$

we are left with proving that $M(a-a_r)v \to 0$ in the scalar case ($N = 1$). Obviously,

$$\begin{aligned}\|M(a - a_r)v\|_2^2 &= \int_B |a - a_r|^2|v|^2 dV \le \|v\|_\infty^2 \int_B |a - a_r|^2 dV \\ &= \|v\|_\infty^2 \int_B |a(z) - a(rz)|^2 dV(z). \qquad (3)\end{aligned}$$

Fix $\varepsilon > 0$. Since a is bounded, there is a compact set $K \subset B$ such that

$$\int_{B\setminus K} |a(z) - a(rz)|^2 dV(z) < \varepsilon/2,$$

and as a is uniformly continuous on K, it follows that

$$\int_K |a(z) - a(rz)|^2 dV(z) < \varepsilon/2$$

whenever r is close enough to 1. This proves the first assertion.

Since $\|M(a_r - a)v\|_2^2$ can also be estimated from above by (3), we get the second assertion. □

Lemma 2.3. *If* $\operatorname{Ind} T(a) = 0$, *then* $T(a_r)$ *is Fredholm and* $\operatorname{Ind} T(a_r) \le 0$ *for all* r *close enough to* 1.

Proof. We employ Lemma 2.1(a) with $A = T(a)$, $F = T(\chi a^{-1})$, $A_r = T(a_r)$, $F_r = T(\chi a_r^{-1})$. Since $a_r|S$ is invertible for r sufficiently close to 1, the operators $T(a_r)$ are Fredholm for these r. The adjoint of $T(\varphi)$ may be identified with $T(\overline{\varphi})$. From Lemma 2.2 we infer that $A_r^* \to A^*$ and $F_r \to F$. Furthermore,

$$F_r A_r = T(\chi) - PM(\chi a_r^{-1})(I - P)M(a_r)P.$$

The operator $T(\chi)$ is I plus a compact operator. We have $a = c + f$ with c in $[C(\overline{B})]^{N\times N}$ and f in $[H^\infty(B)]^{N\times N}$. Consequently, $(I - P)M(a_r)P$ is equal to $(I-P)M(c_r)P$. Taking into account that c is continuous on the closed ball $\overline{B}$, it is easily seen that $M(c_r) \rightrightarrows M(c)$. The operator $(I - P)M(c)P$ is compact. Finally, $M(\chi a_r^{-1}) \to M(\chi a^{-1})$ due to Lemma 2.2. But if $K_r \rightrightarrows K$, K is compact, and $C_r \to C$, then $C_r K_r \rightrightarrows CK$. Thus,

$$PM(\chi a_r^{-1})(I - P)M(a_r)P = PM(\chi a^{-1})(I - P)M(c)P + E_r = L + E_r$$

with a compact operator L and with certain operators E_r such that $E_r \rightrightarrows 0$. Lemma 2.1 now yields the assertion. □

Recall that the Bergman projection $P : L^2(B) \to A^2(B)$ is defined by

$$(Pf)(z) = \int_B \frac{1}{(1-(z,w))^{m+1}} f(w)\, dV(w), \quad z \in B,$$

that the normalized reproducing kernel of $A^2(B)$ is

$$k_z(w) = \frac{(1-|z|^2)^{(m+1)/2}}{(1-(w,z))^{m+1}},$$

and that the Berezin transform $\widetilde{f}$ of $f \in L^\infty(B)$ is the function

$$\widetilde{f}(z) = (fk_z, k_z) = \int_B \frac{(1-|z|^2)^{m+1}}{|1-(w,z)|^{2m+2}} f(w)\, dV(w), \quad z \in B.$$

This is a bounded continuous function in B.

We need the following well-known properties of the Berezin transform. Let $\|\cdot\|$ denote any matrix norm on $\mathbb{C}^{N\times N}$ and let $L_0^\infty(B)$ stand for the functions f in $L^\infty(B)$ for which the essential supremum of $|f(z)|$ on the set $\{z \in B : r < |z| < 1\}$ goes to zero as $r \to 1$. Toeplitz operators with symbols in $[L_0^\infty(B)]^{N\times N}$ are compact. Recall also that f_r is defined by $f_r(z) = f(rz)$ for $z \in B$. A shell is a set of the form $\{z \in B : 1-\delta < |z| < 1\}$ with some $\delta \in (0,1)$.

Lemma 2.4. *Let $f, g \in [L^\infty(B)]^{N\times N}$.*

(a) *If f is continuous on $\overline{B}$, then $\widetilde{f} \in C(\overline{B})$ and $\widetilde{f}|S = f|S$.*
(b) *If f or g is continuous on $\overline{B}$, then $(fg)^\sim - \widetilde{f}\widetilde{g} \in [L_0^\infty(B)]^{N\times N}$.*
(c) *With* sup *meaning the essential supremum,*

$$\sup_{z\in B} \|\widetilde{f}(z)\| \le \sup_{z\in B} \|f(z)\|.$$

(d) *If f is identically zero on some shell, then $\widetilde{f} \in [L_0^\infty(B)]^{N\times N}$.*

Proof. Property (a) can be proved as in [27] using the automorphisms φ_a of B introduced in Section 2.2 of [19]. Property (b) is easily seen using the argument of the proofs of Proposition 3 of [26] or Proposition 6.1.7 of [27]. As for (c), note that

$$\|\widetilde{f}(z)\| = \left\| \int_B f(w)\, |k_z(w)|^2\, dV(w) \right\| \le \int_B \|f(w)\|\, |k_z(w)|^2\, dV(w)$$
$$\le \|f\|_\infty \int_B |k_z(w)|^2\, dV(w) \quad = \|f\|_\infty.$$

Finally, if $f = 0$ on some shell Ω, we can choose a matrix function φ in $[C(\overline{B})]^{N\times N}$ such that $\varphi f = 0$ on B and $\varphi|\Omega_0 = I$ on some shell $\Omega_0 \subset \Omega$. From (b) we infer that

$$(\varphi f)^\sim - \widetilde{\varphi}\widetilde{f} \in [L_0^\infty(B)]^{N\times N}.$$

Since $(\varphi f)^\sim$ is identically zero, it follows that $\widetilde{\varphi}\widetilde{f} \in [L_0^\infty(B)]^{N\times N}$. But property (a) tells us that $\widetilde{\varphi}(z)$ converges uniformly to I as $|z| \to 1$. $\square$

Lemma 2.5. *If* $\xi \in [L_0^\infty(B)]^{N\times N}$, *then* $T(I+\xi_r)$ *is Fredholm of index zero for all* r *sufficiently close to* 1.

Proof. Given any $\varepsilon > 0$, there is an $r_0 \in (0,1)$ such that $\|\xi(z)\| < \varepsilon$ for $r_0 < |z| < 1$. It follows that $\|\xi(rz)\| < \varepsilon$ for $\sqrt{r_0} < r < 1$ and $\sqrt{r_0} < |z| < 1$. Consequently, the essential norm of $T(\xi_r)$, that is, the distance of $T(\xi_r)$ to the compact operators in the operator norm, is smaller than $\|P\|\,\varepsilon$; see, e.g., the proof of [13, Proposition 1.7]. Thus, if $\|P\|\,\varepsilon < 1$, then $T(I+\xi_r) = I + T(\xi_r)$ can be written as $I + A_r + K_r$ with $\|A_r\| < 1$ and with a compact operator K_r. This implies the assertion. □

If $a \in [C(\overline{B})+H^\infty(B)]^{N\times N}$, then $\widetilde{a}-a$ is in $[L_0^\infty(B)]^{N\times N}$ due to Lemma 2.4(a) and due to the fact that $\widetilde{h} = h$ for $h \in H^\infty(B)$. It therefore suffices to prove formula (2) in the case where a coincides with its Berezin transform, $a = \widetilde{a}$.

Recall that $a_S \in [C(S) + H^\infty(S)]^{N\times N}$ is the function given by the radial limits of a. We know that a_S is invertible in $[C(S) + H^\infty(S)]^{N\times N}$. Let b_S be the inverse and define $b \in [C(\overline{B}) + H^\infty(B)]^{N\times N}$ as the Berezin transform of the harmonic extension of b_S. Thus, $b = \widetilde{b}$ on B. Since $ab|S = ba|S = I$, it follows that the operators $T(a)T(b) - I$ and $T(b)T(a) - I$ are compact, and hence $T(b)$ is Fredholm of index zero together with $T(a)$.

Lemma 2.6. *If* $\operatorname{Ind} T(a) = 0$, *then* $\operatorname{Ind} T(a_r) = \operatorname{Ind} T(b_r) = 0$ *for all* r *close enough to* 1.

Proof. Using that $ab = I$ on S and that $\widetilde{h} = h$ for $h \in H^\infty(B)$, we obtain from Lemma 2.4(b) that $I = \widetilde{a}\widetilde{b} + \gamma = ab + \gamma$ with $\gamma \in [L_0^\infty(B)]^{N\times N}$. There are a shell Ω and a constant $M < \infty$ such that a and b are invertible on Ω and $\|a^{-1}\| \le M$ and $\|b^{-1}\| \le M$ on Ω. Hence $I = ab(I + b^{-1}a^{-1}\gamma)$ on Ω, which implies that $I = ab(I+\xi) + \eta$ with $\xi \in [L_0^\infty(B)]^{N\times N}$ and some matrix function $\eta \in [L_0^\infty(B)]^{N\times N}$ which vanishes identically on Ω. We have $I = a_r b_r (I + \xi_r) + \eta_r$ and thus $I = T(a_r)T(b_r)T(I+\xi_r) + T(\eta_r) + K_r$ with some compact operator K_r. Since $T(\eta_r)$ is compact and $T(I+\xi_r)$ is Fredholm of index zero for r close enough to 1 by Lemma 2.5, we conclude that $0 = \operatorname{Ind} T(a_r) + \operatorname{Ind} T(b_r)$ whenever r is sufficiently close to 1. However, we know from Lemma 2.3 that $\operatorname{Ind} T(a_r) \le 0$, and the same lemma applied to $T(b)$ also shows that $\operatorname{Ind} T(b_r) \le 0$. Consequently, both indices must be zero. □

At this point we have proved formula (2) under the extra assumption that $\operatorname{Ind} T(a) = 0$.

Lemma 2.7. *If* $\operatorname{Ind} T(a) = \kappa$, *then* $T(a_r)$ *is Fredholm and* $\operatorname{Ind} T(a_r) = \kappa$ *for all* r *close enough to* 1.

Proof. Choose an integer $k \ge 0$ such that $N + k \ge m$ and consider the $\mathbb{C}^{(N+k)\times(N+k)}$-valued matrix function v given by $v = \operatorname{diag}(a, I_k)$. Clearly $T(v)$ is Fredholm and $\operatorname{Ind} T(v) = \kappa$. Since $N + k \ge m$, we obtain from [25] a matrix function $\psi \in [C(\overline{B})]^{(N+k)\times(N+k)}$ such that $T(\psi)$ is Fredholm and $\operatorname{Ind} T(\psi) = -\kappa$. By Lemma 2.4(a), we may without loss of generality assume that $v = \widetilde{v}$ and

$\psi = \widetilde{\psi}$. Put $w = v\psi$. Then $T(w)$ equals $T(v)T(\psi)$ plus a compact operator, and hence $T(w)$ is Fredholm of index zero. From Lemma 2.6 we deduce that $T(w_r)$ and thus also $T(\widetilde{w}_r)$ is Fredholm of index zero whenever r is close enough to 1. From Lemma 2.4(b) we infer that $\widetilde{w} = v\psi + \gamma$ with $\gamma \in [L_0^\infty(B)]^{N\times N}$. We may write $v\psi = \widetilde{w}(I - \widetilde{w}^{-1}\gamma) = \widetilde{w}(I - \xi) + \eta$ as in the previous proof to obtain that $v_r\psi_r = \widetilde{w}_r(I - \xi_r) + \eta_r$ and thus

$$\operatorname{Ind} T(v_r) + \operatorname{Ind} T(\psi_r) = \operatorname{Ind} T(\widetilde{w}_r) + \operatorname{Ind} T(I + \xi_r) = 0$$

for r sufficiently close to 1. Since

$$\operatorname{Ind} T(\psi_r) = \operatorname{Ind} T(\psi) = -\kappa \quad \text{and} \quad \operatorname{Ind} T(v_r) = \operatorname{Ind} T(a_r),$$

we arrive at the equality $\operatorname{Ind} T(a_r) - \kappa = 0$, as desired. □

The proof of formula (2) is complete. That $\operatorname{Ind} T(a) = 0$ for $N < m$ follows immediately from (2) and the fact that Fredholm Toeplitz operators with continuous matrix symbols always have index zero if $N < m$. This last fact can in turn be proved in several ways. For example, we have Boutet de Monvel's beautiful formula

$$\operatorname{Ind} T(a) = -\frac{1}{(2\pi i)^m} \frac{(m-1)!}{(2m-1)!} \int_S \operatorname{trace}\left((a^{-1}\, da)^{2m-1}\right)$$

for Toeplitz operators with continuous matrix symbols. Proofs can be found in [3], [11], [22]. This formula implies that $\operatorname{Ind} T(a) = 0$ for $N < m$. An argument which is independent of this formula is as follows.

Let $a \in [C(\overline{B})]^{N\times N}$ and suppose $a|S$ is invertible. Then the matrix functions $|a|$ and u in the polar decomposition $a = |a|u$ are continuous on S. We extend $|a|$ and u to functions on B by $|a|(\varrho t) = \varrho|a|(t)$ and $u(\varrho t) = \varrho u(t)$. Then $\operatorname{Ind} T(a) = \operatorname{Ind} T(|a|) + \operatorname{Ind} T(u) = \operatorname{Ind} T(u)$. Note that u is a continuous map of $\mathbb{S}^{2m-1}$ into the group $U(N)$. Suppose we know that the homotopy group $\pi_{2m-1}(U(N))$ is finite. Then $\operatorname{Ind} T(u)$ may assume only finitely many values $\kappa_1, \ldots, \kappa_s$. However,

$$\operatorname{Ind} T(u^n) = \operatorname{Ind}\left(T(u)^n + \text{compact operator}\right) = n \operatorname{Ind} T(u)$$

for every natural number n, and since $\operatorname{Ind} T(u^n)$ must also be one of the numbers $\kappa_1, \ldots, \kappa_s$, the finite set $\{\kappa_1, \ldots, \kappa_s\}$ is invariant under multiplication by an arbitrary natural number. But this is only possible if the set is the singleton $\{0\}$. Consequently, $\operatorname{Ind} T(u) = 0$, as desired.

Finally, it is well known that the homotopy groups $\pi_{2m-1}(U(N))$ are finite for $N < m$. We thank Thomas Püttman of Bochum for acquainting us with the following reasoning. There is a general theorem which says that the rational homotopy groups $\pi_j(G) \otimes \mathbb{Q}$ of a Lie group G of rank r are isomorphic to those of a product of r odd spheres. If $G = U(N)$, then, by [17], these spheres are just the first N odd spheres $\mathbb{S}^1, \mathbb{S}^3, \ldots, \mathbb{S}^{2N-1}$. But if $2k - 1 \le 2N - 1 < 2m - 1$, then $\pi_{2m-1}(\mathbb{S}^{2k-1})$ is finite, and hence $\pi_{2m-1}(\mathbb{S}^1 \times \mathbb{S}^3 \times \cdots \times \mathbb{S}^{2N-1}) \otimes \mathbb{Q} = 0$. It follows that $\pi_{2m-1}(U(N)) \otimes \mathbb{Q} = 0$, which implies that $\pi_{2m-1}(U(N))$ must be finite.

3. μ-ellipticity

We denote the norm of the Bergman projection on $L^p(B)$ by $\|P\|_p$. If $p=2$, then P is an orthogonal projection and hence $\|P\|_2 = 1$. In the case of general $p \in (1,\infty)$ we have $\|P\|_p \geq 1$. Estimates for $\|P\|_p$ can be found in [7] and [28]. From the Riesz-Thorin interpolation theorem we deduce that $p \mapsto \|P\|_p$ is a continuous function (because $\log \|P\|_p$ is a convex function of $1/p$) which is monotonically decreasing on $(1,2]$ and monotonically increasing on $[2,\infty)$.

Let $a \in [L^\infty(B)]^{N\times N}$. From now on we let $\|\cdot\|$ stand for the spectral norm on $\mathbb{C}^{N\times N}$. We also equip $[L^p(B)]^N$ with the norm

$$\|f\| = (\|f_1\|_p^2 + \cdots + \|f_N\|_p^2)^{1/2}.$$

With these two conventions, the norm of the operator $f \mapsto af$ on $[L^p(B)]^N$ is the essential supremum of $\|a(z)\|$, z ranging over B.

Let $\mu \in [2,\infty)$ be a real number. A compact subset $\mathcal{R}$ of $\mathbb{C}^{N\times N}$ is called μ-elliptic if there are invertible matrices $c, d \in \mathbb{C}^{N\times N}$ such that

$$\|I - cwd\| < \sin\frac{\pi}{\mu} \tag{4}$$

for all $w \in \mathcal{R}$. This definition is motivated by the notion of μ-sectoriality, for which see, e.g., [2, p. 221]. Now let $a \in [L^\infty(B)]^{N\times N}$. For a measurable subset U of B with positive measure, we denote by $\mathcal{R}_U(a) \subset \mathbb{C}^{N\times N}$ the essential range of $a|U$. Equivalently, $\mathcal{R}_U(a)$ is the set of all matrices $w \in \mathbb{C}^{N\times N}$ for which the operator of multiplication by $(a-w)|U$ is not invertible on $[L^2(U)]^N$. This implies that $\mathcal{R}_U(a)$ is a compact set. The matrix function a is called μ-elliptic on U if $\mathcal{R}_U(a)$ is a μ-elliptic set. We call a globally μ-elliptic if a is μ-elliptic on some shell.

Proposition 3.1. *Define the number $\mu \in [2,\infty)$ by $1/\sin(\pi/\mu) = \|P\|_p$. If a in $[L^\infty(B)]^{N\times N}$ is μ-elliptic on the entire ball B, then $T(a)$ is invertible on $[A^p(B)]^N$, and if a is globally μ-elliptic, then $T(a)$ is Fredholm of index zero on $[A^p(B)]^N$.*

Proof. Suppose first that a is μ-elliptic on B. Then (4) holds for all w in $\mathcal{R}_B(a)$, which implies that

$$\|I - cad\|_\infty < \sin(\pi/\mu)$$

and hence that

$$\|I - cT(a)dI\| = \|T(I-cad)\| \leq \|P\|_p \, \|I - cad\|_\infty < 1.$$

Consequently, $cT(a)dI$ and thus also $T(a)$ are invertible. Now suppose a is μ-elliptic on a shell. Then (4) holds for all essential values of a on the shell $1-\delta < |z| < 1$. We take any of these values, say w_0, and define a_0 by $a_0(z) = w_0$ for $|z| < 1-\delta$ and $a_0(z) = a(z)$ for $1-\delta < |z| < 1$. Then a_0 is μ-elliptic on B and $T(a) - T(a_0)$ is compact. Since $T(a_0)$ is invertible, we conclude that $T(a)$ is Fredholm of index zero. □

It is readily seen that if $N=1$ and $\mu=2$, then a compact set $\mathcal{R} \subset \mathbb{C}$ is 2-elliptic if and only if $\mathcal{R}$ is contained in an open half-plane whose boundary passes through the origin. This happens in turn if and only if the origin does not belong to the convex hull of $\mathcal{R}$, that is, $0 \notin \operatorname{conv}\mathcal{R}$.

For $N = 1$ and general μ we have the following geometrical description of μ-ellipticity. A $2\pi/\mu$-sector is the open subset in $\mathbb{C}$ lying between two rays which start at the origin and make the angle $2\pi/\mu$ at the origin. Given points $\alpha, \beta \in \mathbb{C}$, we denote by $\mathcal{E}_\mu(\alpha, \beta)$ the boundary and interior of the ellipse with the foci α, β, with the major semi-axis $1/\sin(\pi/\mu)$, and with the minor semi-axis $1/\tan(\pi/\mu)$. Thus, for $\alpha = -1$ and $\beta = 1$,

$$\mathcal{E}_\mu(-1,1) = \left\{ x + iy \in \mathbb{C} : x^2 \sin^2 \frac{\pi}{\mu} + y^2 \tan^2 \frac{\pi}{\mu} \le 1 \right\}.$$

Proposition 3.2. *Let $\mathcal{R}$ be a compact subset of $\mathbb{C}$.*

(a) *The set $\mathcal{R}$ is μ-elliptic if and only if $\mathcal{R}$ is contained in an open disk which is completely contained in a $2\pi/\mu$-sector.*

(b) *If the convex hull of $\mathcal{R}$ is a line segment, $\operatorname{conv}\mathcal{R} = [\alpha, \beta]$, then for $\mathcal{R}$ to be μ-elliptic it is necessary and sufficient that $0 \notin \mathcal{E}_\mu(\alpha, \beta)$.*

Proof. It will be convenient to use the abbreviations

$$s = \sin \frac{\pi}{\mu}, \quad c = \cos \frac{\pi}{\mu}, \quad t = \tan \frac{\pi}{\mu}.$$

(a) Condition (4) amounts to the requirement that $|1 - cwd| < s$. Putting $z_0 = 1/(cd)$, this means that $|w - z_0| < s|z_0|$. The rest is elementary plane geometry.

(b) Since the situation is invariant under homotheties, we may without loss of generality assume that the length of the line segment $[\alpha, \beta]$ is 2. By virtue of (a), we have to show that $[\alpha, \beta]$ is contained in an open disk lying in a $2\pi/\mu$-sector if and only if $0 \notin \mathcal{E}_\mu(\alpha, \beta)$. It is easily seen that $[\alpha, \beta]$ is contained in such a disk if and only if there is a closed disk which has both α and β on its boundary and which is contained in a $2\pi/\mu$-sector. The center z_0 of this disk is on the median of $[\alpha, \beta]$. We parametrize this median by a parameter $\delta \in (-\infty, \infty)$, so that $|\delta|$ is the distance of z_0 to $[\alpha, \beta]$ and the two signs of δ correspond to the two sides of $[\alpha, \beta]$. The radius of this disk is $\sqrt{1 + \delta^2}$. Consequently, the disk is contained in a $2\pi/\mu$-sector if and only if $\sqrt{1 + \delta^2}/|z_0| < s$. This is equivalent to the inequality $\sqrt{1 + \delta^2}/s < |z_0|$, which means that the closed disk $\overline{D}_\delta$ with center z_0 and radius $\sqrt{1 + \delta^2}/s$ does not contain the origin. Thus, $[\alpha, \beta]$ is contained in an open disk lying in a $2\pi/\mu$-sector if and only if there exists a δ such that $0 \notin \overline{D}_\delta$, which happens if and only if $0 \notin \cap \overline{D}_\delta$. The assertion is therefore equivalent to the equality $\cap \overline{D}_\delta = \mathcal{E}_\mu(\alpha, \beta)$. To show this equality we may without loss of generality assume that $[\alpha, \beta] = [-1, 1]$.

We first prove that $\cap \overline{D}_\delta \subset \mathcal{E}_\mu(-1, 1)$. Take a point $x + iy \notin \mathcal{E}_\mu(-1, 1)$ and suppose without loss of generality that $y \le 0$. Take $\delta \ge 0$ as $\delta = -ys^2/c^2$ and consider the disk $\overline{D}_\delta$ with center $i\delta$ and radius $\sqrt{1 + \delta^2}/s$. For the distance d

between $x+iy$ and the center $i\delta$ of this disk we have

$$d^2 = x^2 + (y-\delta)^2 = x^2 + y^2\left(1+\frac{s^2}{c^2}\right)^2 = \frac{1}{s^2}\left(x^2s^2 + y^2\frac{s^2}{c^4}\right)$$
$$= \frac{1}{s^2}\left(x^2s^2 + y^2\frac{s^2}{c^2} + y^2\frac{s^4}{c^4}\right) = \frac{1}{s^2}\left(x^2s^2 + y^2t^2 - y^2\frac{s^4}{c^4}\right),$$

and the radius R od $\overline{D}_\delta$ satisfies

$$R^2 = \frac{1+\delta^2}{s^2} = \frac{1}{s^2}\left(1+y^2\frac{s^4}{c^4}\right).$$

Thus, if $x+iy \notin \mathcal{E}_\mu(-1,1)$ and hence $x^2s^2+y^2t^2>1$, then $d>R$, which implies that $x+iy$ does not belong to $\overline{D}_\delta$ and all the more not to $\cap\overline{D}_\delta$.

To prove that $\mathcal{E}_\mu(-1,1)\subset\cap\overline{D}_\delta$, pick $x+iy\in\mathcal{E}_\mu(-1,1)$. We then have the inequality

$$x^2s^2+y^2t^2\le 1.$$

Let d be the distance between $x+iy$ and the center $i\delta$ of $\overline{D}_\delta$ and let R be the radius of $\overline{D}_\delta$. We must show that $d^2 = x^2+(y-\delta)^2 \le \sqrt{1+\delta^2}/s = R^2$ or equivalently,

$$f(\delta) := 1+\delta^2 - s^2(x^2+(y-\delta)^2)\ge 0$$

for all $\delta\in(-\infty,\infty)$. But the graph of f is a convex parabola, and a straightforward computation reveals that the minimum of $f(\delta)$ is

$$f\left(-y\frac{s^2}{c^2}\right) = 1 - x^2s^2 - y^2t^2 \ge 0.$$

This completes the proof. □

It is well known that the local spectra of Toeplitz operators with piecewise continuous symbols on the Hardy spaces of $\mathbb{S}^1$ are circular arcs. When considering the finite section method for these Toeplitz operators or when passing to such Toeplitz operators with locally sectorial symbols, these circular arcs blow up to lenses. For Toeplitz operators on the Hardy spaces of $\mathbb{S}^1$, μ-sectoriality on $U\subset\mathbb{S}^1$ means that $\mathcal{R}_U(a)$ is contained in a $2\pi/\mu$-sector. This is weaker than μ-ellipticity, which additionally requires that $\mathcal{R}_U(a)$ is contained in a disk lying in a $2\pi/\mu$ sector. The lens $\mathcal{O}_\mu(\alpha,\beta)$ is defined as the set of all points in $\mathbb{C}$ at which the line segment $[\alpha,\beta]$ is seen at an angle of at least $2\pi/\mu$, and for a compact set $\mathcal{R}\subset\mathbb{C}$ with $\operatorname{conv}\mathcal{R}=[\alpha,\beta]$ to be μ-sectorial it is necessary and sufficient that $0\notin\mathcal{O}_\mu(\alpha,\beta)$. This is again weaker than condition (b), because it is easily seen that always $\mathcal{O}_\mu(\alpha,\beta)\subset\mathcal{E}_\mu(\alpha,\beta)$. However note that $\mathcal{O}_\mu(\alpha,\beta)$ touches $\mathcal{E}_\mu(\alpha,\beta)$ from inside at the two points with minimal curvature. Finally, note that if $\mu=2$, then $\mathcal{O}_\mu(\alpha,\beta)$ and $\mathcal{E}_\mu(\alpha,\beta)$ both degenerate to the line segment $[\alpha,\beta]$. Figures 1 and 2 show examples.

The following result provides additional insight in the case where $\mu=2$ and N is arbitrary.

Proposition 3.3. *Let $\mathcal{R}$ be a compact subset of $\mathbb{C}^{N\times N}$. The following are equivalent.*

(i) *$\mathcal{R}$ is 2-elliptic.*
(ii) *There is an invertible matrix $d \in \mathbb{C}^{N\times N}$ such that $\|I - wd\| < 1$ for all $w \in \mathcal{R}$.*
(iii) *There is an invertible matrix $c \in \mathbb{C}^{N\times N}$ such that $\|I - cw\| < 1$ for all $w \in \mathcal{R}$.*
(iv) *There exist invertible matrices $c, d \in \mathbb{C}^{N\times N}$ and a number $\varepsilon > 0$ such that $\mathrm{Re}\,(cwd) \geq \varepsilon I$ for all $w \in \mathcal{R}$.*
(v) *There exist an invertible matrix $d \in \mathbb{C}^{N\times N}$ and a number $\varepsilon > 0$ such that $\mathrm{Re}\,(wd) \geq \varepsilon I$ for all $w \in \mathcal{R}$.*
(vi) *There exist an invertible matrix $c \in \mathbb{C}^{N\times N}$ and a number $\varepsilon > 0$ such that $\mathrm{Re}\,(cw) \geq \varepsilon I$ for all $w \in \mathcal{R}$.*

If, in addition, $\mathrm{conv}\,\mathcal{R}$ is a line segment, that is, a set of the form $\{(1-\xi)\alpha + \xi\beta : \xi \in [0,1]\}$ with certain $\alpha, \beta \in \mathbb{C}^{N\times N}$, then $\mathcal{R}$ is 2-elliptic if and only if $\mathrm{conv}\,\mathcal{R}$ consists of invertible matrices only.

Proof. What we call 2-ellipticity here is analytic sectoriality in the book [2]. The equivalence of (i) to (vi) therefore follows from Lemma 3.6(a),(b) of [2]. The last assertion is due to Clancey [5] and is also proved as Theorem 3.4 in [2]. □

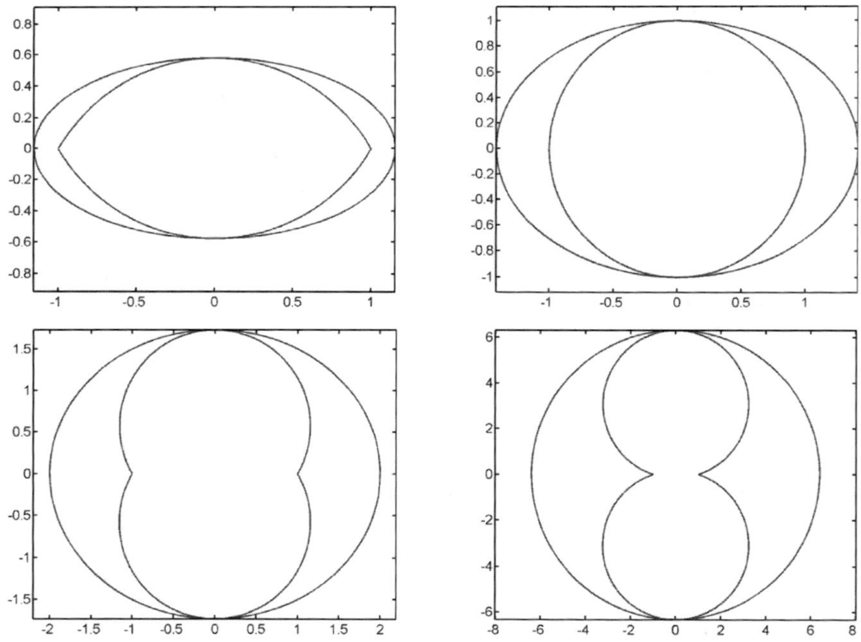

FIGURE 1. The boundaries of the lenses $\mathcal{O}_\mu(-1,1)$ and the ellipses $\mathcal{E}_\mu(-1,1)$ for $\mu = 3, 4, 6, 20$.

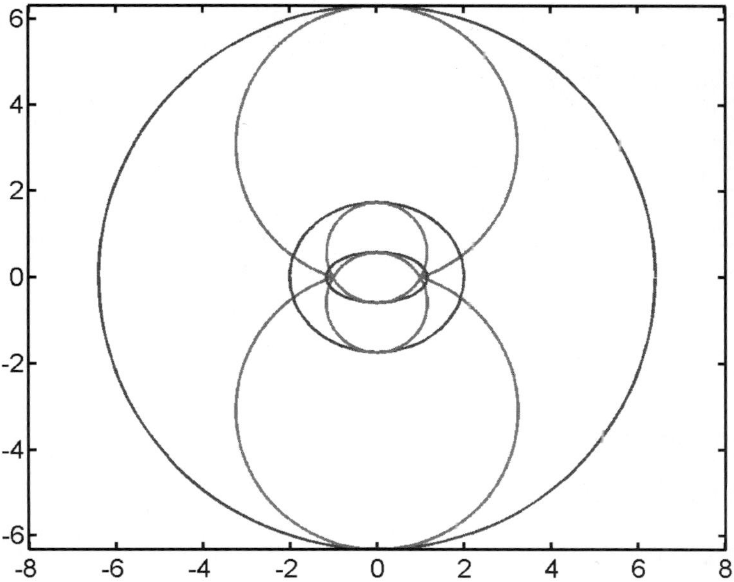

FIGURE 2. Superposition of the boundaries of the lenses $\mathcal{O}_\mu(-1,1)$ and the ellipses $\mathcal{E}_\mu(-1,1)$ for $\mu = 3, 6, 20$.

4. Locally μ-elliptic symbols

We call a matrix function $a \in [L^\infty(B)]^{N\times N}$ locally μ-elliptic if every point $\tau \in S$ has an open neighborhood $U_\tau \subset \mathbb{C}^m$ such that a is μ-elliptic on $U_\tau \cap B$. An equivalent definition is as follows. Given $\tau \in S$, let $\mathcal{R}_\tau(a)$ denote the intersection of the sets $\mathcal{R}_{U\cap B}(a)$ where U ranges over all open neighborhoods $U \subset \mathbb{C}^m$ of τ. Then a is locally μ-elliptic if and only if $\mathcal{R}_\tau(a)$ is a μ-elliptic set for every $\tau \in S$. A standard compactness argument reveals that if a is locally μ-elliptic, then a is even ν-elliptic provided $\nu > \mu$ is sufficiently close to μ.

Obviously, a matrix function $a \in [L^\infty(B)]^{N\times N}$ is locally μ-elliptic if it can be written $a = \varphi\sigma\psi + \eta$ where φ and ψ are matrix functions in $[C(\overline{B})]^{N\times N}$ such that $\varphi|S$ and $\psi|S$ are invertible, $\sigma \in [L^\infty(B)]^{N\times N}$ is globally μ-elliptic, and $\eta \in [L^\infty(B)]^{N\times N}$ vanishes identically on some shell $1-\delta < |z| < 1$.

Proposition 4.1. *Let $a \in [L^\infty(B)]^{N\times N}$ be locally 2-elliptic. Then $a = \sigma\psi + \eta$ where $\psi \in [C(\overline{B})]^{N\times N}$, $\psi|S$ is invertible, $\sigma \in [L^\infty(B)]^{N\times N}$ is 2-elliptic on B, and $\eta \in [L^\infty(B)]^{N\times N}$ vanishes identically on some shell $1-\delta < |z| < 1$.*

Proof. Choose a finite cover $\{U_j\} = \{U_{\tau_j}\}$ of S such that a is 2-elliptic on $U_j \cap B$. By virtue of Proposition 3.3(ii), there are invertible $N \times N$ matrices d_j such that $\|I - a(z)d_j\| < 1$ almost everywhere on $U_j \cap B$. Clearly, $\{U_j\}$ is also a cover of some shell Ω. Now we may proceed exactly as in [18] or the proof of Theorem 3.8 of [2] to obtain a matrix function $d \in [C(\overline{\Omega})]^{N\times N}$ such that $\|I - a(z)d(z)\| < 1$ for almost all $z \in \Omega$. Put $\sigma = ad$ on Ω and $\sigma = w$ on $B \setminus \Omega$, where w is any value in $\mathcal{R}_\Omega(ad)$. Then σ is 2-elliptic on the entire ball B. Since $\|I - ad\| < 1$, the matrix function ad is invertible on Ω, and hence so also is d. Let ψ be any matrix function in $[C(\overline{B})]^{N\times N}$ which coincides with d^{-1} on $\overline{\Omega}$. Thus, $a = \sigma\psi$ on Ω, which implies that $a = \sigma\psi + \eta$ on B with $\eta|\Omega = 0$. □

Theorem 4.2. *Let $1 < p < \infty$ and define $\mu \in [2, \infty)$ by $1/\sin(\pi/\mu) = \|P\|_p$. Suppose $a \in [L^\infty(B)]^{N\times N}$ is locally μ-elliptic. Then $T(a)$ is Fredholm on the space $[A^p(B)]^N$, and there is a number $r_0 \in (0,1)$ such that $T(\widetilde{a}_r)$ is Fredholm for $r_0 < r < 1$ and*

$$\operatorname{Ind} T(a) = \lim_{r\to 1} \operatorname{Ind} T(\widetilde{a}_r).$$

Proof. That $T(a)$ is Fredholm can be proved in a standard way with the help of the local principle of Allan and Douglas, for which see, e.g., [2], [9]. Our assumptions imply that if $p \neq 2$, then $T(a)$ is actually Fredholm on $[A^t(B)]^N$ for all t in some open neighborhood U of the segment $[p, q]$, where $1/p + 1/q = 1$. This can be shown as follows. For the sake of definiteness, let $p > 2$. The invertibility of the local representatives is guaranteed if

$$\|P\|_t \, \|I - c_\tau a d_\tau\|_\infty < 1;$$

recall the proof of Proposition 3.1. But we know that a is locally ν-elliptic for some $\nu > \mu$. Since

$$\|P\|_p = 1/\sin(\pi/\mu) < 1/\sin(\pi/\nu),$$

there is an $\varepsilon > 0$ such that the inequality

$$\|P\|_{p+\varepsilon} < 1/\sin(\pi/\nu)$$

holds. By duality, $\|P\|_q = \|P\|_p$, and hence there is also a $\delta > 0$ such that

$$\|P\|_{q-\delta} < 1/\sin(\pi/\nu).$$

It results that

$$\|P\|_t \le \max\{\|P\|_{p+\varepsilon}, \|P\|_{q-\delta}\} < 1/\sin(\pi/\nu)$$

for $t \in (q - \delta, p + \varepsilon)$.

The operators $T(\widetilde{a}_r)$ have continuous symbols, and hence they are Fredholm on $[A^t(B)]^N$ for all $t \in U$ if they are Fredholm on $[A^2(B)]^N$. The result of Shneiberg [20] implies the index formula for all $t \in U$ once it has been established for $t = 2$. Thus, we are left with proving the theorem for $p = 2$.

Since locally μ-elliptic symbols are locally 2-elliptic, we deduce from Proposition 4.1 that $a = \sigma\psi + \eta$ where ψ is a matrix function in $[C(\overline{B})]^{N\times N}$ such that $\psi|S$ is invertible, $\sigma \in [L^\infty(B)]^{N\times N}$ is 2-elliptic on all of B, and $\eta \in [L^\infty(B)]^{N\times N}$ vanishes identically on some shell $1 - \delta < |z| < 1$. It follows that $T(a) = T(\sigma)T(\psi) + T(\eta)$ plus a compact operator. Moreover, $T(\eta)$ is also compact, and since $T(\sigma)$ is invertible by virtue of Proposition 3.1, we obtain that

$$\operatorname{Ind} T(a) = \operatorname{Ind} T(\psi). \tag{5}$$

We now pass to Berezin transforms. From Lemma 2.4(b) we infer that

$$\widetilde{a} = \widetilde{\sigma}\widetilde{\psi} + \widetilde{\eta} + \gamma$$

with $\gamma \in [L_0^\infty(B)]^{N\times N}$. By Lemma 2.4(d), the Berezin transform of η also belongs to $[L_0^\infty(B)]^{N\times N}$. There are a shell Ω and a constant $M_1 < \infty$ such that $\|\widetilde{\psi}^{-1}\| \le M_1$ on Ω. We have $\|I - c\sigma d\| < 1$ with invertible matrices $c, d \in \mathbb{C}^{N\times N}$ on all of B. From Lemma 2.4(c) we conclude that $\|I - c\widetilde{\sigma} d\| < 1$ on B, which gives $\|\widetilde{\sigma}^{-1}\| \le M_2$ on B with some constant $M_2 < \infty$ and also shows that $T(\widetilde{\sigma}_r)$ is invertible for all r. Thus, on Ω we can write

$$\widetilde{a} = \widetilde{\sigma}\widetilde{\psi}[I + \widetilde{\psi}^{-1}\widetilde{\sigma}^{-1}(\widetilde{\eta} + \gamma)],$$

which implies that on all of B we have

$$\widetilde{a} = \widetilde{\sigma}\widetilde{\psi}[I + \xi] + \theta$$

where ξ and θ are in $[L_0^\infty(B)]^{N\times N}$ and θ vanishes identically on some shell. This gives

$$T(\widetilde{a}_r) = T(\widetilde{\sigma}_r)T(\widetilde{\psi}_r)T(I + \xi_r) + T(\theta_r) + K_r$$

with a compact operator K_r. The operator $T(\theta_r)$ is compact for r sufficiently close to 1, and due to Lemma 2.5 the operator $T(I + \xi_r)$ is Fredholm of index zero for such r. In summary, there is an r_0 such that

$$\operatorname{Ind} T(\widetilde{a}_r) = \operatorname{Ind} T(\widetilde{\sigma}_r) + \operatorname{Ind} T(\widetilde{\psi}_r) + \operatorname{Ind} T(I + \xi_r)$$

where $T(\widetilde{\sigma}_r)$ is invertible and $T(I + \xi_r)$ is Fredholm of index zero. Since

$$\operatorname{Ind} T(\psi) = \lim_{r\to 1} \operatorname{Ind} T(\widetilde{\psi}_r)$$

due to Lemma 2.4(a), we finally see that

$$\lim_{r\to 1} \operatorname{Ind} T(\widetilde{a}_r) = \operatorname{Ind} T(\psi).$$

Comparison with (5) completes the proof. □

We remark that for $N = m = 1$ and $p = 2$ the previous theorem was established in [1]. McDonald [14] considered symbols $a \in L^\infty(B)$ which are uniformly continuous on the two pieces B_+ and B_- which arise when cutting the ball B by a hyperplane of real dimension $2m - 1$. For such symbols, he proved that $T(a)$ is Fredholm on $A^2(B)$ if *and only if* a is locally 2-elliptic. Note that the sets $\mathcal{R}_\tau(a)$ are all either singletons or doubletons, so that $\operatorname{conv}\mathcal{R}_\tau(a)$ is always a line segment in this case. For recent developments in Toeplitz operators with more general piecewise continuous symbols we refer to Loaiza's paper [12] and the nice expositions by Vasilevski in [23] and [24]. However, almost all the available results are for $p = 2$ and in many cases also for $N = m = 1$ only. Fighting with p, N, m at the same time remains a challenge for the future.

References

[1] A. Böttcher, *Toeplitz operators on the disk with locally sctorial symbols.* Rocky Mountain J. Math. **23** (1993), 803–816.

[2] A. Böttcher and B. Silbermann, *Analysis of Toeplitz Operators.* Springer-Verlag, Berlin, Heidelberg, New York 1990.

[3] L. Boutet de Monvel, *On the index of Toeplitz operators of several complex variables.* Invent. Math. **50** (1979), 249–272.

[4] B.R. Choe and Y.J. Lee, *The essential spectra of Toeplitz operators with symbols in* $H^\infty + C$. Math. Japon. **45** (1997), 57–60.

[5] K.F. Clancey, *A local result for system of Riemann-Hilbert barrier problems.* Trans. Amer. Math. Soc. **200** (1974), 315–325.

[6] L.A. Coburn, *Singular integral operators and Toeplitz operators on odd spheres.* Indiana Univ. Math. J. **23** (1973), 433–439.

[7] M.R. Dostanić, *Two sided norm estimate of the Bergman projection on* L^p *spaces.* Czechoslovak Math. J. **58** (2008), 569–575.

[8] R.G. Douglas, *Toeplitz and Wiener-Hopf operators in* $H^\infty + C$. Bull. Amer. Math. Soc. **74** (1968), 895–899.

[9] R.G. Douglas, *Banach Algebra Techniques in Operator Theory.* Academic Press, New York 1972.

[10] R.G. Douglas, *Banach Algebra Techniques in the Theory of Toeplitz Operators.* CBMS Lecture Notes 15, Amer. Math. Soc., Providence, RI 1973.

[11] E. Guentner and N. Higson, *A note on Toeplitz operators.* Internat. J. Math. **7** (1996), 501–513.

[12] M. Loaiza, *On an algebra of Toeplitz operators with piecewise continuous symbols.* Integral Equations Oper. Theory **51** (2005), 141–153.

[13] G. McDonald, *Fredholm properties of a class of Toeplitz operators on the ball.* Indiana Univ. Math. J. **26** (1977), 567–576.

[14] G. McDonald, *Toeplitz operators on the ball with piecewise continuous symbols.* Illinois J. Math. **23** (1979), 286–294.

[15] A. Perälä, J. Taskinen, and J.A. Virtanen, *New results and open problems on Toeplitz operators in Bergman spaces.* New York J. Math. **17a** (2011), 147–164.

[16] A. Perälä and J.A. Virtanen, *A note on the Fredholm properties of Toeplitz operators on weighted Bergman spaces with matrix-valued symbols.* Oper. Matrices **5** (2011), 97–106.

[17] L. Pontryagin, *Sur les nombres de Betti des groups de Lie.* C. R. Acad. Sci. Paris **200** (1935), 1277–1280.

[18] S. Roch, *Locally strongly elliptic singular integral operators.* Wiss. Z. TU Karl-Marx-Stadt **29** (1987), 121–126.

[19] W. Rudin, *Function Theory in the Unit Ball of* $\mathbb{C}^n$. Springer-Verlag, Berlin, Heidelberg, New York 1980.

[20] I.Ya. Shneiberg, *The solvability of linear equations in interpolational families of Banach spaces.* Soviet Math. Dokl. **14** (1973), 1328–1332.

[21] B. Silbermann, *Harmonic approximation of Toeplitz operators and index fomulas.* Integral Equations Oper. Theory **8** (1985), 842–853.

[22] H. Upmeier, *Toeplitz Operators and Index Theory in Several Complex Variables.* Birkhäuser Verlag, Basel 1996.

[23] N.L. Vasilevski, *On the Toeplitz operators with piecewise continuous symbols on the Bergman space.* Oper. Theory: Adv. Appl. **170** (2006), 229–248.

[24] N.L. Vasilevski, *Commutative Algebras of Toeplitz Operators on the Bergman Space.* Birkhäuser Verlag, Basel 2008.

[25] U. Venugopalkrishna, *Fredholm operators associated with strongly pseudoconvex domains in* $\mathbb{C}^n$. J. Funct. Anal. **9** (1972), 349–373.

[26] K. Zhu, *VMO, ESV, and Toeplitz operators on the Bergman space.* Trans. Amer. Math. Soc. **302** (1987), 617–646.

[27] K. Zhu, *Operator Theory in Function Spaces.* Marcel Dekker, New York 1990.

[28] K. Zhu, *A sharp norm estimate for the Bergmann projection on* L^p *spaces.* Contemp. Math. **404** (2006), 199–205.

Albrecht Böttcher
Department of Mathematics
TU Chemnitz
D-09107 Chemnitz, Germany
e-mail: aboettch@mathematik.tu-chemnitz.de

Antti Perälä
Department of Mathematics and Statistics
University of Helsinki
00014 Helsinki, Finland
e-mail: antti.i.perala@helsinki.fi

Operator Theory:
Advances and Applications, Vol. 228, 57–72

The Green Function and Optical Field Enhancement in a Multilayered Microsphere with Metamaterial

Gennadiy Burlak and Vladimir Rabinovich

Abstract. The radiation of a nanosource placed in a coated microsphere with conventional and metamaterial layers having a negative refraction index (NIM) is studied. We consider also that a NIM defect is embedded in such a structure. Our calculations show strong enhancement of the optical field strength assisted by NIM defect. In a resonant case the optical field is almost completely arrested in a vicinity of the defect layer.

Mathematics Subject Classification (2010). Primary 81V80; Secondary 65Q10.

Keywords. Green function, metamaterials, numerical simulations.

1. Introduction

Metamaterials, artificial composite structures with exotic material properties, have emerged as a new frontier of science involving physics, material science and engineering. Nowadays such nanostructured composites are well studied due to intense fundamental and applied research over the past several years [28]–[10]. Next step is a generation of compound structures that can be constructed by alternating of conventional and metamaterial layers. The insertion of metamaterial layers in conventional structures and creation of such a compound environment can open new fundamentals and applied perspectives. In such systems the propagating modes become reconfigurable and can be switched between the left-handed and right-handed modes by changing the position of the metamaterial in the spherical stack[12].

In this paper we study the optical properties of alternating multilayered microspheres with NIM layers. In such a system a basic building block (single-layer unit cell) is a spherical layer that normally has $\lambda/4$ width. An application of the well-known idea about coating by quarter-wave layers opens a possibility to sharply increase the Q factor of such a system up to $Q \sim 10^9$ [5]–[21]. It is important that in such a complex system the remarkable optical phenomena in the interface of

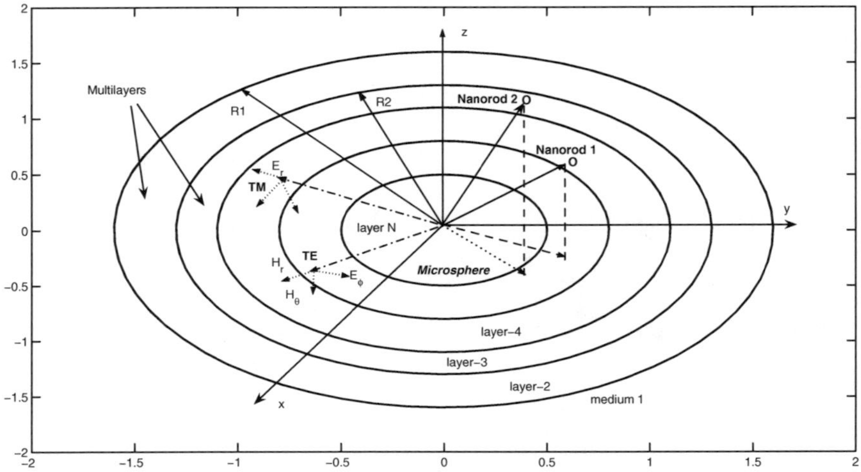

FIGURE 1. Geometry of a multilayered microsphere. A stack of multilayers with embeded NIM layer is deposited on the surface of the microsphere.

conventional and NIM layers [28], [24] can be accumulated. Moreover, the real part of the effective (average) refractive index can become positive, zero or negative along the radial direction. The latter allows imparting new peculiarities to known optical effects [19]. We numerically study the details of spectrum and the optical field distribution of radiating nanoemitter placed in a microsphere coated by alternating conventional and metamaterial layers with negative refraction index (NIM). We consider that a NIM defect is embedded in such a spherical stack. Our calculations show strong enhancement of a nanoemitter field assisted by NIM defect. In resonant case the photon field is almost completely arrested in a vicinity of the defect layer.

2. Basic equations

The spatial scale of the nanoemitter objects ($\sim 1 - 100\,\text{nm}$) is at least one order of magnitude smaller than the spatial scale of microspheres ($\sim 10^3 - 10^4\,\text{nm}$). Therefore in the coated microsphere (Figure 1), we can represent the nanoemitter structure as a point source (object) placed at $\mathbf{r}'$ and having a dipole moment $\mathbf{d}_0$.

It is well known that the solution of the wave equation for the radiated electromagnetic field $\mathbf{E}$ due to a general source $\mathbf{J}(\mathbf{r}')$ is [14], [9]

$$\mathbf{E}(\mathbf{r},\omega) = i\omega\mu_0 \int_V d\mathbf{r}' \widehat{\mathbf{G}}(\mathbf{r},\mathbf{r}',\omega)\widehat{\mu}(\mathbf{r}',\omega)\mathbf{J}(\mathbf{r}'), \tag{2.1}$$

where $\widehat{\mathbf{G}}(\mathbf{r},\mathbf{r}',\omega)$ is the dyadic Green function (DGF), which depends on the type of boundary conditions imposed on $\mathbf{E}(\mathbf{r})$ and contains all the physical information necessary for describing the multilayered structure (the time dependence is assumed to be $e^{i\omega t}$). Eq. (2.1) is complemented by the standard boundary conditions: limitation of the fields in the center of the microsphere and continuity of the tangential components of the fields at the interfaces of layers. We also use Sommerfeld's radiation conditions, where there is only an outgoing wave in the external boundary of the microsphere. In this case, the electromagnetic field $\mathbf{E}$ in the coated structure consists of the sum of the waves radiating in the surrounding medium and the multiple wave reflections due to the interfaces between layers. Substituting the nanoemitter source in the form $\mathbf{J} = i\omega\mathbf{d}$, $\mathbf{d} = \mathbf{d}_0\delta(\mathbf{r}-\mathbf{r}')$ in (2.1), we obtain

$$\mathbf{E}(\mathbf{r},\mathbf{r}',\omega) = -\mathbf{p}_0\widehat{\mathbf{G}}(\mathbf{r},\mathbf{r}',\omega), \tag{2.2}$$

where $\mathbf{p}_0 = (\widehat{\mu}\mathbf{d}_0/\varepsilon_0)\left(\omega^2/c^2\right)$. In such a situation, the nanoemitter frequency spectrum is identical to the dyadic Green function (DGF) spectrum. Thus, the equation of the field generated by a nanoemitter assumes the form of the DGF $\widehat{\mathbf{G}}(\mathbf{r},\mathbf{r}',\omega)$ equation, and is given by [14], [9]

$$\widehat{\mu}(\mathbf{r},\omega)\left[(\nabla\times\widehat{\kappa}(\mathbf{r},\omega))\,\nabla\times\widehat{\mathbf{G}}(\mathbf{r},\mathbf{r}',\omega)-\frac{\omega^2}{c^2}\,\widehat{\varepsilon}(\mathbf{r},\omega)\widehat{\mathbf{G}}(\mathbf{r},\mathbf{r}',\omega)\right] = \delta(\mathbf{r}-\mathbf{r}')\widehat{\mathbf{I}}, \tag{2.3}$$

where $\widehat{\kappa}(\mathbf{r},\omega) = \widehat{\mu}^{-1}(\mathbf{r},\omega)$, and $\mathbf{r}$ is the point where the field is observed, while $\mathbf{r}'$ is the nanoemitter (point source) location, $\widehat{\mathbf{I}}$ is the unit dyadic. For a scalar case with $\widehat{\mu} = \mu_{ik} = \mu\delta_{ik}$, $\widehat{\varepsilon} = \varepsilon_{ik} = \varepsilon\delta_{ik}$ and $\kappa=\mu^{-1}$ from Eq. (2.3) we obtain the DGF equation in well-known form

$$\nabla\times\nabla\times\widehat{\mathbf{G}}(\mathbf{r},\mathbf{r}',\omega)-\frac{\omega^2}{c^2}n^2(\mathbf{r},\omega)\widehat{\mathbf{G}}(\mathbf{r},\mathbf{r}',\omega) = \delta(\mathbf{r}-\mathbf{r}')\widehat{\mathbf{I}}, \tag{2.4}$$

where $n = \pm[\varepsilon\mu]^{1/2}$ is the refraction index that is positive for conventional materials, $n > 0$, and n is negative for metamaterials, $n < 0$. (We note that in some references the equation for DGF Eq. (2.3) is written without $\widehat{\mu}(\mathbf{r},\omega)$ in the left. For a scalar case this leads to a simple renormalization: $\widehat{\mathbf{G}} \to \widehat{\mathbf{G}}/\mu$.) The dyadic Green function $\widehat{\mathbf{G}}(\mathbf{r},\mathbf{r}',\omega)$ in (2.4) satisfies the boundary conditions at the interfaces of spherical layers

$$\widehat{\mathbf{r}}\times\widehat{\mathbf{G}}^{fs} = \widehat{\mathbf{r}}\times\widehat{\mathbf{G}}^{f+1,s}, \qquad \kappa_f(\mathbf{r},\omega)\times\widehat{\mathbf{r}}\times\widehat{\mathbf{G}}^{fs} = \kappa_{f+1}(\mathbf{r},\omega)\times\widehat{\mathbf{r}}\times\widehat{\mathbf{G}}^{fs}. \tag{2.5}$$

Let us consider the multilayered spherical structure: a concentric system of spherical layers contacting with the sphere (concentric stack) deposited onto the surface of the microsphere with a nanoemitter placed in such a structure (see Figure 1). The layers are localized at the distance R_k from the center, where $d_k = R_k - R_{k+1}$ is the width of a kth layer.

Let us first specify some details of the Green function technique for multilayered microspheres and introduce our notations. Following the approach[17], we write down DGF of such a system as follows:

$$\widehat{\mathbf{G}}(\mathbf{r},\mathbf{r}',\omega) = \widehat{\mathbf{G}}^V(\mathbf{r},\mathbf{r}',\omega)\delta_{fs} + \widehat{\mathbf{G}}^{(fs)}(\mathbf{r},\mathbf{r}',\omega), \tag{2.6}$$

where $\widehat{\mathbf{G}}^V(\mathbf{r},\mathbf{r}',\omega)$ represents the contribution of the direct waves from the radiation sources in the unbounded medium, whereas $\widehat{\mathbf{G}}^{(fs)}(\mathbf{r},\mathbf{r}',\omega)$ describes the contribution of the multiple wave reflections and transmissions due to the layer interfaces. The dyadic Green tensor $\widehat{\mathbf{G}}^V(\mathbf{r},\mathbf{r}',\omega)$ in (2.6) is given by

$$\widehat{\mathbf{G}}^V(\mathbf{r},\mathbf{r}',\omega) = \frac{\widehat{\mathbf{r}}\widehat{\mathbf{r}}}{k_s^2}\delta(\mathbf{r}-\mathbf{r}') + \frac{ik_s}{4\pi}\sum_{q=e,o}\sum_{n=1}^{\infty}\sum_{m=0}^{n} C_{nm}^V \widehat{\mathbf{G}}_{q,nm}(\mathbf{r},\mathbf{r}',\omega), \tag{2.7}$$

with

$$C_{nm} = \frac{2n+1}{n(n+1)}\frac{(n-m)!}{(n+m)!}(2-\delta_{0m}), \tag{2.8}$$

where the prime denotes the nanoemitter coordinates $\mathbf{r}' = (r',\theta',\varphi')$, n and m are spherical and azimuthal quantum numbers, respectively, while $k_s = \omega n_s/c$ is the wave number of the medium where the radiated nanoemitters are located. It is worth noting that due to the dyad $\widehat{\mathbf{r}}\widehat{\mathbf{r}}$, the δ-function in (2.7) contributes to the radial (longitudinal) part [9]. Due to the equality $\widehat{\mathbf{r}}\cdot(\theta\widehat{\theta}+\varphi\widehat{\varphi}) = 0$, such a singularity does not contribute to the field (2.2) for the considered case of a dipole, such a case is considered below.

The partial dyadic Green tensor $\widehat{\mathbf{G}}_{q,nm}^V(\mathbf{r},\mathbf{r}',\omega)$ in (2.7) has a form

$$\widehat{\mathbf{G}}_{q,nm}^V(\mathbf{r},\mathbf{r}',\omega) = \begin{pmatrix} \mathbf{M}_{q,nm}^{(1)}(\mathbf{r},k_s)\mathbf{M}_{q,nm}(\mathbf{r}',k_s) + \mathbf{N}_{q,nm}^{(1)}(\mathbf{r},k_s)\mathbf{N}_{q,nm}(\mathbf{r}',k_s),\ \mathbf{r}>\mathbf{r}' \\ \mathbf{M}_{q,nm}(\mathbf{r},k_s)\mathbf{M}_{q,nm}^{(1)}(\mathbf{r}',k_s) + \mathbf{N}_{q,nm}(\mathbf{r},k_1)\mathbf{N}_{q,nm}^{(1)}(\mathbf{r}',k_s)\},\ \mathbf{r}<\mathbf{r}' \end{pmatrix}. \tag{2.9}$$

In Eq. (2.9), vectors $\mathbf{M}$ and $\mathbf{N}$ represent TE- and TM-waves, respectively, where

$$\mathbf{M}_{\substack{e\\o}nm}(k) = \mp\frac{m}{\sin\theta}j_n(kr)P_n^m(\cos\theta)\begin{pmatrix}\sin\\ \cos\end{pmatrix}(m\phi)\mathbf{e}_\theta - j_n(kr)\frac{dP_n^m(\cos\theta)}{d\theta}\begin{pmatrix}\cos\\ \sin\end{pmatrix}(m\phi)\mathbf{e}_\phi\ , \tag{2.10}$$

$$\begin{aligned}\mathbf{N}_{\substack{e\\o},nm}(k) = {}& \frac{n(n+1)}{kr}j_n(kr)P_n^m(\cos\theta)\begin{pmatrix}\cos\\ \sin\end{pmatrix}(m\phi)\mathbf{e}_r \\ &+\frac{1}{kr}\frac{d[rj_n(kr)]}{dr}[\frac{dP_n^m(\cos\theta)}{d\theta}\begin{pmatrix}\cos\\ \sin\end{pmatrix}(m\phi)\mathbf{e}_\theta \\ &\mp\frac{m}{\sin\theta}P_n^m(\cos\theta)\begin{pmatrix}\sin\\ \cos\end{pmatrix}(m\phi)\mathbf{e}_\phi],\end{aligned} \tag{2.11}$$

where $j_n(x)$ and $h_n(x)$ stand for spherical Bessel and Hankel functions [1] respectively, and $P_n^m(x)$ is the associated Legendre function. For the sake of simplicity, we use in (2.10), (2.11) and further on, the standard short notation: $\mathbf{M}_{\substack{e\\o}nm}(k) = \mathbf{M}_{\substack{e\\o}nm}(\mathbf{r},k)$ and $\mathbf{M}'_{\substack{e\\o}nm}(k) = \mathbf{M}_{\substack{e\\o}nm}(\mathbf{r}',k)$. The superscript $^{(1)}$ in

Eqs. (2.9)–(2.17) indicates that in (2.10) and (2.11), the spherical Bessel function $j_n(x)$ has to be replaced by the first-type spherical Hankel function $h_n^{(1)}(x)$ for $r > r'$.

The scattering DGF $\widehat{\mathbf{G}}^{(fs)}(\mathbf{r}, \mathbf{r}', \omega)$ is written as

$$\widehat{\mathbf{G}}^{(fs)}(\mathbf{r}, \mathbf{r}', \omega) = \frac{ik_s}{4\pi} \sum_{q=e,o} \sum_{n=1}^{\infty} \sum_{m=0}^{n} C_{rm}^{(f,s)} \widehat{\mathbf{G}}_{q,nm}(\mathbf{r}, \mathbf{r}', \omega), \tag{2.12}$$

where f and s denote the layers where the field point and source point are located; δ_{fs} is the Kronecker symbol and

$$\begin{aligned}\widehat{\mathbf{G}}_{q,nm}^{(fs)}(\mathbf{r}, \mathbf{r}', \omega) = \Delta_{Nf} &\left(\mathbf{M}_{q,nm}^{(1)}(k_f)\mathbf{P}_M + \mathbf{N}_{q,nm}^{(1)}(k_f)\mathbf{P}_N \right) \\ &+ \Delta_{1f} \left(\mathbf{M}_{q,nm}^{(1)}(k_f)\mathbf{Q}_M + \mathbf{N}_{q,nm}^{(1)}(k_f)\mathbf{Q}_N \right),\end{aligned} \tag{2.13}$$

with

$$\mathbf{P}_M = \Delta_{1s} A_M^{fs} \mathbf{M}'_{q,nm}(k_s) + \Delta_{Ns} B_M^{fs} \mathbf{M}'^{(1)}_{q,nm}(k_s), \tag{2.14}$$

$$\mathbf{P}_N = \Delta_{1s} A_N^{fs} \mathbf{N}'_{q,nm}(k_s) + \Delta_{Ns} B_N^{fs} \mathbf{N}'^{(1)}_{q,nm}(k_s), \tag{2.15}$$

$$\mathbf{Q}_M = \Delta_{1s} C_M^{fs} \mathbf{M}'_{q,nm}(k_s) + \Delta_{Ns} D_M^{fs} \mathbf{M}'^{(1)}_{q,nm}(k_s), \tag{2.16}$$

$$\mathbf{Q}_N = \Delta_{1s} C_M^{fs} \mathbf{N}'_{q,nm}(k_s) + \Delta_{Ns} D_M^{fs} \mathbf{N}'^{(1)}_{q,nm}(k_s), \tag{2.17}$$

where $\Delta_{fs} = 1 - \delta_{fs}$, δ_{fs} is the Kronecker symbol, $k_s = n_s(\omega)\omega/c$, $n(\omega) = \pm\sqrt{\varepsilon_s(\omega)\mu_s(\omega)}$ is the refraction index of the s' layer (see Eq. (3.4) in the next Section). Frequency dependent coefficients $A_k^{fs}(\omega)$, $B_k^{fs}(\omega)$, $C_k^{fs}(\omega)$ and $D_k^{fs}(\omega)$ in (2.14)–(2.17) are defined from the above-mentioned boundary conditions and describe the details of the wave behavior in the interface of the stack layers. The use of boundary conditions yields the relations between these coefficients that can be written in the following matrix form:

$$\mathbf{J}_k^{f+1,s} - \mathbf{I}_k^f \cdot \mathbf{J}_k^{f,s} + \sigma^+ \delta_{,f+1,s} - \mathbf{I}_k^f \cdot \sigma^- \delta_{fs} = 0, \tag{2.18}$$

where $k = M, N$, $f = 1, \ldots, N-1$ and

$$\mathbf{J}_k^{f,s} = \begin{bmatrix} A_k^{f,s} & B_k^{f,s} \\ C_k^{f,s} & D_k^{f,s} \end{bmatrix}, \ \mathbf{I}_k^f = \begin{bmatrix} 1/T_{Ff}^k & R_{Ff}^k/T_{Ff}^k \\ R_{Pf}^k/T_{Pf}^k & 1/T_{Pf}^k \end{bmatrix}, \tag{2.19}$$

$$\sigma^+ = \begin{bmatrix} 1 & 0 \\ 0 & 0 \end{bmatrix}, \ \sigma^- = \begin{bmatrix} 0 & 0 \\ 0 & 1 \end{bmatrix}. \tag{2.20}$$

The reflection R_{jf}^k and the transmittance T_{jf}^k coefficients from Eq. (2.19) are written in Refs. [17], [8], where one can find more details.

We note that the argument of the spherical Bessel $j_n(x)$ or Hankel functions $h_n^{(1,2)}(x)$ ($x = n\omega/c$) in Eqs. (2.13)–(2.19) is positive for conventional layer ($n > 0$) or negative for metamaterial layer ($n > 0$). In general it leads to significant modifications of the scattering coefficients in (2.12). However the situation is more simple in contact of two NIM layers. Really, it is easy to see that due to general properties [1] $j_l(-x) = (-1)^l j_l(x)$ and $h_l^{(1,2)}(-x) = (-1)^l h_l^{(2,1)}(x)$ and since DGF is a bilinear function of the Bessel (or Hankel) functions for points r and r' that are in NIM layers we have $j_l(-x)j_l(-x') = j_l(x)j_l(x')$ and $h_l^{(1)}(-x)h_l^{(1)}(-x') = h_l^{(2)}(x)h_l^{(2)}(x')$. However when the points r and r' belong to different materials the terms $\widehat{\mathbf{G}}_{q,lm}^{(f,s)}(\mathbf{r}, \mathbf{r}', \omega)$ for odd $l = 1, 3, 5, \ldots$ produce the substantial difference in behavior of the DGF for a compound NIM (or conventional) layers stack with respect of conventional stack.

3. Permittivity, permeability, and refractive index of NIP layers

In this section we briefly discuss the properties of NIM layers. Let us consider a causal linear magnetodielectric medium characterized by a (relative) permittivity $\varepsilon(\mathbf{r}, \omega)$ and a (relative) permeability $\mu(\mathbf{r}, \omega)$, both of which are spatially varying, complex functions of frequency satisfying the relations

$$\varepsilon(\mathbf{r}, -\omega^*) = \varepsilon^*(\mathbf{r}, \omega), \quad \mu(\mathbf{r}, -\omega^*) = \mu^*(\mathbf{r}, \omega). \tag{3.1}$$

The relation $n^2(\mathbf{r}, \omega) = \varepsilon(\mathbf{r}, \omega)\mu(\mathbf{r}, \omega)$ formally offers two possibilities for the (complex) refractive index $n(\mathbf{r}, \omega)$

$$n(\mathbf{r}, \omega) = \pm\sqrt{|\varepsilon(\mathbf{r}, \omega)\mu(\mathbf{r}, \omega)|}\, e^{i[\phi_\varepsilon(\mathbf{r},\omega)+\phi_\mu(\mathbf{r},\omega)]/2}, \tag{3.2}$$

where

$$0 < [\phi_\varepsilon(\mathbf{r}, \omega) + \phi_\mu(\mathbf{r}, \omega)]/2 < \pi. \tag{3.3}$$

Further, we follow references [11], [30] that allow us to rewrite Eq. (3.2) as

$$n(\mathbf{r}, \omega) = \sqrt{|\varepsilon(\mathbf{r}, \omega)\mu(\mathbf{r}, \omega)|}\, e^{i[\phi_\varepsilon(\mathbf{r},\omega)+\phi_\mu(\mathbf{r},\omega)]/2}. \tag{3.4}$$

In the following, we refer to the material of a layer as being left-handed (or metamaterial) if the real part of its refractive index is negative. In order to allow a dependence on the frequency of the refractive index, let us restrict our attention to a single-resonance permittivity

$$\varepsilon(\omega) = 1 + \frac{\omega_{\mathrm{P}e}^2}{\omega_{\mathrm{T}e}^2 - \omega^2 - i\omega\gamma_e} \tag{3.5}$$

and a single-resonance permeability

$$\mu(\omega) = 1 + \frac{\omega_{\mathrm{P}m}^2}{\omega_{\mathrm{T}m}^2 - \omega^2 - i\omega\gamma_m}, \tag{3.6}$$

where $\omega_{\mathrm{P}e}$, $\omega_{\mathrm{P}m}$ are the coupling strengths, $\omega_{\mathrm{T}e}$, $\omega_{\mathrm{T}m}$ are the transverse resonance frequencies, and γ_e, γ_m are the absorption parameters. Both the permittivity and the permeability satisfy the Kramers-Kronig relations (see discussion in [26], [15]). Figure 2 shows the NIP refractive index $n(\mathbf{r}, \omega) = \mathrm{Re}\, n(\mathbf{r}, \omega) + i\, \mathrm{Im}\, n(\mathbf{r}, \omega)$ ($\omega = 2\pi f$), with the permittivity $\varepsilon(\omega)$ and the permeability $\mu(\omega)$ being respectively given by Eqs. (3.5) and (3.6) in the frequency interval from 155 THz up to 175 THz. In our case it is used the following typical values: $\omega_{\mathrm{T}m}/2\pi = f_{Tm} = 159.2\,\mathrm{THz}$, $\gamma_m/2\pi = f\gamma = 0.001592\,\mathrm{THz}$, $f_{Te} = 163.9\,\mathrm{THz}$, $f_{Pe} = 119.4\,\mathrm{THz}$,

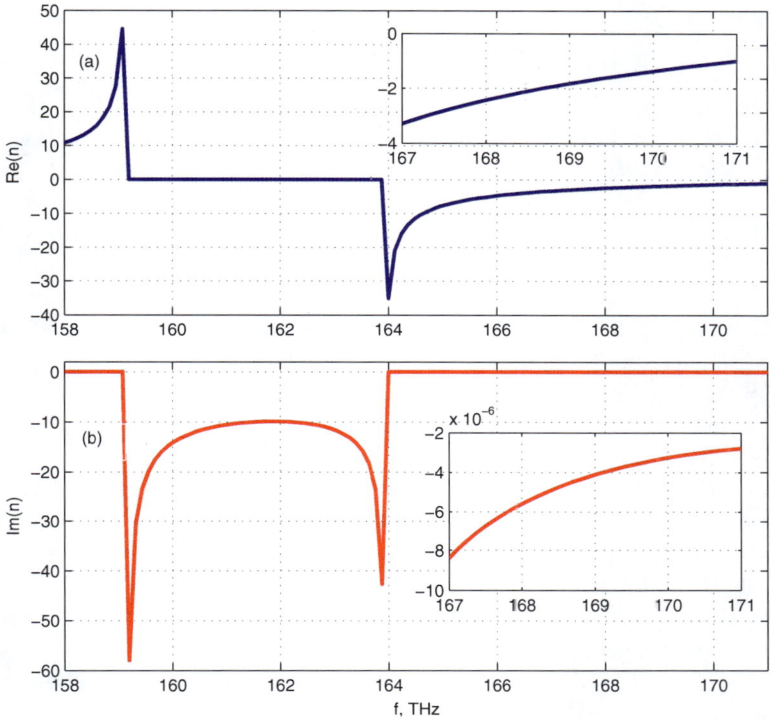

FIGURE 2. Real (a) and imaginary (b) parts of the refractive index $n(\omega)$ ($\omega = 2\pi f$) of LH layer in the stack as functions of frequency $f = \omega/2\pi$, with the permittivity $\varepsilon(\omega)$ and the permeability $\mu(\omega)$ being respectively given by Eqs. (3.5) and (3.6) [$\omega_{\mathrm{T}m} = (10^{16}/2\pi)$ THz, $\omega_{\mathrm{Te}} = 1.03\,\omega_{\mathrm{T}m}$; $\omega_{\mathrm{P}m} = 0.43\,\omega_{\mathrm{T}m}$; $\omega_{\mathrm{P}e} = 0.75\,\omega_{\mathrm{T}m}$; $\gamma_e = \gamma_m = 10^{-7}\,\omega_{\mathrm{T}m}$ (solid lines)]. The values of the parameters have been chosen to be similar to those in Refs. [11]–[23]. Insets show the details of $\mathrm{Re}\,n$ and $\mathrm{Im}\,n$ in the area where $\mathrm{Re}\,n < 0$. See details in text.

$f_{Pm} = 68.44$ THz. In the inset, the details of $n(\mathbf{r}, \omega)$ are shown in the frequency interval from 164 THz up to 175 THz where $\mathrm{Re}\,n(\mathbf{r}, \omega) < 0$. It is worth noting that the negative real part of the refractive index is typically observed together with strong dispersion, so that absorption cannot be disregarded in general. However, in a very recent experiment[29], it was experimentally demonstrated that the incorporation of gain material in a metamaterial makes it possible to fabricate an extremely low-loss and active optical NIM structure. As a result, the original loss-limited negative refractive index can be drastically improved with loss compensation.

4. Numerical results

Analytical solutions to Eq. (2.18) for scattering coefficients $A_k^{f,s}, B_k^{f,s}, C_k^{f,s}$ and $D_k^{f,s}$ for cases of 1 or 2 layers in a spherical stack were derived in Ref. [17]. But corresponding equations are rather laborious, and thus are hardly suitable in practice for studying the frequency spectrum for the cases with more than 2 layers in the stack already. However, namely in such a structure, one can expect physically interesting phenomena due to the wave re-reflections in the layers of the stack. Similarly to the plane case, such phenomena are most pronounced when the thicknesses of the alternating spherical layers are approximately equal to $\lambda/4$ (quarter-wave layers) [4], [27], [6]. In a general case of alternating layers (having small losses), the equality $k_0 |n_k| d_k = \pi/l$, (d_k is width, $k_0 = \omega/c$) is considered, so that $d_k = \pi/lk_0 |n_k|$, where l is integer. In this case, the optical thicknesses of the conventional and NIP material layers are the same $d_1 |n_1| = d_2 |n_2|$.

We consider a spherical stack with $1/k_0 = \Lambda_0/2\pi$, where Λ_0 is the reference wavelength of the structure. The equality $d_k |n_k| = \Lambda_0/4$ corresponds to the quarter-wave case. (Let us remind that the continuity of the fields in the layer interfaces requires the continuity of impedances $Z = (\mu/\varepsilon)^{1/2}$ which is positive for both the NIP and the conventional layers.) Since the amplitude of a spherical electromagnetic wave depends on the distance to the center of the microsphere, such a $\Lambda_0/4$ approximation is only asymptotically close to the plane wave case and can be optimized yet with respect to the local properties of the layers in the stack.

Now we have to identify the nanosource position in a microsphere. If a nanoemitter is placed close to the center of a microsphere, the system is nearly spherically symmetric; therefore the modes with small spherical quantum numbers mainly contribute to the sums in Eqs. (2.7), (2.12). This case is close to a rotational invariant geometry where the dipole moment orientation does not need to be specified. Therefore, we draw more attention to a case where the nanoemitter is placed rather far from the center in one of the layers of the spherical stack. In such a system, the preferred direction (center-source) arises, therefore larger numbers of spherical modes contribute to DGF (2.7), (2.12). As a result the frequency spectrum of DGF becomes richer but more complicated.

In this Section, we numerically explore the details of frequency spectrum and radial dependencies of nanoemitter radiation (the dyadic Green function) for alternating quarter-wave compound stack (Figure 1). The details of the numerical algorithm can be found in Ref. [8]. Since nanoemitters (e.g., nanorods [18]) are highly polarized objects, we further pay more attention to the tangential nanoemitter orientation with $\mathbf{d} = d\widehat{\varphi}$, so only the tangential components of the Green tensor $G_{\varphi\varphi}$ contribute.

The following parameters have been used in our calculations: the geometry of a system is $ABCBCBC..BD$, where the letters A, B, C, D indicate the materials of layers, here A and D are inner and surrounded spherical layers, respectively, while B and C are alternating layers in the spherical stack, $\Lambda_0 = 1.75\,\mu$m ($K_0 = 2\pi f_0/c$, $f_0 = 171.5\,$THz). A bottom microsphere has a refraction index $n_4 = 1.5 + i2 \cdot 10^{-4}$

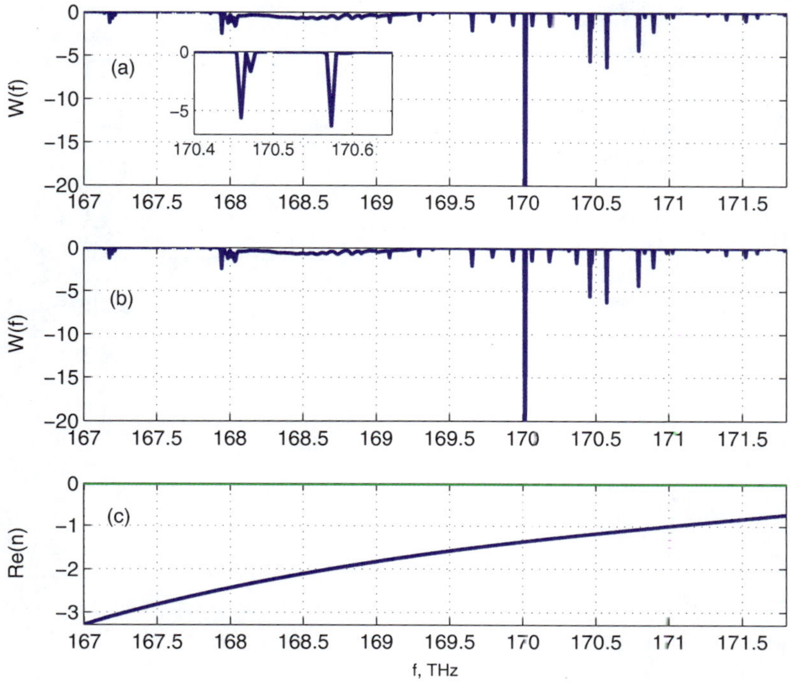

FIGURE 3. The frequency spectrums of $W(r, r', f) = \mathrm{Im}(G_{\varphi\varphi}(r, r', f))$ for (a) $r = r' = 2.5r_1$, and (b) $r = r' = 2.91r_1$, where $r_1 = 1\,\mu$m is the radius of the internal bottom microsphere, panel (c) shows $\mathrm{Re}(n)$, where $n = n(\omega = 2\pi f)$ is the refraction index of NIM layer. Spherical stack consists of 14 layers, the defect LH layer is embedded as 7th layer. See details in text.

(A, glass, radius 1000 nm). The refraction index of the NIP layer is given by Eq. (3.4) (see Figure 2 for details) (B, width 437 nm), $n_2 = 1.46 + i3 \cdot 10^{-3}$ (SiO_2, C, width 300 nm) and $n_1 = 1$ (D, surrounding space). To consider the realistic layers case we added to each n_i a small imaginary part, which corresponds to the material dissipation. To seek for simplicity, further we consider a situation when the embedded NIM defect is the same as other NIM layers in the stack. This allows us to observe the main features of such a compound system. The results of our calculations are shown in Figs. 3–8.

It is known that due to the fluctuation-dissipation theorem the correlation function of the electromagnetic field at points r and r' can be written by means of the macroscopic Green function as follows [16]

$$\langle \mathbf{E}(r)\mathbf{E}(r') \rangle_\omega \sim \mathrm{Im}(\mu^{-1}(\omega)\widehat{\mathbf{G}}(r, r', \omega)),\ \omega = 2\pi f\ . \tag{4.1}$$

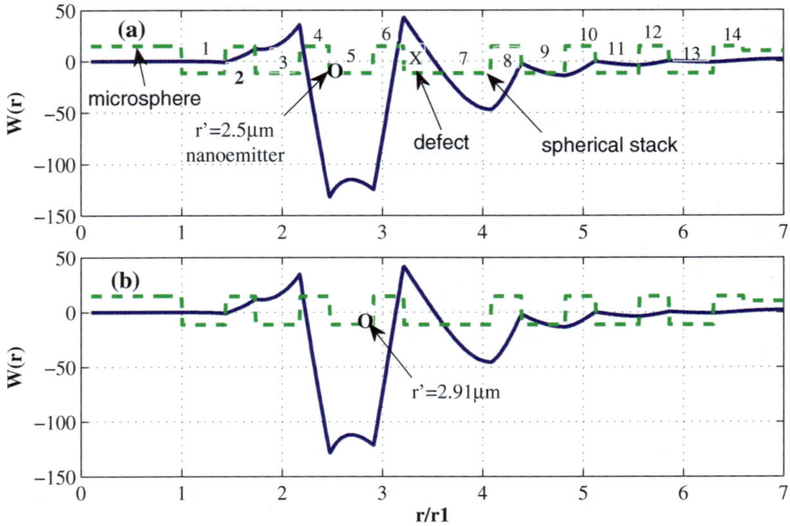

FIGURE 4. Radial distribution of $W(r, r', f_0)$ for a resonance $f_0 = 170.5$ THz and two positions of an emitter (indicated by arrows): (a) $r'/r_1 = 2.5$, and (b) $r'/r_1 = 2.91$, where r_1 is the radius of the internal bottom microsphere ($r/r_1 < 1$). Dash line shows the structure (refractive indices) of the stack. In order to see the field structure clearly the refractive indices are multiplied by 10. Spherical stack has 14 layers, the defect NIM layer is embedded as 7th layer (is marked by symbol X). See details in text.

For case $r = r'$ and small absorbing Eq. (4.1) yields the energy of fluctuating electromagnetic field as $\left\langle \mathbf{E}(r)^2 \right\rangle_\omega \sim \mu^{-1}(\omega)\,\mathrm{Im}(\widehat{\mathbf{G}}(r, r', \omega))$. From the latter we observe that signum $\mathrm{Im}(\widehat{\mathbf{G}}(r, r', \omega)$ must coincide with signum $\mathrm{Re}\,\mu(\omega)$. Therefore $\mathrm{Im}(\widehat{\mathbf{G}}(r, r, \omega))$ in NIP medium (with $\mathrm{Re}\,\mu(\omega) < 0$) is negative.

Further we perform our study as follows. (i) We evaluate the frequency spectrum of the field $W(r, \omega) = \mathrm{Im}(\widehat{\mathbf{G}}(r, r, \omega))$, for fixed r to define the spectral resonances f_i, after that (ii) we evaluate $W(r, r', \omega_0) = \mathrm{Im}(\widehat{\mathbf{G}}(r, r', \omega_0))$ for some resonance ω_0 that allows to study the correlations of field states between various points of the stack r and r', which belong to the same or different layers. The typical structure of the Green function spectrum (that is $G_{\varphi\varphi}(\mathbf{r}, \mathbf{r})$ in our case) is shown in Figure 3.

While calculations we have evaluated complete the complex Green tensor G, however further we will pay attention to the imaginary part $W(r, r', f) = \mathrm{Im}(G_{\varphi\varphi}(r, r', f))$. In a spherical geometry the value of $G_{\varphi\varphi}(r, r', f)$ depends on the source position r' (distance to the center), therefore we start our study for cases when the emitter is placed in the opposite boundaries of a NIM layer. In

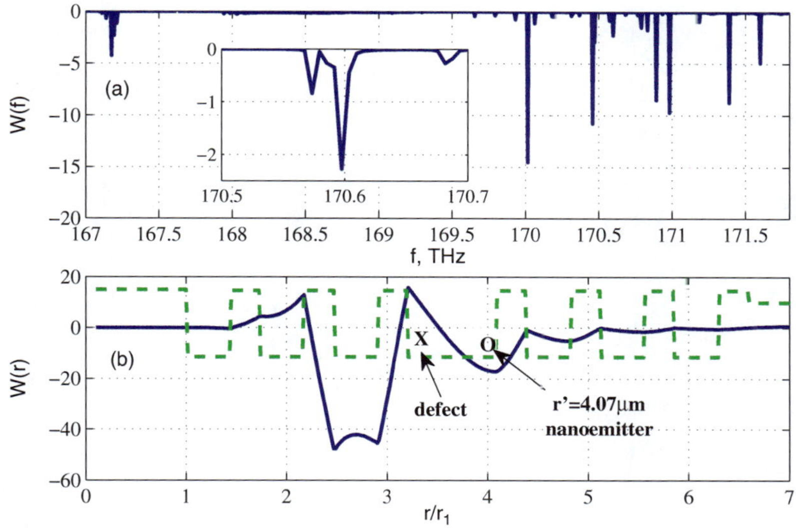

FIGURE 5. (a) The frequency spectrum of the field $W(r,r',f) = \mathrm{Im}(G_{\varphi\varphi}(r,r',f))$ for $r = r' = 4.07$, and (b) radial distribution of the $W(r,r',f_0) = \mathrm{Im}(G_{\varphi\varphi}(r,r',f_0))$ for a resonance $f_0 = 170.5\,\mathrm{THz}$ and the position of the atom: $r'/r_1 = 4.07$, where r_1 is the radius of the internal bottom microsphere. Spherical stack has 14 layers, the defect LH layer is embedded in 7 layer. See details in text.

Figure 3 the frequency spectrum of the field $W(f)$ for (a) $r = r' = 2.5r_1$, and (b) $r = r' = 2.91r_1$, (r_1 is the radius of the internal microsphere) is shown. Such a spectrum consists of various resonances corresponding to eigenfrequencies of such a system. As it is expected, the signum of $\mathrm{Im}(G_{\varphi\varphi}(r,r,f))$ is negative for such a configuration. In Figure 3 panel (c) the frequency dependence of the refraction index of NIM layer $\mathrm{Re}\, n(f)$ is depicted. Spherical stack has 14 layers and a defect NIM layer is embedded as 7th layer in the stack. We observe from Figure 3 (a) and (b) that spectra have similar structure because the width of the NIM layer is small with respect of the distance to center of the microsphere.

After evaluation the structure of resonances for $\mathrm{Im}(G_{\varphi\varphi}(r,r,f))$ we study the spatial (radial) distribution of $W(r,r',f_0) = \mathrm{Im}(G_{\varphi\varphi}(r,r',f_0))$ at fixed source place r' for some resonance $f = f_0$. Such dependencies are shown in Figure 4 for two positions of a source (both in a NIM layer) for a resonance $f_0 = 170.57\,\mathrm{THz}$, see Figure 3 (a). As it was already mentioned in this case the accumulating of partial terms in the sum (2.12) becomes. From Figure 4 (a) and (b) we observe that the field structures in 5th NIM layer are similar but quite different from the field behavior in other layers. It is interesting to see that the field strength W in 5th layer is about 150 (arbitrary units) what is at least in 3–5 times higher than in the other layers. In Figure 4 a dash line shows the structure (refractive indices) of

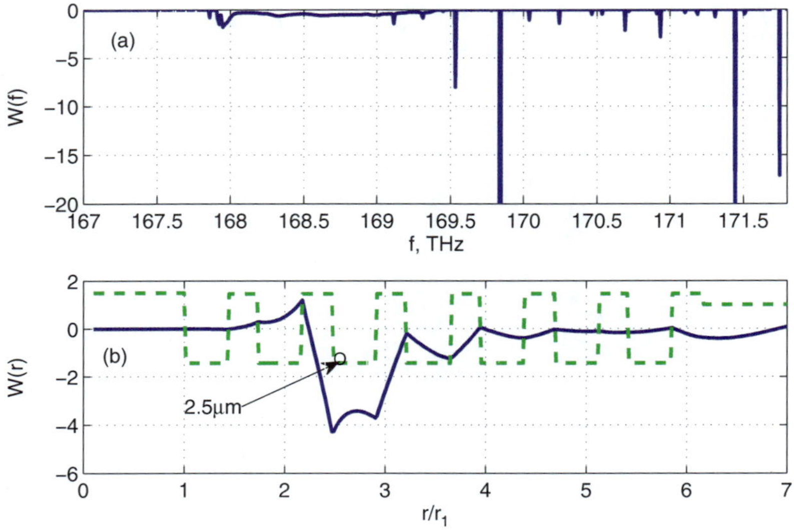

FIGURE 6. Case without a defect in the spherical stack. (a) Frequency spectrum of $W(r, r', f)$ for $r = r' = 2.5r_1$, and (b) radial distribution of the $W(r, r', f_0)$ for a resonance $f_0 = 169.5\,\text{THz}$ and the position of the emitter: $r'/r_1 = 2.5$, where r_1 is the radius of the internal bottom microsphere. Spherical stack has 14 alternating layers. See details in text.

the spherical stack (to see the spatial field structure clearly the refractive indices in Figure 4 (c) are multiplied by 10). Spherical stack has 14 layers, and the defect NIM layer (marked as a symbol X) is embedded as 7th layer in the stack.

To see whether the field shape and strength in 5th layer of Figure 4 is sensitive to the source place we calculate W for other nanoemitter position $r' = 4.07\,\mu\text{m}$ that is farther from the center, but belongs to other NIM layer, see Figure 5. We observe that such a configuration the field amplitude W in 5th layer is strongly (about 3 times) reduced, but the shape of the field remains stable.

Now we investigate whether the field state in 5th layer depends on the presence of the defect in the stack. To do this we calculated the spatial field distribution but without of a defect (indicated as X); in this case spherical stack is exactly periodic one. Figure 6 (a) shows corresponding frequency spectrum with a resonance line closely to 170.5 THz that than was used to evaluate the spatial field distribution shown in Figure 6 (b). We observe from Figure 6 (b) that the shape of W in 5th layer is similar to that is shown in Figure 4 (a), but the amplitude of the field state is considerably less (40 times).

We should explore yet the case when a nanoemitter is placed in conventional material SiO_2 (beyond of NIM layer). Such a configuration is shown in Figure 7. In this layer $\text{Re}(n) > 0$ and, as it is expected in Figure 7 (a) $W(r =$

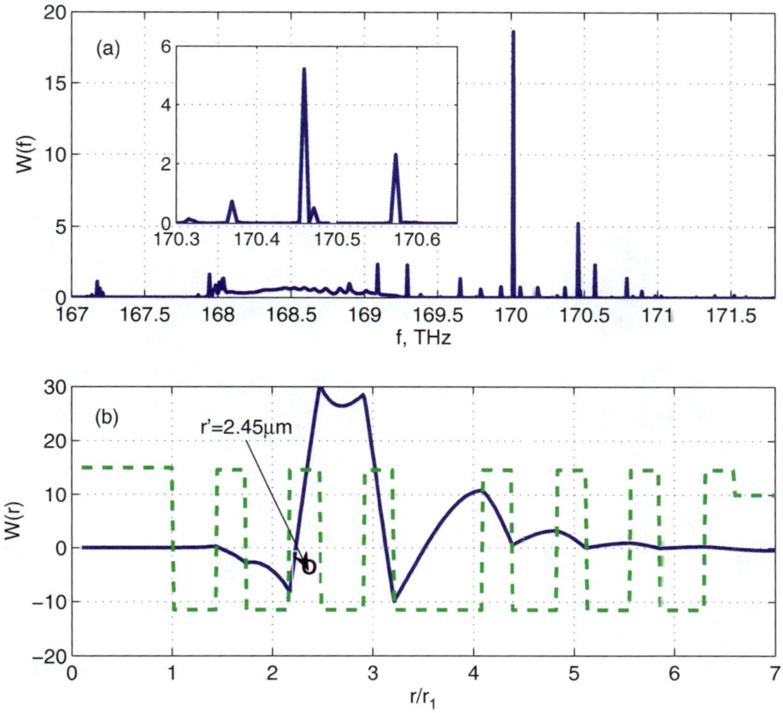

FIGURE 7. (a) The frequency spectrum of the field $W(r, r', f) = \mathrm{Im}(G_{\varphi\varphi}(r, r', f))$ for $r = r' = 2.45r_1$, and (b) radial distribution of the $W(r, r', f_0) = \mathrm{Im}(G_{\varphi\varphi}(r, r', f_0))$ for a resonance $f_0 = 170.5\,\mathrm{THz}$ and the position of the atom: $r'/r_1 = 2.45$ (that is SiO_2 layer), where r_1 is the radius of the internal bottom microsphere. Spherical stack has 14 layers, the defect LH layer is embedded as 7 layer. See details in text.

$r', f) = \mathrm{Im}(G_{\varphi\varphi}(r = r', f))$ is positive. Figure 7 (b) shows the radial dependence $W(r, r', f_0)$ at $r'/r_1 = 2.45$ for the frequency resonance $f_0 = 170.57\,\mathrm{THz}$.

We observe from Figure 7 (b) that the strength W is considerably less with respect to case when a source was placed in NIP layer, Figure 4. This confirms the conclusion that assistance of defect is important to attain the maximum field strength.

In previous figures the frequency spectrum and radial distribution of the field ($\sim \mathrm{Im}(G_{\varphi\varphi}(r, r', f))$) for the multilayered stack were shown. However in experiments it is important to identify the spatial and angular distribution of the optical field, radiated by nanosources located in a coated microsphere. It is of interest to consider the spatial field distribution in a cross-section contained both center of the coated microsphere and nanoemitter for some resonance. Such a distribution is shown in Figure 8 for the resonance $f_0 = 170.57\,\mathrm{THz}$ (see Figure 3(a)). In order to see clearly the field details, Figure 8 shows $-W(r, \varphi)$, where

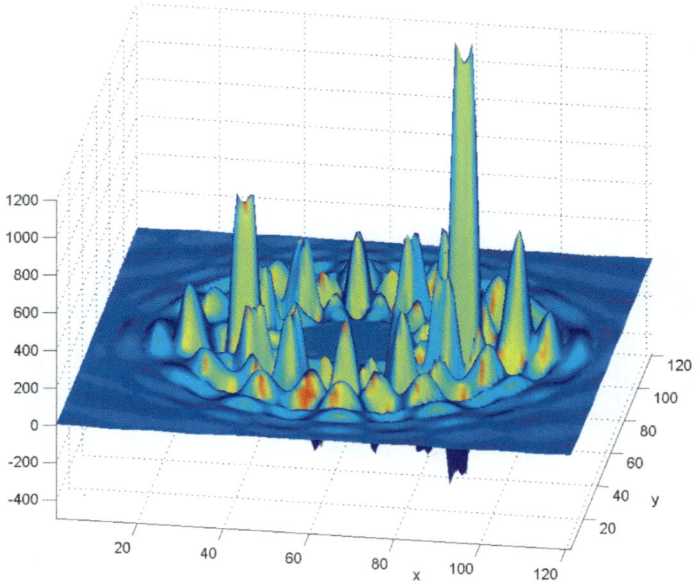

FIGURE 8. The spatial structure $-W(r, \varphi)$ (arbitrary units) where $W(r, \varphi) = \operatorname{Im}(G_{\varphi\varphi}(r, r', \varphi))$ in a cross-section $0 < r < 7\,\mu\text{m}$ and $0 < \varphi < 2\pi$ of the coated microsphere for resonance $f = 170.57\,\text{THz}$. A nanoemitter is placed in NIM layer at point $r' = 2.5\,\mu\text{m}$. One can observe the confinement of field in the vicinity of NIM defect, see Figure 4 (a).

$W(r, \varphi) = \operatorname{Im}(G_{\varphi\varphi}(r, r', \varphi))$. We observe from Figure 8 very sharp field peak in place of the nanosource location $r' = 2.5\,\mu\text{m}$. Such a spatial field structure may be treated as a confinement of the electromagnetic energy inside the NIM layer of coated microsphere. The confinement of the defect optical mode can be explaned as the follows[31]. Once a photon enters the defect region, it encounters two $\lambda/4$ Bragg reflectors (the periodic parts of the stack) before and behind it. This leads that the photon will be strongly reflected back to the defect region and thus remains long time in the defect area. Such a long dwell time results in very high energy field density around the defect. The leakage of photons through such a structure into the outer space obviously is small. We observe from Figure 8 that the field structure inside of multilayered stack is anisotropic and quite intricate, but the field distribution beyond the coated microsphere has a periodic character.

5. Conclusion

We numerically study the details of spectrum and the optical field distribution of nanoemitters placed in a microsphere coated by conventional and metamaterial layers with embedded a NIM defect. By the Green function technique we systematically have investigated the behavior of the nanoemitter fields for the frequency range where the metamaterial has negative index refraction. Our calculations have shown a strong enhancement of the nanoemitter fields correlations assisted by the embedded defect. In resonant case the photon field is almost completely arrested in a NIM layer in vicinity of the defect layer. This allows to confine resonantly the field energy in a multilayered stack in very narrow frequency range in order to create very selective stop-band filters. Incorporating nanoemitters into such structured compound microspheres allow expanding essentially the operational properties of microspheres at engineering of nanometer-sized photon emitters as attractive artificial light sources for advanced optical technologies.

References

[1] Milton Abramowitz, Irene A. Stegun, *Handbook of Mathematical Functions: with Formulas, Graphs, and Mathematical Tables.* Dover Publications, 1965.

[2] S. Ajith Kumar, C.L. Nagendra, H. Ganesh Shanbhogue et al., *Near-infrared band-pass filters from Si/SiO2 multilayer coatings.* Optical Engineering **38** (1999), 368–372.

[3] V. Astratov, *Fundamentals and applications of microsphere resonator circuits.* Springer Series in Optical Sciences **156** (2010), 423–457.

[4] D. Brady, G. Papen, J.E. Sipe, *Spherical distributed dielectric resonators.* Journal of the Optical Society of America B, **10** (1993), 644–649.

[5] V.B. Braginsky, M.L. Gorodetsky, V.S. Ilchenko, *Quality-factor and nonlinear properties of optical whispering-gallery modes.* Physics Letters A **137** (1989), 393–398.

[6] G. Burlak, S. Koshevaya, J. Sanchez-Mondragon et al., *Electromagnetic oscillations in a multilayer spherical stack.* Optics Communications **180** (2000), 49–54.

[7] G. Burlak, *The Classical and Quantum Dynamics of the Multispherical Nanostructures.* Imperial College Press, 2004.

[8] G. Burlak, A.M. Chavez, *Frequency spectrum and confinement of optical radiation of nanorod in multilayered microsphere.* Optics Communications **268** (2006), 330–339.

[9] W. Chew, *Waves, Fields in Inhomogeneous Media.* IEEE Press, New York, 1996.

[10] M. Choi, S.H. Lee, Y. Kim et al., *A terahertz metamaterial with unnaturally high refractive index.* Nature **470** (2011), 369–373.

[11] H.T. Dung, S.Y. Buhmann, L. Knoll et al., *Electromagnetic-field quantization and spontaneous decay in left-handed media.* Phys. Rev. A **68** (2003), 043816–043822.

[12] Fan-Yi Meng, Kuang Zhang, Qun Wu et al., *Reconfigurable composite right/left-handed magnetic-metamaterial waveguide at sub-wavelength scale.* J. Appl. Phys. **109** (2011), 07A309.

[13] I. Gourevich, L.M. Field, Z. Wei et al., *Polymer multilayer particles: A route to spherical dielectric resonators.* Macromolecules **39** (2006), 1449–1455.

[14] J.D. Jackson, *Classical Electrodynamics.* JOHN Willey and Sons, 1975.

[15] P. Kinsler and M.W. McCall, *Causality-Based criteria for a negative refractive index must be used with care.* Phys. Rev. Lett. **101** (2008), 167401–167405.

[16] L.D. Landau and E.M. Lifschitz, *Statistical Physics*, Part 2. Oxford, England: Pergamon Press, 1981.

[17] Le-Wei Li, Pang-Shyan Kooi, Mook-Seng Leong et al., *Electromagnetic dyadic Green's function in spherically multilayered media.* IEEE Trans. Microwave Theory Tech. **42** (1994), 2302–2308.

[18] B. Möller, U. Woggon, M.V. Artemyev et al., *Mode control by nanoengineering light emitters in spherical microcavities.* Appl. Phys. Lett. **83** (2003), 2686–2688.

[19] V.C. Nguyen, L. Chen, K. Halterman, *Total transmission and total reflection by zero index metamaterials with defects.* Phys. Rev. Lett. **105** (2010), 233908–233912.

[20] J.B. Pendry, A.J. Holden, D.J. Robbins, W.J. Stewart, *Magnetism from conductors and enhanced nonlinear phenomena.* IEEE Transactions on Microwave Theory and Techniques **47** (1999), 2075–2080.

[21] A. Petukhova, A.S. Paton, I. Gourevich et al., *Hybrid microspheres with alternating layers of a polymer and metal nanoparticles.* Can. J. Chemistry **88** (2010), 298–304.

[22] V.M. Shalaev, *Optical negative-index metamaterials.* Nature Photonics **1** (2007), 41–48.

[23] R.A. Shelby, D.R. Smith, S. Schultz, *Experimental verification of a negative index of refraction.* Science **292**, No. 5514 (2001), 77–79.

[24] D.R. Smith, J.W. Padilla, D.C. Vier et al., *Composite medium with simultaneously negative permeability and permittivity.* Phys. Rev. Letters **84** (2000), 4184–4187.

[25] I.I. Smolyaninov, Y.J. Hung, C.C. Davis, *Magnifying superlens in the visible frequency range.* Science **315** (2007), 1699–1701.

[26] Mark I. Stockman, *Criterion for negative refraction with low optical losses from a fundamental principle of causality.* Phys. Rev. Lett. **98** (2007), 177404–177408.

[27] K.G. Sullivan, D.G. Hall, *Radiation in spherically symmetric structures.* Phys. Rev. A **50** (1994), 2701–2711.

[28] V.G. Veselago, *The electrodynamics of substances with simultaneously negative values.* Soviet Physics-Uspekhi **10** (1968), 509–514. Uspekhi Fizicheskikh Nauk **92** (1967), 517–522.

[29] S. Xiao, V.P. Drachev, A.V. Kildishev et al., *Loss-free and active optical negative-index metamaterial.* Nature **466** (2010), 735–738.

[30] R.W. Ziolkowski and E. Heyman, *Wave propagation in media having negative permittivity and permeability.* Phys. Rev. E **64** (2001), 056625–056629.

[31] Shi-yao Zhu, Nian-hua Liu, Hang Zheng et al., *Time delay of light propagation through defect modes of one-dimensional photonic band-gap structures.* Optics Communications **174** (2000), 139–145.

Gennadiy Burlak
Centro de Investigación en Ingeniería y Ciencias Aplicadas
Universidad Autónoma del Estado de Morelos, Cuernavaca, Mor., 62209, México.
e-mail: `gburlak@uaem.mx`

Vladimir Rabinovich
Instituto Politécnico Nacional, ESIME Zacatenco, Av. IPN, México D.F. 07738, México
e-mail: `vladimir.rabinovich@gmail.com`

Operator Theory:
Advances and Applications, Vol. 228, 73–93

Potential Methods for Anisotropic Pseudo-Maxwell Equations in Screen Type Problems

O. Chkadua, R. Duduchava and D. Kapanadze

Dedicated to our friend and colleague Vladimir Rabinovich on the occasion of his 70th birthday anniversary

Abstract. We investigate the Neumann type boundary value problems for anisotropic pseudo-Maxwell equations in screen type problems. It is shown that the problem is well posed in tangent Sobolev spaces and unique solvability and regularity results are obtained via potential methods and the coercivity result of Costabel on the bilinear form associated to pseudo-Maxwell equations.

Mathematics Subject Classification (2010). Primary 35J25; Secondary 35C15.

Keywords. Pseudo-Maxwell equations, anisotropic media, uniqueness, existence, integral representation, potential theory, boundary pseudodifferential equation, coerciveness.

1. Introduction

The purpose of the present paper is to investigate the screen-type boundary value problem for pseudo-Maxwell equations

$$\operatorname{curl} \mu^{-1} \operatorname{curl} \boldsymbol{U} - s\, \varepsilon \operatorname{grad} \operatorname{div} (\varepsilon \boldsymbol{U}) - \omega^2 \varepsilon \boldsymbol{U} = 0 \quad \text{in} \quad \Omega, \tag{1.1}$$

where Ω is a bounded or an unbounded domain with boundary, using the potential method.

The present investigation covers the anisotropic case when the matrices

$$\varepsilon = [\varepsilon_{jk}]_{3\times 3}\,, \qquad \mu = [\mu_{jk}]_{3\times 3} \tag{1.2}$$

The investigation was supported by the grant of the Georgian National Science Foundation GNSF/ST07/3-175.

in (1.1) are real valued, constant, symmetric and positive definite, i.e.,

$$\langle\varepsilon\xi,\xi\rangle \geq c|\xi|^2\,, \qquad \langle\mu\xi,\xi\rangle \geq d|\xi|^2\,, \qquad \forall\xi\in\mathbb{R}^3\,,$$

for some positive constants $c>0$, $d>0$, where

$$\langle\eta,\xi\rangle := \sum_{j=1}^{3}\eta_j\overline{\xi}_j, \quad \eta,\ \xi\in\mathbb{C}^3.$$

s is a positive real number and the frequency parameter ω is assumed to be non-zero and complex valued, i.e., $\operatorname{Im}\omega\neq 0$.

The study of boundary value problems in electromagnetism naturally leads us to the pseudo-Maxwell equations inherited with tangent boundary conditions, which are in some sense non-standard for the elliptic equations (1.1), cf. works of Buffa, Costabel, Christiansen, Dauge, Hazard, Lenoir, Mitrea, Nicaise and others. The case with the Dirichlet type boundary condition $\boldsymbol{\nu}\times\boldsymbol{U}$ is mostly investigated by variational methods, here $\boldsymbol{\nu}$ is the unit normal to the boundary $\partial\Omega$. Our goal is to investigate well-posedness of the Neumann type boundary value problems for (1.1) as well as its unique solvability in unbounded domains with screen configuration, i.e.,

$$\Omega = \mathbb{R}^3_{\mathcal{C}} := \mathbb{R}^3\setminus\overline{\mathcal{C}},$$

where $\mathcal{C}$ denotes a smooth open hypersurface with a smooth boundary.

2. Neumann boundary value problems for pseudo-Maxwell equations

From now on throughout the paper, unless stated otherwise, Ω denotes either a bounded $\Omega^+\subset\mathbb{R}^3$ or an unbounded $\Omega^- := \mathbb{R}^3\setminus\overline{\Omega^+}$ domain with smooth boundary $\mathcal{S} := \partial\Omega^+$ and $\boldsymbol{\nu}$ is the outer unit normal vector field to $\mathcal{S}$. Whenever necessary, we will specify the case.

For rigorous formulation of conditions for the unique solvability of the formulated boundary value problems we use the Bessel potential $\mathbb{H}^r(\Omega)$, $\mathbb{H}^r(\mathcal{S})$ spaces. We quote [20] for definitions and properties of these spaces.

By $\mathcal{C}$ we denote an orientable smooth open surface in $\mathbb{R}^3$ (a screen) with boundary $\partial\mathcal{C}$, which has two faces $\mathcal{C}^-$ and $\mathcal{C}^+$ distinguished by the orientation of the normal vector field: $\boldsymbol{\nu}$ is pointing from $\mathcal{C}^+$ to $\mathcal{C}^-$. Moreover, we assume that $\mathcal{C}$ is a part of some smooth and simple (non self intersecting) hypersurface $\mathcal{S}$ that divides the space $\mathbb{R}^3$ into two disjoint domains Ω^+ and $\Omega^- := \mathbb{R}^3\setminus\overline{\Omega^+}$ such that Ω^+ is bounded and $\mathcal{S} = \partial\Omega^\pm$.

The space $\widetilde{\mathbb{H}}^r(\mathcal{C})$ comprises those functions $\varphi\in\mathbb{H}^r(\mathcal{S})$ which are supported in $\overline{\mathcal{C}}$ (functions with the "vanishing traces on the boundary"). For the detailed definitions and properties of these spaces we refer, e.g., to [13, 14, 20]).

We did not distinguish notation for the Banach spaces and their vector analogues unless this does not lead to a confusion. Although we use the boldface letters for vector-functions, in contrast to scalar functions, which are denoted by non-boldface letters.

It is well known that the space $\mathbb{H}^{r-1/2}(\mathcal{S})$ is a trace space for $\mathbb{H}^r(\Omega)$, provided that $r > 1/2$ and the corresponding trace operator is denoted by $\gamma_{\mathcal{S}}$. For the detailed definitions and properties of these spaces we refer, e.g., to [20].

We introduce the following definitions:

$$\mathbb{H}^r_{\varepsilon\boldsymbol{\nu},0}(\mathcal{S}) := \Big\{\boldsymbol{U} \in \mathbb{H}^r(\mathcal{S}) \; : \; \langle\varepsilon\boldsymbol{\nu}, \boldsymbol{U}\rangle = 0\Big\}$$

is a proper linear subspace of $\mathbb{H}^r(\mathcal{S})$. For a constant matrix $\varepsilon = \varepsilon_0 I_3$ the space $\mathbb{H}^r_{\varepsilon\boldsymbol{\nu},0}(\mathcal{S}) = \mathbb{H}^r_{\boldsymbol{\nu},0}(\mathcal{S})$ coincides with the space of tangent vector fields. The operator

$$\pi_{\varepsilon\boldsymbol{\nu}}\boldsymbol{U} := \frac{\varepsilon\boldsymbol{\nu}}{|\varepsilon\boldsymbol{\nu}|} \times \boldsymbol{U} \times \frac{\varepsilon\boldsymbol{\nu}}{|\varepsilon\boldsymbol{\nu}|} = \boldsymbol{U} - \langle\frac{\varepsilon\boldsymbol{\nu}}{|\varepsilon\boldsymbol{\nu}|}, \boldsymbol{U}\rangle\frac{\varepsilon\boldsymbol{\nu}}{|\varepsilon\boldsymbol{\nu}|} = \Big(I - \frac{(\varepsilon\boldsymbol{\nu})(\varepsilon\boldsymbol{\nu})^\top}{|\varepsilon\boldsymbol{\nu}|^2}\Big)\boldsymbol{U},$$

which is actually a multiplication by 3×3 matrix function, is a projection onto the subspace $\pi_{\varepsilon\boldsymbol{\nu}}\mathbb{H}^r(\mathcal{S}) = \mathbb{H}^r_{\varepsilon\boldsymbol{\nu},0}(\mathcal{S})$.

It is easy to see that the operator

$$\pi_{\varepsilon\boldsymbol{\nu}} \; : \; \mathbb{H}^r_{\boldsymbol{\nu},0}(\mathcal{S}) \to \mathbb{H}^r_{\varepsilon\boldsymbol{\nu},0}(\mathcal{S})$$

is continuous and invertible for all $r \in \mathbb{R}$; the inverse mapping is given by the following formula

$$(\pi_{\varepsilon\boldsymbol{\nu}})^{-1}\mathbf{u} = \mathbf{u} - \frac{\langle\boldsymbol{\nu}, \mathbf{u}\rangle}{\langle\boldsymbol{\nu}, \varepsilon\boldsymbol{\nu}\rangle}\,\varepsilon\boldsymbol{\nu}, \quad \mathbf{u} \in \mathbb{H}^r_{\varepsilon\boldsymbol{\nu},0}(\mathcal{S})$$

and we have

$$\pi_{\varepsilon\boldsymbol{\nu}}\boldsymbol{U} - \frac{\langle\boldsymbol{\nu}, \pi_{\varepsilon\boldsymbol{\nu}}\boldsymbol{U}\rangle}{\langle\boldsymbol{\nu}, \varepsilon\boldsymbol{\nu}\rangle}\,\varepsilon\boldsymbol{\nu} = \boldsymbol{U} \qquad \text{for all} \quad \boldsymbol{U} \in \mathbb{H}^r_{\boldsymbol{\nu},0}(\mathcal{S}).$$

We also use the following spaces:

$$\mathbb{H}^1_{\varepsilon\boldsymbol{\nu},0}(\Omega^+) = \Big\{\boldsymbol{U} \in \mathbb{H}^1(\Omega^+) \; : \; \langle\varepsilon\boldsymbol{\nu}, \gamma_{\mathcal{S}}\boldsymbol{U}\rangle = 0 \quad \text{on} \quad \mathcal{S}\Big\},$$

and

$$\mathbb{H}^1_{\varepsilon\boldsymbol{\nu},0}(\mathbb{R}^3_{\mathcal{C}}) = \Big\{\boldsymbol{U} \in \mathbb{H}^1(\mathbb{R}^3_{\mathcal{C}}) \; : \; \langle\varepsilon\boldsymbol{\nu}, \gamma_{\mathcal{C}^\pm}\boldsymbol{U}\rangle = 0 \quad \text{on} \quad \mathcal{C}\Big\}.$$

Theorem 2.1. *The operator in* (1.1)

$$\boldsymbol{A}(D)\boldsymbol{U} := \operatorname{curl}\mu^{-1}\operatorname{curl}\boldsymbol{U} - s\,\varepsilon\operatorname{grad}\operatorname{div}(\varepsilon\boldsymbol{U}) - \omega^2\varepsilon\boldsymbol{U}$$

is elliptic, has a positive definite principal symbol, it is self-adjoint and the following Green's formula holds

$$(\boldsymbol{A}(D)\boldsymbol{U}, \boldsymbol{V})_{\Omega^+} = (\mathfrak{N}(D,\boldsymbol{\nu})\boldsymbol{U}, \boldsymbol{V})_{\mathcal{S}} + \boldsymbol{a}_{\varepsilon,\mu}(\boldsymbol{U}, \boldsymbol{V})_{\Omega^+} - \omega^2(\varepsilon\,\boldsymbol{U}, \boldsymbol{V})_{\Omega^+} \tag{2.1}$$

for all $\boldsymbol{U}, \boldsymbol{V} \in \mathbb{H}^1(\Omega^+)$, *where* $\mathfrak{N}(D,\boldsymbol{\nu})$ *is the Neumann's boundary operator*

$$\mathfrak{N}(D,\boldsymbol{\nu})\boldsymbol{U} := \boldsymbol{\nu} \times \mu^{-1}\operatorname{curl}\boldsymbol{U} - s\operatorname{div}(\varepsilon\boldsymbol{U})\varepsilon\boldsymbol{\nu}, \qquad \boldsymbol{U} \in \mathbb{H}^1(\Omega^+); \tag{2.2}$$

$\boldsymbol{a}_{\varepsilon,\mu}$ *is the natural bilinear differential form associated with the Green formula*

$$\boldsymbol{a}_{\varepsilon,\mu}(\boldsymbol{U}, \boldsymbol{V})_\Omega := (\mu^{-1}\operatorname{curl}\boldsymbol{U}, \operatorname{curl}\boldsymbol{V})_\Omega + s\,(\operatorname{div}(\varepsilon\boldsymbol{U}), \operatorname{div}(\varepsilon\boldsymbol{V}))_\Omega. \tag{2.3}$$

The first part of the result is due to Lemma 3.1 below, while the remaining part is standard and for a similar proof we refer, e.g., to [2].

Based on this fact we obtain that the Neumann's trace $\mathfrak{N}(D,\boldsymbol{\nu})\boldsymbol{U} \in \mathbb{H}^{-\frac{1}{2}}(\mathcal{S})$. Let us mention the well-known fact, that the Neumann boundary value problem

$$\boldsymbol{A}(D)\boldsymbol{U} = 0 \quad \text{in } \Omega^+, \qquad \mathfrak{N}(D,\boldsymbol{\nu})\,\boldsymbol{U} = g \quad \text{on } \mathcal{S}, \qquad g \in \mathbb{H}^{-\frac{1}{2}}(\mathcal{S}),$$

is not an elliptic boundary value problem in the sense of the Shapiro-Lopatinski condition. To overcome this problem we consider the tangent boundary conditions and look for a solution in tangent spaces. First, for any $\boldsymbol{V} \in \mathbb{H}^1_{\varepsilon\boldsymbol{\nu},0}(\Omega^+)$ we have $\pi_{\varepsilon\boldsymbol{\nu}}\boldsymbol{V} = \boldsymbol{V}$ and therefore from (2.2) and (2.3) we obtain

$$(\mathfrak{N}(D,\boldsymbol{\nu})\boldsymbol{U},\boldsymbol{V}) = (\mathfrak{N}(D,\boldsymbol{\nu})\boldsymbol{U},\pi_{\varepsilon\boldsymbol{\nu}}\boldsymbol{V}) = (\pi_{\varepsilon\boldsymbol{\nu}}\mathfrak{N}(D,\boldsymbol{\nu})\boldsymbol{U},\pi_{\varepsilon\boldsymbol{\nu}}\boldsymbol{V}).$$

Thus $\pi_{\varepsilon\boldsymbol{\nu}}\mathfrak{N}(D,\boldsymbol{\nu})\boldsymbol{U}$ is well defined as a functional on $\mathbb{H}^{\frac{1}{2}}_{\varepsilon\boldsymbol{\nu},0}(\mathcal{S})$ and belongs to $\mathbb{H}^{-\frac{1}{2}}_{\varepsilon\boldsymbol{\nu},0}(\mathcal{S})$.

The purpose of the present paper is to investigate the following screen type Neumann boundary value problem (BVP) for pseudo-Maxwell equations:
Find $\boldsymbol{U} \in \mathbb{H}^1_{\varepsilon\boldsymbol{\nu},0}(\mathbb{R}^3_{\mathcal{C}})$ *such that*

$$\begin{cases} \boldsymbol{A}(D)\boldsymbol{U} = \operatorname{curl}\mu^{-1}\operatorname{curl}\boldsymbol{U} - s\,\varepsilon\operatorname{grad}\operatorname{div}(\varepsilon\boldsymbol{U}) - \omega^2\varepsilon\boldsymbol{U} = 0 \quad \text{in} \quad \mathbb{R}^3_{\mathcal{C}}, \\ \gamma^\pm_{\mathcal{C}}\left(\pi_{\varepsilon\boldsymbol{\nu}}\mathfrak{N}(D,\boldsymbol{\nu})\,\boldsymbol{U}\right) = \boldsymbol{g}^\pm \quad \text{on} \quad \mathcal{C}, \end{cases} \tag{2.4}$$

where s *is an arbitrary positive constant and the given data* $\boldsymbol{g}^\pm$ *satisfy the conditions*

$$\boldsymbol{g}^\pm \in \mathbb{H}^{-1/2}_{\varepsilon\boldsymbol{\nu},0}(\mathcal{C}), \quad \boldsymbol{g}^+ - \boldsymbol{g}^- \in r_{\mathcal{C}}\widetilde{\mathbb{H}}^{-1/2}_{\varepsilon\boldsymbol{\nu},0}(\mathcal{C}). \tag{2.5}$$

In Section 5, cf. Theorem 5.3 below, we prove that the screen-type Neumann BVP for pseudo-Maxwell equations (2.4)–(2.5) has a solution which is unique.

3. Vector potentials

We start the section with the following result.

Lemma 3.1. *The basic differential operator* $\boldsymbol{A}(D)$ *in* (2.4) *is elliptic: the principal symbol*

$$\mathcal{A}_{\rm pr}(\xi) := \sigma_{\rm curl}(\xi)\mu^{-1}\sigma_{\rm curl}(\xi) + s\,\varepsilon\left[\xi_j\xi_k\right]_{3\times 3}\varepsilon, \qquad \xi = (\xi_1,\xi_2,\xi_3)^\top \in \mathbb{R}^3, \tag{3.1}$$

where

$$\sigma_{\rm curl}(\xi) := \begin{bmatrix} 0 & i\xi_3 & -i\xi_2 \\ -i\xi_3 & 0 & i\xi_1 \\ i\xi_2 & -i\xi_1 & 0 \end{bmatrix}$$

is non-vanishing, i.e., $\det\mathcal{A}_{\rm pr}(\xi) \neq 0$ *for* $\xi \neq 0$ *and even positive definite, i.e.,*

$$\langle\mathcal{A}_{\rm pr}(\xi)\eta,\eta\rangle \geq c|\xi|^2|\eta|^2 \qquad c = \text{const} > 0, \quad \forall\,\xi \in \mathbb{R}^3, \quad \forall\eta \in \mathbb{C}^3. \tag{3.2}$$

Proof. Let $\mathcal{A}$ be a 3×3 real-valued and symmetric matrix $\mathcal{A}^\top = \mathcal{A}$, positive definite on the Euclidean space $\mathbb{R}^3$

$$\langle \mathcal{A}\xi, \xi\rangle \geq c|\xi|^2\,, \qquad \forall \xi \in \mathbb{R}^3\,.$$

Then $\mathcal{A}$ is positive definite on the complex space $\mathbb{C}^3$

$$\langle \mathcal{A}\eta, \eta\rangle \geq c|\eta|^2\,, \qquad \forall \eta \in \mathbb{C}^3. \tag{3.3}$$

In fact, let $\eta = \eta_r + i\eta_i \in \mathbb{C}^3$, $\eta_r,\ \eta_i \in \mathbb{R}^3$. Then

$$\begin{aligned}\langle \mathcal{A}\eta, \eta\rangle &= \langle \mathcal{A}\eta_r, \eta_r\rangle + \langle \mathcal{A}\eta_i, \eta_i\rangle + i\langle \mathcal{A}\eta_i, \eta_r\rangle - i\langle \mathcal{A}\eta_r, \eta_i\rangle \\ &= \langle \mathcal{A}\eta_r, \eta_r\rangle + \langle \mathcal{A}\eta_i, \eta_i\rangle \geq c|\eta_r|^2 + c|\eta_i|^2 = c|\eta|^2, \qquad \forall \eta \in \mathbb{C}^3,\end{aligned}$$

since $\langle \mathcal{A}\eta_r, \eta_i\rangle = \langle \eta_r, \mathcal{A}\eta_i\rangle = \langle \mathcal{A}\eta_i, \eta_r\rangle$.

If μ is a real, symmetric and positive definite matrix, so is its inverse μ^{-1} and

$$\langle \mu^{-1}\eta, \eta\rangle \geq d_1|\zeta|^2, \qquad d_1 > 0, \quad \forall\, \eta \in \mathbb{R}^3. \tag{3.4}$$

Then the symbol $\mathcal{A}_{\rm pr}(\xi)$ is real valued and symmetric $(\mathcal{A}_{\rm pr})^\top(\xi) = \mathcal{A}_{\rm pr}(\xi)$ and due to (3.3) it suffices to prove the positive definiteness for only real-valued vectors $\eta \in \mathbb{R}^3$.

Applying the first inequality in (1.2) and (3.4) we get:

$$\begin{aligned}\langle \mathcal{A}_{\rm pr}(\xi)\eta, \eta\rangle &= \langle \mu^{-1}\sigma_{\rm curl}(\xi)\eta, \sigma_{\rm curl}(\xi)\eta\rangle + s|\langle \xi, \varepsilon\eta\rangle|^2 \\ &= \langle \mu^{-1}\xi \times \eta, \xi \times \eta\rangle + s|\langle \xi, \varepsilon\eta\rangle|^2 \\ &\geq d_1|\xi \times \eta|^2 + s|\langle \xi, \varepsilon\eta\rangle|^2, \qquad \forall\, \xi,\ \eta \in \mathbb{R}^3.\end{aligned} \tag{3.5}$$

Since the unit sphere in $\mathbb{R}^3$ is compact, it is sufficient to prove that

$$\langle \mathcal{A}_{\rm pr}(\xi)\eta, \eta\rangle > 0, \qquad \forall\, \xi,\ \eta \in \mathbb{R}^3, \quad |\xi| = |\eta| = 1. \tag{3.6}$$

Let us assume the opposite: $\langle \mathcal{A}_{\rm pr}(\xi^0)\eta^0, \eta^0\rangle = 0$ for some $\xi^0 \in \mathbb{R}^3$, $|\xi^0| = 1$ and $\eta^0 \in \mathbb{R}^3$, $|\eta^0| = 1$. Then, due to (3.5),

$$\xi^0 \times \eta^0 = 0, \qquad \langle \xi^0, \varepsilon\eta^0\rangle = 0.$$

The first equality means that the vectors are parallel $\xi^0 = \pm\eta^0$ and, inserted into the second equality, this gives $\langle \varepsilon\eta^0, \eta^0\rangle = 0$. The latter contradicts the inequality (1.2)

$$\langle \varepsilon\eta^0, \eta^0\rangle \geq c|\eta^0| = c > 0.$$

The obtained contradiction verifies (3.6), implies the positive definiteness (3.2) and the ellipticity. □

The elliptic operator $\boldsymbol{A}(D)$ in (2.4) has the fundamental solution (cf. [13])

$$\mathbf{F}_{\boldsymbol{A}}(x) := \mathcal{F}^{-1}_{\xi\to x}\left[\mathcal{A}^{-1}(\xi)\right] = \mathcal{F}^{-1}_{\xi'\to x'}\left[\pm\frac{1}{2\pi}\int_{\mathcal{L}} e^{-i\tau x_3}\mathcal{A}^{-1}(\xi', \tau)d\tau\right],$$
$$\xi' = (\xi_1, \xi_2)^\top \in \mathbb{R}^2, \quad x = (x', x_3) \in \mathbb{R}^3,$$

where $\mathcal{F}^{-1}$ denotes the inverse Fourier transform and $\mathcal{A}(\xi)$ is the full symbol of the operator $\boldsymbol{A}(D)$:

$$\mathcal{A}(\xi) := \sigma_{\rm curl}(\xi)\mu^{-1}\sigma_{\rm curl}(\xi) + s\,\varepsilon\big[\xi_j\xi_k]_{3\times 3}\varepsilon - \omega^2\varepsilon, \quad \xi = (\xi_1, \xi_2, \xi_3)^\top \in \mathbb{R}^3.$$

If $x_3 < 0$ (if, respectively, $x_3 > 0$) we fix the sign "+" (the sign "−") and a contour $\mathcal{L}$ in the upper (in the lower) complex half-plane, which encloses all roots of the polynomial equation $\det \mathcal{A}(\xi) = 0$ in the corresponding half-planes.

Let us consider, respectively, the *single layer* and *double layer* potential operators

$$\mathbf{V}\boldsymbol{U}(x) := \oint_{\mathcal{S}} \mathbf{F}_{\boldsymbol{A}}(x-\tau)\boldsymbol{U}(\tau)\,dS, \tag{3.7}$$

$$\mathbf{W}\boldsymbol{U}(x) := \oint_{\mathcal{S}} [(\mathfrak{N}(D, \boldsymbol{\nu}(\tau))\mathbf{F}_{\boldsymbol{A}})(x-\tau)]^\top \boldsymbol{U}(\tau)\,dS, \qquad x \in \Omega, \tag{3.8}$$

related to pseudo-Maxwell equations in (2.4). Obviously,

$$\boldsymbol{A}(D)\mathbf{V}\boldsymbol{U}(x) = \boldsymbol{A}(D)\mathbf{W}\boldsymbol{U}(x) = 0, \qquad \forall\, \boldsymbol{U} \in \mathbb{L}_1(\mathcal{S}), \quad \forall\, x \in \Omega. \tag{3.9}$$

For the next Propositions 3.2–3.5 and for their proofs we refer, e.g., to [6, 10, 16].

Proposition 3.2. *Let $\Omega \subset \mathbb{R}^3$ be a domain with the smooth boundary $\mathcal{S} = \partial\Omega$.*

The potential operators above map continuously the spaces:

$$\begin{aligned} \mathbf{V} &: \mathbb{H}^r(\mathcal{S}) \to \mathbb{H}^{r+3/2}(\Omega), \\ \mathbf{W} &: \mathbb{H}^r(\mathcal{S}) \to \mathbb{H}^{r+1/2}(\Omega) \qquad \forall\, r \in \mathbb{R}. \end{aligned} \tag{3.10}$$

The direct values $\mathbf{V}_{-1}$, $\mathbf{W}_0$ and $\mathbf{V}_{+1}$ of the potential operators $\mathbf{V}$, $\mathbf{W}$ and $\mathfrak{N}(D, \boldsymbol{\nu})\mathbf{W}$ are pseudodifferential operators of order -1, 0 and 1, respectively, and map continuously the spaces:

$$\begin{aligned} \mathbf{V}_{-1} &: \mathbb{H}^r(\mathcal{S}) \to \mathbb{H}^{r+1}(\mathcal{S}), \\ \mathbf{W}_0 &: \mathbb{H}^r(\mathcal{S}) \to \mathbb{H}^{r}(\mathcal{S}), \\ \mathbf{V}_{+1} &: \mathbb{H}^r(\mathcal{S}) \to \mathbb{H}^{r-1}(\mathcal{S}), \qquad \forall\, r \in \mathbb{R}. \end{aligned} \tag{3.11}$$

Proposition 3.3. *The potential operators on an open, compact, smooth surface $\mathcal{C} \subset \mathbb{R}^3$ have the following mapping properties:*

$$\begin{aligned} \mathbf{V} &: \widetilde{\mathbb{H}}^r(\mathcal{C}) \to \mathbb{H}^{r+3/2}(\mathbb{R}^3_{\mathcal{C}}), \\ \mathbf{W} &: \widetilde{\mathbb{H}}^r(\mathcal{C}) \to \mathbb{H}^{r+1/2}(\mathbb{R}^3_{\mathcal{C}}), \qquad \forall\, r \in \mathbb{R}. \end{aligned} \tag{3.12}$$

The direct values $\mathbf{V}_{-1}$, $\mathbf{W}_0$ and $\mathbf{V}_{+1}$ of the potential operators $\mathbf{V}$, $\mathbf{W}$ and $\mathfrak{N}(D, \boldsymbol{\nu})\mathbf{W}$ are pseudodifferential operators of order -1, 0 and 1, respectively, and have the following mapping properties:

$$\begin{aligned} \mathbf{V}_{-1} &: \widetilde{\mathbb{H}}^r(\mathcal{C}) \to \mathbb{H}^{r+1}(\mathcal{C}), \\ \mathbf{W}_0 &: \widetilde{\mathbb{H}}^r(\mathcal{C}) \to \mathbb{H}^{r}(\mathcal{C}), \\ \mathbf{V}_{+1} &: \widetilde{\mathbb{H}}^r(\mathcal{C}) \to \mathbb{H}^{r-1}(\mathcal{C}), \qquad \forall\, r \in \mathbb{R}. \end{aligned} \tag{3.13}$$

Proposition 3.4. *For the traces of potential operators we have the following Plemelji formulae:*

$$(\gamma_{\mathcal{S}^-}\mathbf{V}\boldsymbol{U})(x) = (\gamma_{\mathcal{S}^+}\mathbf{V}\boldsymbol{U})(x) = \mathbf{V}_{-1}\boldsymbol{U}(x), \tag{3.14}$$

$$(\gamma_{\mathcal{S}^\pm}\mathfrak{N}(D,\boldsymbol{\nu})\mathbf{V}\boldsymbol{U})(x) = \mp\frac{1}{2}\boldsymbol{U}(x) + (\mathbf{W}_0)^*(x,D)\boldsymbol{U}(x), \tag{3.15}$$

$$(\gamma_{\mathcal{S}^\pm}\mathbf{W}\boldsymbol{U})(x) = \pm\frac{1}{2}\boldsymbol{U}(x) + \mathbf{W}_0(x,D)\boldsymbol{U}(x), \tag{3.16}$$

$$(\gamma_{\mathcal{S}^-}\mathfrak{N}(D,\boldsymbol{\nu})\mathbf{W}\boldsymbol{U})(x) = (\gamma_{\mathcal{S}^+}\mathfrak{N}(D,\boldsymbol{\nu})\mathbf{W}\boldsymbol{U})(x) = \mathbf{V}_{+1}\boldsymbol{U}(x), \tag{3.17}$$

$$x \in \mathcal{S}, \quad \boldsymbol{U} \in \mathbb{H}^s(\mathcal{S}),$$

where $(\mathbf{W}_0)^*(x,D)$ *is the adjoint to the pseudodifferential operator* $\mathbf{W}_0(x,D)$*, the direct value of the potential operator* $\mathfrak{N}(D,\boldsymbol{\nu})\mathbf{V}$ *on the boundary* $\mathcal{S}$*.*

Proposition 3.5. *Let the boundary* $\mathcal{S} = \partial\Omega^\pm$ *be a compact smooth surface. Solutions to pseudo-Maxwell equations with anisotropic coefficients* ε *and* μ *are represented as*

$$\boldsymbol{U}(x) = \pm\mathbf{W}(\gamma_{\mathcal{S}^\pm}\boldsymbol{U})(x) \mp \mathbf{V}(\gamma_{\mathcal{S}^\pm}\mathfrak{N}(D,\boldsymbol{\nu})\boldsymbol{U})(x), \qquad x \in \Omega^\pm, \tag{3.18}$$

where $\gamma_{\mathcal{S}^\pm}\mathfrak{N}(D,\boldsymbol{\nu})\Psi$ *is Neumann's trace operator (see* (2.2)*) and* $\gamma_{\mathcal{S}^\pm}\Psi$ *is Dirichlet's trace operator.*

If $\mathcal{C} \subset \mathbb{R}^3$ *is an open compact smooth surface, then a solution to pseudo-Maxwell equations with anisotropic coefficients* ε *and* μ *is represented as*

$$\boldsymbol{U}(x) = \mathbf{W}([\boldsymbol{U}])(x) - \mathbf{V}([\mathfrak{N}(D,\boldsymbol{\nu})\boldsymbol{U}])(x), \qquad x \in \mathbb{R}^3_{\mathcal{C}},$$
$$[\boldsymbol{U}] := \gamma_{\mathcal{C}^+}\boldsymbol{U} - \gamma_{\mathcal{C}^-}\boldsymbol{U}, \quad [\mathfrak{N}(D,\boldsymbol{\nu})\boldsymbol{U}] := \gamma_{\mathcal{C}^+}\mathfrak{N}(D,\boldsymbol{\nu})\boldsymbol{U} - \gamma_{\mathcal{C}^-}\mathfrak{N}(D,\boldsymbol{\nu})\boldsymbol{U}.$$

As a consequence of the representation formula (3.18) we derive the following.

Corollary 3.6. *For a complex-valued frequency a solution to the screen type Neumann BVP for pseudo-Maxwell equations* (2.4)–(2.5) *decays at infinity exponentially, i.e.,*

$$\boldsymbol{U}(x) = \mathcal{O}\left(e^{-\gamma|x|}\right) \qquad \text{as} \quad |x| \to \infty \quad \text{provided that} \quad \operatorname{Im}\omega \neq 0 \tag{3.19}$$

for some $\gamma > 0$*.*

Theorem 3.7. *The screen type Neumann BVP for pseudo-Maxwell equations* (2.4)–(2.5) *has at most one solution in* $\mathbb{H}^1_{\varepsilon\boldsymbol{\nu},0}(\mathbb{R}^3_{\mathcal{C}})$*.*

Proof. Let us consider homogeneous BVP (2.4) with $\boldsymbol{g}^+ = \boldsymbol{g}^- = 0$ (extended by 0 to the complementary surface $\mathcal{C}^c := \mathcal{S} \setminus \overline{\mathcal{C}}$) and apply the Green's formulae with $\boldsymbol{V} = \boldsymbol{U}$; taking into account the boundary conditions in (2.4) and that $\langle \varepsilon\boldsymbol{\nu}, \boldsymbol{U}^\pm\rangle = 0$ and Corollary 3.6, we get

$$0 = \int_{\Omega^\pm} \langle \mu^{-1}\operatorname{curl}\boldsymbol{U}, \operatorname{curl}\boldsymbol{U}\rangle dx + s\int_{\Omega^\pm} |\operatorname{div}\varepsilon\boldsymbol{U}|^2 dx - \omega^2\int_{\Omega^\pm} \langle \varepsilon\boldsymbol{U}, \boldsymbol{U}\rangle dx.$$

Since ε and μ^{-1} are positive definite and $\operatorname{Im}\omega \neq 0$, it follows, that

$$\int_{\mathbb{R}^3} \langle \varepsilon \boldsymbol{U}, \boldsymbol{U} \rangle dx = 0 \quad \text{and, therefore,} \quad \boldsymbol{U} \equiv 0. \qquad \square$$

It is well known that the bilinear differential form $\boldsymbol{a}_{\varepsilon,\mu}(\boldsymbol{U}, \boldsymbol{V})$ in (2.3) is not coercive on $\mathbb{H}^1(\Omega^\pm)$. In the paper [4] M. Costabel suggested the following modified bilinear form (see [17] for an earlier version)

$$\begin{aligned} \boldsymbol{a}^m_{\varepsilon,\mu}(\boldsymbol{U}, \boldsymbol{V})_{\Omega^\pm} &:= (\mu^{-1}\operatorname{curl} \boldsymbol{U}, \operatorname{curl} \boldsymbol{V})_{\Omega^\pm} + s\,(\operatorname{div}(\varepsilon \boldsymbol{U}), \operatorname{div}(\varepsilon \boldsymbol{V}))_{\Omega^\pm} \\ &\quad + (\mathcal{G}rad_{\mathcal{S}} \langle \varepsilon \boldsymbol{\nu}, \boldsymbol{U}^\pm \rangle, \boldsymbol{V}^\pm_\varepsilon)_{\mathcal{S}} - (\mathcal{D}iv_{\mathcal{S}} \boldsymbol{V}^\pm_\varepsilon, \langle \varepsilon \boldsymbol{\nu}, \boldsymbol{U}^\pm \rangle)_{\mathcal{S}} \\ &= (\mu^{-1}\operatorname{curl} \boldsymbol{U}, \operatorname{curl} \boldsymbol{V})_{\Omega^\pm} + s\,(\operatorname{div}(\varepsilon \boldsymbol{U}), \operatorname{div}(\varepsilon \boldsymbol{V}))_{\Omega^\pm} \\ &\quad + 2\operatorname{Re}\,(\mathcal{G}rad_{\mathcal{S}} \langle \varepsilon \boldsymbol{\nu}, \boldsymbol{U}^\pm \rangle, \boldsymbol{V}^\pm_\varepsilon)_{\mathcal{S}}; \end{aligned} \tag{3.20}$$

here $\boldsymbol{U}^\pm := \gamma^\pm_{\mathcal{S}} \boldsymbol{U}$, $\boldsymbol{V}^\pm := \gamma^\pm_{\mathcal{S}} \boldsymbol{V}$ denote the traces of the vector fields on the boundary and

$$\mathcal{X}_\varepsilon := -\frac{s \det \varepsilon}{\langle \varepsilon \boldsymbol{\nu}, \boldsymbol{\nu} \rangle} \boldsymbol{\nu} \times \big(\varepsilon(\mathcal{X} \times \boldsymbol{\nu})\big), \quad \mathcal{X} \in \mathbb{H}^r(\mathcal{S}). \tag{3.21}$$

The surface gradient $\mathcal{G}rad_{\mathcal{S}}$ and the surface divergence $\mathcal{D}iv_{\mathcal{S}}$ are negative adjoint to each-other with respect to the bilinear form on the boundary

$$\mathcal{G}rad_{\mathcal{S}} \varphi := (\mathcal{D}_1\varphi, \mathcal{D}_2\varphi, \mathcal{D}_3\varphi)^\top, \quad \mathcal{D}iv_{\mathcal{S}} \boldsymbol{U} := \mathcal{D}_1 \boldsymbol{U}_1 + \mathcal{D}_2 \boldsymbol{U}_2 + \mathcal{D}_3 \boldsymbol{U}_3,$$

$$\mathcal{D}_j := \partial_j - \nu_j \partial_{\boldsymbol{\nu}}, \quad j = 1, 2, 3,$$

$$(\mathcal{D}iv_{\mathcal{S}} \boldsymbol{U}, \boldsymbol{V})_{\mathcal{S}} = -(\boldsymbol{U}, \mathcal{G}rad_{\mathcal{S}} \boldsymbol{V})_{\mathcal{S}} \qquad \boldsymbol{U}, \boldsymbol{V} \in \mathbb{H}^r(\mathcal{S}).$$

(see [8]).

The following theorem proved in [4] for a bounded domain plays a key role in the present investigation.

Theorem 3.8. *The modified bilinear differential form $\boldsymbol{a}^m_{\varepsilon,\mu}$ in (3.20) is coercive in the space $\mathbb{H}^1(\Omega^\pm)$: there exist positive constants c_1 and c_2 such that*

$$\operatorname{Re} \boldsymbol{a}^m_{\varepsilon,\mu}(\boldsymbol{U}, \boldsymbol{U})_{\Omega^\pm} \geq c_1 \big\| \boldsymbol{U} \big| \mathbb{H}^1(\Omega^\pm) \big\|^2 - c_2 \big\| \boldsymbol{U} \big| \mathbb{L}_2(\Omega^\pm) \big\|^2 \qquad \forall \boldsymbol{U} \in \mathbb{H}^1(\Omega^\pm)$$

provided that

$$\boldsymbol{U}(x) = \mathcal{O}\left(|x|^{-1-\delta}\right) \qquad as \quad |x| \to \infty \tag{3.22}$$

for some $\delta > 0$ if the domain is unbounded.

Proof. As noted already, for a bounded domain Ω^+ the theorem is proved in [4].

Concerning an unbounded domain Ω^-: The coerciveness (3.23) is valid for the domain $\Omega^-_R := \mathbb{S}^3_R \cap \Omega^-$, where $\mathbb{S}^3_R$ is the ball with a sufficiently large radius R, and the bilinear differential form

$$\begin{aligned} \boldsymbol{a}^m_{\varepsilon,\mu,R}(\boldsymbol{U}, \boldsymbol{V}) &:= (\mu^{-1}\operatorname{curl} \boldsymbol{U}, \operatorname{curl} \boldsymbol{V})_{\Omega^-_R} + s\,(\operatorname{div}(\varepsilon \boldsymbol{U}), \operatorname{div}(\varepsilon \boldsymbol{V}))_{\Omega^-_R} \\ &\quad + 2\operatorname{Re}\,(\mathcal{G}rad_{\mathcal{S}} \langle \varepsilon \boldsymbol{\nu}, \boldsymbol{U}^- \rangle, \boldsymbol{V}^-_\varepsilon)_{\mathcal{S}} + 2\operatorname{Re}\,(\mathcal{G}rad_{\mathcal{S}} \langle \varepsilon \boldsymbol{\nu}, \boldsymbol{U}^- \rangle, \boldsymbol{V}^-_\varepsilon)_{\partial \mathbb{S}^3_R}. \end{aligned}$$

The condition (3.22) and the independence of the constants c_1, c_2 from the domain Ω^- configuration, ensures that the limit $R \to \infty$ eliminates the last summand in the form above (the integrals over the surface of the sphere $\partial\mathbb{S}^3_R$) and the coerciveness for the domain Ω^- emerges as the limit case. □

Corollary 3.9. *The quadratic differential form* $\boldsymbol{a}_{\varepsilon,\mu}(\boldsymbol{U},\boldsymbol{U})_{\Omega^+}$ *in* (2.3) *is coercive: there exist positive constants* c_1 *and* c_2 *such that*

$$\operatorname{Re} \boldsymbol{a}_{\varepsilon,\mu}(\boldsymbol{U},\boldsymbol{U})_{\Omega^\pm} \geq c_1\big\|\boldsymbol{U}\big|\mathbb{H}^1(\Omega^\pm)\big\|^2 - c_2\big\|\boldsymbol{U}\big|\mathbb{L}_2(\Omega^\pm)\big\|^2 \tag{3.23}$$

on the space $\mathbb{H}^1_{\varepsilon\boldsymbol{\nu},0}(\Omega^+)$.

The quadratic differential form $\boldsymbol{a}_{\varepsilon,\mu}(\boldsymbol{U},\boldsymbol{U})_{\Omega^-}$ *in* (2.3) *is coercive on the space of* $\mathbb{H}^1_{\varepsilon\boldsymbol{\nu},0}(\Omega^-)$ *of those vector fields which satisfy the condition* (3.22).

Proof. Note that due to definitions (3.21), (3.20) the last summands in the modified form $\boldsymbol{a}^m_{\varepsilon,\mu}(\boldsymbol{U},\boldsymbol{U})$ vanish if either $\langle\varepsilon\boldsymbol{\nu},\boldsymbol{U}^\pm\rangle = 0$, which is the case when $\boldsymbol{U} \in \mathbb{H}^1_{\varepsilon\boldsymbol{\nu},0}(\Omega^-)$, or $\boldsymbol{U}_\varepsilon = 0$ which is the case when $\boldsymbol{U} \in \mathbb{H}^1_{\varepsilon\boldsymbol{\nu}}(\Omega^-)$. Then the modified form coincides with $\boldsymbol{a}_{\varepsilon,\mu}(\boldsymbol{U},\boldsymbol{U})_{\Omega^\pm}$ and the claimed positive definiteness follows from (3.23). □

Corollaries 3.9 and 3.6 imply the following result.

Corollary 3.10. *The quadratic differential form* $\boldsymbol{a}_{\varepsilon,\mu}(\boldsymbol{U},\boldsymbol{V})$ *in* (2.3) *is coercive (satisfies the inequality* (3.23)) *for all vector fields* $\boldsymbol{U} \in \mathbb{H}^1_{\varepsilon\boldsymbol{\nu},0}(\Omega^-)$ *provided they are solutions to pseudo-Maxwell equation.*

Lemma 3.11. *The operator* $\mathbf{V}_{-1}$ *in* (3.7) *is invertible in the following space settings*

$$\mathbf{V}_{-1} \ : \ \mathbb{H}^r(\mathcal{S}) \to \mathbb{H}^{r+1}(\mathcal{S}) \quad \forall r \in \mathbb{R}. \tag{3.24}$$

The principal symbol of the pseudodifferential operator $\mathbf{V}_{-1}$ *is positive definite*

$$\langle V_{-1,\mathrm{pr}}(\mathcal{x},\xi)\eta,\eta\rangle \geq c_0|\eta|^2|\xi|^{-1} \qquad \forall\eta \in C^3, \quad \mathcal{x} \in \mathcal{S}, \quad \xi \in \mathbb{R}^3, \tag{3.25}$$

for some positive constant c_0.

Proof. (see [5, 12, 19] for similar proofs): Let the boundary surface $\mathcal{S} = \partial\Omega^\pm$ be covered by a finite set of open smooth surfaces $\{\mathcal{S}_j\}_{j=1}^M$ and

$$\kappa_j : X_j \mapsto \mathcal{S}_j, \qquad x \in X_j\,, \quad j = 1,\dots,M$$

be diffeomorphisms of open subsets $X_j \subset \mathbb{R}^2$ onto $\mathcal{S}_j$. Let us extend them to the diffeomorphisms of layers:

$$\begin{aligned}
&\varkappa_j \ : \ \widetilde{X}_j \mapsto \widetilde{\mathcal{S}}_j, \qquad \widetilde{X}_j, \widetilde{\mathcal{S}}_j \subset \mathbb{R}^3, \qquad \widetilde{\mathcal{S}}_j \cap \mathcal{S} = \mathcal{S}_j,\\
&\widetilde{X}_j := (-\epsilon,\epsilon)\times X_j, \quad \widetilde{\mathcal{S}}_j := \big\{(\mathcal{x}, t\boldsymbol{\nu}(\mathcal{x})) \ : \ -\epsilon < t < \epsilon, \ \mathcal{x} \in \mathcal{S}_j\big\}\\
&\varkappa_j(x,x_3) := \kappa_j(x) + x_3\boldsymbol{\nu}(\kappa_j(x))\,, \quad x \in X_j\,,\ x_3 \in [-\epsilon,\epsilon]\,, \quad j = 1,\dots,M.
\end{aligned} \tag{3.26}$$

For the principal symbol $V_{-1,\mathrm{pr}}(\mathcal{X},\xi)$ of the operator $\mathbf{V_{-1}}$ we have the formulae

$$V_{-1,\mathrm{pr}}(\kappa_j(x),\xi) = \frac{\mathcal{G}_{\kappa_j}(x)}{2\pi\det\varkappa_j'(x,0)}\int\limits_{-\infty}^{\infty}(\mathcal{A}_{\mathrm{pr}})^{-1}\big([\varkappa_j'(x,0)^\top]^{-1}(\xi,t)\big)\,dt \tag{3.27}$$
$$\text{for}\quad \xi\in\mathbb{R}^2,\quad \mathcal{X}=\kappa_j(x)\mathcal{S}_j,\quad x\in X_j\subset\mathbb{R}^2,$$

where $(\mathcal{A}_{\mathrm{pr}})^{-1}(\xi)$ is the inverse to the principal symbol in (3.1),

$$\varkappa_j'(x,0)=\begin{bmatrix}\partial_1\kappa_1(x) & \partial_2\kappa_1(x) & \nu_1(\varkappa_j(x))\\ \partial_1\kappa_2(x) & \partial_2\kappa_2(x) & \nu_2(\varkappa_j(x))\\ \partial_1\kappa_3(x) & \partial_2\kappa_3(x) & \nu_3(\varkappa_j(x))\end{bmatrix}$$

is the Jacoby matrix of the diffeomorphism in (3.26) and

$$\mathcal{G}_{\kappa_j}:=(\det\|(\partial_k\kappa_j,\partial_l\kappa_j)\|_{2\times2})^{\frac12}\qquad\text{with}\qquad \partial_k\kappa_j:=(\partial_k\kappa_{j1},\partial_k\kappa_{j2})^\top$$

is the square root from the Gram determinant, the surface element on $\mathcal{S}$ (cf., e.g., [5, 12]).

Since $\mathcal{A}_{\mathrm{pr}}$ is positive definite (cf. (3.2)), the same holds for the inverse. In fact, we introduce $\eta'=(\mathcal{A}_{\mathrm{pr}})^{-1}(\xi)\eta\in\mathbb{C}^3$ into (3.2) and proceed as follows:

$$\begin{aligned}\langle(\mathcal{A}_{\mathrm{pr}})^{-1}(\xi)\eta,\eta\rangle &= \langle\eta',\mathcal{A}_{\mathrm{pr}}(\xi)\eta'\rangle\geq c|\eta'|^2|\xi|^2=c|(\mathcal{A}_{\mathrm{pr}})^{-1}(\xi)\eta|^2|\xi|^2\\ &= c_1|(\mathcal{A}_{\mathrm{pr}})^{-1}(|\xi|^{-1}\xi)\eta|^2|\xi|^{-2}\geq c_2|\eta|^2|\xi|^{-2},\quad \forall\xi\in\mathbb{R}^3,\quad\forall\eta\in\mathbb{C}^3,\end{aligned} \tag{3.28}$$

because $(\mathcal{A}_{\mathrm{pr}})^{-1}(\xi)$ is homogeneous of order -2 and, as an invertible matrix, is bounded from below on the unit sphere $|(\mathcal{A}_{\mathrm{pr}})^{-1}(\xi)\eta|\geq c_3>0$ for all $|\xi|=|\eta|=1$. Then, with (3.27) and (3.28) at hand, we derive

$$\begin{aligned}\langle V_{-1,\mathrm{pr}}(\kappa_j(x),\xi)\eta,\eta\rangle &\geq \frac{c_2\mathcal{G}_{\kappa_j}(x)|\eta|^2}{2\pi\det\varkappa_j'(x,0)}\int_{-\infty}^{\infty}|[\varkappa_j'(x,0)^\top]^{-1}(\xi,t)|^{-2}\,dt \qquad (3.29)\\ &\geq c_4|\eta|^2\int_{-\infty}^{\infty}\frac{dt}{t^2+|\xi|^2}=c_0|\eta|^2|\xi|^{-1}\qquad\forall\eta\in\mathbb{C}^3,\quad\forall\xi\in\mathbb{R}^2\end{aligned}$$

and (3.25) is proved.

Since the symbol of the ΨDO $\mathbf{V_{-1}}$ is elliptic (cf. (3.25)) and $\mathcal{S}$ has no boundary, the pseudodifferential operator

$$\boldsymbol{V}_{-1}\;:\;\mathbb{H}^r(\mathcal{S})\longrightarrow\mathbb{H}^{r+1}(\mathcal{S}) \tag{3.30}$$

is Fredholm for arbitrary $r\in\mathbb{R}$ (cf. [11, 13, 14]).

Let us introduce the vectors $\boldsymbol{U}=\boldsymbol{V}=\mathbf{V\Phi}$, $\mathbf{\Phi}\in\mathbb{H}^{-1/2}(\mathcal{S})$ into the Green formula (2.1). Since $\boldsymbol{A}(D)\boldsymbol{U}(x)=\boldsymbol{A}(D)\mathbf{V\Phi}(x)=0$ in $\Omega^\pm$, then by applying the equalities (cf. (3.14), (3.15))

$$\gamma_{\mathcal{S}}^{\pm}\boldsymbol{U}=\gamma_{\mathcal{S}}^{\pm}(\mathbf{V\Phi})=\mathbf{V_{-1}\Phi},\qquad \gamma_{\mathcal{S}}^{\pm}(\mathfrak{N}(D,\boldsymbol{\nu})\mathbf{V\Phi})=\mp\frac12\mathbf{\Phi}+(\mathbf{W_0})^*\mathbf{\Phi}, \tag{3.31}$$

we get

$$\pm\frac{1}{2}(\mathbf{\Phi},\mathbf{V_{-1}\Phi})_{\mathcal{S}}-((\mathbf{W_0})^*\mathbf{\Phi},\mathbf{V_{-1}\Phi})_{\mathcal{S}}=-(\gamma^{\pm}_{\mathcal{S}}(\mathfrak{N}(D,\boldsymbol{\nu})\boldsymbol{U}),\gamma^{\pm}_{\mathcal{S}}\boldsymbol{U})_{\mathcal{S}}$$
$$=\pm(\mu^{-1}\mathrm{curl}\,\boldsymbol{U},\mathrm{curl}\,\boldsymbol{U})_{\Omega^{\pm}}\pm s\,(\mathrm{div}(\varepsilon\boldsymbol{U}),\mathrm{div}(\varepsilon\boldsymbol{U}))_{\Omega^{\pm}}\mp\omega^2(\varepsilon\,\boldsymbol{U},\boldsymbol{U})_{\Omega^{\pm}}.$$

Further, taking the difference of the obtained equalities we find that

$$(\mathbf{V_{-1}\Phi},\mathbf{\Phi})_{\mathcal{S}}=(\mu^{-1}\mathrm{curl}\,\boldsymbol{U},\mathrm{curl}\,\boldsymbol{U})_{\mathbb{R}^3}+s\,(\mathrm{div}(\varepsilon\boldsymbol{U}),\mathrm{div}(\varepsilon\boldsymbol{U}))_{\mathbb{R}^3}-\omega^2(\varepsilon\,\boldsymbol{U},\boldsymbol{U})_{\mathbb{R}^3}. \tag{3.32}$$

Since ω is complex valued, from (3.32) follows that $\mathbf{V_{-1}\Phi}=0$ implies $\boldsymbol{U}\equiv 0$. Then,

$$\mathbf{\Phi}=\gamma_{\mathcal{S}^-}\big(\mathfrak{N}(D,\boldsymbol{\nu})\mathbf{V\Phi}\big)-\gamma_{\mathcal{S}^+}\big(\mathfrak{N}(D,\boldsymbol{\nu})\mathbf{V\Phi}\big)=0,$$

which implies that the kernel of $\mathbf{V_{-1}}$ in $\mathbb{H}^{-1/2}(\mathcal{S})$ is trivial $\mathrm{Ker}\,\mathbf{V_{-1}}=\{0\}$.

Consider the adjoint ΨDO to (3.30)

$$(\mathbf{V_{-1}})^*\ :\ \mathbb{H}^{-\frac{1}{2}}(\mathcal{S})\longrightarrow\mathbb{H}^{\frac{1}{2}}(\mathcal{S}). \tag{3.33}$$

From (3.32) we derive:

$$(\,(\mathbf{V_{-1}})^*\,\mathbf{\Phi},\mathbf{\Phi})_{\mathcal{S}}=\overline{(\mathbf{V_{-1}\Phi},\mathbf{\Phi})_{\mathcal{S}}}$$
$$=(\mu^{-1}\mathrm{curl}\,\boldsymbol{U},\mathrm{curl}\,\boldsymbol{U})_{\mathbb{R}^3}+s\,(\mathrm{div}(\varepsilon\boldsymbol{U}),\mathrm{div}(\varepsilon\boldsymbol{U}))_{\mathbb{R}^3}-\overline{\omega}^2(\varepsilon\,\boldsymbol{U},\boldsymbol{U})_{\mathbb{R}^3}.$$

Repeating the above arguments for the operator $\mathbf{V_{-1}}$ we find out that the operator $(\mathbf{V_{-1}})^*$ in (3.33) has a trivial kernel, i.e., $\mathrm{Ker}\,\mathbf{V_{-1}}=\{0\}$ in $\mathbb{H}^{-1/2}(\mathcal{S})$. The claimed invertibility of the operator $\boldsymbol{V}_{-1}$ in (3.30) follows. □

Lemma 3.12. *The operator* $\mathbf{V}_{-1}$ *is invertible in the following space settings*

$$\boldsymbol{V}_{-1}\ :\ \widetilde{\mathbb{H}}^{r-\frac{1}{2}}(\mathcal{C})\to\mathbb{H}^{r+\frac{1}{2}}(\mathcal{C})\qquad -\frac{1}{2}<r<\frac{1}{2}. \tag{3.34}$$

Proof. In [12, Theorem 2.7] and in [3, Theorem 1.9] it is proved that, since the symbol $V_{-1,\mathrm{pr}}(\mathcal{X},\xi)$ is positive definite, the corresponding operators (3.24) on the manifold (surface) $\mathcal{C}$ with boundary $\partial\mathcal{S}=\emptyset$ is Fredholm for all $r\in\mathbb{R}$. The same operator on the manifold (surface) $\mathcal{C}$ with the boundary $\partial\mathcal{C}\neq\emptyset$ is Fredholm if and only if the condition $|r|<1/2$ holds. In both cases, due to the positive definiteness of the symbol (3.25), the index is trivial $\mathrm{Ind}\,\boldsymbol{V}_{-1}=0$. Indeed, since ω is complex valued, from (3.32) follows that $\boldsymbol{V}_{-1}\mathbf{\Phi}=0$ implies $\boldsymbol{U}\equiv 0$. Then,

$$\mathbf{\Phi}=\gamma_{\mathcal{S}^-}\big(\mathfrak{N}(D,\boldsymbol{\nu})\boldsymbol{V}\mathbf{\Phi}\big)-\gamma_{\mathcal{S}^+}\big(\mathfrak{N}(D,\boldsymbol{\nu})\boldsymbol{V}\mathbf{\Phi}\big)=0,$$

which implies that the kernel of $\boldsymbol{V}_{-1}$ in $\mathbb{H}^{-1/2}(\mathcal{S})$ is trivial $\mathrm{Ker}\,\boldsymbol{V}_{-1}=\{0\}$. Due to the trivial index $\mathrm{Ind}\,\boldsymbol{V}_{-1}=0$ this implies trivial co-kernel

$$\dim\,\mathrm{Coker}\,\boldsymbol{V}_{-1}=\dim\,\mathrm{Ker}\,(\boldsymbol{V}_{-1})^*=0$$

and provides the invertibility of the operator (3.24) for $r=0$.

Concerning a surface $\mathcal{C}$ with boundary $\partial\mathcal{C} \neq \emptyset$: equality (3.32) is valid for an open surface as well

$$\begin{aligned}(\boldsymbol{V}_{-1}\boldsymbol{\Psi},\boldsymbol{\Psi})_{\mathcal{C}} &= (\mu^{-1}\mathrm{curl}\,\boldsymbol{U},\mathrm{curl}\,\boldsymbol{U})_{\mathbb{R}^3} + s\,(\mathrm{div}(\varepsilon\boldsymbol{U}),\mathrm{div}(\varepsilon\boldsymbol{U}))_{\mathbb{R}^3}\\ &\quad - \omega^2(\varepsilon\,\boldsymbol{U},\boldsymbol{U})_{\mathbb{R}^3}, \qquad \boldsymbol{U} = \mathbf{V}\boldsymbol{\Psi}, \qquad \boldsymbol{\Psi} \in \widetilde{\mathbb{H}}^{-1/2}(\mathcal{C}).\end{aligned} \tag{3.35}$$

From (3.35) follows, as for the closed surface, the invertibility of operator (3.34) for $r = 0$.

If a pseudodifferential operators on a manifold with or without boundary is Fredholm in the spaces $\mathbb{H}^s$ for all $s_0 < s < s_1$, it has the same kernel in $\mathbb{H}^s$ for all $s_0 < s < s_1$ (see [9] and also [1, 7, 15] and [10] for similar results). Therefore, the operator $\boldsymbol{V}_{-1}$ in (3.34) is invertible for all $|r| < 1/2$. □

Remark 3.13. For arbitrary complex ω, $\mathrm{Im}\,\omega \neq 0$, the operator $\boldsymbol{V}_{-1}$ is coercive

$$\mathrm{Re}\,(\boldsymbol{V}_{-1}\boldsymbol{\Phi},\boldsymbol{\Phi})_{\mathcal{S}} \geq c_0\big\|\boldsymbol{\Phi}\big|\mathbb{H}^{-1/2}(\mathcal{S})\big\|^2 - c_1\big\|\boldsymbol{\Phi}\big|\mathbb{H}^{-3/2}(\mathcal{S})\big\|^2$$

if $\mathcal{S}$ is closed, and

$$\mathrm{Re}\,(\boldsymbol{V}_{-1}\boldsymbol{\Psi},\boldsymbol{\Psi})_{\mathcal{C}} \geq c_0\big\|\boldsymbol{\Psi}\big|\widetilde{\mathbb{H}}^{-1/2}(\mathcal{C})\big\|^2 - c_1\big\|\boldsymbol{\Psi}\big|\widetilde{\mathbb{H}}^{-3/2}(\mathcal{C})\big\|^2 \tag{3.36}$$

if $\mathcal{C}$ is open, for all $\boldsymbol{\Phi} \in \mathbb{H}^{-1/2}_{\varepsilon\boldsymbol{\nu},0}(\mathcal{S})$, all $\boldsymbol{\Psi} \in \widetilde{\mathbb{H}}^{-1/2}_{\varepsilon\boldsymbol{\nu},0}(\mathcal{C})$, and some positive constants c_0, c_1.

Moreover, for a purely imaginary frequency $\omega = i\beta \neq 0$ the operator $\boldsymbol{V}_{-1}$ is positive definite, i.e.,

$$(\mathbf{V}_{-1}\boldsymbol{\Phi},\boldsymbol{\Phi})_{\mathcal{S}} \geq M\|\boldsymbol{\Phi}\big|\mathbb{H}^{-\frac{1}{2}}(\mathcal{S})\|^2, \quad \boldsymbol{\Phi} \in \mathbb{H}^{-\frac{1}{2}}(\mathcal{S}),$$

if $\mathcal{S}$ is closed, and

$$(\mathbf{V}_{-1}\boldsymbol{\Psi},\boldsymbol{\Psi})_{\mathcal{C}} \geq M\|\boldsymbol{\Psi}\big|\widetilde{\mathbb{H}}^{-\frac{1}{2}}(\mathcal{C})\|^2, \quad \boldsymbol{\Psi} \in \widetilde{\mathbb{H}}^{-\frac{1}{2}}(\mathcal{C}) \tag{3.37}$$

if $\mathcal{C}$ is open, for some $M > 0$.

Indeed, if $\omega = i\beta \neq 0$ then from Green's formula (3.32) we obtain

$$(\mathbf{V}_{-1}\boldsymbol{\Phi},\boldsymbol{\Phi})_{\mathcal{S}} = (\mu^{-1}\mathrm{curl}\,\boldsymbol{U},\mathrm{curl}\,\boldsymbol{U})_{\mathbb{R}^3} + s\,(\mathrm{div}(\varepsilon\boldsymbol{U}),\mathrm{div}(\varepsilon\boldsymbol{U}))_{\mathbb{R}^3} + \beta^2(\varepsilon\boldsymbol{U},\boldsymbol{U})_{\mathbb{R}^3} > 0,$$

if $\boldsymbol{\Phi} \neq 0$ and, therefore, $\boldsymbol{U} \neq 0$ in $\Omega^{\pm}$. Positive and invertible operators are positive definite. For the proof we can lift the operator with the help of Bessel potentials to a positive and invertible operator in $\mathbb{L}_2$-setting, prove the positive definiteness of the lifted operator and, returning to the original operator in the setting (3.24), get the positive definiteness.

The proof of positive definiteness for $\omega = i\beta$ and an open surface $\mathcal{S}$ results from equality (3.35) and is similar to the foregoing case.

The coerciveness (3.36) follows from the positive definiteness (3.37) and the next auxiliary lemma.

Lemma 3.14. *Let $\boldsymbol{A}(D)$ be an elliptic partial differential operator of order 2 with constant $N \times N$ complex-valued matrix coefficients. Let further*

$$\boldsymbol{A}(D)\boldsymbol{\Phi} - \boldsymbol{A}^0(D)\boldsymbol{\Phi} = G\boldsymbol{\Phi}, \qquad \boldsymbol{\Phi} \in \mathbb{H}^1(\Omega)^N,$$

where $G \in \mathbb{C}^{N\times N}(\mathcal{S})$ *is a matrix-function. Let* $\mathbf{V}_{-1}(\mathcal{X},\mathcal{D})$ *be the direct values of the single layer potential operator related to* $\boldsymbol{A}(D)$*, while* $\mathbf{V}^0_{-1}(\mathcal{X},\mathcal{D})$ *be the direct value of the single layer potential operator related to* $\boldsymbol{A}^0(D)$ *(cf. (3.11)).*

Then the difference

$$\boldsymbol{B}_{-3}(\mathcal{X},\mathcal{D}) := \mathbf{V}_{-1}(\mathcal{X},\mathcal{D}) - \mathbf{V}^0_{-1}(\mathcal{X},\mathcal{D}) \tag{3.38}$$

is a pseudodifferential operators of order -3 *(the orders of* ΨDO*s are indicated in the indices).*

Proof. Recall the formulae for the fundamental solution and the direct value of single layer potential for $\boldsymbol{A}(D)$ (cf. (3.7))

$$\mathbf{F}_{\boldsymbol{A}}(x) := \mathcal{F}^{-1}_{\xi\to x}\left[\mathcal{A}^{-1}(\xi)\right], \qquad x \in \Omega,$$
$$\mathbf{V}_{-1}\boldsymbol{\Phi}(\mathcal{X}) := \oint_{\mathcal{S}} F_{\boldsymbol{A}}(\mathcal{X}-y)\boldsymbol{\Phi}(y)\,dS, \quad \mathcal{X}\in\mathcal{S},$$

where $\mathcal{A}(\xi)$ is the full symbol of the operator $\boldsymbol{A}(D)$. Similarly are written the fundamental solution $\mathbf{F}_{\boldsymbol{A}^0}$ and the potential $\mathbf{V}^{\mathbf{0}}_{-\mathbf{1}}$.

The symbol of the pseudodifferential operator (of the difference)

$$\begin{aligned}\boldsymbol{B}_{-3}(\mathcal{X},\mathcal{D})\boldsymbol{\Phi}(\mathcal{X}) &= \mathbf{V}_{-\mathbf{1}}(\mathcal{X},\mathcal{D})\boldsymbol{\Phi}(\mathcal{X}) - \mathbf{V}^{\mathbf{0}}_{-\mathbf{1}}(\mathcal{X},\mathcal{D})\boldsymbol{\Phi}(\mathcal{X})\\ &:= \oint_{\mathcal{S}}\left[\mathbf{F}_{\boldsymbol{A}}(\mathcal{X}-y) - \mathbf{F}_{\boldsymbol{A}^0}(\mathcal{X}-y)\right]\boldsymbol{\Phi}(y)\,dS\end{aligned}$$

is represented locally as

$$\begin{aligned}\mathcal{B}_{-3}(\mathcal{X},\xi) &= V_{-1}(\mathcal{X},\xi) - V^0_{-1}(\mathcal{X},\xi)\\ &= -\frac{\Gamma_{\varkappa_j}(\mathcal{X})}{2\pi\det\widetilde{\varkappa}'_j(\mathcal{X})}\int_{-\infty}^{\infty}\left[\mathcal{A}^{-1}\big(\widetilde{\varkappa}'_j(\mathcal{X})(t,\xi)\big) - (\mathcal{A}^0)^{-1}\big(\widetilde{\varkappa}'_j(\mathcal{X})(t,\xi)\big)\right]dt\\ &= -\frac{\Gamma_{\varkappa_j}(\mathcal{X})}{2\pi\det\widetilde{\varkappa}'_j(\mathcal{X})}\int_{-\infty}^{\infty}\mathcal{A}^{-1}\big(\widetilde{\varkappa}'_j(\mathcal{X})(t,\xi)\big)G(\mathcal{X})(\mathcal{A}^0)^{-1}\big(\widetilde{\varkappa}'_j(\mathcal{X})(t,\xi)\big)\,dt\\ &= \mathcal{O}(|\xi|^{-3}) \qquad \text{as} \quad |\xi|\to\infty, \quad \xi\in\mathbb{R}^2, \quad \mathcal{X}\in\mathcal{S},\end{aligned} \tag{3.39}$$

where $G(\mathcal{X}) := \mathcal{A}^0(\mathcal{X},\xi) - \mathcal{A}(\mathcal{X},\xi)$ is a $N\times N$ matrix-function. The obtained equality (3.39) shows that the pseudodifferential operator $\boldsymbol{B}_{-3}(\mathcal{X},\mathcal{D})$ has order -3 indeed. □

Corollary 3.15. *For a smooth, closed surface* $\mathcal{S}$ *the potential operator* $\mathbf{V}(\mathbf{V}_{-\mathbf{1}})^{-1}$ *has the following mapping properties*

$$\mathbf{P} := \mathbf{V}(\mathbf{V}_{-\mathbf{1}})^{-1} \;:\; \mathbb{H}^{r+\frac12}_{\varepsilon\boldsymbol{\nu},0}(\mathcal{S}) \to \mathbb{H}^{r+1}_{\varepsilon\boldsymbol{\nu},0}(\Omega^{\pm}) \quad \textit{for all} \quad r\in\mathbb{R}, \tag{3.40}$$

while for an open surface $\mathcal{C}$ *it maps*

$$\mathbf{P} := \mathbf{V}(\mathbf{V}_{-\mathbf{1}})^{-1} \;:\; \mathbb{H}^{r+\frac12}_{\varepsilon\boldsymbol{\nu},0}(\mathcal{C}) \to \mathbb{H}^{r+1}_{\varepsilon\boldsymbol{\nu},0}(\mathbb{R}^3_{\mathcal{C}}) \quad \textit{for all} \quad -\tfrac12 < r < \tfrac12.$$

Proof. Since $\gamma_{\mathcal{S}}\mathbf{V}(\mathbf{V}_{-\mathbf{1}})^{-1}\boldsymbol{\Phi} = \boldsymbol{\Phi}$ (see (3.14)) the asserted mapping properties are trivial consequences of the mapping properties of the participating operators $\mathbf{V}$ in (3.10) and of $(\mathbf{V}_{-\mathbf{1}})^{-1}$ in (3.24). □

4. Boundary pseudodifferential equations

First we derive and investigate equivalent boundary pseudodifferential equations for the elliptic Neumann BVP (2.4) in $\Omega^\pm$.

Consider the potential operator

$$\mathbf{P}\boldsymbol{\Phi}(x) := \mathbf{V}(\mathbf{V}_{-1})^{-1}\boldsymbol{\Phi}(x), \qquad \boldsymbol{\Phi} \in \mathbb{H}^{1/2}_{\varepsilon\boldsymbol{\nu},0}(\mathcal{S}), \qquad x \in \Omega, \tag{4.1}$$

where $\Omega = \Omega^\pm$ (cf. (3.40)). Note that $\boldsymbol{U} = \mathbf{P}\boldsymbol{\Phi}(x) = \mathbf{V}\boldsymbol{\Psi}$, $\boldsymbol{\Psi} := (\mathbf{V}_{-1})^{-1}\boldsymbol{\Phi}$, satisfies the basic equation in (2.4) in $\Omega^\pm$ (cf. (3.9)).

By introducing $\boldsymbol{U} = \mathbf{P}\boldsymbol{\Phi}$ from (4.1) into $\gamma_{\mathcal{S}^\pm}(\pi_{\varepsilon\boldsymbol{\nu}}\mathfrak{N}(D,\boldsymbol{\nu})\boldsymbol{U}) = \boldsymbol{g}^\pm$ on $\mathcal{S}$ and using Plemelji's formulae (3.15), we derive the following boundary pseudodifferential equations

$$\mathcal{P}_\pm\boldsymbol{\Phi} = \mp\gamma_{\mathcal{S}^\pm}\pi_{\varepsilon\boldsymbol{\nu}}\mathfrak{N}(D,\boldsymbol{\nu})\mathbf{V}(\mathbf{V}_{-1})^{-1}\boldsymbol{\Phi} = \mp\boldsymbol{g}^\pm,$$

where

$$\mathcal{P}_\pm := \pi_{\varepsilon\boldsymbol{\nu}}\Big(\frac{1}{2}I \mp (\mathbf{W_0})^*\Big)(\mathbf{V}_{-1})^{-1} \tag{4.2}$$

are the *modified Poincaré-Steklov* pseudodifferential operators of order 1.

Lemma 4.1. *The operators*

$$\mathcal{P}_\pm \; : \; \mathbb{H}^{1/2}_{\varepsilon\boldsymbol{\nu},0}(\mathcal{S}) \to \mathbb{H}^{-1/2}_{\varepsilon\boldsymbol{\nu},0}(\mathcal{S}) \tag{4.3}$$

are coercive

$$\mathrm{Re}\,(\mathcal{P}_\pm\boldsymbol{\Phi},\boldsymbol{\Phi})_{\mathcal{S}} \geq c_0\big\|\boldsymbol{\Phi}\big|\mathbb{H}^{1/2}_{\varepsilon\boldsymbol{\nu},0}(\mathcal{S})\big\|^2 - c_1\big\|\boldsymbol{\Phi}\big|\mathbb{L}_{2,\varepsilon\boldsymbol{\nu},0}(\mathcal{S})\big\|^2 \tag{4.4}$$

for some positive constants c_0, c_1 and all $\boldsymbol{\Phi} \in \mathbb{H}^{1/2}_{\varepsilon\boldsymbol{\nu},0}(\mathcal{S})$.

Moreover, the operators have the trivial kernels, i.e., $\mathrm{Ker}\,\mathcal{P}_\pm = \{0\}$ *and are invertible.*

If the frequency is purely imaginary $\omega = i\beta \neq 0$, $\beta \in \mathbb{R}$, the operators $\mathcal{P}_\pm$ are positive definite

$$(\mathcal{P}_\pm\boldsymbol{\Phi},\boldsymbol{\Phi})_{\mathcal{S}} \geq M_\pm\big\|\boldsymbol{\Phi}\big|\mathbb{H}^{1/2}_{\varepsilon\boldsymbol{\nu},0}(\mathcal{S})\big\| \tag{4.5}$$

for some positive constants $M_\pm$.

Proof. By introducing $\widetilde{\boldsymbol{U}} = \mathbf{V}(\mathbf{V}_{-1})^{-1}\boldsymbol{\Phi}$ into the Green formula (2.1) we find out that

$$(\mathcal{P}_\pm\boldsymbol{\Phi},\boldsymbol{\Phi})_{\mathcal{S}} = (\mu^{-1}\mathrm{curl}\,\widetilde{\boldsymbol{U}},\mathrm{curl}\,\widetilde{\boldsymbol{U}})_{\Omega^\pm} + s\,(\mathrm{div}(\varepsilon\widetilde{\boldsymbol{U}}),\mathrm{div}(\varepsilon\widetilde{\boldsymbol{U}}))_{\Omega^\pm} - \omega^2(\varepsilon\,\widetilde{\boldsymbol{U}},\widetilde{\boldsymbol{U}})_{\Omega^\pm}. \tag{4.6}$$

Since $\gamma_{\mathcal{S}}\widetilde{\boldsymbol{U}} = \boldsymbol{\Phi} \in \mathbb{H}^{1/2}_{\varepsilon\boldsymbol{\nu},0}(\mathcal{S})$, due to Corollary 3.9 and Corollary 3.6 the form

$$\boldsymbol{a}_{\varepsilon,\mu}(\widetilde{\boldsymbol{U}},\widetilde{\boldsymbol{U}})_{\Omega^\pm} = (\mu^{-1}\mathrm{curl}\,\widetilde{\boldsymbol{U}},\mathrm{curl}\,\widetilde{\boldsymbol{U}})_{\Omega^\pm} + s\,(\mathrm{div}(\varepsilon\widetilde{\boldsymbol{U}}),\mathrm{div}(\varepsilon\widetilde{\boldsymbol{U}}))_{\Omega^\pm}$$

are coercive: the inequality

$$\boldsymbol{a}_{\varepsilon,\mu}(\widetilde{\boldsymbol{U}},\widetilde{\boldsymbol{U}})_{\Omega^\pm} \geq c_2\big\|\widetilde{\boldsymbol{U}}\big|\mathbb{H}^1_{\varepsilon\boldsymbol{\nu},0}(\Omega^\pm)\big\|^2 - c_3\big\|\widetilde{\boldsymbol{U}}\big|\mathbb{L}_{2,\varepsilon\boldsymbol{\nu},0}(\Omega^\pm)\big\|^2$$

holds for $\widetilde{\boldsymbol{U}} = \mathbf{V}(\mathbf{V}_{-1})^{-1}\boldsymbol{\Phi}$, $\boldsymbol{\Phi} \in \mathbb{H}^{1/2}_{\varepsilon\boldsymbol{\nu},0}(\mathcal{S})$ and some $c_2 > 0$, $c_3 > 0$. From (4.6) we then obtain

$$\operatorname{Re}\,(\mathcal{P}_{\pm}\boldsymbol{\Phi},\boldsymbol{\Phi})_{\mathcal{S}} \geq c_2\big\|\widetilde{\boldsymbol{U}}\big|\mathbb{H}^1_{\varepsilon\boldsymbol{\nu},0}(\Omega^{\pm})\big\|^2 - c_4\big\|\widetilde{\boldsymbol{U}}\big|\mathbb{L}_{2,\varepsilon\boldsymbol{\nu},0}(\Omega^{\pm})\big\|^2$$

for some $c_4 > 0$. Further, invoking the trace theorem (cf. [20]) and the continuity property of the operator $\mathbf{P} = \mathbf{V}(\mathbf{V}_{-1})^{-1}$ (cf. (3.40)) we easily derive the following inequalities

$$\begin{aligned}\big\|\widetilde{\boldsymbol{U}}\big|\mathbb{H}^1_{\varepsilon\boldsymbol{\nu},0}(\Omega^{\pm})\big\| &\geq c_5\big\|\gamma_{\mathcal{S}}\widetilde{\boldsymbol{U}}\ \mathbb{H}^{1/2}_{\varepsilon\boldsymbol{\nu},0}(\mathcal{S})\big\|, \quad c_5 > 0,\\ \big\|\mathbf{V}(\mathbf{V}_{-1})^{-1}\boldsymbol{\Phi}\big|\mathbb{H}^{1/2}_{\varepsilon\boldsymbol{\nu},0}(\Omega^{\pm})\big\| &\leq c_6\big\|\boldsymbol{\Phi}\big|\mathbb{L}_{2,\varepsilon\boldsymbol{\nu},0}(\mathcal{S})\big\|, \quad\ \ c_6 > 0.\end{aligned} \tag{4.7}$$

Applying the inequalities (4.7) we get the estimate with suitable positive constants

$$\begin{aligned}\operatorname{Re}\,(\mathcal{P}_{\pm}\boldsymbol{\Phi},\boldsymbol{\Phi})_{\mathcal{S}} &\geq c_2\big\|\widetilde{\boldsymbol{U}}\big|\mathbb{H}^1_{\varepsilon\boldsymbol{\nu},0}(\Omega^{\pm})\big\|^2 - c_4\big\|\widetilde{\boldsymbol{U}}\big|\mathbb{L}_{2,\varepsilon\boldsymbol{\nu},0}(\Omega^{\pm})\big\|^2\\ &\geq c_7\big\|\gamma_{\mathcal{S}}\widetilde{\boldsymbol{U}}\big|\mathbb{H}^{1/2}_{\varepsilon\boldsymbol{\nu},0}(\mathcal{S})\big\|^2 - c_8\big\|\widetilde{\boldsymbol{U}}\big|\mathbb{H}^{1/2}_{\varepsilon\boldsymbol{\nu},0}(\Omega^{\pm})\big\|^2\\ &= c_7\big\|\boldsymbol{\Phi}\big|\mathbb{H}^{1/2}_{\varepsilon\boldsymbol{\nu},0}(\mathcal{S})\big\|^2 - c_8\big\|\mathbf{V}(\mathbf{V}_{-1})^{-1}\boldsymbol{\Phi}\ \mathbb{H}^{1/2}_{\varepsilon\boldsymbol{\nu},0}(\Omega^{\pm})\big\|^2\\ &\geq c_0\big\|\boldsymbol{\Phi}\big|\mathbb{H}^{1/2}_{\varepsilon\boldsymbol{\nu},0}(\mathcal{S})\big\|^2 - c_1\big\|\boldsymbol{\Phi}\big|\mathbb{L}_{2,\varepsilon\boldsymbol{\nu},0}(\mathcal{S})\big\|^2\end{aligned}$$

for $\boldsymbol{\Phi} \in \mathbb{H}^{1/2}_{\varepsilon\boldsymbol{\nu},0}(\mathcal{S})$. Thus the operator (4.3) is coercive and, therefore, is Fredholm with the index zero. Moreover, it is invertible since it has trivial kernel. Indeed, for $\operatorname{Im}\omega \neq 0$ equating in (4.14) the imaginary part to 0 we get that $(\mathcal{P}_{\pm 1}\boldsymbol{\Phi},\boldsymbol{\Phi})_{\mathcal{S}} = 0$, which implies

$$0 = (\varepsilon\widetilde{\boldsymbol{U}},\widetilde{\boldsymbol{U}})_{\Omega^{\pm}} \geq c\big\|\widetilde{\boldsymbol{U}}\big|\mathbb{L}_2(\Omega^{\pm})\big\|^2 \Longrightarrow \widetilde{\boldsymbol{U}} \equiv 0 \quad \text{in} \quad \Omega^{\pm}.$$

Therefore $\gamma^{\pm}_{\mathcal{S}}\widetilde{\boldsymbol{U}} = \boldsymbol{\Phi} \equiv 0$ on $\mathcal{S}$.

If $\omega = i\beta$ then $\mathcal{P}_{\pm}$ is positive

$$\operatorname{Re}\,(\mathcal{P}_{\pm}\boldsymbol{\Phi},\boldsymbol{\Phi})_{\mathcal{S}} = (\mathcal{P}_{\pm}\boldsymbol{\Phi},\boldsymbol{\Phi})_{\mathcal{S}} = \boldsymbol{a}_{\varepsilon,\mu}(\widetilde{\boldsymbol{U}},\widetilde{\boldsymbol{U}})_{\Omega^{\pm}} + \beta^2(\varepsilon\widetilde{\boldsymbol{U}},\widetilde{\boldsymbol{U}})_{\Omega^{\pm}} > 0$$

if $\widetilde{\boldsymbol{U}} \neq 0$ in $\Omega^{\pm}$ (cf. (4.14)) and, therefore, $\boldsymbol{\Phi} \neq 0$ on $\mathcal{S}$; moreover, $\mathcal{P}_{\pm}$ is coercive

$$(\mathcal{P}_{\pm}\boldsymbol{\Phi},\boldsymbol{\Phi})_{\mathcal{S}} = \operatorname{Re}\,(\mathcal{P}_{\pm}\boldsymbol{\Phi},\boldsymbol{\Phi})_{\mathcal{S}} \geq c_0\big\|\boldsymbol{\Phi}\big|\mathbb{H}^{1/2}_{\varepsilon\boldsymbol{\nu},0}(\mathcal{S})\big\|^2 - c_1\big\|\boldsymbol{\Phi}\big|\mathbb{L}_{2,\varepsilon\boldsymbol{\nu},0}(\mathcal{S})\big\|^2$$

for all $\boldsymbol{\Phi} \in \mathbb{H}^{1/2}_{\varepsilon\boldsymbol{\nu},0}(\mathcal{S})$ (cf. (4.4)). The positive definiteness (4.5) is a consequence of these two properties (see [18, Exercise 2.17]). □

Corollary 4.2. *The operators $\mathcal{P}_{\pm}$ are invertible in the following space settings*

$$\mathcal{P}_{\pm}\ :\ \mathbb{H}^r_{\varepsilon\boldsymbol{\nu},0}(\mathcal{S}) \to \mathbb{H}^{r-1}_{\varepsilon\boldsymbol{\nu},0}(\mathcal{S}), \quad \forall r \in \mathbb{R}. \tag{4.8}$$

Corollary 4.3. *The inverse operators*

$$\mathcal{P}^{-1}_{\pm}\ :\ \mathbb{H}^{-1/2}_{\varepsilon\boldsymbol{\nu},0}(\mathcal{S}) \to \mathbb{H}^{1/2}_{\varepsilon\boldsymbol{\nu},0}(\mathcal{S}), \tag{4.9}$$

which exist due to Lemma 4.1, are coercive

$$\operatorname{Re}\,(\mathcal{P}^{-1}_{\pm}\boldsymbol{\Psi},\boldsymbol{\Psi})_{\mathcal{S}} \geq m_0\big\|\boldsymbol{\Psi}\big|\mathbb{H}^{-1/2}_{\varepsilon\boldsymbol{\nu},0}(\mathcal{S})\big\|^2 - m_1\big\|\boldsymbol{\Psi}\big|\mathbb{H}^{-1}_{\varepsilon\boldsymbol{\nu},0}(\mathcal{S})\big\|^2 \tag{4.10}$$

for some positive constants m_0, m_1 and all $\boldsymbol{\Psi} \in \mathbb{H}^{-1/2}_{\varepsilon\boldsymbol{\nu},0}(\mathcal{S})$.

If the frequency is purely imaginary $\omega = i\beta \neq 0$, $\beta \in \mathbb{R}$, the operators $\mathcal{P}_\pm^{-1}$ are positive definite

$$(\mathcal{P}_\pm^{-1}\boldsymbol{\Psi}, \boldsymbol{\Psi})_{\mathcal{S}} \geq M_\pm \big\| \boldsymbol{\Psi} \big| \mathbb{H}^{-1/2}_{\varepsilon\boldsymbol{\nu},0}(\mathcal{S}) \big\|$$

for some positive constants $M_\pm$.

Proof. Follows from Lemma 4.1 if we introduce $\Phi = \mathcal{P}_\pm^{-1}\boldsymbol{\Psi}$ and recall, that due to the invertibility (4.3) the estimates

$$\frac{1}{m}\big\| \boldsymbol{\Psi} \big| \mathbb{H}^{-1/2}_{\varepsilon\boldsymbol{\nu},0}(\mathcal{S}) \big\|^2 \leqslant \big\| \mathcal{P}_\pm^{-1}\boldsymbol{\Psi} \big| \mathbb{H}^{1/2}_{\varepsilon\boldsymbol{\nu},0}(\mathcal{S}) \big\| \leqslant m \big\| \boldsymbol{\Psi} \big| \mathbb{H}^{-1/2}_{\varepsilon\boldsymbol{\nu},0}(\mathcal{S}) \big\|$$

hold for some $m > 0$. $\square$

Now let us prove analogues of Lemma 4.1 and Corollaries 4.2, 4.3 for a subsurface $\mathcal{C} \subset \mathcal{S}$ with the boundary $\partial\mathcal{C} \neq \emptyset$.

Lemma 4.4. *For an open subsurface $\mathcal{C} \subset \mathcal{S}$ the operators*

$$r_{\mathcal{C}}\mathcal{P}_\pm \,:\, \widetilde{\mathbb{H}}^{1/2}_{\varepsilon\boldsymbol{\nu},0}(\mathcal{C}) \to \mathbb{H}^{-1/2}_{\varepsilon\boldsymbol{\nu},0}(\mathcal{C}) \tag{4.11}$$

are coercive with suitable positive constants c_0, c_1

$$\mathrm{Re}\,(r_{\mathcal{C}}\mathcal{P}_\pm\boldsymbol{\Phi}, \boldsymbol{\Phi})_{\mathcal{C}} \geq c_0 \big\| \boldsymbol{\Phi} \big| \widetilde{\mathbb{H}}^{1/2}_{\varepsilon\boldsymbol{\nu},0}(\mathcal{C}) \big\|^2 - c_1 \big\| \boldsymbol{\Phi} \big| \mathbb{L}_{2,\varepsilon\boldsymbol{\nu},0}(\mathcal{C}) \big\|^2 \tag{4.12}$$

for $\boldsymbol{\Phi} \in \widetilde{\mathbb{H}}^{1/2}_{\varepsilon\boldsymbol{\nu},0}(\mathcal{C})$. Moreover, the operators have the zero kernels $\mathrm{Ker}\, r_{\mathcal{C}}\mathcal{P}_\pm = \{0\}$ *and are invertible.*

If the frequency is purely imaginary $\omega = i\beta \neq 0$, $\beta \in \mathbb{R}$, the operators $r_{\mathcal{C}}\mathcal{P}_\pm$ are positive definite with a suitable positive constants $M_\pm$:

$$(r_{\mathcal{C}}\mathcal{P}_\pm\boldsymbol{\Phi}, \boldsymbol{\Phi})_{\mathcal{C}} \geq M_\pm \big\| \boldsymbol{\Phi} \big| \widetilde{\mathbb{H}}^{1/2}_{\varepsilon\boldsymbol{\nu},0}(\mathcal{C}) \big\|. \tag{4.13}$$

Proof. Using the continuity of the embedding $\widetilde{\mathbb{H}}^{1/2}_{\varepsilon\boldsymbol{\nu},0}(\mathcal{C}) \subset \mathbb{H}^{1/2}_{\varepsilon\boldsymbol{\nu},0}(\mathcal{S})$ and the proved coercivity (4.4), we proceed as follows:

$$\begin{aligned}\mathrm{Re}\,(r_{\mathcal{C}}\mathcal{P}_\pm\boldsymbol{\Phi}, \boldsymbol{\Phi})_{\mathcal{C}} &= \mathrm{Re}\,(\mathcal{P}_\pm\boldsymbol{\Phi}, \boldsymbol{\Phi})_{\mathcal{S}} \geq c_0 \big\| \boldsymbol{\Phi} \big| \mathbb{H}^{1/2}_{\varepsilon\boldsymbol{\nu},0}(\mathcal{S}) \big\|^2 \\ &= c_0 \big\| \boldsymbol{\Phi} \big| \widetilde{\mathbb{H}}^{1/2}_{\varepsilon\boldsymbol{\nu},0}(\mathcal{C}) \big\|^2 - c_1 \big\| \boldsymbol{\Phi} \big| \mathbb{L}_{2,\varepsilon\boldsymbol{\nu},0}(\mathcal{C}) \big\|^2\end{aligned}$$

for all $\boldsymbol{\Phi} \in \widetilde{\mathbb{H}}^{1/2}_{\varepsilon\boldsymbol{\nu},0}(\mathcal{C})$. The coercivity (4.12) is proved. Since $r_{\mathcal{C}}\mathcal{P}_\pm$ is coercive, it is Fredholm and has vanishing index, i.e., $\mathrm{Ind}\,(r_{\mathcal{C}}\mathcal{P}_\pm) = 0$.

Thus, to prove the invertibility we just need to check, that the kernel of the operator in (4.11) is trivial, i.e., $\mathrm{Ker}\, r_{\mathcal{C}}\mathcal{P}_\pm = \{0\}$.

For this purpose we apply the equality

$$\begin{aligned}(\mathcal{P}_\pm\boldsymbol{U}, \boldsymbol{U})_{\mathcal{S}} &= (\mu^{-1}\mathrm{curl}\,\boldsymbol{F}, \mathrm{curl}\,\boldsymbol{F})_{\Omega^\pm} + s\,(\mathrm{div}(\varepsilon\boldsymbol{F}), \mathrm{div}(\varepsilon\boldsymbol{F}))_{\Omega^\pm} \\ &\quad - \omega^2(\varepsilon\,\boldsymbol{F}, \boldsymbol{F})_{\Omega^\pm}\,\boldsymbol{F} = \mathbf{V}(\mathbf{V_{-1}})^{-1}\boldsymbol{U}, \qquad \gamma_{\mathcal{S}}\boldsymbol{F} = \boldsymbol{U} \in \mathbb{H}^{1/2}_{\varepsilon\boldsymbol{\nu},0}(\mathcal{S}),\end{aligned}$$

proved for a surface $\mathcal{S}$ without boundary. By introducing in the above equality the vector $\boldsymbol{U} = \boldsymbol{\Phi} \in \widetilde{\mathbb{H}}^{1/2}_{\varepsilon\boldsymbol{\nu},0}(\mathcal{C}) \subset \mathbb{H}^{1/2}_{\varepsilon\boldsymbol{\nu},0}(\mathcal{S})$, we get

$$\begin{aligned}(r_{\mathcal{C}}\mathcal{P}_{\pm}\boldsymbol{\Phi}, \boldsymbol{\Phi})_{\mathcal{S}} = (\mathcal{P}_{\pm}\boldsymbol{\Phi}, \boldsymbol{\Phi})_{\mathcal{S}} = (\mu^{-1}\mathrm{curl}\,\boldsymbol{F}, \mathrm{curl}\,\boldsymbol{F})_{\Omega^{\pm}} \\ + s\,(\mathrm{div}(\varepsilon\boldsymbol{F}), \mathrm{div}(\varepsilon\boldsymbol{F}))_{\Omega^{\pm}} - \omega^2(\varepsilon\,\boldsymbol{F}, \boldsymbol{F})_{\Omega^{\pm}}\end{aligned} \tag{4.14}$$

for all $\boldsymbol{F} = \mathbf{V}(\mathbf{V}_{-1})^{-1}\boldsymbol{\Phi}$. Since ω is complex valued, from (4.14) follows that the equality $r_{\mathcal{C}}\mathcal{P}_{\pm}\boldsymbol{\Phi} = 0$ implies $\boldsymbol{F} \equiv 0$. Then,

$$(\mathbf{V}_{-1})^{-1}\boldsymbol{\Phi} = \gamma_{\mathcal{S}^-}\big(\mathfrak{N}(D,\boldsymbol{\nu})\boldsymbol{V}(\mathbf{V}_{-1})^{-1}\boldsymbol{\Phi}\big) - \gamma_{\mathcal{S}^+}\big(\mathfrak{N}(D,\boldsymbol{\nu})\boldsymbol{V}(\mathbf{V}_{-1})^{-1}\boldsymbol{\Phi}\big) = 0,$$

which implies, due to the invertibility of $\boldsymbol{V}_{-1}$ (see (3.34)), that $\boldsymbol{\Phi} = 0$ and the kernel is trivial $\mathrm{Ker}\,\boldsymbol{V}_{-1} = \{0\}$. Due to the vanishing index $\mathrm{Ind}\,\boldsymbol{V}_{-1} = 0$ this implies trivial co-kernel

$$\dim\,\mathrm{Coker}\,\boldsymbol{V}_{-1} = \dim\,\mathrm{Ker}\,(\boldsymbol{V}_{-1})^* = 0$$

and provides the invertibility of the operator $r_{\mathcal{C}}\mathcal{P}_{\pm}$ in (4.11).

The positive definiteness (4.13) follows from the positive definiteness (4.5) as in the case of coerciveness. □

Corollary 4.5. *The operators $r_{\mathcal{C}}\mathcal{P}_{\pm}$ are invertible in the following space setting*

$$r_{\mathcal{C}}\mathcal{P}_{\pm}\ :\ \widetilde{\mathbb{H}}^{r+\frac{1}{2}}_{\varepsilon\boldsymbol{\nu},0}(\mathcal{C}) \to \mathbb{H}^{r-\frac{1}{2}}_{\varepsilon\boldsymbol{\nu},0}(\mathcal{C}), \quad -\frac{1}{2} < r < \frac{1}{2}. \tag{4.15}$$

Proof. For a similar proof we refer to the concluding part of the proof of Lemma 3.12. □

5. Proofs of the basic results

Let us look for a solution to the screen-type problem (2.4) in the form

$$\boldsymbol{U}(x) = \begin{cases} \mathbf{V}(\mathbf{V}_{-1})^{-1}\boldsymbol{\Phi}^+(x) & x \in \Omega^+, \\ \mathbf{V}(\mathbf{V}_{-1})^{-1}\boldsymbol{\Phi}^-(x) & x \in \Omega^- \quad \text{for some} \quad \boldsymbol{\Phi}^{\pm} \in \mathbb{H}^{1/2}_{\varepsilon\boldsymbol{\nu},0}(\mathcal{S}). \end{cases} \tag{5.1}$$

Then $\boldsymbol{U}$ satisfies the basic differential equation from BVP (2.4) in the domains $\Omega^{\pm}$ (cf. (3.9)) and, due to the mapping properties of $\mathbf{V}$ we have $\boldsymbol{U} \in \mathbb{H}^1_{\varepsilon\boldsymbol{\nu},0}(\mathbb{R}^3_{\mathcal{C}})$. Further we need to fulfil the boundary conditions (cf. (2.2))

$$r_{\mathcal{C}}\gamma_{\mathcal{S}^{\pm}}\,(\pi_{\varepsilon\boldsymbol{\nu}}\mathfrak{N}(D,\boldsymbol{\nu})\boldsymbol{U}) = \boldsymbol{g}^{\pm} \quad \text{on} \quad \mathcal{C}. \tag{5.2}$$

Due to the Plemelji formulae (3.15) equation (5.2) acquires the form

$$r_{\mathcal{C}}\mathcal{P}_{\pm}\boldsymbol{\Phi}^{\pm} = \boldsymbol{g}^{\pm} \qquad \text{on} \quad \mathcal{C}, \tag{5.3}$$

where $\mathcal{P}_{\pm}$ are the *modified Poincaré-Steklov* pseudodifferential operators of order 1, defined in (4.2).

Let $\ell\boldsymbol{g}^+ \in \mathbb{H}^{-1/2}_{\varepsilon\boldsymbol{\nu},0}(\mathcal{S})$ be a fixed extension of the function $\boldsymbol{g}^+ \in \mathbb{H}^{-1/2}_{\varepsilon\boldsymbol{\nu},0}(\mathcal{C})$ up to the entire closed surface $\mathcal{S}$ and let $\ell_0(\boldsymbol{g}^+ - \boldsymbol{g}^-) \in \mathbb{H}^{-1/2}_{\varepsilon\boldsymbol{\nu},0}(\mathcal{S})$ be an extension by zero of the function $\boldsymbol{g}^+ - \boldsymbol{g}^- \in r_{\mathcal{C}}\widetilde{\mathbb{H}}^{-1/2}_{\varepsilon\boldsymbol{\nu},0}(\mathcal{C})$. Then $\ell\boldsymbol{g}^- := \ell\boldsymbol{g}^+ - \ell_0(\boldsymbol{g}^+ - \boldsymbol{g}^-) \in \mathbb{H}^{-1/2}_{\varepsilon\boldsymbol{\nu},0}(\mathcal{S})$ is an extension of the function $\boldsymbol{g}^- \in \mathbb{H}^{-1/2}_{\varepsilon\boldsymbol{\nu},0}(\mathcal{C})$, i.e.,

$$r_{\mathcal{C}}\ell\boldsymbol{g}^- = \boldsymbol{g}^+ - (\boldsymbol{g}^+ - \boldsymbol{g}^-) = \boldsymbol{g}^- \quad \text{and} \quad r_{\mathcal{C}^c}\ell\boldsymbol{g}^+ = r_{\mathcal{C}^c}\ell\boldsymbol{g}^-. \tag{5.4}$$

Using (5.3) we write the boundary conditions on $\mathcal{S}$ as follows

$$\mathcal{P}_{\pm}\boldsymbol{\Phi}^{\pm} = \mp(\ell\boldsymbol{g}^{\pm} + \boldsymbol{\Psi}^{\pm}),$$

where the functions $\boldsymbol{\Psi}^{\pm} \in \widetilde{\mathbb{H}}^{-1/2}_{\varepsilon\boldsymbol{\nu},0}(\mathcal{C}^c)$ are unknown.

Due to Lemma 4.1 we then obtain

$$\boldsymbol{\Phi}^{\pm} = \mp\mathcal{P}_{\pm}^{-1}\ell\boldsymbol{g}^{\pm} \mp \mathcal{P}_{\pm}^{-1}\boldsymbol{\Psi}^{\pm}. \tag{5.5}$$

From the ellipticity of the differential operator $\boldsymbol{A}(D)$ follows that a generalized solution to the equation $\boldsymbol{A}(D)\boldsymbol{E} = 0$ is analytic in $\mathbb{R}^3_{\mathcal{C}}$ and, therefore, the following continuity conditions

$$\begin{cases} r_{\mathcal{C}^c}\gamma_{\mathcal{S}^+}\boldsymbol{U} - r_{\mathcal{C}^c}\gamma_{\mathcal{S}^-}\boldsymbol{U} = 0\,, \\ r_{\mathcal{C}^c}\gamma_{\mathcal{S}^+}\left(\mathfrak{N}(D,\boldsymbol{\nu})\boldsymbol{U}\right) - r_{\mathcal{C}^c}\gamma_{\mathcal{S}^-}\left(\mathfrak{N}(D,\boldsymbol{\nu})\boldsymbol{U}\right) = 0 \end{cases} \tag{5.6}$$

hold across the complementary surface $\mathcal{C}^c$.

Then taking into the account (5.5) and (5.4) we obtain the following system of equations with respect of the unknown functions $\boldsymbol{\Psi}^{\pm}$:

$$\begin{cases} r_{\mathcal{C}^c}\mathcal{P}_+^{-1}\boldsymbol{\Psi}^+ + r_{\mathcal{C}^c}\mathcal{P}_-^{-1}\boldsymbol{\Psi}^- = -r_{\mathcal{C}^c}\mathcal{P}_+^{-1}\ell\boldsymbol{g}^+ - r_{\mathcal{C}^c}\mathcal{P}_-^{-1}\ell\boldsymbol{g}^- \\ r_{\mathcal{C}^c}\boldsymbol{\Psi}^+ - r_{\mathcal{C}^c}\boldsymbol{\Psi}^- = 0. \end{cases} \tag{5.7}$$

The last equation in (5.7) implies

$$\boldsymbol{\Psi} := \boldsymbol{\Psi}^+ = \boldsymbol{\Psi}^- \in \widetilde{\mathbb{H}}^{-1/2}_{\varepsilon\boldsymbol{\nu},0}(\mathcal{C}^c)$$

and we obtain

$$r_{\mathcal{C}^c}\mathfrak{B}(D)\boldsymbol{\Psi} = \boldsymbol{F}, \tag{5.8}$$

where

$$\mathfrak{B}(D) := \mathcal{P}_+^{-1} + \mathcal{P}_-^{-1}, \tag{5.9}$$

and

$$\boldsymbol{F} := -r_{\mathcal{C}^c}\mathcal{P}_+^{-1}\ell\boldsymbol{g}^+ - r_{\mathcal{C}^c}\mathcal{P}_-^{-1}\ell\boldsymbol{g}^- = -r_{\mathcal{C}^c}\mathfrak{B}(D)\ell\boldsymbol{g}^+ + r_{\mathcal{C}^c}\mathcal{P}_-^{-1}\ell_0(\boldsymbol{g}^+ - \boldsymbol{g}^-).$$

What we obtain is an equivalent pseudodifferential operator to the BVP (2.4) (see the forthcoming Theorem 5.3).

Lemma 5.1. *The operator*

$$r_{\mathcal{C}^c}\mathfrak{B}(D)\ :\ \widetilde{\mathbb{H}}^{-1/2}_{\varepsilon\boldsymbol{\nu},0}(\mathcal{C}^c) \to \mathbb{H}^{1/2}_{\varepsilon\boldsymbol{\nu},0}(\mathcal{C}^c) \tag{5.10}$$

is coercive

$$\operatorname{Re}\left(r_{\mathcal{C}^c}\mathfrak{B}(D)\boldsymbol{\Psi}, \boldsymbol{\Psi}\right)_{\mathcal{C}^c} \geq C_1\left\|\boldsymbol{\Psi}\middle|\widetilde{\mathbb{H}}^{-1/2}_{\varepsilon\boldsymbol{\nu},0}(\mathcal{C}^c)\right\| - C_2\left\|\boldsymbol{\Psi}\middle|\widetilde{\mathbb{H}}^{-1}_{\varepsilon\boldsymbol{\nu},0}(\mathcal{C}^c)\right\|,\ \forall\,\boldsymbol{\Psi} \in \widetilde{\mathbb{H}}^{-1/2}_{\varepsilon\boldsymbol{\nu},0}(\mathcal{C}^c) \tag{5.11}$$

and invertible.

Moreover, if the frequency is purely imaginary $\omega = i\beta \neq 0$, $\beta \in \mathbb{R}$, the operator $\mathfrak{B}(D)$ is positive definite and the inequality

$$(r_{\mathcal{C}^c}\mathfrak{B}(D)\boldsymbol{\Psi}, \boldsymbol{\Psi})_{\mathcal{C}^c} \geq M_0 \big\|\boldsymbol{\Psi}\big|\widetilde{\mathbb{H}}^{-1/2}_{\varepsilon\boldsymbol{\nu},0}(\mathcal{C}^c)\big\|, \qquad \forall\, \boldsymbol{\Psi} \in \widetilde{\mathbb{H}}^{-1/2}_{\varepsilon\boldsymbol{\nu},0}(\mathcal{C}^c) \tag{5.12}$$

holds for some constant $M_0 > 0$.

Proof. Similarly as in Lemma 4.4 the coercivity (5.11) and the positive definiteness (5.12) of the operator $r_{\mathcal{C}^c}\mathfrak{B}(D)$ we obtain from corresponding results for the "non-restricted" operator $\mathfrak{B}(D)$ in (5.9), which follow immediately from similar properties of the summand $\mathcal{P}^{-1}_{\pm}$, established in Corollary 4.3.

From the coercivity (5.11) it follows that the operator in (5.10) is Fredholm and has trivial index, i.e., $\operatorname{Ind} r_{\mathcal{C}^c}\mathfrak{B}(D) = 0$. Then to prove that the operator $r_{\mathcal{C}^c}\mathfrak{B}(D)$ in (5.10) is invertible, it suffices to show that the kernel is trivial, i.e., $\operatorname{Ker} r_{\mathcal{C}^c}\mathfrak{B}(D) = \{0\}$. The latter follows immediately for $\omega = i\beta$ from the positive definiteness (5.12).

By introducing into the Green formula (2.1) the values

$$\boldsymbol{U}^{\pm} = \mathbf{V}(\mathbf{V}_{-1})^{-1}\boldsymbol{\Phi}^{\pm}, \quad \boldsymbol{\Phi}^{\pm} = \mathcal{P}^{-1}_{\pm}\boldsymbol{\Psi}, \quad \boldsymbol{\Psi} \in \widetilde{\mathbb{H}}^{-1/2}_{\varepsilon\boldsymbol{\nu},0}(\mathcal{C}^c)$$

and summing them up, we get

$$\begin{aligned}(\boldsymbol{\Psi}, \mathfrak{B}(D)\boldsymbol{\Psi})_{\mathcal{S}} &= \boldsymbol{a}_{\varepsilon,\mu}(\boldsymbol{U}^+, \boldsymbol{U}^+)_{\Omega^+} - \boldsymbol{a}_{\varepsilon,\mu}(\boldsymbol{U}^-, \boldsymbol{U}^-)_{\Omega^-} \\ &\quad - \omega^2(\varepsilon\,\boldsymbol{U}^+, \boldsymbol{U}^+)_{\Omega^+} - \omega^2(\varepsilon\,\boldsymbol{U}^-, \boldsymbol{U}^-)_{\Omega^-}.\end{aligned} \tag{5.13}$$

Since $\operatorname{Im}\omega \neq 0$, by equating in (5.13) the real and the imaginary parts to 0 we get that $(\boldsymbol{\Psi}, \mathfrak{B}(D)\boldsymbol{\Psi})_{\mathcal{S}} = 0$ implies

$$0 = (\varepsilon\boldsymbol{U}^{\pm}, \boldsymbol{U}^{\pm})_{\Omega^{\pm}} \geq c\big\|\boldsymbol{U}^{\pm}\big|\mathbb{L}_2(\Omega^{\pm})\big\|^2 \Longrightarrow \boldsymbol{U}^{\pm} \equiv 0 \quad \text{in} \quad \Omega^{\pm}.$$

Thus $\gamma^{\pm}_{\mathcal{S}}\boldsymbol{U}^{\pm} = \boldsymbol{\Phi}^{\pm} \equiv 0$ on $\mathcal{S}$ and therefore $\mathcal{P}_{\pm}\boldsymbol{\Phi}^{\pm} = \boldsymbol{\Psi} \equiv 0$ on $\mathcal{S}$. □

Corollary 5.2. *The operator $r_{\mathcal{C}^c}\mathfrak{B}(D)$ is invertible in the following space setting*

$$r_{\mathcal{C}^c}\mathfrak{B}(D)\ :\ \widetilde{\mathbb{H}}^{r-\frac{1}{2}}_{\varepsilon\boldsymbol{\nu},0}(\mathcal{C}^c) \to \mathbb{H}^{r+\frac{1}{2}}_{\varepsilon\boldsymbol{\nu},0}(\mathcal{C}^c), \qquad -\frac{1}{2} < r < \frac{1}{2}.$$

For similar arguments we refer to the concluding part of the proof of Lemma 3.12. The next theorem is the main result of this section.

Theorem 5.3. *Let $0 \leq r < \frac{1}{2}$ and the conditions*

$$\boldsymbol{g}^{\pm} \in \mathbb{H}^{r-1/2}_{\varepsilon\boldsymbol{\nu},0}(\mathcal{C}), \quad \boldsymbol{g}^+ - \boldsymbol{g}^- \in r_{\mathcal{C}}\widetilde{\mathbb{H}}^{r-1/2}_{\varepsilon\boldsymbol{\nu},0}(\mathcal{C}).$$

hold. Let $\ell\boldsymbol{g}^+ \in \mathbb{H}^{r-1/2}_{\varepsilon\boldsymbol{\nu},0}(\mathcal{C})$ be some fixed extension of the data functions $\boldsymbol{g}^+$ up to the entire closed surface $\mathcal{S}$, while $\ell_0(\boldsymbol{g}^+ - \boldsymbol{g}^-) \in \mathbb{H}^{r-1/2}_{\varepsilon\boldsymbol{\nu},0}(\mathcal{S})$ is an extension by zero of the function $\boldsymbol{g}^+ - \boldsymbol{g}^-$.

The elliptic BVP (2.4) has a unique solution $\boldsymbol{U} \in \mathbb{H}^{r+1}_{\varepsilon\boldsymbol{\nu},0}(\mathbb{R}^3_{\mathcal{C}})$ of the form

$$\boldsymbol{U} = \begin{cases} -\mathbf{V}^{\boldsymbol{e}}(\mathbf{V}^{\boldsymbol{e}}_{-1})^{-1}\left[\mathcal{P}^{-1}_{+}\ell\boldsymbol{g}^+ + \mathcal{P}^{-1}_{+}\boldsymbol{\Psi}\right] & \text{in } \Omega^+, \\ \mathbf{V}^{\boldsymbol{e}}(\mathbf{V}^{\boldsymbol{e}}_{-1})^{-1}\left[\mathcal{P}^{-1}_{-}(\ell\boldsymbol{g}^+ - \ell_0(\boldsymbol{g}^+ - \boldsymbol{g}^-)) + \mathcal{P}^{-1}_{-}\boldsymbol{\Psi}\right] & \text{in } \Omega^-, \end{cases} \tag{5.14}$$

where $\boldsymbol{\Psi} \in \widetilde{\mathbb{H}}^{r-\frac{1}{2}}_{\varepsilon\boldsymbol{\nu},0}(\mathcal{C}^c)$ *is a unique solution to the system*

$$\begin{aligned} r_{\mathcal{C}^c}\mathfrak{B}(D)\boldsymbol{\Psi} = \boldsymbol{F} \qquad \text{on } \mathcal{C}^c, \\ \boldsymbol{F} := r_{\mathcal{C}^c}(\mathcal{P}_-)^{-1}\ell_0(\boldsymbol{g}^+ - \boldsymbol{g}^-) - r_{\mathcal{C}^c}\mathfrak{B}(D)\ell\boldsymbol{g}^+, \quad \boldsymbol{F} \in \mathbb{H}^{r+\frac{1}{2}}_{\varepsilon\boldsymbol{\nu},0}(\mathcal{C}^c). \end{aligned} \tag{5.15}$$

The pseudodifferential operator of order -1

$$r_{\mathcal{C}^c}\mathfrak{B}(D) = r_{\mathcal{C}^c}\left[\mathcal{P}_+^{-1} + \mathcal{P}_-^{-1}\right] \;:\; \widetilde{\mathbb{H}}^{r-\frac{1}{2}}_{\varepsilon\boldsymbol{\nu},0}(\mathcal{C}^c) \to \mathbb{H}^{r+\frac{1}{2}}_{\varepsilon\boldsymbol{\nu},0}(\mathcal{C}^c) \tag{5.16}$$

is invertible.

Proof. The proof follows directly from Lemma 5.1 and Corollary 5.2. □

Acknowledgment

The authors thank T. Buchukuri and D. Natroshvili for valuable discussions and remarks.

References

[1] M. Agranovich, *Elliptic singular integrodifferential operators.* Uspekhi Matematicheskikh Nauk **20** (1965), 3–120. (Russian).

[2] T. Buchukuri, R. Duduchava, D. Kapanadze, D. Natroshvili, *On the uniqueness of a solution to an anisotropic Maxwell's equations*, Operator Theory: Advances and Applications **203** (2010), 137–164.

[3] O. Chkadua, R. Duduchava, *Pseudodifferential equations on manifolds with boundary: Fredholm property and asymptotics.* Mathematische Nachrichten **222** (2001), 79–139.

[4] M. Costabel, *A coercive bilinear form for Maxwell's equations.* Journal of Mathematical Analysis and Applications **157** (1991), 527–541.

[5] M. Costabel, W. Wendland, *Strong ellipticity of boundary integral operators.* J. Reine Angew. Mathematik **372** (1986), 34–63.

[6] R. Duduchava, *The Green formula and layer potentials.* Integral Equations and Operator Theory **41** (2001), 127–178.

[7] R. Duduchava, *On singular integral operators on piecewise smooth lines.* Research Notes in Mathematics **8**, "Functional Theoretic Methods in Differential Equations", Pitman, London, (1976), 109–131.

[8] R. Duduchava, D. Mitrea, M. Mitrea, *Differential operators and boundary value problems on surfaces.* Mathematische Nachrichten **279** (2006), 996–1023.

[9] R. Duduchava, D. Natroshvili and E. Shargorodsky, *Boundary value problems of the mathematical theory of cracks.* Proc. I. Vekua Inst. Appl. Math., Tbilisi State University **39** (1990), 68–84.

[10] R. Duduchava, D. Natroshvili, E. Shargorodsky, *Basic boundary value problems of thermoelasticity for anisotropic bodies with cuts* I–II. Georgian Mathematical Journal **2** (1995), 123–140, 259–276.

[11] R. Duduchava, F.-O. Speck, *Pseudodifferential operators on compact manifolds with Lipschitz boundary.* Mathematische Nachrichten **160** (1993), 149–191.

[12] R. Duduchava, W. Wendland, *The Wiener-Hopf method for systems of pseudodifferential equations and its application to crack problems.* Integral Equations and Operator Theory **23** (1995), 294–335,

[13] L. Hörmander, *The Analysis of Linear Partial Differential Operators* **I-IV**, Springer-Verlag, Heidelberg, 1983.

[14] G.C. Hsiao, W.L. Wendland, *Boundary Integral Equations*, Applied Mathematical Sciences, Springer-Verlag Berlin Heidelberg, 2008

[15] R. Kapanadze, *On some properties of singular integral operators in spaces with norms.* Proceedings of Tbilisi State University **129** (1968), 263–277.

[16] V. Kupradze, T. Gegelia, M. Basheleishvili, T. Burchuladze, *Three-Dimensional Problems of the Mathematical Theory of Elasticity and Thermoelasticity.* North-Holland, Amsterdam, 1979.

[17] R. Leis, *Zur Theorie elektromagnetischer Schwingungen in anisotropen inhomogenene Medien.* Mathematische Zeitschrift **106** (1968), 213–224.

[18] W. McLean, *Strongly elliptic systems and boundary integral equations.* Cambridge: Cambridge University Press **XIV**, 2000.

[19] D. Natroshvili, O. Chkadua, E. Shargorodsky, *Mixed problems for homogeneous anisotropic elastic media.* Proceedings I. Vekua Institute of Applied Mathematics, Tblisi State University **39** (1990), 133–181.

[20] H. Triebel, *Interpolation Theory, Function Spaces, Differential Operators.* 2nd edition, Johann Ambrosius Barth Verlag, Heidelberg–Leipzig, 1995.

O. Chkadua
I. Javakhishvili Tbilisi State University
Andrea Razmadze Mathematical Institute
2, University str.
Tbilisi 0186, Georgia

and

Sokhumi State University
9, Jikia str.
Tbilisi 0186, Georgia
e-mail: `chkadua7@yahoo.com`

R. Duduchava and D. Kapanadze
I. Javakhishvili Tbilisi State University
Andrea Razmadze Mathematical Institute
2, University str.
Tbilisi 0186, Georgia
e-mail: `dudu@rmi.ge`
`daka@rmi.ge`

Operator Theory:
Advances and Applications, Vol. 228, 95–125

Functions of Noncommuting Operators in an Asymptotic Problem for a 2D Wave Equation with Variable Velocity and Localized Right-hand Side

Sergey Dobrokhotov, Dmitrii Minenkov,
Vladimir Nazaikinskii and Brunello Tirozzi

Dedicated to Vladimir Rabinovich

Abstract. In the present paper, we use the theory of functions of noncommuting operators, also known as noncommutative analysis (which can be viewed as a far-reaching generalization of pseudodifferential operator calculus), to solve an asymptotic problem for a partial differential equation and show how, starting from general constructions and operator formulas that seem to be rather abstract from the viewpoint of differential equations, one can end up with very specific, easy-to-evaluate expressions for the solution, useful, e.g., in the tsunami wave problem.

Mathematics Subject Classification (2010). Primary 35L05; Secondary 81Q20, 35Q35.

Keywords. Wave equation, localized solutions, asymptotics, noncommutative analysis, Maslov canonical operator.

1. Introduction

In the present paper, we use the theory of functions of noncommuting operators [1–3], aka noncommutative analysis (which can be viewed as a far-reaching generalization of pseudodifferential operator calculus), to solve an asymptotic problem for a partial differential equation and show how, starting from general constructions

Supported by RFBR grants nos. 11-01-00973-a and 11-01-12058-ofi-m-2011 and by the scientific agreement between the Department of Physics, University of Rome "La Sapienza," Rome, and A. Ishlinsky Institute for Problems in Mechanics, Russian Academy of Sciences, Moscow.

and operator formulas that seem to be rather abstract from the viewpoint of differential equations, one can end up with very specific, easy-to-evaluate expressions for the solution, useful, e.g., in the tsunami wave problem.

We consider the Cauchy problem with zero initial data for a 2D wave equation with variable velocity and with right-hand side localized near the origin in space and decaying in time. One physical interpretation of this problem is that it describes, in the linear approximation, the propagation of tsunami waves generated by local vertical displacements of the ocean bottom (see [4–9, 17–19] and also [11–16] and the bibliography therein). Normally, the diameter of the region where these displacements occur (some tens to a hundred of kilometers) is much smaller than the distance traveled by the waves (thousands of kilometers), and their ratio, μ, can serve as a small parameter. Accordingly, we are interested in the asymptotics of the solution as $\mu \to 0$. In the simplest *piston model* of tsunami generation, the bottom displacement occurs instantaneously at $t = 0$. This corresponds to a right-hand side of the form $\delta'(t)v(x)$, where $\delta(t)$ is the Dirac delta function, and the problem is immediately equivalent, via Duhamel's principle, to the Cauchy problem for the homogeneous wave equation with initial data $v(x)$ for the unknown function itself and zero initial data for its t-derivative. Fairly explicit asymptotic solution formulas suitable for easy implementation in *Wolfram Mathematica* [27] were constructed and analyzed for the latter problem in [11–16] on the basis of a generalization of Maslov's canonical operator [1, 20]. Now assume we wish to take into account the fact that the ocean bottom displacement evolves in time rather than happens instantaneously. Then it is natural to consider a right-hand side of the form $g'(t)v(x)$, where $g(t)$ is some smooth approximation to the delta function. An analysis shows that the solution can be represented as the sum of two parts, a propagating part, which travels along the characteristics, and a transient part, which is localized in the vicinity of the origin and decays in time. The propagating part can further be represented as the solution of the Cauchy problem for the homogeneous wave equation with initial data obtained from $v(x)$ by application of certain functions $f(L)$ of the spatial part L of the wave operator, where the corresponding symbols $f(\xi)$ are given by simple formulas expressing them via the Fourier transform of $g(t)$. These initial data, also localized near the origin, will be referred to as the *equivalent source functions*. The transient part of the solution is given by a formula similar to those for the equivalent source functions with the only difference that the function $f(\xi)$ additionally depends on time as a parameter. The transient part is apparently not so important in tsunami wave analysis, but nevertheless it might be useful from the viewpoint of satellite registration of tsunami waves [17–19]. Since, as was mentioned above, the asymptotic formulas for the solution of the Cauchy problem with localized initial data for the homogeneous wave equation are already known from [11–16], we see that the only remaining thing is to compute $f(L)v$ for all these functions $f(\xi)$. It is here that noncommutative analysis comes fully into play. Note that L is an operator with variable coefficients, and so computing the function $f(L)$ efficiently may prove quite a challenging task. However, all we actually need is the asymptotics of $f(L)v$, and methods of noncommutative analysis permit one to prove that $f(L)v = f(L_0)v$ plus an asymptotically small remainder, where L_0 is obtained from L by freezing the coefficients at the origin. Now computing $f(L_0)v$ is a breeze, because $f(L_0)$ is conjugate by the Fourier transform to the operator of multiplication by the function $f(\sigma_{L_0}(p))$, where $\sigma_{L_0}(p)$ is the symbol of L_0.

The one-dimensional counterpart of the problem studied in the present paper was considered in [21]. In the two-dimensional case, the results were announced in [22], where the proofs were partly only sketched and partly absent altogether. Here we develop and refine these results and give complete proofs. Finally, note that we deal with the setting in which the wave propagation velocity is assumed to vanish nowhere. The case it which it vanishes (as it happens on the coastline in the tsunami run-up problem) is much more complicated. The asymptotics of solutions of such degenerate problems in some special cases was considered in the spirit of the approach of [11–16] in [23–25] (see also references therein); in the present paper, we restrict ourselves to wave propagation in open ocean.

The outline of the paper is as follows. In Section 2, we give a detailed statement of the mathematical problem and write out well-known formulas expressing the solution in operator form. Using these formulas, we split the solution into the sum of the propagating and transient parts. Section 3 presents simple formulas for the asymptotics of the solution. The proofs of the theorems stated in this section depend on the results presented in Section 4, which is the most important part of the paper and where the asymptotics of the equivalent source functions and the transient part of the solution are computed with the use of the noncommutative analysis machinery. Finally, Section 5 provides two simple examples; all computations and visualizations in these examples have been done with *Wolfram Mathematica*.

2. Exact solution

2.1. Statement of the problem

Consider the Cauchy problem for the wave equation

$$\frac{\partial^2\eta}{\partial t^2} - \frac{\partial}{\partial x_1}\left(c^2(x)\frac{\partial\eta}{\partial x_1}\right) - \frac{\partial}{\partial x_2}\left(c^2(x)\frac{\partial\eta}{\partial x_2}\right) = Q, \qquad t \geq 0, \tag{2.1}$$

with the initial conditions

$$\eta|_{t=0} = 0, \qquad \eta_t|_{t=0} = 0, \tag{2.2}$$

where $x = (x_1, x_2) \in \mathbf{R}^2$, $\eta = \eta(x,t)$ is the unknown function, $c(x)$ is an everywhere positive smooth function stabilizing at infinity,[1] and the right-hand side $Q = Q(x,t)$ depends on two parameters $\lambda, \mu > 0$ and has the form

$$Q(x,t) = \lambda^2 g_0'(\lambda t)V\left(\frac{x}{\mu}\right), \qquad \text{where } g_0'(\tau) = \frac{dg_0(\tau)}{d\tau}, \tag{2.3}$$

[1]That is, $c(x) = \text{const} > 0$ for sufficiently large $|x|$.

with some smooth real functions $V(y)$, $y \in \mathbf{R}^2$, and $g_0(\tau)$, $\tau \in [0,\infty)$, such that

$$|V^{(\alpha)}(y)| \le C_\alpha(1+|y|)^{-|\alpha|-\varkappa}, \qquad |\alpha| = 0,1,2,\dots, \tag{2.4}$$

$$g_0(0) = 0, \qquad \int_0^\infty g_0(\tau)\,d\tau = 1, \qquad |g_0^{(k)}(\tau)| \le C_k e^{-\nu\tau}, \quad k = 0,1,2,\dots, \tag{2.5}$$

for some $\varkappa > 1$, $\nu > 0$, and positive constants C_α and C_k.

Remark 2.1. One can also consider the case in which $g_0(\tau)$ decays as some (sufficiently large) negative power of τ as $\tau \to \infty$. In this case, the estimates are somewhat more awkward, and we restrict ourselves to the case of the physically natural exponential decay (2.5) for the sake of clarity.

Our aim is to find the asymptotics as $\mu \to 0$ of the solution of problem (2.1) on an arbitrary finite time interval uniformly with respect to λ in the region

$$\lambda\mu > \text{const} > 0. \tag{2.6}$$

This will be done in Sections 3 and 4, and in the present section we deal with the exact solution of the problem.

2.2. Physical interpretation and examples of right-hand sides

First, speaking in terms of the physical interpretation given in the introduction, let us explain the meaning of the parameters λ and μ and condition (2.6). The right-hand side $Q(x,t)$ describes the time evolution (the factor $\lambda^2 g_0'(\lambda t)$) and the spatial shape (the factor $V(x/\mu)$) of the perturbation (the tsunami source). In view of (2.5), λ characterizes the decay rate of the perturbation, so that $1/\lambda \sim t_0$, where t_0 is the mean lifetime of the perturbation. The small parameter μ characterizes the source size r_0, $\mu \sim r_0$. We see that the product $\lambda\mu = r_0/t_0$ has the dimension of velocity and rewrite condition (2.6) in the form

$$\frac{c_0}{\lambda\mu} \equiv \frac{c_0 t_0}{r_0} \le \omega_0, \tag{2.7}$$

where $c_0 = c(0)$, the wave propagation velocity at the origin, is taken to represent the typical wave propagation velocity in the problem and ω_0 is some dimensionless constant. This has a very clear meaning: the waves excited by the perturbation cannot travel too far before the perturbation dies out; they only cover a distance $(c_0 t_0)$ of the same order of magnitude as the diameter r_0 of the perturbation region. We introduce the ratio

$$\omega = \frac{c_0}{\lambda\mu}, \tag{2.8}$$

so that condition (2.7) (and hence (2.6)) becomes

$$\omega < \omega_0. \tag{2.9}$$

Mathematically, condition (2.9) means that the parameter λ is large (at least of the order of μ^{-1}) as $\mu \to 0$. Note that, in view of the first two conditions in (2.5), $\lambda g_0(\lambda t) \to \delta(t)$ and $\lambda^2 g_0'(\lambda t) \to \delta'(t)$ as $\lambda \to \infty$.

In what follows, the dependence on the parameters λ and μ is sometimes not immediately important to the argument, and in such cases we often "hide" these parameters by using the notation

$$g(\tau) = \lambda g_0(\lambda\tau), \quad v(x) = V\left(\frac{x}{\mu}\right), \quad \text{so that} \quad Q(x,t) = g'(t)v(x). \tag{2.10}$$

Next, let us give specific examples of right-hand sides $Q(x,t)$. In practice, the actual ocean bottom displacement is known neither in much detail nor very precisely, because the corresponding measurements are impractical or impossible (cf. [17–19]). This results in certain freedom, which can be turned into an advantage. Namely, when constructing the function $Q(x,t) = g'(t)v(x)$ to be used in the analytical-numerical simulation according to the model (2.1), one should take ansatzes that, on the one hand, fit the general information available about the source shape and evolution and, on the other hand, can be handled efficiently in the computations. (The latter includes the requirement that these functions, as well as their Fourier transforms, be given by closed-form expressions, which permits one to reduce the amount of numerical computations in favor of the less time-consuming analytical transformations.)

A useful class of functions $V(y)$ satisfying (2.4) is described by the expression [10, 14, 16], generalizing [5, 8, 9],

$$V(y) = A\left(1 + \left(\frac{y_1}{b_1}\right)^2 + \left(\frac{y_2}{b_2}\right)^2\right)^{-3/2}, \tag{2.11}$$

where A, b_1, and b_2 are real parameters. The Fourier transform of this function is remarkably simple,

$$\widetilde{V}(p) = Ab_1b_2e^{-\sqrt{b_1^2p_1^2+b_2^2p_2^2}}. \tag{2.12}$$

One can further apply a differential operator

$$\widehat{P} = P\left(\frac{\partial}{\partial y_1}, \frac{\partial}{\partial y_2}\right)$$

with constant coefficients to the function V and then rotate the coordinate system by some angle θ, thus obtaining a broad variety of functions of the form

$$V_{P,\theta}(y) = [\widehat{P}V](T(\theta)y), \qquad T(\theta) \equiv \begin{pmatrix} \cos\theta & \sin\theta \\ -\sin\theta & \cos\theta \end{pmatrix},$$

satisfying (2.4). Such functions model elliptic-shaped sources of various eccentricity and various direction of axes with a wavy relief depending on the differential operator $\widehat{P}$ (see [11, 15]). Figure 1 shows the graph of $V(y)$ rotated by an angle of $\pi/10$ and of its Fourier transform.

Let us also give two examples of functions $g_0(t)$ satisfying (2.5),

$$\text{(a)} \quad g_0(\tau) = ae^{-\tau}(\sin(\alpha\tau + \phi_0) - \sin\phi_0), \qquad \text{(b)} \quad g_0(\tau) = e^{-\tau}P(\tau), \tag{2.13}$$

where $\alpha > 0$ and ϕ are real parameters, $a = (\alpha^2+1)/(\alpha\cos\phi_0 - \alpha^2\sin\phi_0)$ is a normalizing factor, and $P(\tau) = \sum_{k=1}^n (k.)^{-1}P_k\tau^k$ is a polynomial of degree n with $\sum_{k=1}^n P_k = 1$ (see Figure 2).

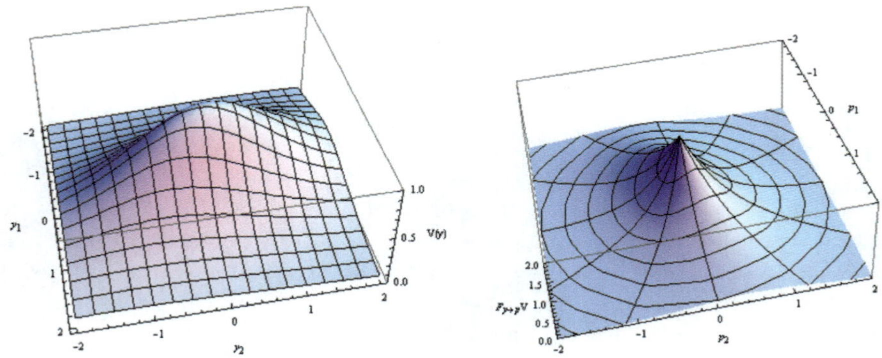

FIGURE 1. The function $V(y)$ with $b_1 = 1$ and $b_2 = 4$ rotated by the angle $\theta = \pi/10$ (left) and its Fourier transform $\widetilde{V}(p)$ (right).

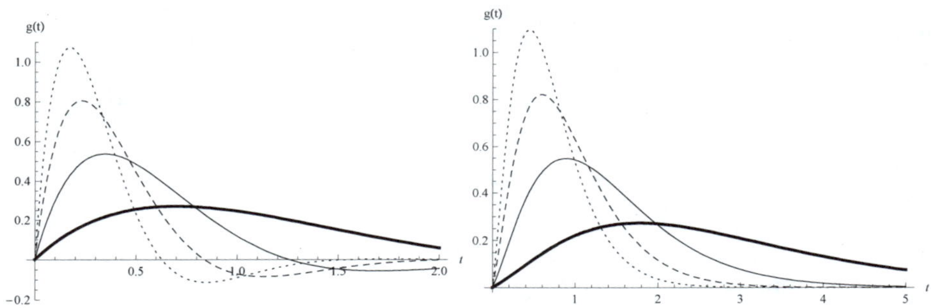

FIGURE 2. Examples of $g(t) = \lambda g_0(\lambda t)$ for $\lambda = 1, 2, 3, 4$: $g_0(\tau) = e^{-\tau}(\sin(\alpha\tau + \varphi_0) - \sin\varphi_0)$ (left diagram); $g_0(\tau) = e^{-\tau}(0.2\tau + 0.4\tau^2)$ (right diagram).

2.3. Operator solution formulas and energy estimates

We denote the spatial part of the wave operator in (2.1) by L; thus,

$$Lu = -\frac{\partial}{\partial x_1}\left(c^2(x)\frac{\partial u}{\partial x_1}\right) - \frac{\partial}{\partial x_2}\left(c^2(x)\frac{\partial u}{\partial x_2}\right) \equiv -\langle\nabla, c^2(x)\nabla\rangle u. \tag{2.14}$$

The operator (2.14) (with domain $W_2^2(\mathbf{R}^2)$) is a nonnegative self-adjoint operator on $L^2(\mathbf{R}^2)$. Let $D = \sqrt{L}$ be the positive square root of L. Problem (2.1) becomes

$$\eta''(t) + D^2\eta(t) = g'(t)v, \qquad \eta|_{t=0} = \eta_t|_{t=0} = 0. \tag{2.15}$$

Duhamel's formula represents the solution of (2.15) as the integral

$$\eta(t) = \int_0^t w(t,\tau)\,d\tau, \tag{2.16}$$

where $w(t,\tau)$ is the solution of the problem[2]

$$w''_{tt}(t,\tau)+D^2w(t,\tau)=0,\quad w|_{t=\tau}=g(\tau)v,\quad w_t|_{t=\tau}=0. \tag{2.17}$$

Indeed, the function (2.16) satisfies (2.15), because

$$\begin{aligned}\eta''(t)+D^2\eta(t)&=\int_0^t\bigl(w''_{tt}(t,\tau)+D^2w(t,\tau)\bigr)\,d\tau+\frac{d}{dt}\bigl(w(t,t)\bigr)+w'_t(t,t)\\&=g'(t)v,\qquad \eta(0)=0,\qquad \eta'(0)=g(0)v=0\end{aligned}$$

in view of (2.5). Now we can use the general solution formula (e.g., see [26, p. 191])

$$u(t)=\cos(Dt)u_0+D^{-1}\sin(Dt)u_1 \tag{2.18}$$

for the abstract hyperbolic Cauchy problem

$$u''(t)+D^2u(t)=0,\qquad u|_{t=0}=u_0,\quad u_t|_{t=0}=u_1 \tag{2.19}$$

and write

$$\eta(t)=\biggl[\int_0^t\cos(D(t-\tau))g(\tau)\,d\tau\biggr]v=\operatorname{Re}\biggl[\int_0^t e^{iD(t-\tau)}g(\tau)\,d\tau\biggr]v. \tag{2.20}$$

Here we have used the fact that $g(\tau)$ is real valued; the real part of an operator A is defined as usual by $\operatorname{Re}A=\frac{1}{2}(A+A^*)$. Formula (2.20) is the desired abstract operator formula for the solution of problem (2.1).

Remark 2.2. Since the operator D is self-adjoint, it follows that the expressions $\cos(Dt)$, $\sin(Dt)/D$, and e^{iDt}, occurring in (2.18) and (2.20), are well defined in the framework of functional calculus for self-adjoint operators as functions $f(D)$ with bounded continuous symbols $f(\xi)=\cos\xi t$, $f(\xi)=\xi^{-1}\sin\xi t$, and $f(\xi)=e^{i\xi t}$, respectively. Moreover, e^{iDt} is none other than the strongly continuous group of unitary operators generated by D, $\operatorname{Re}(f(D))=(\operatorname{Re}f)(D)$, and, for "good" functions $f(\xi)$, the operator $f(D)$ can be defined not only via the integral over the spectral measure but also via the Fourier transform as

$$f(D)u=\frac{1}{\sqrt{2\pi}}\langle\widetilde{f}(\tau),e^{i\tau D}u\rangle,\qquad u\in L^2(\mathbf{R}^2),$$

where $\widetilde{f}$ is the Fourier transform of f and the angle brackets stand for the value of the distribution $\widetilde{f}(\tau)$ on the $L^2(\mathbf{R}^2)$-valued function $e^{i\tau D}u$.

Remark 2.3. The *energy* of the solution of the Cauchy problem (2.19) is defined by the formula [26, p. 191]

$$\mathcal{E}[u](t)=\frac{1}{2}\bigl(\|u'(t)\|^2+\|Du(t)\|^2\bigr)\equiv\frac{1}{2}\bigl(\|u'(t)\|^2+(u(t),Lu(t))\bigr) \tag{2.21}$$

[2]The standard Duhamel formula would give $w|_{t=\tau}=0$ and $w_t|_{t=\tau}=g'(\tau)v$ in (2.17), but we have made use of the special form of the right-hand side.

(where $\|\cdot\|$ stands for the L^2 norm and $(\cdot,\cdot)$ for the L^2 inner product) and is conserved in the course of time. Hence, in view of (2.10) and the estimates (2.4), the solution of problem (2.17) (with τ viewed as a parameter) satisfies

$$\begin{aligned}\mathcal{E}[w](t,\tau) &= \frac{g^2(\tau)}{2}\left\|D\bigl(V(x/\mu)\bigr)\right\|^2 = \frac{g^2(\tau)}{2\mu^2}\int_{\mathbf{R}^2}\left|c(x)\nabla V\left(\frac{x}{\mu}\right)\right|^2 dx_1dx_2\\ &= \frac{g^2(\tau)}{2}\int_{\mathbf{R}^2} c^2(\mu y)|\nabla V(y)|^2\,dy_1dy_2 = \frac{\lambda^2}{2}g_0^2(\lambda\tau)c_0^2\left\|\nabla V\right\|^2(1+O(\mu))\end{aligned}$$

as $\mu\to 0$, where $c_0 = c(0)$. Now it follows from (2.16) that, with some constant C,

$$\begin{aligned}\mathcal{E}[\eta](t) &\le \left\{\int_0^t\sqrt{\mathcal{E}[w](t,\tau)}\,d\tau\right\}^2 + \frac{1}{2}\left\|w(t,t)\right\|^2\\ &\le C\left\{\lambda\int_0^t|g_0(\lambda\tau)|\,d\tau\right\}^2 + \frac{\mu^2\lambda^2}{2}g_0^2(\lambda t)\left\|V\right\|^2\\ &\le C\left\{\int_0^\infty|g_0(\tau)|\,d\tau\right\}^2 + \frac{\mu^2\lambda^2}{2}g_0^2(\lambda t)\left\|V\right\|^2\\ &= O(1)+O(\mu^2\lambda^2e^{-2\nu\lambda t}) = O(1)+O(\omega^{-2}e^{-2\nu\lambda t});\end{aligned}$$

i.e., the energy of the solution is uniformly bounded as $\mu\to 0$ for all $t>\varepsilon>0$. (However, it may have a "spike" of the order of ω^{-2} for $t\sim 1/\lambda$; of course, this is only important if $\omega\ll 1$.) In other words, we have chosen a physically natural normalization of the right-hand side of our problem.

Remark 2.4. For the inhomogeneous wave equation

$$u''(t)+D^2u(t)=F(t), \tag{2.22}$$

one has the energy identity

$$\mathcal{E}[u](t)=\mathcal{E}[u](0)+\operatorname{Re}\int_0^t(F(\tau),u'(\tau))\,d\tau,$$

which implies the well-known estimates[3]

$$\left\|u(t)\right\|_{s+1}+\left\|u'(t)\right\|_s\le C(t)\bigl(\left\|u(0)\right\|_{s+1}+\left\|u'(0)\right\|_s+\sup_{\tau\in[0,t]}\left\|F(\tau)\right\|_s\bigr), \tag{2.23}$$

where $\|\cdot\|_s$ stands for the norm on the Sobolev space $H^s=W_2^s(\mathbf{R}^2)$; in particular, $\|\cdot\|_0=\|\cdot\|$. Of the estimates (2.23), the most important for us is the one with $s=0$ (corresponding to the sum of the energy integral and the L^2 norm), in which the main estimates for the norms of remainders in asymptotic formulas will be obtained. However, occasionally our argument involves estimates with different s.

[3]Their derivation takes into account the fact that the norm $\|u\|_s$ is equivalent to the norm $\|(1+L)^{s/2}u\|$ by virtue of the conditions imposed on the velocity $c(x)$.

2.4. Solution splitting into propagating and transient components

Let us further transform formula (2.20) to reveal the structure of the solution and represent it in a form suitable for subsequent computations. We have

$$\begin{aligned}\int_0^t e^{iD(t-\tau)}g(\tau)\,d\tau &= \int_0^\infty e^{iD(t-\tau)}g(\tau)\,d\tau - \int_t^\infty e^{iD(t-\tau)}g(\tau)\,d\tau \\ &= e^{iDt}\int_0^\infty e^{-iD\tau}g(\tau)\,d\tau - \int_0^\infty e^{-iD\tau}g(\tau+t)\,d\tau.\end{aligned} \tag{2.24}$$

Let $H(\tau)$ be the Heaviside step function ($H(\tau) = 1$ for $\tau \geq 0$ and $H(\tau) = 0$ for $\tau < 0$), and, for $t \geq 0$, let

$$G(\xi,t) = \int_0^\infty e^{-i\xi\tau}g(\tau+t)\,d\tau \equiv \int_{-\infty}^\infty e^{-i\xi\tau}H(\tau)g(\tau+t)\,d\tau$$

be the Fourier transform of $\sqrt{2\pi}H(\tau)g(\tau+t)$ with respect to the variable τ. (Note that $G(\xi,0) = \sqrt{2\pi}\widetilde{g}(\xi)$, where the function $g(\tau)$ is assumed to be extended by zero for the negative values of τ.) Then formula (2.24) can be rewritten as

$$\int_0^t e^{iD(t-\tau)}g(\tau)\,d\tau = e^{iDt}G(D,0) - G(D,t) = \sqrt{2\pi}e^{iDt}\widetilde{g}(D) - G(D,t),$$

and accordingly

$$\eta(t) = \eta_{\text{prop}}(t) + \eta_{\text{trans}}(t), \tag{2.25}$$

where

$$\begin{aligned}\eta_{\text{prop}}(t) &= \sqrt{2\pi}\operatorname{Re}\bigl(e^{iDt}\widetilde{g}(D)\bigr)v \qquad (2.26)\\ &\equiv \sqrt{2\pi}\cos(Dt)\operatorname{Re}\widetilde{g}(D)v - \sqrt{2\pi}\sin(Dt)\operatorname{Im}\widetilde{g}(D)v,\\ \eta_{\text{trans}}(t) &= -\operatorname{Re}(G(D,t))v. \qquad (2.27)\end{aligned}$$

The function $\eta_{\text{prop}}(t)$ given by (2.26) is the solution of the Cauchy problem for the homogeneous wave equation

$$u''(t) + D^2u(t) = 0 \tag{2.28}$$

with the initial data

$$u_0 = \sqrt{2\pi}\operatorname{Re}\widetilde{g}(D)v, \qquad u_1 = -\sqrt{2\pi}\operatorname{Im}\widetilde{g}(D)Dv. \tag{2.29}$$

(This follows from the comparison of (2.26) with (2.28).) Hence this function will be called the *propagating component* of the solution, and the initial data (2.29) for $\eta_{\text{prop}}(t)$ will be called the *equivalent source functions*. We shall see in Section 3.3 that, exactly as one should expect, $\eta_{\text{prop}}(t)$ propagates along the characteristics.

The function $\eta_{\text{trans}}(t)$ given by (2.27) will be called the *transient component* of the solution, because it exponentially decays as $\lambda t \to \infty$, as shown by the following proposition. (We shall also see in Section 3.1 that $\eta_{\text{trans}}(t)$ always remains localized near the origin.)

Proposition 2.5. *As $\mu \to 0$, the propagating component satisfies the estimates*

$$\|\eta_{\text{prop}}(t)\|_1 = O(1), \qquad \big\|\eta'_{\text{prop}}(t)\big\| = O(1),$$

and the transient component satisfies the estimates

$$\|\eta_{\text{trans}}(t)\|_1 = O(e^{-\nu\lambda t}), \qquad \|\eta'_{\text{trans}}(t)\| = O(\omega^{-1}e^{-\nu\lambda t}),$$

where ν is the constant in condition (2.5).

Proof. We will estimate the transient part (2.27) of the solution directly and the propagating part (2.26) via the Cauchy data (2.29) by using the energy estimates. Formulas (2.27) and (2.29) involve the real and imaginary parts of the operator $G(D,t)$ applied to the original right-hand side source function v. (Recall that $\widetilde{g}(D)$ is a special case of $G(D,t)$ for $t=0$.) Thus, we need to estimate the operator $G(D,t)$. Note that, for an arbitrary bounded measurable function $f(\xi)$, one has

$$\big\|f(D)\colon H^0 \to H^0\big\| \le \sup_{\xi\in\mathbf{R}}|f(\xi)|, \quad \big\|f(D)\colon H^1 \to H^1\big\| \le C\sup_{\xi\in\mathbf{R}}|f(\xi)| \tag{2.30}$$

with some constant C independent of f.[4] Thus, we need estimates for the function $G(\xi,t)$. Since $g(\tau)=\lambda g_0(\lambda\tau)$, we have

$$G(\xi,t) = G_0(\xi/\lambda, \lambda t), \tag{2.31}$$

where

$$G_0(\xi,t) = \int_0^\infty e^{-i\xi\tau} g_0(\tau+t)\,d\tau \tag{2.32}$$

is the Fourier transform of the function $\sqrt{2\pi}H(\tau)g_0(\tau+t)$ with respect to τ. By Lemma 2.6 below, we have

$$|\sqrt{2\pi}\widetilde{g}(\xi)| = |G_0(\xi/\lambda,0)| \le C_{00}$$

and hence, by (2.29) and (2.30),

$$\|u_0\|_1 = \sqrt{2\pi}\,\|\operatorname{Re}\widetilde{g}(D)v\|_1 \le CC_{00}\,\|v\|_1 = O(1),$$
$$\|u_1\| = \sqrt{2\pi}\,\|\operatorname{Im}\widetilde{g}(D)Dv\| \le C_{00}\,\|Dv\| \le \widetilde{C}\,\|v\|_1 = O(1),$$

because $v = V(x/\mu)$ and hence $\|v\|_1 = O(1)$ (cf. the computation in Remark 2.3). Now the energy estimates (2.23) for $s=0$ give the desired estimates for $\eta_{\text{prop}}(t)$. The estimates for the transient part go as follows, again with the use of Lemma 2.6:

$$\begin{aligned}\|\eta_{\text{trans}}(t)\|_1 = \|\operatorname{Re}G(D,t)v\|_1 &\le C\sup_{\xi}|G_0(\xi/\lambda,\lambda t)|\,\|v\|_1\\ &\le CC_{00}e^{-\nu\lambda t}\,\|v\|_1 = O(e^{-\nu\lambda t}),\end{aligned}$$

[4] Indeed, the first estimate is obvious, because the operator D is self-adjoint on $H^0 = L_2(\mathbf{R}^2)$. To obtain the second estimate, we replace the norm on H^1 by the equivalent Hilbert norm $\|u\| = (u,(1+L)u)^{1/2}$ (cf. Remark 2.4); then the operator D becomes self-adjoint on H^1, and the second estimate follows.

$$\|\eta'_{\rm trans}(t)\| = \left\|\operatorname{Re}\frac{\partial G}{\partial t}(D,t)v\right\| \le \sup_\xi \left|\lambda \frac{\partial G_0}{\partial t}\left(\frac{\xi}{\lambda},\lambda t\right)\right| \|v\|$$
$$\le C_{01}\lambda e^{-\nu\lambda t}\|v\| = O(\mu\lambda e^{-\nu\lambda t}) = O(\omega^{-1}e^{-\nu\lambda t}),$$

because $v = V(x/\mu)$ and hence $\|v\| = O(\mu)$. This completes the proof. □

The following lemma establishes the estimates for $G_0(\xi,t)$ used in the proof given above and also estimates that will be useful below.

Lemma 2.6. *The function $G_0(\xi,t)$ satisfies the estimates*

$$\left|\frac{\partial^{m+k}G_0}{\partial t^m\partial\xi^k}(\xi,t)\right| \le C_{km}e^{-\nu t}(1+|\xi|)^{-k-1}, \qquad k,m = 0,1,2,\dots, \tag{2.33}$$

with some constants C_{km}. For $t=0$ and $m=0$, one has the better estimates

$$\left|\frac{\partial^k G_0}{\partial\xi^k}(\xi,0)\right| \le C_{k0}(1+|\xi|)^{-k-2}, \qquad k=0,1,2,\dots. \tag{2.34}$$

Proof. First, let us prove the estimates (2.33) and (2.34) for $|\xi|\le 1$. Then we have

$$\left|\frac{\partial^{m+k}G_0}{\partial t^m\partial\xi^k}(\xi,t)\right| = \left|\int_0^\infty (-i\tau)^k e^{-i\xi\tau} g_0^{(m)}(\tau+t)\,d\tau\right| \le C_m e^{-\nu t}\int_0^\infty \tau^k e^{-\nu\tau}\,d\tau$$

by (2.5), whence the claim follows. Now let $|\xi|>1$. Then we write

$$\frac{\partial^m G_0}{\partial t^m}(\xi,t) = \left(\frac{i}{\xi}\right)^N \int_0^\infty \left[\frac{d^N}{d\tau^N}(e^{-i\xi\tau})\right] g_0^{(m)}(\tau+t)\,d\tau$$

for some integer $N>k+1$ and then integrate by parts N times, thus obtaining

$$\frac{\partial^m G_0}{\partial t^m}(\xi,t) = \sum_{l=1}^N (i\xi)^{-l} g_0^{(m+l-1)}(t) + (i\xi)^{-N}\int_0^\infty e^{-i\xi\tau} g_0^{(m+N)}(\tau+t)\,d\tau. \tag{2.35}$$

Next, we differentiate both sides of (2.35) k times with respect to ξ, which gives

$$\frac{\partial^{m+k}G_0}{\partial t^m\partial\xi^k}(\xi,t) = i^{-k}\sum\nolimits_{l=1}^N \frac{(l+k-1)!}{(l-1)!}(i\xi)^{-l-k} g_0^{(m+l-1)}(t)$$
$$+\, i^{-k}\sum\nolimits_{s=0}^k \binom{k}{s}\frac{(l+s-1)!}{(l-1)!}(i\xi)^{-N-s}\int_0^\infty \tau^{k-s}e^{-i\xi\tau}g_0^{(m+N)}(\tau+t)\,d\tau.$$

Here all factors $g_0^{(m+l-1)}(t)$ and the integral are bounded in modulus by const $\cdot\, e^{-\nu t}$ by virtue of (2.5), and the smallest power of ξ^{-1} on the right-hand side is ξ^{-k-1}, which implies the estimate (2.33). For $t=0$ and $m=0$, the smallest power of ξ^{-1} on the right-hand side is ξ^{-k-2}, since $g_0(0)=0$, and we have the estimate (2.34). □

3. Asymptotics of the solution

In this section, we describe the asymptotics as $\mu \to 0$ of the solution $\eta(t) = \eta_{\mathrm{prop}}(t)+\eta_{\mathrm{trans}}(t)$ of problem (2.1), (2.2). In all theorems in this section, we assume that all conditions stated in Section 2.1 are satisfied. Recall that the problem also contains the large parameter λ, which is related to μ by the condition $\omega < \omega_0$ (see (2.9)), where $\omega = c_0(\lambda\mu)^{-1}$ (see (2.8)). If ω can be treated as a second small parameter (i.e., the distance traveled by the waves in the lifetime of the source is much smaller than the source diameter), then additional Taylor series expansions in ω lead to further simplifications in the asymptotic formulas.

3.1. Asymptotics of the transient component

The asymptotics of the transient component $\eta_{\mathrm{trans}}(t)$ of the solution as $\mu \to 0$ is given by the following theorem.

Theorem 3.1. *One has*

$$\eta_{\mathrm{trans}}(x,t) = -\frac{1}{2\pi}\iint \operatorname{Re} G_0(\omega|p|,\lambda t)\widetilde{V}(p)e^{ipx/\mu}dp_1\,dp_2 + R(t), \tag{3.1}$$

or, in the polar coordinates (r,φ), $x = r\mathbf{n}(\varphi)$, *where* $\mathbf{n}(\varphi) = (\cos\varphi, \sin\varphi)$,

$$\begin{aligned}\eta_{\mathrm{trans}}(r\mathbf{n}(\varphi)) = -\frac{1}{2\pi}\int_0^{2\pi}\int_0^{\infty} &\rho \operatorname{Re} G_0(\omega\rho,\lambda t)\\ &\times \widetilde{V}\big(\rho\mathbf{n}(\psi)\big)e^{ir\rho\cos(\psi-\varphi)/\mu}d\rho\,d\psi + R(t),\end{aligned} \tag{3.2}$$

where the remainder $R(t)$ *satisfies the estimates*

$$\|R(t)\|_1 = O(\mu e^{-\nu\lambda t}), \qquad \|R'(t)\| = O(\mu\omega^{-1}e^{-\nu\lambda t}), \qquad \mu\to 0. \tag{3.3}$$

Proof. Consider the operators

$$L^{(0)} = -c_0^2\nabla^2, \qquad D^{(0)} = (L^{(0)})^{1/2}. \tag{3.4}$$

Thus, $L^{(0)}$ is obtained by freezing the coefficients of the operator L at the origin, and $D^{(0)}$ is just the positive square root of the positive self-adjoint operator $L^{(0)}$.

Lemma 3.2. *One has*

$$\big[\operatorname{Re} G(D^{(0)},t) - \operatorname{Re} G(D,t)\big]V\left(\frac{x}{\mu}\right) = R(t), \tag{3.5}$$

where $R(t)$ *satisfies the estimates* (3.3).

The proof of this lemma will be given in Section 4. Thus, the operator D in the expression (2.27) for $\eta_{\mathrm{trans}}(x,t)$ can be replaced by the operator $D^{(0)}$ with constant coefficients. Now it remains to compute $\operatorname{Re} G(D^{(0)},t)V(x/\mu)$. Since $D^{(0)}$ is an operator with constant coefficients and with symbol $c_0|p|$, it follows that

$$F(D^{(0)}) = \mathcal{F}^{-1}\circ F(c_0|p|)\circ\mathcal{F} \tag{3.6}$$

for any function $F(\xi)$, where $\mathcal{F}$ is the Fourier transform and the middle factor on the right-hand side is the operator of multiplication by $F(c_0|p|)$. Thus we obtain

$$G(D^{(0)},t)V\left(\frac{x}{\mu}\right)=\frac{1}{2\pi}\iint \operatorname{Re} G_0(\omega|p|,\lambda t)\widetilde{V}(p)e^{ipx/u}\,dp_1\,dp_2$$

(we have used the formula for $G(\xi,t)$ and made obvious changes of variables), which proves Theorem 3.1. □

3.2. Asymptotics of the equivalent source functions

Now we proceed to the computation of the propagating component of the solution. It satisfies the Cauchy problem (2.28), (2.29), and so a good starting point would be to compute the equivalent source functions (2.29). Once we compute them (asymptotically) and prove that they are localized near the origin, we can use the methods developed in [11–16] to obtain the asymptotics of the propagating solution component. However, we would like to apply ready-to-use formulas from these papers rather than to write out new formulas based on the same ideas. The formulas in [11–16] were obtained for the case in which u_1, the initial data for the t-derivative of the solution, is zero. So we resort to the following trick.

Proposition 3.3. *The propagating solution component $\eta_{\text{prop}}(t)$ can be represented in the form*

$$\eta_{\text{prop}}(t)=\eta_1(t)+\eta_2'(t), \tag{3.7}$$

where $\eta_1(t)$ and $\eta_2(t)$ are the solutions of the Cauchy problems

$$\eta_1''(t)+D^2\eta_1=0,\quad \eta_1|_{t=0}=\sqrt{2\pi}\operatorname{Re}\widetilde{g}(D)v,\qquad \eta_1'|_{t=0}=0, \tag{3.8}$$

$$\eta_2''(t)+D^2\eta_2=0,\quad \eta_2|_{t=0}=\sqrt{2\pi}D^{-1}\operatorname{Im}\widetilde{g}(D)v,\quad \eta_2'|_{t=0}=0. \tag{3.9}$$

Proof. The sum (3.7) obviously satisfies the wave equation (2.28). Next,

$$\eta_2'|_{t=0}=0,\qquad (\eta_2')'|_{t=0}=\eta_2''|_{t=0}=-D^2\eta_2|_{t=0}=-\sqrt{2\pi}D\operatorname{Im}\widetilde{g}(D)v,$$

which shows that the initial conditions (2.29) are satisfied and hence completes the proof. □

Thus, let us compute the asymptotics of the new equivalent source functions

$$\eta_{10}=\sqrt{2\pi}\operatorname{Re}\widetilde{g}(D)v,\qquad \eta_{20}=\sqrt{2\pi}D^{-1}\operatorname{Im}\widetilde{g}(D)v. \tag{3.10}$$

Theorem 3.4. *The equivalent source functions* (3.10) *have the following asymptotics as $\mu\to 0$:*

$$\eta_{10}=U_1\left(\frac{x}{\mu}\right)+R_1,\qquad \eta_{20}=U_2\left(\frac{x}{\mu}\right)+R_2, \tag{3.11}$$

where the Fourier transforms of the functions $U_1(y)$ and $U_2(y)$ are given by the formulas

$$\widetilde{U}_1(p)=\sqrt{2\pi}\operatorname{Re}\widetilde{g}_0(\omega|p|)\widetilde{V}(p),\qquad \widetilde{U}_2(p)=\sqrt{2\pi}\lambda^{-1}\operatorname{Im}\frac{\widetilde{g}_0(\omega|p|)}{\omega|p|}\widetilde{V}(p) \tag{3.12}$$

and the remainders satisfy the estimates

$$\|R_1\|_1 = O(\mu), \qquad \|R_2\|_2 = O(\mu). \tag{3.13}$$

Proof. The proof goes along the same lines as that of Theorem 3.1. Namely, we prove that the operator D in formulas (3.10) can asymptotically be replaced by $D^{(0)}$ and then compute the Fourier transforms of U_1 and U_2 using formula (3.6). The latter computation is trivial, and we omit it altogether. As for the first part, it is given by the following lemma, which will be proved together with Lemma 3.2 in Section 4.

Lemma 3.5. *One has*

$$\begin{aligned} &\sqrt{2\pi}\big[\operatorname{Re}\widetilde{g}(D) - \operatorname{Re}\widetilde{g}(D^{(0)})\big]V\left(\frac{x}{\mu}\right) = R_1, \\ &\sqrt{2\pi}\big[D^{-1}\operatorname{Im}\widetilde{g}(D) - (D^{(0)})^{-1}\operatorname{Im}\widetilde{g}(D^{(0)})\big]V\left(\frac{x}{\mu}\right) = R_2, \end{aligned} \tag{3.14}$$

where R_1 and R_2 satisfy the estimates (3.13).

This completes the proof of Theorem 3.4. □

Remark 3.6. If we replace η_{10} and η_{20} in the Cauchy problems for η_1 and η_2 by $U_1(x/\mu)$ and $U_2(x/\mu)$, respectively, then the resulting error $\delta(t)$ in the computation of $\eta_{\text{prop}}(t)$ will satisfy the estimates

$$\|\delta(t)\|_1 = O(\mu), \qquad \|\delta'(t)\| = O(\mu), \qquad \mu \to 0,$$

uniformly on any finite time interval. Indeed, let us write $\delta(t) = \delta_1(t) + \delta_2'(t)$, where δ_1 and δ_2 are the errors in η_1 and η_2, respectively. Then, by virtue of the energy estimates (2.23), we have

$$\begin{aligned} \|\delta_1(t)\|_1 + \|\delta_1'(t)\| &\le C(t)\,\|\delta_1(0)\|_1 = C(t)\,\|R_1\|_1 = O(\mu), \\ \|\delta_2'(t)\|_1 + \|\delta_2''(t)\| &\le C(t)\,\|\delta_2''(0)\| = C(t)\,\big\|D^2\delta_2(0)\big\| \\ &= C(t)\,\big\|D^2 R_2\big\| \le C_1(t)\,\|R_2\|_2 = O(\mu). \end{aligned}$$

Thus, the accuracy provided by Theorem 3.4 permits computing the propagating part of the solution modulo $O(\mu)$ in the energy norm.

3.3. Asymptotics of the propagating part

Remark 3.6 shows that, to compute the asymptotics of the propagating part of the solution of problem (2.1) modulo $O(\mu)$ in the energy norm, it suffices to solve problems (3.8) and (3.9) asymptotically with the initial data replaced by the functions $U_1(x/\mu)$ and $U_2(x/\mu)$ indicated in Theorem 3.4. Thus, we need to solve the problems

$$\eta_1''(t) + D^2\eta_1 = 0, \quad \eta_1|_{t=0} = U_1(x/\mu), \quad \eta_1'|_{t=0} = 0, \tag{3.15}$$

$$\eta_2''(t) + D^2\eta_2 = 0, \quad \eta_2|_{t=0} = U_2(x/\mu), \quad \eta_2'|_{t=0} = 0. \tag{3.16}$$

(We denote the new unknown functions by the same letters $\eta_{1,2}$; this will not lead to a misunderstanding.) The initial data in these problems are localized near the origin, and hence the asymptotics of solutions of these problems modulo $O(\mu)$ in all spaces H^s can be obtained with the use of the approach developed in [11–14] and based on the Maslov canonical operator [1, 20]. Let us briefly recall this construction.

3.3.1. Bicharacteristics, canonical operator and solution formulas. In the phase space $\mathbf{R}^4_{x,p}$ with the coordinates $(x,p) = (x_1, x_2, p_1, p_2)$, consider the Hamiltonian system

$$\dot{p} = -\frac{\partial \mathcal{H}}{\partial x}, \qquad \dot{x} = \frac{\partial \mathcal{H}}{\partial p} \tag{3.17}$$

corresponding to the Hamiltonian function $\mathcal{H} = |p|c(x)$. This system determines the Hamiltonian phase flow $g^t_{\mathcal{H}}$. Let $\mathbf{n}(\psi) = {}^t(\cos\psi, \sin\psi)$. Consider the Lagrangian manifold $\Lambda_0 = \{p = \mathbf{n}(\psi), x = \alpha\mathbf{n}(\psi)\}$, isomorphic to the two-dimensional cylinder, where $\psi \in [0, 2\pi)$ and $\alpha \in \mathbf{R}$ are coordinates on Λ_0. By shifting this manifold along the flow $g^t_{\mathcal{H}}$, we obtain the family of Lagrangian manifolds $\Lambda_t = g^t_{\mathcal{H}}\Lambda_0$, each of which is equipped with the same coordinate system (ψ, α) as Λ_0. We take the point with coordinates $(\psi, \alpha) = (0, 0)$ for the distinguished point on Λ_0 and construct the Maslov canonical operator $K^h_{\Lambda_t}$ [1, 20] on each of the manifolds Λ_t. (Here $h \to 0$ is the small parameter occurring in the construction of the canonical operator; all Jacobians in the definition of $K^h_{\Lambda_t}$ are taken with respect to the coordinates (ψ, α).)

It follows from the results in [11–14] that the asymptotics of the solutions $\eta_{1,2}$ of problems (3.15) and (3.16) can be obtained as follows. Using the Fourier transforms (3.12) of the equivalent source functions computed in Theorem 3.4, we introduce the following two smooth functions on Λ_t, independent of t and α but depending on the coordinate ψ and an additional parameter ρ:

$$\varphi_1(\psi, \rho) = \widetilde{U}_1(\rho\mathbf{n}(\psi)) = \sqrt{2\pi}\,\mathrm{Re}\,\widetilde{g}_0(\omega\rho)\widetilde{V}(\rho\mathbf{n}(\psi)),$$

$$\varphi_2(\psi, \rho) = \widetilde{U}_2(\rho\mathbf{n}(\psi)) = \sqrt{2\pi}\lambda^{-1}\,\mathrm{Im}\,\frac{\widetilde{g}_0(\omega\rho)}{\omega\rho}\widetilde{V}(\rho\mathbf{n}(\psi)).$$

Then the formulas in [11, 13] give

$$\eta_{1,2}(t) = \sqrt{\frac{\mu}{2\pi}}\,\mathrm{Re}\Big(e^{-i\pi/4}\int_0^\infty K^{\mu/\rho}_{\Lambda_t}(\sqrt{\rho}\,\varphi_{1,2}(\psi, \rho))\,d\rho\Big) + O(\mu). \tag{3.18}$$

Let us find the derivative $\eta_2'(t)$. By the commutation formula [20] for the canonical operator, we have

$$\begin{aligned}\eta_2'(t) &= \sqrt{\frac{\mu}{2\pi}}\,\mathrm{Re}\Big(e^{-i\pi/4}\int_0^\infty \frac{\partial}{\partial t}K^{\mu/\rho}_{\Lambda_t}(\sqrt{\rho}\,\varphi_2(\psi, \rho))\,d\rho\Big)\\ &= \sqrt{\frac{\mu}{2\pi}}\,\mathrm{Re}\Big(e^{-i\pi/4}\int_0^\infty K^{\mu/\rho}_{\Lambda_t}\Big[-\frac{i\rho}{\mu}\mathcal{H}|_{\Lambda_t}\sqrt{\rho}\,\varphi_2(\psi, \rho)\Big]\,d\rho\Big) + O(\mu).\end{aligned}$$

But the Hamiltonian $\mathcal{H}$ is preserved along the trajectories of the Hamiltonian system, and hence $\mathcal{H}\big|_{\Lambda_t} = \mathcal{H}\big|_{\Lambda_0} = c(\alpha\mathbf{n}(\psi))$. It was shown in [11] that, modulo lower-order terms, one can set $\alpha = 0$ in the functions on Λ_t. Taking into account the definition of φ_2, we obtain

$$\eta_2'(t) = \sqrt{\mu}\,\mathrm{Re}\biggl(e^{-\frac{i\pi}{4}} \int_0^\infty K_{\Lambda_t}^{\mu/\rho}\bigl[\sqrt{\rho}\bigl(-i\,\mathrm{Im}\,\widetilde{g}_0(\omega\rho)\widetilde{V}(\rho\mathbf{n}(\psi))\bigr)\bigr]\,d\rho\biggr) + O(\mu).$$

Finally, we use the formula $\eta_{\mathrm{prop}}(t) = \eta_1(t) + \eta_2'(t)$ and arrive at the following theorem.

Theorem 3.7. *The propagating part of the solution has the following asymptotics*:

$$\eta_{\mathrm{prop}}(t) = \sqrt{\mu}\,\mathrm{Re}\biggl(e^{-\frac{i\pi}{4}} \int_0^\infty K_{\Lambda_t}^{\mu/\rho}\bigl[\sqrt{\rho}\,\overline{\widetilde{g}_0}\,(\omega\rho)\widetilde{V}(\rho\mathbf{n}(\psi))\bigr]\,d\rho\biggr) + R(t), \qquad (3.19)$$

where the bar stands for complex conjugation and the remainder satisfies the estimates

$$\|R(t)\|_1 = O(\mu), \qquad \|R'(t)\| = O(\mu) \qquad (3.20)$$

uniformly on any finite interval of time t.

3.3.2. Asymptotics near the front. Now let us compute the propagating part (3.19) of the solution in more explicit terms. To this end, we need some geometry. Let $(P(t,\psi), X(t,\psi))$, $\psi \in [0.2\pi)$, be the family of solutions of the Hamiltonian system (3.17) with the initial conditions

$$p|_{t=0} = \mathbf{n}(\psi), \quad x|_{t=0} = 0. \qquad (3.21)$$

For each t, the equations $p = P(t,\psi)$, $x = X(t,\psi)$, $\psi \in [0.2\pi)$, define a smooth closed curve Γ_t in the four-dimensional phase space $\mathbf{R}^4_{x,p}$; this curve is called the *wave front* in $\mathbf{R}^4_{x,p}$. The projection $\gamma_t = \{x = X(t,\psi)\colon \psi \in [0.2\pi)\}$ of Γ_t into $\mathbf{R}^2_x$ is called the *front in the configuration space.* In contrast to Γ_t, the curve γ_t may well be nonsmooth; namely, it may have turning (or focal) points (in this case, $X_\psi = 0$ for some ψ) and points of self-intersection. Moreover, the front γ_0 at the initial time $t = 0$ is just the point $x = 0$.

For each t, the function (3.19) is localized in a neighborhood of the front γ_t [11–16]. Formula (3.19) provides the global asymptotics of the propagating part of the solution; i.e., this formula holds both near regular and near focal points of the front. The formula can be simplified in a neighborhood of any point of the front, but the simplified expression depends on whether the point is regular or focal. Here we restrict ourselves to the case of a neighborhood of a regular point.

Take some time t and angle ψ^0 and assume that the point $X(t,\psi^0) \in \gamma_t$ is not focal; i.e., $X_\psi(t,\psi^0) \neq 0$. In some neighborhood of $X(t,\psi^0)$, we can introduce the local coordinates (ψ, y), where $y = y(x,t)$ is the (signed) distance between the point x and the front and $\psi = \psi(x,t)$ is determined by the condition that the vector $x - X(t,\psi(x,t))$ is orthogonal to the vector tangent to the wave front at the point $X(t,\psi(x,t))$; in other words

$$\langle x - X(t,\psi(x,t)), X_\psi(t,\psi(x,t))\rangle = 0.$$

Set

$$S(x,t) = \langle P(t,\psi(x,t)), x - X(t,\psi(x,t))\rangle.$$

Next, we introduce the *Morse index* $m(t,\psi^0)$ of the trajectory $X(\tau,\psi^0)$, $\tau\in(0,t]$, which is the number of zeros of the function $|X_\psi(\tau,\psi^0)|$ on the half-open interval $\tau\in(0,t]$ [20].

It may happen that some region of points x where we intend to write out the asymptotics simultaneously belongs to several neighborhoods of the above-mentioned type, where the corresponding points $X(t,\psi^0)$ lie on several (but finitely many!) distinct arcs of the front γ_t. (For example, this is the case if we study the asymptotics near a point of self-intersection of the front γ_t.) Then all these arcs contribute to the asymptotics at such points x, and we use an additional subscript j to distinguish these neighborhoods as well as all associated objects (ψ^0, $\psi(x,t)$, $S(x,t)$, Morse index, etc.). Now from the results in [11, 12, 15, 16] we obtain the following theorem.

Theorem 3.8. *In a neighborhood of the front γ_t but outside a neighborhood of the focal points, the asymptotic formula* (3.19) *for the propagating part of the solution can be rewritten in the form*

$$\eta_{\text{prop}}(t)=\sqrt{\mu}\,\mathrm{Re}\sum_j\left[\frac{e^{-i\pi m(\psi_j^0,t)/2}}{\sqrt{|X_\psi(\psi,t)|}}\sqrt{\frac{c_0}{c(X(\psi,t))}}F\left(\frac{S_j(x,t)}{\mu},\psi\right)\right]_{\psi=\psi_j(x,t)}+R(t),\tag{3.22}$$

where

$$F(z,\psi)=e^{-i\pi/4}\int_0^\infty\sqrt{\rho}\,\bar{\tilde{g}}_0(\omega\rho)\widetilde{V}(\rho\mathbf{n}(\psi))e^{iz\rho}\,d\rho,\tag{3.23}$$

$R(t)$ satisfies the estimate (3.20), *and the sum with respect to j is taken over all distinct arcs of γ_t contributing to the asymptotics at x.*[5]

Remark 3.9. The factor

$$\frac{1}{\sqrt{|X_\psi(\psi,t)|}}\sqrt{\frac{c_0}{c(X(\psi,t))}}$$

includes the two-dimensional analog of the so-called Green law and the trajectory divergence related to the velocity $c(x)$ (with height $c^2(x)$ describing the bottom topography). The function F depends on the time and space shape of the source generating the waves [11–16]. Formulas (3.19), (3.22), and (3.23) apply to any localized perturbation.

4. Obtaining asymptotic expansions by noncommutative analysis

The aim of this section is to prove Lemmas 3.2 and 3.5. Vaguely speaking, these lemmas state that the replacement of the operator D by the operator $D^{(0)}$ (3.4) with constant coefficients in certain expressions results in an $O(\mu)$ error. However, it is much easier to deal with functions of the differential operators L and $L^{(0)}$ than with functions of their square roots, the pseudodifferential operators $D=\sqrt{L}$ and

[5] More formally, for example, fix an $\varepsilon>0$; the intersection of γ_t with the ε-neighborhood of x can be covered by finitely many arcs of length $\le\varepsilon$; take the contributions of all these arcs.

$D^{(0)} = \sqrt{L^{(0)}}$. Hence in Section 4.1 we represent the latter functions via the former and accordingly restate the lemmas. In Section 4.2, we make all noncommutative computations.

4.1. Eliminating the square roots

Lemma 4.1. *The functions* $\operatorname{Re} G_0(\xi, t)$ *and* $\xi^{-1} \operatorname{Im} G_0(\xi, t)$ *are smooth even functions of* ξ *and hence smooth functions of* ξ^2.

Proof. The function $g_0(\tau)$ is real valued, and $\overline{G_0(\xi, t)} = G_0(-\xi, t)$ by (2.32). Thus, $\operatorname{Re} G_0(-\xi, t) = \operatorname{Re} G_0(\xi, t)$ and $\operatorname{Im} G_0(-\xi, t) = -\operatorname{Im} G_0(\xi, t)$; i.e., the real part of G_0 is an even function of ξ, and the imaginary part of G_0 is an odd function of ξ. Hence the desired claim follows. □

Now let us introduce the functions

$$\begin{aligned} f_1(\xi) &= \operatorname{Re} G_0(\xi^{1/2}, 0), \\ f_2(\xi) &= \xi^{-1/2} \operatorname{Im} G_0(\xi^{1/2}, 0), \\ f_3(\xi, t) &= \operatorname{Re} G_0(\xi^{1/2}, t). \end{aligned} \tag{4.1}$$

By Lemma 4.1, these functions are smooth for all ξ, including $\xi = 0$. Formulas (3.10) for the equivalent sources and (2.27) for the transient solution component can now be rewritten as

$$\begin{aligned} \eta_{10} &= f_1(\lambda^{-2} L) V\left(\frac{x}{\mu}\right), \\ \eta_{20} &= \lambda^{-1} f_2(\lambda^{-2} L) V\left(\frac{x}{\mu}\right), \\ \eta_{\text{trans}}(t) &= -f_3(\lambda^{-2} L, \lambda t) V\left(\frac{x}{\mu}\right). \end{aligned} \tag{4.2}$$

Indeed, for example,

$$\begin{aligned} \lambda^{-1} f_2(\lambda^{-2} L) &= \lambda^{-1} (\lambda^{-2} L)^{-1/2} \operatorname{Im} G_0\big((\lambda^{-2} L)^{1/2}, 0\big) \\ &= D^{-1} \operatorname{Im} G_0(\lambda^{-1} D, 0) = D^{-1} \operatorname{Im} G(D, 0) = \sqrt{2\pi} D^{-1} \operatorname{Im} \widetilde{g}(D). \end{aligned}$$

The following theorem is an equivalent restatement of Lemmas 3.2 and 3.5 in terms of functions of L and $L^{(0)}$. (We write $R_3(t) = -R(t)$ to unify the notation.)

Theorem 4.2. *One has*

$$\begin{aligned} f_1(\lambda^{-2} L) V\left(\frac{x}{\mu}\right) &= f_1(\lambda^{-2} L^{(0)}) V\left(\frac{x}{\mu}\right) + R_1, \\ \lambda^{-1} f_2(\lambda^{-2} L) V\left(\frac{x}{\mu}\right) &= \lambda^{-1} f_2(\lambda^{-2} L^{(0)}) V\left(\frac{x}{\mu}\right) + R_2, \\ f_3(\lambda^{-2} L, \lambda t) V\left(\frac{x}{\mu}\right) &= f_3(\lambda^{-2} L^{(0)}, \lambda t) V\left(\frac{x}{\mu}\right) + R_3(t), \end{aligned} \tag{4.3}$$

where the remainders satisfy the estimate

$$\begin{aligned} \|R_1\|_1 &= O(\mu), & \|R_2\|_2 &= O(\mu), \\ \|R_3(t)\|_1 &= O(\mu e^{-\nu\lambda t}), & \|R_3'(t)\| &= O(\mu^2 \lambda e^{-\nu\lambda t}). \end{aligned} \tag{4.4}$$

The proof will be given below in Section 4.2.

We need some estimates for the symbols (4.1). These are provided by the following lemma.

Lemma 4.3. *The following estimates hold for the functions* (4.1):

$$\begin{gathered} |f_1^{(k)}(\xi)| \le C_{k0}(1+|\xi|)^{-1-k}, \qquad |f_2^{(k)}(\xi)| \le C_{km}(1+|\xi|)^{-3/2-k}, \\ \left| \frac{\partial^{k+m} f_3(\xi,t)}{\partial x^k \partial t^m} \right| \le C_{km} e^{-\nu t}(1+|\xi|)^{-1/2-k}, \end{gathered} \tag{4.5}$$

$k = 0, 1, 2, \dots$, *where the* C_{km} *are some constants* (*in general, different from those introduced earlier*).

Proof. For $k = 0$, the desired estimates (4.5) readily follow from (2.33) and (2.34); it suffices to replace ξ by $\xi^{1/2}$ (and use the fact that the functions f_j given by (4.1) are smooth and in particular continuous at $\xi = 0$). Next, note that if $f(\xi) = F(\xi^{1/2})$, where $F(\zeta)$ is a smooth even function, then $f'(\xi) = \Psi(\xi^{1/2})$, where $\Psi(\zeta) = \frac{1}{2}F'(\zeta)/\zeta$ is again a smooth even function. Thus, it suffices to prove that if a smooth even function F satisfies estimates of the form

$$|F^{(k)}(\zeta)| \le d_k(1+|\zeta|)^{-k-k_0}, \qquad k = 0, 1, 2, \dots,$$

for some k_0, then Ψ satisfies the same estimates but with k_0 increased by 2 and with new constants d_k, each of which is a finite linear combination of the old ones. This is trivial for $|\zeta| \ge 1$, and in the region $|\zeta| < 1$ one can use the identity

$$\zeta^{-1}F'(\zeta) = \zeta^{-1}\big(F'(\zeta) - F'(0)\big) = \int_0^1 F''(\vartheta\zeta)\, d\theta. \qquad \square$$

4.2. Computation of the transient part and the equivalent sources

Now we will prove Lemmas 3.2 and 3.5 by proving the equivalent Theorem 4.2. Let $f(\xi)$ be any of the functions $f_1(\xi)$, $f_2(\xi)$, and $f_3(\xi, t)$ given by (4.1) or the function $f_4(\xi, t) = \partial f_3(\xi, t)/\partial t$. We need to compute the difference

$$\mathcal{R} \equiv \mathcal{R}(x, \lambda, \mu) = \big(f(\lambda^{-2}L) - f(\lambda^{-2}L^{(0)})\big) V\left(\frac{x}{\mu}\right) \tag{4.6}$$

and estimate it in an appropriate norm. Let us make the change of variables $x = \mu y$. In the new variables, (4.6) becomes

$$\mathcal{R} = \big(f(L_y) - f(L_y^{(0)})\big) V(y), \tag{4.7}$$

where

$$L_y = -\omega^2 \nabla_y \frac{c^2(\mu y)}{c_0^2} \nabla_y, \qquad L_y^{(0)} = -\omega^2 \nabla_y^2, \qquad \nabla_y = \left(\frac{\partial}{\partial y_1}, \frac{\partial}{\partial y_2}\right),$$

and $\omega = c_0/(\lambda\mu)$ is bounded by condition (2.6).

To compute this difference, we use the machinery of noncommutative analysis. We refer the reader to [1, 2] for details concerning the definition and properties of functions of noncommuting operators and only recall that a function $F(\overset{1}{A}_1,\dots,\overset{n}{A}_n)$ of (possibly, noncommuting) operators $A_1,\dots,A_n$ can be defined as follows in the particular case where the A_j are the generators of uniformly bounded strongly continuous one-parameter operator groups e^{iA_jt}, $t\in\mathbf{R}$, on a Hilbert space H:

$$F(\overset{1}{A}_1,\dots,\overset{n}{A}_n)u=\frac{1}{(2\pi)^{n/2}}\int_{\mathbf{R}^n}\widetilde{F}(t_1,t_2,\dots,t_n)e^{iA_nt_n}\cdots e^{iA_1t_1}u\,dt_1\cdots dt_n,$$

$u\in H$, where $\widetilde{F}$ is the Fourier transform of the symbol F, which is assumed to satisfy certain conditions (e.g., see [2]) guaranteeing that the integral on the right-hand side is well defined. The numbers (Feynman indices) over operators indicate the order of their action: of any two operators, the operator with the smaller Feynman index stands to the right of the operators with the larger Feynman index in products.

It follows by the zero-order Newton formula of noncommutative analysis (see [1] and [2, Theorem I.8]) that

$$f(L_y)-f(L_y^{(0)})=\frac{\delta f}{\delta\xi}(\overset{3}{L}_y,\overset{1}{L_y^{(0)}})\overset{2}{\overline{(L_y-L_y^{(0)})}}=\frac{\delta f}{\delta\xi}(\overset{3}{L}_y,\overset{1}{L_y^{(0)}})\overset{2}{T},\tag{4.8}$$

where

$$\frac{\delta f}{\delta\xi}(\xi_1,\xi_2)=\frac{f(\xi_1)-f(\xi_2)}{\xi_1-\xi_2}$$

is the first difference quotient of f and

$$T=L_y-L_y^{(0)}=\omega^2\Big\langle\nabla_y,\Big(1-\frac{c^2(\mu y)}{c_0^2}\Big)\nabla_y\Big\rangle\equiv\omega^2\sum_{j=1}^2\frac{\partial}{\partial y_j}\phi(\mu y)\frac{\partial}{\partial y_j}.$$

Here we have denoted

$$\phi(z)=1-\frac{c^2(z)}{c_0^2};$$

this function is uniformly bounded together with all of its derivatives for $z\in\mathbf{R}^2$, and $\phi(0)=0$.

We further transform the right-hand side of (4.8) as follows.

Proposition 4.4.

$$\frac{\delta f}{\delta\xi}(\overset{3}{L}_y,\overset{1}{L_y^{(0)}})\overset{2}{T}=\frac{\delta f}{\delta\xi}(\overset{3}{L}_y,\overset{2}{L_y^{(0)}})\overset{1}{T}+\frac{\delta^2 f}{\delta\xi^2}(\overset{4}{L}_y,\overset{3}{L_y^{(0)}},\overset{1}{L_y^{(0)}})\overset{2}{\overline{[T,L_y^{(0)}]}},\tag{4.9}$$

where

$$\frac{\delta^2 f}{\delta\xi^2}(\xi_1,\xi_2,\xi_3)=\frac{\frac{\delta f}{\delta\xi}(\xi_1,\xi_2)-\frac{\delta f}{\delta\xi}(\xi_1,\xi_3)}{\xi_2-\xi_3}$$

is the second difference quotient of f and

$$[T, L_y^{(0)}] = TL_y^{(0)} - L_y^{(0)}T$$

is the commutator of T and $L_y^{(0)}$.

Proof. The proof of (4.9) mimics the derivation of the general commutation formula [2, Proposition I.3]

$$\overset{2}{A}f(\overset{1}{B}) - \overset{1}{A}f(\overset{2}{B}) = \frac{\delta f}{\delta \xi}(\overset{1}{B}, \overset{3}{B})\overline{[A, B]}^{\,2} \tag{4.10}$$

of noncommutative analysis, with T playing the role of A and $L_y^{(0)}$ playing the role of B. Recall this derivation (e.g., see [2, pp. 52–53]). We need to compute

$$[A, f(B)] \equiv Af(B) - f(B)A = \overset{2}{A}f(\overset{1}{B}) - \overset{1}{A}f(\overset{2}{B}).$$

The Feynman indices can be chosen independently for either summand on the right, and we can write

$$[A, f(B)] = \overset{2}{A}f(\overset{1}{B}) - \overset{2}{A}f(\overset{3}{B}) = \overset{2}{A}(f(\overset{1}{B}) - f(\overset{3}{B})) = \overset{2}{A}(\overset{1}{B} - \overset{3}{B})\frac{\delta f}{\delta \xi}(\overset{1}{B}, \overset{3}{B}).$$

(Here we have used the identity $f(x) - f(\xi) = (x - \xi)\frac{\delta f}{\delta \xi}(x, \xi)$, which is in fact the definition of $\delta f/\delta \xi$.) Next, we move apart the Feynman indices over the B's, thus obtaining

$$\overset{2}{A}(\overset{1}{B} - \overset{3}{B})\frac{\delta f}{\delta \xi}(\overset{1}{B}, \overset{3}{B}) = \overset{2}{A}(\overset{1}{B} - \overset{3}{B})\frac{\delta f}{\delta \xi}(\overset{0}{B}, \overset{4}{B}) = \overline{[A, B]}^{\,2}\frac{\delta f}{\delta \xi}(\overset{0}{B}, \overset{4}{B}) = \overline{[A, B]}^{\,2}\frac{\delta f}{\delta \xi}(\overset{1}{B}, \overset{3}{B}).$$

(In the middle, we have written $\overset{2}{A}(\overset{1}{B} - \overset{3}{B}) = \overline{[A, B]}^{\,2}$ using the fact that no other operators in the expression have Feynman indices in the interval $[1, 3]$.) Thus, we arrive at the desired commutation formula (4.10).

The derivation of (4.9) differs from this only in that now, instead of $f(\overset{1}{B})$, we have $\frac{\delta f}{\delta \xi}(\overset{3}{L_y}, \overset{1}{L_y^{(0)}})$; i.e., there is an additional operator argument, $\overset{3}{L_y}$, but this argument does not invalidate the computation, because its Feynman number does not lie between those of $A = T$ and $B = L_y^{(0)}$. □

Let us evaluate the commutator $[T, L_y^{(0)}]$.

Proposition 4.5. *One has*

$$[T, L_y^{(0)}] = \mu T_1, \quad \textit{where} \quad \|T_1 \colon H^s(\mathbf{R}_y^2) \longrightarrow H^{s-3}(\mathbf{R}_y^2)\| \le C_s$$

for all s with some constants C_s independent of μ as $\mu \to 0$.

Proof. We have

$$\begin{aligned}[T, L_y^{(0)}] &= \omega^4 \langle \nabla_y, [\nabla_y^2, \phi(\mu y)] \\ &= \mu\omega^4 \sum_{j=1}^{2} \Big\langle \nabla_y, \Big(2\phi'_{z_j}(\mu y) \frac{\partial}{\partial y_j} + \mu \phi''_{z_j z_j}(\mu y) \Big) \nabla_y \Big\rangle,\end{aligned}$$

and it remains to recall that $\phi(z)$ is uniformly bounded together with all derivatives. □

By Propositions 4.4 and 4.5, we can write

$$f(L_y) - f(L_y^{(0)}) = \frac{\delta f}{\delta \xi}(\overset{3}{L_y}, \overset{2}{L_y^{(0)}})\overset{1}{T} + \mu \frac{\delta^2 f}{\delta \xi^2}(\overset{4}{L_y}, \overset{3}{L_y^{(0)}}, \overset{1}{L_y^{(0)}})\overset{2}{T_1}.$$

Accordingly,

$$\mathcal{R} = \big(f(L_y) - f(L_y^{(0)})\big)V = AW + \mu BV, \tag{4.11}$$

where

$$W = TV, \qquad A = \frac{\delta f}{\delta \xi}(\overset{2}{L_y}, \overset{1}{L_y^{(0)}}), \qquad B = \frac{\delta^2 f}{\delta \xi^2}(\overset{4}{L_y}, \overset{3}{L_y^{(0)}}, \overset{1}{L_y^{(0)}})\overset{2}{T_1}. \tag{4.12}$$

Let us estimate the expression (4.11) for $f = f_j$, $j = 1, 2, 3, 4$.

Proposition 4.6. *One has $V \in H^s(\mathbf{R}_y^2)$ for every s.*

Proof. This follows from the estimates (2.4). □

Proposition 4.7. *For every s, one has $W \in H^s(\mathbf{R}_y^2)$ and*

$$\|W\|_{H^s(\mathbf{R}_y^2)} = O(\mu), \qquad \mu \to 0.$$

Proof. We have

$$\phi(\mu y) = \mu \langle F(\mu y), y \rangle,$$

where the vector function

$$F(z) = \int_0^1 \frac{\partial \phi}{\partial z}(\theta z)\, d\theta$$

is bounded together with all derivatives, and hence for the function $W = TV$ we obtain

$$W(y) = \mu\omega^2 \bigg(\frac{\partial}{\partial y_1} \langle F(\mu y), y \rangle \frac{\partial V(y)}{\partial y_1} + \frac{\partial}{\partial y_2} \langle F(\mu y), y \rangle \frac{\partial V(y)}{\partial y_2} \bigg).$$

Since, by virtue of the estimates (2.4), the function $y_j \partial V(y)/\partial y_k$ lies in $H^s(\mathbf{R}_y^2)$ for every s, we arrive at the desired assertion. □

Proposition 4.8. *Let $f = f_j$, $j = 1, 2, 3, 4$. Then for each $s \in \mathbf{R}$ there exists a constant C_s independent of $\mu \to 0$ such that*

$$\big\|A\colon H^s(\mathbf{R}_y^2) \to H^s(\mathbf{R}_y^2)\big\| \le C_s, \qquad \big\|B\colon H^s(\mathbf{R}_y^2) \to H^{s-3}(\mathbf{R}_y^2)\big\| \le C_s$$

for $j = 1, 2$,

$$\left\|A\colon H^s(\mathbf{R}_y^2) \to H^s(\mathbf{R}_y^2)\right\| \le C_s e^{-\nu t}, \ \left\|B\colon H^s(\mathbf{R}_y^2) \to H^{s-3}(\mathbf{R}_y^2)\right\| \le C_s e^{-\nu t}$$

for $j = 3, 4$.

Proof. We make use of the following representation of the kth difference quotient:

$$\begin{aligned}\frac{\delta^k f}{\delta \xi^k}(\xi_1, \dots, \xi_{k+1}) &= \int_{\Delta_k} f^{(k)}(\theta_1\xi_1 + \cdots + \theta_{k+1}\xi_{k+1})\, d\theta_1 \cdots d\theta_k \\ &= \frac{1}{\sqrt{2\pi}} \int_{\Delta_k} \left(\int_{-\infty}^{\infty} \widetilde{f^{(k)}}(p) e^{ip(\theta_1\xi_1 + \cdots + \theta_{k+1}\xi_{k+1})} dp \right) d\theta_1 \cdots d\theta_k,\end{aligned}$$

where $\widetilde{f^{(k)}}(p)$ is the Fourier transform of the kth derivative $f^{(k)}(\xi)$ and

$$\Delta_k = \{(\theta_1, \dots, \theta_{k+1} \in \mathbf{R}^{k+1} \colon \theta_1 + \cdots + \theta_{k+1} = 1,\ \theta_j \ge 0,\ j = 1, \dots, k+1\}$$

is the standard k-simplex. Hence

$$\begin{aligned}\frac{\delta f}{\delta \xi}(\overset{2}{L_y}, \overset{1}{L_y^{(0)}}) &= \frac{1}{\sqrt{2\pi}} \int_{\Delta_1} \left(\int_{-\infty}^{\infty} \widetilde{f}'(p) e^{ip\theta_1 L_y} e^{ip\theta_2 L_y^{(0)}} dp \right) d\theta_1, \\ \frac{\delta^2 f}{\delta \xi^2}(\overset{4}{L_y}, \overset{3}{L_y^{(0)}}, \overset{1}{L_y^{(0)}})\overset{2}{T_1} & \\ &= \frac{1}{\sqrt{2\pi}} \int_{\Delta_2} \left(\int_{-\infty}^{\infty} \widetilde{f}''(p) e^{ip\theta_1 L_y} e^{ip\theta_2 L_y^{(0)}} T_1 e^{ip\theta_3 L_y^{(0)}} dp \right) d\theta_1 d\theta_2.\end{aligned} \tag{4.13}$$

Let us estimate the operators (4.13). To this end, we use the following lemma.

Lemma 4.9. *For each* s, *there exists a constant* $\widetilde{C}_s$ *independent of* $\mu \to 0$ *such that*

$$\left\|e^{itL_y} \colon H^s(\mathbf{R}_y^2) \to H^s(\mathbf{R}_y^2)\right\| \le \widetilde{C}_s, \qquad \left\|e^{itL_y^{(0)}} \colon H^s(\mathbf{R}_y^2) \to H^s(\mathbf{R}_y^2)\right\| \le \widetilde{C}_s$$

for all $t \in \mathbf{R}$.

Proof. For $s = 0$, the claim is obvious, because the operators $L_y^{(0)}$ and L_y are self-adjoint in $L^2(\mathbf{R}_y)$. For other values of s, one equips $H^s(\mathbf{R}_y^2)$ with the equivalent norm $\left\|(1 + L_y)^{s/2} u\right\|$, so that the operator L_y becomes self-adjoint. This norm depends on the parameter μ, but it is not hard to prove (for positive integer s by a straightforward computation, and for other s by duality and interpolation) that the constants in the inequalities specifying the equivalence of norms remain bounded as $\mu \to 0$. The argument for $L_y^{(0)}$ is simpler, because the parameter μ is not involved. The proof of Lemma 4.9 is complete. □

Now we can finish the proof of Proposition 4.8. If $f = f_1, f_2, f_3$, or f_4, then it follows from Lemma 4.3 that the Fourier transforms of f' and f'' belong to $L^1(\mathbf{R})$, and in the case of f_3 and f_4 the L^1-norm decays as $e^{-\nu t}$. By combining this with Lemma 4.9 and with the estimate for T_1 in Proposition 4.5, we arrive at the assertion of Proposition 4.8. □

By applying Propositions 4.6, 4.7, and 4.8 to formulas (4.11) and (4.12), we find that $\mathcal{R} = O(\mu)$ in all $H^s(\mathbf{R}^2_y)$ for $f = f_1$ and $f = f_2$ and $\mathcal{R} = O(\mu e^{-\nu t})$ in all $H^s(\mathbf{R}^2_y)$ for $f = f_3$ and $f = f_4$. Let us finally estimate the remainders R_j in (4.3). We should take into account the additional factor λ^{-1} for $j = 2$ and pass from the variables y to the original variables $x = \mu y$. Since

$$\|u\|_s \equiv \|u\|_{H^s(\mathbf{R}^2_x)} \leq \mu^{1-s} \|u\|_{H^s(\mathbf{R}^2_y)} \qquad \text{for } \mu \leq 1 \text{ and } s > 0, \tag{4.14}$$

we arrive at the desired estimates (4.4). For example, for R_2 we obtain

$$\|R_2\|_2 \leq C\mu\lambda^{-1}\mu^{1-2} = C\lambda^{-1} \leq \frac{C\omega}{c_0}\mu$$

(where the factor λ^{-1} comes from (4.2) and the factor $\mu^{1-2} = \mu^{-1}$ from (4.14) for $s = 2$). The estimates for R_1 and R_3 are similar. The proof of Theorem 4.2 and hence of Lemmas 3.2 and 3.5 is complete. □

5. Examples

In conclusion, let us present two simple examples in which the asymptotics of the solution of the Cauchy problem (2.1), (2.2) with a special right-hand side will be demonstrated. Namely, we use the right-hand side (2.3), $Q(x,t) = \lambda^2 g_0'(\lambda t)V(x/\mu)$, where $V(y) = A(1 + (y_1/b_1)^2 + (y_2/b_2)^2)^{-3/2}$ is the simplest spatial shape factor (2.11) and the function $g_0(\tau)$ is given by one of formulas (a) (a *sine source*) and (b) (a *polynomial source*) in Eq. (2.13).

Recall that the asymptotics of the solution is given by Theorem 3.1, Eqs. (3.1) and (3.2) (the transient solution component) and by Theorem 3.8, Eq. (3.22) (the propagating solution component away from the focal points). The transient component $\eta_{\text{trans}}(x,t)$ and the wave profile $F(z,\psi)$ (see (3.23)) of the propagating component depend only on the right-hand side and on the parameters λ, μ, and ω; they are represented by integrals which, for our choice of the right-hand side, can be evaluated (or, in the case of the transient component, considerably simplified) analytically. The other ingredients of the asymptotic formula (3.22) for the propagating component (the phase functions $S_j(x,t)$, the Lagrangian coordinates $\psi_j(x,t)$, the Morse index $m(\psi_j^0,t)$, and the factors responsible for the Green law and for the trajectory divergence) depend on the solution of the Cauchy problem (3.21) for the Hamiltonian system (3.17), which, except for the simplest cases, should be solved numerically.

Accordingly, our exposition in both examples is as follows. First, we find the function $G_0(\xi,t)$ (2.32), which plays a crucial role in all subsequent calculations. Then we write out the wave profile $F(z,\psi)$ and finally present the expression for the transient component $\eta_{\text{trans}}(x,t)$ of the solution. In the second example, we also numerically compute the trajectories and display snapshots of the solution obtained with the use of *Wolfram Mathematica*.

The calculations are mostly carried out in polar coordinates, so let us rewrite formula (2.12) for the Fourier transform of V in the polar coordinates (ρ, ψ), where $p = \rho\mathbf{n}(\psi)$ with $\mathbf{n}(\psi) = (\cos\psi, \sin\psi)$:

$$\widetilde{V}\big(\rho\mathbf{n}(\psi)\big) = Ab_1b_2e^{-\rho\beta(\psi)}, \quad \text{where } \beta(\psi) \equiv \sqrt{b_1^2\cos^2\psi + b_2^2\sin^2\psi}. \tag{5.1}$$

5.1. The case of a sine source

Let

$$g_0(\tau) = ae^{-\tau}(\sin(\alpha\tau + \phi_0) - \sin\phi_0),$$

where $a = (\alpha^2+1)/(\alpha\cos\phi_0 - \alpha^2\sin\phi_0)$ is a normalizing factor. By evaluating the integral in (2.32), we obtain

$$G_0(\xi, t) = ae^{-t}\Big(\frac{ie^{-i(\alpha t+\phi_0)}/2}{1+i\alpha+i\xi} - \frac{ie^{i(\alpha t+\phi_0)}/2}{1-i\alpha+i\xi} - \frac{\sin\phi_0}{1+i\xi}\Big). \tag{5.2}$$

We see that $G_0(\xi,t)$ is a rational function of ξ. Moreover, a routine computation (which we omit) shows that it can be represented in the form

$$G_0(\xi,t) = \sum_m q_m(t)\big(R_m(\xi^2) + i\xi Q_m(\xi^2)\big), \tag{5.3}$$

where $R_m(\zeta)$ and $Q_m(\zeta)$ are rational functions with real coefficients and with denominators nonvanishing for $\zeta \ge 0$. This is, of course, consistent with the assertion in Lemma 4.1 concerning the parity of the real and imaginary parts of G_0. As to $\widetilde{g}_0(\xi)$, we have

$$\widetilde{g}_0(\xi) = \frac{1}{\sqrt{2\pi}}G_0(\xi,0) = \frac{a}{\sqrt{2\pi}}\Big(\frac{ie^{-i\phi_0}/2}{1+i\alpha+i\xi} - \frac{ie^{i\phi_0}/2}{1-i\alpha+i\xi} - \frac{\sin\phi_0}{1+i\xi}\Big). \tag{5.4}$$

To evaluate the wave profile $F(z,\psi)$ of the propagating solution component, we substitute the functions (5.1) and (5.4) into formula (3.23) and obtain

$$\begin{aligned} F(z,\psi) &= \frac{aAb_1b_2e^{-i\pi/4}}{\sqrt{2\pi}\omega^{3/2}} \\ &\quad\times \int_0^\infty \sqrt{\rho}\Big(\frac{ie^{-i\phi_0}/2}{1+i\alpha-i\rho} - \frac{ie^{i\phi_0}/2}{1-i\alpha-i\rho} - \frac{\sin\phi_0}{1-i\rho}\Big)e^{-\rho\omega^{-1}(\beta(\psi)-iz)}\,d\rho \\ &= \frac{aAb_1b_2e^{-i\pi/4}}{\sqrt{2\pi}\omega^{3/2}}\Big[\frac{i}{2}e^{-i\phi_0}\mathrm{I}_0\big(\omega^{-1}(\beta(\psi)-iz), 1+i\alpha\big) \\ &\quad - \frac{i}{2}e^{i\phi_0}\mathrm{I}_0\big(\omega^{-1}(\beta(\psi)-iz), 1-i\alpha\big) - \mathrm{I}_0\big(\omega^{-1}(\beta(\psi)-iz), 1\big)\sin\phi_0\Big], \end{aligned}$$

where the integral

$$\mathrm{I}_0(C_1, C_2) = \int_0^\infty \frac{\sqrt{\rho}e^{-C_1\rho}\,d\rho}{C_2 - i\rho}, \quad C_1, C_2 \in \mathbf{C},\ \operatorname{Re}C_1 > 0,\ \arg C_2 \ne \frac{\pi}{2}, \tag{5.5}$$

can be expressed via the complementary error function

$$\operatorname{erfc}(w) = \frac{2}{\sqrt{\pi}}\int_w^\infty e^{-v^2}\,dv$$

by the formula

$$\mathrm{I}_0(C_1, C_2) = \frac{i\sqrt{\pi}}{\sqrt{C_1}} + e^{-i\pi/4}\pi\sqrt{C_2}e^{iC_1C_2}\operatorname{erfc}\left(e^{i\pi/4}\sqrt{C_1C_2}\right).$$

To evaluate the transient term of the solution, we substitute the functions (5.1) and (5.2) into (3.2) and obtain

$$\begin{aligned}
&\eta_{\mathrm{trans}}(r\mathbf{n}(\varphi))\\
&= \frac{aAb_1b_2e^{-\lambda t}}{2\pi\omega^2}\int_0^{2\pi} \mathrm{Re}\Big[\frac{i}{z}(\sin(\alpha\lambda t+\phi_0)-\sin\phi_0)+\sin\phi_0\int_0^\infty \frac{e^{-\rho z}d\rho}{\rho-i}\\
&\quad+\frac{\alpha-i}{2}e^{-i(\alpha t+\phi_0)}\int_0^\infty \frac{e^{-\rho z}d\rho}{\rho+\alpha-i}\\
&\quad+\frac{\alpha+i}{2}e^{i(\alpha t+\phi_0)}\int_0^\infty \frac{e^{-\rho z}d\rho}{\rho-\alpha-i}\Big]d\psi+O(\mu)\\
&= \frac{aAb_1b_2e^{-\lambda t}}{2\pi\omega^2}\int_0^{2\pi} \mathrm{Re}\Big[\frac{i}{z}(\sin(\alpha\lambda t+\phi_0)-\sin\phi_0)\\
&\quad+\sin\phi_0\frac{i}{2}e^{-iz}\left(\pi+2i\mathrm{Ci}(z)-2\mathrm{Si}(z)\right)\\
&\quad+\frac{\alpha-i}{2}e^{-i(\alpha t+\phi_0)}e^{(\alpha-i)z}E_1\left((\alpha-i)z\right)\\
&\quad+\frac{\alpha+i}{2}e^{i(\alpha t+\phi_0)}e^{-(\alpha+i)z}E_1\left(-(\alpha+i)z\right)\Big]d\psi+O(\mu),
\end{aligned}$$

where $z = z(r,\varphi,\psi) = \omega^{-1}\left(\beta(\psi) - ir\mu^{-1}\cos(\psi-\varphi)\right)$, $\mathrm{Re}(z) > 0$, and

$$E_1(z) \equiv \int_z^{+\infty}\frac{e^{-t}}{t}dt, \qquad \mathrm{Ci}\,(z) \equiv -\int_z^\infty \frac{\cos t}{t}dt, \quad \mathrm{Si}\,(z) \equiv \int_0^z \frac{\sin t}{t}dt.$$

5.2. The case of a polynomial source

Now let

$$g_0(\tau) = e^{-\tau}P(\tau),$$

where

$$P(\tau) = \sum_{k=1}^n \tfrac{P_k}{k!}\tau^k$$

is a polynomial of degree n with coefficients P_k such that $P_0 = 0$ and $\sum_{k=1}^n P_k = 1$. Let us use formula (2.32) for $G_0(\xi,\tau)$. Since

$$\int_0^\infty e^{-t-\tau-i\xi\tau}(t+\tau)^k\,d\tau = e^{-t}\left(t+i\frac{\partial}{\partial\xi}\right)^k\frac{1}{1+i\xi},$$

it follows that

$$G_0(\xi,t) = e^{-t}P\left(t+i\frac{\partial}{\partial\xi}\right)\frac{1}{1+i\xi}, \quad \widetilde{g}_0(\xi) = \frac{1}{\sqrt{2\pi}}P\left(i\frac{\partial}{\partial\xi}\right)\frac{1}{1+i\xi}, \tag{5.6}$$

and we see that $G_0(\xi, t)$ again has the form (5.3). Using (5.1), (5.6) and (3.23), we evaluate the wave profile of the propagating part of the solution as follows:

$$F(z,\psi) = \frac{Ab_1b_2e^{-i\pi/4}}{\sqrt{2\pi}\omega^{3/2}}\left[P\Big(-\frac{\partial}{\partial C_2}\Big)\mathrm{I}_0(\rho(\beta(\psi)-iz)/\omega, C_2)\right]\Big|_{C_2=1}$$

$$= -i\frac{Ab_1b_2\sqrt{\pi}}{\sqrt{2}\omega^{3/2}}e^{iC_1}\left[P\Big(-\overset{2}{C_1}\Big(\overset{1}{\overline{i+\frac{1}{2C_1}+\frac{d}{dC_1}}}\Big)\Big)\mathrm{erfc}(\sqrt{iC_1})\Big)\right]\Big|_{C_1=\frac{\beta(\psi)-iz}{\omega}},$$

where $\mathbf{I}_0(C_1, C_2)$ is the integral (5.5).

Remark 5.1. In both examples, one can prove that the following asymptotic formulas hold for the functions $F(z,\psi)$ for small ω:

$$F(z,\psi) = \frac{ib_1b_2}{2\sqrt{2}(z+i\beta(\psi))^{3/2}} + O(\omega).$$

This means that for small ω the solution of the inhomogeneous problem (corresponding to "sources stretched in time") passes into the solution of the homogeneous problem (corresponding to "instantaneous sources").

Let us compute the transient term of the solution for the case in which $P(\tau)$ is a second-order polynomial; then

$$G_0(\xi,t) = e^{-t}\Big(\frac{P_2t^2/2+(P_1-P_2)t-P_1}{1+\xi^2} + \frac{2P_2t+2P_1-3P_2}{(1+\xi^2)^2} + \frac{4P_2}{(1+\xi^2)^3}\Big)$$
$$- i\xi e^{-t}\Big(\frac{P_2t^2/2+P_1t}{1+\xi^2} + \frac{2P_2t+2P_1-P_2}{(1+\xi^2)^2} + \frac{4P_2}{(1+\xi^2)^3}\Big).$$

For the transient term, we find

$$\eta_{\text{trans}} = -\lambda^2e^{-\lambda t}\Big[\Big(P_2\lambda^2t^2/2+(P_1-P_2)\lambda t-P_1\Big)\Theta_1\Big(\frac{x}{\mu}\Big)$$
$$+\Big(2P_2\lambda^3t+(2P_1-3P_2)\lambda^2\Big)\Theta_2\Big(\frac{x}{\mu}\Big)+4P_2\lambda^4\Theta_3\Big(\frac{x}{\mu}\Big)\Big],$$

where

$$\Theta_k(y,\mu) = \frac{Ab_1b_2}{2\pi\lambda^{2k}}\int_{\mathbb{R}^2}\frac{e^{i\langle p,y\rangle}e^{-\sqrt{(b_1p_1)^2+(b_2p_2)^2}}}{(1+(\omega|p|)^2)^k}\,dp_1dp_2.$$

If we pass to the polar coordinates by setting $y = r\mathbf{n}(\varphi)$ and $p = \rho\mathbf{n}(\psi)$, then we obtain

$$\Theta_k(r\mathbf{n}(\varphi),\mu) = \frac{Ab_1b_2}{2\pi\lambda^{2k}}\int_0^\infty\int_0^{2\pi}\frac{\rho e^{-\rho(\beta(\psi)-ir\cos(\psi-\varphi))}}{(1+\omega^2\rho^2)^k}\,d\rho d\psi.$$

Here one can evaluate the integral over ρ. For $k = 1, 2, 3$, we obtain

$$\Theta_1(r\mathbf{n}(\varphi),\mu) = \frac{Ab_1b_2}{2\pi\lambda^2\omega^2}\int_0^{2\pi}d\psi\Big(-\cos(z)\mathrm{Ci}\,(z)+\frac{1}{2}\sin(z)\big(\pi-2\mathrm{Si}\,(z)\big)\Big),$$

$$\Theta_2(r\mathbf{n}(\varphi),\mu) = \frac{Ab_1b_2}{8\pi\lambda^2\omega^2}\int_0^{2\pi} d\psi\Big(2 - 2z\sin(z)\mathrm{Ci}(z) - z\cos(z)\big(\pi - 2\mathrm{Si}(z)\big)\Big),$$

$$\Theta_3(r\mathbf{n}(\varphi),\mu) = \frac{Ab_1b_2}{32\pi\lambda^2\omega^2}\int_0^{2\pi} d\psi\Big(4 - z\sin(z)\big(\pi z + 2\mathrm{Ci}(z) - 2z\mathrm{Si}(z)\big) + z\cos(z)(-\pi + 2z\mathrm{Ci}(z) + 2\mathrm{Si}(z))\Big),$$

where $z(\psi) = \omega^{-1}(\beta - ir\cos(\psi - \varphi))$.

An illustration of the solution given by the sum of propagating and transient terms in the second example is shown in Figure 3. Here the propagating part is calculated for the constant velocity $c(x) \equiv c_0 = 1$, and other constants are $b_1 = 1, b_2 = 2, \Lambda = 1, \mu = 0.1, P_1 = 0, P_2 = 1$. The first four snapshots are taken at small times $t = 0.3, 0.7, 1.0, 1.5$ to show how the transient term behaves, and the last three snapshots are taken at large times $t = 1.5, 4.0, 6.5$. At $t = 6.5$, the transient term practically disappears, while the propagating part continues its motion.The function g_0 and the wave profile for $P_1 = -2, P_2 = 3$, and various λ are compared in Figure 4. For small λ, the wave profile has the form that "reproduces" the shape of the function g_0, while for large λ the wave profile is almost the same as for $g_0 = \delta(t)$.

References

[1] V.P. Maslov, *Operational Methods.* Mir, 1973.

[2] V.E. Nazaikinskii, B.Yu. Sternin, and V.E. Shatalov, *Methods of Noncommutative Analysis.* Walter de Gruyter, 1996.

[3] V.P. Maslov and M.V. Karasev, *Nonlinear Poisson Brackets, Geometry and Quantization.* Amer. Math. Soc., 1993.

[4] C.C. Mei, *The Applied Dynamics of Ocean Surface Waves.* World Scientific, 1989.

[5] E.N. Pelinovski, *Hydrodynamics of Tsunami Waves.* Nizhni Novgorod, 1996.

[6] Yu.I. Shokin, L.B. Chubarov, A.G. Marchuk, and K.V. Simonov, *Numerical Experiment in Tsunami Problem.* Nauka, Siberian Branch, 1989.

[7] B.V. Levin, *Tsunami and Seequake in the Ocean.* Nature **5** (1996), 48–61.

[8] S.F. Dotsenko, B.Yu. Sergievskii, and L.V. Cherkasov, *Space Tsunami Waves Generated by Alternating Displacement of the Ocean Surface.* Tsunami Research **1** (1986), 7–14.

[9] S. Wang, *The Propagation of the Leading Wave*, ASCE Specialty Conference on Coastal Hydrodynamics, University of Delaware, June 29–July 1, 1987, pp. 657–670.

[10] S.Ya. Sekerzh-Zenkovich, *Simple Asymptotic Solution to the Cauchy–Poisson Problem for Leading Waves.* Russ. J. Math. Phys. **16**:2 (2009), 215–222.

[11] S. Dobrokhotov, S. Sekerzh-Zenkovich, B. Tirozzi, and T. Tudorovski, *Description of Tsunami Propagation Based on the Maslov Canonical Operator.* Doklady Math. **74**:1 (2006), 592–596.

[12] S.Yu. Dobrokhotov, A.I. Shafarevich, and B. Tirozzi, *Localized Wave and Vortical Solutions to Linear Hyperbolic Systems and Their Application to the Linear Shallow Water Equations.* Russ. J. Math. Phys. **15**:2 (2008), 192–221.

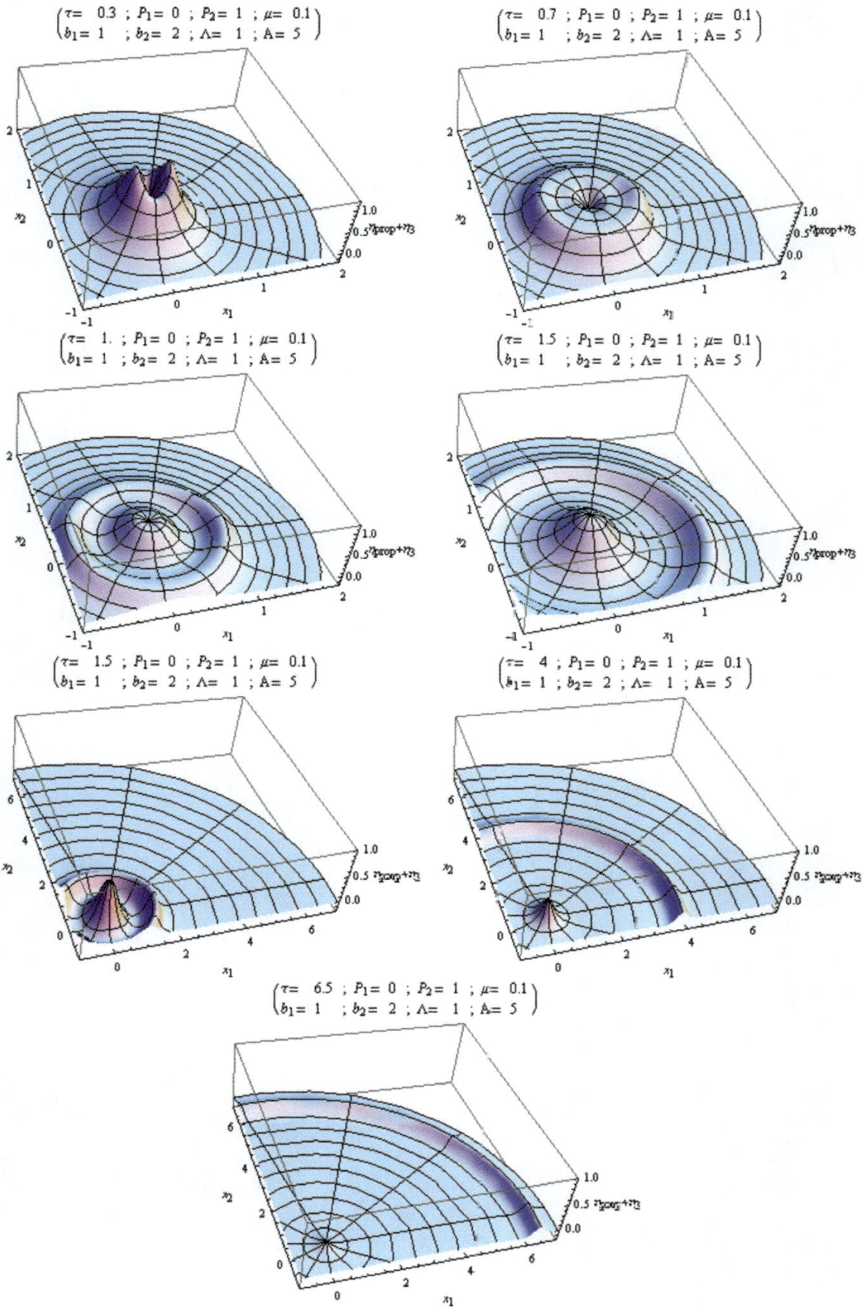

FIGURE 3. Sum of waves $\eta_{\text{prop}} + \eta_{\text{trans}}$.

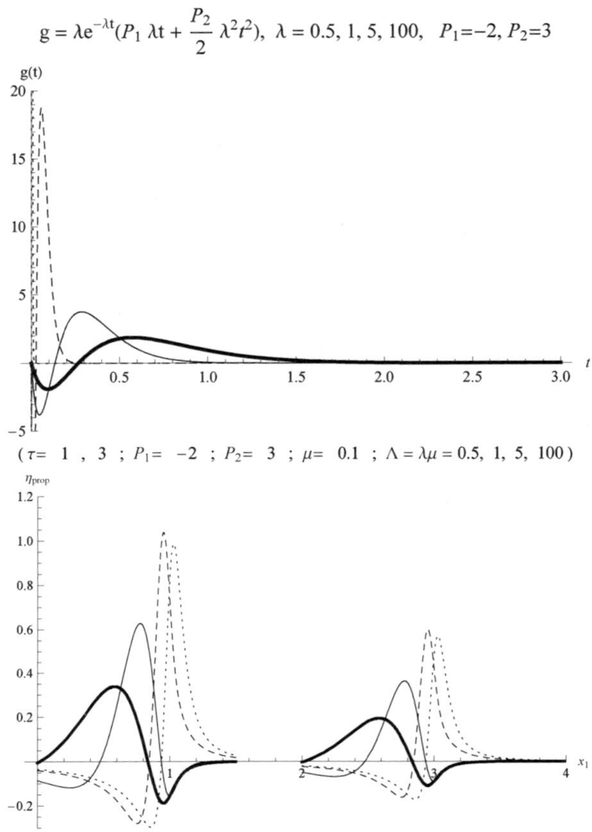

FIGURE 4. Examples of profiles of propagating waves.

[13] S.Yu. Dobrokhotov, B. Tirozzi, and C.A. Vargas, *Behavior Near the Focal Points of Asymptotic Solutions to the Cauchy Problem for the Linearized Shallow Water Equations with Initial Localized Perturbations.* Russ. J. Math. Phys. **16**:2 (2009), 228–245.

[14] S.Yu. Dobrokhotov, R. Nekrasov, and B. Tirozzi, *Asymptotic Solutions of the Linear Shallow-Water Equations with Localized Initial Data.* J. Engng. Math. **69**:2 (2011), 225–242.

[15] S.Yu. Dobrokhotov, S.Ya Sekerzh-Zenkovich, B. Tirozzi, and B. Volkov, *Explicit Asymptotics for Tsunami Waves in Framework of the Piston Model.* Russ. J. Earth Sci. **8** (ES403), 1–12 (2006).

[16] S.Yu. Dobrokhotov, B.I. Volkov, S.Ya. Sekerzh-Zenkovich, and B. Tirozzi, *Asymptotic description of tsunami waves in a frame of piston model: General construction and explicitly solvable models.* Fund. Applied Geophysics **2**:4 (2009), 15–29.

[17] B. Levin, V. Kaistrenko, A. Kharlamov, M. Chepareva, and V. Kryshny, *Physical Processes in the Ocean as Indicators for Direct Tsunami Registration from Satellite.*

Tsunami '93, Wakayama, Japan, August 23–27, 1993. In *Proceedings of the IUGG / IOC International Tsunami Sumposium*, Wakayama, Japan, 1993, pp. 309–320.

[18] B.W. Levin and V.M. Kaistrenko, *Recent Tsunamis and Seaquakes, its Physical Modelling and Satellite Registration.* In *Intern. Union of Geodesy and Geophys. XXI General Assembly. Boulder, Colorado, July* 2–14, 1995*: Abstr. Week A.*, Boulder, 1995, p. A337.

[19] H. Yeh, V. Titov, V. Gusyakov, E. Pelinovsky, V. Khramushib, and V. Kaistrenko, *The* 1994 *Shikotan Earthquake Tsunamis.* PAGEOPH. **144**:3–4 (1995), 856–874.

[20] V.P. Maslov and M.V. Fedoriuk, *Semi-Classical Approximation in Quantum Mechanics.* Reidel, 1981.

[21] D. Bianchi, S. Dobrokhotov, and B. Tirozzi, *Asymptotics of Localized Solutions of the One-Dimensional Wave Equation with Variable Velocity. II: Taking into Account a Source on the Right-Hand Side and a Weak Dispersion.* Russ. J. Math. Phys. **15**:4 (2008), 427–446.

[22] S.Yu. Dobrokhotov, V.E. Nazaikinskii, and B. Tirozzi, *Asymptotic Solutions of 2D Wave Equations with Variable Velocity and Localized Right-Hand Side.* Russ. J. Math. Phys. **17**:1 (2010), 66–76.

[23] S.Yu. Dobrokhotov, V.E. Nazaikinskii, and B. Tirozzi, *Asymptotic Solution of the One-Dimensional Wave Equation with Localized Initial Data and with Degenerating Velocity: I.* Russ. J. Math. Phys. **17**:4 (2010), 428–444.

[24] S.Yu. Dobrokhotov, V.E. Nazaikinskii, and B. Tirozzi, *Asymptotic Solutions of the Wave Equation with Degenerating Velocity and Localized Initial Data (Two-Dimensional Case).* Algebra Analiz **22**:6 (2010), 67–90.

[25] S.Yu. Dobrokhotov and B. Tirozzi, *Localized Solutions of the One-Dimensional Nonlinear System of Shallow Water Equations with Velocity* $c = \sqrt{x}$. Uspekhi Mat. Nauk **65**:1 (391) (2010), 185–186.

[26] M.Sh. Birman and M.Z. Solomyak, *Spectral Theory of Self-Adjoint Operators in Hilbert Space.* Reidel, 1987.

[27] *Wolfram Mathematica®*, www.wolfram.com.

Sergey Dobrokhotov, Dmitrii Minenkov and Vladimir Nazaikinskii
A. Ishlinsky Institute for Problems in Mechanics
Russian Academy of Sciences, Moscow
Moscow Institute of Physics and Technology
e-mail: dobr@ipmnet.ru
minenkov_ds@gmail.com
nazay@ipmnet.ru

Brunello Tirozzi
Department of Physics, University "La Sapienza," Rome
e-mail: brunello.tirozzi@roma1.infn.it

Operator Theory:
Advances and Applications, Vol. 228, 127–150

On Toeplitz and Hankel Operators with Oscillatory Symbols Containing Blaschke Products and Applications to the KdV Equation

Sergei Grudsky and Alexei Rybkin

To Vladimir Rabinovich on the occasion of his 70th birthday

Abstract. We derive an asymptotic formula for the argument of a Blaschke product in the upper half-plane with purely imaginary zeros. We then use this formula to find conditions for the quotient of two such Blaschke products to be continuous on the real line. These results are applied to certain Hankel and Toeplitz operators arising in the Cauchy problem for the Korteweg-de Vries equation. Our main theorems include certain compactness conditions for Hankel operators and invertibility conditions for Toeplitz operators with oscillating symbols containing such quotients. As a by-product we obtain a well-posedness result for the Korteweg-de Vries equation.

Mathematics Subject Classification (2010). 30D30, 45E05, 34B20, 37K15, 47B35.

Keywords. Blaschke product, Hankel operator, Toeplitz operator, KdV equation.

1. Introduction

The theory of Toeplitz operators on Hardy spaces with symbols having discontinuities of the second kind has been in focus of one of the authors (see, e.g., [2–5], [9], [14, 15] and the literature cited therein). The range of symbols under consideration is quite large and varies from discontinuities with rapidly oscillating

The first author is partially supported by PROMEP (México) via "Proyecto de Redes", by the CONACYT grant 102800 (México), and by grant 2012-1.5-12-000-1004-8501 (Russia).

The second author is partially supported by the NSF under grant DMS 1009673.

behavior (oscillations of power, exponential and super-exponential types) to slowly oscillating (e.g., logarithmic). A large variety of generalizations of classical almost periodic symbols has been considered. For example, the so-called α-almost periodic and α-semi-almost periodic symbols have been studied in great detail [3] (see also [4, 6, 7] for matrix-valued analogs). We note that those generalizations are highly non trivial. The main problem is that, as opposed to traditional symbols (continuous or with at most jump discontinuities), the Toeplitz operators with those more general symbols need not be Fredholm, i.e., the kernels and co-kernels may be infinitely dimensional. This raises serious problems: finding criteria for one-sided and generalized invertibility, construction of bases in kernels and co-kernels, to name just two. Addressing these issues has required developing new methods (see monographs [7, 9]). We mention here only the method of the so-called "u-periodic factorizations of symbols". Further development of the theory of Toeplitz and Hankel operators with such symbols would therefore be interesting in its own right due to the nontriviality of its methods.

What is perhaps even more important is that, while the symbols above may look a bit artificially complicated, there are some problems of mathematical physics and partial differential equations where such symbols naturally appear. In particular, a symbol with a cubic oscillation of its argument is a main player in the study of the Cauchy problem for the Korteweg-de Vries (KdV) equation [18–20].

In the present paper we consider Toeplitz and Hankel operators with symbols which besides the cubic oscillation contain quotients of Blaschke products with zeros on the imaginary line. We obtain asymptotics of such Blaschke products and then use them to find some sufficient conditions for continuity of their quotients. We then apply these results to study one-sided invertibility of the corresponding Toeplitz operator and compactness of the Hankel operator. We emphasize that our interest to this circle of problems was stimulated by certain well-posedness issues more related to the Cauchy problem for the KdV equation.

Let us describe our main objects in detail. Consider the Blaschke product in the upper half-plane $\mathbb{C}_+ := \{z \in \mathbb{C} | \operatorname{Im} z > 0\}$

$$B(z) = \prod_{n=1}^{\infty} \frac{z - i\kappa_n}{z + i\kappa_n}, \tag{1.1}$$

with purely imaginary simple zeros such that

$$\kappa_n > \kappa_{n+1} > 0 \quad \text{and} \quad \lim \kappa_n = 0, \quad n \to \infty. \tag{1.2}$$

Such Blaschke products are of course very specific but they do arise in the spectral and scattering theories for Schrödinger operators (see, e.g., [17]). Typically, $i\kappa_n = \sqrt{E_n}$ where E_n is the (negative) nth bound state of a Schrödinger operator.

It is well known (see [10, 16]) that $B(z)$ is convergent for any $z \in \overline{\mathbb{C}}_+ \setminus \{0\}$ if and only if

$$\sum_{n=1}^{\infty} \kappa_n < \infty. \tag{1.3}$$

Of course $B(z)$ is analytic in any neighborhood of a real point x not containing 0. We are specifically concerned with the asymptotic behavior of suitably defined $\arg B(x)$ as $x \to 0$ and conditions providing continuity at $x = 0$ of

$$Q(x) := \frac{B_1(x)}{B_2(x)}, \tag{1.4}$$

where $B_{1,2}(x)$ are two Blaschke products given by (1.1). The results obtained are then applied to the study of Toeplitz and Hankel operators with symbols

$$a(x) = D(x)\, Q(x), \tag{1.5}$$

where either $D \in H_+^\infty + C(\dot{\mathbb{R}})$ or $D \in \overline{H_+^\infty} + C(\dot{\mathbb{R}})$. We recall that H_+^∞ stands for the Hardy space of analytic and bounded functions in the upper half-plane $\mathbb{C}_+$ and $C(\dot{\mathbb{R}})$ is the space of functions continuous on the one point compactification of the real axis $\mathbb{R}$. The class of operators with such symbols is quite broad (see (4.10) below) and includes the Hankel and Toeplitz operators arising in the initial value problem for the Korteweg-de Vries (KdV) equation. We use our results on Hankel and Toeplitz operators to describe some subtle properties of solutions to the KdV equation which we believe cannot be achieved by usual PDE methods. We emphasize that although Hankel operators naturally appear in many other (if not every) so-called completely integrable systems of nonlinear PDEs (see, e.g., [1]), not much from the theory of Hankel and Toeplitz operators have been actually used there so far. We believe that the language of Hankel and Toeplitz operators is quite adequate in the setting of completely integrable systems and the theory of those operators will find more useful applications in integrable systems.

This work is organized as follows. In Section 2 we derive an asymptotic formula for the argument of the Blaschke product (1.1). The sufficient conditions of continuity of the function $Q(x)$ (1.4) at the point $x = 0$ are given in Section 3. Applications to the theory of Toeplitz and Hankel operators with oscillating symbol are considered in Section 4. In Section 5 we apply our results to the theory of the KdV equation.

2. Argument of Blaschke products

Let $B(x)$ be of the form (1.1)–(1.3) and let the branch of $\arctan x$ be chosen such that $\arctan x \in \left(-\frac{\pi}{2}, \frac{\pi}{2}\right)$ for $x \in \mathbb{R}$. We define the Blaschke product (1.1) under conditions (1.2)–(1.3) such that the function

$$B : \dot{\mathbb{R}} \setminus \{0\} \to \mathbb{C}, \quad x \mapsto B(x)$$

is continuous, $B(\infty) = 1$ and $|B(x)| = 1$ for all $x \in \dot{\mathbb{R}} \setminus \{0\}$. So we can choose a branch of $\arg B$ such that $\arg B(x)$ is continuous on $\dot{\mathbb{R}} \setminus \{0\}$ and $\arg B(\infty) = 0$. The following statement is elementary.

Theorem 2.1. *The function* $\arg B(x)$ *is continuously increasing on* $\mathbb{R} \setminus \{0\}$,

$$\arg B(x) = -2 \sum_{n=1}^{\infty} \arctan \frac{\kappa_n}{x}, \quad x \neq 0 \tag{2.1}$$

and

$$\arg B(x) = -\arg B(-x), \quad x \in \mathbb{R}, \tag{2.2}$$

$$\lim_{x \to \pm 0} \arg B(x) = \mp\infty. \tag{2.3}$$

Proof. Since for $\pm x > 0$

$$\arg \frac{x - i\kappa_n}{x + i\kappa_n} = 2 \arg(x - i\kappa_n) = -2 \arctan \frac{\kappa_n}{x},$$

we immediately have (2.1) and (2.2). The series is convergent due to the Blaschke condition (1.3). It follows from

$$\sum_{n=1}^{\infty} \left| \arctan \frac{\kappa_n}{x} \right| > \sum_{\kappa_n > |x|} \left| \arctan \frac{\kappa_n}{x} \right| > \sum_{\kappa_n > |x|} \frac{\pi}{4}$$

that (2.3) holds. The function $-2 \arctan \frac{\kappa_n}{x}$ is clearly increasing on $\mathbb{R}_+ := (0, +\infty)$ and $\mathbb{R}_- := (-\infty, 0)$ respectively and so is $\arg B(x)$. □

With each Blaschke product B of the type (1.1) we associate a function f constructed as follows. Fix a point $\kappa_{1/2} > \kappa_1$ and define f

$$f : [1/2, \infty) \to (0, \kappa_{1/2}], \quad x \mapsto f(x)$$

as a continuous strictly decreasing function that interpolates the points $\{(1/2, \kappa_{1/2}), (1, \kappa_1), (2, \kappa_{2,}), \dots\}$. That is

$$f(1/2) = \kappa_{1/2}, \quad f(n) = \kappa_n, \quad n = 1, 2, \dots. \tag{2.4}$$

We call such f a function associated with a Blaschke product B of the type (1.1). Similarly, given a continuous suitably decreasing function f, we call a Blaschke product B of the type (1.1) satisfying (2.4) a Blaschke product associated with f.

Hypothesis 2.2. Let $B(z)$ be a Blaschke product of the form (1.1)–(1.3) such that:

i) its zeros $\{i\kappa_n\}$ satisfy

$$\lim_{n \to \infty} \frac{\kappa_n - \kappa_{n+1}}{\kappa_n} = 0; \tag{2.5}$$

ii) there exists a continuously differentiable associated function $f(x)$ such that

$$\lim_{n \to \infty} \sup_{-1/2 \leq s \leq 1/2} \frac{|f(n+s) - f(n) + s(\kappa_n - \kappa_{n+1})|}{(\kappa_n - \kappa_{n+1})} = 0. \tag{2.6}$$

Theorem 2.3. *Under Hypothesis* 2.2

$$\arg B(x) = -2 \int_{1/2}^{\infty} \arctan \frac{f(u)}{x} du + o(1), \quad x \to 0. \tag{2.7}$$

Proof. Since the function $\arg B(x)$ is odd, it is enough to consider the case $x > 0$. Let $\epsilon_n(x)$ be the difference

$$\epsilon_n(x) := \arctan \frac{\kappa_n}{x} - \int\limits_{n-1/2}^{n+1/2} \arctan \frac{f(u)}{x} du.$$

It is easy to see that

$$\begin{aligned}
\epsilon_n(x) &= \int\limits_0^{1/2} \Bigg[\left(\arctan \frac{f(n)}{x} - \arctan \frac{f(n+s)}{x} \right) \\
&\quad + \left(\arctan \frac{f(n)}{x} - \arctan \frac{f(n-s)}{x} \right) \Bigg] ds \\
&= \int\limits_0^{1/2} \left[\arctan \frac{x(f(n)-f(n+s))}{x^2 + f(n)f(n+s)} + \arctan \frac{x(f(n)-f(n-s))}{x^2+f(n)f(n-s)} \right] ds \\
&= \int\limits_0^{1/2} [\arctan \delta_n(s,x) + \arctan \delta_n(-s,x)] \, ds,
\end{aligned}$$

where

$$\delta_n(s,x) := \frac{x(f(n)-f(n+s))}{x^2+f(n)f(n+s)}, \quad s \in [-1/2, 1/2].$$

By a direct computation

$$\epsilon_n(x) = \int\limits_0^{1/2} \arctan \frac{\delta_n(s,x) + \delta_n(-s,x)}{1 - \delta_n(s,x)\delta_n(-s,x)} ds.$$

Since $-\delta_n(s,x)\delta_n(-s,x) > 0$, we have

$$|\epsilon_n(x)| \le \int\limits_0^{1/2} |\delta_n(s,x) + \delta_n(-s,x)| ds.$$

For $s \in [-1/2, 1/2]$, we set

$$\begin{aligned}
\Delta_n &:= \kappa_n - \kappa_{n+1}, \\
\Delta_n^{(1)}(s) &:= f(n+s) - f(n), \\
\Delta_n^{(2)}(s) &:= \Delta_n^{(1)}(s) - s\Delta_n^{(1)}(1).
\end{aligned}$$

Note that $\Delta_n^{(1)}(1) = -\Delta_n, \ \Delta_n^{(2)}(1) = 0$ and

$$\Delta_n^{(2)}(s) = f(n+s) - f(n) + s(\kappa_n - \kappa_{n+1}). \tag{2.8}$$

Let us evaluate now

$$\begin{aligned}\delta_n(s,x)+\delta_n(-s,x) = & -x\left\{\frac{\Delta_n^{(2)}(s)}{x^2+f(n)f(n+s)}+\frac{\Delta_n^{(2)}(-s)}{x^2+f(n)f(n-s)}\right.\\ & \left.+\frac{s\Delta_n(f(n+s)-f(n-s))f(n)}{(x^2+f(n)f(n+s))(x^2+f(n)f(n-s))}\right\}\\ = & -x\left\{\frac{\Delta_n^{(2)}(s)}{x^2+f(n)f(n+s)}+\frac{\Delta_n^{(2)}(-s)}{x^2+f(n)f(n-s)}\right.\\ & \left.+\frac{s\Delta_n(\Delta_n^{(1)}(s)-\Delta_n^{(1)}(-s))f(n)}{(x^2+f(n)f(n+s))(x^2+f(n)f(n-s))}\right\}.\end{aligned}$$

Consider two cases: $f(n)\geq x$ and $f(n)<x$. If $f(n)\geq x$, then

$$\begin{aligned}|\delta_n(s,x)+\delta_n(-s,x)|\leq & \frac{x\left|\Delta_n^{(2)}(s)\right|}{f(n)f(n+s)}+\frac{x\left|\Delta_n^{(2)}(-s)\right|}{f(n)f(n-s)}\\ & +\frac{x\,|s|\;\Delta_n(|\Delta_n^{(1)}(s)|+|\Delta_n^{(1)}(-s)|)}{f(n)f(n+s)f(n-s)}.\end{aligned}$$

Since for $s\in[-1/2,1/2]$

$$f(n+s)>f(n+1)$$

and

$$|\Delta_n^{(1)}(|s|)|<\Delta_n,\quad \Delta_n^{(1)}(-|s|)<\Delta_{n-1},$$

one has

$$|\delta_n(s,x)+\delta_n(-s,x)|<\left\{\frac{|\Delta_n^{(2)}(s)|+|\Delta_n^{(2)}(-s)|}{\Delta_n}+\frac{\Delta_n+\Delta_{n-1}}{2f(n)}\right\}\frac{x\Delta_n}{f(n)f(n+1)}.$$

Recalling (2.8), it follows from (2.6) and (2.5) that[1]

$$|\delta_n(s,x)+\delta_n(-s,x)|\lesssim\alpha_n\frac{x\Delta_n}{f(n)f(n+1)},$$

where α_n is independent of s, and $\lim_{n\to\infty}\alpha_n=0$.

If $f(n)<x$ then

$$\begin{aligned}&|\delta_n(s,x)+\delta_n(-s,x)|\\ &\lesssim\frac{|\Delta_n^{(2)}(s)|}{x}+\frac{|\Delta_n^{(2)}(-s)|}{x}+\frac{s\,\Delta_n\left\{|\Delta_n^{(1)}(s)|+|\Delta_n^{(1)}(-s)|\right\}f(n)}{x^3}.\end{aligned}$$

It follows from (2.6) that

$$\sup_{-1/2\leq s\leq 1/2}\frac{|\Delta_n^{(1)}(s)|}{\Delta_n}=\sup_{-1/2\leq s\leq 1/2}\frac{|\Delta_n^{(2)}(s)-s\Delta_n|}{\Delta_n}$$

[1]We write $f\lesssim g$ if $f\leq Cg$ with some $C>0$ independent of arguments of the functions f and g.

is bounded with respect to n and hence

$$\left\{|\Delta_n^{(1)}(s)| + |\Delta_n^{(1)}(-s)|\right\} \frac{sf(n)\Delta_n}{x^3} \lesssim \frac{|\Delta_n^{(1)}(s)| + |\Delta_n^{(1)}(-s)|}{\Delta_n} \left(\frac{\Delta_n}{x}\right)^2 \lesssim \left(\frac{\Delta_n}{x}\right)^2.$$

Therefore

$$|\delta_n(s,x) + \delta_n(-s,x)| \lesssim \frac{|\Delta_n^{(2)}(s)| + |\Delta_n^{(2)}(-s)|}{\Delta_n} \cdot \frac{\Delta_n}{x} + \left(\frac{\Delta_n}{x}\right)^2 \lesssim \beta_n \frac{\Delta_n}{x},$$

where β_n is independent of s and $\lim_{n\to\infty} \beta_n = 0$, and we finally have

$$|\epsilon_n(x)| \lesssim \begin{cases} \alpha_n \dfrac{x\Delta_n}{f(n)f(n+1)}, & f(n) \geq x \\ \beta_n \dfrac{\Delta_n}{x}, & f(n) < x. \end{cases}$$

We now estimate the remainder $\delta(x) := \arg B(x) + 2 \int\limits_{1/2}^{\infty} \arctan \dfrac{f(u)}{x}\, du$

for $x > 0$ small enough. We have

$$\begin{aligned} |\delta(x)| &\lesssim \sum_{n=1}^{\infty} |\epsilon_n(x)| \leq \left\{ \sum_{f(n)\geq\sqrt{x}} + \sum_{x\leq f(n)<\sqrt{x}} + \sum_{f(n)<x} \right\} |\epsilon_n(x)| \\ &\lesssim \sum_{f(n)\geq\sqrt{x}} x \left\{\frac{1}{f(n+1)} - \frac{1}{f(n)}\right\} + \sum_{x\leq f(n)<\sqrt{x}} \sigma_1(x) x \left\{\frac{1}{f(n+1)} - \frac{1}{f(n)}\right\} \\ &\quad + \sum_{f(n)<x} \frac{\sigma_2(x)}{x} \{f(n) - f(n+1)\}, \end{aligned}$$

where

$$\sigma_1(x) := \sup\left\{\alpha_n : x \leq f(n) < \sqrt{x}\right\}, \quad \sigma_2(x) := \sup\left\{\beta_n : f(n) < x\right\}.$$

Thus, we have

$$\begin{aligned} |\delta(x)| &\lesssim x \left(\frac{1}{f(n_1+1)} - \frac{1}{f(1)}\right) \\ &\quad + x\sigma_1(x) \left(\frac{1}{f(n_2+1)} - \frac{1}{f(n_1+1)}\right) + \frac{\sigma_2(x)}{x} f(n_2+1), \end{aligned}$$

where

$$n_1 = \max\left\{n : f(n) \geq \sqrt{x}\right\}, \quad n_2 = \max\left\{n : f(n) \geq x\right\}.$$

It is easy to see that

$$\lim_{x\to 0} \sigma_1(x) = \lim_{x\to 0} \sigma_2(x) = \lim_{x\to 0} \frac{x}{f(n_1+1)} = 0$$

and

$$\lim_{x\to 0} \frac{x}{f(n_2+1)} = 1.$$

Hence $\lim_{x\to 0} \delta(x) = 0$, and the theorem is proved. □

Theorem 2.4. *Under Hypothesis* 2.2

$$\arg B(x) = \frac{\pi}{2}\operatorname{sgn}(x) - 2x\int_0^1 \frac{f^{-1}(v)}{x^2+v^2}dv + o(1), \quad x\to 0, \tag{2.9}$$

where $f^{-1} : (0, \kappa_{1/2}] \to [1/2, \infty)$ *is the inverse function of* f.

Proof. As above we may assume $x > 0$. By Theorem 2.3 for $x \to 0$ one has

$$\begin{aligned}\arg B(x) &= -2\int_{1/2}^{\infty} \arctan\frac{f(u)}{x}du + o(1)\\ &= -2u\arctan\frac{f(u)}{x}\bigg|_{1/2}^{\infty} + 2x\int_{1/2}^{\infty}\frac{uf'(u)du}{x^2+f^2(u)} + o(1)\\ &= \arctan\frac{\kappa_{1/2}}{x} + 2x\int_{1/2}^{\infty}\frac{ud\,f(u)}{x^2+f^2(u)} + o(1).\end{aligned}$$

Here we have used

$$\lim_{u\to\infty} u\arctan\left(\frac{f(u)}{x}\right) = 0, \tag{2.10}$$

that can be easily shown by contradiction. If (2.10) does not hold, then there exists a sequence of positive numbers $\{u_n\}_{n=1}^{\infty}$ such that

$$\lim_{n\to\infty} u_n = \infty \tag{2.11}$$

and

$$u_n f(u_n) \geq \delta > 0. \tag{2.12}$$

By the definition of the function $f(u)$, the integral

$$I := \int_{u_1}^{\infty} f(u)\,du$$

is finite. Since $f(u)$ is a continuous strictly decreasing function it follows from (2.12) that

$$I = \sum_{n=1}^{\infty} \int_{u_n}^{u_{n+1}} f(u)\, du \geq \delta \sum_{n=1}^{\infty} \frac{u_{n+1} - u_n}{u_{n+1}}$$

$$\gtrsim \delta \sum_{n=1}^{\infty} \left| \ln \left(1 - \frac{u_{n+1} - u_n}{u_{n+1}} \right) \right|$$

$$= \delta \ln \prod_{n=1}^{\infty} \frac{u_{n+1}}{u_n}.$$

Thus the last infinite product is convergent, and hence

$$\lim_{N \to \infty} \prod_{n=1}^{N} \frac{u_{n+1}}{u_n} = \lim_{N \to \infty} \frac{u_{N+1}}{u_1}$$

must be finite, which contradicts (2.11). Thus (2.10) holds true.

Changing the variable $v = f(u)$ we continue

$$\arg B(x) = \arctan \frac{\kappa_{1/2}}{x} - 2x \int_0^{\kappa_{1/2}} \frac{f^{-1}(v)dv}{x^2 + v^2} + o(1)$$

$$= -2x \int_0^1 \frac{f^{-1}(v)dv}{x^2 + v^2} + 2x \int_{\kappa_{1/2}}^1 \frac{f^{-1}(v)dv}{x^2 + v^2} + \arctan \left(\frac{\kappa_{1/2}}{x} \right) + o(1).$$

Due to $\lim_{x \to 0} \arctan \frac{\kappa_{1/2}}{x} = \frac{\pi}{2}$ and $\sup \left\{ f^{-1}(v) : v \in [\kappa_{1/2}, 1] \right\} < \infty$ we have

$$2x \left| \int_{\kappa_{1/2}}^1 \frac{f^{-1}(v)dv}{x^2 + v^2} \right| \lesssim x \left| \int_{\kappa_{1/2}}^1 \frac{dv}{x^2 + v^2} \right| = \left| \arctan \frac{1}{x} - \arctan \frac{\kappa_{1/2}}{x} \right|.$$

That is

$$\lim_{x \to 0} 2x \int_{\kappa_{1/2}}^1 \frac{f^{-1}(v)}{x^2 + v^2} dv = 0$$

and (2.9) follows. □

In place of Hypothesis 2.2 we can state somewhat stronger.

Hypothesis 2.5. Let $B(z)$ be a Blaschke product of the form (1.1)–(1.3) that has an associated function $f(x)$ such that $|f'(x)|$ is decreasing and

$$\lim_{n \to \infty} \frac{f^{(l)}(n) - f^{(l)}(n+1)}{f^{(l)}(n)} = 0, \quad l = 0, 1. \tag{2.13}$$

Hypothesis 2.5 implies Hypothesis 2.2. For $l = 0$ condition (2.13) is the same as i) of Hypothesis 2.2 and one only needs to show that (2.13) for $l = 1$ implies (2.6). Indeed, it is easy to see that

$$\Delta_n^{(2)}(s) = f'(n_0)s - f'(n_1)s = (f'(n_0) - f'(n_1))s,$$

with some n_0 and n_1 from $[n, n+1]$ and $[n, n+s]$ respectively. One has

$$\left|\frac{\Delta_n^{(2)}(s)}{\Delta_n}\right| = \left|\frac{f'(n_0) - f'(n_1)}{f'(n_1)}\right| s \leq \left|\frac{f'(n-1) - f'(n+1)}{f'(n+1)}\right|$$
$$\leq \left|\frac{f'(n-1) - f'(n)}{f'(n-1)}\right| \left|\frac{f'(n-1)}{f'(n+1)}\right| + \left|\frac{f'(n) - f'(n+1)}{f'(n)}\right| \left|\frac{f'(n)}{f'(n+1)}\right|.$$

Since $f'(x)$ satisfies (2.13) we immediately conclude that (2.6) holds.

Hypothesis 2.5 is of course much easier to verify and a simple example is in order.

Example. Take

$$f(x) = x^{-\alpha} \ln^{\beta} x, \tag{2.14}$$

where β is any real number if $\alpha > 1$ and $\beta < -1$ if $\alpha = 1$. It follows from

$$f'(x) = -\frac{f(x)}{x}\left(\alpha - \frac{\beta}{\ln x}\right),$$

that $f(x)$ is continuous and decreasing for x large enough. Moreover for some $n_0 \in [n, n+1]$

$$\left|\frac{f(n) - f(n+1)}{f(n)}\right| = \left|\frac{f'(n_0)}{f(n)}\right| \to 0$$

and condition (2.13) for $l = 0$ holds. Similarly using the second derivative of $f(x)$ one verifies that (2.13) holds also for $l = 1$. Therefore any Blaschke product associated with the function (2.14) satisfies Hypothesis 2.5.

Let us demonstrate now how Theorem 2.4 applies in the case of (2.14) with $\alpha > 1$ and $\beta = 0$.

Example. Take

$$f(x) = x^{-\alpha}, \quad \alpha > 1,$$

then $f^{-1}(v) = v^{-1/\alpha}$ and by (2.9) for $x > 0$ we have

$$\arg B(x) = \frac{\pi}{2} - 2x \int_0^1 \frac{v^{-1/\alpha}}{x^2 + v^2}\, dv + o(1) = \frac{\pi}{2} - 2x^{-1/\alpha} \int_0^{1/x} \frac{u^{-1/\alpha}\, du}{1 + u^2} + o(1), \quad x \to 0.$$

Due to the symmetry of $\arg B(x)$ we finally obtain

$$\arg B(x) = \left(\frac{\pi}{2} - c|x|^{-\frac{1}{\alpha}}\right) \operatorname{sgn}(x) + o(1), \quad x \to 0,$$

where $c := 2 \int_0^{\infty} \frac{u^{-1/\alpha}}{1 + u^2}\, du$.

3. Quotient of Blaschke products

In this section we consider the continuity of the quotient $Q(x) = B_1(x)/B_2(x)$ of Blaschke products $B_{1,2}(x)$ subject to Hypothesis 2.2. More specifically, we study conditions on $B_{1,2}$ providing continuity of $\arg Q(x)$ as $x = 0$. The following statement is the main result of this section.

Theorem 3.1. *Let $B_{1,2}$ be subject to Hypothesis 2.2 and $f_{1,2}$ be associated with $B_{1,2}$ functions. Set*

$$r(v) := f_1^{-1}(v) - f_2^{-1}(v).$$

The function $\arg Q(x)$ is continuos at $x = 0$ if at least one of the following holds:

i) $\lim_{v\to 0} r(v)$ *exists;*

ii) *there exists $c_1 \in \mathbb{C}$, such that* $\int_0^v r(s)ds - c_1 v = o(v)$.

Proof. Let $x > 0$ and assume condition i). Then by Theorem 2.4 we have

$$\arg Q(x) = -2x \int_0^1 \frac{r(v)}{x^2 + v^2} dv + o(1), \quad x \to 0.$$

Introduce a function $O_1(v) := r(v) - r_0$ where $r_0 = \lim_{v\to 0} r(v)$. Then, by i), we get

$$\begin{aligned}\arg Q(x) &= -2r_0 \int_0^1 \frac{x\,dv}{x^2+v^2} - 2x \int_0^1 \frac{O_1(v)dv}{x^2+v^2} + o(1)\\ &= -2r_0 \arctan\frac{1}{x} - 2x \int_0^1 \frac{O_1(v)dv}{x^2+v^2} + o(1).\end{aligned}$$

Estimate the integral in the last equation:

$$\begin{aligned}\left| x \int_0^1 \frac{O_1(v)}{x^2+v^2} dv \right| &\lesssim \alpha(x) \int_0^{\sqrt{x}} \frac{x\,dv}{x^2+v^2} + \int_{\sqrt{x}}^1 \frac{x\,dv}{x^2+v^2}\\ &= \alpha(x) \arctan x^{-1/2} + (\arctan x^{-1} - \arctan x^{-1/2}),\end{aligned}$$

where $\alpha(x) = \sup\{|O_1(v)| : v \in [0, \sqrt{x}]\}$. Thus we obtain

$$\lim_{x\to 0} x \int_0^1 \frac{O_1(v)dv}{x^2+v^2} = 0$$

and the theorem is proven under condition i).

Assume that ii) is satisfied. Denoting $F(v) := \int_0^v r(s)ds$ we have

$$\begin{aligned}\arg Q(x) &= -2x \int_0^1 \frac{dF(v)}{x^2+v^2} + o(1)\\ &= -2x\, \frac{F(v)}{x^2+v^2}\Bigg|_{\nu=0}^{1} - 4x\int_0^1 \frac{v\,F(v)dv}{(x^2+v^2)^2} + o(1)\\ &= -\frac{2x\,F(1)}{x^2+1} - 4xc_1\int_0^1 \frac{v^2\,dv}{(x^2+v^2)^2} - 4x\int_0^1 \frac{v\,O_2(v)\,dv}{(x^2+v^2)^2} + o(1),\end{aligned}$$

where $O_2(v) := F(v) - c_1 v$. Consider the last integrals:

$$x\int_0^1 \frac{v^2\,dv}{(x^2+v^2)^2} = \int_0^{1/x} \frac{s^2 ds}{(1+s^2)^2} = \int_0^{\infty} \frac{s^2 ds}{(1+s^2)^2} + o(1), \tag{3.1}$$

and

$$\begin{aligned}\left| x\int_0^1 \frac{vO_2(v)\,dv}{(x^2+v^2)^2}\right| &\lesssim |x|\left\{\beta(x)\int_0^{\sqrt{x}} \frac{v^2 dv}{(x^2+v^2)^2} + \int_{\sqrt{x}}^1 \frac{v^2 dv}{(x^2+v^2)^2}\right\}\\ &\lesssim \beta(x)\int_0^{1/\sqrt{x}} \frac{s^2\,ds}{(1+s^2)^2} + \int_{1/\sqrt{x}}^{1/x} \frac{s^2\,ds}{(1+s^2)^2},\end{aligned}$$

where $\beta(x) = \sup\left\{\frac{|O_2(v)|}{v} : v \in (0,\sqrt{x})\right\}$. Hence we have

$$\lim_{x\to 0} x\int_0^1 \frac{vO_2(v)dv}{(x^2+v)^2} = 0. \tag{3.2}$$

Taking into account (3.1) and (3.2), the assertion is proven under condition ii). □

Example. Consider a Blaschke product B_1 satisfying Hypothesis 2.2. Let f_1 be a function associated with B_1 set $f_1 =: f$. Next let $\alpha(x)$ be a continuous function such that $\lim_{x\to\infty} \alpha(x) = 0$ and $\beta(x) := x + \alpha(x)$ is monotonically increasing. Define $f_2(x) := f(\beta(x)+c)$, where c is a real constant. Then $f_2^{-1}(v) = f^{-1}(v) - c - \alpha_1(v)$, where $\lim_{v\to 0} \alpha_1(v) = 0$, and hence

$$r(v) = f^{-1}(v) - f_2^{-1}(v) = c + \alpha_1(v) \to c, \quad v \to 0.$$

By Theorem 3.1 $\lim_{x\to 0} \arg B_1(x)/B_2(x)$ exists.

Example. Consider a more delicate case of Theorem 3.1 (part ii)). Let $\beta(x) = x + p(x)$, where $p(x)$ is a periodic continuous function such that $\beta(x)$ is increasing on $[1/2, \infty)$. Then the inverse function has the form

$$\beta^{-1}(v) = v - q(v),$$

where $q(v)$ is a periodic continuous function. As in the previous example, let us construct two Blaschke products B_1 and B_2 with the associated functions f_1 and f_2. Let $f_1(x) = f(x)$, where f satisfies Hypothesis 2.2 and such that $f'(x)$ is monotonic function and $f''(x)$ is bounded. Set $f_2(x) = f(\beta(x))$. Then $f_2^{-1}(v) = \beta^{-1}(f^{-1}(v)) = f^{-1}(v) - q(f^{-1}(v))$ and $r(v) = q(f^{-1}(v))$. Consider

$$F(v) = \int_0^v q(f^{-1}(u))du = \int_0^v q_0 du + \int_0^v q_1(f^{-1}(u))du,$$

where q_0 is the zero Fourier coefficient of $q(v)$ and $q_1(v) = q(v) - q_0$. Then

$$F(v) = q_0 v - \int_{f^{-1}(v)}^{\infty} q_1(u) f'(u)du.$$

Let $F_1(v)$ be an antiderivative of $q_1(v)$. That is $F_1'(v) = q_1(v)$. Then

$$\begin{aligned} F(v) &= q_0 v - F_1(u)f'(u)\Big|_{f^{-1}(v)}^{\infty} + \int_{f^{-1}(v)}^{\infty} F_1(u)f''(u)du \\ &= q_0 v + \frac{F_1(f^{-1}(v))}{(f^{-1}(v))'} + \int_{f^{-1}(v)}^{\infty} F_1(u)f''(u)du. \end{aligned}$$

Since $f'(f^{-1}(v)) = \dfrac{1}{(f^{-1}(v))'}$ and $F_1(f^{-1}(v))$ is bounded, one has

$$|F(v) - q_0 v| \lesssim \left(\left| \frac{v}{v\,(f^{-1}(v))'} \right| + \left| \int_{f^{-1}(v)}^{\infty} f''(u)du \right| \right).$$

Let $v = f(x)$, then

$$|F(v) - q_0 v| \lesssim \left(v \left| \frac{f'(x)}{f(x)} \right| + \left| f'(x) \right| \right) \lesssim v \left| \frac{f'(x)}{f(x)} \right|.$$

Condition (2.5) and the monotonicity of the function $f'(x)$ imply the condition ii) of Theorem 3.1 and $\arg Q(x)$ approaches a finite limit as $x \to 0$.

Theorem 3.1 has a consequence which will be crucial in the last section.

Corollary 3.2. *Let*

$$B_1(z)=\prod_{n=1}^{\infty}\frac{z-i\nu_n}{z+i\nu_n}\quad \textit{and}\quad B_2(z)=\prod_{n=1}^{\infty}\frac{z-i\kappa_n}{z+i\kappa_n}$$

be two Blaschke products subject to Hypothesis 2.2 *with interlacing zeros (i.e.,* $\kappa_n>\nu_n>\kappa_{n+1}$ *for any* $n\in\mathbb{N}$*) and associated functions* f_1 *and* f_2*. If there exists a real continuously differentiable function* f *such that*

$$f\,(2x-1)=f_1\,(x)\,,\qquad f\,(2x)\;=f_2\,(x)\,,$$

and

$$f\,(n)=\begin{cases}\kappa^{(1)}_{\frac{n+1}{2}}, & n \;\; \textit{is odd}\\ \nu^{(2)}_{\frac{n}{2}}, & n\;\;\textit{is even}\end{cases},$$

then $\arg B_1(x)/B_2(x)$ *is continuous on the real line.*

Proof. Indeed

$$f_1^{-1}(v)-f_2^{-1}(v)=\frac{f^{-1}(v)+1}{2}-\frac{f^{-1}(v)}{2}=\frac{1}{2}$$

and Theorem 3.1 now applies. □

4. Toeplitz and Hankel operators

Let $H^2_\pm$ be the usual Hardy space of the upper and lower half-planes. By the Paley-Wiener theorem

$$H^2_\pm=\left\{f:f(x)=\int_0^\infty g(t)e^{\pm itx}dt,\;\; x\in\mathbb{R},\;\; g\in L_2(\mathbb{R}_+)\right\}.$$

Let $P^\pm$ be the orthogonal projector of $L_2(\mathbb{R})$ onto $H^2_\pm(\mathbb{R})$. The operators $P^\pm$ can be written as follows

$$P^\pm=\frac{1}{2}(I\pm S),$$

where

$$(Sf)(x):=\frac{1}{\pi i}\int_{\mathbb{R}}\frac{f(\tau)}{\tau-x}\,d\tau\;:\;L_2(\mathbb{R})\to L_2(\mathbb{R}),$$

with the singular integral understood in the sense of the Cauchy principal value.

The Toeplitz operator with a symbol[2] $a\,(x)\in L_\infty(\mathbb{R})$ is defined by

$$T(a)f:=P^+\,af:H^2_+\to H^2_+. \tag{4.1}$$

Let

$$(Jf)(x)=f(-x):L_2(\mathbb{R})\to L_2(\mathbb{R}) \tag{4.2}$$

---

[2] $L_\infty(\mathbb{R})$ is the usual space of functions essentially bounded on $\mathbb{R}$.

be the reflection operator. The Hankel operator with the symbol a is given by the formula

$$(\mathbb{H}(a)f)(x) := (JP^{-}af)(x) : H^2_+ \to H^2_+. \tag{4.3}$$

The theory of Toeplitz and Hankel operators is given, e.g., in [8, 12, 13]. Recall a few more definitions.

Definition 4.1. A bounded linear operator A acting in a Banach space B is called left (right) invertible if there exists a bounded in B operator A_ℓ^{-1} (A_r^{-1}) such that

$$A_\ell^{-1}A = I \quad (AA_r^{-1} = I),$$

where I is the identity operator on B.

Definition 4.2. A bounded linear operator A is called Fredholm if

$$\operatorname{Im} A = \overline{\operatorname{Im} A}, \quad \dim\ker A < \infty, \text{ and } \dim(B/\operatorname{Im} A) < \infty.$$

The number

$$\operatorname{ind}(A) := \dim\ker A - \dim(B/\operatorname{Im} A)$$

is called the index of the operator A.

Define the distance between a function $a \in L_\infty(\mathbb{R})$ and a subset $M \subset L_\infty(\mathbb{R})$ as

$$\operatorname{dist}(a, M) := \inf_{m\in M} \operatorname*{ess\,sup}_{x\in\mathbb{R}} |a(x) - m(x)|$$

Introduce

$$H^\infty_+ + C(\dot{\mathbb{R}}) := \{f + g : f \in H^\infty_+, \quad g \in C(\dot{\mathbb{R}})\}.$$

This space is a closed subspace (and even a closed subalgebra) of $L_\infty(\mathbb{R})$ and is particularly important in the theory of Toeplitz and Hankel operators. We will use the following well-known results.

Theorem 4.3 (Widom-Devinatz, see [8], p. 59). *Let $a(x)$ be a unimodular function (that is $|a(x)| = 1$ for almost all $x \in \mathbb{R}$). Then the operator $T(a)$ defined by* (4.1)

i) *is left invertible if and only if* $\operatorname{dist}(a, H^\infty_+) < 1$;
ii) *is right invertible if and only if* $\operatorname{dist}(a, \overline{H^\infty_+}) < 1$;
iii) *is invertible if and only if* $\operatorname{dist}(a, GH^\infty_+) < 1$,
where $GH^\infty_+ \subset H^\infty_+$ is the set of all invertible in H^∞_+ elements.

Theorem 4.4 (I. Gohberg, see [8], [12, 13]). *Let $a(x) \in C(\dot{\mathbb{R}})$, then the operator $T(a)$ is Fredholm if and only if $a(x) \neq 0$ for all $x \in \mathbb{R}$. Moreover*

$$\operatorname{ind}(T(a)) = -\operatorname{wind} a,$$

where $\operatorname{wind} a$ *is the number of rotations which the point $z = a(x)$ makes around the origin in the complex plane (when x moves along $\mathbb{R}$ from $-\infty$ to $+\infty$).*

Theorem 4.5 ([8], Ch.2, [9], Theorem 2.7). *Let $a(x) \in L_\infty(\mathbb{R})$ and* $\operatorname{ess\,inf}\{|a(x)| : x \in \mathbb{R}\} > 0$. *Then*

i) *if $a(x) \in H^\infty_+ + C(\dot{\mathbb{R}})$ but $1/a(x) \notin H^\infty_+ + C(\dot{\mathbb{R}})$ then $T(a)$ is left invertible;*

ii) *if* $a(x) \in \overline{H_+^\infty} + C(\dot{\mathbb{R}})$ *but* $1/a(x) \notin \overline{H_+^\infty} + C(\dot{\mathbb{R}})$ *then* $T(a)$ *is right invertible;*
iii) *if* $a(x) \in (H_+^\infty + C(\dot{\mathbb{R}})) \cap (\overline{H_+^\infty} + C(\dot{\mathbb{R}}))$ *then* $T(a)$ *is Fredholm.*

Theorem 4.6 ([8], [12, 13]). *Let* $a(x) \in L_\infty(\mathbb{R})$. *Then*

$$\| \mathbb{H}(a) \| \leq \| a \|_{L_\infty}$$

and the Hankel operator (4.3) *is compact if and only if*

$$a(x) \in H_+^\infty + C(\dot{\mathbb{R}}).$$

Note that if $h(x) \in H_+^\infty$ then $\mathbb{H}(h) = 0$ and consequently

$$\mathbb{H}(a) = \mathbb{H}(a - h). \tag{4.4}$$

Consider now

$$a(x) = D\,(x)\, B_1(x)/B_2(x), \tag{4.5}$$

where $D\,(x)$ is a unimodular function and $B_{1,2}(x)$ are Blaschke products satisfying the conditions of Theorem 3.1. Then Theorems 3.1, 4.4 and 4.5 imply the following result.

Theorem 4.7. *Let* a *have the form* (4.5).

i) *If* $D \in H_+^\infty + C(\dot{\mathbb{R}})$ $\quad (D \in \overline{H_+^\infty} + C(\dot{\mathbb{R}}))$ *and* $1/D \notin H_+^\infty + C(\dot{\mathbb{R}})$ $\quad (1/D \notin \overline{H_+^\infty} + C(\dot{\mathbb{R}}))$ *then* $T(a)$ *is left (right) invertible.*
ii) *If* $D \in (H_+^\infty + C(\dot{\mathbb{R}})) \cap (\overline{H_+^\infty} + C(\dot{\mathbb{R}}))$ *then* $T(a)$ *is Fredholm.*
iii) *If* $D \in C(\dot{\mathbb{R}})$ *then* $a \in C(\dot{\mathbb{R}})$ *and* $T(a)$ *is Fredholm and*

$$\operatorname{ind}(T(a)) = -\operatorname{wind} a(x).$$

We will also need

Theorem 4.8. *Let a function* a *have the form* (4.5) *with some* $D \in H_+^\infty + C(\dot{\mathbb{R}})$ *and* $1/D \notin H_+^\infty(\mathbb{R}) + C(\dot{\mathbb{R}})$. *Then the Hankel operator* $\mathbb{H}(a)$ *is compact,* $\| \mathbb{H}(a) \| < 1$ *and hence the operator* $I + \mathbb{H}(a)$ *is invertible.*

Proof. The compactness of the operator $\mathbb{H}(a)$ is a direct consequence of Theorem 4.6. Turn to the invertibility of $I + \mathbb{H}(a)$. By Theorem 4.7, the operator $T(a)$ is left invertible and thus by Theorem 4.3 (i) there exists a function $h(x)$ from H_+^∞ such that $\| a - h \|_{L_\infty} < 1$. By (4.4), $\mathbb{H}(a) = \mathbb{H}(a - h)$ and hence by Theorem 4.6

$$\| \mathbb{H}(a) \| \leq \| a - h \|_{L_\infty} < 1 \tag{4.6}$$

and operator $I + \mathbb{H}(a)$ is invertible. □

The symbol

$$\phi(x) = e^{i\left(tx^3 + cx\right)} D\,(x)\,, \quad t > 0, c \in \mathbb{R} \tag{4.7}$$

arises in the inverse scattering transform method for the Korteweg-de Vries (KdV) equation (see [18, 19]). The form of the unimodular function $D\,(x)$ depends on the properties of the initial data in the Cauchy problem for the KdV equation. In certain particular cases discussed in the next section the function $D(x)$ is of the form

$$D(x) = \frac{B_1(x)}{B_2(x)} I(x), \tag{4.8}$$

where $B_{1,2}(x)$ are Blaschke products with zeros converging to 0 along the imaginary axis and $I(x)$ is an inner function ($I(x) \in H_+^\infty$ and $|I(x)| = 1$ a.e. on $\mathbb{R}$). To apply Theorem 4.8 to the case of (4.7)–(4.8) we need one result from [3, 9].

Definition 4.9. Let Δ be a real-valued function defined for all sufficiently large $x > 0$. The function Δ is called regular if it is strictly monotonically increasing, twice continuously differentiable and satisfies

$$\liminf_{x\to\infty} \frac{x\Delta''(x)}{\Delta'(x)} > -2,$$

$$\lim_{x\to\infty} \frac{x\Delta''(x)}{\Delta'(x)^2} = 0,$$

$$\lim_{x\to\infty} \frac{\sqrt{x}\Delta''(x)}{\Delta'(x)^{3/2}} = 0.$$

Theorem 4.10 ([3], [9], Ch. 5). *If the homeomorphism $\delta(x) : \mathbb{R} \to \mathbb{R}$ is a regular function and $\delta(-x) = -\delta(x)$ for sufficiently large $x > 0$, then*

$$\exp\{i\xi\delta(x)\} \in H_+^\infty + C(\dot{\mathbb{R}})$$

for all $\xi > 0$. Moreover the following representation holds

$$\exp\{i\xi\delta(x)\} = B_\xi(x)C_\xi(x), \tag{4.9}$$

where $B_\xi(x)$ is a Blaschke product with an infinite number of zeros with no accumulation points at a finite distance and $C_\xi(x)$ is a unimodular function from $C(\dot{\mathbb{R}})$.

The following theorem is one of the main results of this paper.

Theorem 4.11. *Let $B_{1,2}(x)$ be Blaschke products of the form (1.1) with zeros satisfying the conditions of Theorem 2.4 and Theorem 3.1 and let $I(x)$ be an inner function. Consider*

$$\phi(x) = e^{i(tx^3+cx)} \frac{B_1(x)}{B_2(x)} I(x), \quad t > 0, \quad c \in \mathbb{R}. \tag{4.10}$$

Then the Toeplitz operator $T(\phi) : H_+^2 \to H_+^2$ is left invertible, the Hankel operator $\mathbb{H}(\phi) : H_+^2 \to H_+^2$ is compact and the operator $I + \mathbb{H}(\phi) : H_+^2 \to H_+^2$ is invertible.

Proof. By Theorem 3.1

$$Q(x) = \frac{B_1(x)}{B_2(x)} \in C(\dot{\mathbb{R}}).$$

It follows from Theorem 4.10 that

$$e^{i(tx^3+cx)} \in H_+^\infty + C(\dot{\mathbb{R}})$$

(it is easy to check that function $\delta(x) := tx^3 + cx$ is regular). Since the set $H_+^\infty + C(\dot{\mathbb{R}})$ is an algebra we have

$$\phi(x) \in H_+^\infty + C(\dot{\mathbb{R}}). \tag{4.11}$$

It remains to demonstrate that

$$1/\phi(x) \notin H_+^\infty(\mathbb{R}) + C(\dot{\mathbb{R}}). \tag{4.12}$$

To this end consider

$$1/\phi(x) = \overline{B_\xi(x)}\, d(x),$$

where $B_\xi(x)$ is as in (4.9) $d(x) \in C(\dot{\mathbb{R}})$ and $|d(x)| = 1$ for all $x \in \mathbb{R}$. Since the Blaschke product $B_\xi(x)$ has an infinite number of zeros, we conclude that $\dim\ker T(1/\phi) = \infty$ (see, e.g., [9], p. 24) and hence the operator $T(1/\phi)$ cannot be Fredholm. On the other hand if (4.12) doesn't hold, i.e., $1/\phi \in H_+^\infty + C(\dot{\mathbb{R}})$ (and (4.11) also holds), then ([8, 12, 13]) $T(1/\phi)$ must be Fredholm. This contradiction proves (4.12). □

5. Applications to the Korteweg-de Vries equation

In this section we apply the results obtained in the previous sections to soliton theory (see, e.g., the book [1] by Ablowitz-Clarkson). We do not assume that the reader is familiar with this theory and therefore present here some background information. Consider the initial value (Cauchy) problem for the Korteweg-de Vries (KdV) equation

$$\frac{\partial u(x,t)}{\partial t} - 6u(x,t)\frac{\partial u(x,t)}{\partial x} + \frac{\partial^3 u(x,t)}{\partial x^3} = 0, \quad t \geq 0, x \in \mathbb{R}. \tag{5.1}$$

$$u(x,0) = q(x). \tag{5.2}$$

This equation is arguably the most celebrated nonlinear partial differential equations. It was derived by Korteweg and de Vries in 1895 as a model for describing shallow water but remained essentially unused until the 50s when it was found to be particularly important in plasma physics. In 1955, Fermi, Pasta, and Ulam took a chain of harmonic oscillators coupled with a quadratic nonlinearity and investigated how the energy in one mode would spread to the rest. (One of the first dynamics calculations carried out on a computer.) They found that the system cycled periodically and never came to the rest. This was a striking phenomenon which back then had no explanation. Although Fermi, Pasta, and Ulam never published their observation, the equation drew attention of mathematicians and theoretical physicists. The breakthrough occurred in the mid 60s when Gardner, Greene, Kruskal, and Miura found a truly ingenious way to linearize it. Their method, now called the inverse scattering transform (IST), is a major achievement of the 20th century mathematics and with its help we have learned an incredible amount about the KdV equation and physical systems described by it[3]. We have given here only a small part of the fascinating story behind the KdV equation. The interested reader can learn more about the history in [1] or any other book on soliton theory.

Conceptually, the IST is similar to the Fourier transform and consists, as the standard Fourier transform method, of the following three steps:

1. the direct transform mapping the (real) initial data $q(x)$ to a new set of variables S_0 in which (5.1) turns into a very simple first-order linear ordinary equation for $S(t)$ with the initial condition $S(0) = S_0$;
2. solve then this linear ordinary differential equation for $S(t)$;
3. apply the inverse transform to find $u(x,t)$ from $S(t)$.

In its original edition due to Gardner-Greene-Kruskal-Miura (see, e.g., [1]), S_0 was the set of the so-called scattering data associated with the pair of Schrödinger operators $H_0 = -d^2/dx^2$ and $H_q = -d^2/dx^2 + q\,(x)$ on $L_2\,(\mathbb{R})$. Moreover, this procedure comes with a beautiful formula

$$u\,(x,t) = -2\frac{\partial^2}{\partial x^2} \log\det\left(I + \mathbb{M}_{x,t}\right), \tag{5.3}$$

where $\mathbb{M}_{x,t} : L_2\,(0,\infty) \to L_2\,(0,\infty)$ is a two parametric family of integral operators

$$\left(\mathbb{M}_{x,t} f\right)(y) = \int_0^\infty M_{x,t}\,(y+s)\, f\,(s)\, ds, \quad f \in L_2\,(0,\infty)\,, \tag{5.4}$$

explicitly constructed in terms of $S(t)$.

One immediately sees that the operator defined by (5.4) is Hankel. We describe this operator following [18, 19]. The operator (5.4) is unitary equivalent to

$$\mathbb{H}_{x,t} := \mathbb{H}^{(1)}_{x,t} + \mathbb{H}^{(2)}_{x,t}. \tag{5.5}$$

The first operator on the right-hand side of (5.5) is the Hankel operator defined by (4.3) with the symbol $R_{x,t}$ given by

$$R_{x,t}\,(\lambda) = e^{2i\lambda\left(4\lambda^2 t - x\right)} R\,(\lambda)\,,$$

where $R\,(\lambda)$ is the so-called reflection coefficient corresponding to the pair of Schrödinger operators H_0, H_q. We can easily do without presenting its formal definition by stating its properties. For a.e. real λ

$$R\,(-\lambda) = \overline{R\,(\lambda)}, \quad |R\,(\lambda)| \leq 1. \tag{5.6}$$

Note that (5.6) implies that $\mathbb{H}\,(R(x,t))$ is self-adjoint.

The other operator $\mathbb{H}^{(2)}_{x,t}$ on the right-hand side of (5.5) is also a Hankel operator corresponding to the measure

$$d\rho_{x,t}(\alpha) := e^{2\alpha\left(4\alpha^3 t - x\right)} d\rho(\alpha),$$

[3]Similar methods have also been developed for many other physically important evolution nonlinear partial differential equations (PDE), which are typically referred to as completely integrable.

where $\rho(\alpha)$ is a measure subject to

$$\operatorname{Supp}\rho \subseteq [0,a], \quad d\rho \geq 0, \quad \int_0^a d\rho < \infty. \tag{5.7}$$

The measure ρ is related to the negative spectrum of H_q but its explicit expression in terms of H_q is not essential in our consideration. What we need is the following relation between the support of ρ and the negative spectrum of H_q:

$$\alpha \in \operatorname{Supp}\,\rho \iff -\alpha^2 \in \operatorname{Spec}(H_q) \cap \mathbb{R}_-.$$

More specifically, the operator $\mathbb{H}_{x,t}^{(2)}$ is unitarily equivalent to $\chi_{\mathbb{R}_+}\widehat{\rho}_{x,t}\mathcal{F}$, where $\chi_{\mathbb{R}_+}$ is the Heaviside function of $\mathbb{R}_+$, $\mathcal{F}$ is the Fourier operator

$$(\mathcal{F}f)(\lambda) = \frac{1}{\sqrt{2\pi}} \int_{-\infty}^{\infty} e^{i\lambda x} f(x)dx,$$

and $\widehat{\rho}_{x,t}$ is the Fourier transform of the measure[4] $\rho_{x,t}$.

The pair of functions $(R_{x,t}, \rho_{x,t})$ is called the scattering data and we view $\mathbb{H}_{x,t}$ as the Hankel operator associated with $(R_{x,t}, \rho_{x,t})$.

It is quite easy to see that the Hankel operator $\chi_{\mathbb{R}_+}\widehat{\rho}_{x,t}\mathcal{F}$ is (self-adjoint) non-negative. The operator $\mathbb{H}_{x,t}^{(2)}$ then is also non-negative for any real x and $t \geq 0$. That is

$$\mathbb{H}_{x,t}^{(2)} \geq 0 \tag{5.8}$$

and it is all we can say so far about $\mathbb{H}_{x,t}$ based upon (5.6) and (5.7). Besides the full line Schrödinger operator H_q, introduce $H_q^D = -d^2/dx^2 + q(x)$ defined on $L_2(\mathbb{R}_-)$ with the Dirichlet boundary condition $u(0) = 0$. We label quantities related to H_q^D with a superscript D. We are now able to state the main result of this section.

Theorem 5.1. *Assume that the initial profile $q(x)$ in (5.2) is real, locally integrable, supported on $(-\infty, 0)$ and such that*

$$\inf \operatorname{Spec}(H_q) = -a^2 > -\infty. \tag{5.9}$$

Then the Cauchy problem for the KdV equation (5.1)–(5.2) has a unique solution $u(x,t)$ which is a meromorphic function in x on the whole complex plane with no real poles for any $t > 0$ if at least one of the following conditions holds:

1. *The operator H_q^D has a non-empty absolutely continuous spectrum;*
2. *The set $i\operatorname{Supp}\rho$ is a set of uniqueness of H_+^∞ functions;*
3. *The sets $\operatorname{Supp}\rho^D = \{\nu_n\}_{n\geq 1}$ and $\operatorname{Supp}\rho = \{\kappa_n\}_{n\geq 1}$ satisfy the Blaschke condition and the corresponding Blaschke products are subject to the conditions of Corollary 3.2.*

[4] We recall $\widehat{\mu}(\lambda) := \frac{1}{\sqrt{2\pi}} \int_{-\infty}^{\infty} e^{i\lambda x} d\mu(x)$.

Proof. Under conditions 1 and/or 2, the theorem is already proven in [18, 19] and it remains to show that the conclusion of theorem also holds under condition 3. Moreover, the arguments of [18, 19] (see also [20]) based upon (5.3) can be easily adjusted to handle condition 3 if we prove that the operator $I + \mathbb{H}_{x,t}$ is invertible under this condition.

Without loss of generality, we may assume that the operator H_q^D has an empty absolutely continuous spectrum (otherwise we are under condition 1). The structure of the reflection coefficient $R(\lambda)$ is studied in [17] where it is shown that $R(\lambda)$ admits the following factorization

$$R(\lambda) = \lim_{m\to\infty} \left\{ \left(\prod_{n=1}^{m} \frac{\lambda - i\nu_n}{\lambda + i\nu_n} \right) \left(\prod_{n=1}^{m} \frac{\lambda - i\kappa_n}{\lambda + i\kappa_n} \right)^{-1} \right\} S(\lambda), \quad \lambda \in \mathbb{C}_+, \tag{5.10}$$

where $S \in H_+^\infty$ and S is contractive on $\mathbb{C}_+$ (i.e., $|S(\lambda)| \leq 1, \quad \lambda \in \mathbb{C}_+$) and the sequence $\{\nu_n\}_{n\geq 1}$ is such that

$$\{-\nu_n^2\}_{n\geq 1} = \operatorname{Spec}\left(H_q^D\right) \cap \mathbb{R}_-,$$

(the negative spectrum of the half-line Dirichlet Schrödinger operator), and the sequence $\{\kappa_n\}_{n\geq 1}$ is such that

$$\{-\kappa_n^2\}_{n\geq 1} = \operatorname{Spec}(H_q) \cap \mathbb{R}_-,$$

(the negative spectrum of the full-line Schrödinger operator). Moreover these sequences are interlacing, i.e.,

$$\kappa_n > \nu_n > \kappa_{n+1} \text{ for any } n \in \mathbb{N}. \tag{5.11}$$

Since we have assumed that the operator H_q^D has no absolutely continuous spectrum, $|S(\lambda)| = 1$ for a.e. real λ (see, e.g., [17]) and hence $S(\lambda) = I(\lambda)$ where $I(\lambda)$ is an inner function of $\mathbb{C}_+$.

Note next that

$$\lim_{m\to\infty} \left\{ \left(\prod_{n=1}^{m} \frac{\lambda - i\nu_n}{\lambda + i\nu_n} \right) \left(\prod_{n=1}^{m} \frac{\lambda - i\kappa_n}{\lambda + i\kappa_n} \right)^{-1} \right\} = \left(\prod_{n=1}^{\infty} \frac{\lambda - i\nu_n}{\lambda + i\nu_n} \right) \left(\prod_{n=1}^{\infty} \frac{\lambda - i\kappa_n}{\lambda + i\kappa_n} \right)^{-1}$$
$$=: B_1(\lambda) B_2(\lambda)^{-1},$$

where

$$B_1(\lambda) = \prod_{n=1}^{\infty} \frac{\lambda - i\nu_n}{\lambda + i\nu_n} \quad \text{and} \quad B_2(\lambda) = \prod_{n=1}^{\infty} \frac{\lambda - i\kappa_n}{\lambda + i\kappa_n}.$$

We have thus arrived at the factorization

$$R(\lambda) = \frac{B_1(\lambda)}{B_2(\lambda)} I(\lambda), \quad \lambda \in \mathbb{C}_+,$$

and hence for every $x \in \mathbb{R}$ and $t > 0$ the function

$$R_{x,t}(\lambda) = e^{2i\lambda\left(4\lambda^2 t - x\right)} R(\lambda),$$

by Corollary 3.2, satisfies the conditions of Theorem 4.11 and hence

$$\|\mathbb{H}(R_{x,t})\| < 1.$$

This immediately implies that

$$\left\|\mathbb{H}^{(1)}_{x,t}\right\| = \|\mathbb{H}(R_{x,t})\| < 1.$$

Therefore $I + \mathbb{H}^{(1)}_{x,t} \geq 0$ and is boundedly invertible. Due to (5.8)

$$I + \mathbb{H}^{(1)}_{x,t} + \mathbb{H}^{(2)}_{x,t} = I + \mathbb{H}^{(1)}_{x,t} \geq 0$$

is also boundedly invertible and the theorem is proven. □

Note that Theorem 5.1 represents an existence and uniqueness result for the KdV equation in a very strong sense. We refer the interested reader to [18, 19] for detailed discussions of statements like Theorem 5.1 and the extensive recent literature on the subject cited therein.

Let us discuss what the conditions of Theorem 5.1 actually mean in terms of the initial profile $q(x)$ in (5.2). Condition (5.9) means that the spectrum of H_q is bounded from below, which (see, e.g., [11]) is satisfied if

$$\sup_x \int_{x-1}^{x} \max(-q, 0) < \infty. \tag{5.12}$$

The condition (5.12) becomes also necessary for (5.9) if q is negative. Note that (5.9) imposes no restriction on the positive part $\max(q, 0)$ of $q(x)$ (e.g., it can grow arbitrarily fast at $-\infty$ or look like the stock market) but H_q still satisfies (5.9).

Condition 1 means that $q(x)$ has a certain pattern of behavior at $-\infty$. The precise statement is rather complicated but particular examples are easy. Condition 1 is satisfied if, for example, q is quasi-periodic on $(-\infty, 0)$ or approaches a constant as $x \to -\infty$ sufficiently fast.

Condition 2 means that the negative spectrum of H_q is, in a way, rich enough. Condition 2 holds if, loosely speaking, $\max(-q, 0)$ (the negative part of q) is large. A typical example would be $q(x) \to -c^2$ as $x \to -\infty$ for some real c (so-called step like initial profiles).

Condition 3 is much trickier as the problem of the negative spectrum distribution for the Schrödinger operator is notoriously difficult. In fact, besides the Lieb-Thirring estimate [21]

$$\sum_{n\geq 1} \kappa_n \lesssim \int_{\mathbb{R}} \max(-q, 0), \tag{5.13}$$

nothing is known about the distribution of $\{\kappa_n\}$ in general. The reason for that is a poor understanding of how individual eigenvalues $-\kappa_n^2$ of H_q depend on q and even (5.13) was a good open problem for quite some time. By the same token constructing a nontrivial explicit example of $q(x)$ subject to condition 3 but not condition 1 appears to be a real challenge. Note that one can always start with a desired spectrum and then work backwards to an essentially non-computable (and quite pathological) $q(x)$ via the Gelfand-Levitan-Marchenko inverse method.

The following statement is important.

Corollary 5.2. *The conclusions of Theorem* 5.1 *hold if* $q(x)$ *in* (5.2) *is real, locally integrable, supported on* $\mathbb{R}_-$ *and such that*

$$\int_{-\infty}^{0} |x| \max(-q(x), 0)\, dx < \infty. \tag{5.14}$$

Proof. The condition (5.14) clearly implies (5.12). Furthermore, it is well known that the negative spectra of H_q and H_q^D are finite under the condition (5.14). Hence $\{\kappa_n\}$ and $\{\nu_n\}$ are also finite and Corollary 3.2 clearly applies. We are then under Condition 3. □

We emphasize that even Corollary 5.2 is new and nontrivial as it cannot be achieved by usual PDEs techniques. We however conjecture that the condition (5.9) alone will be sufficient for Theorem 5.1 to hold. We are not sure if condition (5.9) implies that $I + \mathbb{H}(R_{x,t})$ is boundedly invertible but there are some strong reasons to believe that $I + \mathbb{H}_{x,t}$ has this property.

References

[1] M.J. Ablowitz and P.A. Clarkson, *Solitons, Nonlinear Evolution Equations and Inverse Scattering.* **149,** Cambridge University Press, Cambridge, 1991.

[2] A. Böttcher and S. Grudsky, *On the composition of Muckenhoupt weights and inner functions.* J. London Math. Soc. (2) **58** (1998), no. 1, 172–184.

[3] A. Böttcher, S. Grudsky, and I. Spitkovsky, *Toeplitz operators with frequency modulated semi-almost periodic symbols.* J. Fourier Anal. Appl. **7** (2001), no. 5, 523–535.

[4] A. Böttcher, S. Grudsky, and I. Spitkovsky, *Block Toeplitz operators with frequency-modulated semi-almost periodic symbols.* Int. J Math. Sci. (2003), no. 34, 2157–2176.

[5] A. Böttcher and S.M. Grudsky, *Toeplitz operators with discontinuous symbols: phenomena beyond piecewise continuity.* Singular integral operators and related topics (Tel Aviv, 1995), Oper. Theory Adv. Appl., vol. 90, Birkhäuser, Basel, 1996, pp. 55–118.

[6] A. Böttcher, S.M. Grudsky, and I.M. Spitkovsky, *Matrix functions with arbitrarily prescribed left and right partial indices.* Integral Equations Operator Theory **36** (2000), no. 1, 71–91.

[7] A. Böttcher, Y.I. Karlovich, and I.M. Spitkovsky, *Convolution Operators and Factorization of Almost Periodic Matrix Functions.* Oper. Theory Adv. Appl., vol. 131, Birkhäuser Verlag, Basel, 2002.

[8] A. Böttcher and B. Silbermann, *Analysis of Toeplitz Operators.* Second ed., Springer Monographs in Mathematics, Springer-Verlag, Berlin, 2006.

[9] V. Dybin and S.M. Grudsky, *Introduction to the Theory of Toeplitz Operators with Infinite Index.* Oper. Theory Adv. Appl., vol. 137, Birkhäuser Verlag, Basel, 2002.

[10] J.B. Garnett, *Bounded Analytic Functions.* Pure and Applied Mathematics, vol. 96, Academic Press Inc. [Harcourt Brace Jovanovich Publishers], New York, 1981.

[11] I.M. Glazman, *Direct Methods of Qualitative Spectral Analysis of Singular Differential Operators.* Translated from the Russian by the IPST staff, Israel Program for Scientific Translations, Jerusalem, 1965, 1966.

[12] I. Gohberg and N. Krupnik, *One-dimensional Linear Singular Integral Equations. Vol. I. Introduction.* Oper. Theory Adv. Appl., vol. 53, Birkhäuser Verlag, Basel, 1992.

[13] I. Gohberg and N. Krupnik, *One-dimensional Linear Singular Integral Equations. Vol. II. General Theory and Applications.* Oper. Theory Adv. Appl., vol. 54, Birkhäuser Verlag, Basel, 1992.

[14] S. Grudsky and E. Shargorodsky, *Spectra of Toeplitz operators and compositions of Muckenhoupt weights with Blaschke products.* Integral Equations Operator Theory **61** (2008), no. 1, 63–75.

[15] S.M. Grudsky, *Toeplitz operators and the modelling of oscillating discontinuities with the help of Blaschke products*, Problems and methods in mathematical physics (Chemnitz, 1999), Oper. Theory Adv. Appl., vol. 121, Birkhäuser, Basel, 2001, pp. 162–193.

[16] K. Hoffman, *Banach Spaces of Analytic Functions.* Prentice-Hall Series in Modern Analysis, Prentice-Hall Inc., Englewood Cliffs, N.J., 1962.

[17] A. Rybkin, *The analytic structure of the reflection coefficient, a sum rule and a complete description of the Weyl m-function of half-line Schrödinger operators with L_2-type potentials.* Proc. Roy. Soc. Edinburgh Sect. A **136** (2006), no. 3, 615–632.

[18] A. Rybkin, *Meromorphic solutions to the KdV equation with non-decaying initial data supported on a left half line.* Nonlinearity **23** (2010), no. 5, 1143–1167.

[19] A. Rybkin, *The Hirota τ-function and well-posedness of the KdV equation with an arbitrary step-like initial profile decaying on the right half-line.* Nonlinearity **24** (2011), no. 10, 2953–2990.

[20] A. Rybkin, *Spatial analyticity of solutions to integrable systems. I. The KdV case.* Preprint, 2011.

[21] T. Weidl, *On the Lieb-Thirring constants $L_{\gamma,1}$ for $\gamma \geq 1/2$.* Comm. Math. Phys. **178** (1996), 135–146.

Sergei Grudsky
Departamento de Matematicas
CINVESTAV del I.P.N.
Aportado Postal 14-740
07360 Mexico, D.F., Mexico
e-mail: `grudsky@math.cinvestav.mx`

Alexei Rybkin
Department of Mathematics and Statistics
University of Alaska Fairbanks
PO Box 756660
Fairbanks, AK 99775, USA
e-mail: `arybkin@alaska.edu`

Operator Theory:
Advances and Applications, Vol. 228, 151–172

Mellin Quantisation in Corner Operators

N. Habal and B.-W. Schulze

Dedicated to the 70th anniversary of V. Rabinovich

Abstract. We construct Mellin quantisations or operator conventions, applied to corner-degenerate pseudo-differential symbols, referring to geometric corners of singularity order $k \in \mathbb{N}$, and we obtain holomorphic Mellin symbols in $z \in \mathbb{C}^k$, also with a corresponding degenerate behaviour.

Mathematics Subject Classification (2010). Primary 35S35; Secondary 35J70, 47G30, 58J40.

Keywords. Holomorphic Mellin symbols, corner-degenerate symbols, kernel cut-off, operators on corner singularities, Mellin quantisations.

1. Introduction

This investigation is devoted to the construction of operator conventions or quantisations that are associated with corner degenerate operators within a suitable pseudo-differential algebra. The terminology corner-degenerate refers to configurations, locally described by $(\mathbb{R}_+)^k \times \mathbb{R}^n \times \mathbb{R}^q \ni (r, x, y)$, with $r = (r_1, \dots, r_k)$ being a tuple of half-axis variables and $y = (y^1, \dots, y^k)$ constituted by "higher" edge variables, $y^j \in \mathbb{R}^{q_j}$, $q = \sum_{j=1}^k q_j$. A symbol $p(r, x, y, \rho, \xi, \eta)$ in the variables $(r, x, y) \in \mathbb{R}^k_+ \times \mathbb{R}^n \times \mathbb{R}^q$ and covariables (ρ, ξ, η) is called corner degenerate if there is a symbol $\tilde{p}(r, x, y, \tilde{\rho}, \xi, \tilde{\eta}) \in S^\mu((\overline{\mathbb{R}}_+)^k \times \mathbb{R}^n \times \mathbb{R}^q \times \mathbb{R}^{k+n+q}_{\tilde{\rho},\xi,\tilde{\eta}})$ in the standard symbol class of order $\mu \in \mathbb{R}$ (to be formulated below) which is smooth in r up to $r = 0$ such that for $\rho = (\rho_1, \dots, \rho_k)$, $\eta = (\eta^1, \dots, \eta^k)$ we have

$$p(r, x, y, \rho, \xi, \eta) \qquad (1.1)$$
$$= \tilde{p}(r, x, y, r_1\rho_1, r_1r_2\rho_2, \dots, r_1r_2 \dots r_k\rho_k, \xi, r_1\eta^1, r_1r_2\eta^2, \dots, r_1r_2 \dots r_k\eta^k).$$

Degenerate symbols of the above-mentioned kind appear as symbols of degenerate differential operators, and we obtain pseudo-differential symbols when we pass to an operator algebra that contains the degenerate differential operators together with the parametrices of elliptic elements. If the singularities are of conical or edge type, we talk about Fuchs-type or edge-degenerate symbols, otherwise, for higher k about corner-degenerate symbols of the respective singularity order. From the case $k = 1, 2$ it is well known that in the process of building up pseudo-differential algebras of the above type there appear so-called Mellin operator conventions, also called quantisations. Those rephrase operators in r referring to the Fourier transform to actions based on the Mellin transform, cf. [1], [2], [4], [8], [9], [14], [23], [25]. In operator algebras that reflect asymptotics of solutions it is also essential to specify the Mellin symbols to be holomorphic (or meromorphic) in the respective Mellin covariables. The program of this article is to establish such quantisations for arbitrary $k \in \mathbb{N}$ and to characterise natural classes of Mellin symbols in the respective complex covariables, again with a corresponding degenerate behaviour.

In order to make our iterative construction work with increasing k we have to reproduce the process once again for $k = 1, 2$ in more detailed form than carried out before in [24]. In addition we generalise here the degenerate symbols to the case of a higher edge-degenerate dependence on several edge covariables.

In the present article the main focus is to illustrate the iterative process in this form for the step from $k = 1$ in Section 2 to $k = 2$ in Section 3 and to observe a number of new structures such as kernel cut-offs in different covariables and asymptotic summations that avoid destroying holomorphy in the covariables. After these preparations, the iterative process really works, and the corresponding result is formulated in Section 4.

Holomorphic Mellin symbols and associated operators only furnish some part of the higher corner algebras. The asymptotic parts of those algebras form "complementary" ingredients, cf. the papers [7], [28]. Those are not studied in this article. Similarly as in cone and edge theories for $k = 1$ it will be interesting to study further special cases and applications. Let us also point out that our corner-degenerate operators correspond to corners as described in [26] but different to spaces like $\mathbb{R}^k_+ \times \mathbb{R}^q$ with some other metric. Theories with complete metrics usually give rise to degenerate behaviour of simpler structure such as of "multi-Fuchs" type which is not the topic here.

PDE problems on manifolds with corner singularities or also with non-compact "ends up to infinity" have attracted many mathematicians since a long time. The field of singular PDEs in that sense remained an active area of research. The motivation lies in numerous applications and new challenges, see [9], [10] or [26].

Let us finally give a list of references to illustrate the long history of the singular analysis and the variety of different aspects of the analysis that employs Mellin techniques, Dauge [3], Eskin [5], Komech [11], Kondratyev [12], Pham The Lai [15], Plamenevskij [16–18], Rabinovich [19, 20], Rempel and Schulze [21, 22].

2. Mellin quantisation for singularities of first order

In this section we briefly sketch the well-known approach to construct holomorphic Mellin symbols in the case of edge-degenerate symbols.

Let $S^\mu(U \times \mathbb{R}^n)$ for $\mu \in \mathbb{R}$ and an open set $U \subseteq \mathbb{R}^m$ denote the space of all (so-called symbols) $a(x,\xi) \in C^\infty(U \times \mathbb{R}^n)$ satisfying the symbolic estimates

$$|D_x^\alpha D_\xi^\beta a(x,\xi)| \leq c\,\langle\xi\rangle^{\mu-|\beta|}$$

for all $(x,\xi) \in K\times\mathbb{R}^n$, $K \Subset U$, $\alpha \in \mathbb{N}^m$, $\beta \in \mathbb{N}^n$, with constants $c = c(\alpha,\beta,K) > 0$; $\mathbb{N} = \{0,1,2,\ldots\}$. Moreover, by $S^\mu_{\mathrm{cl}}(U \times \mathbb{R}^n)$ we denote the subspace of classical elements $a(x,\xi)$, i.e., with an asymptotic expansion $a \sim \sum_{j=0}^\infty a_{\mu-j}$, where $a_{\mu-j}$ are homogeneous of order $\mu-j$ in the sense $a_{\mu-j}(x,\lambda\xi) = \lambda^{\mu-j}a_{\mu-j}(x,\xi)$ for $\lambda \geq 1$, $|\xi| \geq \text{const} > 0$, for all j. Recall that $S^\mu(U \times \mathbb{R}^n)$ and $S^\mu_{\mathrm{cl}}(U \times \mathbb{R}^n)$ are Fréchet in a canonical way. We write $S^\mu(\mathbb{R}^n)$ and $S^\mu_{\mathrm{cl}}(\mathbb{R}^n)$ for the subsets of x-independent symbols; those are closed subspaces of $S^\mu(U \times \mathbb{R}^n)$ and $S^\mu_{\mathrm{cl}}(U \times \mathbb{R}^n)$, respectively. We set $S^{-\infty}(U \times \mathbb{R}^n) = \bigcap_{\mu\in\mathbb{R}} S^\mu(U \times \mathbb{R}^n)$ which is equal to $\mathcal{S}(\mathbb{R}^n, C^\infty(U))$, the Schwartz space of $C^\infty(U)$-valued functions.

We will employ several variants of such symbol spaces, in particular, with holomorphic dependence on some covariables. If E is a Fréchet space and $G \subseteq \mathbb{C}$ an open set, by $\mathcal{A}(G,E)$ we denote the space of all holomorphic functions in G with values in E, in the topology of uniform convergence on compact subsets. For instance, it will be important to possess the space $S^\mu_{\mathcal{O}}(U \times \mathbb{R}^n)$ of all $h(x,z,\xi) \in \mathcal{A}(\mathbb{C}_z, S^\mu(U_x \times \mathbb{R}^n_\xi))$ such that

$$h(x,\beta+i\rho,\xi) \in S^\mu(U \times \mathbb{R}^{1+n}_{\rho,\xi})$$

for every $\beta \in \mathbb{R}$, such that $h(x,\beta+i\rho,\xi)$ is a bounded set in $S^\mu(U\times\mathbb{R}^{1+n}_{\rho,\xi})$ when β varies over a compact interval. We will employ below an alternative terminology, namely, with

$$\Gamma_\beta := \{z \in \mathbb{C} : \operatorname{Re} z = \beta\}$$

and denote by $S^\mu(U\times\Gamma_\beta\times\mathbb{R}^n_\xi)$ the space of all symbols in the covariables (z,ξ) with z varying on Γ_β. Then it is more intuitive to distinguish the spaces for different β and to say that

$$h(x,z,\xi)|_{\Gamma_\beta} \in S^\mu(U \times \Gamma_\beta \times \mathbb{R}^n)$$

holds uniformly in finite β-intervals.

In the case $U = \Sigma \times \Sigma$ for an open set $\Sigma \subseteq \mathbb{R}^n$ we also write (x,x') rather than x. For a symbol $a(x,x',\xi) \in S^\mu(\Sigma \times \Sigma \times \mathbb{R}^n)$ we set

$$\mathrm{Op}(a)u(x) := \iint e^{i(x-x')\xi}a(x,x',\xi)u(x')dx'đ\xi, \tag{2.1}$$

$đ\xi := (2\pi)^{-n}d\xi$, $u \in C_0^\infty(\Sigma)$, which is the pseudo-differential operator associated with a via the Fourier transform. Instead of Op we also write Op_x in order to point out the action with respect to the x-variable.

Given an open Riemannian manifold X by $L^\mu(X;\mathbb{R}^l)$ we denote the space of pseudo-differential operators on X of order μ where the local amplitude functions $a(x,\xi,\lambda)$ are symbols in $S^\mu(\Sigma \times \mathbb{R}^{n+l})$ for $\Sigma \subseteq \mathbb{R}^n$ open, $n = \dim X$, modulo smoothing operators. The latter are identified with $\mathcal{S}(\mathbb{R}^l, C^\infty(X \times X)) = \mathcal{S}(\mathbb{R}^l, L^{-\infty}(X))$, where the identification $L^{-\infty}(X) = C^\infty(X \times X)$ refers to the measure dx on X, i.e., $c(x,x') \in C^\infty(X \times X)$ corresponds to the operator $C_0^\infty(X) \ni u \to \int c(x,x')u(x')dx'$. The subspace with $a(x,\xi,\lambda) \in S^\mu_{\mathrm{cl}}(\Sigma \times \mathbb{R}^{n+l})$ will be denoted by $L^\mu_{\mathrm{cl}}(X;\mathbb{R}^l)$. We systematically employ $L^\mu(X;\mathbb{R}^l)$ or $L^\mu_{\mathrm{cl}}(X;\mathbb{R}^l)$ in canonical Fréchet topologies. More details on this kind of notation may be found in [24]. An $A(\lambda) \in L^\mu(X;\mathbb{R}^l)$ is said to be parameter-dependent elliptic if for the local amplitude functions $a(x,\xi,\lambda)$ there are $a^{(-1)}(x,\xi,\lambda) \in S^{-\mu}(\Sigma \times \mathbb{R}^{n+l})$ such that $(aa^{(-1)})(x,\xi,\lambda) - 1 \in S^{-1}(\Sigma \times \mathbb{R}^{n+l})$. In that case there is an $A^{(-1)}(\lambda) \in L^{-\mu}(X;\mathbb{R}^l)$ such that $(AA^{(-1)})(\lambda) \in L^{-\infty}(X;\mathbb{R}^l)$. We call $A^{(-1)}$ a parameter-dependent parametrix of A. In our applications the parameter λ is often splitted up into (z,η) for $z \in \Gamma_\beta$, $\eta \in \mathbb{R}^q$. In that case we also write $L^\mu(X;\Gamma_\beta \times \mathbb{R}^q)$ for the corresponding space of parameter-dependent operators.

Singular spaces contain half-axis variables, and in an intrinsic description of operators it is natural to employ the Mellin transform. We set

$$Mu(z) := \int_0^\infty r^{z-1}u(r)dr$$

which is a function in $\mathcal{A}(\mathbb{C})$ when $u \in C_0^\infty(\mathbb{R}_+)$. Recall that the weighted Mellin transform $M_\gamma u(z) := Mu(z)|_{\Gamma_{1/2-\gamma}}$ extends to an isomorphism

$$M_\gamma : r^\gamma L^2(\mathbb{R}_+) \to L^2(\Gamma_{1/2-\gamma}),$$

and the inverse is $(M_\gamma^{-1}g)(r) = \int_{\Gamma_{1/2-\gamma}} r^{-z}g(z)đz$, $đz := (2\pi i)^{-1}dz$. For $\gamma = 0$ we also write M rather than M_0.

Let us define

$$\mathrm{op}_M^\gamma(f)u := M_\gamma^{-1} f M_\gamma u \tag{2.2}$$

for a symbol $f(r,z) \in C^\infty(\mathbb{R}_+, S^\mu(\Gamma_{1/2-\gamma}))$ which is a pseudo-differential operator on $\mathbb{R}_+$, based on the weighted Mellin transform. Clearly, analogously as (2.1) in (2.2) we also may admit symbols depending on $(r,r') \in \mathbb{R}_+ \times \mathbb{R}_+$. Note that the substitution $r = e^{-t} =: \chi(t)$, $\chi : \mathbb{R} \to \mathbb{R}_+$, yields the relation

$$\begin{aligned} \mathrm{op}_M^{1/2}(f)u(r) &= \int_{\mathbb{R}}\int_0^\infty \left(\frac{r}{r'}\right)^{-i\rho} f(r,i\rho)u(r')\frac{dr'}{r'}đ\rho \\ &= (\chi^*)^{-1} \iint e^{i(t-t')\rho} f(e^{-t},i\rho)v(t')dt'đ\rho \end{aligned} \tag{2.3}$$

for $v(t) = u(e^{-t}) = (\chi^* u)(t)$.

In the following we apply the Mellin operator convention (2.2) also to symbols taking values in some vector space, e.g., a space of operators.

As noted before, one of the major issues in our consideration is to pass from the Fourier to the Mellin operator convention and to achieve Mellin symbols that are holomorphic in the covariable z. In this context we always accept remainders that may be ignored in the (non-canonical) quantisation.

Let us consider parameter-dependent symbols of the kind

$$p(r,x,\rho,\xi,\eta) = \tilde{p}(r,x,r\rho,\xi,r\eta)$$

for $\tilde{p}(r,x,\tilde{\rho},\xi,\tilde{\eta}) \in S^\mu(\overline{\mathbb{R}}_+ \times \Sigma \times \mathbb{R}^{1+n+q}_{\tilde{\rho},\xi,\tilde{\eta}})$. Note that the smoothness up to $r=0$ can be defined in different equivalent ways. We may ask the symbolic estimates for functions in $C^\infty(\overline{\mathbb{R}}_+ \times \Sigma \times \mathbb{R}^{1+n+q}_{\tilde{\rho},\xi,\tilde{\eta}})$ uniformly in $(r,x) \in [0,R) \times K$ for any $R>0$ and $K \Subset \Sigma$. Another way is to start with symbols over $\mathbb{R} \times \Sigma$ and restrict them to $\overline{\mathbb{R}}_+ \times \Sigma$. The equivalence follows from a variant of Seeley's theorem.

For a symbol $f(r,x,z,\xi) \in S^\mu(\overline{\mathbb{R}}_+ \times \Sigma \times \Gamma_0 \times \mathbb{R}^n)$ we form

$$k(f)(r,x,\theta,\xi) := \int_{\Gamma_0} \theta^{-z} f(r,x,z,\xi) đz = \big(M^{-1}_{1/2,z\to\theta} f\big)(r,x,\theta,\xi) \tag{2.4}$$

which gives us a $S^\mu(\overline{\mathbb{R}}_+ \times \Sigma \times \mathbb{R}^n)$-valued distribution on $\mathbb{R}_{+,\theta}$. Then, for any $\psi(\theta) \in C_0^\infty(\mathbb{R}_+)$ that is equal to 1 in a neighbourhood of $\theta=1$ we set

$$\begin{aligned} \mathcal{V}_\psi(f)(r,x,z,\xi) &:= \int_{\mathbb{R}_+} \theta^{z-1}\psi(\theta) k(f)(r,x,\theta,\xi) d\theta \\ &= \big(M_{1/2,\theta\to z}\psi(\theta) k(f)\big)(r,x,z,\xi). \end{aligned} \tag{2.5}$$

This belongs to $S^\mu_{\mathcal{O}}(\overline{\mathbb{R}}_+ \times \Sigma \times \mathbb{R}^n)$ and $\mathcal{V}_\psi(\cdot)$ represents a continuous operator

$$\mathcal{V}_\psi : S^\mu(\overline{\mathbb{R}}_+ \times \Sigma \times \Gamma_0 \times \mathbb{R}^n) \to S^\mu_{\mathcal{O}}(\overline{\mathbb{R}}_+ \times \Sigma \times \mathbb{R}^n)$$

where

$$\mathcal{V}_\psi(f)|_{\Gamma_0} = f \bmod S^{-\infty}(\overline{\mathbb{R}}_+ \times \Sigma \times \Gamma_0 \times \mathbb{R}^n). \tag{2.6}$$

The operator $\mathcal{V}_\psi$ is referred to as a kernel cut-off operator, here with respect to the Mellin transform. The Fourier version of the kernel cut-off is discussed, for instance, [23] or [24].

Theorem 2.1. *Let $p(r,x,y,\rho,\xi,\tilde{\eta}) := \tilde{p}(r,x,y,r\rho,\xi,\tilde{\eta})$ for $\tilde{p}(r,x,y,\tilde{\rho},\xi,\tilde{\eta}) \in S^\mu(\overline{\mathbb{R}}_+ \times \Sigma \times \Omega \times \mathbb{R}^{1+q}_{\tilde{\rho},\xi,\tilde{\eta}})$, $\Sigma \subseteq \mathbb{R}^n$, $\Omega \subseteq \mathbb{R}^q$ open. Then there exists an*

$$h(r,x,y,z,\xi,\tilde{\eta}) \in S^\mu_{\mathcal{O}}(\overline{\mathbb{R}}_+ \times \Sigma \times \Omega \times \mathbb{R}^q_{\tilde{\eta}})$$

such that

$$\operatorname{Op}_{r,x}(p)(y,\tilde{\eta}) = \operatorname{op}^\gamma_M \operatorname{Op}_x(h)(y,\tilde{\eta}) \bmod C^\infty(\Omega, L^{-\infty}(\mathbb{R}_+ \times \Sigma; \mathbb{R}^q_{\tilde{\eta}})), \tag{2.7}$$

for every $\gamma \in \mathbb{R}$.

The (non-canonical) map

$$p(r,x,y,\rho,\xi,\tilde{\eta}) \to h(r,x,y,z,\xi,\tilde{\eta}) \tag{2.8}$$

defined by Theorem 2.1 may be interpreted as a change of the quantisation rule for amplitude functions; operators for $p(r,x,y,\rho,\xi,\tilde{\eta})$ refer to the Fourier transform in r, operators associated with $h(r,x,y,z,\xi,\tilde{\eta})$ refer to the Mellin transform in r. To have a convenient notation we will also call (2.8) a Mellin quantisation (on the level of symbols) and the correspondence in opposite direction the inverse Mellin quantisation. Theorem 2.1 may be found in [24, Theorem 3.2.7]. Concerning alternative arguments, cf. [13].

In order to treat the higher singular case below we have to go back once again to the method of [24] and give a more transparent proof. The dependence on y is not the essential point. Therefore, it will often be dropped.

Lemma 2.2. *Let p and $\tilde{p}$ as in Theorem 2.1 and*

$$b(r,x,\tau,\xi,\tilde{\eta}) := \tilde{p}(r,x,\tau,\xi,\tilde{\eta}). \tag{2.9}$$

Set $f_0(r,x,i\tau,\xi,\tilde{\eta}) := b(r,x,-\tau,\xi,\tilde{\eta}) \in S^\mu(\overline{\mathbb{R}}_+ \times \Sigma \times \Gamma_0 \times \mathbb{R}^{n+q}_{\xi,\tilde{\eta}})$. Then we have

$$\mathrm{Op}_{r,x}(p)(\tilde{\eta}) = \mathrm{op}^{1/2}_{M_r}\mathrm{Op}_x(f_0)(\tilde{\eta}) + \mathrm{Op}_{r,x}(p_1)(\tilde{\eta}) \bmod L^{-\infty}(\mathbb{R}_+ \times \Sigma; \mathbb{R}^q_{\tilde{\eta}}) \tag{2.10}$$

where

$$\mathrm{op}^{1/2}_M(f_0)(x,\xi,\tilde{\eta})u(r) = \int_{-\infty}^{\infty}\int_0^\infty \left(\frac{r}{r'}\right)^{-i\tau} f_0(r,x,i\tau,\xi,\tilde{\eta})u(r')\frac{dr'}{r'}đ\tau,$$

and

$$p_1(r,x,\rho,\xi,\tilde{\eta}) := \tilde{p}_1(r,x,r\rho,\xi,\tilde{\eta}) \text{ for } \tilde{p}_1(r,x,\tilde{\rho},\xi,\tilde{\eta}) \in S^{\mu-1}(\overline{\mathbb{R}}_+ \times \Sigma \times \mathbb{R}^{1+n+q}_{\tilde{\rho},\xi,\tilde{\eta}}).$$

Proof. We consider the diffeomorphism $\chi : \mathbb{R}_t \to \mathbb{R}_{+,r}$, $\chi(t) := e^{-t} = r$, and form

$$\mathrm{Op}_{t,x}(a)(\tilde{\eta})v(t) = \mathrm{Op}_x\Big(\iint e^{i(t-t')\tau} f_0(e^{-t},x,i\tau,\xi,\tilde{\eta})v(t')dt'đ\tau\Big)$$

for $a(t,x,\tau,\xi,\tilde{\eta}) = f_0(e^{-t},x,i\tau,\xi,\tilde{\eta})$. Then we have

$$\mathrm{op}^{1/2}_{M_r}\mathrm{Op}_x(f_0)(\tilde{\eta})u(r) = (\chi^*)^{-1}\mathrm{Op}_{t,x}(a)(\tilde{\eta})v(t)$$

for $v(t) = (\chi^* u)(t) = u(e^{-t})$, i.e., when χ_* denotes the operator push forward under χ it follows that $\mathrm{op}^{1/2}_{M_r}\mathrm{Op}_x(f_0)(\tilde{\eta}) = \chi_*\mathrm{Op}_{t,x}(a)(\tilde{\eta})$. In abuse of notation we often suppress the x-variable; basically we consider the diffeomorphism $\chi \times \mathrm{id}_\Sigma : \mathbb{R} \times \Sigma \to \mathbb{R}_+ \times \Sigma$. As such it is a pseudo-differential operator on $\mathbb{R}_+ \ni r$ based on the Fourier transform, i.e., there is a $c(r,x,\rho,\xi,\tilde{\eta}) \in S^\mu(\mathbb{R}_+ \times \Sigma \times \mathbb{R}^{1+n+q}_{\rho,\xi,\tilde{\eta}})$ such that

$$\mathrm{op}^{1/2}_{M_r}\mathrm{Op}_x(f_0)(\tilde{\eta}) = \mathrm{Op}_{r,x}(c)(\tilde{\eta}) \bmod L^{-\infty}(\mathbb{R}_+ \times \Sigma; \mathbb{R}^q_{\tilde{\eta}}).$$

The well-known asymptotic formula for symbols under operator push forwards tells us that

$$c(r,x,\rho,\xi,\tilde{\eta})|_{r=\chi(t),\rho=\tau} \sim \sum_{j=0}^\infty \frac{1}{j!}\big(\partial^j_\tau a\big)(t,x,(d\chi(t))\tau,\xi,\tilde{\eta})\Phi_j(t,\tau) \tag{2.11}$$

where $d\chi(t) = -e^{-t}$, $\Phi_j(t,\tau) = D^j_{t'}e^{i\delta(t,t')\tau}|_{t'=t}$ for $\delta(t,t') = \chi(t') - \chi(t) - d\chi(t)(t' - t)$, and $\Phi_0 \equiv 1$. We have $\Phi_j(t,\tau)|_{t=-\log r,\tau=\rho} = \Psi_j(r,r\rho)$ where $\Psi_j(r,\tilde{\rho})$ is a polynomial of degree $\leq j/2$ with coefficients in $C^\infty(\overline{\mathbb{R}}_+)$, cf. [24, Lemma 3.2.9]. Let us now verify that

$$c(r,x,\rho,\xi,\tilde{\eta}) = \tilde{p}(r,x,r\rho,\xi,\tilde{\eta}) + \tilde{p}_1(r,x,r\rho,\xi,\tilde{\eta}) \bmod S^{-\infty}(\mathbb{R}_+ \times \Sigma \times \mathbb{R}^{1+n+q}_{\rho,\xi,\tilde{\eta}}) \tag{2.12}$$

for a symbol $\tilde{p}_1(r,x,\tilde{\rho},\xi,\tilde{\eta}) \in S^{\mu-1}(\overline{\mathbb{R}}_+ \times \Sigma \times \mathbb{R}^{1+n+q}_{\tilde{\rho},\xi,\tilde{\eta}})$. The characterisation of the first summand on the right of (2.12) comes from

$$\begin{aligned} & a(t,x,(d\chi(t))\tau,\xi,\tilde{\eta})|_{t=\chi^{-1}(r),\tau=\rho} \\ &= f_0(e^{-t},x,-i(d\chi(t))\tau,\xi,\tilde{\eta})|_{t=\chi^{-1}(r),\tau=\rho} \\ &= b(r,x,r\rho,\xi,\tilde{\eta}) = \tilde{p}(r,x,r\rho,\xi,\tilde{\eta}). \end{aligned} \tag{2.13}$$

Here we employ the formula (2.9), the definition of f_0 in terms of b and the relation

$$a(t,x,\tau,\xi,\tilde{\eta}) = f_0(e^{-t},x,-i\tau,\xi,\tilde{\eta}).$$

The other summand $\tilde{p}_1$ in (2.12) will be obtained by an asymptotic summation. First analogously as (2.13) we find

$$\frac{1}{j!}(\partial_\tau^j a)(t,x,(d\chi(t))\tau,\xi,\tilde{\eta})|_{t=\chi^{-1}(r),\tau=\rho} =: \tilde{g}_j'(r,x,r\rho,\xi,\tilde{\eta})$$

for $\tilde{g}_j'(r,x,\tilde{\rho},\xi,\tilde{\eta}) \in S^{\mu-j}(\overline{\mathbb{R}}_+ \times \Sigma \times \mathbb{R}^{1+n+q}_{\tilde{\rho},\xi,\tilde{\eta}})$, for any fixed $j \geq 1$. This entails

$$\begin{aligned} \frac{1}{j!}(\partial_\tau^j a)(t,x,(d\chi(t))\tau,\xi,\tilde{\eta})|_{t=\chi^{-1}(r),\tau=\rho} &= \tilde{g}_j'(r,x,r\rho,\xi,\tilde{\eta})\Psi_j(r,r\rho) \\ &=: \tilde{g}_j(r,x,r\rho,\xi,\tilde{\eta}) \end{aligned} \tag{2.14}$$

for $\tilde{g}_j(r,x,\tilde{\rho},\xi,\tilde{\eta}) \in S^{\mu-j/2}(\overline{\mathbb{R}}_+ \times \Sigma \times \mathbb{R}^{1+n+q}_{\tilde{\rho},\xi,\tilde{\eta}})$. Then we define

$$\tilde{p}_1(r,x,\tilde{\rho},\xi,\tilde{\eta}) \sim -\sum_{j=1}^{\infty} \tilde{g}_j(r,x,\tilde{\rho},\xi,\tilde{\eta})$$

where the asymptotic sum is carried out in $S^{\mu-1}(\overline{\mathbb{R}}_+ \times \Sigma \times \mathbb{R}^{1+n+q}_{\tilde{\rho},\xi,\tilde{\eta}})$ (i.e., in the class with smoothness in r up to 0). The minus sign on the right is chosen for algebraic reasons as we shall see in the iterative construction below. This yields altogether the formula (2.12), or equivalently,

$$c(r,x,\rho,\xi,\tilde{\eta}) = p(r,x,\rho,\xi,\tilde{\eta}) + p_1(r,x,\rho,\xi,\tilde{\eta}) \bmod S^{-\infty}(\mathbb{R}_+ \times \Sigma \times \mathbb{R}^{1+n+q}_{\rho,\xi,\tilde{\eta}})$$

for $p(r,x,\rho,\xi,\tilde{\eta}) = \tilde{p}(r,x,r\rho,\xi,\tilde{\eta})$ as before, $p_1(r,x,\rho,\xi,\tilde{\eta}) = \tilde{p}_1(r,x,r\rho,\xi,\tilde{\eta})$. Thus we have

$$\mathrm{op}_{M_r}^{1/2}\mathrm{Op}_x(f_0)(\tilde{\eta}) = \mathrm{Op}_{r,x}(p)(\tilde{\eta}) + \mathrm{Op}_{r,x}(p_1)(\tilde{\eta}) \tag{2.15}$$

modulo an operator family in $L^{-\infty}(\mathbb{R}_+ \times \Sigma; \mathbb{R}^q_{\tilde{\eta}})$. □

Proof of Theorem 2.1. In order to prove the formula (2.7) for arbitrary weight γ it suffices to observe that because of the holomorphic dependence of h on z we have

$$\mathrm{op}_M^\gamma \mathrm{Op}_x(h)(y,\eta) = \mathrm{op}_M^{1/2}\mathrm{Op}_x(h)(y,\eta), \tag{2.16}$$

on functions with compact support in $r \in \mathbb{R}_+$. The argument is Cauchy's theorem, cf. also [24, Proposition 2.3.69]. Applying Lemma 2.2 for p_1 rather than p it follows that there are

$$\begin{aligned} & f_1(r,x,i\tau,\xi,\tilde{\eta}) \in S^{\mu-1}(\overline{\mathbb{R}}_+ \times \Sigma \times \Gamma_0 \times \mathbb{R}^{n+q}_{\xi,\tilde{\eta}}), \\ & \tilde{p}_2(r,x,\tilde{\rho},\xi,\tilde{\eta}) \in S^{\mu-2}(\overline{\mathbb{R}}_+ \times \Sigma \times \mathbb{R}^{1+n-q}_{\tilde{\rho},\xi,\tilde{\eta}}) \end{aligned}$$

such that for $p_2(r,x,\rho,\xi,\tilde{\eta}) := \tilde{p}_2(r,x,r\rho,\xi,\tilde{\eta})$ we have

$$\mathrm{Op}_{r,x}(p_1)(\tilde{\eta}) = \mathrm{op}_{M_r}^{1/2}\mathrm{Op}_x(f_1)(\tilde{\eta}) + \mathrm{Op}_{r,x}(p_2)(\tilde{\eta}) \bmod L^{-\infty}(\mathbb{R}_+ \times \Sigma; \mathbb{R}^q_{\tilde{\eta}}).$$

This gives us

$$\mathrm{Op}_{r,x}(p)(\tilde{\eta}) = \sum_{j=0}^{1} \mathrm{op}_{M_r}^{1/2}\mathrm{Op}_x(f_j)(\tilde{\eta}) + \mathrm{Op}_{r,x}(p_2)(\tilde{\eta}).$$

By iterating the arguments we obtain a recursive process which yields

$$p_j(r,x,\rho,\xi,\tilde{\eta}) = \tilde{p}_j(r,x,r\rho,\xi,\tilde{\eta})$$

for some $\tilde{p}_j(r,x,\tilde{\rho},\xi,\tilde{\eta}) \in S^{\mu-j}(\overline{\mathbb{R}}_+ \times \Sigma \times \mathbb{R}^{1+n+q}_{\tilde{\rho},\xi,\tilde{\eta}})$ for every $j \in \mathbb{N}$ and resulting symbols $f_j(r,x,i\tau,\xi,\tilde{\eta}) \in S^{\mu-j}(\overline{\mathbb{R}}_+ \times \Sigma \times \Gamma_0 \times \mathbb{R}^{n+q}_{\xi,\tilde{\eta}})$. Then the asymptotic sum

$$f(r,x,i\tau,\xi,\tilde{\eta}) \sim \sum_{j=0}^{\infty} f_j(r,x,i\tau,\xi,\tilde{\eta}) \tag{2.17}$$

carried out in the space $S^{\mu}(\overline{\mathbb{R}}_+ \times \Sigma \times \Gamma_0 \times \mathbb{R}^{n+q}_{\xi,\tilde{\eta}})$ has the property

$$\mathrm{Op}_{r,x}(p)(\tilde{\eta}) = \mathrm{op}_{M}^{1/2}\mathrm{Op}_x(f)(\tilde{\eta}) \bmod L^{-\infty}(\mathbb{R}_+ \times \Sigma; \mathbb{R}^q_{\tilde{\eta}}). \tag{2.18}$$

We now apply (2.5) and set $h(r,x,z,\xi,\tilde{\eta}) := \mathcal{V}_\psi(f)(r,x,z,\xi,\tilde{\eta})$ in the version with $(\xi,\tilde{\eta})$ instead of ξ. Taking into account (2.6) and (2.18) it follows that

$$\mathrm{Op}_{r,x}(p)(\tilde{\eta}) = \mathrm{op}_{M}^{1/2}\mathrm{Op}_x(h)(\tilde{\eta}) \bmod L^{-\infty}(\mathbb{R}_+ \times \Sigma; \mathbb{R}^q_{\tilde{\eta}}). \tag{2.19}$$

□

Definition 2.3. Let $M^{\mu}_{\mathcal{O}}(X;\mathbb{R}^q)$ denote the subspace of all operator families

$$h(z,\eta) \in \mathcal{A}(\mathbb{C}_z, L^{\mu}(X;\mathbb{R}^q_\eta))$$

such that $h(\beta+i\rho,\eta) \in L^{\mu}(X;\Gamma_\beta \times \mathbb{R}^q_\eta)$ for every $\beta \in \mathbb{R}$, uniformly in $c \le \beta \le c'$ for all reals $c \le c'$. For $q=0$ we simply write $M^{\mu}_{\mathcal{O}}(X)$.

Theorem 2.4. *For every $f(r,y,z,\eta) \in C^\infty(\overline{\mathbb{R}}_+ \times \Omega, L^{\mu}(X;\Gamma_\beta \times \mathbb{R}^q))$ there exists an $h(r,y,z,\eta) \in C^\infty(\overline{\mathbb{R}}_+ \times \Omega, M^{\mu}_{\mathcal{O}}(X;\mathbb{R}^q))$, namely, $h(r,y,z,\eta) = \mathcal{V}_\psi(f)(r,y,z,\eta)$, with obvious meaning of notation, analogous to (2.5), such that*

$$h|_{\Gamma_\beta} = f \bmod C^\infty(\Omega, L^{-\infty}(\mathbb{R}_+ \times X; \mathbb{R}^q_\eta))$$

and h is unique modulo $C^\infty(\mathbb{R}_+ \times \Omega, M^{-\infty}_{\mathcal{O}}(X;\mathbb{R}^q_\eta))$.

Recall that Theorem 2.4 is a consequence of the construction on the kernel cut-off at the beginning which implies, in particular, that $h(r,y,z,\eta) \in C^\infty(\overline{\mathbb{R}}_+ \times \Omega, M^{\mu}_{\mathcal{O}}(X;\mathbb{R}^q))$, $h(r,y,z,\eta)|_{\Gamma_\beta} \in C^\infty(\overline{\mathbb{R}}_+ \times \Omega, L^{\mu-1}(X;\Gamma_\beta \times \mathbb{R}^q))$ for some fixed $\beta \in \mathbb{R}$ implies $h(r,y,z,\eta) \in C^\infty(\overline{\mathbb{R}}_+ \times \Omega, M^{\mu-1}_{\mathcal{O}}(X;\mathbb{R}^q))$, cf. the arguments for [9, Remark 6.1.6].

As a consequence of the definition we have in $M^{\mu}_{\mathcal{O}}(X;\mathbb{R}^q)$ a natural semi-norm system which turns it to a Fréchet space. For completeness we recall the following result, cf. [24, Theorem 3.2.7].

Corollary 2.5. *Let*

$$p(r,y,\rho,\tilde{\eta}) := \tilde{p}(r,y,r\rho,\tilde{\eta}) \text{ for } \tilde{p}(r,y,\tilde{\rho},\tilde{\eta}) \in C^\infty(\overline{\mathbb{R}}_+ \times \Omega, L^\mu(X;\mathbb{R}^{1+q}_{\tilde{\rho},\tilde{\eta}})).$$

Then there exists an

$$h(r,y,z,\tilde{\eta}) \in C^\infty(\overline{\mathbb{R}}_+ \times \Omega, M^\mu_{\mathcal{O}}(X;\mathbb{R}^q_{\tilde{\eta}}))$$

such that

$$\mathrm{op}^\gamma_M(h)(y,\tilde{\eta}) = \mathrm{Op}_r(p)(y,\tilde{\eta}) \bmod C^\infty(\Omega, L^{-\infty}(\mathbb{R}_+ \times X;\mathbb{R}^q_{\tilde{\eta}}))$$

for every $\gamma \in \mathbb{R}$.

Corollary 2.6. *Let*

$$p(r,y,\rho,\eta) := \tilde{p}(r,y,r\rho,r\eta) \text{ for } \tilde{p}(r,y,\tilde{\rho},\tilde{\eta}) \in C^\infty(\overline{\mathbb{R}}_+ \times \Omega, L^\mu(X;\mathbb{R}^{1+q}_{\tilde{\rho},\tilde{\eta}})).$$

Then there exists an

$$h(r,y,z,\tilde{\eta}) \in C^\infty(\overline{\mathbb{R}}_+ \times \Omega, M^\mu_{\mathcal{O}}(X;\mathbb{R}^q_{\tilde{\eta}}))$$

such that for $h(r,y,z,\eta) = \tilde{h}(r,y,z,r\eta)$ *we have*

$$\mathrm{op}^\gamma_M(h)(y,\eta) = \mathrm{Op}_r(p)(y,\eta) \bmod C^\infty(\Omega, L^{-\infty}(\mathbb{R}_+ \times X;\mathbb{R}^q_\eta)) \tag{2.20}$$

for every $\gamma \in \mathbb{R}$.

The proof follows from a simple modification of Theorem 2.1. The extra r-factor at η is regarded as an r-dependence of coefficients, operating from the left on functions in $r \in \mathbb{R}_+$.

Remark 2.7.

(i) The operator functions both in the versions of Corollaries 2.5 and 2.6 belong to $C^\infty(\Omega, L^\mu(\mathbb{R}_+ \times X;\mathbb{R}^q_{\tilde{\eta}}))$, $C^\infty(\Omega, L^\mu(\mathbb{R}_+ \times X;\mathbb{R}^q_\eta))$, respectively, and the claimed relations refer to the respective families of mappings $C_0^\infty(\mathbb{R}_+ \times X) \to C^\infty(\mathbb{R}_+ \times X)$. The role of the obtained left-hand sides of (2.20) is, that the operators in Mellin quantised form admit continuous extensions to weighted Sobolev spaces, in contrast to the Fourier representations of the respective degenerate operators on the right.

(ii) In the "full" cone and edge algebras that contain Mellin operator families as in Corollary 2.6 it is advisable to take classical symbols. The constructions so far restrict to the classical case.

3. The case of second-order corners

Symbols of the kind (1.1) for $k = 2$ have the form

$$p(r,x,y,\rho,\xi,\eta) = \tilde{p}(r,x,y,r_1\rho_1,r_1r_2\rho_2,\xi,r_1\eta^1,r_1r_2\eta^2) \tag{3.1}$$

for $r := (r_1,r_2)$, $y := (y^1,y^2) \in \mathbb{R}^q$ for $q := q_1 + q_2$, and we intend to show an iterated Mellin quantisation result analogously as Theorem 2.1.

Let us give a brief motivation of symbols like (3.1). If we consider, for instance, a Riemannian metric on $\mathbb{R}_+ \times \mathbb{R}_+ \times \mathbb{R}^n_x \times \mathbb{R}^{q_1} \times \mathbb{R}^{q_1} \ni (r_1, r_2, x, y^1, y^2)$ of the form

$$dr_2^2 + r_2^2\big(dr_1^2 + r_1^2 dx^2 + (dy^1)^2\big) + (dy^2)^2.$$

Then the Laplace-Beltrami operator has a symbol like $r_1^{-2} r_2^{-2} p(r, x, y, \rho, \xi, \eta)$ with p being of the form (3.1) and $\tilde{p}(r, x, y, \tilde{\rho}, \xi, \tilde{\eta})$ elliptic of order 2 with respect to the covariables $(\tilde{\rho}, \xi, \tilde{\eta})$, up to $r = 0$.

Another motivation of such symbols comes from the analysis of elliptic (non-degenerate) differential operators of order μ in a corner configuration, say, embedded in an Euclidean space. Then symbols in degenerate form (3.1) appear by iteratively introducing polar coordinates into the operator, here twice, and together with a weight factor $r_1^{-\mu} r_2^{-\mu}$. In a similar manner there occur degenerate symbols (1.1) when we have corners of orders k. From the spaces of Definition 2.3 we can pass to certain derived spaces, for instance $C^\infty(\overline{\mathbb{R}}_+ \times \overline{\mathbb{R}}_+ \times \mathbb{R}^q, M^\mu_{\mathcal{O}_{z_1}}(X; \Gamma_\delta \times \mathbb{R}^q_{\tilde{\eta}}))$ for Γ_δ in the variable z_2, $\delta \in \mathbb{R}$.

Definition 3.1. Let $M^\mu_{\mathcal{O}_{z_1}, \mathcal{O}_{z_2}}(X; \mathbb{R}^q_\eta)$ defined to be the space of all $h(z_1, z_2, \eta) \in \mathcal{A}(\mathbb{C}_{z_2}, M^\mu_{\mathcal{O}_{z_1}}(X; \mathbb{R}^q_\eta))$ such that

$$h(z_1, z_2, \eta)|_{\Gamma_\delta} \in M^\mu_{\mathcal{O}_{z_1}}(X; \Gamma_\delta \times \mathbb{R}^q_\eta), \tag{3.2}$$

for every $\delta \in \mathbb{R}$, uniformly in compact δ-intervals.

This space is Fréchet in a natural way. In Definition 3.1 we may interchange the role of z_1 and z_2 and get the same space, in other words $M^\mu_{\mathcal{O}_{z_1}, \mathcal{O}_{z_2}}(X; \mathbb{R}^q_\eta) \cong M^\mu_{\mathcal{O}_{z_2}, \mathcal{O}_{z_1}}(X; \mathbb{R}^q_\eta)$. Concerning generalities on holomorphic functions in several complex variables, see [6]. The following result tells us that the spaces of Definition 3.1 are very rich for every $\mu \in \mathbb{R}$.

Theorem 3.2.

(i) *For every*

$$f(r_1, r_2, y, z_1, z_2, \eta) \in C^\infty(\overline{\mathbb{R}}_+ \times \overline{\mathbb{R}}_+ \times \mathbb{R}^q, M^\mu_{\mathcal{O}_{z_1}}(X; \Gamma_\delta \times \mathbb{R}^q))$$

there exists an $h(r_1, r_2, y, z_1, z_2, \eta) \in C^\infty(\overline{\mathbb{R}}_+ \times \overline{\mathbb{R}}_+ \times \mathbb{R}^q, M^\mu_{\mathcal{O}_{z_1}, \mathcal{O}_{z_2}}(X; \mathbb{R}^q))$ *such that*

$$h|_{\Gamma_\delta} = f \bmod C^\infty(\mathbb{R}_+ \times \mathbb{R}_+ \times \mathbb{R}^q, M^{-\infty}_{\mathcal{O}_{z_1}}(X; \Gamma_\delta \times \mathbb{R}^q_\eta))$$

and h *is unique modulo* $C^\infty(\mathbb{R}_+ \times \mathbb{R}_+ \times \mathbb{R}^q, M^{-\infty}_{\mathcal{O}_{z_1}, \mathcal{O}_{z_2}}(X; \mathbb{R}^q_\eta))$.

(ii) *For every* $m(r_1, r_2, y, z_1, z_2, \eta) \in C^\infty(\overline{\mathbb{R}}_+ \times \overline{\mathbb{R}}_+ \times \mathbb{R}^q, L^\mu(X; \Gamma_\beta \times \Gamma_\delta \times \mathbb{R}^q_\eta))$ *with fixed* $\beta, \delta \in \mathbb{R}$ *there exists an*

$$h(r_1, r_2, y, z_1, z_2, \eta) \in C^\infty(\overline{\mathbb{R}}_+ \times \overline{\mathbb{R}}_+ \times \mathbb{R}^q, M^\mu_{\mathcal{O}_{z_1}, \mathcal{O}_{z_2}}(X; \mathbb{R}^q_\eta))$$

such that

$$h|_{\Gamma_\beta \times \Gamma_\delta} = m \bmod C^\infty(\mathbb{R}_+ \times \mathbb{R}_+ \times \mathbb{R}^q, L^{-\infty}(X; \Gamma_\beta \times \Gamma_\delta \times \mathbb{R}^q)), \tag{3.3}$$

and h *is unique modulo* $C^\infty(\mathbb{R}_+ \times \mathbb{R}_+ \times \mathbb{R}^q, M^{-\infty}_{\mathcal{O}_{z_1}, \mathcal{O}_{z_2}}(X; \mathbb{R}^q_\eta))$.

Proof. Let us consider the assertion (i). The proof of (ii) then follows from a slight modification of Theorem 2.4 with respect to z_1 combined with (i). In order to find h we apply the kernel cut-off formula (2.5), modified for global operator functions along X and z_1 restricted for a moment to Γ_β for a fixed β. Without loss of generality we assume $\delta = 0$, since a translation in the complex plane gives rise to the construction for any other δ. Then if ψ is a cut-off function on $\mathbb{R}_{+,\theta}$ we may set

$$\mathcal{V}_\psi(f)(r_1, r_2, y, z_1, z_2, \eta) = \int_{\mathbb{R}_+} \theta^{z_2-1}\psi(\theta)k(f)(r_1, r_2, y, z_1, \theta, \eta)d\theta$$

for

$$k(f)(r_1, r_2, y, z_1, \theta, \eta) := \int_{\Gamma_0} \theta^{-z_2} f(r_1, r_2, y, z_1, z_2, \eta)đz_2.$$

The variables r_1, r_2 and y do not affect the process, similarly as in the considerations around the formula (2.5). So we drop these variables here. From the same source we see that if z_1 is varying on Γ_β we obtain

$$\mathcal{V}_\psi(f)(z_1, z_2, \eta) \in L^\mu(X; \Gamma_\beta \times \Gamma_0 \times \mathbb{R}^q_\eta). \tag{3.4}$$

Another information in this context is that (3.4) holds uniformly in compact β-intervals. Now the application of $\mathcal{V}_\psi$ in z_2 and for fixed β gives us a holomorphic dependence on z_2 with the symbol property (after x-localisation) in (x, z_1, z_2, ξ, η) for $z_1 \in \Gamma_\beta$ and $z_2 \in \Gamma_\delta$ for every δ, uniformly in finite δ-intervals. This uniformity then holds for (β, δ) varying in compact sets in $\mathbb{R}^2$. At the same time we have holomorphic dependence separately in z_1 and z_2 for $z_1 \in \Gamma_\beta$ and $z_2 \in \Gamma_\delta$, respectively. A classical lemma of Osgood implies that then that $h := \mathcal{V}_\psi(f)(z_1, z_2, \eta)$ is jointly holomorphic in $(z_1, z_2) \in \mathbb{C} \times \mathbb{C}$ with values in $L^\mu(X; \mathbb{R}^q_\eta)$. Thus the properties required in Definition 3.1 are satisfied. □

Remark 3.3.

(i) An equivalent definition of $M^\mu_{\mathcal{O}_{z_1},\mathcal{O}_{z_2}}(X; \mathbb{R}^q)$ is that this space consists of all $h(z_1, z_2, \eta) \in \mathcal{A}(\mathbb{C}_{z_1} \times \mathbb{C}_{z_2}, L^\mu(X; \mathbb{R}^q_\eta))$ such that

$$h(z_1, z_2, \eta)|_{\Gamma_\beta \times \Gamma_\delta} \in L^\mu(X; \Gamma_\beta \times \Gamma_\delta \times \mathbb{R}^q)$$

for every $(\beta, \delta) \in \mathbb{R}^2$, uniformly in compact subsets of $\mathbb{R}^2$. The iterated application of $\mathcal{V}_{\psi,z_1}\mathcal{V}_{\psi,z_2}$ gives rise to a continuous operator

$$\mathcal{V}_{\psi,z_1}\mathcal{V}_{\psi,z_2} : L^\mu(X; \Gamma_\beta \times \Gamma_\delta \times \mathbb{R}^q) \to M^\mu_{\mathcal{O}_{z_1},\mathcal{O}_{z_2}}(X; \mathbb{R}^q)$$

for any fixed $\beta, \delta \in \mathbb{R}$, and this is the same as $\mathcal{V}_{\psi,z_2}\mathcal{V}_{\psi,z_1}$.

(ii) Let $h(z_1, z_2, \eta) \in M^\mu_{\mathcal{O}_{z_1},\mathcal{O}_{z_2}}(X; \mathbb{R}^q)$, $h(z_1, z_2, \eta)|_{\Gamma_\delta} \in M^{\mu-1}_{\mathcal{O}_{z_1}}(X; \Gamma_\delta \times \mathbb{R}^q)$ for some fixed $\delta \in \mathbb{R}$. Then we have $h(z_1, z_2, \eta) \in M^{\mu-1}_{\mathcal{O}_{z_1},\mathcal{O}_{z_2}}(X; \mathbb{R}^q)$.

An analogous property holds when we interchange the role of z_1 and z_2.

Theorem 3.4. *Let $h_j \in C^\infty(\overline{\mathbb{R}}_+ \times \overline{\mathbb{R}}_+ \times \mathbb{R}^q, M^{\mu_j}_{\mathcal{O}_{z_1},\mathcal{O}_{z_2}}(X; \mathbb{R}^q))$, $j \in \mathbb{N}$, be an arbitrary sequence with $\mu_{j+1} < \mu_j$ for all j, $\mu_j \to -\infty$ as $j \to \infty$, then there is an $h \in$*

$C^\infty(\overline{\mathbb{R}}_+ \times \overline{\mathbb{R}}_+ \times \mathbb{R}^q, M^\mu_{\mathcal{O}_{z_1},\mathcal{O}_{z_2}}(X;\mathbb{R}^q))$, $\mu := \mu_0$, *unique modulo* $C^\infty(\overline{\mathbb{R}}_+ \times \overline{\mathbb{R}}_+ \times \mathbb{R}^q, M^{-\infty}_{\mathcal{O}_{z_1},\mathcal{O}_{z_2}}(X;\mathbb{R}^q))$, *such that*

$$h - \sum_{j=0}^{N} h_j \in C^\infty(\overline{\mathbb{R}}_+ \times \overline{\mathbb{R}}_+ \times \mathbb{R}^q, M^{\mu_{N+1}}_{\mathcal{O}_{z_1},\mathcal{O}_{z_2}}(X;\mathbb{R}^q))$$

for every $N \in \mathbb{N}$.

Proof. Without loss of generality we consider the case of (r_1, r_2, y)-independent h_j. For an asymptotic summation of $L^\mu(X)$-valued operator function we usually go back to the local symbols that we denote by

$$h_j(r_1, r_2, x, z_1, z_2, \xi, \eta). \tag{3.5}$$

Then it suffices to carry out the asymptotic sum over j and return to global operator functions along X, using a partition of unity, etc.

The symbols (3.5) belong to the space $S^\mu_{\mathcal{O}_{z_1},\mathcal{O}_{z_2}}(\Sigma \times \mathbb{R}^{n+q}_{\xi,\eta})$ which is defined to be the set of all $g(x, z_1, z_2, \eta) \in \mathcal{A}(\mathbb{C}_{z_2}, S^\mu_{\mathcal{O}_{z_1}}(\Sigma_x \times \mathbb{R}^{n+q}_{\xi,\eta}))$ such that $g|_{\Gamma_\delta} \in S^\mu_{\mathcal{O}_{z_1}}(\Sigma \times \Gamma_\delta \times \mathbb{R}^{n+q}_{\xi,\eta})$ for every $\delta \in \mathbb{R}$, uniformly in compact δ-intervals. The space $S^\mu_{\mathcal{O}_{z_1},\mathcal{O}_{z_2}}(\Sigma \times \mathbb{R}^{n+q}_{\xi,\eta})$ is Fréchet in a canonical way.

If we try to carry out the asymptotic sum of $h_j \in S^{\mu_j}_{\mathcal{O}_{z_1},\mathcal{O}_{z_2}}(\Sigma \times \mathbb{R}^{n+q})$ we have to take into account that the standard way of performing a convergent sum $\sum_{j=0}^\infty \chi_j(z_1, z_2, \xi, \eta) h_j(x, z_1, z_2, \xi, \eta)$ with j-depending excision functions χ_j will not produce an element in $S^\mu_{\mathcal{O}_{z_1},\mathcal{O}_{z_2}}(\Sigma \times \mathbb{R}^{n+q})$. In fact, the summands are not holomorphic in the complex variables. However, we may form the asymptotic sum in the covariables $(z_1, z_2, \xi, \eta) \in \Gamma_0 \times \Gamma_0 \times \mathbb{R}^{n+q}$ which gives us a symbol $f \in S^\mu(\Sigma \times \Gamma_0 \times \Gamma_0 \times \mathbb{R}^{n+q})$ and then apply the kernel cut-off operator $\mathcal{V}_{\psi_1}$ to f, first with respect to $z_1 \in \Gamma_0$ which gives us

$$\mathcal{V}_{\psi_1}(f)(x, z_1, z_2, \xi, \eta) \in S^\mu_{\mathcal{O}_{z_1}}(\Sigma \times \Gamma_0 \times \mathbb{R}^{n+q})$$

according to (2.5). Then, applying analogously $\mathcal{V}_{\psi_2}$ with respect to $z_2 \in \Gamma_0$ yields

$$\mathcal{V}_{\psi_2}\mathcal{V}_{\psi_1}(f)(x, z_1, z_2, \xi, \eta) \in S^\mu_{\mathcal{O}_{z_1},\mathcal{O}_{z_2}}(\Sigma \times \mathbb{R}^{n+q}).$$

These constructions work because of the well-known properties of kernel cut-off operators and the property that

$$h(x, z_1, z_2, \xi, \eta) \in S^\mu_{\mathcal{O}_{z_1},\mathcal{O}_{z_2}}(\Sigma \times \mathbb{R}^{n+q})$$

and

$$h(x, \beta + i\rho, \delta + i\tau, \xi, \eta) \in S^{\mu-1}(\Sigma \times \Gamma_\beta \times \Gamma_\delta \times \mathbb{R}^{n+q}_{\xi,\eta}) \text{ for some fixed } \beta, \delta \in \mathbb{R}$$

entails $h(x, z_1, z_2, \xi, \eta) \in S^{\mu-1}_{\mathcal{O}_{z_1},\mathcal{O}_{z_2}}(\Sigma \times \mathbb{R}^{n+q})$. □

Remark 3.5. In asymptotic summations we employed repeatedly a combination of standard asymptotic summation combined with kernel cut-off, in order to restore the holomorphic dependence on the complex covariables. There is also an alternative way of carrying out asymptotic sums by applying a summation on the kernel side, similarly as in [23, Section 3.2.2, Proposition 3; Section 3.2.3, Theorem 4]. This idea preserves holomorphy in the variables that are not touched by the partial kernel cut-off; see also [13] for a similar argument in terms of Volterra symbols.

Theorem 3.6. *For every*

$$\boldsymbol{p}(r_1, r_2, y, z_1, \rho_2, \tilde{\eta}) = \tilde{\boldsymbol{p}}(r_1, r_2, y, z_1, r_2\rho_2, \tilde{\eta})$$

for

$$\tilde{\boldsymbol{p}}(r_1, r_2, y, z_1, \tilde{\rho}_2, \tilde{\eta}) \in C^\infty(\overline{\mathbb{R}}_+ \times \overline{\mathbb{R}}_+ \times \mathbb{R}^q, M^\mu_{\mathcal{O}_{z_1}}(X; \mathbb{R}_{\tilde{\rho}_2} \times \mathbb{R}^q_{\tilde{\eta}}))$$

there exists an

$$\boldsymbol{h}(r_1, r_2, y, z_1, z_2, \tilde{\eta}) \in C^\infty(\overline{\mathbb{R}}_+ \times \overline{\mathbb{R}}_+ \times \mathbb{R}^q, M^\mu_{\mathcal{O}_{z_1}, \mathcal{O}_{z_2}}(X; \mathbb{R}^q_{\tilde{\eta}}))$$

such that

$$\mathrm{Op}_{r_2}\mathrm{op}^{\beta_1}_{M_{r_1}}(\boldsymbol{p})(y, \tilde{\eta}) = \mathrm{op}^{\beta_2}_{M_{r_2}}\mathrm{op}^{\beta_1}_{M_{r_1}}(\boldsymbol{h})(y, \tilde{\eta}) \tag{3.6}$$

modulo $C^\infty(\mathbb{R}^q, L^{-\infty}(\mathbb{R}_+ \times \mathbb{R}_+ \times X; \mathbb{R}^q_{\tilde{\eta}}))$ *for any reals* β_1, β_2.

For the proof we prepare the following lemma, and we drop the y-variables again.

Lemma 3.7. *Let* $\mathbf{p}$ *and* $\tilde{\mathbf{p}}$ *be as in the assumptions of Theorem* 3.6 *and*

$$C^\infty(\overline{\mathbb{R}}_+ \times \overline{\mathbb{R}}_+, M^\mu_{\mathcal{O}_{z_1}}(X; \mathbb{R}^{1+q}_{\tau,\tilde{\eta}})) \ni \boldsymbol{b}(r_1, r_2, z_1, \tau, \tilde{\eta}) := \tilde{\boldsymbol{p}}(r_1, r_2, z_1, \tau, \tilde{\eta}).$$

Set

$$C^\infty(\overline{\mathbb{R}}_+ \times \overline{\mathbb{R}}_+, M^\mu_{\mathcal{O}_{z_1}}(X; \Gamma_0 \times \mathbb{R}^q)) \ni \boldsymbol{f}_0(r_1, r_2, z_1, i\tau, \tilde{\eta}) := \boldsymbol{b}(r_1, r_2, z_1, -\tau, \tilde{\eta}).$$

Then we have

$$\mathrm{Op}_{r_2}(\boldsymbol{p})(r_1, z_1, \tilde{\eta})|_{\Gamma_\beta} = \mathrm{op}^{1/2}_{M_{r_2}}(\boldsymbol{f}_0)(r_1, z_1, \tilde{\eta})|_{\Gamma_\beta} + \mathrm{Op}_{r_2}(\boldsymbol{p}_1)(r_1, z_1, \tilde{\eta})|_{\Gamma_\beta} \tag{3.7}$$

modulo $C^\infty(\mathbb{R}_{+,r_1}, L^{-\infty}(\mathbb{R}_{+,r_2} \times X; \Gamma_\beta \times \mathbb{R}^q_{\tilde{\eta}}))$ *for every fixed real* β, *where*

$$\boldsymbol{p}_1(r_1, r_2, z_1, \rho_2, \tilde{\eta}) = \tilde{\boldsymbol{p}}_1(r_1, r_2, z_1, r_2\rho_2, \tilde{\eta})$$

for $\tilde{\boldsymbol{p}}_1(r_1, r_2, z_1, \tilde{\rho}_2, \tilde{\eta}) \in C^\infty(\overline{\mathbb{R}}_+ \times \overline{\mathbb{R}}_+, M^{\mu-1}_{\mathcal{O}_{z_1}}(X; \mathbb{R}_{\tilde{\rho}_2} \times \mathbb{R}^q_{\tilde{\eta}}))$.

Proof. We first let z_1 vary on Γ_0. Then we apply the operator push forward under $\chi : \mathbb{R}_t \to \mathbb{R}_{+,r_2}$, $\chi(t) := e^{-t}$, to $\mathrm{op}^{1/2}_{M_{r_2}}(\boldsymbol{f}_0)(r_1, z_1, \tilde{\eta})$. This gives us

$$\mathrm{op}^{1/2}_{M_{r_2}}(\boldsymbol{f}_0)(r_1, z_1, \tilde{\eta}) = \chi_*\mathrm{Op}_t(\boldsymbol{a})(r_1, z_1, \tilde{\eta})$$

for $\boldsymbol{a}(r_1, t, z_1, \tau, \tilde{\eta}) = \boldsymbol{f}_0(r_1, e^{-t}, z_1, i\tau, \tilde{\eta})$. Thus, according to the transformation rule of pseudo-differential operators under diffeomorphisms it follows that

$$\mathrm{op}^{1/2}_{M_{r_2}}(\boldsymbol{f}_0)(r_1, z_1, \tilde{\eta}) = \mathrm{Op}_{r_2}(\mathbf{c})(r_1, z_1, \tilde{\eta}) \bmod C^\infty(\mathbb{R}_{+,r_1}, L^{-\infty}(\mathbb{R}_+ \times X; \mathbb{R}^q_{\tilde{\eta}})),$$

for a function $\boldsymbol{c}(r_1, r_2, z_1, \rho_2, \tilde{\eta}) \in C^\infty(\mathbb{R}_+ \times \mathbb{R}_+, L^\mu(X; \Gamma_{0,z_1} \times \mathbb{R}_{\rho_2} \times \mathbb{R}^q_{\tilde{\eta}}))$ which has an asymptotic expansion

$$\boldsymbol{c}(r_1, r_2, z_1, \rho_2, \tilde{\eta})|_{r_2=\chi(t), \rho_2=\tau} \sim \sum_{j=0}^{\infty} \frac{1}{j!} (\partial_\tau^j \boldsymbol{a})(r_1, t, z_1, (d\chi(t))\tau, \tilde{\eta}) \Phi_j(t, \tau) \quad (3.8)$$

cf. (2.11), for $\Phi_j(t,\tau)$ only depending on χ. Now the term on the right of (3.8) for $j = 0$ just coincides with $\tilde{\boldsymbol{p}}(r_1, r_2, z_1, r_2\rho_2, \tilde{\eta})$ and hence

$$\boldsymbol{c}(r_1, r_2, z_1, \rho_2, \tilde{\eta}) = \tilde{\boldsymbol{p}}(r_1, r_2, z_1, r_2\rho_2, \tilde{\eta}) + \tilde{\boldsymbol{p}}_1(r_1, r_2, z_1, r_2\rho_2, \tilde{\eta}) \quad (3.9)$$

modulo $C^\infty(\mathbb{R}_+ \times \mathbb{R}_+, L^{-\infty}(X; \Gamma_0 \times \mathbb{R}^{1+q}_{\rho_2, \tilde{\eta}}))$ for some

$$\tilde{\boldsymbol{p}}_1(r_1, r_2, z_1, \tilde{\rho}_2, \tilde{\eta}) \in C^\infty(\overline{\mathbb{R}}_+ \times \overline{\mathbb{R}}_+, M^{\mu-1}_{\mathcal{O}_{z_1}}(X; \mathbb{R}_{\tilde{\rho}_2} \times \mathbb{R}^q_{\tilde{\eta}})). \quad (3.10)$$

In (3.10) we indicated holomorphic dependence of $\tilde{\boldsymbol{p}}_1$ on $z_1 \in \mathbb{C}$, although we first interpreted z_1 as a variable on Γ_0. However, the asymptotic sum for $\tilde{\boldsymbol{p}}_1$ may be produced in combination with a kernel cut-off step in z_1, and this yields $\tilde{\boldsymbol{p}}_1$ as a holomorphic function. Since $\tilde{\boldsymbol{p}}$ is holomorphic also in z_1, the equivalence (3.9) modulo $C^\infty(\mathbb{R}_+ \times \mathbb{R}_+, L^{-\infty}(X; \Gamma_\beta \times \mathbb{R}^{1+q}_{\rho_2, \tilde{\eta}}))$ holds under restriction to $\Gamma_\beta \ni z_1$. In the asymptotic summation for $\tilde{\boldsymbol{p}}_1$ we employed the fact that the smoothness in r_1 and r_2 up to 0 remains under control indeed, and that the holomorphy in z_1 is guaranteed when we combine the asymptotic process in the covariables in $\Gamma_0 \times \mathbb{R}^{1+q}_{\tilde{\rho}_2, \tilde{\eta}}$ with a kernel cut off step with respect to z_1. The relation (3.7) now contains holomorphic functions in z_1 and the smoothing remainders appear first for $z_1 \in \Gamma_0$. Then, when we take $z_1 \in \Gamma_\beta$ for any other real β the remainder is again of such a quality, as indicated in (3.7). □

Proof of Theorem 3.6. Similarly as in (2.16) it suffices to show the assertion (3.6) for some specific β_1, β_2, e.g., $\beta_1 = \beta_2 = 1/2$ which is again a consequence of Cauchy's theorem. By iterating the result of Lemma 3.7 we produce in a recursive manner

$$\boldsymbol{p}_j(r_1, r_2, z_1, \rho_2, \tilde{\eta}) = \tilde{\boldsymbol{p}}_j(r_1, r_2, z_1, r_2\rho_2, \tilde{\eta})$$

for $\tilde{\boldsymbol{p}}_j(r_1, r_2, z_1, \tilde{\rho}_2, \tilde{\eta}) \in C^\infty(\overline{\mathbb{R}}_+ \times \overline{\mathbb{R}}_+, M^{\mu-j}_{\mathcal{O}_{z_1}}(X; \mathbb{R}_{\tilde{\rho}_2} \times \mathbb{R}^q_{\tilde{\eta}}))$. Analogously as (3.7) we obtain equivalences

$$\operatorname{Op}_{r_2}(\boldsymbol{p}_j)(r_1, z_1, \tilde{\eta}) = \operatorname{op}^{1/2}_{M_{r_2}}(\boldsymbol{f}_j)(r_1, z_1, \tilde{\eta}) + \operatorname{Op}_{r_2}(\boldsymbol{p}_{j+1})(r_1, z_1, \tilde{\eta}),$$

for $\boldsymbol{f}_j(r_1, r_2, z_1, i\tau, \tilde{\eta}) \in C^\infty(\overline{\mathbb{R}}_+ \times \overline{\mathbb{R}}_+, M^{\mu-j}_{\mathcal{O}_{z_1}}(X; \Gamma_0 \times \mathbb{R}^{1+q}_{\tau, \tilde{\eta}}))$ where $\boldsymbol{p}_0 := \boldsymbol{p}$ for all $z_1 \in \mathbb{C}$, with remainders in $C^\infty(\mathbb{R}_+, L^{-\infty}(\mathbb{R}_+ \times X; \Gamma_\beta \times \mathbb{R}^q_{\tilde{\eta}}))$ when restricted to Γ_β in z_1. Thus we have for every $l \in \mathbb{N}$

$$\operatorname{Op}_{r_2}(\boldsymbol{p})(r_1, z_1, \tilde{\eta}) = \operatorname{op}^{1/2}_{M_{r_2}}\Big(\sum_{j=0}^{l} \boldsymbol{f}_j\Big)(r_1, z_1, \tilde{\eta}) + \operatorname{Op}_{r_2}(\boldsymbol{p}_{l+1})(r_1, z_1, \tilde{\eta})$$

modulo a reminder of the above-mentioned kind.

The analogue of the step (2.17) cannot be done immediately in the framework of operator functions that are holomorphic in z_1, since the asymptotic summation with excision functions in the covariables $(z_1, z_2, \tilde{\eta}) \in \Gamma_\beta \times \Gamma_0 \times \mathbb{R}^q_{\tilde{\eta}}$ destroys the holomorphic dependence on z_1. However, we may fix β and first treat z_1 as a covariable on Γ_β. Without loss of generality we take $\beta = 1/2$. Then the asymptotic sum gives us

$$m(r_1, r_2, i\alpha, i\tau, \tilde{\eta}) \sim \sum_{j=0}^{\infty} \boldsymbol{f}_j(r_1, r_2, i\alpha, i\tau, \tilde{\eta}), \tag{3.11}$$

for $m(r_1, r_2, i\alpha, i\tau, \tilde{\eta}) \in C^\infty(\overline{\mathbb{R}}_+ \times \overline{\mathbb{R}}_+, L^\mu(X; \Gamma_0 \times \Gamma_0 \times \mathbb{R}^q_{\tilde{\eta}}))$.
Applying now the kernel cut-off operator $\mathcal{V}_\psi$ as in Theorem 2.4 with respect to $z_1 \in \Gamma_0$ we obtain an $\boldsymbol{m}(r_1, r_2, z_1, i\tau, \tilde{\eta}) \in C^\infty(\overline{\mathbb{R}}_+ \times \overline{\mathbb{R}}_+, M^\mu_{\mathcal{O}_{z_1}}(X; \Gamma_0 \times \mathbb{R}^q_{\tilde{\eta}}))$ such that

$$m(r_1, r_2, i\alpha, i\tau, \tilde{\eta}) = \boldsymbol{m}(r_1, r_2, z_1, i\tau, \tilde{\eta})|_{z_1 \in \Gamma_0}$$

modulo $C^\infty(\mathbb{R}_+ \times \mathbb{R}_+, L^{-\infty}(X; \Gamma_0 \times \Gamma_0 \times \mathbb{R}^q_{\tilde{\eta}}))$. Moreover, applying $\mathcal{V}_\psi$ with respect to $z_2 \in \Gamma_0$ as in Theorem 3.2 (i) yields a function

$$\boldsymbol{h}(r_1, r_2, z_1, z_2, \tilde{\eta}) \in C^\infty(\overline{\mathbb{R}}_+ \times \overline{\mathbb{R}}_+, M^\mu_{\mathcal{O}_{z_1}, \mathcal{O}_{z_2}}(X; \mathbb{R}^q_{\tilde{\eta}})) \tag{3.12}$$

such that

$$\boldsymbol{m}(r_1, r_2, z_1, i\tau, \tilde{\eta}) = \boldsymbol{h}(r_1, r_2, z_1, z_2, \tilde{\eta})|_{z_2 \in \Gamma_0}.$$

It follows that

$$\mathrm{Op}_{r_2}(\boldsymbol{p})(r_1, z_1, \tilde{\eta}) = \mathrm{op}^{1/2}_{M_{r_2}}(\boldsymbol{h})(r_1, z_1, \tilde{\eta})$$

modulo $C^\infty(\mathbb{R}_{+, r_1}, L^{-\infty}(\mathbb{R}_+ \times X; \Gamma_0 \times \mathbb{R}^q_{\tilde{\eta}}))$. This gives us finally (3.6) for $\beta_1 = \beta_2 = 1/2$ after applying $\mathrm{op}^{1/2}_{M_{r_1}}$ on both sides. □

Let us now turn to the Mellin quantisation of corner-degenerate symbols which is the main issue of this section. Those symbols can be written as (3.1), now for

$$\tilde{p}(r, x, y, \tilde{\rho}_1, \tilde{\tilde{\rho}}_2, \xi, \tilde{\eta}^1, \tilde{\tilde{\eta}}^2) \in S^\mu(\overline{\mathbb{R}}_+ \times \overline{\mathbb{R}}_+ \times \mathbb{R}^{n+q}_{x,y} \times \mathbb{R}^{2+n+q}_{\tilde{\rho}_1, \tilde{\tilde{\rho}}_2, \xi, \tilde{\eta}^1, \tilde{\tilde{\eta}}^2}). \tag{3.13}$$

Similarly as before we pass to globalized corner-degenerate operator functions along X, i.e., we start with functions

$$p(r_1, r_2, y, \rho_1, \rho_2, \eta^1, \eta^2) = \tilde{p}(r_1, r_2, y, r_1\rho_1, r_1 r_2 \rho_2, r_1 \eta^1, r_1 r_2 \eta^2) \tag{3.14}$$

for

$$\tilde{p}(r, y, \tilde{\rho}_1, \tilde{\tilde{\rho}}_2, \tilde{\eta}^1, \tilde{\tilde{\eta}}^2) \in C^\infty(\overline{\mathbb{R}}_+ \times \overline{\mathbb{R}}_+ \times \mathbb{R}^q, L^\mu(X; \mathbb{R}^{2+q}_{\tilde{\rho}_1, \tilde{\tilde{\rho}}_2, \tilde{\eta}^1, \tilde{\tilde{\eta}}^2})). \tag{3.15}$$

Theorem 3.8. *For every p of the form (3.14) with (3.15) there exists an $\tilde{h}(r, y, z_1, \tilde{z}_2, \tilde{\eta}^1, \tilde{\tilde{\eta}}^2) \in C^\infty(\overline{\mathbb{R}}_+ \times \overline{\mathbb{R}}_+ \times \mathbb{R}^q_y, M^\mu_{\mathcal{O}_{z_1}, \mathcal{O}_{\tilde{z}_2}}(X; \mathbb{R}^q_{\tilde{\eta}^1, \tilde{\tilde{\eta}}^2}))$ such that for*

$$h(r, y, z_1, z_2, \eta^1, \eta^2) := \tilde{h}(r, y, z_1, r_1 z_2, r_1 \eta^1, r_1 r_2 \eta^2) \tag{3.16}$$

we have

$$\mathrm{Op}_{r_2, r_1}(p)(y, \eta) = \mathrm{op}^{\beta_2}_{M_{r_2}} \mathrm{op}^{\beta_1}_{M_{r_1}}(h)(y, \eta) \tag{3.17}$$

modulo $C^\infty(\mathbb{R}^q, L^{-\infty}(\mathbb{R}_+ \times \mathbb{R}_+ \times X; \mathbb{R}^q))$ for any reals β_1, β_2.

Proof. The proof will be given in several steps. First we apply Theorem 2.1 in global form with respect to X. Then the result reads as follows. For every

$$p^1(r_1, y^1, \rho_1, \tilde{\eta}^1) = \tilde{p}^1(r_1, y^1, r_1\rho_1, \tilde{\eta}^1)$$

where $\tilde{p}^1(r_1, y^1, \tilde{\rho}_1, \tilde{\eta}^1) \in C^\infty(\overline{\mathbb{R}}_+ \times \mathbb{R}^{q_1}, L^\mu(X; \mathbb{R}^{1+q_1}_{\tilde{\rho}_1, \tilde{\eta}^1}))$ there exists an

$$h^1(r_1, y^1, z_1, \tilde{\eta}^1) \in C^\infty(\overline{\mathbb{R}}_+ \times \mathbb{R}^{q_1}, M^\mu_{\mathcal{O}_{z_1}}(X; \mathbb{R}^{q_1}_{\tilde{\eta}^1}))$$

such that

$$\mathrm{Op}_{r_1}(p^1)(y^1, \tilde{\eta}^1) = \mathrm{op}^{\beta_1}_{M_{r_1}}(h^1)(y^1, \tilde{\eta}^1)$$

for every $\beta_1 \in \mathbb{R}$, modulo smoothing remainders. The assertion extends in an obvious manner to the case when we let p^1 and h^1 depend on further variables and covariables $(r_2, y^2, \tilde{\rho}_2, \tilde{\eta}^2) \in \overline{\mathbb{R}}_+ \times \mathbb{R}^{q_2} \times \mathbb{R}^{1+q_2}_{\tilde{\rho}_2, \tilde{\eta}^2}$. Taking into account these additional variables we get an analogue of Theorem 2.1 that states the Mellin quantisation in the r_1-variable, namely a correspondence

$$p^1(r_1, r_2, y, \rho_1, \tilde{\rho}_2, \tilde{\eta}) \rightsquigarrow h^1(r_1, r_2, y, z_1, \tilde{\rho}_2, \tilde{\eta})$$

with the replacement of notation for $y = (y^1, y^2)$ and $\tilde{\eta} = (\tilde{\eta}^1, \tilde{\eta}^2) \in \mathbb{R}^q$.

$$h^1 \rightsquigarrow \tilde{\boldsymbol{p}},\ \tilde{\boldsymbol{p}}(r_1, r_2, y, \tilde{\rho}_2, \tilde{\eta}) \in C^\infty(\overline{\mathbb{R}}_+ \times \mathbb{R}^{q_2} \times \mathbb{R}^q, M^\mu_{\mathcal{O}_{z_1}}(X; \mathbb{R}^{1+q}_{\tilde{\rho}_2, \tilde{\eta}}))$$

we are in the situation of Theorem 3.6 and we can produce an

$$\boldsymbol{h}(r_1, r_2, y, z_1, z_2, \tilde{\eta}) \in C^\infty(\overline{\mathbb{R}}_+ \times \mathbb{R}^{q_2} \times \mathbb{R}^q, M^\mu_{\mathcal{O}_{z_1}, \mathcal{O}_{z_2}}(X; \mathbb{R}^q_{\tilde{\eta}}))$$

such that in the notation of Theorem 3.6

$$\mathrm{Op}_{r_2}\mathrm{op}^{\beta_1}_{M_{r_1}}(\boldsymbol{p})(y, \tilde{\eta}) = \mathrm{op}^{\beta_2}_{M_{r_2}}\mathrm{op}^{\beta_1}_{M_{r_1}}(\boldsymbol{h})(y, \tilde{\eta})$$

modulo $C^\infty(\mathbb{R}^q, L^{-\infty}(\mathbb{R}_+ \times \mathbb{R}_+ \times X; \mathbb{R}^q_{\tilde{\eta}}))$. From now on we drop again the variables y. We establish a modification of Theorem 3.6. More precisely we apply Theorem 3.6 to families of operator functions of the same nature, with parameter $c \in \mathbb{R}_+$. Starting with $\boldsymbol{p}, \tilde{\boldsymbol{p}}$ as in Theorem 3.6 we form

$$\boldsymbol{p}^{\mathrm{c}}(r_1, r_2, z_1, \rho_2, \tilde{\eta}) := \tilde{\boldsymbol{p}}(r_1, r_2, z_1, cr_2\rho_2, \tilde{\eta}^1, c\tilde{\eta}^2)$$

where

$$\tilde{\boldsymbol{p}}^{\mathrm{c}}(r_1, r_2, z_1, \tilde{\rho}_2, \tilde{\eta}^1, \tilde{\eta}^2) := \tilde{\boldsymbol{p}}(r_1, r_2, z_1, c\tilde{\rho}_2, \tilde{\eta}^1, c\tilde{\eta}^2)$$

belongs to $C^\infty(\overline{\mathbb{R}}_+ \times \overline{\mathbb{R}}_+, M^\mu_{\mathcal{O}_{z_1}}(X; \mathbb{R}_{\tilde{\rho}_2} \times \mathbb{R}^q_{\tilde{\eta}}))$ for every fixed $c \in \mathbb{R}_+$; $\tilde{\eta} = (\tilde{\eta}^1, \tilde{\eta}^2)$. An inspection of the proof of Theorem 3.6 shows that the resulting Mellin symbol $\boldsymbol{h}^{\mathrm{c}}$ is of the form

$$\boldsymbol{h}^{\mathrm{c}}(r_1, r_2, z_1, z_2, \tilde{\eta}) = \tilde{\boldsymbol{h}}(c, r_1, r_2, z_1, cz_2, \tilde{\eta}^1, c\tilde{\eta}^2)$$

for an $\tilde{\boldsymbol{h}}(c, r_1, r_2, z_1, \tilde{z}_2, \tilde{\eta}^1, \tilde{\tilde{\eta}}^2) \in C^\infty(\overline{\mathbb{R}}_{+,c} \times \overline{\mathbb{R}}_+ \times \overline{\mathbb{R}}_+, M^\mu_{\mathcal{O}_{z_1}, \mathcal{O}_{\tilde{z}_2}}(X; \mathbb{R}^q_{\tilde{\eta}^1, \tilde{\tilde{\eta}}^2}))$, i.e., we have

$$\mathrm{Op}_{r_2}\mathrm{op}^{\beta_1}_{M_{r_1}}(\boldsymbol{p}^{\mathrm{c}})(\tilde{\eta}) = \mathrm{op}^{\beta_2}_{M_{r_2}}\mathrm{op}^{\beta_1}_{M_{r_1}}(\boldsymbol{h}^{\mathrm{c}})(\tilde{\eta}) \bmod L^{-\infty}(\mathbb{R}_+ \times \mathbb{R}_+ \times X; \mathbb{R}^q_{\tilde{\eta}}).$$

Our point is to verify that, in the c-version of the relation (3.7) we simply replace $\tilde{\eta}$ by $(\tilde{\eta}^1, c\tilde{\eta}^2)$ and $i\tau$ by $ic\tau$. All the smoothing remainders will depend on c, however, this does not affect the final result, since such remainders are accepted in the quantisation. The relation (3.7) takes the form

$$\mathrm{Op}_{r_2}(\boldsymbol{p}^{\mathrm{c}})(r_1,z_1,\tilde{\eta})|_{\Gamma_\beta} = \mathrm{op}_{M_{r_2}}^{1/2}(\boldsymbol{f}_0^{\mathrm{c}})(r_1,z_1,\tilde{\eta})|_{\Gamma_\beta} + \mathrm{Op}_{r_2}(\boldsymbol{p}_1^{\mathrm{c}})(r_1,z_1,\tilde{\eta})|_{\Gamma_\beta} \tag{3.18}$$

modulo a remainder as in (3.7) for a $\boldsymbol{p}_1^{\mathrm{c}}$ of analogous structure as $\boldsymbol{p}^{\mathrm{c}}$, but of order $\mu-1$ and $\boldsymbol{f}_0^{\mathrm{c}}(r_1,r_2,z_1,i\tau,\tilde{\eta}) = \boldsymbol{f}_0(r_1,r_2,z_1,ic\tau,\tilde{\eta}^1,c\tilde{\eta}^2)$. In addition in this computation we generate $\boldsymbol{p}_1^{\mathrm{c}}$ with an extra smooth dependence on c, i.e., we get

$$\boldsymbol{p}_1^{\mathrm{c}}(r_1,r_2,z_1,r_2\rho_2,\tilde{\eta}) = \tilde{\boldsymbol{p}}_1(c,r_1,r_2,z_1,cr_2\rho_2,\tilde{\eta}^1,c\tilde{\eta}^2). \tag{3.19}$$

The same iteration process as before gives us functions

$$\boldsymbol{f}_j^{\mathrm{c}}(r_1,r_2,z_1,i\tau,\tilde{\eta}) = \boldsymbol{f}_j(c,r_1,r_2,z_1,ic\tau,\tilde{\eta}^1,c\tilde{\eta}^2)$$

for every $j\geq 1$. The asymptotic sum analogously as (3.11) yields an m^{c} that can be carried out in such a way that $i\tau$ and $\tilde{\eta}^2$ contains the factor c, and the subsequent z_1-kernel cut-off preserves this structure as well. This gives us finally $\boldsymbol{h}^{\mathrm{c}}$ of the desired structure. Summing up we proved the following result, with slightly modified meaning of $\tilde{\boldsymbol{p}}^{\mathrm{c}}$ now with an extra c-dependence. For every

$$\boldsymbol{p}^{\mathrm{c}}(r_1,r_2,z_1,\rho_2,\tilde{\eta}) := \tilde{\boldsymbol{p}}(c,r_1,r_2,z_1,cr_2\rho_2,\tilde{\eta}^1,c\tilde{\eta}^2)$$

for

$$\tilde{\boldsymbol{p}}(c,r_1,r_2,z_1,\tilde{\rho}_2,\tilde{\eta}^1,\tilde{\tilde{\eta}}^2) \in C^\infty(\overline{\mathbb{R}}_{+,c}\times\overline{\mathbb{R}}_{+,r_1}\times\overline{\mathbb{R}}_{+,r_2}, M^{\mu}_{\mathcal{O}_{z_1}}(X;\mathbb{R}^{1+q}_{\tilde{\rho}_2,\tilde{\eta}^1,\tilde{\tilde{\eta}}^2}))$$

there exists an

$$\boldsymbol{h}^{\mathrm{c}}(r_1,r_2,z_1,z_2,\tilde{\eta}) := \tilde{\boldsymbol{h}}(c,r_1,r_2,z_1,cz_2,\tilde{\eta}^1,c\tilde{\eta}^2)$$

for

$$\tilde{\boldsymbol{h}}(c,r_1,r_2,z_1,\tilde{z}_2,\tilde{\eta}^1,\tilde{\tilde{\eta}}^2) \in C^\infty(\overline{\mathbb{R}}_+\times\overline{\mathbb{R}}_+\times\overline{\mathbb{R}}_+, M^{\mu}_{\mathcal{O}_{z_1},\mathcal{O}_{\tilde{z}_2}}(X;\mathbb{R}^{q}_{\tilde{\eta}^1,\tilde{\tilde{\eta}}^2}))$$

such that

$$\begin{aligned}\mathrm{Op}_{r_2}&\mathrm{op}_{M_{r_1}}^{\beta_1}(\boldsymbol{p}^{\mathrm{c}})(\tilde{\eta}) = \\ &\mathrm{op}_{M_{r_2}}^{\beta_2}\mathrm{op}_{M_{r_1}}^{\beta_1}(\boldsymbol{h}^{\mathrm{c}})(\tilde{\eta}) \bmod C^\infty(\mathbb{R}_{+,c},L^{-\infty}(\mathbb{R}_+\times\mathbb{R}_+\times X;\mathbb{R}^q_{\tilde{\eta}})).\end{aligned} \tag{3.20}$$

What concerns the part of the proof which generates $\boldsymbol{p}_1^{\mathrm{c}}$ we also see that the factor c remains at the covariables $i\tau$ and $\tilde{\eta}^2$. In computing $\boldsymbol{p}_1^{\mathrm{c}}$ we also apply an analogue of what we did in (2.11). It follows analogously as (3.8) a function

$$\boldsymbol{c}^{\mathrm{c}}(r_1,r_2,z_1,\rho_2,\tilde{\eta}^1,\tilde{\eta}^2) = \boldsymbol{c}(c,r_1,r_2,z_1,c\rho_2,\tilde{\eta}^1,c\tilde{\eta}^2),$$

via

$$\begin{aligned}\boldsymbol{c}^{\mathrm{c}}(r_1,r_2,z_1,\rho_2,&\tilde{\eta}^1,\tilde{\eta}^2)|_{r_2=\chi(t),\rho_2=\tau} \\ &\sim \sum_{j=0}^{\infty}\frac{1}{j!}(\partial_\tau^j\boldsymbol{a})(r_1,t,z_1,(d\chi(t))c\tau,\tilde{\eta}^1,c\tilde{\eta}^2)\Phi_j(t,\tau).\end{aligned} \tag{3.21}$$

Let us explain the sense of carrying out the asymptotic sum (3.21). The contribution of

$$\boldsymbol{a}^{\mathrm{c}}(r_1,t,z_1,\tau,\tilde{\eta}^1,\tilde{\eta}^2) := \boldsymbol{a}(c,r_1,t,z_1,c\tau,\tilde{\eta}^1,c\tilde{\eta}^2)$$

for $\boldsymbol{a}(c, r_1, t, z_1, c\tau, \tilde{\eta}^1, c\tilde{\eta}^2) = \boldsymbol{f}_0(c, r_1, e^{-t}, z_1, ic\tau, \tilde{\eta}^1, c\tilde{\eta}^2)$ contains in its argument the right combination of $i\tau$ and $\tilde{\eta}^2$ with the factor c, namely, $ic\tau$ and $c\tilde{\eta}^2$. Although $\Phi_j(t, \tau)$ is independent of c, we gain through the τ-differentiation of $\boldsymbol{a}^{\mathrm{c}}$ in the c-dependent version of (3.8) also an extra factor c^j. The function $\Phi_j(t, \tau)$ has the form

$$\Phi_j(t, \tau) = \Psi_j(r_2, \tilde{\tau})|_{\tilde{\tau}=e^{-t}\tau, r_2=e^{-t}}$$

for the above-mentioned polynomial $\Psi_j(r_2, \tilde{\tau})$ of degree $\leq j/2$ in $\tilde{\tau}$, with smooth dependence in r_2 up to $r_2 = 0$. In addition we have

$$c^j \Psi_j(r_2, \tilde{\tau}) =: \psi_j(c, r_2, \tilde{\tilde{\tau}}),$$

$\tilde{\tilde{\tau}} = c\tilde{\tau}$, with a function $\psi_j(c, r_2, \tilde{\tilde{\tau}})$ being a polynomial in $\tilde{\tilde{\tau}}$ of degree $\leq j/2$ and smooth in $(c, r_1, r_2) \in \overline{\mathbb{R}}_+ \times \overline{\mathbb{R}}_+ \times \overline{\mathbb{R}}_+$ up to $(0, 0, 0)$. In order to see how to carry out the asymptotic sum (3.21) with a control of smoothness in $(c, r_1, r_2) \in \overline{\mathbb{R}}_+ \times \overline{\mathbb{R}}_+ \times \overline{\mathbb{R}}_+$ up to $(0, 0, 0)$ we replace the summands by

$$\tilde{\boldsymbol{g}}_j(c, r_1, r_2, z_1, \tilde{\tilde{\tau}}, \tilde{\eta}) := \boldsymbol{a}^{(j)}(c, r_1, r_2, z_1, \tilde{\tilde{\tau}}, \tilde{\eta})\psi_j(c, r_2, \tilde{\tilde{\tau}})$$

for

$$\boldsymbol{a}^{(j)}(c, r_1, r_2, z_1, \tilde{\tilde{\tau}}, \tilde{\eta}) := \frac{1}{j!}(\partial_\tau^j \boldsymbol{a})(c, r_1, -\log r_2, z_1, \tilde{\tilde{\tau}}, \tilde{\eta})$$

where $\tilde{\boldsymbol{g}}_j(c, r_1, r_2, z_1, \tilde{\tilde{\tau}}, \tilde{\eta}) \in C^\infty(\overline{\mathbb{R}}_+ \times \overline{\mathbb{R}}_+ \times \overline{\mathbb{R}}_+, M_{\mathcal{O}_{z_1}}^{\mu-j/2}(X; \mathbb{R}^{1+q}_{\tilde{\tilde{\tau}}, \tilde{\eta}}))$. We form

$$\tilde{\boldsymbol{p}}_1(c, r_1, r_2, z_1, \tilde{\tilde{\tau}}, \tilde{\eta}) = -\sum_{j=1}^{\infty} \tilde{\boldsymbol{g}}_j(c, r_1, r_2, z_1, \tilde{\tilde{\tau}}, \tilde{\eta}).$$

This yields

$$\begin{aligned} &\boldsymbol{c}^{\mathrm{c}}(r_1, r_2, z_1, \rho_2, \tilde{\eta}) \\ &\quad = \tilde{\boldsymbol{p}}(c, r_1, r_2, z_1, cr_2\rho_2, \tilde{\eta}^1, c\tilde{\eta}^2) + \tilde{\boldsymbol{p}}_1(c, r_1, r_2, z_1, cr_2\rho_2, \tilde{\eta}^1, c\tilde{\eta}^2) \end{aligned}$$

modulo $C^\infty(\mathbb{R}_+ \times \mathbb{R}_+ \times \mathbb{R}_+, L^{-\infty}(X; \Gamma_0 \times \mathbb{R}^{1+q}_{\rho_2, \tilde{\eta}}))$ for some $\tilde{\boldsymbol{p}}_1(c, r_1, r_2, z_1, \tilde{\rho}_2, \tilde{\eta}) \in C^\infty(\overline{\mathbb{R}}_{+,c} \times \overline{\mathbb{R}}_{+,r_1} \times \overline{\mathbb{R}}_{+,r_2} \times \mathbb{R}^q_{\tilde{\eta}}, M_{\mathcal{O}}^{\mu-1}(X; \mathbb{R}_{\tilde{\rho}_2} \times \mathbb{R}^q_{\tilde{\eta}}))$.This gives us (2.19) for (3.19) which is the c-variant of (3.7).

In the final step of the proof we simply interpret c as $r_1 \in \overline{\mathbb{R}}_+$ and $\tilde{\eta}$ as $(r_1\eta^1, r_1 r_2\eta^2)$. Then the assertion of the theorem is an immediate consequence of (3.20). This is an admissible argument, since the r_1 and r_2 variables are involved as variables in left symbols, i.e., they act in the respective pseudo-differential operators as multiplications from the left. All the c dependent smoothing remainders occurring in the first part of the proof after replacements $c \rightsquigarrow r_1$, $\tilde{\eta} \rightsquigarrow (r_1\eta^1, r_1 r_2\eta^2)$ turn again to smoothing remainders of analogous quality. $\square$

4. Mellin symbols for higher-order corners

The method of proving Theorem 3.8 is iterative. It is nearly straightforward to generalise the respective operator-valued symbol spaces with holomorphic dependence on several complex variables. Therefore, we briefly formulate the result and content ourselves with a few remarks.

First we have the following analogue of Definition 3.1. As before let X be a closed C^∞ manifold.

Definition 4.1. The space $M^\mu_{\mathcal{O}_z}(X;\mathbb{R}^q_\eta)$ for $z=(z_1,\dots,z_k)$, $\eta\in\mathbb{R}^q$, $k\geq 2$, $\mu\in\mathbb{R}$, is defined to be the set of all $h(z',z_k,\eta)$, $z'=(z_1,\dots,z_{k-1})$, such that $h(z',z_k,\eta)\in\mathcal{A}(\mathbb{C}_{z_k},M^\mu_{\mathcal{O}_{z'}}(X;\mathbb{R}^q_\eta))$ with

$$h(z',z_k,\eta)\in M^\mu_{\mathcal{O}_{z'}}(X;\Gamma_\delta\times\mathbb{R}^q_\eta)$$

for every $\delta\in\mathbb{R}$, uniformly in compact δ-intervals. We also write $M^\mu_{\mathcal{O}}(X;\mathbb{R}^q)$ rather than $M^\mu_{\mathcal{O}_z}(X;\mathbb{R}^q)$.

Here we inductively assume that the spaces $M^\mu_{\mathcal{O}_{z'}}(X;\mathbb{R}^l_\lambda)$ are already defined for every $l\in\mathbb{N}$, with a Fréchet topology that follows in a natural way from the definition. As before when we write Γ_δ as a component of a space of parameters we mean the real variable $\operatorname{Im} z_k$ with z_k varying on Γ_δ. The space $M^\mu_{\mathcal{O}}(X;\mathbb{R}^q)$ can be defined in many equivalent ways; e.g., analogously as in Remark 3.3(i). In addition there is an analogue of Remark 3.3(ii). Moreover, we have continuous embeddings $M^{\mu'}_{\mathcal{O}}(X;\mathbb{R}^q)\hookrightarrow M^\mu_{\mathcal{O}}(X;\mathbb{R}^q)$ for any $\mu'\leq\mu$.

Observe that there are analogues of Theorems 3.2 and 3.4 also for arbitrary k. Those play a similar role for higher corner operators as the above theorems for the case $k=2$.

Consider an operator function

$$p(r,y,\rho,\eta)=\tilde{p}(r,y,r_1\rho_1,r_1r_2\rho_2,\dots,r_1\dots r_k\rho_k,r_1\eta^1,r_1r_2\eta^2,\dots,r_1\dots r_k\eta^k) \tag{4.1}$$

for

$$\tilde{p}(r,y,\tilde{\rho},\tilde{\eta})\in C^\infty((\overline{\mathbb{R}}_+)^k\times\Omega,L^\mu(X;\mathbb{R}^{k+q}_{\tilde{\rho},\tilde{\eta}})), \tag{4.2}$$

$\Omega\subseteq\mathbb{R}^q$ open, $(\overline{\mathbb{R}}_+)^k=\overline{\mathbb{R}}_+\times\cdots\times\overline{\mathbb{R}}_+$ (k factors). Then our main result is as follows.

Theorem 4.2. *For every p as in (4.1) with (4.2) there exists an $\tilde{h}(r,y,\tilde{z},\tilde{\eta})\in C^\infty((\overline{\mathbb{R}}_+)^k\times\Omega,M^\mu_{\mathcal{O}_{\tilde{z}}}(X;\mathbb{R}^q_{\tilde{\eta}}))$, $\tilde{z}=(\tilde{z}_1,\dots,\tilde{z}_k)$, such that for $h(r,y,z,\eta)$ defined by*

$$\tilde{h}(r,y,z_1,r_1z_2,r_1r_2z_3,\dots,r_1r_2\dots r_{k-1}z_k,r_1\eta^1,r_1r_2\eta^2,\dots,r_1r_2\dots r_k\eta^k)$$

we have

$$\operatorname{Op}_{r_k,\dots,r_1}(p)(y,\eta)=\operatorname{op}^{\beta_k}_{M_{r_k}}\dots\operatorname{op}^{\beta_1}_{M_{r_1}}(h)(y,\eta)\ \mathrm{mod}\ C^\infty(\Omega,L^{-\infty}(\mathbb{R}^k_+\times X;\mathbb{R}^q))$$

for any reals $\beta_1,\dots,\beta_k$.

Remark 4.3. Theorem 4.2 refers to an equivalence of (y,η)-depending families of operators $C_0^\infty(\mathbb{R}_+^k \times X) \to C^\infty(\mathbb{R}_+^k \times X)$ modulo smoothing operators. The role of the Mellin quantisation is to modify the original operator by removing some smoothing (possibly singular) error and to obtain someone that admits a continuous extension between suitable weighted Sobolev spaces. This aspect cannot be discussed in detail here; it requires more voluminous considerations.

Remark 4.4. The Mellin quantisation result of Theorem 4.2 may be specified for classical symbols. Then, starting with classical p we obtain classical h in the sense that everywhere in the definition of holomorphic Mellin symbols we have L_{cl}^μ instead of L^μ. This aspect is of relevance in the higher corner pseudo-differential calculus for analogous reasons as in cone and edge algebras corresponding to the case $k = 1$.

Let us finally observe that the spaces $M_{\mathcal{O}}^\mu(X;\mathbb{R}^q)$ admit a concept of ellipticity and parametrices that is also a part of the elliptic theory in higher corner operators. First note that

$$h_j(z,\eta) \in M_{\mathcal{O}}^{\mu_j}(X;\mathbb{R}^q), \ j = 1, 2,$$

implies $h_1(z,\eta)h_2(z,\eta) \in M_{\mathcal{O}}^{\mu_1+\mu_2}(X;\mathbb{R}^q)$ for arbitrary $\mu_1, \mu_2 \in \mathbb{R}$.

An element $h(z,\eta) \in M_{\mathcal{O}}^\mu(X;\mathbb{R}^q)$ is called elliptic if there is a tuple $\beta = (\beta_1, \dots, \beta_k) \in \mathbb{R}^k$ such that

$$h(z,\eta)|_{\Gamma_\beta} \in L^\mu(X;\Gamma_\beta \times \mathbb{R}^q) \text{ for } \Gamma_\beta := \Gamma_{\beta_1} \times \cdots \times \Gamma_{\beta_k}$$

is parameter-dependent elliptic of order μ with the parameters $(z_1, \dots, z_k, \eta) \in \Gamma_\beta \times \mathbb{R}^q$. This definition is independent of the choice of β, i.e., the condition is equivalent to the one with respect to $\Gamma_{\beta'}$ for any other $\beta' \in \mathbb{R}^k$.

Theorem 4.5. *Let $h(z,\eta) \in M_{\mathcal{O}}^\mu(X;\mathbb{R}^q)$ be elliptic; then there exists an*

$$h^{(-1)}(z,\eta) \in M_{\mathcal{O}}^{-\mu}(X;\mathbb{R}^q)$$

such that

$$h(z,\eta)h^{(-1)}(z,\eta) = 1, \ h^{(-1)}(z,\eta)h(z,\eta) = 1 \bmod M_{\mathcal{O}}^{-\infty}(X;\mathbb{R}^q). \tag{4.3}$$

Proof. By assumption $f(z,\eta) := h(z,\eta)|_{\Gamma_\beta} \in L^\mu(X;\Gamma_\beta \times \mathbb{R}^q)$ is parameter-dependent elliptic. Let $f^{(-1)}(z,\eta) \in L^{-\mu}(X;\Gamma_\beta \times \mathbb{R}^q)$ be a parameter-dependent parametrix. Then we may set

$$h^{(-1)}(z,\eta) := \big(\mathcal{V}_{\psi,z} f^{(-1)}\big)(z,\eta)$$

where $\mathcal{V}_{\psi,z}$ is the multiple kernel cut-off operator $\mathcal{V}_{\psi,z} := \mathcal{V}_{\psi,z_k} \dots \mathcal{V}_{\psi,z_1}$. For simplicity we took the same ψ with respect to different variables; we could distinguish with ψ's in different variables as well. The property (4.3) then follows from a corresponding relation between $f(z,\eta)$, $f^{(-1)}(z,\eta)$ and the corresponding analogues of the above-mentioned properties of the kernel cut-off operator. □

References

[1] D. Calvo, C.-I. Martin, and B.-W. Schulze, *Symbolic structures on corner manifolds*, RIMS Conf. dedicated to L. Boutet de Monvel on "Microlocal Analysis and Asymptotic Analysis", Kyoto, August 2004, Keio University, Tokyo, 2005, pp. 22–35.

[2] D. Calvo and B.-W. Schulze, *Edge symbolic structures of second generation*, Math. Nachr. **282** (2009), 348–367.

[3] M. Dauge, *Elliptic Boundary Value Problems on Corner Domains*, Lecture Notes in Mathematics **1341**, Springer-Verlag, Berlin, Heidelberg, 1988.

[4] N. Dines, *Elliptic Operators on Corner Manifolds*, Ph.D. thesis University of Potsdam, 2006.

[5] G.I. Eskin, *Boundary Value Problems for Elliptic Pseudo-Differential Equations*, Transl. of Nauka, Moskva, 1973, Math. Monographs, Amer. Math. Soc. **52**, Providence, Rhode Island 1980.

[6] R.C. Gunning and H. Rossi, *Analytic Functions of Several Complex Variables*, American Mathematical Soc., Rhode Island, 1965.

[7] N. Habal and B.-W. Schulze, *Holomorphic corner symbols*, J. Pseudo-Differ. Oper. Appl. **2**, 4 (2011), 419–465.

[8] G. Harutyunyan and B.-W. Schulze, *The relative index for corner singularities*, Integr. Equ. Oper. Theory **54**, 3 (2006), 385–426.

[9] G. Harutyunyan and B.-W. Schulze, *Elliptic Mixed, Transmission and Singular Crack Problems*, European Mathematical Soc., Zürich, 2008.

[10] D. Kapanadze and B.-W. Schulze, *Crack Theory and Edge Singularities*, Kluwer Academic Publ., Dordrecht, 2003.

[11] A.I. Komech, *Elliptic boundary problems for pseudo-differential operators on manifolds with conical points*, Mat. Sb. **86**, 2 (1971), 268–298.

[12] V.A. Kondratyev, *Boundary value problems for elliptic equations in domains with conical points*, Trudy Mosk. Mat. Obshch. **16** (1967), 209–292.

[13] T. Krainer, *Parabolic Pseudo-Differential Operators and Long-time Asymptotics of Solutions*, Ph.D. thesis, University of Potsdam, 2000.

[14] L. Maniccia and B.-W. Schulze, *An algebra of meromorphic corner symbols*, Bull. des Sciences Math. **127**, 1 (2003), 55–99.

[15] Pham The Lai, *Problème de Dirichlet dans un cône avec paramètre spectral pour une classe d'espaces de Sobolev à poids*, Comm. Part. Diff. Equ. **4** (1979), 389–445.

[16] B.A. Plamenevskij, *On the boundedness of singular integrals in spaces with weight*, Mat. Sb. **25**, 4 (1968), 573–592.

[17] B.A. Plamenevskij, *Algebras of Pseudodifferential Operators*, Nauka: Moscow, 1986.

[18] B.A. Plamenevskij and V.N. Senichkin, *On the representation of algebras of pseudo-differential operators on manifolds with discontinuous symbols* (Russian), Izv. Akad. Nauk SSSR **51**, 4 (1987), 833–859.

[19] V.S. Rabinovich, *Pseudo-differential operators in non-bounded domains with conical structure at infinity*, Mat. Sb. **80** (1969), 77–97.

[20] V.S. Rabinovich, *Mellin pseudo-differential operators with operator symbols and their applications*, Operator Theory: Adv. and Appl., Birkhäuser Verlag, Basel, **78** (1995), 271–279.

[21] S. Rempel and B.-W. Schulze, *Parametrices and boundary symbolic calculus for elliptic boundary problems without transmission property*, Math. Nachr. **105** (1982), 45–149.

[22] S. Rempel and B.-W. Schulze, *Complete Mellin and Green symbolic calculus in spaces with conormal asymptotics*, Ann. Glob. Anal. Geom. **4**, 2 (1986), 137–224.

[23] B.-W. Schulze, *Pseudo-Differential Operators on Manifolds with Singularities*, North-Holland, Amsterdam, 1991.

[24] B.-W. Schulze, *Boundary Value Problems and Singular Pseudo-Differential Operators*, J. Wiley, Chichester, 1998.

[25] B.-W. Schulze, *Operators with symbol hierarchies and iterated asymptotics*, Publications of RIMS, Kyoto University **38**, 4 (2002), 735–802.

[26] B.-W. Schulze, *The iterative structure of the corner calculus*, Oper. Theory: Adv. Appl. **213**, Pseudo-Differential Operators: Analysis, Application and Computations (L. Rodino et al. eds.), Birkhäuser Verlag, Basel, 2011, 79–103.

[27] B.-W. Schulze and J. Seiler, *The edge algebra structure of boundary value problems*, Ann. Glob. Anal. Geom. **22** (2002), 197–265.

[28] B.-W. Schulze and M.W. Wong, *Mellin and Green operators of the corner calculus*, J. Pseudo-Differ. Oper. Appl. **2**, 4 (2011), 467–507.

[29] J. Seiler, *Pseudodifferential Calculus on Manifolds with Non-Compact Edges*, Ph.D. thesis, University of Potsdam, 1997.

[30] I. Witt, *Explicit algebras with the Leibniz-Mellin translation product*, Math. Nachr. **280**, 3 (2007), 326–337.

N. Habal and B.-W. Schulze
Am Neuen Palais 10
D-14469 Potsdam, Germany
e-mail: habal@math.uni-potsdam.de
schulze@math.uni-potsdam.de

Operator Theory:
Advances and Applications, Vol. 228, 173–183

Pseudodifferential Operators on Variable Lebesgue Spaces

Alexei Yu. Karlovich and Ilya M. Spitkovsky

To Professor Vladimir Rabinovich on the occasion of his 70th birthday

Abstract. Let $\mathcal{M}(\mathbb{R}^n)$ be the class of bounded away from one and infinity functions $p:\mathbb{R}^n\to[1,\infty]$ such that the Hardy-Littlewood maximal operator is bounded on the variable Lebesgue space $L^{p(\cdot)}(\mathbb{R}^n)$. We show that if a belongs to the Hörmander class $S^{n(\rho-1)}_{\rho,\delta}$ with $0<\rho\le 1$, $0\le\delta<1$, then the pseudodifferential operator $\mathrm{Op}(a)$ is bounded on $L^{p(\cdot)}(\mathbb{R}^n)$ provided that $p\in\mathcal{M}(\mathbb{R}^n)$. Let $\mathcal{M}^*(\mathbb{R}^n)$ be the class of variable exponents $p\in\mathcal{M}(\mathbb{R}^n)$ represented as $1/p(x)=\theta/p_0+(1-\theta)/p_1(x)$ where $p_0\in(1,\infty)$, $\theta\in(0,1)$, and $p_1\in\mathcal{M}(\mathbb{R}^n)$. We prove that if $a\in S^0_{1,0}$ slowly oscillates at infinity in the first variable, then the condition

$$\lim_{R\to\infty}\inf_{|x|+|\xi|\ge R}|a(x,\xi)|>0$$

is sufficient for the Fredholmness of $\mathrm{Op}(a)$ on $L^{p(\cdot)}(\mathbb{R}^n)$ whenever $p\in\mathcal{M}^*(\mathbb{R}^n)$. Both theorems generalize pioneering results by Rabinovich and Samko [24] obtained for globally log-Hölder continuous exponents p, constituting a proper subset of $\mathcal{M}^*(\mathbb{R}^n)$.

Mathematics Subject Classification (2010). Primary 47G30; Secondary 42B25, 46E30.

Keywords. Pseudodifferential operator, Hörmander symbol, slowly oscillating symbol, variable Lebesgue space, Hardy-Littlewood maximal operator, Fefferman-Stein sharp maximal operator, Fredholmness.

1. Introduction

We denote the usual operators of first-order partial differentiation on $\mathbb{R}^n$ by $\partial_{x_j}:=\partial/\partial_{x_j}$. For every multi-index $\alpha=(\alpha_1,\dots,\alpha_n)$ with non-negative integers α_j, we write $\partial^\alpha:=\partial^{\alpha_1}_{x_1}\dots\partial^{\alpha_n}_{x_n}$. Further, $|\alpha|:=\alpha_1+\dots+\alpha_n$, and for each vector $\xi=(\xi_1,\dots,\xi_n)\in\mathbb{R}^n$, define $\xi^\alpha:=\xi_1^{\alpha_1}\dots\xi_n^{\alpha_n}$ and $\langle\xi\rangle:=(1+|\xi|_2^2)^{1/2}$ where $|\xi|_2$ stands for the Euclidean norm of ξ.

Let $C_0^\infty(\mathbb{R}^n)$ denote the set of all infinitely differentiable functions with compact support. Recall that, given $u \in C_0^\infty(\mathbb{R}^n)$, a pseudodifferential operator $\mathrm{Op}(a)$ is formally defined by the formula

$$(\mathrm{Op}(a)u)(x) := \frac{1}{(2\pi)^n} \int_{\mathbb{R}^n} d\xi \int_{\mathbb{R}^n} a(x,\xi)u(y)e^{i\langle x-y,\xi\rangle} dy,$$

where the symbol a is assumed to be smooth in both the spatial variable x and the frequency variable ξ, and satisfies certain growth conditions (see, e.g., [26, Chap. VI]). An example of symbols one might consider is the class $S^m_{\rho,\delta}$, introduced by Hörmander [13], consisting of $a \in C^\infty(\mathbb{R}^n \times \mathbb{R}^n)$ with

$$|\partial_\xi^\alpha \partial_x^\beta a(x,\xi)| \le C_{\alpha,\beta} \langle \xi \rangle^{m-\rho|\alpha|+\delta|\beta|} \quad (x,\xi \in \mathbb{R}^n),$$

where $m \in \mathbb{R}$ and $0 \le \delta, \rho \le 1$ and the positive constants $C_{\alpha,\beta}$ depend only on α and β.

The study of pseudodifferential operators $\mathrm{Op}(a)$ with symbols in $S^0_{1,0}$ on so-called variable Lebesgue spaces was started by Rabinovich and Samko [24, 25].

Let $p : \mathbb{R}^n \to [1,\infty]$ be a measurable a.e. finite function. By $L^{p(\cdot)}(\mathbb{R}^n)$ we denote the set of all complex-valued functions f on $\mathbb{R}^n$ such that

$$I_{p(\cdot)}(f/\lambda) := \int_{\mathbb{R}^n} |f(x)/\lambda|^{p(x)} dx < \infty$$

for some $\lambda > 0$. This set becomes a Banach space when equipped with the norm

$$\|f\|_{p(\cdot)} := \inf \big\{ \lambda > 0 : I_{p(\cdot)}(f/\lambda) \le 1 \big\}.$$

It is easy to see that if p is constant, then $L^{p(\cdot)}(\mathbb{R}^n)$ is nothing but the standard Lebesgue space $L^p(\mathbb{R}^n)$. The space $L^{p(\cdot)}(\mathbb{R}^n)$ is referred to as a *variable Lebesgue space.*

Lemma 1.1. (see, e.g., [15, Thm. 2.11] or [10, Thm. 3.4.12]) *If $p : \mathbb{R}^n \to [1,\infty]$ is an essentially bounded measurable function, then $C_0^\infty(\mathbb{R}^n)$ is dense in $L^{p(\cdot)}(\mathbb{R}^n)$.*

We will always suppose that

$$1 < p_- := \operatorname*{ess\,inf}_{x \in \mathbb{R}^n} p(x), \quad \operatorname*{ess\,sup}_{x \in \mathbb{R}^n} p(x) =: p_+ < \infty. \tag{1.1}$$

Under these conditions, the space $L^{p(\cdot)}(\mathbb{R}^n)$ is separable and reflexive, and its dual space is isomorphic to $L^{p'(\cdot)}(\mathbb{R}^n)$, where

$$1/p(x) + 1/p'(x) = 1 \quad (x \in \mathbb{R}^n)$$

(see, e.g., [15] or [10, Chap. 3]).

Given $f \in L^1_{\mathrm{loc}}(\mathbb{R}^n)$, the Hardy-Littlewood maximal operator is defined by

$$Mf(x) := \sup_{Q \ni x} \frac{1}{|Q|} \int_Q |f(y)| dy$$

where the supremum is taken over all cubes $Q \subset \mathbb{R}^n$ containing x (here, and throughout, cubes will be assumed to have their sides parallel to the coordinate axes). By $\mathcal{M}(\mathbb{R}^n)$ denote the set of all measurable functions $p:\mathbb{R}^n \to [1,\infty]$ such that (1.1) holds and the Hardy-Littlewood maximal operator is bounded on $L^{p(\cdot)}(\mathbb{R}^n)$.

Assume that (1.1) is fulfilled. Diening [7] proved that if p satisfies

$$|p(x) - p(y)| \le \frac{c}{\log(e + 1/|x-y|)} \quad (x, y \in \mathbb{R}^n) \tag{1.2}$$

and p is constant outside some ball, then $p \in \mathcal{M}(\mathbb{R}^n)$. Further, the behavior of p at infinity was relaxed by Cruz-Uribe, Fiorenza, and Neugebauer [5, 6], where it was shown that if p satisfies (1.2) and there exists a $p_\infty > 1$ such that

$$|p(x) - p_\infty| \le \frac{c}{\log(e + |x|)} \quad (x \in \mathbb{R}^n), \tag{1.3}$$

then $p \in \mathcal{M}(\mathbb{R}^n)$. Following [10, Section 4.1], we will say that if conditions (1.2)–(1.3) are fulfilled, then p is *globally log-Hölder continuous.*

Conditions (1.2) and (1.3) are optimal for the boundedness of M in the pointwise sense; the corresponding examples are contained in [21] and [5]. However, neither (1.2) nor (1.3) is necessary for $p \in \mathcal{M}(\mathbb{R}^n)$. Nekvinda [19] proved that if p satisfies (1.1)–(1.2) and

$$\int_{\mathbb{R}^n} |p(x) - p_\infty| c^{1/|p(x)-p_\infty|}\, dx < \infty \tag{1.4}$$

for some $p_\infty > 1$ and $c > 0$, then $p \in \mathcal{M}(\mathbb{R}^n)$. One can show that (1.3) implies (1.4), but the converse, in general, is not true. The corresponding example is constructed in [3]. Nekvinda further relaxed condition (1.4) in [20]. Lerner [16] (see also [10, Example 5.1.8]) showed that there exist discontinuous at zero or/and at infinity exponents, which nevertheless belong to $\mathcal{M}(\mathbb{R}^n)$. We refer to the recent monograph [10] for further discussions concerning the class $\mathcal{M}(\mathbb{R}^n)$.

Our first main result is the following theorem concerning the boundedness of pseudodifferential operators on variable Lebesgue spaces.

Theorem 1.2. *Let $0 < \rho \le 1$, $0 \le \delta < 1$, and $a \in S^{n(\rho-1)}_{\rho,\delta}$. If $p \in \mathcal{M}(\mathbb{R}^n)$, then* $\operatorname{Op}(a)$ *extends to a bounded operator on the variable Lebesgue space $L^{p(\cdot)}(\mathbb{R}^n)$.*

The respective result for $a \in S^0_{1,0}$ and p satisfying (1.1)–(1.3) was proved by Rabinovich and Samko [24, Theorem 5.1].

Following [24, Definition 4.5], a symbol $a \in S^m_{1,0}$ is said to be *slowly oscillating at infinity* in the first variable if

$$|\partial_\xi^\alpha \partial_x^\beta a(x,\xi)| \le C_{\alpha\beta}(x) \langle \xi \rangle^{m-|\alpha|},$$

where

$$\lim_{x\to\infty} C_{\alpha\beta}(x) = 0 \tag{1.5}$$

for all multi-indices α and $\beta \ne 0$. We denote by SO^m the class of all symbols slowly oscillating at infinity. Finally, we denote by SO^m_0 the set of all symbols $a \in SO^m$, for which (1.5) holds for all multi-indices α and β. The classes SO^m and SO^m_0 were introduced by Grushin [12].

We denote by $\mathcal{M}^*(\mathbb{R}^n)$ the set of all variable exponents $p \in \mathcal{M}(\mathbb{R}^n)$ for which there exist constants $p_0 \in (1,\infty)$, $\theta \in (0,1)$, and a variable exponent $p_1 \in \mathcal{M}(\mathbb{R}^n)$ such that

$$\frac{1}{p(x)} = \frac{\theta}{p_0} + \frac{1-\theta}{p_1(x)}$$

for almost all $x \in \mathbb{R}^n$. Rabinovich and Samko observed in the proof of [24, Theorem 6.1] that if p satisfies (1.1)–(1.3), then $p \in \mathcal{M}^*(\mathbb{R}^n)$. It turns out that the class $\mathcal{M}^*(\mathbb{R}^n)$ contains many interesting exponents which are not globally log-Hölder continuous (see [14]). In particular, there exists $\varepsilon > 0$ such that for every α, β satisfying $0 < \beta < \alpha \le \varepsilon$ the function

$$p(x) = 2 + \alpha + \beta \sin\big(\log(\log|x|)\chi_{\{x\in\mathbb{R}^n:|x|\ge e\}}(x)\big) \quad (x \in \mathbb{R}^n)$$

belongs to $\mathcal{M}^*(\mathbb{R}^n)$.

As usual, we denote by I the identity operator on a Banach space. Recall that a bounded linear operator A on a Banach space is said to be Fredholm if there is an (also bounded linear) operator B such that the operators $AB - I$ and $BA - I$ are compact. In that case the operator B is called a *regularizer* for the operator A.

Our second main result is the following sufficient condition for the Fredholmness of pseudodifferential operators on variable Lebesgue spaces.

Theorem 1.3. *Suppose* $p \in \mathcal{M}^*(\mathbb{R}^n)$ *and* $a \in SO^0$. *If*

$$\lim_{R\to\infty} \inf_{|x|+|\xi|\ge R} |a(x,\xi)| > 0, \tag{1.6}$$

then the operator $\operatorname{Op}(a)$ *is Fredholm on the variable Lebesgue space* $L^{p(\cdot)}(\mathbb{R}^n)$.

As it was the case with Theorem 1.2, for p satisfying (1.1)–(1.3) this result was established by Rabinovich and Samko [24, Theorem 6.1]. Notice that for such p condition (1.6) is also necessary for the Fredholmness (see [24, Theorems 6.2 and 6.5]). Whether or not the necessity holds in the setting of Theorem 1.3, remains an open question.

The paper is organized as follows. In Section 2.2, the Diening-Růžička generalization (see [11]) of the Fefferman-Stein sharp maximal theorem to the variable exponent setting is stated. Further, Diening's results [8] on the duality and left-openness of the class $\mathcal{M}(\mathbb{R}^n)$ are formulated. In Section 2.4 we discuss a pointwise estimate relating the Fefferman-Stein sharp maximal operator of $\operatorname{Op}(a)u$ and $M_q u := M(|u|^q)^{1/q}$ for $q \in (1,\infty)$ and $u \in C_0^\infty(\mathbb{R}^n)$. Such an estimate for the range of parameters ρ, δ, and $m = n(\rho - 1)$ as in Theorem 1.2 was recently obtained by Michalowski, Rule, and Staubach [17]. Combining this key pointwise estimate with the sharp maximal theorem and taking into account that M_q is bounded on $L^{p(\cdot)}(\mathbb{R}^n)$ for some $q \in (1,\infty)$ whenever $p \in \mathcal{M}(\mathbb{R}^n)$, we give the proof of Theorem 1.2 in Section 2.5.

Section 3 is devoted to the proof of the sufficient condition for the Fredholmness of a pseudodifferentail operator with slowly oscillating symbol. In Section 3.1, we state analogues of the Riesz-Thorin and Krasnoselskii interpolation theorems for variable Lebesgue spaces. Section 3.2 contains the composition formula for pseudodifferential operators with slowly oscillating symbols and the compactness result for pseudodifferential operators with symbols in SO_0^{-1}. Both results are essentially due to Grushin [12]. Section 3.3 contains the proof of Theorem 1.3. Its outline is as follows. From (1.6) it follows that there exist symbols $b_R \in SO^0$ and $\varphi_R + c \in SO_0^{-1}$ such that $I - \mathrm{Op}(a)\,\mathrm{Op}(b_R) = \mathrm{Op}(\varphi_R + c)$. Since $\varphi_R + c \in SO_0^{-1}$, the operator $\mathrm{Op}(\varphi_R + c)$ is compact on all standard Lebesgue spaces. Its compactness on the variable Lebesgue space $L^{p(\cdot)}(\mathbb{R}^n)$ is proved by interpolation, since it is bounded on the variable Lebesgue space $L^{p_1(\cdot)}(\mathbb{R}^n)$, where p_1 is the variable exponent from the definition of the class $\mathcal{M}^*(\mathbb{R}^n)$. Actually, the class $\mathcal{M}^*(\mathbb{R}^n)$ is introduced exactly for the purpose to perform this step. Therefore $\mathrm{Op}(b_R)$ is a right regularizer for $\mathrm{Op}(a)$ on $L^{p(\cdot)}(\mathbb{R}^n)$. In the same fashion it can be shown that $\mathrm{Op}(b_R)$ is a left regularizer for $\mathrm{Op}(a)$. Thus $\mathrm{Op}(a)$ is Fredholm.

2. Boundedness of the operator $\mathrm{Op}(a)$

2.1. Lattice property of variable Lebesgue spaces

We start with the following simple but important property of variable Lebesgue spaces. Usually it is called the lattice property or the ideal property.

Lemma 2.1. (see, e.g., [10, Thm. 2.3.17]) *Let $p : \mathbb{R}^n \to [1, \infty]$ be a measurable a.e. finite function. If $g \in L^{p(\cdot)}(\mathbb{R}^n)$, f is a measurable function, and $|f(x)| \le |g(x)|$ for a.e. $x \in \mathbb{R}^n$, then $f \in L^{p(\cdot)}(\mathbb{R}^n)$ and $\|f\|_{p(\cdot)} \le \|g\|_{p(\cdot)}$.*

2.2. The Fefferman-Stein sharp maximal function

Let $f \in L^1_{\mathrm{loc}}(\mathbb{R}^n)$. For a cube $Q \subset \mathbb{R}^n$, put

$$f_Q := \frac{1}{|Q|} \int_Q f(x)dx.$$

The Fefferman-Stein sharp maximal function is defined by

$$M^{\#} f(x) := \sup_{Q \ni x} \frac{1}{|Q|} \int_Q |f(x) - f_Q| dx,$$

where the supremum is taken over all cubes Q containing x.

It is obvious that $M^{\#} f$ is pointwise dominated by Mf. Hence, by Lemma 2.1,

$$\|M^{\#} f\|_{p(\cdot)} \le \mathrm{const} \|f\|_{p(\cdot)} \quad \text{for} \quad f \in L^{p(\cdot)}(\mathbb{R}^n)$$

whenever $p \in \mathcal{M}(\mathbb{R}^n)$. The converse is also true. For constant p this fact goes back to Fefferman and Stein (see, e.g., [26, Chap. IV, Section 2.2]). The variable exponent analogue of the Fefferman-Stein theorem was proved by Diening and Růžička [11].

Theorem 2.2. (see [11, Thm. 3.6] or [10, Thm. 6.2.5]) *If $p, p' \in \mathcal{M}(\mathbb{R}^n)$, then there exists a constant $C_{\#}(p) > 0$ such that for all $f \in L^{p(\cdot)}(\mathbb{R}^n)$,*

$$\|f\|_{p(\cdot)} \le C_{\#}(p) \|M^{\#} f\|_{p(\cdot)}.$$

2.3. Duality and left-openness of the class $\mathcal{M}(\mathbb{R}^n)$

Let $1 \le q < \infty$. Given $f \in L^q_{\rm loc}(\mathbb{R}^n)$, the *qth maximal operator* is defined by

$$M_q f(x) := \sup_{Q \ni x} \left(\frac{1}{|Q|} \int_Q |f(y)|^q dy \right)^{1/q},$$

where the supremum is taken over all cubes $Q \subset \mathbb{R}^n$ containing x. For $q = 1$ this is the usual Hardy-Littlewood maximal operator. Diening [8] established the following deep duality and left-openness result for the class $\mathcal{M}(\mathbb{R}^n)$.

Theorem 2.3. (see [8, Thm. 8.1] or [10, Thm. 5.7.2]) *Let $p : \mathbb{R}^n \to [1, \infty]$ be a measurable function satisfying* (1.1). *The following statements are equivalent:*

(a) *M is bounded on $L^{p(\cdot)}(\mathbb{R}^n)$;*
(b) *M is bounded on $L^{p'(\cdot)}(\mathbb{R}^n)$;*
(c) *there exists an $s \in (1/p_-, 1)$ such that M is bounded on $L^{sp(\cdot)}(\mathbb{R}^n)$;*
(d) *there exists a $q \in (1, \infty)$ such that M_q is bounded on $L^{p(\cdot)}(\mathbb{R}^n)$.*

2.4. The crucial pointwise estimate

One of the main steps in the proof of Theorem 1.2 is the following pointwise estimate.

Theorem 2.4. (see [17, Thm. 3.3]) *Let $1 < q < \infty$ and $a \in S^m_{\rho,\delta}$ with $0 < \rho \le 1$, $0 \le \delta < 1$, and $m = n(\rho - 1)$. For every $u \in C_0^\infty(\mathbb{R}^n)$,*

$$M^\#(\mathrm{Op}(a)u)(x) \le C(q, a) M_q u(x) \quad (x \in \mathbb{R}^n),$$

where $C(q, a)$ is a positive constant depending only on q and the symbol a.

This theorem generalizes the pointwise estimate by Miller [18, Theorem 2.8] for $a \in S^0_{1,0}$ and by Álvarez and Hounie [1, Theorem 4.1] for $a \in S^m_{\rho,\delta}$ with the parameters satisfying $0 < \delta \le \rho \le 1/2$ and $m \le n(\rho - 1)$.

Let $0 < s < 1$. One of the main steps in the Rabinovich and Samko's proof [24] of the boundedness on $L^{p(\cdot)}(\mathbb{R}^n)$ of the operator $\mathrm{Op}(a)$ with $a \in S^0_{1,0}$ is another pointwise estimate

$$M^\#(|\,\mathrm{Op}(a)u|^s)(x) \le C[Mu(x)]^s \quad (x \in \mathbb{R}^n)$$

for all $u \in C_0^\infty(\mathbb{R}^n)$, where C is a positive constant independent of u. It was proved in [24, Corollary 3.4] following the ideas of Álvarez and Pérez [2], where the same estimate is obtained for the Calderón-Zygmund singular integral operator in place of the pseudodifferential operator $\mathrm{Op}(a)$.

2.5. Proof of Theorem 1.2

Suppose $p \in \mathcal{M}(\mathbb{R}^n)$. Then, by Theorem 2.3, $p' \in \mathcal{M}(\mathbb{R}^n)$ and there exists a number $q \in (1, \infty)$ such that M_q is bounded on $L^{p(\cdot)}(\mathbb{R}^n)$. In other words, there exists a positive constant $\widetilde{C}(p, q)$ depending only on p and q such that for all $u \in L^{p(\cdot)}(\mathbb{R}^n)$,

$$\|M_q u\|_{p(\cdot)} \le \widetilde{C}(p, q) \|u\|_{p(\cdot)}. \tag{2.1}$$

From Theorem 2.2 it follows that there exists a constant $C_\#(p)$ such that for all $u \in C_0^\infty(\mathbb{R}^n)$,

$$\|\operatorname{Op}(a)u\|_{p(\cdot)} \le C_\#(p)\|M^\#(\operatorname{Op}(a)u)\|_{p(\cdot)}. \tag{2.2}$$

On the other hand, from Theorem 2.4 and Lemma 2.1 we obtain that there exists a positive constant $C(q,a)$, depending only on q and a, such that

$$\|M^\#(\operatorname{Op}(a)u)\|_{p(\cdot)} \le C(q,a)\|M_q u\|_{p(\cdot)}. \tag{2.3}$$

Combining (2.1)–(2.3), we arrive at

$$\|\operatorname{Op}(a)u\|_{p(\cdot)} \le C_\#(p)C(q,a)\widetilde{C}(p,q)\|u\|_{p(\cdot)}$$

for all $u \in C_0^\infty(\mathbb{R}^n)$. It remains to recall that $C_0^\infty(\mathbb{R}^n)$ is dense in $L^{p(\cdot)}(\mathbb{R}^n)$ (see Lemma 1.1). □

3. Fredholmness of the operator $\operatorname{Op}(a)$

3.1. Interpolation theorem

For a Banach space X, let $\mathcal{B}(X)$ and $\mathcal{K}(X)$ denote the Banach algebra of all bounded linear operators and its ideal of all compact operators on X, respectively.

Theorem 3.1. *Let $p_j : \mathbb{R}^n \to [1,\infty]$, $j = 0,1$, be a.e. finite measurable functions, and let $p_\theta : \mathbb{R}^n \to [1,\infty]$ be defined for $\theta \in [0,1]$ by*

$$\frac{1}{p_\theta(x)} = \frac{\theta}{p_0(x)} + \frac{1-\theta}{p_1(x)} \quad (x \in \mathbb{R}^n).$$

Suppose A is a linear operator defined on $L^{p_0(\cdot)}(\mathbb{R}^n) \cup L^{p_1(\cdot)}(\mathbb{R}^n)$.

(a) *If $A \in \mathcal{B}(L^{p_j(\cdot)}(\mathbb{R}^n))$ for $j = 0,1$, then $A \in \mathcal{B}(L^{p_\theta(\cdot)}(\mathbb{R}^n))$ for all $\theta \in [0,1]$ and*

$$\|A\|_{\mathcal{B}(L^{p_\theta(\cdot)}(\mathbb{R}^n))} \le 4\|A\|^{\theta}_{\mathcal{B}(L^{p_0(\cdot)}(\mathbb{R}^n))}\|A\|^{1-\theta}_{\mathcal{B}(L^{p_1(\cdot)}(\mathbb{R}^n))}.$$

(b) *If $A \in \mathcal{K}(L^{p_0(\cdot)}(\mathbb{R}^n))$ and $A \in \mathcal{B}(L^{p_1(\cdot)}(\mathbb{R}^n))$, then $A \in \mathcal{K}(L^{p_\theta(\cdot)}(\mathbb{R}^n))$ for all $\theta \in (0,1)$.*

Part (a) is proved in [10, Corollary 7.1.4] under the more general assumption that p_j may take infinite values on sets of positive measure (and in the setting of arbitrary measure spaces). Part (b) was proved in [24, Proposition 2.2] under the additional assumptions that p_j satisfy (1.1)–(1.3). It follows without these assumptions from a general interpolation theorem by Cobos, Kühn, and Schonbeck [4, Theorem 3.2] for the complex interpolation method for Banach lattices satisfying the Fatou property. Indeed, the complex interpolation space $[L^{p_0(\cdot)}(\mathbb{R}^n), L^{p_1(\cdot)}(\mathbb{R}^n)]_{1-\theta}$ is isomorphic to the variable Lebesgue space $L^{p_\theta(\cdot)}(\mathbb{R}^n)$ (see [10, Theorem 7.1.2]), and $L^{p_j(\cdot)}(\mathbb{R}^n)$ have the Fatou property (see [10, p. 77]).

3.2. Calculus of pseudodifferential operators

Let $m \in \mathbb{Z}$ and $OPSO^m$ be the class of all pseudodifferential operators $\operatorname{Op}(a)$ with $a \in SO^m$. By analogy with [12, Section 2] one can get the following *composition formula* (see also [22, Theorem 6.2.1] and [23, Chap. 4]).

Proposition 3.2. *If* $\mathrm{Op}(a_1) \in OPSO^{m_1}$ *and* $\mathrm{Op}(a_2) \in OPSO^{m_2}$, *then their product* $\mathrm{Op}(a_1)\,\mathrm{Op}(a_2) = \mathrm{Op}(\sigma)$ *belongs to* $OPSO^{m_1+m_2}$ *and its symbol* σ *is given by*

$$\sigma(x,\xi) = a_1(x,\xi)a_2(x,\xi) + c(x,\xi), \quad x,\xi \in \mathbb{R}^n,$$

where $c \in SO_0^{m_1+m_2-1}$.

Proposition 3.3. *Let* $1 < q < \infty$. *If* $c \in SO_0^{-1}$, *then* $\mathrm{Op}(c) \in \mathcal{K}(L^q(\mathbb{R}^n))$.

Proof. From Theorem 1.2 it follows that $\mathrm{Op}(c) \in \mathcal{B}(L^q(\mathbb{R}^n))$ for all constant exponents $q \in (1,\infty)$. By [12, Theorem 3.2], $\mathrm{Op}(c) \in \mathcal{K}(L^2(\mathbb{R}^n))$. Hence, by the Krasnoselskii interpolation theorem (Theorem 3.1(b) for constant p_j with $j = 0,1$), $\mathrm{Op}(c) \in \mathcal{K}(L^q(\mathbb{R}^n))$ for all $q \in (1,\infty)$. □

3.3. Proof of Theorem 1.3

The idea of the proof is borrowed from [12, Theorem 3.4] and [24, Theorem 6.1]. Let $\varphi \in C_0^\infty(\mathbb{R}^n \times \mathbb{R}^n)$ be such that $\varphi(x,\xi) = 1$ if $|x| + |\xi| \le 1$ and $\varphi(x,\xi) = 0$ if $|x| + |\xi| \ge 2$. For $R > 0$, put

$$\varphi_R(x,\xi) = \varphi(x/R, \xi/R), \quad x,\xi \in \mathbb{R}^n.$$

From (1.6) it follows that there exists an $R > 0$ such that

$$\inf_{|x|+|\xi|\ge R} |a(x,\xi)| > 0.$$

Then it is not difficult to check that

$$b_R(x,\xi) := \begin{cases} \dfrac{1-\varphi_R(x,\xi)}{a(x,\xi)} & \text{if} \quad |x|+|\xi| \ge R, \\ 0 & \text{if} \quad |x|+|\xi| < R, \end{cases}$$

belongs to SO^0. It is also clear that $\varphi_R \in SO^0$.

From Proposition 3.2 it follows that there exists a function $c \in SO_0^{-1}$ such that

$$\mathrm{Op}(ab_R) - \mathrm{Op}(a)\,\mathrm{Op}(b_R) = \mathrm{Op}(c). \tag{3.1}$$

On the other hand, since

$$a(x,\xi)b_R(x,\xi) = 1 - \varphi_R(x,\xi), \quad x,\xi \in \mathbb{R}^n,$$

we have

$$\mathrm{Op}(ab_R) = \mathrm{Op}(1-\varphi_R) = I - \mathrm{Op}(\varphi_R). \tag{3.2}$$

Combining (3.1)–(3.2), we get

$$I - \mathrm{Op}(a)\,\mathrm{Op}(b_R) = \mathrm{Op}(\varphi_R) + \mathrm{Op}(c) = \mathrm{Op}(\varphi_R + c). \tag{3.3}$$

Since $p \in \mathcal{M}^*(\mathbb{R}^n)$, there exist $p_0 \in (1,\infty)$, $\theta \in (0,1)$, and $p_1 \in \mathcal{M}(\mathbb{R}^n)$ such that

$$\frac{1}{p(x)} = \frac{\theta}{p_0} + \frac{1-\theta}{p_1(x)} \quad (x \in \mathbb{R}^n).$$

From Theorem 1.2 we conclude that all pseudodifferential operators considered above are bounded on $L^{p_0}(\mathbb{R}^n)$, $L^{p(\cdot)}(\mathbb{R}^n)$, and $L^{p_1(\cdot)}(\mathbb{R}^n)$. Since $\varphi_R + c \in SO_0^{-1}$, from Proposition 3.3 it follows that $\mathrm{Op}(\varphi_R + c) \in \mathcal{K}(L^{p_0}(\mathbb{R}^n))$. Then, by Theorem 3.1(b), $\mathrm{Op}(\varphi_R + c) \in \mathcal{K}(L^{p(\cdot)}(\mathbb{R}^n))$. Therefore, from (3.3) it follows that $\mathrm{Op}(b_R)$ is a right regularizer for $\mathrm{Op}(a)$. Analogously it can be shown that $\mathrm{Op}(b_R)$ is also a left regularizer for $\mathrm{Op}(a)$. Thus $\mathrm{Op}(a)$ is Fredholm on $L^{p(\cdot)}(\mathbb{R}^n)$. □

4. Addendum

After the paper was accepted for publication, Lars Diening communicated to us [9] a short (but nontrivial!) proof of the inclusion $\mathcal{M}(\mathbb{R}^n) \subseteq \mathcal{M}^*(\mathbb{R}^n)$. Thus, the following result holds.

Theorem 4.1 (Diening). *We have* $\mathcal{M}(\mathbb{R}^n) = \mathcal{M}^*(\mathbb{R}^n)$.

Proof. By definition, $\mathcal{M}^*(\mathbb{R}^n) \subseteq \mathcal{M}(\mathbb{R}^n)$. Let us show the reverse inclusion. Suppose that $p \in \mathcal{M}(\mathbb{R}^n)$. By Theorem 2.3(c), there exists a constant $r \in (1,\infty)$ such that $p/r \in \mathcal{M}(\mathbb{R}^n)$. Then, in view of Theorem 2.3(b), $(p/r)' \in \mathcal{M}(\mathbb{R}^n)$. Applying Theorem 2.3(c) once again, we see that there is a constant $s \in (1,\infty)$ such that $\frac{1}{s}\left(\frac{p}{r}\right)' \in \mathcal{M}(\mathbb{R}^n)$. Therefore, by Theorem 2.3(b),

$$p_1 := \left(\frac{1}{s}\left(\frac{p}{r}\right)'\right)' \in \mathcal{M}(\mathbb{R}^n).$$

Simple calculations show that

$$\frac{1}{p_1(x)} = 1 - s + \frac{rs}{p(x)} \quad (x \in \mathbb{R}^n). \tag{4.1}$$

To prove that $p \in \mathcal{M}^*(\mathbb{R}^n)$, we have to find $p_0 \in (1,\infty)$ and $\theta \in (0,1)$ such that

$$\frac{1}{p(x)} = \frac{\theta}{p_0} + \frac{1-\theta}{p_1(x)} \quad (x \in \mathbb{R}^n). \tag{4.2}$$

Equalities (4.1) and (4.2) give

$$\frac{1}{p(x)} = \frac{\theta}{p_0} + (1-\theta)(1-s) + \frac{(1-\theta)rs}{p(x)} \quad (x \in \mathbb{R}^n).$$

The choice $\theta := 1 - \frac{1}{rs} \in (0,1)$ leads to

$$\frac{1}{p(x)} = \frac{1-\frac{1}{rs}}{p_0} + \frac{1-s}{rs} + \frac{1}{p(x)} \quad (x \in \mathbb{R}^n).$$

So, necessarily,

$$p_0 = \frac{rs-1}{s-1} \in (1,\infty).$$

Thus, $p \in \mathcal{M}^*(\mathbb{R}^n)$, which finishes the proof. □

Consequently, our Theorem 1.3 can be restated as follows.

Theorem 4.2. *Suppose* $p \in \mathcal{M}(\mathbb{R}^n)$ *and* $a \in SO^0$. *If*

$$\lim_{R\to\infty} \inf_{|x|+|\xi|\ge R} |a(x,\xi)| > 0,$$

then the operator $\mathrm{Op}(a)$ *is Fredholm on the variable Lebesgue space* $L^{p(\cdot)}(\mathbb{R}^n)$.

On the other hand, Theorem 4.1 immediately implies that there exist functions in $\mathcal{M}^*(\mathbb{R}^n)$ different from globally log-Hölder continuous exponents, since the latter constitute a proper subclass of $\mathcal{M}(\mathbb{R}^n)$. Therefore, the results of our preprint [14], while being formally correct, are redundant.

Acknowledgement.

We would like to thank Lars Diening for his kind permission to present his Theorem 4.1 in our paper.

References

[1] J. Álvarez and J. Hounie, *Estimates for the kernel and continuity properties of pseudo-differential operators.* Ark. Mat. **28** (1990), 1–22.

[2] J. Álvarez and C. Pérez, *Estimates with A_∞ weights for various singular integral operators.* Boll. Un. Mat. Ital. A (7) **8** (1994), 123–133.

[3] C. Capone, D. Cruz-Uribe, and A. Fiorenza, *The fractional maximal operator and fractional integrals on variable L^p spaces.* Rev. Mat. Iberoamericana **23** (2007), 743–770.

[4] F. Cobos, T. Kühn, and T. Schonbek, *One-sided compactness results for Aronszajn-Gagliardo functors.* J. Funct. Analysis **106** (1992), 274–313.

[5] D. Cruz-Uribe, A. Fiorenza, and C.J. Neugebauer, *The maximal function on variable L^p spaces.* Ann. Acad. Sci. Fenn. Math. **28** (2003), 223–238.

[6] D. Cruz-Uribe, A. Fiorenza, and C.J. Neugebauer, *Corrections to: "The maximal function on variable L^p spaces".* Ann. Acad. Sci. Fenn. Math. **29** (2004), 247–249.

[7] L. Diening, *Maximal function on generalized Lebesgue spaces $L^{p(\cdot)}$.* Math. Inequal. Appl. **7** (2004), 245–253.

[8] L. Diening, *Maximal function on Musielak-Orlicz spaces and generlaized Lebesgue spaces.* Bull. Sci. Math. **129** (2005), 657–700.

[9] L. Diening, *Private communication*, 29th of October, 2011.

[10] L. Diening, P. Harjulehto, P. Hästö, and M. Růžička, *Lebesgue and Sobolev Spaces with Variable Exponents.* Lecture Notes in Mathematics **2017**, 2011.

[11] L. Dieinig and M. Růžička, *Calderón-Zygmund operators on generalized Lebesgue spaces $L^{p(\cdot)}$ and problems related to fluid dynamics.* J. Reine Angew. Math., **563** (2003), 197–220.

[12] V.V. Grushin, *Pseudodifferential operators on $\mathbb{R}^n$ with bounded symbols.* Funct. Anal. Appl. **4** (1970), 202–212.

[13] L. Hörmander, *Pseudo-differential operators and hypoelliptic equations.* In: "Singular integrals (Proc. Sympos. Pure Math., Vol. X, Chicago, Ill., 1966)", pp. 138–183. Amer. Math. Soc., Providence, R.I., 1967.

[14] A.Yu. Karlovich and I.M. Spitkovsky, *On an interesting class of variable exponents,* arXiv:1110.0299v1.

[15] O. Kováčik and J. Rákosník, *On spaces $L^{p(x)}$ and $W^{k,p(x)}$.* Czechoslovak Math. J. **41(116)** (1991), no. 4, 592–618.

[16] A.K. Lerner, *Some remarks on the Hardy-Littlewood maximal function on variable L^p spaces.* Math. Z. **251** (2005), no. 3, 509–521.

[17] N. Michalowski, D. Rule, and W. Staubach, *Weighted L^p boundedness of pseudodifferential operators and applications.* Canad. Math. Bull. **55** (2012), 555–570.

[18] N. Miller, *Weighted Sobolev spaces and pseudodifferential operators with smooth symbols.* Trans. Amer. Math. Soc. **269** (1982), 91–109.

[19] A. Nekvinda, *Hardy-Littlewood maximal operator on $L^{p(x)}(\mathbb{R}^n)$.* Math. Inequal. Appl. **7** (2004), 255–265.

[20] A. Nekvinda, *Maximal operator on variable Lebesgue spaces for almost monotone radial exponent.* J. Math. Anal. Appl. **337** (2008), 1345–1365.

[21] L. Pick and M. Růžička, *An example of a space of Lp(x) on which the Hardy-Littlewood maximal operator is not bounded.* Expo. Math. **19** (2001), 369–371.

[22] V.S. Rabinovich, *An itroductory course on pseudodifferential operators.* Textos de Matemática, Instituto Superior Técnico, Lisboa, 1998.

[23] V.S. Rabinovich, S. Roch, and B. Silbermann, *Limit Operators and Their Applications in Operator Theory.* Operator Theory: Advances and Applications, vol. 150. Birkhäuser, Basel, 2004.

[24] V.S. Rabinovich and S.G. Samko, *Boundedness and Fredholmness of pseudodifferential operators in variable exponent spaces.* Integral Equations Operator Theory **60** (2008), 507–537.

[25] V.S. Rabinovich and S.G. Samko, *Pseudodifferential operators approach to singular integral operators in weighted variable exponent Lebesgue spaces on Carleson curves.* Integral Equations Operator Theory **69** (2011), 405–444.

[26] E. Stein, *Harmonic Analysis: Real-Variable Methods, Orthogonality, and Oscillatory Integrals.* Princeton Uinversity Press, Princeton, NJ, 1993.

Alexei Yu. Karlovich
Departamento de Matemática, Faculdade de Ciências e Tecnologia
Universidade Nova de Lisboa, Quinta da Torre
2829–516 Caparica, Portugal
e-mail: `oyk@fct.unl.pt`

Ilya M. Spitkovsky
Department of Mathematics, College of William & Mary
Williamsburg, VA, 23187-8795, USA
e-mail: `ilya@math.wm.edu`

Operator Theory:
Advances and Applications, Vol. 228, 185–207

On Convolution Type Operators with Piecewise Slowly Oscillating Data

Yuri I. Karlovich and Iván Loreto Hernández

To Vladimir Rabinovich on the occasion of his 70th birthday

Abstract. Applying the theory of pseudodifferential and Calderón-Zygmund operators, we study the compactness of commutators of multiplication operators aI and convolution operators $W^0(b)$ on weighted Lebesgue spaces $L^p(\mathbb{R}, w)$ with $p \in (1, \infty)$ and Muckenhoupt weights w for some classes of piecewise slowly oscillating functions $a \in PSO^\diamond$ and $b \in PSO^\diamond_{p,w}$ on the real line $\mathbb{R}$. Then we study the Banach algebra $\mathcal{Z}_{p,w}$ generated by the operators $aW^0(b)$ with functions $a \in SO^\diamond$ and $b \in SO^\diamond_{p,w}$ admitting slowly oscillating discontinuities at every point $\lambda \in \mathbb{R} \cup \{\infty\}$. Applying the method of limit operators under some condition on Muckenhoupt weights w, we describe the maximal ideal space of the commutative quotient Banach algebra $\mathcal{Z}^\pi_{p,w} = \mathcal{Z}_{p,w}/\mathcal{K}_{p,w}$ where $\mathcal{K}_{p,w}$ is the ideal of compact operators on $L^p(\mathbb{R}, w)$, define the Gelfand transform for $\mathcal{Z}^\pi_{p,w}$ and establish the Fredholmness for the operators $A \in \mathcal{Z}_{p,w}$.

Mathematics Subject Classification (2010). Primary 47G10; Secondary 45E10, 46J10, 47A53, 47B47.

Keywords. Convolution type operator, piecewise slowly oscillating function, *BMO* and *VMO* functions, commutator, maximal ideal space, Fredholmness.

1. Introduction

Let $\mathcal{B}(X)$ denote the Banach algebra of all bounded linear operators acting on a Banach space X, let $\mathcal{K}(X)$ be the closed two-sided ideal of all compact operators in $\mathcal{B}(X)$, and let $\mathcal{B}^\pi(X) = \mathcal{B}(X)/\mathcal{K}(X)$ be the Calkin algebra of the cosets $A^\pi := A + \mathcal{K}(X)$ where $A \in \mathcal{B}(X)$. An operator $A \in \mathcal{B}(X)$ is said to be *Fredholm*, if its image is closed and the spaces $\ker A$ and $\ker A^*$ are finite-dimensional (see, e.g., [10]).

Partially supported by the SEP-CONACYT Project No. 168104 (México) and by PROMEP (México) via "Proyecto de Redes".

A measurable function $w : \mathbb{R} \to [0, \infty]$ is called a weight if the preimage $w^{-1}(\{0, \infty\})$ of the set $\{0, \infty\}$ has measure zero. For $1 < p < \infty$, a weight w belongs to the *Muckenhoupt class* $A_p(\mathbb{R})$ if

$$c_{p,w} := \sup_I \left(\frac{1}{|I|} \int_I w^p(x)dx \right)^{1/p} \left(\frac{1}{|I|} \int_I w^{-q}(x)dx \right)^{1/q} < \infty,$$

where $1/p + 1/q = 1$, and supremum is taken over all intervals $I \subset \mathbb{R}$ of finite length $|I|$.

In what follows we assume that $1 < p < \infty$ and $w \in A_p(\mathbb{R})$, and consider the weighted Lebesgue space $L^p(\mathbb{R}, w)$ equipped with the norm

$$\|f\|_{L^p(\mathbb{R},w)} := \left(\int_{\mathbb{R}} |f(x)|^p w^p(x)dx \right)^{1/p}.$$

As is known (see, e.g., [12]), the Cauchy singular integral operator $S_{\mathbb{R}}$ given by

$$(S_{\mathbb{R}} f)(x) = \lim_{\varepsilon \to 0} \frac{1}{\pi i} \int_{\mathbb{R} \setminus (x-\varepsilon, x+\varepsilon)} \frac{f(t)}{t - x} \, dt, \quad x \in \mathbb{R}, \tag{1.1}$$

is bounded on every space $L^p(\mathbb{R}, w)$ with $1 < p < \infty$ and $w \in A_p(\mathbb{R})$.

Let $\mathcal{F} : L^2(\mathbb{R}) \to L^2(\mathbb{R})$ denote the *Fourier transform*,

$$(\mathcal{F}f)(x) := \widehat{f}(x) := \int_{\mathbb{R}} f(t)e^{itx}dt, \ x \in \mathbb{R},$$

A function $a \in L^\infty(\mathbb{R})$ is called a *Fourier multiplier* on $L^p(\mathbb{R}, w)$ if the convolution operator $W^0(a) := \mathcal{F}^{-1} a \mathcal{F}$ maps the dense subset $L^2(\mathbb{R}) \cap L^p(\mathbb{R}, w)$ of $L^p(\mathbb{R}, w)$ into itself and extends to a bounded linear operator on $L^p(\mathbb{R}, w)$. Let $M_{p,w}$ stand for the Banach algebra of all Fourier multipliers on $L^p(\mathbb{R}, w)$ equipped with the norm $\|a\|_{M_{p,w}} := \|W^0(a)\|_{\mathcal{B}(L^p(\mathbb{R},w))}$.

Setting $\mathcal{B}_{p,w} := \mathcal{B}(L^p(\mathbb{R}, w))$ and $\mathcal{K}_{p,w} := \mathcal{K}(L^p(\mathbb{R}, w))$ for $p \in (1, \infty)$ and $w \in A_p(\mathbb{R})$, we consider the Banach subalgebra

$$\mathfrak{A}_{p,w} := \operatorname{alg}\big(aI, W^0(b) : \ a \in PSO^\diamond, \ b \in PSO^\diamond_{p,w}\big) \tag{1.2}$$

of $\mathcal{B}_{p,w}$ generated by all multiplication operators aI ($a \in PSO^\diamond$) and all convolution operators $W^0(b)$ ($b \in PSO^\diamond_{p,w}$), and the Banach subalgebra

$$\mathcal{Z}_{p,w} := \operatorname{alg}\big(aI, W^0(b) : \ a \in SO^\diamond, \ b \in SO^\diamond_{p,w}\big) \tag{1.3}$$

of $\mathfrak{A}_{p,w}$ generated by all the operators $aW^0(b)$ with $a \in SO^\diamond$ and $b \in SO^\diamond_{p,w}$, where the algebras $PSO^\diamond \subset L^\infty(\mathbb{R})$ and $PSO^\diamond_{p,w} \subset M_{p,w}$ of piecewise slowly oscillating functions on the real line $\mathbb{R}$ and the algebras $SO^\diamond \subset L^\infty(\mathbb{R})$ and $SO^\diamond_{p,w} \subset M_{p,w}$ of slowly oscillating functions admitting slowly oscillating discontinuities at arbitrary points $\lambda \in \mathbb{R} \cup \{\infty\}$ are defined in Section 2.

In the present paper, applying the theory of pseudodifferential and Calderón-Zygmund operators, we study the compactness of the commutators

$$[aI, W^0(b)] := aW^0(b) - W^0(b)aI$$

on the weighted Lebesgue spaces $L^p(\mathbb{R}, w)$ with $p \in (1, \infty)$ and $w \in A_p(\mathbb{R})$ for some classes of functions $a \in PSO^\diamond$ and $b \in PSO^\diamond_{p,w}$. Obtained results extend those in [11, Lemmas 7.1–7.4] and [1, Theorem 4.2, Corollary 4.3] to the weighted Lebesgue spaces $L^p(\mathbb{R}, w)$ with general Muckenhoupt weights w and to wider classes of data functions a, b. In addition, this implies that the quotient Banach algebra $\mathcal{Z}^\pi_{p,w} := \mathcal{Z}_{p,w}/\mathcal{K}_{p,w}$ is a central subalgebra of the Banach algebra $\mathfrak{A}^\pi_{p,w} := \mathfrak{A}_{p,w}/\mathcal{K}_{p,w}$. Then, under some condition on Muckenhoupt weights $w \in A_p(\mathbb{R})$, we describe the maximal ideal space of the commutative Banach algebra $\mathcal{Z}^\pi_{p,w}$, define the Gelfand transform for $\mathcal{Z}^\pi_{p,w}$ and establish the Fredholmness for the operators $A \in \mathcal{Z}_{p,w}$. To this end we use the method of limit operators, which was essentially developed and applied to different classes of integral and pseudodifferential operators by V.S. Rabinovich and his co-authors (see, e.g., [20], [7], [23] and the references therein).

The paper is organized as follows. In Section 2 we introduce the Banach algebras of slowly oscillating and piecewise slowly oscillating functions. In Section 3 we describe the maximal ideal spaces of the commutative Banach algebras $SO^\diamond_{p,w}$. In Section 4 we study the compactness of commutators of convolution type operators with piecewise slowly oscillating data. Section 5 is devoted to applications of limit operators. Finally, in Section 6, using the results of Section 5, we describe the maximal ideal space of the commutative Banach algebra $\mathcal{Z}^\pi_{p,w}$ and study the Fredholmness of operators $A \in \mathcal{Z}_{p,w}$.

2. Algebras of piecewise slowly oscillating functions

2.1. The C^*-algebra $SO^\diamond$

Let Γ be the unit circle $\mathbb{T} = \{z \in \mathbb{C} : |z| = 1\}$ or the one-point compactification $\dot{\mathbb{R}} := \mathbb{R} \cup \{\infty\}$ of the real line $\mathbb{R}$. For a bounded measurable function $f : \Gamma \to \mathbb{C}$ and a set $I \subset \Gamma$, let

$$\operatorname{osc}(f, I) = \operatorname{ess\,sup}\big\{|f(t) - f(s)| : t, s \in I\big\}.$$

Following [3, Section 4], we say that a function $f \in L^\infty(\Gamma)$ is called *slowly oscillating at a point* $\eta \in \Gamma$ if for every $r \in (0,1)$ or, equivalently, for some $r \in (0,1)$,

$$\lim_{\varepsilon \to 0} \operatorname{osc}\big(f, \Gamma_{r\varepsilon,\varepsilon}(\eta)\big) = 0 \quad \text{for} \quad \eta \neq \infty,$$
$$\lim_{\varepsilon \to \infty} \operatorname{osc}\big(f, \Gamma_{r\varepsilon,\varepsilon}(\eta)\big) = 0 \quad \text{for} \quad \eta = \infty,$$

where

$$\Gamma_{r\varepsilon,\varepsilon}(\eta) := \begin{cases} \big\{z \in \Gamma : \ r\varepsilon \le |z - \eta| \le \varepsilon\big\} & \text{if } \eta \neq \infty, \\ \big\{z \in \Gamma : \ r\varepsilon \le |z| \le \varepsilon\big\} & \text{if } \eta = \infty. \end{cases}$$

For each $\eta \in \Gamma$, let $SO_\eta(\Gamma)$ denote the C^*-subalgebra of $L^\infty(\Gamma)$ defined by

$$SO_\eta(\Gamma) := \Big\{f \in C_b(\Gamma \setminus \{\eta\}) : \ f \text{ slowly oscillates at } \eta\Big\},$$

where $C_b(\Gamma \setminus \{\eta\}) := C(\Gamma \setminus \{\eta\}) \cap L^\infty(\Gamma)$. Hence, setting $SO_\lambda := SO_\lambda(\dot{\mathbb{R}})$ for all $\lambda \in \dot{\mathbb{R}}$, we conclude that

$$\begin{aligned} SO_\infty &= \Big\{ f \in C_b(\dot{\mathbb{R}} \setminus \{\infty\}) : \lim_{x \to +\infty} \operatorname{osc}\big(f, [-x, -x/2] \cup [x/2, x]\big) = 0 \Big\}, \\ SO_\lambda &= \Big\{ f \in C_b(\dot{\mathbb{R}} \setminus \{\lambda\}) : \lim_{x \to +0} \operatorname{osc}\big(f, \lambda + ([-x, -x/2] \cup [x/2, x])\big) = 0 \Big\} \end{aligned} \tag{2.1}$$

for $\lambda \in \mathbb{R}$. Let $SO^\diamond$ be the minimal C^*-subalgebra of $L^\infty(\mathbb{R})$ that contains all the C^*-algebras SO_λ with $\lambda \in \dot{\mathbb{R}}$. In particular, $SO^\diamond$ contains $C(\dot{\mathbb{R}})$.

Lemma 2.1. *Let $\lambda \in \dot{\mathbb{R}}$, $a \in SO_\lambda$, and let $\gamma : \mathbb{T} \to \dot{\mathbb{R}}$ be the homeomorphism given by $\gamma(t) = i(1+t)/(1-t)$. Then $a \circ \gamma \in SO_\eta(\mathbb{T})$ where $\eta := \gamma^{-1}(\lambda)$.*

Proof. First, let $\lambda \in \mathbb{R}$ and hence $\eta \in \mathbb{T} \setminus \{1\}$. Fix $\delta < |\eta - 1|$ and put $\mathbb{T}_\delta(\eta) := \big\{ t \in \mathbb{T} : \ 0 < |t - \eta| \le \delta \big\}$. Since $\gamma'(\eta) = 2i/(1-\eta)^2 \ne 0$, we conclude that

$$0 < m := \inf_{t \in \mathbb{T}_\delta(\eta)} \left| \frac{\gamma(t) - \lambda}{t - \eta} \right| < M := \sup_{t \in \mathbb{T}_\delta(\eta)} \left| \frac{\gamma(t) - \lambda}{t - \eta} \right| < \infty. \tag{2.2}$$

Then, taking $\varepsilon \in (0, \delta)$ and setting $r := m/(2M) \in (0,1)$ and $\varepsilon' := \varepsilon M$, we infer from (2.2) by analogy with [3, Lemma 4.2] that if $\varepsilon/2 \le |t - \eta| \le \varepsilon$ then

$$\begin{aligned} m|t - \eta| \le |\gamma(t) - \lambda| \le M|t - \eta| &\Rightarrow m\varepsilon/2 \le |\gamma(t) - \lambda| \le M\varepsilon \\ &\Leftrightarrow r\varepsilon' \le |\gamma(t) - \lambda| \le \varepsilon'. \end{aligned}$$

Hence $\gamma(\mathbb{T}_{\varepsilon/2,\varepsilon}(\eta)) \subset \dot{\mathbb{R}}_{r\varepsilon',\varepsilon'}(\lambda)$, which implies that $a \circ \gamma \in SO_\eta(\mathbb{T})$ for every $a \in SO_\lambda$.

Let now $\lambda = \infty$ and hence $\eta = 1$. Fix $\delta \in (0,2)$. As $\lim_{t \to 1} \big((t-1)\gamma(t)\big) = 2i$, we again get

$$0 < m := \min_{t \in \mathbb{T}_\delta(1)} \big|(t-1)\gamma(t)\big| < M := \max_{t \in \mathbb{T}_\delta(1)} \big|(t-1)\gamma(t)\big| < \infty. \tag{2.3}$$

Then for $\varepsilon/2 \le |t - 1| \le \varepsilon < \delta$ we deduce from (2.3) that

$$\begin{aligned} m|t-1|^{-1} \le |\gamma(t)| \le M|t-1|^{-1} &\Rightarrow m\varepsilon^{-1} \le |\gamma(t)| \le M2\varepsilon^{-1} \\ &\Leftrightarrow r\varepsilon' \le |\gamma(t)| \le \varepsilon', \end{aligned}$$

where $r = m/(2M) \in (0,1)$ and $\varepsilon' := 2\varepsilon^{-1}M$. Consequently, $\gamma(\mathbb{T}_{\varepsilon/2,\varepsilon}(1)) \subset \dot{\mathbb{R}}_{r\varepsilon',\varepsilon'}(\infty)$, which means that $a \circ \gamma \in SO_1(\mathbb{T})$ if $a \in SO_\infty$. □

Corollary 2.2. *For every $\lambda \in \mathbb{R}$, the mapping $a \mapsto a \circ \beta_\lambda$ defined by the homeomorphism*

$$\beta_\lambda : \ \dot{\mathbb{R}} \to \dot{\mathbb{R}}, \quad x \mapsto \frac{\lambda x - 1}{x + \lambda} \tag{2.4}$$

is an isometric isomorphism of the C^*-algebra SO_λ onto the C^*-algebra SO_∞.

Proof. Obviously, the C^*-algebras $SO_\eta(\mathbb{T})$ for all $\eta \in \mathbb{T}$ are isometrically isomorphic. In particular, setting $\alpha_\eta(t) = \eta t$ for all $t \in \mathbb{T}$, we conclude that the map $a \mapsto a \circ \alpha_\eta$ is an isomorphism of $SO_\eta(\mathbb{T})$ onto $SO_1(\mathbb{T})$. Applying Lemma 2.1, we infer that the map $a \mapsto a \circ \beta_\lambda$, where $\beta_\lambda = \gamma \circ \alpha_\eta \circ \gamma^{-1}$ (see (2.4)) and $\lambda = \gamma(\eta)$, is an isometric isomorphism $SO_\lambda \to SO_\eta(\mathbb{T}) \to SO_1(\mathbb{T}) \to SO_\infty$. □

2.2. Fourier multipliers

Let $C^n(\mathbb{R})$ be the set of all n times continuously differentiable functions $a : \mathbb{R} \to \mathbb{C}$, and let $V(\mathbb{R})$ be the Banach algebra of all functions $a : \mathbb{R} \to \mathbb{C}$ with finite total variation

$$V(a) := \sup\Big\{\sum_{i=1}^{n} |a(t_i) - a(t_{i-1})| : -\infty < t_0 < t_1 < \dots < t_n < +\infty,\ n \in \mathbb{N}\Big\}$$

where the supremum is taken over all finite partitions of the real line $\mathbb{R}$ and the norm in $V(\mathbb{R})$ is given by $\|a\|_V = \|a\|_{L^\infty(\mathbb{R})} + V(a)$. As is known (see, e.g., [15, Chapter 9]), every function $a \in V(\mathbb{R})$ has finite one-sided limits at every point $t \in \dot{\mathbb{R}}$.

Let PC be the C^*-algebra of all functions on $\mathbb{R}$ having finite one-sided limits at every point $t \in \dot{\mathbb{R}}$. If $a \in PC$ has finite total variation, then $a \in M_{p,w}$ for all $p \in (1, \infty)$ and all $w \in A_p(\mathbb{R})$ according to Stechkin's inequality

$$\|a\|_{M_{p,w}} \le \|S_\mathbb{R}\|_{\mathcal{B}(L^p(\mathbb{R},w))}\big(\|a\|_{L^\infty(\mathbb{R})} + V(a)\big) \tag{2.5}$$

(see, e.g., [11, Theorem 2.11] and [9]), where the Cauchy singular integral operator $S_\mathbb{R}$ is given by (1.1).

The following result obtained in [18, Corollary 2.10] supply us with another class of Fourier multipliers in $M_{p,w}$.

Theorem 2.3. *If $a \in C^3(\mathbb{R} \setminus \{0\})$ and $\|D^k a\|_{L^\infty(\mathbb{R})} < \infty$ for all $k = 0, 1, 2, 3$, where $(Da)(x) = xa'(x)$ for $x \in \mathbb{R}$, then the convolution operator $W^0(a)$ is bounded on every weighted Lebesgue space $L^p(\mathbb{R}, w)$ with $1 < p < \infty$ and $w \in A_p(\mathbb{R})$, and*

$$\|a\|_{M_{p,w}} \le C_{p,w} \max\big\{\|D^k a\|_{L^\infty(\mathbb{R})} :\ k = 0, 1, 2, 3\big\} < \infty,$$

where the constant $C_{p,w} \in (0, \infty)$ depends only on p and w.

2.3. Banach algebras $SO^\diamond_{p,w}$

For $\lambda \in \dot{\mathbb{R}}$, we consider the commutative Banach algebras

$$SO^3_\lambda := \Big\{a \in SO_\lambda \cap C^3(\mathbb{R} \setminus \{\lambda\}) : \lim_{x \to \lambda} (D^k_\lambda a)(x) = 0,\ k = 1, 2, 3\Big\}$$

equipped with the norm

$$\|a\|_{SO^3_\lambda} := \max\big\{\|D^k_\lambda a\|_{L^\infty(\mathbb{R})} :\ k = 0, 1, 2, 3\big\},$$

where $(D_\lambda a)(x) = (x - \lambda)a'(x)$ for $\lambda \in \mathbb{R}$ and $(D_\lambda a)(x) = xa'(x)$ if $\lambda = \infty$. By Theorem 2.3, $SO^3_\lambda \subset M_{p,w}$. Let $SO_{\lambda,p,w}$ denote the closure of SO^3_λ in $M_{p,w}$, and let $SO^\diamond_{p,w}$ be the Banach subalgebra of $M_{p,w}$ generated by all the algebras $SO_{\lambda,p,w}$ $(\lambda \in \dot{\mathbb{R}})$. Because $M_{p,w} \subset M_2 = L^\infty(\mathbb{R})$, we conclude that $SO^\diamond_{p,w} \subset SO^\diamond$.

2.4. Banach algebras $PSO^\diamond_{p,w}$

We denote by $C_{p,w}(\dot{\mathbb{R}})$ (resp., $C_{p,w}(\overline{\mathbb{R}})$, $PC_{p,w}$) the closure in $M_{p,w}$ of the set of all functions $a \in C(\dot{\mathbb{R}})$ (resp., $a \in C(\overline{\mathbb{R}})$, $a \in PC$) of finite total variation. Obviously, $C_{p,w}(\dot{\mathbb{R}})$, $C_{p,w}(\overline{\mathbb{R}})$ and $PC_{p,w}$ are Banach subalgebras of $M_{p,w}$, and

$$C_{p,w}(\dot{\mathbb{R}}) \subset C(\dot{\mathbb{R}}), \quad C_{p,w}(\overline{\mathbb{R}}) \subset C(\overline{\mathbb{R}}), \quad PC_{p,w} \subset PC.$$

Let $PSO^\diamond$ be the C^*-subalgebra of $L^\infty(\mathbb{R})$ generated by the C^*-algebras $SO^\diamond$ and PC, and let $PSO^\diamond_{p,w}$ be the Banach subalgebra of $M_{p,w}$ generated by the Banach algebras $SO^\diamond_{p,w}$ and $PC_{p,w}$.

3. The maximal ideal space of the Banach algebra $SO^\diamond_{p,w}$

In what follows, let $M(\mathcal{A})$ denote the maximal ideal space of a commutative Banach algebra $\mathcal{A}$. If $\mathcal{C}$ is a Banach subalgebra of $\mathcal{A}$ and $\lambda \in M(\mathcal{C})$, then the set $M_\lambda(\mathcal{A}) := \{\xi \in M(\mathcal{A}) : \xi|_{\mathcal{C}} = \lambda\}$ is called the fiber of $M(\mathcal{A})$ over λ. Hence for every Banach algebra $\mathcal{A} \subset L^\infty(\mathbb{R})$ with $M(C(\dot{\mathbb{R}}) \cap \mathcal{A}) = \dot{\mathbb{R}}$ and every $\lambda \in \dot{\mathbb{R}}$, the fiber $M_\lambda(\mathcal{A})$ denotes the set of all characters (multiplicative linear functionals) of $\mathcal{A}$ that annihilate the set $\{f \in C(\dot{\mathbb{R}}) \cap \mathcal{A} : f(\lambda) = 0\}$.

Identifying the points $\lambda \in \dot{\mathbb{R}}$ with the evaluation functionals δ_λ on $\dot{\mathbb{R}}$, $\delta_\lambda(f) = f(\lambda)$ for $f \in C(\dot{\mathbb{R}})$, we infer that the maximal ideal space $M(SO^\diamond)$ of $SO^\diamond$ is of the form

$$M(SO^\diamond) = \bigcup_{\lambda \in \dot{\mathbb{R}}} M_\lambda(SO^\diamond) \tag{3.1}$$

where $M_\lambda(SO^\diamond) := \big\{\xi \in M(SO^\diamond) : \xi|_{C(\dot{\mathbb{R}})} = \delta_\lambda\big\}$ are fibers of $M(SO^\diamond)$ over $\lambda \in \dot{\mathbb{R}}$. Applying Corollary 2.2 and [4, Proposition 5], we infer that for every $\lambda \in \dot{\mathbb{R}}$,

$$M_\lambda(SO^\diamond) = M_\lambda(SO_\lambda) = M_\infty(SO_\infty) = (\mathrm{clos}_{SO^*_\infty} \mathbb{R}) \setminus \mathbb{R}, \tag{3.2}$$

where $\mathrm{clos}_{SO^*_\infty} \mathbb{R}$ is the weak-star closure of $\mathbb{R}$ in SO^*_∞, the dual space of SO_∞.

The fiber $M_\infty(SO_\infty)$ is related to the partial limits of a function $a \in SO_\infty$ at infinity as follows (see [7, Corollary 4.3] and [1, Corollary 3.3]).

Proposition 3.1. *If $\{a_k\}_{k=1}^\infty$ is a countable subset of SO_∞ and $\xi \in M_\infty(SO_\infty)$, then there exists a sequence $\{g_n\} \subset \mathbb{R}_+$ such that $g_n \to \infty$ as $n \to \infty$, and for every $t \in \mathbb{R} \setminus \{0\}$ and every $k \in \mathbb{N}$, $\lim_{n\to\infty} a_k(g_n t) = \xi(a_k)$.*

Let $1 < p < \infty$ and $w \in A_p(\mathbb{R})$. For every $\lambda \in \dot{\mathbb{R}}$ we consider three unital commutative Banach algebras with the same unit which are homomorphically embedded one into another:

$$SO^3_\lambda \subset SO_{\lambda,p,w} \subset SO_\lambda. \tag{3.3}$$

To study the relations between their maximal ideal spaces, by analogy with [1] and [18], we use the following result (see [30, Theorem 3.10]).

Theorem 3.2. *Let B_i $(i = 1, 2, 3)$ be commutative Banach algebras with the same unit which are homomorphically inclosed one into another, $B_1 \subset B_2 \subset B_3$. Suppose that B_1 is dense in B_2 and every multiplicative linear functional defined on B_1 extends to a multiplicative linear functional on B_3. Then every multiplicative linear functional on B_2 also extends to a multiplicative linear functional on B_3.*

Given $\lambda \in \dot{\mathbb{R}}$, we consider the commutative algebra $C(\dot{\mathbb{R}}) \cap SO^3_\lambda \subset SO^3_\lambda$. As $M(C(\dot{\mathbb{R}}) \cap SO^3_\lambda) = \dot{\mathbb{R}}$, we see that for every $t \in \dot{\mathbb{R}}$ the set

$$M_t(SO^3_\lambda) = \{\xi \in M(SO^3_\lambda) : \xi|_{C(\dot{\mathbb{R}}) \cap SO^3_\lambda} = \delta_t\}$$

is the fiber of $M(SO^3_\lambda)$ over the point t. Then

$$M(SO^3_\lambda) = \bigcup_{t \in \dot{\mathbb{R}}} M_t(SO^3_\lambda), \tag{3.4}$$

where $M_t(SO^3_\lambda) = \{t\}$ for all $t \in \dot{\mathbb{R}} \setminus \{\lambda\}$.

Modifying the proof of [4, Proposition 5], we obtain the following.

Lemma 3.3. *If $\lambda \in \dot{\mathbb{R}}$, then the fiber $M_\lambda(SO^3_\lambda)$ has the form*

$$M_\lambda(SO^3_\lambda) = \left(\mathrm{clos}_{(SO^3_\lambda)^*}(\dot{\mathbb{R}} \setminus \{\lambda\})\right) \setminus (\dot{\mathbb{R}} \setminus \{\lambda\})$$

where $\mathrm{clos}_{(SO^3_\lambda)^*}(\dot{\mathbb{R}} \setminus \{\lambda\})$ is the weak-star closure of $\dot{\mathbb{R}} \setminus \{\lambda\}$ in $(SO^3_\lambda)^*$, the dual space of SO^3_λ.

Proof. First, let us prove that

$$M(SO^3_\lambda) \subset \mathrm{clos}_{(SO^3_\lambda)^*}(\dot{\mathbb{R}} \setminus \{\lambda\}). \tag{3.5}$$

Fix $\xi \in M(SO^3_\lambda)$. Any $(SO^3_\lambda)^*$-neighborhood of ξ is of the form

$$U := U_{a_1,\dots,a_n;\varepsilon}(\xi) = \{\eta \in (SO^3_\lambda)^* : |\eta(a_i) - \xi(a_i)| < \varepsilon,\ i = 1, \dots, n\},$$

where $\varepsilon > 0$ and $a_1, \dots, a_n \in SO^3_\lambda$. We must show that there is a $t_0 \in \dot{\mathbb{R}} \setminus \{\lambda\}$ such that $\delta_{t_0} \in U$. Put $a := |a_1 - \xi(a_1)| + \dots + |a_n - \xi(a_n)|$. According to [13, § 13, Theorem 1], $a \in SO^3_\lambda$ and then $\xi(a) = 0$. Therefore, a is not invertible in SO^3_λ and, hence, there is a sequence $\{t_n\} \subset \dot{\mathbb{R}} \setminus \{\lambda\}$ such that $\lim_{n\to\infty} a(t_n) = 0$. Since $|a_i(t) - \xi(a_i)| \le a(t)$ for all $t \in \dot{\mathbb{R}} \setminus \{\lambda\}$ and each i, we infer that there exists a $t_0 \in \dot{\mathbb{R}} \setminus \{\lambda\}$ such that $|a_i(t_0) - \xi(a_i)| \le a(t_0) < \varepsilon$ for each i. Thus, $\delta_{t_0} \in U$, which implies (3.5).

It is clear that $\delta_t \notin M_\lambda(SO^3_\lambda)$ for $t \in \dot{\mathbb{R}} \setminus \{\lambda\}$ because there is a function $b \in C(\dot{\mathbb{R}}) \cap SO^3_\lambda$ such that $b(t) \neq 0$ for $t \in \dot{\mathbb{R}} \setminus \{\lambda\}$ and $b(\lambda) = 0$. Therefore, by (3.5), $M_\lambda(SO^3_\lambda) \subset \left(\mathrm{clos}_{(SO^3_\lambda)^*}(\dot{\mathbb{R}} \setminus \{\lambda\})\right) \setminus (\dot{\mathbb{R}} \setminus \{\lambda\})$.

Conversely, let $\xi \in \left(\mathrm{clos}_{(SO^3_\lambda)^*}(\dot{\mathbb{R}} \setminus \{\lambda\})\right) \setminus (\dot{\mathbb{R}} \setminus \{\lambda\})$, let $a, b \in SO^3_\lambda$, and let $\varepsilon > 0$. Then choose $t \in \dot{\mathbb{R}} \setminus \{\lambda\}$ so that $\delta_t \in U_{a,b,ab;\varepsilon}(\xi)$. We have

$$\begin{aligned} |\xi(ab) - \xi(a)\xi(b)| &\le |\xi(ab) - a(t)b(t)| + |a(t) - \xi(a)||b(t)| + |\xi(a)||b(t) - \xi(b)| \\ &\le \varepsilon + \varepsilon|b(t)| + |\xi(a)|\varepsilon. \end{aligned}$$

Since $\varepsilon > 0$ can be chosen arbitrarily, we get $\xi(ab) = \xi(a)\xi(b)$, that is, $\xi \in M(SO^3_\lambda)$. But as $\xi \notin \dot{\mathbb{R}} \setminus \{\lambda\} = \bigcup_{t\in\dot{\mathbb{R}}\setminus\{\lambda\}} M_t(SO^3_\lambda)$, we conclude that actually $\xi \in M_\lambda(SO^3_\lambda)$. ☐

Lemma 3.4. *If $\lambda \in \dot{\mathbb{R}}$, then every multiplicative linear functional defined on SO^3_λ extends to a multiplicative linear functional on SO_λ.*

Proof. Fix $\lambda \in \dot{\mathbb{R}}$. Since $M_t(SO^3_\lambda) = \{\delta_t\}$ for all $t \in \dot{\mathbb{R}} \setminus \{\lambda\}$ and the evaluation functionals δ_t identified with the points $t \in \dot{\mathbb{R}} \setminus \{\lambda\}$ belong to $M(SO_\lambda)$, it remains to prove the existence of the required extensions for the multiplicative linear functionals $\xi \in M_\lambda(SO^3_\lambda)$.

By Lemma 3.3, every functional $\xi \in M_\lambda(SO^3_\lambda)$ is the limit of a net $\{t_\alpha\} \subset \dot{\mathbb{R}} \setminus \{\lambda\}$ that does not converge to functionals $t \in \dot{\mathbb{R}} \setminus \{\lambda\}$, that is,

$$\xi(a) = \lim_\alpha t_\alpha(a) \quad \text{for every} \quad a \in SO^3_\lambda,$$

where $t_\alpha(a) = a(t_\alpha)$ for $a \in SO^3_\lambda$ and $t_\alpha \in \dot{\mathbb{R}} \setminus \{\lambda\}$. Then, for every $a \in SO^3_\lambda$, $\{t_\alpha(a)\}$ is a Cauchy net in $\mathbb{C}$, that is, for every $\varepsilon > 0$ there exists $\gamma > 0$ such that $|t_\alpha(a) - t_\beta(a)| < \varepsilon$ if $\alpha, \beta \succ \gamma$. Given $b \in SO_\lambda$, there is a sequence $\{a_n\} \subset SO^3_\lambda$ such that $\lim_{n\to\infty} \|b - a_n\|_{L^\infty(\mathbb{R})} = 0$. Then from the relations

$$\begin{aligned}|t_\alpha(b) - t_\beta(b)| &\le |t_\alpha(b - a_n)| + |t_\alpha(a_n) - t_\beta(a_n)| + |t_\beta(a_n - b)| \\ &\le \|b - a_n\|_{L^\infty(\mathbb{R})} + |t_\alpha(a_n) - t_\beta(a_n)| + \|b - a_n\|_{L^\infty(\mathbb{R})}\end{aligned}$$

it follows that $\{t_\alpha(b)\}$ is a Cauchy net in $\mathbb{C}$ for every $b \in SO_\lambda$, and hence this net converges in $\mathbb{C}$ and its limit is unique.

For each $b \in SO_\lambda$, we define $\tilde{\xi}(b) := \lim_\alpha t_\alpha(b)$. Then $\tilde{\xi}$ is a multiplicative linear functional on SO_λ (and therefore bounded of norm 1 by [25, Proposition 10.6, Theorem 10.7]) because

$$\tilde{\xi}(bd) = \lim_\alpha t_\alpha(bd) = \lim_\alpha t_\alpha(b) \lim_\alpha t_\alpha(d) = \tilde{\xi}(b)\tilde{\xi}(d) \quad \text{for all} \quad b, d \in SO_\lambda.$$

Thus, $\tilde{\xi}$ is a required extension of $\xi \in M(SO^3_\lambda)$ to SO_λ. ☐

Note that the proof of Lemma 3.4 in case $\lambda = \infty$ improves that in [18, Lemma 3.3].

Applying (3.3), the density of SO^3_λ in $SO_{\lambda,p,w}$ in the norm of $M_{p,w}$ and Lemma 3.4, we conclude that for every $\lambda \in \dot{\mathbb{R}}$ the commutative Banach algebras

$$B_1 = SO^3_\lambda, \quad B_2 = SO_{\lambda,p,w}, \quad B_3 = SO_\lambda$$

satisfy the conditions of Theorem 3.2. By Theorem 3.2, every multiplicative linear functional on $SO_{\lambda,p,w}$ extends to a multiplicative linear functional on SO_λ, and hence $M(SO_{\lambda,p,w}) \subset M(SO_\lambda)$. On the other hand, $M(SO_\lambda) \subset M(SO_{\lambda,p,w})$ because $SO_{\lambda,p,w} \subset SO_\lambda$. Thus we get the following result.

Lemma 3.5. *If $1 < p < \infty$, $w \in A_p(\mathbb{R})$ and $\lambda \in \dot{\mathbb{R}}$, then the maximal ideal spaces of $SO_{\lambda,p,w}$ and SO_λ coincide as sets, that is, $M(SO_{\lambda,p,w}) = M(SO_\lambda)$.*

Fix $p \in (1, \infty)$ and $w \in A_p(\mathbb{R})$. Lemma 3.5 and relations (3.2) imply that

$$M_\lambda(SO^\diamond_{p,w}) = M_\lambda(SO_{\lambda,p,w}) = M_\lambda(SO_\lambda) = M_\infty(SO_\infty) \tag{3.6}$$

for every $\lambda \in \dot{\mathbb{R}}$. Analogously to (3.1) and (3.4) we obtain

$$M(SO^\diamond_{p,w}) = \bigcup_{\lambda \in \dot{\mathbb{R}}} M_\lambda(SO^\diamond_{p,w}). \tag{3.7}$$

Applying (3.7), (3.6) and (3.1) we arrive at the following result.

Theorem 3.6. *If $1 < p < \infty$ and $w \in A_p(\mathbb{R})$, then the maximal ideal spaces of $SO^\diamond_{p,w}$ and $SO^\diamond$ coincide as sets, $M(SO^\diamond_{p,w}) = M(SO^\diamond)$.*

4. Compactness of commutators of convolution type operators

4.1. $SO^\diamond$ and VMO

Let $\Gamma \in \{\mathbb{R}, \mathbb{T}\}$. Given a locally integrable function $f \in L^1_{\mathrm{loc}}(\Gamma)$ and a finite interval I on Γ, let $|I|$ denote the length of I and let

$$I(f) := |I|^{-1} \int_I f(t)dt$$

denote the average of f over I. For $a > 0$, consider the quantities

$$M_a(f) := \sup_{|I| \le a} |I|^{-1} \int_I |f(t) - I(f)|dt,$$

$$M_0(f) := \lim_{a \to 0} M_a(f), \quad \|f\|_* := \lim_{a \to \infty} M_a(f). \tag{4.1}$$

The function $f \in L^1_{\mathrm{loc}}(\Gamma)$ is said to have bounded mean oscillation, $f \in BMO(\Gamma)$, if $\|f\|_* < \infty$. The space $BMO(\Gamma)$ is a Banach space under the norm $\|\cdot\|_*$, provided that two functions differing by a constant are identified. A function $f \in BMO(\Gamma)$ is said to have vanishing mean oscillation, $f \in VMO(\Gamma)$, if $M_0(f) = 0$. As is well known (see, e.g., [26]), $VMO(\Gamma)$ is a closed subspace of $BMO(\Gamma)$.

Consider the homeomorphism $\gamma : \mathbb{T} \to \dot{\mathbb{R}}$, $\gamma(t) = i(1+t)/(1-t)$. By [12, Chapter VI, Corollary 1.3], $f \in BMO(\mathbb{R})$ if and only if $f \circ \gamma \in BMO(\mathbb{T})$, and the norms of these functions are equivalent. On the other hand,

$$VMO := \{f \circ \gamma^{-1} : f \in VMO(\mathbb{T})\} \tag{4.2}$$

is a proper closed subspace of $VMO(\mathbb{R})$. Since $VMO(\mathbb{T})$ is the closure of $C(\mathbb{T})$ in $BMO(\mathbb{T})$ (see, e.g., [12, p. 253]), (4.2) implies the following property of VMO.

Proposition 4.1. *VMO is the closure in $BMO(\mathbb{R})$ of the set $C(\dot{\mathbb{R}})$.*

Let H^∞ be the closed subalgebra of $L^\infty(\mathbb{R})$ that consists of all functions being non-tangential limits on $\mathbb{R}$ of bounded analytic functions on the upper half-plane.

Theorem 4.2. *The C^*-algebra $SO^\diamond$ is contained in the C^*-algebra QC of quasi-continuous functions on $\dot{\mathbb{R}}$, where*

$$QC := (H^\infty + C(\dot{\mathbb{R}})) \cap (\overline{H^\infty} + C(\dot{\mathbb{R}})) = VMO \cap L^\infty(\mathbb{R}). \tag{4.3}$$

Proof. If $a \in SO_\infty$, then $\widetilde{a} := a \circ \gamma \in SO_1(\mathbb{T})$ in view of Lemma 2.1. By [27, Lemma 1] and [22, Lemma A4], every function $\widetilde{a} \in SO_1(\mathbb{T})$ belongs to the set $VMO(\mathbb{T}) \cap L^\infty(\mathbb{T})$. According to [26], $VMO(\mathbb{T}) \cap L^\infty(\mathbb{T}) = QC(\mathbb{T})$, where $QC(\mathbb{T}) := (H^\infty(\mathbb{T}) + C(\mathbb{T})) \cap (\overline{H^\infty(\mathbb{T})} + C(\mathbb{T}))$ is the C^*-algebra of quasicontinuous functions on $\mathbb{T}$, and $H^\infty(\mathbb{T})$ is the closed subalgebra of $L^\infty(\mathbb{T})$ consisting of all functions being non-tangential limits on $\mathbb{T}$ of bounded analytic functions on $\mathbb{D} := \{z \in \mathbb{C} : |z| < 1\}$. Thus, $\widetilde{a} \in QC(\mathbb{T})$, and therefore $a = \widetilde{a} \circ \gamma^{-1} \in QC$, where QC is given by (4.3). Applying (4.3) and Corollary 2.2, we infer that for all $\lambda \in \dot{\mathbb{R}}$ the C^*-algebras SO_λ, and hence $SO^\diamond$, are contained in QC. □

4.2. Compactness of commutators

Given $1 < p < \infty$ and $w \in A_p(\mathbb{R})$, we consider the Banach algebra $\mathcal{B}_{p,w}$ and its ideal of compact operators $\mathcal{K}_{p,w}$. In case $w \equiv 1$ we abbreviate $\mathcal{B}_{p,1}$ and $\mathcal{K}_{p,1}$ to $\mathcal{B}_p$ and $\mathcal{K}_p$, respectively. The notation $C_p(\dot{\mathbb{R}})$, $C_p(\overline{\mathbb{R}})$, PC_p and $SO_{\infty,p}$ is understood analogously.

For two algebras $\mathcal{A}$ and $\mathcal{B}$ contained in a Banach algebra $\mathcal{C}$, we denote by $\mathrm{alg}\,(\mathcal{A}, \mathcal{B})$ the Banach subalgebra of $\mathcal{C}$ generated by the algebras $\mathcal{A}$ and $\mathcal{B}$.

First we recall two known results on the compactness of commutators.

Lemma 4.3. [11, Lemmas 7.1–7.4] *Let* $1 < p < \infty$.

(a) *If* $a \in PC$, $b \in PC_p$, *and* $a(\pm\infty) = b(\pm\infty) = 0$, *then* $aW^0(b), W^0(b)aI \in \mathcal{K}_p$.

(b) *If* $a \in C(\dot{\mathbb{R}})$ *and* $b \in PC_p$, *or* $a \in PC$ *and* $b \in C_p(\dot{\mathbb{R}})$, *then* $[aI, W^0(b)] \in \mathcal{K}_p$.

(c) *If* $a \in C(\overline{\mathbb{R}})$ *and* $b \in C_p(\overline{\mathbb{R}})$, *then* $[aI, W^0(b)] \in \mathcal{K}_p$.

Theorem 4.4. [1, Theorem 4.2, Corollary 4.3] *If* $1 < p < \infty$ *and either* $a \in \mathrm{alg}\,(SO_\infty, PC)$ *and* $b \in SO_{\infty,p}$, *or* $a \in SO_\infty$ *and* $b \in \mathrm{alg}\,(SO_{\infty,p}, PC_p)$, *or* $a \in \mathrm{alg}\,(SO_\infty, C(\overline{\mathbb{R}}))$ *and* $b \in \mathrm{alg}\,(SO_{\infty,p}, C_p(\overline{\mathbb{R}}))$, *then* $[aI, W^0(b)] \in \mathcal{K}_p$.

The use of a weighted analogue [16] of the Krasnoselskii theorem [19, Theorem 3.10] on interpolation of compactness, which follows from the Stein-Weiss interpolation theorem (see, e.g., [5, Corollary 5.5.4]), leads to the following compactness result.

Lemma 4.5. [16, Corollary 5.3] *If a linear operator T is bounded on every weighted Lebesgue space $L^p(\mathbb{R}, w)$ $(1 < p < \infty,\ w \in A_p(\mathbb{R}))$ and T is compact on the space $L^2(\mathbb{R})$, then T is compact on every space $L^p(\mathbb{R}, w)$.*

Applying the theory of pseudodifferential and Calderón-Zygmund operators and Theorem 4.2, we establish the following compactness result for weighted Lebesgue spaces.

Theorem 4.6. *Let* $1 < p < \infty$ *and* $w \in A_p(\mathbb{R})$. *If* $a \in PSO^\diamond$ *and* $b \in SO^\diamond_{p,w}$, *or* $a \in SO^\diamond$ *and* $b \in PSO^\diamond_{p,w}$, *or* $a \in \mathrm{alg}\,(SO_\infty, C(\overline{\mathbb{R}}))$ *and* $b \in \mathrm{alg}\,(SO_{\infty,p,w}, C_{p,w}(\overline{\mathbb{R}}))$, *then the commutator* $[aI, W^0(b)]$ *is compact on the space* $L^p(\mathbb{R}, w)$.

Proof. By definition, $PSO^\diamond_{p,w} = \text{alg}\,(SO^\diamond_{p,w}, PC_{p,w}) \subset M_{p,w}$, where $SO^\diamond_{p,w}$ is the Banach subalgebra of $M_{p,w}$ generated by all the algebras $SO_{\lambda,p,w}$ $(\lambda \in \dot{\mathbb{R}})$, $SO_{\lambda,p,w}$ is the closure of SO^3_λ in $M_{p,w}$, and $PC_{p,w}$ is the closure in $M_{p,w}$ of the set of piecewise continuous functions of finite total variation, and hence $PC_{p,w}$ is the closure in $M_{p,w}$ of the set of all piecewise constant functions with finite sets of discontinuities (see, e.g., [11, Remark 2.12]). In addition, $C_{p,w}(\overline{\mathbb{R}})$ is the closure of $C(\overline{\mathbb{R}}) \cap V(\mathbb{R})$ in $M_{p,w}$. Consequently, it is sufficient to prove the compactness of the commutator $[aI, W^0(b)]$ in the following four cases for each $\lambda, \mu \in \dot{\mathbb{R}}$:

1) $a \in SO_\lambda$ and $b \in SO^3_\mu$,
2) $a \in SO_\lambda$ and $b(x) = \text{sgn}(x-\mu)$,
3) $a(x) = \text{sgn}(x-\lambda)$ and $b \in SO^3_\mu$,
4) $a \in C(\overline{\mathbb{R}})$ and $b \in C(\overline{\mathbb{R}}) \cap V(\mathbb{R})$.

Under these conditions on functions a and b, the commutators $[aI, W^0(b)]$ are bounded linear operators on every Lebesgue space $L^p(\mathbb{R}, w)$ with $1 < p < \infty$ and $w \in A_p(\mathbb{R})$. Hence, according to Lemma 4.5, it is sufficient to prove the compactness of the commutator $[aI, W^0(b)]$ only on the space $L^2(\mathbb{R})$, which implies its compactness on all the spaces $L^p(\mathbb{R}, w)$. Thus, in the case of $L^2(\mathbb{R})$ we may replace $b \in SO^3_\mu$ by $b \in SO_\mu$. Then the case 3) is reduced to the case 2) under the transform $A \mapsto \mathcal{F}A\mathcal{F}^{-1}$. Indeed, $\mathcal{F}a\mathcal{F}^{-1} = W^0(\widetilde{b})$ and $\mathcal{F}W^0(b)\mathcal{F}^{-1} = \widetilde{a}I$ where $\widetilde{b}(x) = a(-x)$ and $\widetilde{a} = b$. Because the assertion in the case 4) for the space $L^2(\mathbb{R})$ follows from Lemma 4.3(c), it only remains to consider cases 1) and 2).

1) Let $a \in SO_\lambda$ and $b \in SO_\mu$ $(\lambda, \mu \in \dot{\mathbb{R}})$. If $\lambda = \mu = \infty$, then the compactness of the commutator $[aI, W^0(b)]$ follows from Theorem 4.4. Let $\lambda \in \mathbb{R}$ and $\mu = \infty$. In this case we assume without loss of generality that $b \in SO^3_\infty$. Then from [18, Lemma 2.2] it follows that the distribution $K = \mathcal{F}^{-1}b$ agrees with a function $K(\cdot)$ differentiable on $\mathbb{R} \setminus \{0\}$ and such that

$$\big|K(x)\big| \le A_0|x|^{-1}, \quad \big|K'(x)\big| \le A_1|x|^{-2} \quad \text{for all} \quad x \in \mathbb{R} \setminus \{0\}, \tag{4.4}$$

where the constants A_α $(\alpha = 0, 1)$ are estimated by

$$A_\alpha \le C_\alpha \max\big\{\|D^k b\|_{L^\infty(\mathbb{R})} : \ k = 0, 1, 2, 3\big\},$$

$(Db)(x) = xb'(x)$ for $x \in \mathbb{R}$ and the constants $C_\alpha \in (0, \infty)$ depend only on α. Hence $K(\cdot)$ is a classical Calderón-Zygmund kernel, and the convolution operator $W^0(b)$ can be considered as the Calderón-Zygmund operator given by

$$(Tf)(x) = \text{v.p.} \int_{\mathbb{R}} K(x-y)f(y)dy \quad \text{for } x \in \mathbb{R}, \tag{4.5}$$

where T is bounded on every weighted Lebesgue space $L^p(\mathbb{R}, w)$ with $p \in (1, \infty)$ and $w \in A_p(\mathbb{R})$ (see, e.g., Theorem 2.3). In particular, the second condition in (4.4) implies that there is a constant $A_2 \in (0, \infty)$ such that

$$|K(x-y) - K(x)| \le A_2|y|^\delta |x|^{-1-\delta} \quad \text{for } |x| \ge 2|y| > 0, \tag{4.6}$$

where $\delta \in (0,1)$. Moreover, because the convolution operator $W^0(b)$ is bounded on the space $L^2(\mathbb{R})$, we conclude from [29, p. 291, Proposition 2] that

$$\sup_{0<r<R<\infty} \left| \int_{r<|x|<R} K(x)dx \right| < \infty. \tag{4.7}$$

Since conditions (4.4), (4.6) and (4.7) for the operator $T = W^0(b)$ represented in the form (4.5) are fulfilled, we infer from [14, Theorem 7.5.6] that there exists a constant $C \in (0,\infty)$ such that

$$\big\| [aI, W^0(b)] \big\|_{\mathcal{B}_2} \le C\|a\|_* \tag{4.8}$$

for every $a \in BMO(\mathbb{R})$, where $\mathcal{B}_2 = \mathcal{B}(L^2(\mathbb{R}))$ and $\|\cdot\|_*$ is given by (4.1). On the other hand, by Theorem 4.2, the function $a \in SO_\lambda$ belongs to the Banach space VMO. Hence, by Proposition 4.1, for every $a \in SO_\lambda$ there exists a sequence $\{a_n\} \in C(\dot{\mathbb{R}})$ such that $\lim_{n\to\infty} \|a - a_n\|_* = 0$, and therefore, by (4.8),

$$\lim_{n\to\infty} \big\| [aI, W^0(b)] - [a_nI, W^0(b)] \big\|_{\mathcal{B}_2} = \lim_{n\to\infty} \big\| [(a-a_n)I, W^0(b)] \big\|_{\mathcal{B}_2} = 0. \tag{4.9}$$

But $[a_nI, W^0(b)] \in \mathcal{K}_2$ for all $a_n \in C(\dot{\mathbb{R}})$ and $b \in SO_\infty$ in virtue of Theorem 4.4. Thus, we deduce from (4.9) that the commutator $[aI, W^0(b)]$ is compact on the space $L^2(\mathbb{R})$ for every $a \in SO_\lambda$ and every $b \in SO_\infty$, which proves the assertion for $\lambda \in \mathbb{R}$ and $\mu = \infty$.

The case $\lambda = \infty$ and $\mu \in \mathbb{R}$ is reduced to the previous one under the transform $A \mapsto \mathcal{F}A\mathcal{F}^{-1}$. Let now $\lambda, \mu \in \mathbb{R}$. Then

$$[aI, W^0(b)] = [(a - a(\infty))I, W^0(b - b(\infty))], \tag{4.10}$$

and there exist functions $c, d \in C(\dot{\mathbb{R}})$ that vanish at ∞ and functions $\widetilde{a}, \widetilde{b} \in L^\infty(\mathbb{R})$ such that $a - a(\infty) = \widetilde{a}c$ and $b - b(\infty) = d\widetilde{b}$. Since the operators $cW^0(d)$ and $W^0(d)cI$ are compact on the space $L^2(\mathbb{R})$ due to Lemma 4.3(a), we infer from (4.10) and the equality

$$[(a - a(\infty))I, W^0(b - b(\infty))] = \widetilde{a}c\, W^0(d)W^0(\widetilde{b}) - W^0(\widetilde{b})W^0(d)\, c\widetilde{a}I$$

that $[aI, W^0(b)] \in \mathcal{K}_2$ for all $\lambda, \mu \in \mathbb{R}$, which completes the proof of compactness of the commutator $[aI, W^0(b)]$ in case 1).

2) Let now $a \in SO_\lambda$ and $b(x) = \operatorname{sgn}(x - \mu)$ where $\lambda, \mu \in \dot{\mathbb{R}}$. Clearly, we may exclude $\mu = \infty$. Since $a \in QC = (H^\infty + C(\dot{\mathbb{R}})) \cap (\overline{H^\infty} + C(\dot{\mathbb{R}}))$ in view of Theorem 4.2, it immediately follows from the Hartman compactness result (see, e.g., [8, Theorem 2.18]) that $[aI, S_\mathbb{R}] \in \mathcal{K}_2$ (see also [21, Section 4]). Applying then the equality $W^0(b) = -e_{-\mu}S_\mathbb{R}e_\mu I$ where $e_\mu(x) = e^{i\mu x}$ for $\mu, x \in \mathbb{R}$, we infer that the commutator $[aI, W^0(b)] = -e_{-\mu}[aI, S_\mathbb{R}]e_\mu I$ is compact on the space $L^2(\mathbb{R})$, which completes the proof in case 2). □

5. Applications of limit operators

Fix $w \in A_p(\mathbb{R})$. Then $v = \log w$ is a BMO function on $\mathbb{R}$ (see, e.g., [12, Chapter 6]). Hence, from [12, Chapter 6, Theorem 1.2] it follows the existence of two continuous functions

$$v_\pm(x) := x \int_x^{\pm\infty} \frac{v(\tau)}{\tau^2}\, d\tau \quad \text{for} \quad \pm x > 0. \tag{5.1}$$

The functions $v_\pm$ are differentiable almost everywhere on $\mathbb{R}_\pm = \{x : \pm x > 0\}$, and

$$xv'_\pm(x) = v_\pm(x) - v(x) \quad \text{for almost all } x \in \mathbb{R}_\pm. \tag{5.2}$$

In what follows we assume that at least one of the functions $x \mapsto xv'_\pm(x)$ belongs to L^∞ at a neighborhood $U_\pm \subset \mathbb{R}_\pm$ of $\pm\infty$, respectively. By (5.2), this is equivalent to the condition

$$v_- - v \in L^\infty(U_-) \quad \text{or} \quad v_+ - v \in L^\infty(U_+). \tag{5.3}$$

We say that a weight w is locally equivalent to a weight W at a neighborhood $U_\pm$ of $\pm\infty$ if $w/W,\ W/w \in L^\infty(U_\pm)$. Thus, in view of (5.2), the weight $w = e^v$ is locally equivalent to the weight $w_\pm = e^{v_\pm}$ at a neighborhood $U_\pm$ of $\pm\infty$, respectively. Therefore, the weights $\widetilde{w}_\pm$ that coincide with w on $\mathbb{R} \setminus U_\pm$ and with $w_\pm$ on $U_\pm$, belong to $A_p(\mathbb{R})$ along with w, and $A \in \mathcal{B}(L^p(\mathbb{R}, w))$ if and only if $A \in \mathcal{B}(L^p(\mathbb{R}, \widetilde{w}_\pm))$.

Let $e_\lambda(x) = e^{i\lambda x}$ for all $\lambda, x \in \mathbb{R}$, and let $U_\lambda = W^0(e_\lambda)$ is the translation operator acting by the rule $(U_\lambda f)(x) = f(x - \lambda)$ for $x \in \mathbb{R}$. Let $SO_p^\diamond := SO_{p,1}^\diamond$. As usual, for all $a \in SO^\diamond$ and all $\xi \in M(SO^\diamond)$ we put $a(\xi) := \xi(a)$.

Lemma 5.1. *If $1 < p < \infty$, $w = e^v \in A_p(\mathbb{R})$, $a \in SO^\diamond$, $b \in SO_{p,w}^\diamond$, and (5.3) holds, then for every $\xi \in M_\infty(SO^\diamond)$ there is a sequence $\{h_n\} \subset (0,\infty)$ such that $h_n \to +\infty$ as $n \to \infty$, $\lim_{n\to\infty} a(h_n) = a(\xi)$, $\lim_{n\to\infty} b(h_n) = b(\xi)$, and*

$$\operatorname*{s-lim}_{n\to\infty} \left(e_{h_n}(aI)e_{h_n}^{-1}I\right) = aI, \quad \operatorname*{s-lim}_{n\to\infty} \left(e_{h_n}W^0(b)e_{h_n}^{-1}I\right) = b(\xi)I, \tag{5.4}$$

on the space $L^p(\mathbb{R}, w)$,

$$\begin{aligned} \operatorname*{s-lim}_{n\to\infty} \left(U_{-h_n}\widetilde{w}_+ a\widetilde{w}_+^{-1}U_{h_n}\right) &= a(\xi)I, \\ \operatorname*{s-lim}_{n\to\infty} \left(U_{-h_n}\widetilde{w}_+ W^0(b)\widetilde{w}_+^{-1}U_{h_n}\right) &= W^0(b), \end{aligned} \tag{5.5}$$

on the space $L^p(\mathbb{R})$ if $v_+ - v \in L^\infty(U_+)$, and

$$\begin{aligned} \operatorname*{s-lim}_{n\to\infty} \left(U_{h_n}\widetilde{w}_- a\widetilde{w}_-^{-1}U_{-h_n}\right) &= a(\xi)I, \\ \operatorname*{s-lim}_{n\to\infty} \left(U_{h_n}\widetilde{w}_- W^0(b)\widetilde{w}_-^{-1}U_{-h_n}\right) &= W^0(b), \end{aligned} \tag{5.6}$$

on the space $L^p(\mathbb{R})$ if $v_- - v \in L^\infty(U_-)$.

Proof. By definition (see Subsection 2.3), every function $b \in SO_{p,w}^\diamond$ is approximated in the Banach algebra $M_{p,w}$ by a sequence of functions $b_m = \sum_{\lambda\in F_m} b_{m,\lambda}$ where $b_{m,\lambda} \in SO_\lambda^3$ for all $m \in \mathbb{N}$ and all $\lambda \in \dot{\mathbb{R}}$, and F_m are finite subsets of $\dot{\mathbb{R}}$ that contain ∞. Hence $b_m \in SO_{p,w}^\diamond \cap SO_p^\diamond$.

Fix $\xi \in M_\infty(SO^\diamond)$. Since the set $\{b_{m,\infty} \in SO^3_\infty : m \in \mathbb{N}\}$ is at most countable, from Proposition 3.1 it follows that there exists a sequence $\{h_n\} \subset (0,\infty)$ such that $h_n \to +\infty$ as $n \to \infty$ and

$$\xi(b_{m,\infty}) = \lim_{n\to\infty} b_{m,\infty}(h_n) \quad \text{for all} \quad m \in \mathbb{N}. \tag{5.7}$$

Then we infer from (5.7) and the estimate

$$\begin{aligned} \big|b_{m,\infty}(x+h_n) - b_{m,\infty}(h_n)\big| &= \left| \int_{h_n}^{x+h_n} t b'_{m,\infty}(t)\, \frac{dt}{t} \right| \\ &\le \sup_{|t-h_n|\le |x|} \big|t b'_{m,\infty}(t)\big| \left| \ln\left(\frac{x+h_n}{h_n}\right)\right| \end{aligned} \tag{5.8}$$

that

$$\xi(b_{m,\infty}) = \lim_{n\to\infty} b_{m,\infty}(x+h_n) \tag{5.9}$$

for every $x \in \mathbb{R}$ and all $m \in \mathbb{N}$. Because the functions $b_{m,\lambda}$ are continuous at ∞ for all $m \in \mathbb{N}$ and all $\lambda \in \mathbb{R}$, we conclude that for these m and λ,

$$\lim_{n\to\infty} b_{m,\lambda}(x+h_n) = b_{m,\lambda}(\infty) = \xi(b_{m,\lambda}). \tag{5.10}$$

As the limit

$$b = \lim_{m\to\infty} \sum_{\lambda\in F_m} b_{m,\lambda}, \tag{5.11}$$

is uniform in the norm of $M_{p,w}$, we deduce from (5.9), (5.10) and (5.11), that for every $x \in \mathbb{R}$,

$$\begin{aligned} \lim_{n\to\infty} b(x+h_n) &= \lim_{n\to\infty}\lim_{m\to\infty} \sum_{\lambda\in F_m} b_{m,\lambda}(x+h_n) \\ &= \lim_{m\to\infty}\lim_{n\to\infty} \sum_{\lambda\in F_m} b_{m,\lambda}(x+h_n) \\ &= \lim_{m\to\infty} \sum_{\lambda\in F_m} \xi(b_{m,\lambda}) = \xi(b). \end{aligned} \tag{5.12}$$

Moreover, in view of (2.1) one can easily prove that the convergence in (5.12) is uniform on compacts of $\mathbb{R}$.

On the other hand, on the space $L^p(\mathbb{R},w)$ we also have the following:

$$\operatorname*{s-lim}_{n\to\infty} \left(e_{h_n} W^0(b) e_{h_n}^{-1}\right) = \operatorname*{s-lim}_{n\to\infty} W^0(b(\cdot+h_n)) = \operatorname*{s-lim}_{n\to\infty} W^0(b(h_n)). \tag{5.13}$$

Indeed, according to [17, Lemma 2.5] established by analogy with [28, Section 3.2], for every $f \in L^p(\mathbb{R},w)$ and every $\varphi \in L^1(\mathbb{R})$ with $\int_{\mathbb{R}} \varphi(x)dx = 1$, we have

$$\lim_{\varepsilon\to 0} \|f * \varphi_\varepsilon - f\|_{L^p(\mathbb{R},w)} = 0, \tag{5.14}$$

where $\varphi_\varepsilon(x) = \varepsilon^{-1}\varphi(x/\varepsilon),\ x \in \mathbb{R},\ \varepsilon > 0$. Choosing now rapidly decreasing functions $\varphi \in \mathcal{S}(\mathbb{R})$ whose Fourier transforms $\mathcal{F}\varphi$ have compact supports in $\mathbb{R}$, we derive from (5.14) that the set Φ of functions in $L^p(\mathbb{R},w)$ whose Fourier transforms have compact supports in $\mathbb{R}$ is dense in $L^p(\mathbb{R}, w)$. Therefore, for every function $f \in \Phi$ there is a function $\psi \in C^\infty(\mathbb{R})$ with a compact support in $\mathbb{R}$ such that

$$\big(W^0[b(\cdot + h_n) - b(h_n)]\big) f = \mathcal{F}^{-1}[b(\cdot + h_n) - b(h_n)]\psi\mathcal{F}f. \tag{5.15}$$

By analogy with (5.12), from (5.11) it follows that

$$\begin{aligned}
&\lim_{n\to\infty} \big\| \big[b(\cdot + h_n) - b(h_n)\big]\psi \big\|_{M_{p,w}} \\
&\quad = \lim_{n\to\infty} \lim_{m\to\infty} \Big\| \sum_{\lambda\in F_m} \big[b_{m,\lambda}(\cdot + h_n) - b_{m,\lambda}(h_n)\big]\psi \Big\|_{M_{p,w}} \\
&\quad = \lim_{m\to\infty} \lim_{n\to\infty} \Big\| \sum_{\lambda\in F_m} \big[b_{m,\lambda}(\cdot + h_n) - b_{m,\lambda}(h_n)\big]\psi \Big\|_{M_{p,w}} \\
&\quad \le \lim_{m\to\infty} \sum_{\lambda\in F_m} \lim_{n\to\infty} \big\| \big[b_{m,\lambda}(\cdot + h_n) - b_{m,\lambda}(h_n)\big]\psi \big\|_{M_{p,w}}.
\end{aligned} \tag{5.16}$$

In view of Theorem 2.3, we obtain

$$\begin{aligned}
&\Big\| \big[b_{m,\lambda}(\cdot + h_n) - b_{m,\lambda}(h_n)\big]\psi \Big\|_{M_{p,w}} \\
&\quad \le C_{p,w} \max_{k=0,1,2,3} \Big\| D^k \big(\big[b_{m,\lambda}(\cdot + h_n) - b_{m,\lambda}(h_n)\big]\psi \big) \Big\|_{L^\infty(\mathbb{R})},
\end{aligned} \tag{5.17}$$

where $(Da)(x) = xa'(x)$.

Let $K := \operatorname{supp} \psi$. Since K is a compact subset of $\mathbb{R}$ and

$$\begin{aligned}
&\lim_{n\to\infty} \max_{x\in K} \big| \big[b_{m,\lambda}(x + h_n) - b_{m,\lambda}(h_n)\big] \big| = 0, \\
&\lim_{n\to\infty} \max_{x\in K} \big| (D^k b_{m,\lambda})(x + h_n) \big| = 0 \quad (k = 1, 2, 3)
\end{aligned}$$

(see [7, Section 4]), we infer from (5.17) and the relations

$$\begin{aligned}
&D^k \big(\big[b_{m,\lambda}(\cdot + h_n) - b_{m,\lambda}(h_n)\big]\psi \big) \\
&\quad = \sum_{\nu=0}^{k} \binom{k}{\nu} \big(D^\nu \big[b_{m,\lambda}(\cdot + h_n) - b_{m,\lambda}(h_n)\big]\big)(D^{k-\nu}\psi),
\end{aligned}$$

$\max_{k=0,1,2,3} \|D^k\psi\|_{L^\infty(\mathbb{R})} < \infty$, and

$$\begin{aligned}
D\big[b_{m,\lambda}(x + h_n) - b_{m,\lambda}(h_n)\big] &= \frac{x}{x + h_n}\,(Db_{m,\lambda})(x + h_n), \\
D^2\big[b_{m,\lambda}(x + h_n) - b_{m,\lambda}(h_n)\big] &= \frac{x^2}{(x + h_n)^2}\,(D^2 b_{m,\lambda})(x + h_n) \\
&\quad + \frac{x h_n}{(x + h_n)^2}\,(Db_{m,\lambda})(x + h_n),
\end{aligned}$$

$$D^3\big[b_{m,\lambda}(x+h_n)-b_{m,\lambda}(h_n)\big] = \frac{x^3}{(x+h_n)^3}\,(D^3 b_{m,\lambda})(x+h_n) + \frac{3x^2h_n}{(x+h_n)^3}\,(D^2 b_{m,\lambda})(x+h_n) + \frac{xh_n^2-x^2h_n}{(x+h_n)^3}\,(Db_{m,\lambda})(x+h_n),$$

that

$$\lim_{n\to\infty}\big\|\big[b_{m,\lambda}(\cdot+h_n)-b_{m,\lambda}(h_n)\big]\psi\big\|_{M_{p,w}}=0. \tag{5.18}$$

Then from (5.16) and (5.18) it follows that

$$\lim_{n\to\infty}\big\|\big[b(\cdot+h_n)-b(h_n)\big]\psi\big\|_{M_{p,w}}=0,$$

which together with (5.15) implies (5.13). Hence,

$$\operatorname*{s-lim}_{n\to\infty}\big(e_{h_n}W^0(b)e_{h_n}^{-1}\big)=\operatorname*{s-lim}_{n\to\infty}W^0(b(h_n))=\operatorname*{s-lim}_{n\to\infty}b(h_n)I=b(\xi)I.$$

Thus, we obtain (5.4) because the first equality in (5.4) is evident.

It remains to prove (5.5) because the proof of (5.6) is similar. Since the weight $w_+=e^{v_+}$ is equivalent to the weight $w=e^v$ at a neighborhood of $+\infty$, we conclude that the operators $\widetilde{w}_+ a\widetilde{w}_+^{-1}I$ and $\widetilde{w}_+W^0(b)\widetilde{w}_+^{-1}I$ for $a\in SO^\diamond$ and $b\in SO^\diamond_{p,w}$ are bounded on the space $L^p(\mathbb{R})$. Because the first equality in (5.5) is evident, let us prove the second equality there.

First, suppose that $b\in SO^\diamond_{p,w}\cap SO^\diamond_p$. Because the function $x\mapsto xv_+'(x)$ belongs to $L^\infty(U_+)$, we infer by analogy with (5.8) that for all sufficiently large $h_n>0$,

$$\left|\int_{h_n}^{x+h_n} tv_+'(t)\,\frac{dt}{t}\right|\le \operatorname*{ess\,sup}_{t\in U_+}\big|tv_+'(t)\big|\,\left|\ln\Big(\frac{x+h_n}{h_n}\Big)\right|,$$

which implies that

$$\lim_{n\to\infty}\left(\int_{h_n}^{x+h_n} tv_+'(t)\,\frac{dt}{t}\right)=0$$

uniformly on compacts of $\mathbb{R}$. Therefore,

$$\operatorname*{s-lim}_{n\to\infty}\Big[\exp\big(v_+(x+h_n)-v_+(h_n)\big)I\Big]=\operatorname*{s-lim}_{n\to\infty}\left[\exp\left(\int_{h_n}^{x+h_n} tv_+'(t)\,\frac{dt}{t}\right)I\right]=I.$$

Consequently, for $b\in M_{p,w}\cap M_p$, we get

$$\begin{aligned}\operatorname*{s-lim}_{n\to\infty}\big(U_{-h_n}\widetilde{w}_+W^0(b)\widetilde{w}_+^{-1}U_{h_n}\big)&=\operatorname*{s-lim}_{n\to\infty}\Big(e^{v_+(x+h_n)}W^0(b)e^{-v_+(x+h_n)}I\Big)\\&=\operatorname*{s-lim}_{n\to\infty}\Big(\exp\big(v_+(x+h_n)-v_+(h_n)\big)W^0(b)\exp\big(v_+(h_n)-v_+(x+h_n)\big)I\Big)\\&=W^0(b),\end{aligned} \tag{5.19}$$

which completes the proof of (5.5) for $b\in SO^\diamond_{p,w}\cap SO^\diamond_p$.

For all $b\in SO^\diamond_{p,w}\cap SO^\diamond_p$ from (5.19) it follows that

$$\big\|W^0(b)\big\|_{\mathcal{B}(L^p(\mathbb{R}))}\le\big\|\widetilde{w}_+W^0(b)\widetilde{w}_+^{-1}I\big\|_{\mathcal{B}(L^p(\mathbb{R}))}\le\big\|W^0(b)\big\|_{\mathcal{B}(L^p(\mathbb{R},\widetilde{w}_+))},$$

and therefore, for these b,

$$\|b\|_{M_p} \le \|b\|_{M_{p,\widetilde{w}_+}}. \tag{5.20}$$

Since every function $b \in SO^\diamond_{p,w}$ is approximated in the Banach algebra $M_{p,w}$ by a sequence of functions $b_m \in SO^\diamond_{p,w} \cap SO^\diamond_p$ and since the norms $\|\cdot\|_{M_{p,\widetilde{w}_+}}$ and $\|\cdot\|_{M_{p,w}}$ are equivalent, we infer from (5.20) that $SO^\diamond_{p,w} \subset SO^\diamond_p$ for all considered weights $w \in A_p(\mathbb{R})$, which implies (5.5) for every $b \in SO^\diamond_{p,w}$. □

In particular, the proof of Lemma 5.1 gives the following result.

Corollary 5.2. *If $1 < p < \infty$ and $w \in A_p(\mathbb{R})$ satisfies (5.3), then $SO^\diamond_{p,w} \subset SO^\diamond_p$.*

6. Fredholm study of the Banach algebra $\mathcal{Z}_{p,w}$

Along with the Banach algebra $\mathcal{Z}_{p,w} \subset \mathcal{B}_{p,w}$ given by (1.3), we consider the Banach subalgebra

$$\widehat{\mathcal{Z}}_{p,w} := \operatorname{alg}\big(aI, W^0(b) :\ a \in C(\dot{\mathbb{R}}),\ b \in C_{p,w}(\dot{\mathbb{R}})\big) \tag{6.1}$$

of $\mathcal{Z}_{p,w}$ generated by the operators aI and $W^0(b)$ with $a \in C(\dot{\mathbb{R}})$ and $b \in C_{p,w}(\dot{\mathbb{R}})$.

By analogy with [24, Proposition 5.8.1], we obtain the following.

Lemma 6.1. *If $1 < p < \infty$ and $w \in A_p(\mathbb{R})$, then the Banach algebra $\widehat{\mathcal{Z}}_{p,w}$ contains the ideal $\mathcal{K}_{p,w}$ of compact operators in $\mathcal{B}_{p,w}$.*

Proof. As is well known, every compact operator on the space $L^p(\mathbb{R}, w)$ can be uniformly approximated in $\mathcal{B}_{p,w}$ by a finite sum of rank one operators of the form

$$(T\varphi)(x) = a(x) \int_{\mathbb{R}} b(y)\, \varphi(y)\, dy \quad (x \in \mathbb{R}), \tag{6.2}$$

where $a \in L^p(\mathbb{R}, w)$, $b \in L^q(\mathbb{R}, w^{-1})$ and $1/p + 1/q = 1$. Because the set $C_0(\mathbb{R})$ of continuous functions on $\mathbb{R}$ with compact support is dense in $L^p(\mathbb{R}, w)$ and $L^q(\mathbb{R}, w^{-1})$, we can take $a, b \in C_0(\mathbb{R})$ in (6.2). Then there is a number $M > 0$ such that the set $\{x - y : x \in \operatorname{supp} a,\ y \in \operatorname{supp} b\}$ is contained in the segment $[-M, M]$. Choose now a function

$$k(x) := \begin{cases} 1 - \exp\big((x+M)^{-3}\big) & \text{if } x < -M, \\ 1 & \text{if } x \in [-M, M], \\ 1 - \exp\big(-(x-M)^{-3}\big) & \text{if } x > M. \end{cases}$$

Then (6.2) can be rewritten in the form

$$(T\varphi)(x) = a(x) \int_{\mathbb{R}} k(x-y) b(y)\, \varphi(y)\, dy = \big[aW^0(\widehat{k}) b\varphi\big](x) \quad (x \in \mathbb{R}).$$

It remains to prove that $\widehat{k} \in C_{p,w}(\dot{\mathbb{R}})$ because then $T \in \widehat{\mathcal{Z}}_{p,w}$ in view of (6.1).

Obviously, $k \in C(\dot{\mathbb{R}})$ and $\lim_{x\to\pm\infty}[x^3k(x)] = \pm 1$. It is easily seen that the functions $k(x), k'(x), xk(x), xk'(x), xk''(x)$ belong to the space $L^1(\mathbb{R})$. Hence, $\widehat{k}, \widehat{k}\,' \in C(\dot{\mathbb{R}})$ and $\widehat{k}(\infty) = \widehat{k}\,'(\infty) = 0$. Moreover,

$$\begin{aligned} x^2\widehat{k}\,'(x) &= \int_{\mathbb{R}} x^2(iy)e^{ixy}k(y)dy \\ &= -\int_{\mathbb{R}} xe^{ixy}[k(y) + yk'(y)]dy \\ &= -i\int_{\mathbb{R}} e^{ixy}[2k'(y) + yk''(y)]dy, \end{aligned}$$

and therefore $\widehat{k}\,' \in L^1(\mathbb{R})$, which implies that the function $\widehat{k} \in C(\dot{\mathbb{R}})$ is of bounded total variation. Hence, by (2.5), $\widehat{k} \in C_{p,w}(\dot{\mathbb{R}})$ for all $p \in (1,\infty)$ and all $w \in A_p(\mathbb{R})$. ☐

By Lemma 6.1, $\mathcal{K}_{p,w} \subset \widehat{\mathcal{Z}}_{p,w} \subset \mathcal{Z}_{p,w} \subset \mathfrak{A}_{p,w}$, where $\mathfrak{A}_{p,w}$ is given by (1.2). Then from Theorem 4.6 it follows that the commutative Banach algebra $\mathcal{Z}^\pi_{p,w} = \mathcal{Z}_{p,w}/\mathcal{K}_{p,w}$ is a central subalgebra of the Banach algebra $\mathfrak{A}^\pi_{p,w} = \mathfrak{A}_{p,w}/\mathcal{K}_{p,w}$.

Theorem 6.2. *If $1 < p < \infty$, $w = e^v \in A_p(\mathbb{R})$, and (5.3) holds with $v_\pm$ given by (5.1), then the maximal ideal space $M\big(\mathcal{Z}^\pi_{p,w}\big)$ of the algebra $\mathcal{Z}^\pi_{p,w}$ is homeomorphic to the set*

$$\begin{aligned} \Omega := &\Big(\bigcup_{t\in\mathbb{R}} M_t(SO^\diamond) \times M_\infty(SO^\diamond)\Big) \cup \Big(M_\infty(SO^\diamond) \times \bigcup_{t\in\mathbb{R}} M_t(SO^\diamond)\Big) \\ &\cup \Big(M_\infty(SO^\diamond) \times M_\infty(SO^\diamond)\Big) \end{aligned} \tag{6.3}$$

equipped with topology induced by the product topology of $M(SO^\diamond) \times M(SO^\diamond)$, and the Gelfand transform $\Gamma : \mathcal{Z}^\pi_{p,w} \to C(\Omega)$, $A^\pi \mapsto \mathcal{A}(\cdot,\cdot)$ is defined on the generators $A^\pi = (aW^0(b))^\pi$ of the algebra $\mathcal{Z}^\pi_{p,w}$, where $a \in SO^\diamond$ and $b \in SO^\diamond_{p,w}$, by $\mathcal{A}(\xi,\eta) = a(\xi)b(\eta)$ for all $(\xi,\eta) \in \Omega$.

Proof. Note that if J is a maximal ideal of $\mathcal{Z}^\pi_{p,w}$, then

$$J \cap \big\{aI + \mathcal{K}_{p,w} : a \in SO^\diamond\big\} \quad\text{and}\quad J \cap \mathrm{clos}\big\{W^0(b) + \mathcal{K}_{p,w} : b \in SO^\diamond_{p,w}\big\}$$

are maximal ideals of the commutative Banach algebras

$$\big\{aI + \mathcal{K}_{p,w} : a \in SO^\diamond\big\} \quad\text{and}\quad \mathrm{clos}\big\{W^0(b) + \mathcal{K}_{p,w} : b \in SO^\diamond_{p,w}\big\}, \tag{6.4}$$

respectively (see [10, Lemma 1.33]). Therefore, taking into account the relations

$$\begin{aligned} M\big(\big\{aI + \mathcal{K}_{p,w} : a \in SO^\diamond\big\}\big) &= M(SO^\diamond), \\ M\big(\mathrm{clos}\big\{W^0(b) + \mathcal{K}_{p,w} : b \in SO^\diamond_{p,w}\big\}\big) &= M(SO^\diamond_{p,w}), \end{aligned}$$

and the fact that $M(SO^\diamond_{p,w}) = M(SO^\diamond)$ due to Theorem 3.6, we conclude that for every point $(\xi,\eta) \in M(SO^\diamond) \times M(SO^\diamond)$ there exists the closed two-sided (not necessarily maximal) ideal $\mathcal{I}^\pi_{\xi,\eta}$ of the Banach algebra $\mathcal{Z}^\pi_{p,w}$ generated by the maximal ideals

$$\begin{gathered}\Big\{aI+\mathcal{K}_{p,w}:\ a\in SO^\diamond,\ \xi(a)=0\Big\}\quad\text{and}\\ \operatorname{clos}\Big\{W^0(b)+\mathcal{K}_{p,w}:\ b\in SO^\diamond_{p,w},\ \eta(b)=0\Big\}\end{gathered}\tag{6.5}$$

of the commutative Banach algebras (6.4), respectively. Thus, by virtue of (3.1), the maximal ideal space of $\mathcal{Z}^\pi_{p,w}$ can be identified with a subset of

$$\begin{aligned}&M(SO^\diamond)\times M(SO^\diamond)\\ &\quad=\Big(\bigcup_{t\in\mathbb{R}}M_t(SO^\diamond)\cup M_\infty(SO^\diamond)\Big)\times\Big(\bigcup_{t\in\mathbb{R}}M_t(SO^\diamond)\cup M_\infty(SO^\diamond)\Big).\end{aligned}$$

Fix $(\xi,\eta)\in\bigcup_{t\in\mathbb{R}}M_t(SO^\diamond)\times\bigcup_{t\in\mathbb{R}}M_t(SO^\diamond)$. Given $a\in SO^\diamond$ and $b\in SO^\diamond_{p,w}$ choose functions $a_1\in C(\dot{\mathbb{R}})$ and $b_1\in C_{p,w}(\dot{\mathbb{R}})$ such that $a(\xi)=a_1(\xi)$, $b(\eta)=b_1(\eta)$, and $a_1(\infty)=b_1(\infty)=0$. Then

$$aW^0(b)=T_1+T_2+T_3\tag{6.6}$$

where

$$T_1=(a-a_1)W^0(b),\quad T_2=a_1W^0(b-b_1),\quad T_3=a_1W^0(b_1).$$

The operator T_3 is compact by Lemma 4.3(a), and the cosets T_1^π,T_2^π belong to the ideal $I^\pi_{\xi,\eta}$. Thus, the smallest closed two-sided ideal of $\mathcal{Z}^\pi_{p,w}$ which corresponds to the point $(\xi,\eta)\in\bigcup_{t\in\mathbb{R}}M_t(SO^\diamond)\times\bigcup_{t\in\mathbb{R}}M_t(SO^\diamond)$ coincides with the whole algebra $\mathcal{Z}^\pi_{p,w}$. So, the maximal ideals of the algebra $\mathcal{Z}^\pi_{p,w}$ can only correspond to points $(\xi,\eta)\in\Omega$, where Ω is given by (6.3).

It remains to show that for all $(\xi,\eta)\in\Omega$, the closed two-sided ideals $\mathcal{I}^\pi_{\xi,\eta}$ generated by the maximal ideals (6.5) are maximal ideals of the commutative Banach algebra $\mathcal{Z}^\pi_{p,w}$.

First, let us prove that these ideals are proper. To this end we need to show that for all $(\xi,\eta)\in\Omega$ the ideals $\mathcal{I}^\pi_{\xi,\eta}$ do not contain the coset $I^\pi=I+\mathcal{K}_{p,w}$. Clearly, the ideals $\mathcal{I}^\pi_{\xi,\eta}$ are closures in $\mathcal{B}^\pi_{p,w}=\mathcal{B}^\pi_{p,w}/\mathcal{K}_{p,w}$ of the sets

$$\Big\{\sum_{n=1}^N[a_nI]^\pi A_n^\pi+\sum_{m=1}^M[W^0(b_m)]^\pi B_m^\pi\Big\}\tag{6.7}$$

where $a_n\in SO^\diamond$, $\xi(a_n)=0$, $b_m\in SO^\diamond_{p,w}$, $\eta(b_m)=0$, and $A_n,B_m\in\mathcal{Z}_{p,w}$.

Given $t\in\dot{\mathbb{R}}$, let $(\xi,\eta)\in M_t(SO^\diamond)\times M_\infty(SO^\diamond)$. Assume that $I^\pi\in\mathcal{I}^\pi_{\xi,\eta}$. Hence, by (6.7), there is a sequence of operators of the form

$$C_k=\sum_{n=1}^{N_k}a_{n,k}A_{n,k}+\sum_{m=1}^{M_k}W^0(b_{m,k})B_{m,k}\tag{6.8}$$

with $a_{n,k}\in SO^\diamond$, $\xi(a_{n,k})=0$, $b_{m,k}\in SO^\diamond_{p,w}$, $\eta(b_{m,k})=0$ and $A_{n,k},B_{m,k}\in\mathcal{Z}_{p,w}$, and there is a sequence of compact operators $K_k\in\mathcal{K}_{p,w}$ such that $C_k^\pi\in\mathcal{I}^\pi_{\xi,\eta}$ and $\lim_{k\to\infty}\|C_k+K_k-I\|=0$. Since for every $\eta\in M_\infty(SO^\diamond)$ and every countable set $\{b_k\} \subset SO^\diamond_{p,w}$ there is a sequence $h_\nu \to +\infty$ in $\mathbb{R}$ such that $\lim_{\nu\to\infty} b_k(x+h_\nu) = \eta(b_k)$ for all $x \in \mathbb{R}$ (see [7, Corollary 4.3]), we conclude that there exists a sequence $\{h_\nu\} \subset \mathbb{R}_+$ such that $\lim_{\nu\to\infty} h_\nu = +\infty$ and, for all $m = 1,2,\dots,M_k$ and all $k \in \mathbb{N}$, $\lim_{\nu\to\infty} b_{m,k}(h_\nu) = 0$ and therefore, by (5.4) in Lemma 5.1, $\text{s-lim}_{n\to\infty}\big(e_{h_\nu} W^0(b_{m,k}) e_{-h_\nu} I\big) = 0$. Moreover, from (6.8), the algebraic properties of limit operators (see [7, Proposition 6.1]) and [17, Lemma 3.8] it follows that we can choose the sequence $\{h_\nu\}$ in such a way that there exists the strong limit

$$\widetilde{C}_k := \operatorname*{s-lim}_{\nu\to\infty}\big(e_{h_\nu}(C_k+K_k)e_{-h_\nu} I\big) = \sum_{n=1}^{N_k} a_{n,k}\widetilde{A}_{n,k} \ \in \mathcal{B}_{p,w},$$

where $\widetilde{A}_{n,k} = \widetilde{a}_{n,k} I$ and $\widetilde{a}_{n,k} \in SO^\diamond$. Hence, $\lim_{k\to\infty} \widetilde{C}_k = I$, which is impossible because the operators $\widetilde{C}_k$ belong to the maximal ideal $\{aI : \ a \in SO^\diamond, \ \xi(a) = 0\}$ of the C^*-algebra $\{aI : \ a \in SO^\diamond\}$.

Given $t \in \mathbb{R}$, let now $(\xi,\eta) \in M_\infty(SO^\diamond) \times M_t(SO^\diamond)$, and we again assume that $I^\pi \in \mathcal{I}^\pi_{\xi,\eta}$. For definiteness, suppose that $v_+ - v \in L^\infty(U_+)$ in (5.3). Then analogously to the previous case there exists a sequence of operators C_k of the form (6.8), where $a_{n,k} \in SO^\diamond$, $\xi(a_{n,k}) = 0$, $b_{m,k} \in SO^\diamond_{p,w}$, $\eta(b_{m,k}) = 0$ and $A_{n,k}, B_{m,k} \in \mathcal{Z}_{p,w}$, and there exists a sequence of compact operators $K_k \in \mathcal{K}_{p,w}$ such that $C_k^\pi \in \mathcal{I}^\pi_{\xi,\eta}$ and $\lim_{k\to\infty} \|C_k + K_k - I\| = 0$. Consequently, on the space $L^p(\mathbb{R})$ we get

$$\lim_{k\to\infty}\left(\sum_{n=1}^{N_k} a_{n,k}\widetilde{w}_+ A_{n,k}\widetilde{w}_+^{-1} I + \sum_{m=1}^{M_k} \widetilde{w}_+ W^0(b_{m,k}) B_{m,k}\widetilde{w}_+^{-1} I + \widetilde{w}_+ K_k \widetilde{w}_+^{-1} I\right) = I,$$

where $\widetilde{w}_+$ is the weight constructed in Section 5 and equivalent to the weight w at a neighborhood U_+ of $+\infty$. Since for every $\xi \in M_\infty(SO^\diamond)$ and every countable set $\{a_k\} \subset SO^\diamond$ there is a sequence $h_\nu \to +\infty$ in $\mathbb{R}$ such that $\lim_{\nu\to\infty} a_k(x+h_\nu) = \xi(a_k)$ for all $x \in \mathbb{R}$ (see [7, Corollary 4.3]), we conclude that there exists a sequence $\{h_\nu\} \subset \mathbb{R}_+$ such that $\lim_{\nu\to\infty} h_\nu = +\infty$ and for all $n = 1,2,\dots,N_k$ and all $k \in \mathbb{N}$, $\lim_{\nu\to\infty} a_{n,k}(h_\nu) = 0$ and therefore $\text{s-lim}_{\nu\to\infty}(U_{-h_\nu} a_{n,k} U_{h_\nu}) = 0$, where $U_{h_\nu} = W^0(e_{h_\nu})$ is a translation operator. Because the function $x \mapsto x v_+'(x)$ belongs to $L^\infty(U_+)$, we infer from (5.5) in Lemma 5.1 that

$$\operatorname*{s-lim}_{\nu\to\infty}\big(U_{-h_\nu}\widetilde{w}_+ W^0(b_{m,k})\widetilde{w}_+^{-1} U_{h_\nu}\big) = W^0(b_{m,k}).$$

Using then (6.8), the algebraic properties of limit operators (see [7, Proposition 6.1]) and [8, Lemma 18.9], we can choose the sequence $\{h_\nu\}$ in such a way that there exists the strong limit

$$\widehat{C}_k := \operatorname*{s-lim}_{\nu\to\infty}\big(U_{-h_\nu}\widetilde{w}_+(C_k+K_k)\widetilde{w}_+^{-1} U_{h_\nu}\big) = \sum_{m=1}^{M_k} W^0(b_{m,k})\widehat{B}_{m,k} \ \in \mathcal{B}_p,$$

where $\widehat{B}_{m,k} = W^0(\widehat{b}_{m,k})$ and $\widehat{b}_{m,k} \in SO^\diamond_{p,w} \subset SO^\diamond_p$ for considered weights w. Hence, $\lim_{k\to\infty} \widehat{C}_k = I$ in the norm of $\mathcal{B}_p$, which is impossible because the operators $\widehat{C}_k$ belong to the maximal ideal $\operatorname{clos}\{W^0(b) : b \in SO^\diamond_{p,w}, \eta(b) = 0\}$ of the Banach algebra $\operatorname{clos}\{W^0(b) : b \in SO^\diamond_{p,w}\}$.

Thus, for all $(\xi,\eta) \in \Omega$ the ideals $\mathcal{I}^\pi_{\xi,\eta}$ do not contain the unit coset I^π, and hence these ideals are proper. Suppose, contrary to our claim on the maximality of the ideal $\mathcal{I}^\pi_{\xi,\eta}$, that for a point $(\xi,\eta) \in \Omega$ there is a proper closed two-sided ideal $\widetilde{\mathcal{I}}^\pi_{\xi,\eta}$ of the algebra $\mathcal{Z}^\pi_{p,w}$ that properly contains the ideal $\mathcal{I}^\pi_{\xi,\eta}$. Then there is a coset $A^\pi \in \mathcal{Z}^\pi_{p,w}$ which belongs to $\widetilde{\mathcal{I}}^\pi_{\xi,\eta} \setminus \mathcal{I}^\pi_{\xi,\eta}$. Since in view of (6.6),

$$(aW^0(b))^\pi - (a(\xi)W^0(b(\eta)))^\pi = (aW^0(b))^\pi - (a(\xi)b(\eta)I)^\pi \in \mathcal{I}^\pi_{\xi,\eta} \tag{6.9}$$

for all $a \in SO^\diamond$ and all $b \in SO^\diamond_{p,w}$, and since $A^\pi \notin \mathcal{I}^\pi_{\xi,\eta}$, there exists a complex number $c \neq 0$ such that $A^\pi - (cI)^\pi \in \mathcal{I}^\pi_{\xi,\eta}$. Hence $(cI)^\pi \in \widetilde{\mathcal{I}}^\pi_{\xi,\eta}$ because $A^\pi \in \widetilde{\mathcal{I}}^\pi_{\xi,\eta}$ and $\mathcal{I}^\pi_{\xi,\eta} \subset \widetilde{\mathcal{I}}^\pi_{\xi,\eta}$. But the coset $(cI)^\pi$ is invertible in the algebra $\mathcal{Z}^\pi_{p,w}$, which implies that the ideal $\widetilde{\mathcal{I}}^\pi_{\xi,\eta}$ coincides with the whole algebra $\mathcal{Z}^\pi_{p,w}$. Thus the ideal $\widetilde{\mathcal{I}}^\pi_{\xi,\eta}$ is not proper, a contradiction. Consequently, all the ideals $\mathcal{I}^\pi_{\xi,\eta}$ for $(\xi,\eta) \in \Omega$ are maximal, and therefore $M(\mathcal{Z}^\pi_{p,w})$ can be identified with Ω given by (6.3).

Furthermore, by (6.9), the value of the Gelfand transform of the coset $A^\pi = (aW^0(b))^\pi$ at a point $(\xi,\eta) \in \Omega$ equals $a(\xi)b(\eta)$ for each choice of $a \in SO^\diamond$ and $b \in SO^\diamond_{p,w}$. This defines the Gelfand transform for the whole algebra $\mathcal{Z}^\pi_{p,w}$. □

Corollary 6.3. *If $1 < p < \infty$, $w = e^v \in A_p(\mathbb{R})$, and (5.3) holds, then the operator $A \in \mathcal{Z}_{p,w}$ is Fredholm on the space $L^p(\mathbb{R}, w)$ if and only if the Gelfand transform of the coset A^π is invertible, that is, if $\mathcal{A}(\xi,\eta) \neq 0$ for all $(\xi,\eta) \in \Omega$.*

Since $\mathcal{Z}^\pi_{p,w}$ is a central subalgebra of the Banach algebra $\mathfrak{A}^\pi_{p,w}$, applying the Allan-Douglas local principle (see, e.g., [10]) and the two idempotents theorem (see, e.g., [6]), one can construct a Fredholm theory for the Banach algebra $\mathfrak{A}_{p,w}$ (cf. [1]–[2]). We will consider this question in a forthcoming paper.

References

[1] M.A. Bastos, A. Bravo, and Yu.I. Karlovich, *Convolution type operators with symbols generated by slowly oscillating and piecewise continuous matrix functions.* Operator Theory: Advances and Applications **147** (2004), 151–174.

[2] M.A. Bastos, A. Bravo, and Yu.I. Karlovich, *Symbol calculus and Fredholmness for a Banach algebra of convolution type operators with slowly oscillating and piecewise continuous data.* Math. Nachr. **269–270** (2004), 11–38

[3] M.A. Bastos, C.A. Fernandes, and Yu.I. Karlovich, *C^*-algebras of integral operators with piecewise slowly oscillating coefficients and shifts acting freely.* Integral Equations and Operator Theory **55** (2006), 19–67.

[4] M.A. Bastos, Yu.I. Karlovich, and B. Silbermann, *Toeplitz operators with symbols generated by slowly oscillating and semi-almost periodic matrix functions.* Proc. London Math. Soc. (3) **89** (2004), 697–737.

[5] J. Bergh and J. Löfström, *Interpolation Spaces. An Introduction.* Springer, Berlin, 1976.

[6] A. Böttcher and Yu.I. Karlovich, *Carleson Curves, Muckenhoupt Weights, and Toeplitz Operators.* Progress in Mathematics **154**, Birkhäuser, Basel, 1997.

[7] A. Böttcher, Yu.I. Karlovich, and V.S. Rabinovich, *The method of limit operators for one-dimensional singular integrals with slowly oscillating data.* J. Operator Theory **43** (2000), 171–198.

[8] A. Bötcher, Yu.I. Karlovich, and I.M. Spitkovsky, *Convolution Operators and Factorization of Almost Periodic Matrix Functions.* Operator Theory: Advances and Applications **131**, Birkhäuser, Basel, 2002.

[9] A. Bötcher and I.M. Spitkovsky, *Wiener-Hopf integral operators with PC symbols on spaces with Muckenhoupt weight.* Revista Matemática Iberoamericana **9** (1993), 257–279.

[10] A. Böttcher and B. Silbermann, *Analysis of Toeplitz Operators.* 2nd edition, Springer, Berlin, 2006.

[11] R.V. Duduchava, *Integral Equations with Fixed Singularities.* Teubner, Leipzig, 1979.

[12] J.B. Garnett, *Bounded Analytic Functions.* Academic Press, New York, 1981.

[13] I.M. Gelfand, D.A. Raikov, and G.E. Shilov, *Commutative Normed Rings.* Fizmatgiz, Moscow, 1960 (Russian); English transl.: Chelsea, New York, 1964.

[14] L. Grafakos, *Classical and Modern Fourier Analysis.* Pearson/Prentice Hall, Upper Saddle River, NJ, 2004.

[15] N.B. Haaser and J.A. Sullivan, *Real Analysis.* Dover Publications, New York, 1991.

[16] Yu.I. Karlovich, *Boundedness and compactness of pseudodifferential operators with non-regular symbols on weighted Lebesgue spaces*, Integral Equations and Operator Theory **73** (2012), 217–254.

[17] Yu.I. Karlovich and J. Loreto Hernández, *Wiener-Hopf operators with semi-almost periodic matrix symbols on weighted Lebesgue spaces.* Integral Equations and Operator Theory **62** (2008), 85–128.

[18] Yu.I. Karlovich and J. Loreto Hernández, *Wiener-Hopf operators with slowly oscillating matrix symbols on weighted Lebesgue spaces.* Integral Equations and Operator Theory **64** (2009), 203–237.

[19] M.A. Krasnoselskii, P.P. Zabreiko, E.I. Pustylnik, and P.E. Sobolevskii, *Integral Operators in Spaces of Summable Functions.* Nauka, Moscow, 1966 (Russian); English transl.: Noordhoff I.P., Leyden, 1976.

[20] B.V. Lange and V.S. Rabinovich, *Pseudo-differential operators on* $\mathbb{R}^n$ *and limit operators.* Math. USSR, Sb. **57** (1987), 183–194.

[21] S.C. Power, *Fredholm Toeplitz operators and slow oscillation.* Can. J. Math. **32** (1980), 1058–1071.

[22] S.C. Power, *Hankel Operators on Hilbert Space.* Pitman Research Notes in Math. **64**, Pitman, Boston, 1982.

[23] V.S. Rabinovich, S. Roch, and B. Silbermann, *Limit Operators and Their Applications in Operator Theory.* Operator Theory: Advances and Applications **150**, Birkhäuser, Basel, 2004.

[24] S. Roch, P.A. Santos, and B. Silbermann, *Non-commutative Gelfand Theories. A Tool-kit for Operator Theorists and Numerical Analysts.* Springer, London, 2011.

[25] W. Rudin, *Functional Analysis.* 2nd edition, McGraw-Hill Inc., New York, 1991.

[26] D. Sarason, *Functions of vanishing mean oscillation.* Trans. Amer. Math. Soc. **207** (1975), 391–405.

[27] D. Sarason, *Toeplitz operators with piecewise quasicontinuous symbols.* Indiana Univ. Math. J. **26** (1977), 817–838.

[28] E.M. Stein, *Singular Integrals and Differentiability Properties of Functions.* Princeton Univ. Press, Princeton, NJ, 1970.

[29] E.M. Stein, *Harmonic Analysis: Real-Variable Methods, Orthogonality, and Oscillatory Integrals.* Princeton Univ. Press, Princeton, NJ, 1993.

[30] I.B. Simonenko and Chin Ngok Min, *Local Method in the Theory of One-Dimensional Singular Integral Equations with Piecewise Continuous Coefficients. Noetherity.* Rostov Univ. Press, Rostov on Don, 1986 (Russian).

Yuri I. Karlovich
Facultad de Ciencias
Universidad Autónoma del Estado de Morelos
Av. Universidad 1001, Col. Chamilpa
C.P. 62209 Cuernavaca, Morelos, México
e-mail: `karlovich@uaem.mx`

Iván Loreto Hernández
Instituto de Matemáticas
Universidad Nacional Autónoma de México
Av. Universidad s/n, Col. Lomas de Chamilpa
C.P. 62210 Cuernavaca, Morelos, México
e-mail: `ivan@matcuer.unam.mx`

Operator Theory:
Advances and Applications, Vol. 228, 209–238

Transmutations and Spectral Parameter Power Series in Eigenvalue Problems

Vladislav V. Kravchenko and Sergii M. Torba

Dedicated to 70th birthday anniversary of Prof. Dr. Vladimir S. Rabinovich.

Abstract. We give an overview of recent developments in Sturm-Liouville theory concerning operators of transmutation (transformation) and spectral parameter power series (SPPS). The possibility to write down the dispersion (characteristic) equations corresponding to a variety of spectral problems related to Sturm-Liouville equations in an analytic form is an attractive feature of the SPPS method. It is based on a computation of certain systems of recursive integrals. Considered as families of functions these systems are complete in the L_2-space and result to be the images of the nonnegative integer powers of the independent variable under the action of a corresponding transmutation operator. This recently revealed property of the Delsarte transmutations opens the way to apply the transmutation operator even when its integral kernel is unknown and gives the possibility to obtain further interesting properties concerning the Darboux transformed Schrödinger operators.

We introduce the systems of recursive integrals and the SPPS approach, explain some of its applications to spectral problems with numerical illustrations, give the definition and basic properties of transmutation operators, introduce a parametrized family of transmutation operators, study their mapping properties and construct the transmutation operators for Darboux transformed Schrödinger operators.

Mathematics Subject Classification (2010). Primary 34B24, 34L16, 65L15, 81Q05, 81Q60; Secondary 34L25, 34L40.

Keywords. Sturm-Liouville operator, Sturm-Liouville problem, complex eigenvalue, transmutation operator, transformation operator, Schrödinger operator, spectral parameter power series, Darboux transformation, quantum well, scalar potential.

Research was supported by CONACYT, Mexico. Research of second named author was supported by DFFD, Ukraine (GP/F32/030) and by SNSF, Switzerland (JRP IZ73Z0 of SCOPES 2009–2012).

1. Introduction

Transmutation operators also called operators of transformation are a widely used tool in the theory of linear differential equations (see, e.g., [3], [10], [49], [51], [63] and the recent review [61]). It is well known that under certain quite general conditions the transmutation operator transmuting the operator $A = -\frac{d^2}{dx^2} + q(x)$ into $B = -\frac{d^2}{dx^2}$ is a Volterra integral operator with good properties. Its integral kernel can be obtained as a solution of a certain Goursat problem for the Klein-Gordon equation with a variable coefficient. In particular, the elementary solutions of the equation $Bv = \lambda v$ are transformed into the solutions of the equation $Au = \lambda u$. If the integral kernel of the transmutation operator is unknown, and usually this is the case, there is no way to apply it to an arbitrary smooth function. This obstacle strongly restricts the application of the transmutation operators confining it to purely theoretical purposes.

Recently, in [9] a relation of the transmutation operators with another fundamental object of the Sturm-Liouville theory was revealed. Sometimes this object is called the L-basis [24] where L refers to a corresponding linear ordinary differential operator. The L-basis is an infinite family of functions $\{\varphi_k\}_{k=0}^{\infty}$ such that $L\varphi_k = 0$ for $k = 0, 1$, $L\varphi_k = k(k-1)\varphi_{k-2}$, for $k = 2, 3, \dots$ and all φ_k satisfy certain prescribed initial conditions. In [41], [42], [45] it was shown that the L-basis naturally arises in a representation of the solutions of the Sturm-Liouville equation in terms of powers of the spectral parameter. The approach based on such representation is called the spectral parameter power series (SPPS) method. The functions φ_k which constitute the L-basis appear as the expansion coefficients in the SPPS. In [41], [42] and [45] convenient representations for their practical computation were proposed which converted the SPPS method into an efficient and highly competitive technique for solving a variety of spectral and scattering problems related to Sturm-Liouville equations (see [12], [13], [37], [39], [45], [47]). The above-mentioned relation between the transmutation operators and the functions φ_k called in the present paper the recursive integrals consists in the fact established in [9] that for every system $\{\varphi_k\}_{k=0}^{\infty}$ there exists a transmutation operator $\mathbf{T}$ such that $\mathbf{T}[x^k] = \varphi_k$, i.e., the functions φ_k are the images of the usual powers of the independent variable. Moreover, it was shown how this operator can be constructed and how it is related to the "canonical" transformation operator considered, e.g., in [51, Chapter 1]. This result together with the practical formulas for calculating the functions φ_k makes it possible to apply the transmutation technique even when the integral kernel of the operator is unknown. Indeed, now it is easy to apply the transmutation operator to any function approximated by a polynomial.

Deeper understanding of the mapping properties of the transmutation operators led us in [46] to the explicit construction of the transmutation operator for a Darboux transformed Schrödinger operator by a known transmutation operator for the original Schrödinger operator as well as to several interesting relations between the two transmutation operators. These relations also allowed us to prove the main theorem on the transmutation operators under a weaker condition than it was known before (not requiring the continuous differentiablity of the potential in the Schrödinger operator).

In the present paper we overview the recent results related to the SPPS approach explaining and illustrating its main advantage, the possibility to write down in an analytic form the characteristic equation of the spectral problem. This equation can be approximated in different ways, and its solutions give us the eigenvalues of the problem. In other words the eigenvalue problem reduces to computation of zeros of a certain complex analytic function given by its Taylor series whose coefficients are obtained as simple linear combinations of the values of the functions φ_k at a given point. We discuss different applications of the SPPS method and give the results of some comparative numerical calculations.

Following [9] and [46] we introduce a parametrized family of transmutation operators and study their mapping properties,we give an explicit representation for the kernel of the transmutation operator corresponding to the Darboux transformed potential in terms of the transmutation kernel for its superpartner (Theorem 6.2). Moreover, this result leads to interesting commutation relations between the two transmutation operators (Corollary 6.6) which in their turn allow us to obtain a transmutation operator for the one-dimensional Dirac system with a scalar potential as well as to prove the main property of the transmutation operator under less restrictive conditions than it has been proved until now. We give several examples of explicitly constructed kernels of transmutation operators. It is worth mentioning that in the literature there are very few explicit examples and even in the case when q is a constant such kernel was presented recently in [9]. The results discussed in the present paper allow us to enlarge considerably the list of available examples and give a relatively simple tool for constructing Darboux related sequences of the transmutation kernels.

2. Recursive integrals: a question on the completeness

Let $f \in C^2(a,b) \cap C^1[a,b]$ be a complex-valued function and $f(x) \neq 0$ for any $x \in [a,b]$. The interval (a,b) is assumed being finite. Let us consider the following functions

$$X^{(0)}(x) \equiv 1, \qquad X^{(n)}(x) = n \int_{x_0}^{x} X^{(n-1)}(s) \left(f^2(s)\right)^{(-1)^n} \mathrm{d}s, \qquad x_0 \in [a,b], \quad n = 1,2,\ldots. \tag{2.1}$$

We pose the following questions. **Is the family of functions $\{X^{(n)}\}_{n=0}^{\infty}$ complete let us say in $L_2(a,b)$? What about the completeness of $\{X^{(2n)}\}_{n=0}^{\infty}$ or $\{X^{(2n+1)}\}_{n=0}^{\infty}$?**

The following example shows that both questions are meaningful and natural.

Example 2.1. Let $f \equiv 1$, $a = 0$, $b = 1$. Then it is easy to see that choosing $x_0 = 0$ we have $X^{(0)}(x) = 1$, $X^{(1)}(x) = x$, $X^{(2)}(x) = x^2$, $X^{(3)}(x) = x^3, \ldots$. Thus, the family of functions $\{X^{(n)}\}_{n=0}^{\infty}$ is complete in $L_2(0,1)$. Moreover, both $\{X^{(2n)}\}_{n=0}^{\infty}$ and $\{X^{(2n+1)}\}_{n=0}^{\infty}$ are complete in $L_2(0,1)$ as well.

If instead of $a=0$ we choose $a=-1$ then $\{X^{(n)}\}_{n=0}^{\infty}$ **is still complete in** $L_2(-1,1)$ **but neither** $\{X^{(2n)}\}_{n=0}^{\infty}$ **nor** $\{X^{(2n+1)}\}_{n=0}^{\infty}$.

Together with the family of functions $\{X^{(n)}\}_{n=0}^{\infty}$ we consider also another similarly defined family of functions $\{\widetilde{X}^{(n)}\}_{n=0}^{\infty}$,

$$\widetilde{X}^{(0)} \equiv 1, \qquad \widetilde{X}^{(n)}(x) = n\int_{x_0}^{x} \widetilde{X}^{(n-1)}(s)\left(f^2(s)\right)^{(-1)^{n-1}} \mathrm{d}s,$$
$$x_0 \in [a,b], \quad n=1,2,\dots. \tag{2.2}$$

Remark 2.2. As we show below the introduced families of functions are closely related to the one-dimensional Schrödinger equations of the form $u''-qu=\lambda u$ where q is a complex-valued continuous function. Slightly more general families of functions can be studied in relation to Sturm-Liouville equations of the form $(py')'+qy=\lambda ry$. Their definition based on a corresponding recursive integration procedure is given in [42], [45], [37].

We introduce the infinite system of functions $\{\varphi_k\}_{k=0}^{\infty}$ defined as follows

$$\varphi_k(x) = \begin{cases} f(x)X^{(k)}(x), & k \text{ odd},\\ f(x)\widetilde{X}^{(k)}(x), & k \text{ even}. \end{cases} \tag{2.3}$$

The system (2.3) is closely related to the notion of the L-basis introduced and studied in [24]. Here the letter L corresponds to a linear ordinary differential operator.

Together with the system of functions (2.3) we define the functions $\{\psi_k\}_{k=0}^{\infty}$ using the "second half" of the recursive integrals (2.1) and (2.2),

$$\psi_k(x) = \begin{cases} \dfrac{\widetilde{X}^{(k)}(x)}{f(x)}, & k \text{ odd},\\ \dfrac{X^{(k)}(x)}{f(x)}, & k \text{ even}. \end{cases} \tag{2.4}$$

The following result obtained in [41] (for additional details and simpler proof see [42] and [45]) establishes the relation of the system of functions $\{\varphi_k\}_{k=0}^{\infty}$ and $\{\psi_k\}_{k=0}^{\infty}$ to the Sturm-Liouville equation.

Theorem 2.3 ([41]). *Let q be a continuous complex-valued function of an independent real variable $x\in[a,b]$ and λ be an arbitrary complex number. Suppose there exists a solution f of the equation*

$$f''-qf=0 \tag{2.5}$$

on (a,b) such that $f\in C^2(a,b)\cap C^1[a,b]$ and $f(x)\neq 0$ for any $x\in[a,b]$. Then the general solution $u\in C^2(a,b)\cap C^1[a,b]$ of the equation

$$u''-qu=\lambda u \tag{2.6}$$

on (a,b) *has the form*

$$u = c_1 u_1 + c_2 u_2$$

where c_1 *and* c_2 *are arbitrary complex constants,*

$$u_1 = \sum_{k=0}^{\infty} \frac{\lambda^k}{(2k)!} \varphi_{2k} \qquad \text{and} \qquad u_2 = \sum_{k=0}^{\infty} \frac{\lambda^k}{(2k+1)!} \varphi_{2k+1} \tag{2.7}$$

and both series converge uniformly on $[a,b]$ *together with the series of the first derivatives which have the form*

$$u_1' = f' + \sum_{k=1}^{\infty} \frac{\lambda^k}{(2k)!} \left(\frac{f'}{f} \varphi_{2k} + 2k\, \psi_{2k-1} \right) \qquad \text{and}$$

$$u_2' = \sum_{k=0}^{\infty} \frac{\lambda^k}{(2k+1)!} \left(\frac{f'}{f} \varphi_{2k+1} + (2k+1)\, \psi_{2k} \right). \tag{2.8}$$

The series of the second derivatives converge uniformly on any segment $[a_1, b_1] \subset (a,b)$.

The representation (2.7) offers the linearly independent solutions of (2.6) in the form of spectral parameter power series (SPPS). The possibility to represent solutions of the Sturm-Liouville equation in such form is by no means a novelty, though it is not a widely used tool (in fact, besides the work reviewed below and in [37] we are able to mention only [4, Sect. 10], [23] and the recent paper [40]) and to our best knowledge for the first time it was applied for solving spectral problems in [45]. The reason of this underuse of the SPPS lies in the form in which the expansion coefficients were sought. Indeed, in previous works the calculation of coefficients was proposed in terms of successive integrals with the kernels in the form of iterated Green functions (see [4, Sect. 10]). This makes any computation based on such representation difficult, less practical and even proofs of the most basic results like, e.g., the uniform convergence of the spectral parameter power series for any value of $\lambda \in \mathbb{C}$ (established in Theorem 2.3) are not an easy task. For example, in [4, p. 16] the parameter λ is assumed to be small and no proof of convergence is given.

The way of how the expansion coefficients in (2.7) are calculated according to (2.1), (2.2) is relatively simple and straightforward, this is why the estimation of the rate of convergence of the series (2.7) presents no difficulty, see [45]. Moreover, in [7] a discrete analogue of Theorem 2.3 was established and the discrete analogues of the series (2.7) resulted to be finite sums.

Another crucial feature of the introduced representation of the expansion coefficients in (2.7) consists in the fact that not only these coefficients (denoted by φ_k in (2.3)) are required for solving different spectral problems related to the Sturm-Liouville equation. Indeed, the functions $\widetilde{X}^{(2k+1)}$ and $X^{(2k)}$, $k = 0, 1, 2, \ldots$ do not participate explicitly in the representation (2.7). Nevertheless, together with the functions φ_k they appear in the representation (2.8) of the derivatives of the solutions and therefore also in characteristic equations corresponding to the spectral problems.

In the present work we also overview another approach developed in [43], [8], [9] and [46] where the formal powers (2.1) and (2.2) were considered as infinite families of functions intimately related to the corresponding Sturm-Liouville operator. As we show this leads to a deeper understanding of the transmutation operators [3], [10] also known as transformation operators [49], [51]. Indeed, the functions $\varphi_k(x)$ result to be the images of the powers x^k under the action of a corresponding transmutation operator [9]. This makes it possible to apply the transmutation operator even when the operator itself is unknown (and this is the usual situation – very few explicit examples are available) due to the fact that its action on every polynomial is known. This result was used in [8] and [9] to prove the completeness (Runge-type approximation theorems) for families of solutions of two-dimensional Schrödinger and Dirac equations with variable complex-valued coefficients.

Remark 2.4. It is easy to see that by definition the solutions u_1 and u_2 from (2.7) satisfy the following initial conditions

$$u_1(x_0) = f(x_0), \qquad u_1'(x_0) = f'(x_0),$$
$$u_2(x_0) = 0, \qquad u_2'(x_0) = 1/f(x_0).$$

Remark 2.5. It is worth mentioning that in the regular case the existence and construction of the required f presents no difficulty. Let q be real valued and continuous on $[a, b]$. Then (2.5) possesses two linearly independent regular solutions v_1 and v_2 whose zeros alternate. Thus one may choose $f = v_1 + iv_2$. Moreover, for the construction of v_1 and v_2 in fact the same SPPS method may be used [45].

Theorem 2.3 together with the results on the completeness of Sturm-Liouville eigenfunctions and generalized eigenfunctions [51] implies the validity of the following two statements. For their detailed proofs we refer to [43] and [44] respectively.

Theorem 2.6 ([43]). *Let (a, b) be a finite interval and $f \in C^2(a, b) \cap C^1[a, b]$ be a complex-valued function such that $f(x) \neq 0$ for any $x \in [a, b]$.*

If $x_0 = a$ (or $x_0 = b$) then each of the four systems of functions $\{X^{(2n)}\}_0^\infty$, $\{X^{(2n+1)}\}_0^\infty$, $\{\widetilde{X}^{(2n)}\}_0^\infty$, $\{\widetilde{X}^{(2n+1)}\}_0^\infty$ is complete in $L_2(a, b)$.

If x_0 is an arbitrary point of the interval (a, b) then each of the following two combined systems of functions $\{\widetilde{X}^{(2n)}\}_{n=0}^\infty \cup \{X^{(2n+1)}\}_{n=0}^\infty$ and $\{\widetilde{X}^{(2n+1)}\}_{n=0}^\infty \cup \{X^{(2n)}\}_{n=0}^\infty$ is complete in $L_2(a, b)$.

Theorem 2.7 ([44]). *Let f satisfy the conditions of the preceding theorem and $\{\varphi_k\}_{k=0}^\infty$ be the system of functions defined by (2.3) with x_0 being an arbitrary point of the interval $[a, b]$. Then for any complex-valued continuous, piecewise continuously differentiable function h defined on $[a, b]$ and for any $\varepsilon > 0$ there exists such $N \in \mathbb{N}$ and such complex numbers α_k, $k = 0, 1, \ldots, N$ that*

$$\max_{x\in[a,b]} \left| h(x) - \sum_{k=0}^{N} \alpha_k \varphi_k(x) \right| < \varepsilon.$$

3. Dispersion relations for spectral problems and approximate solutions

The SPPS representation (2.7) for solutions of the Sturm-Liouville equation (2.6) is very convenient for writing down the dispersion (characteristic) relations in an analytical form. This fact was used in [13], [37], [39], [45], [47] for approximating solutions of different eigenvalue problems. Here in order to explain this we consider two examples: the Sturm-Liouville problem and the quantum-mechanical eigenvalue problem. As the performance of the SPPS method in application to classical (regular and singular) Sturm-Liouville problems was studied in detail in [45] here we consider the Sturm-Liouville problems with boundary conditions which depend on the spectral parameter λ. This situation occurs in many physical models (see, e.g., [5, 14, 19, 20, 27, 64] and references therein) and is considerably more difficult from the computational point of view. Moreover, as we show in this section the SPPS method is applicable to models admitting complex eigenvalues – an important advantage in comparison with the best purely numerical techniques all of them being based on the shooting method.

Consider the equation $u'' - qu = \lambda u$ together with the boundary conditions

$$u(a)\cos\alpha + u'(a)\sin\alpha = 0, \tag{3.1}$$

$$\beta_1 u(b) - \beta_2 u'(b) = \phi(\lambda)\big(\beta_1' u(b) - \beta_2' u'(b)\big), \tag{3.2}$$

where α is an arbitrary complex number, ϕ is a complex-valued function of the variable λ and β_1, β_2, β_1', β_2' are complex numbers. For some special forms of the function ϕ such as $\phi(\lambda) = \lambda$ or $\phi(\lambda) = \lambda^2 + c_1\lambda + c_2$, results were obtained [19], [64] concerning the regularity of the problem; we will not dwell upon the details. Notice that the SPPS approach is applicable as well to a more general Sturm-Liouville equation $(pu')' + qu = \lambda ru$. For the corresponding details we refer to [37] and [45].

For simplicity, let us suppose that $\alpha = 0$ and hence the condition (3.1) becomes $u(a) = 0$. Then choosing the initial integration point in (2.1) and (2.2) as $x_0 = a$ and taking into account Remark 2.4 we obtain that if an eigenfunction exists it necessarily coincides with u_2 up to a multiplicative constant. In this case condition (3.2) becomes equivalent to the equality [45], [37]

$$\big(f(b)\phi_1(\lambda) - f'(b)\phi_2(\lambda)\big)\sum_{k=0}^{\infty}\frac{\lambda^k}{(2k+1)!}X^{(2k+1)}(b) - \frac{\phi_2(\lambda)}{f(b)}\sum_{k=0}^{\infty}\frac{\lambda^k}{(2k)!}X^{(2k)}(b) = 0, \tag{3.3}$$

where $\phi_{1,2}(\lambda) = \beta_{1,2} - \beta_{1,2}'\phi(\lambda)$. This is the characteristic equation of the considered spectral problem. Calculation of eigenvalues given by (3.3) is especially simple in the case of ϕ being a polynomial of λ. Precisely this particular situation was considered in all of the above-mentioned references concerning Sturm-Liouville problems with spectral parameter dependent boundary conditions. In any case the knowledge of an explicit characteristic equation (3.3) for the spectral problem makes possible its accurate and efficient solution. For this the infinite sums in (3.3) are truncated after a certain $N \in \mathbb{N}$. The paper [45] contains several numerical tests corresponding to a variety of computationally difficult problems. All they reveal an excellent performance of the SPPS method. We do not review them here referring the interested reader to [45]. Instead we consider another interesting example from [37], a Sturm-Liouville problem admitting complex eigenvalues.

Example 3.1. Consider the equation (2.6) with $q \equiv 0$ on the interval $(0, \pi)$ with the boundary conditions $u(0) = 0$ and $u(\pi) = -\lambda^2 u(\pi)$. The exact eigenvalues of the problem are $\lambda_n = n^2$ together with the purely imaginary numbers $\lambda_\pm = \pm i$. Application of the SPPS method with $N = 100$ and 3000 interpolating points (used for representing the integrands as splines) delivered the following results $\lambda_1 = 1$, $\lambda_2 = 4.0000000000007$, $\lambda_3 = 9.00000000001$, $\lambda_4 = 15.99999999996$, $\lambda_5 = 25.000000002$, $\lambda_6 = 35.99999997$, $\lambda_7 = 49.0000004$, $\lambda_8 = 63.9999994$, $\lambda_9 = 80.9996$, $\lambda_{10} = 100.02$ and $\lambda_\pm = \pm i$. Thus, the complex eigenvalues are as easily and accurately detected by the SPPS method as the real eigenvalues. Note that for a better accuracy in calculation of higher eigenvalues of a Sturm-Liouville problem an additional simple shifting procedure described in [45] and based on the representation of solutions not as series in powers of λ but in powers of $(\lambda - \lambda_0)$ is helpful. We did not apply it here and hence the accuracy of the calculated value of λ_{10} is considerably worse than the accuracy of the first calculated eigenvalues which in general can be improved by means of the mentioned shifting procedure.

Figures 1–3 give us an idea about the stability of the computed eigenvalues when N increases. In Figure 1 we plot λ_1 and λ_2 computed with $N = 14, 16, \ldots, 120$. Figure 2 shows λ_3 computed with $N = 24, 30, \ldots, 140$ and Figure 3 shows λ_4 computed with $N = 40, 50, \ldots, 140$ Similar plots can be done for calculated higher eigenvalues. In all cases the computed eigenvalues reveal a remarkable stability when N increases.

An attractive feature of the SPPS method is the possibility to easily plot the characteristic relation. In Figure 4 we show the absolute value of the expression from the left-hand side of (3.3) as a function of the complex variable λ for the considered example. Its zeros coincide with the eigenvalues of the problem. It is important to mention that such plot is obtained in a fraction of a second. This is due to the fact that once the required formal powers $X^{(n)}$ are computed (and this takes several seconds) the calculation of the characteristic relation involves only simple algebraic operations.

Let us consider the one-dimensional Schrödinger equation

$$Hu(x) = -u''(x) + Q(x)u(x) = \lambda u(x), \quad x \in \mathbb{R}, \tag{3.4}$$

where

$$Q(x) = \begin{cases} \alpha_1, & x < 0, \\ q(x), & 0 \le x \le h, \\ \alpha_2, & x > h, \end{cases} \tag{3.5}$$

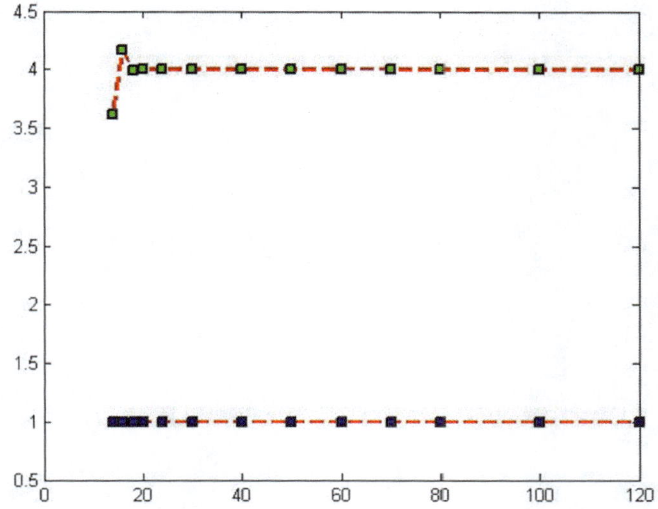

FIGURE 1. The approximate eigenvalues λ_1 and λ_2 from Example 3.1 computed using different number N of formal powers.

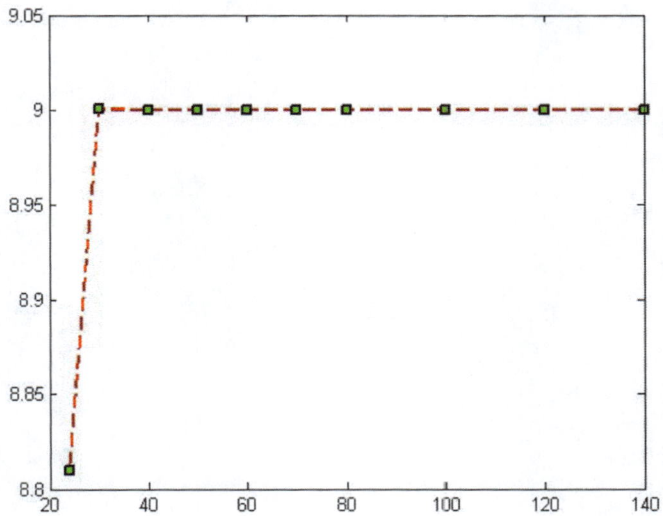

FIGURE 2. The approximate values of λ_3 from Example 3.1 computed using different number N of formal powers.

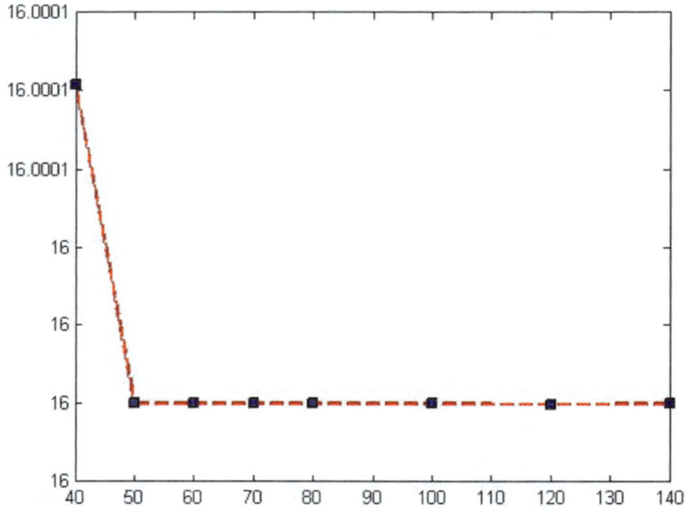

FIGURE 3. The approximate values of λ_4 from Example 3.1 computed using different number N of formal powers.

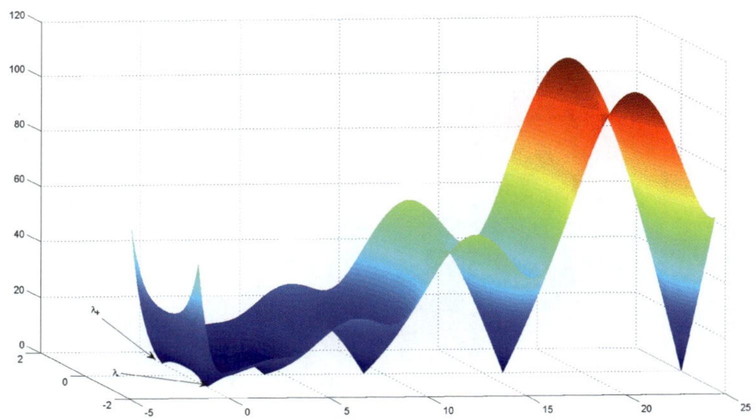

FIGURE 4. The absolute value of the expression from the left-hand side of (3.3) as a function of the complex variable λ for the considered example. With the arrows we indicate the calculated complex eigenvalues $\lambda_{\pm}$. The other zeros of the graph correspond to the first five real eigenvalues of the problem.

α_1 and α_2 are complex constants and q is a continuous complex-valued function defined on the segment $[0, h]$. Thus, outside a finite segment the potential Q admits constant values, and at the end points of the segment the potential may have discontinuities. We are looking for such values of the spectral parameter $\lambda \in \mathbb{C}$ for which the Schrödinger equation possesses a solution u belonging to the Sobolev space $H^2(\mathbb{R})$ which in the case of the potential of the form (3.5) means that we are looking for solutions exponentially decreasing at $\pm\infty$. This eigenvalue problem is one of the central in quantum mechanics for which H is a self-adjoint operator in $L^2(\mathbb{R})$ with the domain $H^2(\mathbb{R})$. It implies that Q is a real-valued function. In this case the operator H has a continuous spectrum $\left[\min\{\alpha_1, \alpha_2\}, +\infty\right)$ and a discrete spectrum located on the set

$$\left[\min_{x\in[0,h]} q(x), \min\{\alpha_1, \alpha_2\}\right).$$

Computation of energy levels of a quantum well described by the potential Q is a problem of physics of semiconductor nanostructures (see, e.g., [31]). Other important models which reduce to the spectral problem (3.4) arise in studying the electromagnetic and acoustic wave propagation in inhomogeneous waveguides (see for instance [2], [16], [25], [17], [6], [56], [53]).

A characteristic equation for this spectral problem in terms of spectral parameter power series was obtained in [13] (see also [37]) where a simple numerical algorithm based on the approximation of the characteristic equation was implemented and compared to other known numerical techniques. Here we only give an example from [13].

The usual approach to numerical solution of the considered eigenvalue problem consists in applying the shooting method (see, e.g., [31]) which is known to be unstable, relatively slow and to the difference of the SPPS approach does not offer any explicit equation for determining eigenvalues and eigenfunctions. In [30] another method based on approximation of the potential by square wells was proposed. It is limited to the case of symmetric potentials. The approach based on the SPPS is completely different and does not require any shooting procedure, approximation of the potential or numerical differentiation. Derived from the exact characteristic equation its approximation is considered, and in fact numerically the problem is reduced to finding zeros of a polynomial $\sum_{k=0}^{N} a_k \mu^k$ in the interval $\left[\min q(x), 0\right)$, $(\mu^2 = -\lambda)$.

As an example, consider the potential Q defined by the expression $Q(x) = -\upsilon \operatorname{sech}^2 x$, $x \in (-\infty, \infty)$. It is not of a finite support, nevertheless its absolute value decreases rapidly when $x \to \pm\infty$. The original problem is approximated by a problem with a finite support potential $\widehat{Q}$ defined by the equality

$$\widehat{Q}(x) = \begin{cases} 0, & x < -a \\ -\upsilon \operatorname{sech}^2 x, & -a \le x \le a \\ 0, & x > a. \end{cases}$$

An attractive feature of the potential Q is that its eigenvalues can be calculated explicitly (see, e.g., [26]). In particular, for $v = m(m+1)$ the eigenvalue λ_n is given by the formula $\lambda_n = -(m-n)^2$, $n = 0, 1, \ldots$.

The results of application of the SPPS method for $v = 12$ are given in Table 1 in comparison with the exact values and the results from [30].

n	Exact values	Num.res. from [30]	Num.res. using SPPS ($N = 180$)
0	-9	-9.094	-8.999628656
1	-4	-4.295	-3.999998053
2	-1	-0.885	-0.999927816

TABLE 1. Approximations of λ_n of the Hamiltonian $H = -D^2 - 12\operatorname{sech}^2 x$.

The results obtained by means of SPPS are considerably more accurate, and as was pointed out above the application of the SPPS method has much less restrictions.

4. Transmutation operators

We slightly modify here the definition given by Levitan [49] adapting it to the purposes of the present work. Let E be a linear topological space and E_1 its linear subspace (not necessarily closed). Let A and B be linear operators: $E_1 \to E$.

Definition 4.1. A linear invertible operator T defined on the whole E such that E_1 is invariant under the action of T is called a transmutation operator for the pair of operators A and B if it fulfills the following two conditions.

1. Both the operator T and its inverse T^{-1} are continuous in E;
2. The following operator equality is valid
$$AT = TB \tag{4.1}$$
or which is the same
$$A = TBT^{-1}.$$

Very often in literature the transmutation operators are called the transformation operators. Here we keep ourselves to the original term coined by Delsarte and Lions [23]. Our main interest concerns the situation when $A = -\frac{d^2}{dx^2} + q(x)$, $B = -\frac{d^2}{dx^2}$, and q is a continuous complex-valued function. Hence for our purposes it will be sufficient to consider the functional space $E = C[a, b]$ with the topology of uniform convergence and its subspace E_1 consisting of functions from $C^2[a, b]$. One of the possibilities to introduce a transmutation operator on E was considered by Lions [50] and later on in other references (see, e.g., [51]), and consists in constructing a Volterra integral operator corresponding to a midpoint of the segment of interest. As we begin with this transmutation operator it is convenient to consider a symmetric segment $[-a, a]$ and hence the functional space $E = C[-a, a]$. It is worth mentioning that other well-known ways to construct the transmutation operators (see, e.g., [49], [63]) imply imposing initial conditions on the functions and consequently lead to transmutation operators satisfying (4.1) only on subclasses of E_1.

Thus, we consider the space $E = C[-a, a]$ and an operator of transmutation for the defined above A and B can be realized in the form (see, e.g., [49] and [51]) of a Volterra integral operator

$$Tu(x) = u(x) + \int_{-x}^{x} K(x,t)u(t)dt \tag{4.2}$$

where $K(x,t) = H\left(\frac{x+t}{2}, \frac{x-t}{2}\right)$ and H is the unique solution of the Goursat problem

$$\frac{\partial^2 H(u,v)}{\partial u\, \partial v} = q(u+v)H(u,v), \tag{4.3}$$

$$H(u,0) = \frac{1}{2}\int_0^u q(s)\, ds, \qquad H(0,v) = 0 \tag{4.4}$$

If the potential q is continuously differentiable, the kernel K itself is the solution of the Goursat problem

$$\left(\frac{\partial^2}{\partial x^2} - q(x)\right) K(x,t) = \frac{\partial^2}{\partial t^2} K(x,t),$$

$$K(x,x) = \frac{1}{2}\int_0^x q(s)\, ds, \qquad K(x,-x) = 0.$$

If the potential q is n times continuously differentiable, the kernel $K(x,t)$ is $n+1$ times continuously differentiable with respect to both independent variables (see [51]).

An important property of this transmutation operator consists in the way how it maps solutions of the equation

$$v'' + \omega^2 v = 0 \tag{4.5}$$

into solutions of the equation

$$u'' - q(x)u + \omega^2 u = 0 \tag{4.6}$$

where ω is a complex number. Denote by $e_0(i\omega, x)$ the solution of (4.6) satisfying the initial conditions

$$e_0(i\omega, 0) = 1 \qquad \text{and} \qquad e_0'(i\omega, 0) = i\omega.$$

The subindex "0" indicates that the initial conditions correspond to the point $x = 0$ and the letter "e" reminds us that the initial values coincide with the initial values of the function $e^{i\omega x}$.

The transmutation operator (4.2) maps $e^{i\omega x}$ into $e_0(i\omega, x)$,

$$e_0(i\omega, x) = T[e^{i\omega x}] \tag{4.7}$$

(see [51, Theorem 1.2.1]).

Following [51] we introduce the following notations

$$K_c(x,t;h) = h + K(x,t) + K(x,-t) + h\int_t^x \{K(x,\xi) - K(x,-\xi)\}d\xi$$

where h is a complex number, and

$$K_s(x,t;\infty) = K(x,t) - K(x,-t).$$

Theorem 4.2 ([51]). *Solutions $c(\omega,x;h)$ and $s(\omega,x;\infty)$ of equation (4.6) satisfying the initial conditions*

$$c(\omega,0;h) = 1, \qquad c'_x(\omega,0;h) = h \tag{4.8}$$

$$s(\omega,0;\infty) = 0, \qquad s'_x(\omega,0;\infty) = 1 \tag{4.9}$$

can be represented in the form

$$c(\omega,x;h) = \cos\omega x + \int_0^x K_c(x,t;h)\cos\omega t\,dt \tag{4.10}$$

and

$$s(\omega,x;\infty) = \frac{\sin\omega x}{\omega} + \int_0^x K_s(x,t;\infty)\frac{\sin\omega t}{\omega}\,dt. \tag{4.11}$$

Denote by

$$T_c u(x) = u(x) + \int_0^x K_c(x,t;h)u(t)dt \tag{4.12}$$

and

$$T_s u(x) = u(x) + \int_0^x K_s(x,t;\infty)u(t)dt \tag{4.13}$$

the corresponding integral operators. As was pointed out in [9], they are not transmutations on the whole subspace E_1, they even do not map all solutions of (4.5) into solutions of (4.6). For example, as we show below

$$\left(-\frac{d^2}{dx^2} + q(x)\right) T_s[1] \neq T_s\left[-\frac{d^2}{dx^2}(1)\right] = 0$$

when q is constant.

Example 4.3. Transmutation operator for operators $A := \frac{d^2}{dx^2} + c$, c is a constant, and $B := \frac{d^2}{dx^2}$. According to (4.3) and (4.4), finding the kernel of transmutation operator is equivalent to finding the function $H(s,t) = K(s+t, s-t)$ satisfying the Goursat problem

$$\frac{\partial^2 H(s,t)}{\partial s\partial t} = -cH(s,t), \quad H(s,0) = -\frac{cs}{2}, \quad H(0,t) = 0.$$

The solution of this problem is given by [28, (4.85)]

$$H(s,t) = -\frac{c}{2}\int_0^s J_0\big(2\sqrt{ct(s-\xi)}\big)\,d\xi = -\frac{\sqrt{cst}J_1(2\sqrt{cst})}{2t},$$

where J_0 and J_1 are the Bessel functions of the first kind, and the formula is valid even if the radicand is negative. Hence,

$$K(x,y) = H\left(\frac{x+y}{2}, \frac{x-y}{2}\right) = -\frac{1}{2}\frac{\sqrt{c(x^2-y^2)}J_1\left(\sqrt{c(x^2-y^2)}\right)}{x-y}. \tag{4.14}$$

From (4.14) we get the 'sine' kernel

$$K_s(x,t;\infty) = -\frac{t\sqrt{c(x^2-t^2)}J_1\left(\sqrt{c(x^2-t^2)}\right)}{x^2-t^2},$$

and can check the above statement about the operator T_s,

$$T_s[1](x) = 1 - \int_0^x \frac{t\sqrt{c(x^2-t^2)}J_1\left(\sqrt{c(x^2-t^2)}\right)}{x^2-t^2}\,dt = J_0(x\sqrt{c}),$$

$$\left(\frac{d^2}{dx^2}+c\right)T_s[1] = \frac{\sqrt{c}J_1(x\sqrt{c})}{x} \neq 0.$$

For the rest of this section suppose that f is a solution of (2.5) fulfilling the condition of Theorem 2.3 on a finite segment $[-a,a]$. We normalize f in such a way that $f(0)=1$ and let $f'(0)=h$ where h is some complex constant. Define the system of functions $\{\varphi_k\}_{k=0}^{\infty}$ by this function f with the use of (2.1), (2.2) and (2.3). The system of functions $\{\varphi_k\}_{k=0}^{\infty}$ is related to the transmutation operators T_c (with the same parameter h in the kernel) and T_s in a way that it is the union of functions which are the result of acting of operator T_s on the odd powers of independent variable and of operator T_c on the even powers of independent variable. The following theorem holds, see [9] for the details of the proof.

Theorem 4.4 ([9]). *Let q be a continuous complex-valued function of an independent real variable $x \in [-a,a]$, and f be a particular solution of (2.5) such that $f \in C^2(-a,a)$, $f \neq 0$ on $[-a,a]$ and normalized as $f(0)=1$. Let φ_k, $k \in \mathbb{N}_0 := \mathbb{N} \cup \{0\}$ are functions defined by (2.3). Then the following equalities are valid*

$$\varphi_k = T_c[x^k] \quad \textit{when } k \in \mathbb{N}_0 \textit{ is even or equal to zero}$$

and

$$\varphi_k = T_s[x^k] \quad \textit{when } k \in \mathbb{N} \textit{ is odd.}$$

As for the transmutation operator T, it does not map all powers of the independent variable into the functions φ_k. Instead, the following theorem holds.

Theorem 4.5 ([9]). *Under the conditions of Theorem 4.4 the following equalities are valid*

$$\varphi_k = T[x^k] \quad \textit{when } k \textit{ is odd} \tag{4.15}$$

and

$$\varphi_k - \frac{h}{k+1}\varphi_{k+1} = T[x^k] \quad \textit{when } k \in \mathbb{N}_0 \textit{ is even or equal to zero} \tag{4.16}$$

where by h we denote $f'(0) \in \mathbb{C}$.

Taking into account the first of former relations the second can be written also as follows

$$\varphi_k = T\left[x^k + \frac{h}{k+1}x^{k+1}\right] \quad \text{when } k \in \mathbb{N}_0 \text{ is even or equal to zero.}$$

Remark 4.6. Let f be the solution of (2.5) satisfying the initial conditions

$$f(0) = 1, \quad \text{and} \quad f'(0) = 0. \tag{4.17}$$

If it does not vanish on $[-a, a]$ then from Theorem 4.5 we obtain that $\varphi_k = T[x^k]$ for any $k \in \mathbb{N}_0$. In general, of course there is no guaranty that the solution satisfying (4.17) have no zero on $[-a, a]$. Hence the operator T transmutes the powers of x into φ_k whose construction is based on the solution f satisfying (4.17) only in some neighborhood of the origin. In the next section we show how to change the operator T so that the new operator map x^k into $\varphi_k(x)$ on the whole segment $[-a, a]$.

Note that in Theorem 4.5 the operator T does not depend on the function f, so the right-hand sides of the equalities (4.15) and (4.16) do not change with the change of f. Consider two non-vanishing solutions f and g of (2.5) normalized as $f(0) = g(0) = 1$ and let φ_k^f and φ_k^g be the functions obtained from f and g respectively by means of (2.1), (2.2) and (2.3). The relation between φ_k^f and φ_k^g are given by the following proposition which may be easily deduced from equalities (4.15) and (4.16).

Proposition 4.7. *The following equalities hold*

$$\varphi_k^f = \varphi_k^g \quad \textit{when } k \in \mathbb{N} \textit{ is odd,}$$

and

$$\varphi_k^f = \varphi_k^g + \frac{h_f - h_g}{k+1}\varphi_{k+1}^g \quad \textit{when } k \in \mathbb{N}_0 \textit{ is even,}$$

where $h_f = f'(0)$ *and* $h_g = g'(0)$.

5. A parametrized family of transmutation operators

In [9] a parametrized family of operators $\mathbf{T}_h$, $h \in \mathbb{C}$ was introduced, given by the integral expression

$$\mathbf{T}_h u(x) = u(x) + \int_{-x}^{x} \mathbf{K}(x, t; h)u(t)dt \tag{5.1}$$

where

$$\mathbf{K}(x, t; h) = \frac{h}{2} + K(x, t) + \frac{h}{2}\int_t^x \big(K(x, s) - K(x, -s)\big)\, ds. \tag{5.2}$$

They are related to operators T_s and T_c (with the parameter h in the kernel of the latter operator) by

$$\mathbf{T}_h = T_c P_e + T_s P_o, \tag{5.3}$$

where $P_e f(x) = \big(f(x)+f(-x)\big)/2$ and $P_o f(x) = \big(f(x)-f(-x)\big)/2$ are projectors onto even and odd functions, respectively. In this section we show that the operators $\mathbf{T}_h$ are transmutations, summarize their properties and in Theorem 5.8 we show how they act on powers of x.

Let us notice that $\mathbf{K}(x,t;0) = K(x,t)$ and that the expression

$$\begin{aligned}\mathbf{K}(x,t;h) - \mathbf{K}(x,-t;h) &= K(x,t) - K(x,-t) - \frac{h}{2}\int_{-t}^{t} (K(x,s) - K(x,-s))\,ds\\ &= K(x,t) - K(x,-t)\end{aligned}$$

does not depend on h. Thus, it is possible to compute $\mathbf{K}(x,t;h)$ for any h by a given $\mathbf{K}(x,t;h_1)$ for some particular value h_1. We formulate this result as the following statement.

Theorem 5.1 ([9]). *The integral kernels $\mathbf{K}(x,t;h)$ and $\mathbf{K}(x,t;h_1)$ are related by the expression*

$$\mathbf{K}(x,t;h) = \frac{h-h_1}{2} + \mathbf{K}(x,t;h_1) + \frac{h-h_1}{2}\int_t^x \big(\mathbf{K}(x,s;h_1) - \mathbf{K}(x,-s;h_1)\big)\,ds. \tag{5.4}$$

The operator $\mathbf{T}_h$ may be expressed in terms of another operator $\mathbf{T}_{h_1}$ and in particular, in terms of the operator T. The following proposition holds.

Proposition 5.2. *The operators $\mathbf{T}_{h_1}$ and $\mathbf{T}_{h_2}$ are related by the expression*

$$\mathbf{T}_{h_2}u = \mathbf{T}_{h_1}\left[u(x) + \frac{h_2-h_1}{2}\int_{-x}^{x} u(t)\,dt\right] \tag{5.5}$$

valid for any $u\in C[-a,a]$. In particular,

$$\mathbf{T}_h u = T\left[u(x) + \frac{h}{2}\int_{-x}^{x} u(t)\,dt\right]. \tag{5.6}$$

Proof. Using formulas (5.1) and (5.2) we obtain

$$\begin{aligned}\mathbf{T}_h u = Tu + \frac{h}{2}\int_{-x}^{x} u(t)\,dt + \frac{h}{2}\int_{-x}^{x} u(t)\int_t^x K(x,s)\,ds\,dt\\ - \frac{h}{2}\int_{-x}^{x} u(t)\int_{-t}^{x} K(x,s)\,ds\,dt,\end{aligned}$$

and after changing the order of integration in the last two integrals, we have

$$\begin{aligned}T_h u = Tu &+ \frac{h}{2}\int_{-x}^{x} u(t)\,dt + \frac{h}{2}\int_{-x}^{x} K(x,s)\int_{-x}^{s} u(t)\,dt\,ds\\ &- \frac{h}{2}\int_{-x}^{x} K(x,s)\int_{-x}^{-s} u(t)\,dt\,ds\end{aligned}$$

$$
\begin{aligned}
&= Tu + \frac{h}{2}\int_{-x}^{x} u(t)\,dt + \frac{h}{2}\int_{-x}^{x} K(x,s)\left[\int_{-x}^{0} + \int_{0}^{s} u(t)\,dt\right] ds \\
&\quad - \frac{h}{2}\int_{-x}^{x} K(x,s)\left[\int_{-x}^{0} - \int_{-s}^{0} u(t)\,dt\right] ds \\
&= Tu + \frac{h}{2}\int_{-x}^{x} u(t)\,dt + \frac{h}{2}\int_{-x}^{x} K(x,s)\int_{-s}^{s} u(t)\,dt\,ds = T\left[u(x) + \frac{h}{2}\int_{-x}^{x} u(t)\,dt\right].
\end{aligned}
$$

Since $\int_{-x}^{x}\int_{-t}^{t} u(s)\,ds\,dt = 0$ for any function $u \in C[-a,a]$, we have from (5.6) that

$$
\begin{aligned}
&\mathbf{T}_{h_1}\left[u(x) + \int_{-x}^{x} u(t)\,dt\right] \\
&= T\left[u(x) + \frac{h_2 - h_1}{2}\int_{-x}^{x} u(t)\,dt + \frac{h_1}{2}\int_{-x}^{x}\left(u(t) + \frac{h_2 - h_1}{2}\int_{-t}^{t} u(s)\,ds\right) dt\right] \\
&= T\left[u(x) + \frac{h_2}{2}\int_{-x}^{x} u(t)\,dt\right] = \mathbf{T}_{h_2} u. \qquad \square
\end{aligned}
$$

Using (4.8)–(4.13) and (5.3) it is possible to check how the operators $\mathbf{T}_h$ act on solutions of (4.5).

Proposition 5.3 ([46]). *The operator $\mathbf{T}_h$ maps a solution v of an equation $v'' + \omega^2 v = 0$, where ω is a complex number, into a solution u of the equation $u'' - q(x)u + \omega^2 u = 0$ with the following correspondence of the initial values*

$$
u(0) = v(0), \qquad u'(0) = v'(0) + hv(0). \tag{5.7}
$$

Remark 5.4. Formulas (5.7) are valid for any function $v \in C^1[-a,a]$.

We know that the integral kernel of the transmutation operator T is related to the solution of the Goursat problem (4.3)–(4.4). A similar result holds for the operators $\mathbf{T}_h$.

Theorem 5.5 ([46]). *In order for the function $K(x,t;h)$ to be the kernel of a transmutation operator acting as described in Proposition 5.3, it is necessary and sufficient that $H(u,v;h) := K(u+v, u-v;h)$ be a solution of the Goursat problem*

$$
\frac{\partial^2 H(u,v;h)}{\partial u\,\partial v} = q(u+v)H(u,v;h),
$$

$$
H(u,0;h) = \frac{h}{2} + \frac{1}{2}\int_0^u q(s)\,ds, \qquad H(0,v;h) = \frac{h}{2}.
$$

If the potential q is continuously differentiable, the function $K(x,t;h)$ itself must be the solution of the Goursat problem

$$
\left(\frac{\partial^2}{\partial x^2} - q(x)\right) K(x,t;h) = \frac{\partial^2}{\partial t^2} K(x,t;h), \tag{5.8}
$$

$$
K(x,x;h) = \frac{h}{2} + \frac{1}{2}\int_0^x q(s)\,ds, \qquad K(x,-x;h) = \frac{h}{2}. \tag{5.9}
$$

Under some additional requirements on the potential q the operators $\mathbf{T}_h$ are transmutations in the sense of Definition 4.1. The following theorem generalizes the results obtained in [46].

Theorem 5.6. *Suppose the potential q satisfies either of the following two conditions.*

- $q \in C^1[-a,a]$;
- $q \in C[-a,a]$ *and there exists a particular complex-valued solution g of (2.5) non-vanishing on $[-a,a]$.*

Then the operator $\mathbf{T}_h$ given by (5.1) satisfies the equality

$$\left(-\frac{d^2}{dx^2}+q(x)\right)\mathbf{T}_h[u]=\mathbf{T}_h\left[-\frac{d^2}{dx^2}(u)\right] \tag{5.10}$$

for any $u \in C^2[-a,a]$.

Proof. In [46] the theorem was proved for the case $q \in C^1[-a,a]$ and for the case when the particular solution g from the statement satisfies the conditions $g(0)=1$ and $g'(0)=h$.

We may normalize the particular solution g as $g(0)=1$. Suppose that $g'(0)=h_1$. We know already that (5.10) holds for the operator $\mathbf{T}_{h_1}$. To finish the proof, we use (5.5) and obtain

$$\begin{aligned}\left(-\frac{d^2}{dx^2}+q(x)\right)\mathbf{T}_h[u]&=\left(-\frac{d^2}{dx^2}+q(x)\right)\mathbf{T}_{h_1}\left[u(x)+\frac{h-h_1}{2}\int_{-x}^{x}u(t)\,dt\right]\\&=-\mathbf{T}_{h_1}\left[u''(x)+\frac{h-h_1}{2}\frac{d^2}{dx^2}\int_{-x}^{x}u(t)\,dt\right]\\&=-\mathbf{T}_{h_1}\left[u''(x)+\frac{h-h_1}{2}\int_{-x}^{x}u''(t)\,dt\right]=\mathbf{T}_h\left[-\frac{d^2}{dx^2}(u)\right].\end{aligned}$$ □

Remark 5.7. As was pointed out in Remark 2.5, in the regular case the non-vanishing solution g of (2.5) exists due to the alternation of zeroes of two linearly independent solutions. Of course, it would be interesting to prove that the operators $\mathbf{T}_h$ are transmutations in the general case of complex-valued potentials $q \in C[-a,a]$ without any additional assumption.

Suppose now that a function f is a solution of (2.5), non-vanishing on $[-a,a]$ and normalized as $f(0)=1$. Let $h := f'(0)$ be some complex constant. Define as before the system of functions $\{\varphi_k\}_{k=0}^{\infty}$ by this function f and by (2.3). The following theorem states that the operator $\mathbf{T}_h$ transmutes powers of x into the functions φ_k.

Theorem 5.8 ([9]). *Let q be a continuous complex-valued function of an independent real variable $x \in [-a,a]$, and f be a particular solution of (2.5) such that $f \in C^2(-a,a)$ together with $1/f$ are bounded on $[-a,a]$ and normalized as $f(0)=1$, and let $h := f'(0)$, where h is a complex number. Then the operator (5.1) with the kernel defined by (5.2) transforms x^k into $\varphi_k(x)$ for any $k \in \mathbb{N}_0$.*

Thus, we clarified that the system of functions $\{\varphi_k\}$ may be obtained as the result of the Volterra integral operator acting on powers of the independent variable. As was mentioned before, this offers an algorithm for transmuting functions in the situation when $\mathbf{K}(x,t;h)$ is unknown. Moreover, properties of the Volterra integral operator such as boundedness and bounded invertibility in many functional spaces gives us a tool to prove the completeness of the system of function $\{\varphi_k\}$ in various situations.

Example 5.9. Consider a function $k(x,t) = \frac{t-1}{2(x+1)}$ (later, in Example 6.8 it is explained how it can be obtained). We have

$$(\partial_x^2 - \partial_t^2)k(x,t) = \frac{t-1}{(x+1)^3} = \frac{2}{(x+1)^2} \cdot \frac{t-1}{2(x+1)},$$

$k(x,-x) = \frac{-x-1}{2(x+1)} = -\frac{1}{2}$ and $k(x,x) = \frac{x-1}{2(x+1)} = -\frac{1}{2} + \frac{1}{2}\int_0^x \frac{2}{(s+1)^2}\,ds$, thus the function $k(x,t)$ satisfies the Goursat problem (5.8)–(5.9) with $q(x) = 2/(x+1)^2$ and $h = -1$ and by Theorem 5.5 is the kernel of the transmutation operator $\mathbf{T}_{-1}$.

Consider the function $f = \mathbf{T}_{-1}[1] = \frac{1}{x+1}$ as a solution of (2.5) such that $f(0) = 1$ and $f'(0) = h = -1$, nonvanishing on any $[-a,a] \subset (-1,1)$. The first 3 functions φ_k are given by

$$\varphi_0 = f = \frac{1}{x+1}, \quad \varphi_1 = \frac{x^3+3x^2+3x}{3(x+1)}, \quad \varphi_2 = \frac{2x^3+3x^2}{3(x+1)}.$$

It can be easily checked that

$$\mathbf{T}_{-1}x = x + \int_{-x}^{x} \frac{(t-1)t}{2(x+1)}\,dt = \frac{x^3+3x^2+3x}{3(x+1)} = \varphi_1,$$

$$\mathbf{T}_{-1}x^2 = x^2 + \int_{-x}^{x} \frac{(t-1)t^2}{2(x+1)}\,dt = \frac{2x^3+3x^2}{3(x+1)} = \varphi_2.$$

We can calculate the kernel K of the original operator T by (5.4), it is given by

$$K(x,t) = \frac{2x+2t+x^2-t^2}{4(x+1)}$$

and we can check that $T[x] = \varphi_1$ and $T[1] = \frac{x^3+3x^2+3x+3}{3(x+1)} = \varphi_0 + \varphi_1$ in accordance with Theorem 4.5.

6. Transmutation operators and Darboux-transformed equations

To construct the system of functions $\{\varphi_k\}_{k=0}^\infty$ we use the half of the functions $\{X^{(k)}, \widetilde{X}^{(k)}\}_{k=0}^\infty$. What about the second half? Note that starting with the function $1/f$ we obtain the same system of functions $\{X^{(k)}, \widetilde{X}^{(k)}\}_{k=0}^\infty$ with the only change that $X_f^{(k)}$ becomes $\widetilde{X}_{1/f}^{(k)}$ and $\widetilde{X}_f^{(k)}$ becomes $X_{1/f}^{(k)}$. Hence the "second half" of the functions $\{X^{(k)}, \widetilde{X}^{(k)}\}_{k=0}^\infty$ from (2.3) is used. The function $1/f$ is continuous complex valued and non-vanishing, and is a solution of the equation $u'' - q_2 u = 0$, where $q_2 = 2\left(f'/f\right)^2 - q$. The last equation is known as the Darboux transformation of the original equation. The Darboux transformation is closely related to the factorization of the Schrödinger equation, and nowadays it is used in dozens of works, see, e.g., [18, 29, 52, 58] in connection with solitons and integrable systems, e.g., [1, 32, 55, 57] and the review [59] of applications to quantum mechanics.

For the convenience denote the potential of the original equation by q_1 and the corresponding Sturm-Liouville operator by $A_1 := \frac{d^2}{dx^2} - q_1(x)$. Suppose a solution f of the equation $A_1 f = 0$ is given such that $f(x) \neq 0$, $x \in [-a, a]$, it is normalized as $f(0) = 1$ and $h := f'(0)$ is some complex number. Denote the Darboux-transformed operator by $A_2 := \frac{d^2}{dx^2} - q_2(x)$, where $q_2(x) = 2\left(\frac{f'(x)}{f(x)}\right)^2 - q_1(x)$.

From the previous section we know that there exists a transmutation operator $\mathbf{T}_{1;h}$ for the original equation (2.6) with the potential q_1 and such that

$$\mathbf{T}_{1;h} x^k = \varphi_k, \quad k \in \mathbb{N}_0. \tag{6.1}$$

The subindex "1" in the notation $\mathbf{T}_{1;h}$ indicates that the transmutation operator corresponds to A_1.

Similarly, there exists a transmutation operator $\mathbf{T}_{2;-h}$ for the Darboux-transformed operator A_2 such that

$$\mathbf{T}_{2;-h} x^k = \psi_k, \quad k \in \mathbb{N}_0, \tag{6.2}$$

where the family of functions $\{\psi_k\}_{k=0}^{\infty}$ is defined by (2.4).

It is interesting to obtain some relations between the operators $\mathbf{T}_{1;h}$ and $\mathbf{T}_{2;-h}$ and between their integral kernels $\mathbf{K}_1$ and $\mathbf{K}_2$. In this section we explain how to construct the integral kernel $\mathbf{K}_2$ by the known integral kernel $\mathbf{K}_1$ and show that the operators $\mathbf{T}_{1;h}$ and $\mathbf{T}_{2;-h}$ satisfy certain commutation relations with the operator of differentiation.

We remind some well-known facts about the Darboux transformation. First, $1/f$ is a solution of $A_2 u = 0$. Second, it is closely related to the factorization of Sturm-Liouville and one-dimensional Schrödinger operators. Namely, we have

$$A_1 = \frac{d^2}{dx^2} - q_1(x) = \left(\partial_x + \frac{f'}{f}\right)\left(\partial_x - \frac{f'}{f}\right) = \frac{1}{f}\partial_x f^2 \partial_x \frac{1}{f}\cdot, \tag{6.3}$$

$$A_2 = \frac{d^2}{dx^2} - q_2(x) = \left(\partial_x - \frac{f'}{f}\right)\left(\partial_x + \frac{f'}{f}\right) = f\partial_x \frac{1}{f^2}\partial_x f\,\cdot. \tag{6.4}$$

Suppose that u is a solution of the equation $A_1 u = \omega u$ for some $\omega \in \mathbb{C}$. Then the function $v = \left(\partial_x - \frac{f'}{f}\right)u = \left(f\partial_x \frac{1}{f}\right)u$ is a solution of the equation $A_2 v = \omega v$, and vice versa, given a solution v of $A_2 v = \omega v$, the function $u = \left(\partial_x + \frac{f'}{f}\right)v = \left(\frac{1}{f}\partial_x f\right)v$ is a solution of $A_1 u = \omega u$.

Suppose that the operator $\mathbf{T}_1 := \mathbf{T}_{1;h}$ which transmutes the operator A_1 into the operator $B = d^2/dx^2$ and the powers x^k into the functions φ_k is known in the sense that its kernel $\mathbf{K}_1(x, t; h)$ is given. As before $h = f'(0)$. Then the function $1/f$ is the non-vanishing solution of the equation $A_2 u = 0$ satisfying $1/f(0) = 1$ and $(1/f)'(0) = -h$. Hence we are looking for the operator $\mathbf{T}_2 := \mathbf{T}_{2;-h}$ transmuting the operator A_2 into the operator B and the powers x^k into the functions ψ_k.

Let us explain the idea for obtaining the operator $\mathbf{T}_2$. We want to find an operator transforming solutions of the equation $Bu + \omega^2 u = 0$ into solutions of the equation $A_2 u + \omega^2 u = 0$, see the first diagram below. Starting with a solution σ of the equation $(\partial_x^2 + \omega^2)\sigma = 0$, by application of $\mathbf{T}_1$ we get a solution of $(A_1 + \omega^2)u = 0$, and the expression $\big(f\partial_x \frac{1}{f}\big)\mathbf{T}_1\sigma$ is a solution of $(A_2 + \omega^2)v = 0$. But the operator $\big(f\partial_x \frac{1}{f}\big)\mathbf{T}_1$ is unbounded and hence cannot coincide with the operator $\mathbf{T}_2$. In order to find the required bounded operator we may consider the second copy of the equation $(\partial_x^2 + \omega^2)u = 0$, which is a result of the Darboux transformation applied to $(\partial_x^2 + \omega^2)\sigma = 0$ with respect to the particular solution $g \equiv 1$ and construct the operator $\mathbf{T}_2$ by making the second diagram commutative. In order to obtain a bounded operator $\mathbf{T}_2$, instead of using $f\partial_x \frac{1}{f}$ for the last step, we will use the inverse of $\frac{1}{f}\partial_x f$, i.e., $\frac{1}{f}\big(\int_0^x f(s)\cdot ds + C\big)$.

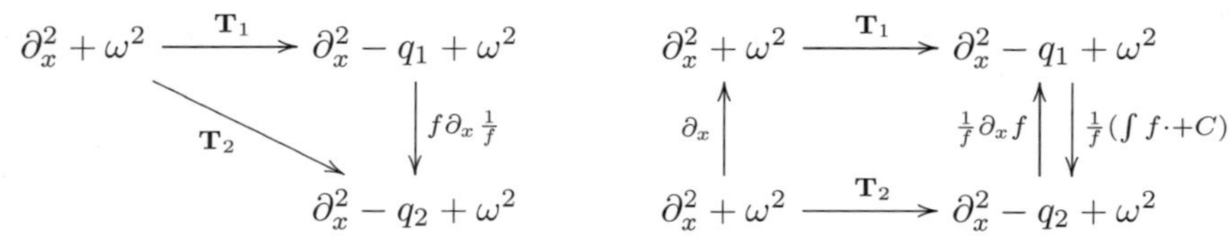

That explains how to obtain the following theorem.

Theorem 6.1 ([46]). *The operator T_2, acting on solutions u of equations $(\partial_x^2 + \omega^2)u = 0$, $\omega \in \mathbb{C}$ by the rule*

$$T_2[u](x) = \frac{1}{f(x)}\left(\int_0^x f(\eta)\mathbf{T}_1[u'](\eta)\,d\eta + u(0)\right) \tag{6.5}$$

coincides with the transmutation operator $\mathbf{T}_{2;-h}$.

Now we show that the operator T_2 can be written as a Volterra integral operator and, as a consequence, extended by continuity to a wider class of functions. To obtain simpler expression for the integral kernel $\mathbf{K}_2(x,t;-h)$ we have to suppose that the original integral kernel $\mathbf{K}_1(x,t;h)$ is known in the larger domain than required by definition (5.1). Namely, suppose that the function $\mathbf{K}_1(x,t;h)$ is known and is continuously differentiable in the domain $\overline{\Pi}: -a \le x \le a, -a \le t \le a$. We refer the reader to [46] for further details.

Theorem 6.2 ([46]). *The operator T_2 admits a representation as the Volterra integral operator*

$$T_2[u](x) = u(x) + \int_{-x}^x \mathbf{K}_2(x,t;-h)u(t)\,dt, \tag{6.6}$$

with the kernel

$$\mathbf{K}_2(x,t;-h) = -\frac{1}{f(x)}\left(\int_{-t}^x \partial_t\mathbf{K}_1(s,t;h)f(s)\,ds + \frac{h}{2}f(-t)\right). \tag{6.7}$$

Such representation is valid for any function $u \in C^1[-a,a]$.

By Theorems 6.1 and 6.2 the Volterra operators T_2 and $\mathbf{T}_2$ coincide on the set of finite linear combinations of solutions of the equations $(\partial_x^2 + \omega^2)u = 0$, $\omega \in \mathbb{C}$. Since this set is dense in $C[-a,a]$, by continuity of T_2 and $\mathbf{T}_2$ we obtain the following corollaries.

Corollary 6.3 ([46]). *The operator* T_2 *given by* (6.6) *with the kernel* (6.7) *coincides with* $\mathbf{T}_2$ *on* $C[-a,a]$.

Corollary 6.4 ([46]). *The operator* T_2 *given by* (6.5) *coincides with* $\mathbf{T}_2$ *on* $C^1[-a,a]$.

Operator A_1 is the Darboux transformation of the operator A_2 with respect to the solution $1/f$, hence we may obtain another relation between the operators $\mathbf{T}_1$ and $\mathbf{T}_2$.

Corollary 6.5 ([46]). *For any function* $u \in C^1[-a,a]$ *the equality*

$$\mathbf{T}_1[u](x) = f(x)\left(\int_0^x \frac{1}{f(\eta)}\mathbf{T}_2[u'](\eta)\,d\eta + u(0)\right) \tag{6.8}$$

is valid.

From the second commutative diagram at the beginning of this subsection we may deduce some commutation relations between the operators $\mathbf{T}_1$, $\mathbf{T}_2$ and d/dx. The proof immediately follows from (6.5) and (6.8).

Corollary 6.6 ([46]). *The following operator equalities hold on* $C^1[-a,a]$:

$$\partial_x f \mathbf{T}_2 = f\mathbf{T}_1\partial_x \tag{6.9}$$

$$\partial_x \frac{1}{f}\mathbf{T}_1 = \frac{1}{f}\mathbf{T}_2\partial_x. \tag{6.10}$$

In [44] the following notion of generalized derivatives was introduced. Consider a function g assuming that both f and g possess the derivatives of all orders up to the order n on the segment $[-a,a]$. Then in $[-a,a]$ the following generalized derivatives are defined

$$\gamma_0(g)(x) = g(x),$$

$$\gamma_k(g)(x) = \left(f^2(x)\right)^{(-1)^{k-1}}\left(\gamma_{k-1}(g)\right)'(x)$$

for $k = 1,2,\dots,n$.

Let a function u be defined by the equality

$$g = \frac{1}{f}\mathbf{T}_1 u,$$

and assume that $u \in C^n[-a,a]$. Note that below we do not necessarily require that the functions f and g be from $C^n[-a,a]$. With the use of (6.9) and (6.10) we have

$$\gamma_1(g) = f^2 \cdot \left(\frac{1}{f}\mathbf{T}_1 u\right)' = f^2 \cdot \frac{1}{f}\mathbf{T}_2 u' = f\mathbf{T}_2 u',$$

$$\gamma_2(g) = \frac{1}{f^2}\cdot\left(f\mathbf{T}_2 u'\right)' = \frac{1}{f^2}\cdot f\mathbf{T}_1 u'' = \frac{1}{f}\mathbf{T}_1 u''.$$

By induction we obtain the following corollary.

Corollary 6.7 ([46]). *Let* $u \in C^n[-a,a]$ *and* $g = \frac{1}{f}\mathbf{T}_1 u$. *Then*

$$\gamma_k(g) = f\mathbf{T}_2 u^{(k)} \qquad \text{if } k \text{ is odd}, \quad k \le n,$$

and

$$\gamma_k(g) = \frac{1}{f}\mathbf{T}_1 u^{(k)} \qquad \text{if } k \text{ is even}, \ k \le n.$$

Example 6.8. We start with the operator $A_0 = d^2/dx^2$. We have to pick up such a solution f of the equation $A_0 f = 0$ that $f'/f \ne 0$. This is in order to obtain an operator $A_1 \ne A_0$ as a result of the Darboux transformation of A_0. For such solution f consider, e.g., $f_0(x) = x+1$. Both f_0 and $1/f_0$ are bounded on any segment $[-a,a] \subset (-1;1)$ and the Darboux transformed operator has the form $A_1 = \frac{d^2}{dx^2} - \frac{2}{(x+1)^2}$.

The transmutation operator T for A_0 is obviously an identity operator and $K_0(x,t;0) = 0$. Since $f_0'(0) = 1$, we look for the parametrized operator $\mathbf{T}_{0;1}$. Its kernel is given by (5.4): $\mathbf{K}_0(x,t;1) = 1/2$. From Theorem 6.2 we obtain the transmutation kernel for the operator A_1

$$\mathbf{K}_1(x,t;-1) = -\frac{1}{x+1}\cdot\frac{1-t}{2} = \frac{t-1}{2(x+1)}, \tag{6.11}$$

the kernel from Example 5.9.

To obtain a less trivial example consider again the operator $A_1 = \frac{d^2}{dx^2} - \frac{2}{(x+1)^2}$ and the function $f_1(x) = (x+1)^2$ as a solution of $A_1 f = 0$. Since $h = f_1'(0) = 2$, we compute $\mathbf{K}_1(x,t;2)$ from (6.11) using (5.4)

$$\mathbf{K}_1(x,t;2) = \frac{3x^2 + 6x + 4 - 3t^2 + 2t}{4(x+1)}.$$

The Darboux transformation of the operator A_1 with respect to the solution f_1 is the operator $A_2 = \frac{d^2}{dx^2} - \frac{6}{(x+1)^2}$ and by Theorem 6.2 the transmutation operator $\mathbf{T}_{2;-2}$ for A_2 is given by the Volterra integral operator (5.1) with the kernel

$$\begin{aligned}\mathbf{K}_2(x,t;-2) &= -\frac{1}{(x+1)^2}\left(\int_{-t}^{x} \frac{-3t+1}{2(s+1)}(s+1)^2\,ds + (1-t)^2\right)\\ &= \frac{(3t-1)(x+1)^2 - 3(t-1)^2(t+1)}{4(x+1)^2}.\end{aligned}$$

This procedure may be continued iteratively. Consider the operators

$$A_n := \frac{d^2}{dx^2} - \frac{n(n+1)}{(x+1)^2}.$$

The function $f_n(x) = (x+1)^{n+1}$ is a solution of the equation $A_n f = 0$. The Darboux transformation of the operator A_n with respect to the solution f_n is the operator

$$\frac{d^2}{dx^2} - 2\left(\frac{f_n'(x)}{f_n(x)}\right)^2 + \frac{n(n+1)}{(x+1)^2} = \frac{d^2}{dx^2} - \frac{(n+1)(n+2)}{(x+1)^2},$$

i.e., exactly the operator A_{n+1}. If we know $\mathbf{K}_n(x,t;-n)$ for the operator A_n, by (5.4) we compute the kernel $\mathbf{K}_n(x,t;n+1)$ corresponding to the solution $f_n(x)$ and by Theorem 6.2 we may calculate the kernel $\mathbf{K}_{n+1}(x,t;-n-1)$. Careful analysis shows that we have to integrate only polynomials in all integrals involved, so the described procedure can be performed up to any fixed n.

Example 6.9. Consider the Schrödinger equation

$$u'' + 2\,\mathrm{sech}^2(x)\,u = u. \tag{6.12}$$

This equation appears in soliton theory and as an example of a reflectionless potential in the one-dimensional quantum scattering theory (see. e.g., [48]). Equation (6.12) can be obtained as a result of the Darboux transformation of the equation $u'' = u$ with respect to the solution $f(x) = \cosh x$. The transmutation operator for the operator $A_1 = \partial_x^2 - 1$ was calculated in [9, Example 3]. Its kernel is given by the expression

$$\mathbf{K}_1(x,t;0) = \frac{1}{2}\frac{\sqrt{x^2-t^2}\,I_1(\sqrt{x^2-t^2})}{x-t},$$

where I_1 is the modified Bessel function of the first kind. Hence from Theorem 6.2 we obtain the transmutation kernel for the operator $A_2 = \partial_x^2 + 2\,\mathrm{sech}^2\,x - 1$

$$\mathbf{K}_2(x,t;0) = \frac{1}{2\cosh(x)}\int_{-t}^{x}\left(\frac{I_0(\sqrt{s^2-t^2})t}{s-t} - \frac{\sqrt{s^2-t^2}\,I_1(\sqrt{s^2-t^2})}{(s-t)^2}\right)\cosh s\,ds.$$

7. Transmutation operator for the one-dimensional Dirac equation with a Lorentz scalar potential

One-dimensional Dirac equations with Lorentz scalar potentials are widely studied (see, for example, [11, 15, 33–36, 38, 54, 60, 62] and [55] for intertwining techniques for them).

According to [54] the Dirac equation in one space dimension with a Lorentz scalar potential can be written as

$$(\partial_x + m + S(x))\Psi_1 = E\Psi_2, \tag{7.1}$$

$$(-\partial_x + m + S(x))\Psi_2 = E\Psi_1, \tag{7.2}$$

where m ($m > 0$) is the mass and $S(x)$ is a Lorentz scalar. Denote $\eta = m + S$ and write the system (7.1), (7.2) in a matrix form as

$$\begin{pmatrix} \partial_x + \eta & 0 \\ 0 & \partial_x - \eta \end{pmatrix}\begin{pmatrix} \Psi_1 \\ \Psi_2 \end{pmatrix} = E\begin{pmatrix} 0 & 1 \\ -1 & 0 \end{pmatrix}\begin{pmatrix} \Psi_1 \\ \Psi_2 \end{pmatrix}.$$

In order to apply the results on the transmutation operators and factorizations (6.3), (6.4) we consider a function f such that

$$\frac{f'(x)}{f(x)} = -\eta = -m - S(x).$$

We can take $f(x) = \exp\left(-\int_0^x (m + S(s))\,ds\right)$, then $f(0) = 1$ and f does not vanish. Suppose the operators $\mathbf{T}_1$ and $\mathbf{T}_2$ are transmutations for the operators $A_1 = \left(\partial_x + \frac{f'}{f}\right)\left(\partial_x - \frac{f'}{f}\right)$ and $A_2 = \left(\partial_x - \frac{f'}{f}\right)\left(\partial_x + \frac{f'}{f}\right)$ respectively (corresponding to functions f and $1/f$ in the sense of Proposition 5.3). As was shown in [46] with the use of commutation relations (6.9) and (6.10), the operator

$$\mathbf{T} = \begin{pmatrix} \mathbf{T}_1 & 0 \\ 0 & \mathbf{T}_2 \end{pmatrix}$$

transmutes any solution $\begin{pmatrix} u_1 \\ u_2 \end{pmatrix}$ of the system

$$u_1' = E u_2 \tag{7.3}$$

$$u_2' = -E u_1 \tag{7.4}$$

into the solution $\begin{pmatrix} \Psi_1 \\ \Psi_2 \end{pmatrix}$ of the system (7.1),(7.2) with the initial conditions $\Psi_1(0) = u_1(0)$, $\Psi_2(0) = u_2(0)$. And vice versa if $\begin{pmatrix} \Psi_1 \\ \Psi_2 \end{pmatrix}$ is a solution of the system (7.1), (7.2), then the operator $\begin{pmatrix} \mathbf{T}_1^{-1} & 0 \\ 0 & \mathbf{T}_2^{-1} \end{pmatrix}$ transmutes it into the solution $\begin{pmatrix} u_1 \\ u_2 \end{pmatrix}$ of (7.3),(7.4) such that $u_1(0) = \Psi_1(0)$, $u_2(0) = \Psi_2(0)$.

Consider two systems of functions $\{\varphi_k\}_{k=0}^\infty$ and $\{\psi_k\}_{k=0}^\infty$ constructed from the function f by (2.3) and (2.4). The general solution of the system (7.3),(7.4) is given by

$$u_1 = C_1 v_1 + C_2 v_2$$
$$u_2 = C_2 v_1 - C_1 v_2,$$

where C_1 and C_2 are arbitrary constants and

$$v_1(x) = \cos Ex = \sum_{k=0}^\infty \frac{(-1)^k E^{2k}}{(2k)!} x^{2k},$$

$$v_2(x) = \sin Ex = \sum_{k=0}^\infty \frac{(-1)^k E^{2k+1}}{(2k+1)!} x^{2k+1}.$$

From (6.1) and (6.2) we see that the general solution of the one-dimensional Dirac system (7.1),(7.2) has the form

$$\Psi_1 = C_1 \sum_{k=0}^{\infty} \frac{(-1)^k E^{2k}}{(2k)!} \varphi_{2k} + C_2 \sum_{k=0}^{\infty} \frac{(-1)^k E^{2k+1}}{(2k+1)!} \varphi_{2k+1},$$

$$\Psi_2 = C_2 \sum_{k=0}^{\infty} \frac{(-1)^k E^{2k}}{(2k)!} \psi_{2k} - C_1 \sum_{k=0}^{\infty} \frac{(-1)^k E^{2k+1}}{(2k+1)!} \psi_{2k+1}.$$

Remark 7.1. It is possible to consider the two- or three-dimensional Dirac system and to construct the transmutation operator for it under some conditions on the potential. But the techniques involved, such as bicomplex numbers, pseudoanalytic function theory, Vekua equation and formal powers go well beyond the scope of the present article. We refer interested readers to the recent paper [8].

References

[1] V.G. Bagrov and B.F. Samsonov, *Darboux transformation, factorization, and supersymmetry in one-dimensional quantum mechanics.* Teoret. Mat. Fiz. **104** (1995), no. 2, 356–367 (in Russian); translation in Theoret. and Math. Phys. **104** (1995), no. 2, 1051–1060.

[2] C.A. Balanis, *Advanced Engineering Electromagnetics.* John Wiley & Sons, 1989.

[3] H. Begehr and R. Gilbert, *Transformations, transmutations and kernel functions,* vol. 1–2. Longman Scientific & Technical, Harlow, 1992.

[4] R. Bellman, *Perturbation techniques in mathematics, engineering and physics.* Dover Publications, 2003.

[5] J. Ben Amara and A.A. Shkalikov, *A Sturm-Liouville problem with physical and spectral parameters in boundary conditions.* Mathematical Notes **66** (1999), no. 2, 127–134.

[6] L.M. Brekhovskikh, *Waves in layered media.* New York, Academic Press, 1960.

[7] H. Campos and V.V. Kravchenko, *A finite-sum representation for solutions for the Jacobi operator.* Journal of Difference Equations and Applications **17** (2011) No. 4, 567–575.

[8] H. Campos, V.V. Kravchenko and L. Mendez, *Complete families of solutions for the Dirac equation using bicomplex function theory and transmutations.* Adv. Appl. Clifford Algebras (2012), Published online. DOI: 10.1007/s00006-012-0349-1.

[9] H. Campos, V.V. Kravchenko and S. Torba, *Transmutations, L-bases and complete families of solutions of the stationary Schrödinger equation in the plane.* J. Math. Anal. Appl. **389** (2012), No. 2, 1222–1238.

[10] R.W. Carroll, *Transmutation theory and applications.* Mathematics Studies, Vol. 117, North-Holland, 1985.

[11] J. Casahorrán, *Solving smultaneously Dirac and Ricatti equations.* Journal of Nonlinear Mathematical Physics **5** (1985), No. 4, 371–382.

[12] R. Castillo, K.V. Khmelnytskaya, V.V. Kravchenko and H. Oviedo, *Efficient calculation of the reflectance and transmittance of finite inhomogeneous layers.* J. Opt. A: Pure and Applied Optics **11** (2009), 065707.

[13] R. Castillo R, V.V. Kravchenko, H. Oviedo and V.S. Rabinovich, *Dispersion equation and eigenvalues for quantum wells using spectral parameter power series.* J. Math. Phys., **52** (2011), 043522 (10 pp.)

[14] B. Chanane, *Sturm-Liouville problems with parameter dependent potential and boundary conditions.* J. Comput. Appl. Math. **212** (2008), 282–290.

[15] C.-Y. Chen, *Exact solutions of the Dirac equation with scalar and vector Hartmann potentials.* Physics Letters A. **339** (2005), 283–287.

[16] A.H. Cherin, *An introduction to Optical Fibers.* McGraw-Hill, 1983.

[17] W.C. Chew, *Waves and fields in inhomogeneous media.* Van Nostrand Reinhold, New York, 1990.

[18] J.L. Cieśliński, *Algebraic construction of the Darboux matrix revisited.* J. Phys. A: Math. Theor. **42** (2009), 404003.

[19] W.J. Code and P.J. Browne, *Sturm-Liouville problems with boundary conditions depending quadratically on the eigenparameter.* J. Math. Anal. Appl. **309** (2005), 729–742.

[20] H. Coşkun and N. Bayram, *Asymptotics of eigenvalues for regular Sturm-Liouville problems with eigenvalue parameter in the boundary condition.* J. Math. Anal. Appl. 306 (2005), no. 2, 548–566.

[21] J. Delsarte, *Sur une extension de la formule de Taylor.* J Math. Pures et Appl. **17** (1938), 213–230.

[22] J. Delsarte, *Sur certaines transformations fonctionnelles relatives aux équations linéaires aux dérivées partielles du second ordre.* C. R. Acad. Sc. **206** (1938), 178–182.

[23] J. Delsarte and J.L. Lions, *Transmutations d'opérateurs différentiels dans le domaine complexe.* Comment. Math. Helv. **32** (1956), 113–128.

[24] M.K. Fage and N.I. Nagnibida. *The problem of equivalence of ordinary linear differential operators.* Novosibirsk: Nauka, 1987 (in Russian).

[25] L.B. Felsen and N. Marcuvitz, *Radiation and Scattering of Waves.* IEEE Press, New York, 1994.

[26] S. Flügge, *Practical Quantum Mechanics.* Berlin: Springer-Verlag, 1994.

[27] Ch.T. Fulton, *Two-point boundary value problems with eigenvalue parameter contained in the boundary conditions.* Proc. Roy. Soc. Edinburgh Sect. A **77** (1977), no. 3–4, 293–308.

[28] P.R. Garabedian, *Partial differential equations.* New York–London: John Willey and Sons, 1964.

[29] C. Gu, H. Hu, and Z. Zhou, *Darboux Transformations in Integrable Systems,* Springer-Verlag, Berlin, 2005.

[30] R.L. Hall, *Square-well representations for potentials in quantum mechanics.* J. Math. Phys. **33** (1992), 3472–3476.

[31] P. Harrison, *Quantum Wells, Wires and Dots: Theoretical and Computational Physics of Semiconductor Nanostructures.* Chichester: Wiley, 2010.

[32] A.D. Hemery and A.P. Veselov, *Whittaker-Hill equation and semifinite-gap Schrödinger operators.* J. Math. Phys. **51** (2010), 072108; doi:10.1063/1.3455367.

[33] J.R. Hiller, *Solution of the one-dimensional Dirac equation with a linear scalar potential.* Am. J. Phys. **70(5)** (2002), 522–524.

[34] C.-L. Ho, *Quasi-exact solvability of Dirac equation with Lorentz scalar potential.* Ann. Physics **321** (2006), No. 9, 2170–2182.

[35] R. Jackiw and S.-Y. Pi, *Persistence of zero modes in a gauged Dirac model for bilayer graphene.* Phys. Rev. B **78** (2008), 132104.

[36] N. Kevlishvili, G. Piranishvili, *Klein paradox in modified Dirac and Salpeter equations.* Fizika **9** (2003), No. 3,4, 57–61.

[37] K.V. Khmelnytskaya, V.V. Kravchenko and H.C. Rosu, *Eigenvalue problems, spectral parameter power series, and modern applications.* Submitted, available at arXiv:1112.1633.

[38] K.V. Khmelnytskaya and H.C. Rosu, *An amplitude-phase (Ermakov–Lewis) approach for the Jackiw–Pi model of bilayer graphene.* J. Phys. A: Math. Theor. **42** (2009), 042004.

[39] K.V. Khmelnytskaya and H.C. Rosu, *A new series representation for Hill's discriminant.* Annals of Physics **325** (2010), 2512–2521.

[40] A. Kostenko and G. Teschl, *On the singular Weyl–Titchmarsh function of perturbed spherical Schrödinger operators.* J. Differential Equations **250** (2011), 3701–3739.

[41] V.V. Kravchenko, *A representation for solutions of the Sturm-Liouville equation.* Complex Variables and Elliptic Equations **53** (2008), 775–789.

[42] V.V. Kravchenko, *Applied pseudoanalytic function theory.* Basel: Birkhäuser, Series: Frontiers in Mathematics, 2009.

[43] V.V. Kravchenko, *On the completeness of systems of recursive integrals.* Communications in Mathematical Analysis, Conf. **03** (2011), 172–176.

[44] V.V. Kravchenko, S. Morelos and S. Tremblay, *Complete systems of recursive integrals and Taylor series for solutions of Sturm-Liouville equations.* Mathematical Methods in the Applied Sciences, **35** (2012), 704–715.

[45] V.V. Kravchenko and R.M. Porter, *Spectral parameter power series for Sturm-Liouville problems.* Mathematical Methods in the Applied Sciences **33** (2010), 459–468.

[46] V.V. Kravchenko and S. Torba, *Transmutations for Darboux transformed operators with applications.* J. Phys. A: Math. Theor. **45** (2012), # 075201 (21 pp.).

[47] V.V. Kravchenko and U. Velasco-García, *Dispersion equation and eigenvalues for the Zakharov-Shabat system using spectral parameter power series.* J. Math. Phys. **52** (2011), 063517.

[48] G.L. Lamb, *Elements of soliton theory.* John Wiley & Sons, New York, 1980.

[49] B.M. Levitan, *Inverse Sturm-Liouville problems.* VSP, Zeist, 1987.

[50] J.L. Lions, *Solutions élémentaires de certains opérateurs différentiels à coefficients variables.* Journ. de Math. **36** (1957), Fasc 1, 57–64.

[51] V.A. Marchenko, *Sturm-Liouville operators and applications.* Basel: Birkhäuser, 1986.

[52] V. Matveev and M. Salle, *Darboux transformations and solitons.* New York, Springer, 1991.

[53] H. Medwin and C.S. Clay, *Fundamentals of Oceanic Acoustics.* Academic Press, Boston, San Diego, New York, 1997.

[54] Y. Nogami and F.M. Toyama, *Supersymmetry aspects of the Dirac equation in one dimension with a Lorentz scalar potential.* Physical Review A. **47** (1993), no. 3, 1708–1714.

[55] L.M. Nieto, A.A. Pecheritsin and B.F. Samsonov, *Intertwining technique for the one-dimensional stationary Dirac equation*, Annals of Physics **305** (2003), 151–189.

[56] O.A. Obrezanova and V.S. Rabinovich, *Acoustic field, generated by moving source in stratified waveguides.* Wave Motion **27** (1998), 155–167.

[57] A.A. Pecheritsin, A.M. Pupasov and B.F. Samsonov, *Singular matrix Darboux transformations in the inverse-scattering method*, J. Phys. A: Math. Theor. **44** (2011), 205305.

[58] C. Rogers and W.K. Schief, *Backlund and Darboux transformations: geometry and modern applications in soliton theory.* Cambridge University Press, 2002.

[59] H. Rosu, *Short survey of Darboux transformations*, Proceedings of "Symmetries in Quantum Mechanics and Quantum Optics", Burgos, Spain, 1999, 301–315.

[60] R.K. Roychoudhory and Y.P. Varshni, *Shifted $1/N$ expansion and scalar potential in the Dirac equation.* J. Phys. A: Math. Gen. **20** (1987), L1083–L1087.

[61] S.M. Sitnik, *Transmutations and applications: a survey.* `arXiv:1012.3741v1 [math.CA]`, originally published in the book: "*Advances in Modern Analysis and Mathematical Modeling*" Editors: Yu.F. Korobeinik, A.G. Kusraev, Vladikavkaz: Vladikavkaz Scientific Center of the Russian Academy of Sciences and Republic of North Ossetia–Alania, 2008, 226–293.

[62] R. Su, Yu Zhong and S. Hu, *Solutions of Dirac equation with one-dimensional scalar-like potential.* Chinese Phys.Lett. **8** (1991), no. 3, 114–117.

[63] K. Trimeche. *Transmutation operators and mean-periodic functions associated with differential operators.* London: Harwood Academic Publishers, 1988.

[64] J. Walter, *Regular eigenvalue problems with eigenvalue parameter in the boundary condition.* Math. Z. **133** (1973), 301–312.

Vladislav V. Kravchenko and Sergii M. Torba
Department of Mathematics, CINVESTAV del IPN, Unidad Queretaro
Libramiento Norponiente No. 2000, Fracc. Real de Juriquilla
Queretaro, Qro. C.P. 76230 MEXICO
e-mail: `vkravchenko@math.cinvestav.edu.mx`
`storba@math.cinvestav.edu.mx`

Operator Theory:
Advances and Applications, Vol. 228, 239–259

Estimate from Below for the Growth of Solution to the Navier-Stokes Equation When the Solution Blows Up

Valeri V. Kucherenko

Dedicated to the 70th birthday of the excellent scientist and great friend Vladimir Rabinovich

Abstract. Estimates for the growth of solution from below to the Navier-Stokes equation are provided.

Mathematics Subject Classification (2010). 35Q30.

Keywords. Navier-Stokes equation, estimates from below, existence of solution.

1. Introduction

The 3D Navier-Stokes equations describe the motion of a viscous incompressible fluid in $\mathbb{R}^3$. The equations need to be solved for an unknown divergence-free velocity vector $u(x,t) = (u_i(x,t))_{i=1}^3$ and pressure $p(x,t)$ [2], [5]. Here we consider the case when the fluid is filling the domain Ω and $\overline{\Omega}$ is a compact set in $\mathbb{R}^3$ with C^∞ boundary $\partial\Omega$. Let $\mathbb{Q}_T := \Omega \times [0,T)$, $\mathbb{Q}_\infty := \Omega \times [0,+\infty)$. The Navier-Stokes equations in dimensionless coordinates have the form

$$\frac{\partial u_i}{\partial t} + \sum_{j=1}^{3} u_j \frac{\partial u_i}{\partial x_j} = \nu \Delta u_i - \frac{\partial p}{\partial x_i} + f_i(x,t), \ (x.t) \in \mathbb{Q}_\infty, \tag{1.1}$$

$$\text{div } u = \sum_{j=1}^{3} \frac{\partial u_j}{\partial x_j} = 0, \ (x,t) \in \mathbb{Q}_\infty, \tag{1.2}$$

This research was partially supported by the CONACYT Project No. 80504 (Mexico) and by the SIP – IPN Project No. 20121028.

with the initial condition

$$u(x,0) = u^0(x), \ x \in \Omega, \tag{1.3}$$

and the boundary condition

$$u(x,t)|_{\partial\Omega} = 0, \ t \geq 0. \tag{1.4}$$

Notation and preliminary results. We denote by $J^{\cdot}(\Omega)$ the set of all sufficiently smooth solenoidal vectors with compact support in the domain Ω, and by $H(\Omega)$ the completion of $J^{\cdot}(\Omega)$ in the norm $W_0^{1,2}(\Omega)$. Let $J^{\circ}(\Omega)$ be the completion of the set $J^{\cdot}(\Omega)$ in $L_2(\Omega)$ and denote by P an orthogonal projection (Leray's projection) of the Hilbert space $L_2(\Omega)$ onto the subspace $J^{\circ}(\Omega)$. The norm of a real vector function f in the space $L_2(\Omega)$ is defined as

$$\|f(\cdot,t)\| := \left\{ \int_{\Omega} |f(x,t)|^2 \, dx \right\}^{1/2},$$

where $|f(x,t)|$ is magnitude of the $\mathbb{R}^3$ vector $f(x,t)$. The scalar product of vectors f, g in $\mathbb{R}^3$ is denoted by fg, and the scalar product in the space $L_2(\Omega)$ is denoted by $(\cdot,\cdot)$. The norm of a function $u(\cdot,t)$ in the Sobolev space $W_0^{1,2}(\Omega)$ is defined as

$$\|u(\cdot,t)\|_{1,2} := \left\{ \int_{\Omega} \sum_{k=1}^{3} |u_{x_k}(x,t)|^2 \, dx \right\}^{1/2}; \tag{1.5}$$

and the norm of a function $u(\cdot,t)$ in the Sobolev space $W^{k,2}(\Omega)$ is denoted by $\|u(\cdot,t)\|_{k,2}$. The scalar product in the Hilbert space $W^{k,2}(\Omega)$ is denoted by $(f,g)_{k,2}$; and the norm in the space $L_p(\Omega)$ is denoted as $\|\cdot\|_p$ (for $p \neq 2$). The subspace of $L_2(\mathbb{Q}_T)$ such that $u(\cdot,t)$ for all fixed t belongs to $J^{\circ}(\Omega)$ is denoted by $L_2^{\circ}(\mathbb{Q}_T)$. In the case when we use the spaces $W^{k,2}(\Omega)$ for several domains, we include the domains in the notation (for example, $\|u(\cdot,t)\|_{k,2}(\Omega)$, $\|f(\cdot,t)\|(\Omega)$, etc.).

The following problem was considered in [5]:

$$\begin{aligned} \nu\Delta u &= -\operatorname{grad} p + f(x), \ f \in L_2(\Omega); \\ \operatorname{div} u &= 0, \ u|_{\partial\Omega} = 0. \end{aligned} \tag{1.6}$$

A generalized solution of problem (1.6) is defined as a function $u \in H(\Omega)$ that satisfies the identity

$$\nu \int_{\Omega} \sum_{k=1}^{3} u_{x_k} \Phi_{x_k} dx = -\int_{\Omega} f\Phi dx$$

for any $\Phi \in H(\Omega)$. A generalized solution exists and it is unique [5]. In [10], [11], [5, p. 67] the following estimate for the generalized solution of problem (1.6) is provided:

$$\|u\|_{2,2} + \|\operatorname{grad} p\| \leq c\,\|f\|. \tag{1.7}$$

Applying the projection P to equation (1.6) we obtain a symmetric in $J^{\circ}(\Omega)$ operator $\widetilde{\Delta}$ given by

$$\widetilde{\Delta} u := P\Delta u \tag{1.8}$$

for all $u \in W^{2,2}(\Omega) \cap H(\Omega)$. The operator $\widetilde{\Delta}$ was considered in [5, pp. 44–47] and it was proved that there exists a self-adjoint extension of the operator $\widetilde{\Delta}$ as an operator in $\dot{J}^{\circ}(\Omega)$. This extension $\widetilde{\Delta}$ maps its domain of definition $D(\widetilde{\Delta})$ onto $J^{\circ}(\Omega)$ and there exists an operator $\widetilde{\Delta}^{-1}$.

It follows from inequality (1.7) that $D(\widetilde{\Delta}) = W^{2,2}(\Omega) \cap H(\Omega)$ and, therefore, the operator $P\Delta u$ is self-adjoint. From equation (1.6) we infer that

$$-\int_{\Omega} \sum_{k=1}^{3} u_{x_k} w_{x_k} dx = \int_{\Omega} w(\widetilde{\Delta} u) dx \tag{1.9}$$

for functions $u \in W^{2,2}(\Omega) \cap H(\Omega)$ and $w \in H(\Omega)$. Inequality (1.7) implies that the norms $\|\Delta u\|$ and $\|\widetilde{\Delta} u\|$ of the functions $u \in W^{2,2}(\Omega) \cap H(\Omega)$ are equivalent [5],

$$\|\widetilde{\Delta} u\| \leq \|\Delta u\| \leq a_1 \|\widetilde{\Delta} u\|. \tag{1.10}$$

In the Galerkin approximation to solutions of the Navier-Stokes problem we use orthogonal in $L_2(\Omega)$ eigenfunctions of the operator $\widetilde{\Delta}$ such that

$$\widetilde{\Delta} a^k(x) = \lambda_k a^k(x), \quad a^k(x) \in H(\Omega) \cap W^{2,2}(\Omega), \tag{1.11}$$

that is,

$$\Delta a^k = \lambda_k a^k - \operatorname{grad} p_k,$$

$$\operatorname{div} a^k(x) = 0, \quad a^k\big|_{\partial\Omega} = 0.$$

In what follows we use several formulas for the projection P. For a vector function $w \in W^{1,2}(\Omega)$, the following relations hold

$$Pw = w - \operatorname{grad} p, \tag{1.12}$$

$$\Delta p = \operatorname{div} w, \quad \left.\frac{\partial p}{\partial n}\right|_{\partial\Omega} = wn, \tag{1.13}$$

where n is the normal unit vector to the boundary $\partial\Omega$.

For this Neumann problem with the boundary $\partial\Omega \in C^\infty$ it was proved that the kernel of the problem is one-dimensional and consists of constants. In the case when $w|_{\partial\Omega} = 0$ and the pressure p is orthogonal to the constants in Ω, in [7] it was shown that

$$\|p\|_{2,2} \leq c \|\operatorname{div} w\|. \tag{1.14}$$

Let now $u \in W^{2,2}(\Omega) \cap W_0^{1,2}(\Omega)$. We will show in Lemma 2.1 below that the function $(u, \nabla)u := \sum_{i=1}^3 u_i \partial_{x_i} u$ belongs to $L_2(\Omega)$. By definition of Leray's projection, we have

$$P(u, \nabla)u = (u, \nabla)u - \operatorname{grad} p, \tag{1.15}$$

$$\Delta p = \sum_{1 \leq i,j \leq 3} \frac{\partial u_j}{\partial x_i} \frac{\partial u_i}{\partial x_j}, \quad \left.\frac{\partial p}{\partial n}\right|_{\partial\Omega} = 0. \tag{1.16}$$

In what follows we consider a generalized solution to the Navier-Stokes problem.

Definition 1.1. A generalized solution of the Navier-Stokes problem (abbreviation GSNS) (1.1), (1.2), (1.3) in the cylinder $\mathbb{Q}_T$ with the initial data

$$u^0(\cdot) \in H(\Omega), \tag{1.17}$$

and the right-hand side $f(\cdot,\cdot) \in L_2(\mathbb{Q}_T)$, with $\sup_{t\in[0,T]} \|f(\cdot,t)\| < \infty$, is a vector function $(x,t) \mapsto u(x,t)$ such that: 1) $u(\cdot,t) \in H(\Omega)$ for $t \in [0,T)$; the function $u(\cdot,t)$ is strongly continuous in $t \in [0,T)$ as a function with values in $H(\Omega)$ and $\|u(\cdot,t) - u^0(\cdot)\|_{1,2} \to 0$ as $t \to 0$; 2) generalized derivatives $u_t, u_{x_k}, u_{x_m x_k}, p_{x_j}$ are in the space $L_2(\mathbb{Q}_t)$ for $t \in [0,T)$ and satisfy equation (1.1). Then $u_t(\cdot,t) \in J^\circ(\Omega)$ for almost all $t \in [0,T)$.

Lemma 1.2. *If the GSNS exists in $\mathbb{Q}_T$, then it is unique.*

Proof. Evidently, a GSNS is a generalized solution in the integral sense [5]. It was proved that the generalized solution in the integral sense on $[0,t)$ is unique if $\sup_{\tau\in[0,t)} \|u(\cdot,\tau)\|_{1,2} < \infty$. As the norm $\|u(\cdot,\tau)\|_{1,2}$ of the GSNS is bounded on any segment $[0,t]$ with $t < T$, the GSNS is unique in the cylinder $\mathbb{Q}_T$. □

The GSNS solutions have important energy integral estimates [5], [8] for $t < \infty$:

$$\|u(\cdot,t)\| \le \|u(\cdot,0)\| + \int_0^t \|f(\cdot,\tau)\|\, d\tau, \tag{1.18}$$

$$\begin{aligned} &\frac{1}{2}\|u(\cdot,t)\|^2 + \nu \int_0^t \sum_{i=1}^3 \|u_{x_i}(\cdot,\tau)\|^2\, d\tau \\ &\le \frac{1}{2}\|u(\cdot,0)\|^2 + \frac{1}{2}\int_0^t \|f(\cdot,\tau)\|^2\, d\tau + \frac{1}{2}\int_0^t \left\{\|u(\cdot,0)\| + \int_0^s \|f(\cdot,\tau)\|\, d\tau\right\}^2 ds. \end{aligned} \tag{1.19}$$

Let $u(x,t) \in C^2(\mathbb{Q}_T)$ be a classical solution to the Navier-Stokes equation, hence $\partial_t u \in H(\Omega)$. Applying projection P to the Navier-Stokes equations (1.1) we obtain

$$\frac{\partial u}{\partial t} - \nu\widetilde{\Delta} u = -P(u,\nabla)u + Pf. \tag{1.20}$$

Now, let $u \in W^{2,2}(\Omega) \cap H(\Omega)$, so taking scalar square in $L_2(\Omega)$ on the left- and right-hand side of equality (1.20) we obtain the second integral inequality for the parabolic equation [6] in the form

$$\begin{aligned} &\nu\frac{d}{dt}\|u(\cdot,t)\|_{1,2}^2 + \|u_t(\cdot,t)\|^2 + \nu^2\big\|\widetilde{\Delta} u(\cdot,t)\big\|^2 \\ &\le 2\big\|(u,\nabla)u(\cdot,t)\big\|^2 + 2\|f(\cdot,t)\|^2, \quad u \in W^{2,2}(\Omega) \cap H(\Omega). \end{aligned} \tag{1.21}$$

Due to equivalence of the norms (1.10) we can substitute the term $\big\|\widetilde{\Delta} u(\cdot,t)\big\|^2$ in the left-hand side of (1.21) by the term $c\|\Delta u(\cdot,t)\|^2$.

Below we show that the energy integral estimates (1.19) and inequality (1.21) can be applied to the Galerkin approximations $u^n(x,t)$ for the GSNS

$$u^n(x,t) := \sum_{k=1}^{n} c_{kn}(t)a^k(x). \tag{1.22}$$

In 2D case the existence and uniqueness of GSNS is proved globally in $\mathbb{Q}_\infty$ by the Galerkin method on the basis of inequality (1.19) and the following multiplicative inequality for a real functions in the space $W_0^{1,2}(\Omega)$, $\Omega \in \mathbb{R}^2$ [5]:

$$\int_\Omega v^4 dx_1 dx_2 \le c \int_\Omega v^2 dx_1 dx_2 \int_\Omega |\text{grad } v|^2 \, dx_1 dx_2. \tag{1.23}$$

In 3D case multiplicative inequality (1.23) does not hold, therefore, up to now the existence and uniqueness of 3D GSNS can be proved only locally in $\mathbb{Q}_{T_l}$; the time T_l depends on the initial data [5], [8].

By virtue of the imbedding theorem, the 3D GSNS belongs to $C(\Omega)$ for almost all t in $[0, T_l)$. We set

$$y_u(t) := \max_{x\in\Omega} |u(x,t)| \,.$$

Below it is proved that on the interval $[0, T_l)$ where the GSNS solution $u(x,t)$ exists, the following inequality holds:

$$\|u(\cdot,t)\|_{1,2}^2 \le \|u(\cdot,0)\|_{1,2}^2 \exp\left\{\frac{4}{\nu}\int_0^t y_u^2(\tau)d\tau\right\} + 2\int_0^t \exp\left\{\frac{4}{\nu}\int_s^t y_u^2(\tau)d\tau\right\} \|f\|^2 \, ds. \tag{1.24}$$

Therefore, the condition

$$\sup_{t\in[0,T]} \|u(\cdot,t)\|_{1,2}^2 < \infty \quad \text{for all} \quad 0 \le T < T_l \tag{1.25}$$

is necessary for the existence of classical solution in the interval $[0, T_l)$.

In the case when for the classical solution to the Navier-Stokes problem

$$\sup_{t\in[0,T_l)} \|u(\cdot,t)\|_{1,2}^2 = \infty, \tag{1.26}$$

in Lemma 3.2 we establish the following estimate from below for the norm $\|u(\cdot,t)\|_{1,2}$ of this solution in the interval $[0, T_l)$ for $T_l < +\infty$:

$$\|u(\cdot,t)\|_{1,2}^2 \ge \frac{a}{\sqrt{T_l - t}} - b, \tag{1.27}$$

where a, b are some positive constants.

We obtain the GSNS as a limit of the Galerkin approximations $u^n(x,t)$. Observe that the Galerkin approximations are defined for all $t \in [0, \infty)$. Let

$$S_g := \left\{T \ge 0 : \sup_n \sup_{t\in[0,T]} \|u^n(\cdot,t)\|_{1,2}^2 < \infty\right\} \tag{1.28}$$

and let

$$T_g := \sup_{T\in S_g} T. \tag{1.29}$$

Below in Theorem 3.3 we prove that $T_g > 0$, and if $T_g = \infty$, then evidently a GSNS exists and it is unique in $[0, \infty)$. Further, if $T_g < \infty$, then

$$\sup_n \sup_{t \in [0,T_g)} \|u^n(\cdot,t)\|_{1,2}^2 = \infty, \tag{1.30}$$

and the GSNS exists only in the interval $[0, T_g)$ where the norm $\|u(\cdot,t)\|_{1,2}$ of the GSNS can be estimated from below as

$$\|u(\cdot,t)\|_{1,2}^2 \geq \frac{a}{\sqrt{T_g - t}} - b. \tag{1.31}$$

Evidently the classical solution such that $u \in C^2(\mathbb{Q}_T)$ is also the GSNS, therefore, this classical solution exists only in the time interval $[0, T_g)$, that is, $T_l = T_g$.

2. Priory estimates for the classical solutions

Estimates for the norm $\|u(\cdot,t)\|_{1,2}$ of classical solution can be derived from inequality (1.21) on the basis of the following lemma.

Lemma 2.1. *If the components of a real vector function v are in $W^{2,2}(\Omega)$, then:*

1) *in* 2D *case*

$$\int_\Omega \left(v_k \frac{\partial v_i}{\partial x_j}\right)^2 dx \tag{2.1}$$
$$\leq c\left\{\int_\Omega v^2 dx\right\}^{1/2}\left\{\int_\Omega \left(\left|\frac{\partial v}{\partial x_j}\right|^2 + v^2\right)dx\right\}\left\{\int_\Omega \left(|\Delta v|^2 + v^2\right)dx\right\}^{1/2};$$

2) *in* 3D *case*

$$\int_\Omega \left(v_k \frac{\partial v_i}{\partial x_j}\right)^2 dx \leq c\left\{\int_\Omega \left(\sum_{l=1}^3 \left|\frac{\partial v}{\partial x_l}\right|^2 + v^2\right)dx\right\}^{3/2}\left\{\int_\Omega \left(|\Delta v|^2 + v^2\right)dx\right\}^{1/2}. \tag{2.2}$$

3) *If vector functions $v, w \in W^{2,2}(\Omega) \cap W_0^{1,2}(\Omega)$, then in* 2D *case*

$$\int_\Omega \left(v_k \frac{\partial v_i}{\partial x_j}\right)^2 dx \leq c\left\{\int_\Omega v^2 dx\right\}^{1/2}\left\{\int_\Omega \sum_{m=1}^2 \left|\frac{\partial v}{\partial x_m}\right|^2 dx\right\}\left\{\int_\Omega |\Delta v|^2 dx\right\}^{1/2}, \tag{2.3}$$

and in 3D *case*

$$\int_\Omega \left(v_k \frac{\partial v_i}{\partial x_j}\right)^2 dx \leq c\left\{\int_\Omega \sum_{j=1}^3 \left|\frac{\partial v}{\partial x_j}\right|^2 dx\right\}^{3/2}\left\{\int_\Omega |\Delta v|^2 dx\right\}^{1/2}; \tag{2.4}$$

$$\int_\Omega w^2 v_{x_i}^2 dx \leq c\left\{\int_\Omega \sum_{k=1}^3 |w_{x_k}|^2 dx\right\}\left\{\int_\Omega |v_{x_i}|^2 dx\right\}^{1/2}\left\{\int_\Omega |\Delta v|^2 dx\right\}^{1/2}. \tag{2.5}$$

Proof. 1) Let us consider the 2D case. Performing the standard extension of functions $w, v \in W^{2,2}(\Omega)$ onto some cube Π, as functions in $W_0^{2,2}(\Pi)$, such that the norms of extended functions $\widetilde{w}, \widetilde{v}$, in the space $W_0^{2,2}(\Pi)$ are equivalent to the norms of the functions w, v in $W^{2,2}(\Omega)$ [1]. So the linear operator of extension $A : \widetilde{w} = Aw$, satisfies the following estimates [1]

$$\|Aw\|_{k,2}(\Pi) \leq c\,\|w\|_{k,2}(\Omega), \quad k = 0, 1, 2. \tag{2.6}$$

Furthermore, applying to the functions from the spaces $W_0^{1,2}(\Pi), W_0^{2,2}(\Pi)$ the multiplicative inequality [5], [3], [9] we obtain that

$$\begin{gathered}\left\{\int_\Pi \widetilde{w}^4 dx\right\}^{1/2} \leq c\left\{\int_\Pi \widetilde{w}^2 dx\right\}^{1/2}\left\{\int_\Pi \sum_{i=1}^2 |\widetilde{w}_{x_i}|^2\, dx\right\}^{1/2}, \\ \int_\Pi \widetilde{v}_{x_i}^4\, dx \leq c\left\{\int_\Pi \widetilde{v}_{x_i}^2\, dx\right\}^{1/2}\left\{\int_\Pi \sum_{j=1}^2 |\widetilde{v}_{x_i x_j}|^2\, dx\right\}^{1/2}.\end{gathered} \tag{2.7}$$

Combining the Cauchy inequality, inequalities (2.7) and estimates (2.6), we get

$$\begin{aligned}\int_\Omega w^2 v_{x_i}^2\, dx &\leq \left\{\int_\Pi \widetilde{w}^4 dx\right\}^{1/2}\left\{\int_\Pi \widetilde{v}_{x_i}^4\, dx\right\}^{1/2} \leq c\left\{\int_\Omega w^2 dx\right\}^{1/2} \\ &\times \left\{\int_\Omega v_{x_i}^2\, dx\right\}^{1/2}\left\{\int_\Omega\left(\sum_{i=1}^2 |w_{x_i}|^2 + w^2\right) dx\right\}^{1/2}\left\{\int_\Omega (|\Delta v|^2 + v^2)\, dx\right\}^{1/2}.\end{aligned} \tag{2.8}$$

Inequality (2.1) is a direct consequence of inequality (2.8).

2) Let us now consider the 3D case. The Hölder inequality provides that

$$\begin{aligned}\int_\Omega w^2 v_{x_i}^2\, dx &\leq \left\{\int_\Omega |w|^6\, dx\right\}^{1/3}\left\{\int_\Omega |v_{x_i}|^3\, dx\right\}^{2/3} \\ &\leq \left\{\int_\Omega |w|^6\, dx\right\}^{1/3}\left\{\int_\Omega |v_{x_i}|^2\, dx\right\}^{1/3}\left\{\int_\Omega |v_{x_i}|^4\, dx\right\}^{1/3}.\end{aligned} \tag{2.9}$$

Then, applying the imbedding theorem and the multiplicative inequalities to the extended functions $\widetilde{w}, \widetilde{v}$ we obtain the estimates:

$$\begin{gathered}\left\{\int_\Pi |\widetilde{w}|^6\, dx\right\}^{1/3} \leq c\left\{\int_\Pi \sum_{i=1}^3 |\widetilde{w}_{x_i}|^2\, dx\right\}, \\ \left\{\int_\Pi |\widetilde{v}_{x_i}|^4\, dx\right\}^{1/3} \leq c\left\{\int_\Pi |\widetilde{v}_{x_i}|^2\, dx\right\}^{1/6}\left\{\int_\Pi \sum_{j=1}^3 |\widetilde{v}_{x_j x_i}|^2\, dx\right\}^{1/2}.\end{gathered} \tag{2.10}$$

Substituting this estimates in inequality (2.9) we get

$$\int_\Pi \widetilde{w}^2 \widetilde{v}_{x_i}^2\, dx \le c \left\{ \int_\Pi \sum_{k=1}^3 |\widetilde{w}_{x_k}|^2\, dx \right\} \left\{ \int_\Pi |\widetilde{v}_{x_i}|^2\, dx \right\}^{1/2} \left\{ \int_\Pi \sum_{j=1}^3 \left|\widetilde{v}_{x_j x_i}\right|^2 dx \right\}^{1/2}. \tag{2.11}$$

Inequality (2.2) is a direct consequence of inequality (2.11) and estimates of the norms (2.6) for the extended functions $\widetilde{w}$, $\widetilde{v}$.

3) Functions $u \in W_0^{1,2}(\Omega) \cap W^{2,2}(\Omega)$ satisfy the inequalities [5]:

$$\|u\|_{2,2} \le c\,\|\Delta u\|\,, \quad \|u\| \le c\,\|\Delta u\|_{1,2}\,. \tag{2.12}$$

Estimates (2.3), (2.4) are direct consequences of estimates (2.1), (2.2) and estimates (2.12). Estimate (2.5) follows from estimates (2.11) and (2.12). □

Now, in what follows some estimates for the norm $\|u(\cdot,t)\|_{1,2}$ of solution to the Navier-Stokes problem are deduced. Below in the inequalities by c, c_i are denoted constants that depend on the domain Ω.

Lemma 2.2. *The $C^2(\mathbb{Q}_T)$ solution to the Navier-Stokes problem* (1.1)–(1.4) *satisfies the following differential inequalities.*

1° *In the* 2D *case*

$$\begin{aligned} &\frac{d}{dt}\|u(\cdot,t)\|_{1,2}^2 + \nu^{-1}\|u_t(\cdot,t)\|^2 + \nu c^2 \|\Delta u(\cdot,t)\|^2 \\ &\quad \le \frac{c_1^2}{\nu^3}\|u(\cdot,t)\|^2 \|u(\cdot,t)\|_{1,2}^4 + \frac{4}{\nu}\|f(\cdot,t)\|^2. \end{aligned} \tag{2.13}$$

2° *In the* 3D *case*

$$\begin{aligned} &\frac{d}{dt}\|u(\cdot,t)\|_{1,2}^2 + \nu^{-1}\|u_t(\cdot,t)\|^2 + \nu c_2^2 \|\Delta u(\cdot,t)\|^2 \\ &\quad \le \frac{c_3}{\nu^3}\|u(\cdot,t)\|_{1,2}^6 + \frac{c_4}{\nu}\|f(\cdot,t)\|^2 \le \frac{c_6}{\nu^3}\left[\|u(\cdot,t)\|_{1,2}^2 + \nu\|f(\cdot,t)\|^{2/3}\right]^3. \end{aligned} \tag{2.14}$$

3° *In both the* 2D *and* 3D *cases*

$$\frac{d}{dt}\|u(\cdot,t)\|_{1,2}^2 \le \frac{36}{\nu} y_u^2(t)\,\|u(\cdot,t)\|_{1,2}^2 + \frac{2}{\nu}\|f(\cdot,t)\|^2\,. \tag{2.15}$$

4° *In the* 3D *case there exists such a constant c_0 that the norm $\|u(\cdot,t)\|_{1,2}$ satisfies the inequality*

$$\|u(\cdot,t)\|_{1,2}^2 < \frac{\nu^2}{c_0} \quad \textit{on} \quad [0,T], \tag{2.16}$$

when

$$\frac{4}{\nu}\int_0^T \|f(\cdot,t)\|^2\, dt + \|u(\cdot,0)\|_{1,2}^2 < \frac{\nu^2}{c_0}. \tag{2.17}$$

Proof. 1° Let us consider the 2D case. For $u \in W_0^{1,2}(\Omega) \cap W^{2,2}(\Omega)$ we can estimate the norm $\|(u,\nabla)u(x,t)\|^2$ through inequality (2.3) as:

$$\int_\Omega |(u,\nabla)u|^2\,dx \le c_2\,\|u\|\,\|u\|_{1,2}^2\,\|\Delta u\|\,. \tag{2.18}$$

Then applying the inequality $|ab| \le \frac{a^2}{2\nu^2} + \frac{\nu^2 b^2}{2}$ and equivalence (1.10) to inequality (2.18), we obtain

$$\int_\Omega |(u,\nabla)u|^2\,dx \le \frac{c_3}{\nu^2}\,\|u\|^2\,\|u\|_{1,2}^4 + \frac{\nu^2}{2}\left\|\widetilde{\Delta} u\right\|^2. \tag{2.19}$$

Furthermore, substituting estimate (2.19) in the right-hand side of inequality (1.21), subtracting the term $\frac{\nu^2}{2}\left\|\widetilde{\Delta} u(\cdot,t)\right\|^2$ from the right- and left-hand sides of inequality (1.21) and using equivalence of the norms (1.10), we deduce differential inequality (2.13) with some constant c, c_1.

2° Now, let us consider the 3D case. For $u \in W_0^{1,2}(\Omega) \cap W^{2,2}(\Omega)$ we can estimate the norm $\|(u,\nabla)u(x,t)\|^2$ by using inequality (2.4):

$$\int_\Omega |(u,\nabla)u|^2\,dx \le c\,\|u\|_{1,2}^{3/2}\,\|u\|_{2,2}\,.$$

Substituting this estimate in the right-hand side of inequality (1.21), applying the inequality $|ab| \le \frac{a^2}{2\nu^2} + \frac{\nu^2 b^2}{2}$, subtracting the term $\frac{\nu^2}{2}\left\|\widetilde{\Delta} u(x,t)\right\|^2$ from the right- and left-hand sides of inequality (1.21) and using equivalence of the norms (1.10), we deduce differential inequality (2.14) with some constants c_i.

3° Evidently, the following inequality holds:

$$\|(u,\nabla)u(\cdot,t)\|^2 \le 9y_u^2(t)\,\|u(\cdot,t)\|_{1,2}^2 \tag{2.20}$$

Substituting estimate (2.20) in the right-hand side of inequalities (1.21) we obtain inequality (2.15).

4° It follows from inequality (2.4) and estimate (2.12) for $u \in W_0^{1,2}(\Omega) \cap W^{2,2}(\Omega)$ that

$$\int_\Omega |(u,\nabla)u|^2\,dx \le c^2 c_0\,\|u\|_{1,2}^2\,\|u\|_{2,2}^2\,.$$

Now, we substitute this estimate in inequality (1.21), apply equivalence (1.10) and subtract the term $c^2 c_0\,\|u\|_{1,2}^2\,\|\Delta u\|^2$. As a result, we obtain

$$\nu\frac{d}{dt}\|u(\cdot,t)\|_{1,2}^2 + \|u_t(\cdot,t)\|^2 + c^2\left(\nu^2 - c_0\,\|u(\cdot,t)\|_{1,2}^2\right)\|\Delta u(\cdot,t)\|^2 \le 2\,\|f(\cdot,t)\|^2\,. \tag{2.21}$$

Evidently, by condition (2.17) inequality (2.16) is satisfied in some interval $[0,t)$. Suppose t^* is the minimum point in $[0,T]$, where $\|u(x,t^*)\,|_{1,2}^2 = \frac{\nu^2}{c_0}$, and $t^* < T$; hence $\|u(\cdot,t^*)\|_{1,2}^2 < \frac{\nu^2}{c_0}$ in $[0,t^*)$. Therefore, from inequality (2.21) it follows that

$$\nu\frac{d}{dt}\|u(\cdot,t)\|_{1,2}^2 \le 2\,\|f(\cdot,t)\|^2,$$

and by inequality (2.17) we conclude that

$$\|u(\cdot,t)\|_{1,2}^2 \le \frac{4}{\nu}\int_0^{t^*} \|f(\cdot,t)\|^2\, dt + \|u(\cdot,0)\|_{1,2}^2 < \frac{\nu^2}{c_0}$$

in $[0,t^*]$. This contradiction proves inequality (2.16). □

Conclusion 2.3.

1° *Estimate* (1.24) *is a direct consequence of differential inequality* (2.15).

2° *Assume that the integral* $\int_0^t \|f(\cdot,\tau)\|^2\, d\tau$ *is bounded for all* $t \ge 0$. *Then in the* 2D *case we have the following.*

a) *The function* $\|u(\cdot,t)\|^2$ *is bounded for all* $t \ge 0$ *by inequality* (1.18) *and the integral* $\int_0^t \|u(\cdot,\tau)\|_{1,2}^2\, d\tau$ *is bounded for all* $t \ge 0$ *by inequality* (1.19). *Hence, the integral* $\int_0^t w(\tau)d\tau$ *of the function*

$$w(\tau) := \frac{c}{\nu^3}\left\{\left(\|u(\cdot,\tau)\|^2\right)\|u(\cdot,\tau)\|_{1,2}^2\right\},$$

is also bounded for all $t \ge 0$ *by inequalities* (1.18), (1.19).

b) *Due to inequality* (2.13) *and Gronwall's lemma, we obtain*

$$\begin{aligned}\|u(\cdot,t)\|_{1,2}^2 &\le \exp\left\{\int_0^t w(\tau)d\tau\right\}\|u^0(\cdot)\|_{1,2}^2 \\ &+ \frac{4}{\nu}\int_0^t \exp\left\{\int_s^t w(\tau)d\tau\right\}\|f(\cdot,s)\|^2\, ds.\end{aligned} \tag{2.22}$$

As the integral $\int_0^t w(\tau)d\tau$ *is bounded for all* $t \ge 0$, *the function* $\|u(\cdot,t)\|_{1,2}^2$ *is bounded for all* $t \ge 0$ *by a constant depending on* $\|u^0\|$ *and* $\int_0^t \|f(\cdot,\tau)\|^2\, d\tau$.

3. The Galerkin approximations and existence of the GSNS

1° *Let* $a^k(x) \in H(\Omega)$ be *vector functions chosen to be the eigenfunctions of problem* (1.11). The system $\{a^k(x)\}$ is orthonormal in $L_2(\Omega)$ and dense in $J^\circ(\Omega)$.

Let $u^0(x) \in H(\Omega)$. Evidently,

$$u^0(x) = \sum_{k=1}^{\infty} c_k a^k(x). \tag{3.1}$$

We search the Galerkin approximate solutions $u^n(x,t)$ of problem (1.1) in the form

$$u^n(x,t) = \sum_{k=1}^{n} c_{kn}(t) a^k(x). \tag{3.2}$$

The functions $c_{kn}(t)$ are determined by the initial data

$$c_{kn}(0) = c_k, \quad k = 1,\dots,n,$$

and the Galerkin conditions

$$(u_t^n - f, a^j) + \sum_{i=1}^{3} \left\{ \nu(u_{x_i}^n, a_{x_i}^j) - (u_i^n u^n, a_{x_i}^j) \right\} = 0. \tag{3.3}$$

Conditions (3.3) were obtained formally from system (1.1) substituting u by u^n, multiplying by the function a^j and integrating over Ω. The Galerkin conditions (3.3) represent a system of ordinary differential equations of the form

$$\frac{dc_{jn}}{dt} - \nu \sum_{k=1}^{n} a_{jk} c_{kn} + \sum_{p,k=1}^{n} a_{jpk} c_{pn} c_{kn} = f_j, \tag{3.4}$$

where a_{jk}, a_{jpk} are constants and $f_j = (f, a^j)$.

If we multiply relations (3.3) by $c_{jn}(t)$ and sum them up with respect to the index j from 1 to n, we obtain

$$\frac{1}{2} \frac{d}{dt} \|u^n(\cdot, t)\|^2 + \nu \sum_{i=1}^{3} \left\| u_{x_i}^n(\cdot, t) \right\|^2 = (f, u^n). \tag{3.5}$$

To derive equation (3.5), we use the following identity

$$\left(\sum_{i=1}^{3} u_i u_{x_i}, u \right) = 0, \tag{3.6}$$

which holds for functions in $H(\Omega)$. Then, from equation (3.5) and the Cauchy inequality it follows that

$$\frac{d}{dt} \|u^n(\cdot, t)\| \leq \|f(\cdot, t)\|,$$

whence

$$\|u^n(\cdot, t)\| \leq \|u^n(\cdot, 0)\| + \int_0^t \|f(\cdot, \tau)\| \, d\tau. \tag{3.7}$$

Therefore, substituting estimate (3.7) into the right-hand side of inequality (3.5) and integrating the resulting inequality by t we infer that the basic estimates (1.19) are true for the Galerkin approximations $u^n(x,t)$ as well.

As the functions $a^k(x)$ are orthonormal in $L_2(\Omega)$, it follows from inequality (3.7) that

$$\sum_{k=1}^{n} c_{kn}^2(t) \leq \left(\|u^n(\cdot, 0)\| + \int_0^t \|f(\cdot, \tau)\| \, d\tau \right)^2.$$

Hence, if $\int_0^t \|f(\cdot, \tau)\| \, d\tau < \infty$ *for all* $t \geq 0$, *then system* (3.4) *for the coefficients* c_{jn} *has a solution in the interval* $[0, +\infty)$.

2° *Now we verify that inequality* (1.21) *holds for the Galerkin approximations as well.* As the eigenfunctions $a^k(x)$ of problem (1.11) belong to $W^{2,2}(\Omega) \cap H(\Omega)$ [5], we can rewrite equations (3.3) in an equivalent form

$$\left\{ (u_t^n, a^j) - \nu(\tilde{\Delta} u^n, a^j) \right\} = \left(-\sum_{i=1}^{3} u_i^n u_{x_i}^n + f, a^j \right), \; j = 1, \ldots, n. \tag{3.8}$$

Multiplying these equations by $\frac{dc_{jn}}{dt} - \nu\lambda_j c_{jn}$ and summing them up with respect to the index j from 1 to n, after some simple standard transformation, we obtain

$$\nu\frac{d}{dt}\sum_{i=1}^{3}\left\|u^n_{x_i}(\cdot,t)\right\|^2 + \|u^n_t(\cdot,t)\|^2 + \nu^2\left\|\widetilde{\Delta}u^n(\cdot,t)\right\|^2 = \left(-\sum_{i=1}^{3}u^n_i u^n_{x_i} + f, u^n_t - \nu\widetilde{\Delta}u^n\right). \tag{3.9}$$

Leray's projection P in the bases $\{a^k(x) : k = 1, \dots\}$ has the form $Pf = \sum_{k=1}^{\infty}(f, a^k)a^k$. Let $P_n f$ be

$$P_n f := \sum_{k=1}^{n}(f, a^k)a^k.$$

Hence, as $P_n u^n_t = u^n_t$, $P_n\widetilde{\Delta}u^n = \widetilde{\Delta}u^n$, we can rewrite equations (3.8) in the form

$$\left(u^n_t - \nu\widetilde{\Delta}u^n, P_n g\right) = \left(-\sum_{i=1}^{3}u^n_i u^n_{x_i} + f, P_n g\right) \tag{3.10}$$

for any function $g(x,t) \in L_2(\mathbb{Q}_\infty)$. We set $g = -\sum_{i=1}^{3}u^n_i u^n_{x_i} + f$, and, therefore, the right-hand side of equation (3.9) can be presented in the form

$$\left(-\sum_{i=1}^{3}u^n_i u^n_{x_i} + f, u^n_t - \nu\widetilde{\Delta}u^n\right) = \left(P_n\Big[-\sum_{i=1}^{3}u^n_i u^n_{x_i} + f\Big], u^n_t - \nu\widetilde{\Delta}u^n\right) = \left\|P_n\Big[-\sum_{i=1}^{3}u^n_i u^n_{x_i} + f\Big]\right\|^2. \tag{3.11}$$

Now, we substitute equality (3.11) into the right-hand side of inequality (3.9) and obtain the inequality

$$\nu\frac{d}{dt}\sum_{i=1}^{3}\left\|u^n_{x_i}(\cdot,t)\right\|^2 + \|u^n_t(\cdot,t)\|^2 + \nu^2\left\|\widetilde{\Delta}u^n(\cdot,\tau)\right\|^2 \le 2\left\|\sum_{i=1}^{3}u^n_i u^n_{x_i}\right\|^2 + 2\|f\|^2, \tag{3.12}$$

which means that inequality (1.21) also holds for the Galerkin approximations $u^n(x,t)$.

Lemma 3.1. *The Galerkin approximations $u^n(x,t)$ satisfy inequalities* (2.13)–(2.15) *of Lemma* 2.2.

Proof. The proof of Lemma 2.2 is based on the inequalities of Lemma 2.1 and inequality (1.21). As the Galerkin approximations also satisfy all these inequalities, then inequalities of Lemma 2.2 also hold for the Galerkin approximations. □

The global existence of GSNS in the 2D case is well known, in particular, this follows from inequality (2.22).

Now, we consider the 3D case and inequality (2.14).

Lemma 3.2.

1) *Let $b \geq 0$ and $z(t)$ be a function in $C^1([\tau, T))$, $z(t) > 0$. Suppose that the function $z(t)$ on the interval $[\tau, T)$ satisfies the inequality*

$$\frac{d}{dt} z(t) \leq \frac{\gamma}{\alpha}\big(z(t) + b\big)^{1+\alpha}, \quad \alpha > 0. \tag{3.13}$$

We put

$$T^*(\tau) := \gamma^{-1}(z(\tau) + b)^{-\alpha}, \quad T_m := \min(T, T^*(\tau) + \tau). \tag{3.14}$$

Then the following estimate holds on the interval $[\tau, T_m)$:

$$z(t) \leq \frac{\gamma^{-1/\alpha}}{(T^*(\tau) - (t - \tau))^{1/\alpha}} - b. \tag{3.15}$$

2) *Consider a function $z(t)$ in $C^1([0, T_l))$ such that $z(t) > 0$, inequality (3.13) is satisfied and $\overline{\lim}_{t \to T_l}\, z(t) = +\infty$. Then the following inequality holds on the interval $[0, T_l)$:*

$$z(t) \geq \frac{\gamma^{-1/\alpha}}{(T_l - t)^{1/\alpha}} - b. \tag{3.16}$$

3) *If $\sup_{t \in [0,T_l]} \|f(\cdot, t)\| < \infty$ and $\sup_{t \in [0,T_l)} \|u(\cdot, t)\|_{1,2}^2 = +\infty$, then for the classical solution to the Navier-Stokes problem on the interval $[0, T_l)$, $T_l < +\infty$, estimate (1.27) holds for some positive constants a, b.*

Proof. 1) Dividing inequality (3.13) by $\big(z(t) + b\big)^{1+\alpha}$ and integrating the result in the interval $[\tau, t]$, we obtain

$$-(z(t) + b)^{-\alpha} + (z(\tau) + b)^{-\alpha} \leq \gamma(t - \tau). \tag{3.17}$$

Therefore,

$$(z(\tau) + b)^{-\alpha} - \gamma(t - \tau) \leq (z(t) + b)^{-\alpha},$$

and

$$(z(t) + b)^{\alpha} \leq 1/\{(z(\tau) + b)^{-\alpha} - \gamma(t - \tau)\}. \tag{3.18}$$

Inequality (3.15) is a direct consequence of inequality (3.18) and definition (3.14).

2) Suppose now that in inequality (3.17) we take $t \to T_l$. Then the condition $\overline{\lim}_{t \to T_l}\, z(t) = +\infty$ implies that

$$(z(\tau) + b)^{-\alpha} \leq \gamma(T_l - \tau) \tag{3.19}$$

for all $\tau \in [0, T_l)$. From inequality (3.19) it follows that for all $\tau \in [0, T_l)$ we have

$$(z(\tau) + b)^{\alpha} \geq \frac{1}{\gamma(T_l - \tau)}$$

and, therefore, inequality (3.16) is satisfied.

3) Consider inequality (2.14) for a classical solution. We set

$$z(t) := \|u(\cdot,t)\|_{1,2}^2\,,$$

$$b := c_6\nu^{2/3} \sup_{t\in[0,T_l]} \|f(\cdot,t)\|^{2/3}\,, \tag{3.20}$$

$$\gamma := \frac{2c_6}{\nu^3}. \tag{3.21}$$

With such notation inequality (2.14) has the form (3.13) for $\alpha = 2$. Hence, by part 2) of the lemma, we obtain estimate (1.27) with the constant $a = \left(\frac{2c_6}{\nu^3}\right)^{-1/2}$. □

If the initial data and the integral $\int_0^\infty \|f(\cdot,t)\|^2\,dt$ are small enough, i.e.,

$$\frac{4}{\nu}\int_0^\infty \|f(\cdot,t)\|^2\,dt + \|u(\cdot,0)\|_{1,2}^2 < \frac{\nu^2}{c_0},$$

then the global existence of the GSNS to the Navier-Stokes problem in the cylinder $\mathbb{Q}_\infty = \Omega\times[0,+\infty)$ easily follows from estimates (2.16), (2.17).

Now, denote by $W(T)$ the Banach space obtained as the completion of the set of functions

$$\left\{g(x,t) : g = \sum_{k=1}^{N} f_k(t)\varphi_k(x);\, f_k(t)\in C^1([0,T]), \varphi_k \in W^{1,2}(\Omega)\cap H(\Omega), N<\infty\right\}$$

in the norm $\|\cdot\|_{W(T)}$, where, with $c_2^2>0$,

$$\|u\|_{W(T)}^2 := \max_{t\in[0,T]} \|u(\cdot,t)\|_{1,2}^2 + \int_0^T \left\{\nu^{-1}\|\partial_t u(\cdot,t)\|^2 + \nu c_2^2 \|\Delta u(\cdot,t)\|^2\right\} dt. \tag{3.22}$$

Evidently, by definition the GSNS belongs to the space $W(T)$. Below we prove the convergence of the Galerkin approximations in the space $W(T)$ to the GSNS. In the following theorem we use the notations S_g and T_g from (1.28) and (1.29), respectively.

Theorem 3.3. (3D case) *Suppose that the initial data and the right-hand side $f(x,t)$ of the Navier-Stokes problem* (1.1)–(1.4) *in a domain $\Omega\in\mathbb{R}^3$ with compact closure and a C^∞ boundary $\partial\Omega$ satisfy the conditions*

$$u^0(\cdot)\in H(\Omega);\ \sup_{t\in[0,\infty]} \|f(\cdot,t)\| < \infty. \tag{3.23}$$

Let b be the constant defined by equality (3.20). *Then the following assertions are fulfilled.*

1) *The GSNS exists and it is unique on the interval $[0,T_g)$ for $T_g \ge T^*$, where*

$$T^* = \gamma^{-1}\left(\|u^0\|_{1,2}^2 + b\right)^{-2}. \tag{3.24}$$

The norm $\|u(\cdot,t)\|_{1,2}$ *of the GSNS on the interval* $[0,T^*)$ *satisfies the inequality*

$$\|u(\cdot,t)\|_{1,2}^2 \leq \frac{\gamma^{-1/2}}{\sqrt{T^*-t}} - b. \tag{3.25}$$

2) *If* $T_g < \infty$, *then*

$$\sup_n \sup_{t\in[0,T_g)} \|u^n(\cdot,t)\|_{1,2}^2 = \infty, \tag{3.26}$$

and the GSNS on the interval $[0,T_g)$ *satisfies the inequality*

$$\|u(\cdot,t)\|_{1,2}^2 \geq \frac{\gamma^{-1/2}}{\sqrt{T_g-t}} - b. \tag{3.27}$$

If $T_g = \infty$, *then the GSNS exists on* $[0,+\infty)$.

3) *The Galerkin approximations* u^n *converge to the GSNS in the norm* $\|\cdot\|_{W(T)}$ *for all intervals* $[0,T]$ *and all* $T < T_g$.

Proof. 1a) The existence of the GSNS is proved by the Galerkin method. As the Galerkin approximations $u^n(x,t)$ satisfy inequality (2.14), by Lemma 3.2 we obtain the estimate

$$\|u^n(\cdot,t)\|_{1,2}^2 \leq \frac{\gamma^{-1/2}}{\sqrt{T_n^*-t}} - b. \tag{3.28}$$

where $T_n^* = \gamma^{-1}\big(\|u_n^0\|_{1,2}^2 + b\big)^{-2}$ and $u_n^0(x) := \sum_{k=1}^n a^k(x)(u^0,a^k)$. For a function $u^0 \in H(\Omega)$, we get $\|u_n^0 - u^0\|_{1,2} \to 0$ as $n\to\infty$ [5], and, therefore, $T_n^* \to T^* = \gamma^{-1}\big(\|u^0\|_{1,2}^2 + b\big)^{-2}$.

Substituting the estimate (3.28) into the right-hand side of the first inequality in (2.14) for the function $u^n(x,t)$, we get

$$\frac{d}{dt}\|u(\cdot,t)\|_{1,2}^2 + \nu^{-1}\|u_t(\cdot,t)\|^2 + \nu c_2^2\|\Delta u(\cdot,t)\|^2 \leq \frac{c}{(T_n^*-t)^{3/2}} + \frac{c_4}{\nu}\|f(\cdot,t)\|^2.$$

Integrating the latter inequality by t, we infer that

$$\begin{aligned}\|u(\cdot,t)\|_{1,2}^2 &+ \int_0^t \big(\nu^{-1}\|u_t(\cdot,\tau)\|^2 + \nu c_2^2\|\Delta u(\cdot,\tau)\|^2\big)d\tau \\ &\leq \|u^0\|_{1,2}^2 + 2c\left\{\frac{1}{(T_n^*-t)^{1/2}} - \frac{1}{(T_n^*)^{1/2}}\right\} + \frac{c_4}{\nu}\int_0^t \|f(\cdot,\tau)\|^2 d\tau.\end{aligned} \tag{3.29}$$

Since the right-hand side of inequality (3.29) monotonically increases in t, we conclude that

$$\max_{\tau\in[0,t]}\|u(\cdot,\tau)\|_{1,2}^2 \leq \|u^0\|_{1,2}^2 + 2c\left\{\frac{1}{(T_n^*-t)^{1/2}} - \frac{1}{(T_n^*)^{1/2}}\right\} + \frac{c_4}{\nu}\int_0^t \|f(\cdot,\tau)\|^2 d\tau. \tag{3.30}$$

Summing inequalities (3.29) and (3.30) and taking $t = T$, we obtain the inequality

$$\|u^n\|^2_{W(T)} \leq 2\left\|u^0\right\|^2_{1,2} + a_1\left\{\frac{1}{(T^*_n - T)^{1/2}} - \frac{1}{(T^*_n)^{1/2}}\right\} + a_2 \tag{3.31}$$

with some positive constants a_i, $i = 1, 2$. Hence $T_g \geq T^* > 0$.

Furthermore, as inequality (2.14) holds for the Galerkin approximations $u^n(x,t)$ on the intervals $[0,T]$, $T < T_g$, integrating again the first inequality in (2.14) for $u^n(x,t)$ by $t \in [0,T]$ and taking into account (1.28), we conclude that the Galerkin approximations also satisfy the inequality

$$\|u^n\|^2_{W(T)} \leq 2\left\|u^0\right\|^2_{1,2} + \frac{a_3 T}{\nu^3}\left(\sup_n \sup_{t\in[0,T]} \|u^n(\cdot,t)\|^2_{1,2}\right)^3 + \frac{a_4}{\nu}\int_0^T \|f(\cdot,t)\|^2\, dt. \tag{3.32}$$

Therefore we can choose a subsequence $u^{n_q}(x,t)$ such that the sequences $u^{n_q}(x,t)$, $u^{n_q}_t(x,t)$, $u^{n_q}_{x_m}(x,t)$, $u^{n_q}_{x_i x_j}(x,t)$ are weakly converging in $L_2(\mathbb{Q}_T)$ for all $0 < T < T_g$.

Let us prove that the sequences $u^{n_q}_{x_m}(x,t)$, $u^{n_q}(x,t)$ *strongly converge in* $L_2(\mathbb{Q}_T)$ (for all $0 < T < T_g$) applying the Friedrich inequality. This inequality asserts that for any function u in $W^{1,2}_0(\Omega)$ and any $\varepsilon > 0$ there exist N_ε functions ω_j, $j = 1, \dots, N_\varepsilon$, such that

$$\int_\Omega u^2(x)dx \leq \sum_{j=1}^{N_\varepsilon}\left(\int_\Omega u\omega_j dx\right)^2 + \varepsilon\int_\Omega (\text{grad } u)^2 dx. \tag{3.33}$$

We prove that estimate (3.33) is also valid for all the function u from $W^{1,2}(\Omega)$. Let S be a rectangle that contains some vicinity of $\overline{\Omega}$, and let A be a bounded operator of extension [1],

$$A : W^{1,2}(\Omega) \to W^{1,2}_0(S),\ (Au)|_\Omega = u$$

satisfying the inequalities

$$\|Au\|_{k,2}(S) \leq c\,\|u\|_{k,2}(\Omega),\ k = 0, 1. \tag{3.34}$$

Actually, the unique analytical expression for the extension operator A generates two bounded operators $A_0 : L_2(\Omega) \to L_2(\Omega)$ and $A_1 : W^{1,2}(\Omega) \to W^{1,2}_0(S)$ with the norms $\|A_0\|$, $\|A_1\|$.

Consider the Friedrich inequality in rectangle S, and let ω_j, $j = 1, 2, \dots, N_\varepsilon$ be functions in Friedrich's inequality for $W^{1,2}_0(S)$. We can take for ω_j, $j = 1, \dots, N_\varepsilon$, the orthogonal trigonometric system in the rectangle S. The Friedrich inequality in the rectangle S implies that

$$\int_S (A_0 u)^2 dx \leq \sum_{j=1}^{N_\varepsilon}\left(\int_S (A_0 u)\omega_j dx\right)^2 + \varepsilon\int_S (\text{grad}\,(A_1 u))^2 dx. \tag{3.35}$$

By virtue of equality $\int_S (A_0 u)\omega_j dx = \int_\Omega u(A_0^*\omega_j)dx$ and estimate (3.34) for $k=1$, from (3.35) we get

$$\begin{aligned}\int_\Omega u^2(x)dx &\le \int_S (A_0 u)^2 dx\\ &\le \sum_{j=1}^{N_\varepsilon}\left(\int_\Omega u(A_0^*\omega_j)dx\right)^2 + \varepsilon\,\|A_1\|^2 \int_\Omega \left\{(\text{grad } u)^2 + u^2\right\}dx,\end{aligned} \tag{3.36}$$

whence

$$\int_\Omega u^2(x)dx \le \frac{1}{1-\varepsilon\,\|A_1\|^2}\sum_{j=1}^{N_\varepsilon}\left(\int_\Omega u(A_0^*\omega_j)dx\right)^2 + \frac{\varepsilon\,\|A_1\|^2}{1-\varepsilon\,\|A_1\|^2}\int_\Omega (\text{grad } u)^2 dx. \tag{3.37}$$

We use the above inequality for $u = \partial_{x_k}(u^{n_i} - u^{n_j})$ and integrate it with respect to t from 0 to T obtaining:

$$\begin{aligned}&\int_0^T \int_\Omega \left|\partial_{x_k}(u_l^{n_i} - u_l^{n_j})\right|^2 dxdt\\ &\le \frac{1}{1-\varepsilon\,\|A_1\|^2}\sum_{j=1}^{N_\varepsilon}\int_0^T \left[\int_\Omega \left\{\partial_{x_k}(u_l^{n_i} - u_l^{n_j})\right\}(A_0^*\omega_j)dx\right]^2 dt\\ &+ \frac{\varepsilon\,\|A_1\|^2}{1-\varepsilon\,\|A_1\|^2}\int_0^T \int_\Omega \sum_{m=1}^{3}\left|\partial^2_{x_k x_m}(u_l^{n_i} - u_l^{n_j})\right|^2 dxdt.\end{aligned} \tag{3.38}$$

Observe that the Galerkin approximations u^{n_i} satisfy inequality (3.32) and $\|u^{n_i}\|_{2,2} \le c\,\|\Delta u^{n_i}\|$ as $u^{n_i} \in H(\Omega)\cap W^{2,2}(\Omega)$. Therefore, the last integral in the right-hand side of inequality (3.38) does not exceed a fixed constant multiplied by ε. The first integral in the right-hand side of inequality (3.38) can be considered arbitrarily small for sufficiently large n_i, n_j, as the sequence $\{u_{x_k}^{n_q}(x,t)\}$ converges weakly in $L_2(\mathbb{Q}_T)$ and in $L_2(\Omega)$ to the function weakly continuous in t [5], and hence the integral

$$\int_0^T \left[\int_\Omega \{\partial_{x_k}(u_l^{n_i} - u_l^{n_j})\}(A_0^*\omega_j)dx\right]^2 dt \to 0$$

as $n_i, n_j \to \infty$. Thus, the right-hand side of (3.38) can be considered arbitrarily small for sufficiently large indices n_i, n_j. This proves that the sequences $\{u_{x_k}^{n_i}\}$, $k = 1,2,3$ converge strongly in $L_2(\mathbb{Q}_T)$. Estimate $\|u\|^2 \le c\,\|u\|_{1,2}^2$ is valid for $u \in H(\Omega)$, and we obtain that the sequence $\{u^{n_i}\}$ also converges in $L_2(\mathbb{Q}_T)$.

1b) *Now we prove that the sequence* $\{(u^{n_i},\nabla)u^{n_i}\}$ *strongly converges in* $L_2(\mathbb{Q}_T)$. With this goal in mind we employ inequality (2.5) and set $w = u_l^{n_i} - u_l^{n_j}$,

$v = u_k^{n_i}$. Thus, we get

$$\int_0^T \int_\Omega \left|(u_l^{n_i} - u_l^{n_j})\partial_{x_l} u_k^{n_i}\right|^2 dx \leq c \max_{t\in[0,T]} \|u_k^{n_i}(\cdot,t)\|_{1,2} \tag{3.39}$$
$$\times \left\{ \max_{t\in[0,T]} \|u_l^{n_i}(\cdot,t)\|_{1,2} + \max_{t\in[0,T]} \left\|u_l^{n_j}(\cdot,t)\right\|_{1,2} \right\} \int_0^T \left\|u_l^{n_i} - u_l^{n_j}\right\|_{1,2} \|\Delta u_k^{n_i}\| \, dt.$$

Due to inequality (3.32) the numbers

$$\max_{t\in[0,T]} \|u_l^{n_i}(\cdot,t)\|_{1,2}, \quad \max_{t\in[0,T]} \left\|u_l^{n_j}(\cdot,t)\right\|_{1,2}, \quad \max_{t\in[0,T]} \|u_k^{n_i}(\cdot,t)\|_{1,2}$$

are bounded in the interval $[0,T]$, $0 < T < T_g$ by some constant $C(T)$ uniformly with respect to n_i, n_j, l. Hence, applying the Cauchy inequality to the right-hand side of (3.39) we obtain

$$\int_0^T \int_\Omega \left|(u_l^{n_i} - u_l^{n_j})\partial_{x_l} u_k^{n_i}\right|^2 dx$$
$$\leq C_1(T) \left\{ \int_0^T \|\Delta u_k^{n_i}\|^2 dt \right\}^{1/2} \left\{ \int_0^T \|u_l^{n_i} - u_l^{n_j}\|_{1,2}^2 dt \right\}^{1/2}. \tag{3.40}$$

By virtue of inequality (3.32) the numbers $\left\{ \int_0^T \|\Delta u_k^{n_i}\|^2 dt \right\}^{1/2}$ are uniformly bounded by some constant $C_2(T)$ in $[0,T]$, $0 < T < T_g$, and in part 1a) it was proved that $\left\{ \int_0^T \left\|u_l^{n_i} - u_l^{n_j}\right\|_{1,2}^2 dt \right\} \to 0$ as n_i, $n_j \to \infty$. Therefore, the right-hand side in inequality (3.40) can be considered arbitrarily small as n_i, $n_j \to \infty$.

We consider the following inequality in a similar way:

$$\int_0^T \int_\Omega \left|u_k^{n_i} \partial_{x_k} (u_l^{n_i} - u_l^{n_j})\right|^2 dx$$
$$\leq c \left(\max_{t\in[0,T]} \|u_k^{n_i}(\cdot,t)\|_{1,2}^2 \right) \int_0^T \left\|u_l^{n_i} - u_l^{n_j}\right\|_{1,2} \left\|\Delta(u_l^{n_i} - u_l^{n_j})\right\| dt$$
$$\leq C(T) \left\{ \int_0^T \left\|\Delta(u_l^{n_i} - u_l^{n_j})\right\|^2 dt \right\}^{1/2} \left\{ \int_0^T \left\|u_l^{n_i} - u_l^{n_j}\right\|_{1,2}^2 dt \right\}^{1/2} \tag{3.41}$$
$$\leq C_1(T) \left\{ \int_0^T \left\|u_l^{n_i} - u_l^{n_j}\right\|_{1,2}^2 dt \right\}^{1/2}.$$

From inequality (3.41) it follows the convergence:

$$\int_0^T \left|u_k^{n_i} \partial_{x_k} (u_l^{n_i} - u_l^{n_j})\right|^2 dt \to 0 \quad \text{as} \quad n_i, n_j \to \infty.$$

Combining inequalities (3.40), (3.41) we infer that the sequence $\{(u^{n_i}, \nabla)u^{n_i}\}$ strongly converges in $L_2(\mathbb{Q}_T)$ to some function ψ,

$$\psi := \lim_{n_j\to\infty} (u^{n_j}, \nabla) u^{n_j}. \tag{3.42}$$

1c) Equality (3.8) can be rewritten in the following equivalent form:

$$u_t^n - \nu\widetilde{\Delta}u^n = -P_n(u^n,\nabla)u^n + P_n f. \tag{3.43}$$

Therefore, we obtain

$$\begin{aligned}&\partial_t(u^n - u^m) - \nu\widetilde{\Delta}(u^n - u^m) \\ &\quad = -(P_n - P_m)(u^n,\nabla)u^n + (P_n - P_m)f + P_m\{(u^m,\nabla)u^m - (u^n,\nabla)u^n\}; \\ &(u^n - u^m)|_{t=0} = (P_n - P_m)u^0.\end{aligned} \tag{3.44}$$

Using the equivalence

$$a_1\int_0^T \left\|\widetilde{\Delta}(u^n - u^m)\right\|^2 dt \le \int_0^T \|\Delta(u^n - u^m)\|^2 dt \le a_2\int_0^T \left\|\widetilde{\Delta}(u^n - u^m)\right\|^2 dt,$$

by standard calculations from (3.44) we get the inequality

$$\begin{aligned}&\max_{[0,T]}\|(u^n - u^m)(\cdot,t)\|_{1,2}^2 + \int_0^T \left\{\|\partial_t(u^n - u^m)\|^2 + \nu c\,\|\Delta(u^n - u^m)\|^2\right\} dt \\ &\quad \le \left\|(P_n - P_m)u^0\right\|_{1,2}^2 + c\int_0^T \Big\{\|(P_n - P_m)(u^n,\nabla)u^n\|^2 \\ &\qquad + \|(P_n - P_m)f\|^2 + \|(u^m,\nabla)u^m - (u^n,\nabla)u^n\|^2\Big\}dt.\end{aligned} \tag{3.45}$$

From [5, pp. 44–46] it follows that $\left\|(P_n - P_m)u^0\right\|_{1,2}^2 \to 0$ and $\|(P_n - P_m)f\| \to 0$ as $n,\ m \to \infty$. We proved above that $\int_0^T \|(u^{n_j},\nabla)u^{n_j} - (u^{n_i},\nabla)u^{n_i}\|^2 dt \to 0$ as $i, j \to \infty$. Evidently,

$$\begin{aligned}&\int_0^T \left\|(P_{n_j} - P_{n_i})(u^{n_j},\nabla)u^{n_j}\right\|^2 dt \\ &\quad \le 4\int_0^T \|(u^{n_j},\nabla)u^{n_j} - \psi\|^2 dt + 2\int_0^T \left\|(P_{n_j} - P_{n_i})\psi\right\|^2 dt,\end{aligned} \tag{3.46}$$

where $\psi := \lim_{n_j\to\infty}(u^{n_j},\nabla)u^{n_j}$. Consequently, the right-hand side of inequality (3.46) tends to zero as $n_j \to \infty$. Hence, the right-hand side of inequality (3.45) for $n = n_j$, $m = n_i$ tends to zero as $i, j \to \infty$, that is, the Galerkin approximations $\{u^{n_j}\}$ converge in the norm $\|\cdot\|_{W(T)}$ to the function

$$u := \lim_{j\to\infty} u^{n_j} \in W(T). \tag{3.47}$$

If we substitute the expressions $(u_l^{n_i} - u_l^{n_j})\partial_{x_l}u_k^{n_i}$ by $(u_l^{n_i} - u_l)\partial_{x_l}u_k^{n_i}$ and $u_k^{n_i}\partial_{x_k}(u_l^{n_i} - u_l^{n_j})$ by $u_k\partial_{x_k}(u_l^{n_i} - u_l)$ in inequalities (3.40) and (3.41), then, similarly to 1b), we obtain the following convergence in $L_2(\mathbb{Q}_T)$:

$$\lim_{n_j\to\infty}(u^{n_j},\nabla)u^{n_j} = (u,\nabla)u = \psi.$$

Note that linear combinations of the functions $a^j, j = 1,\dots,$ with time-dependent coefficients $d_j(t)$ are dense in $L_2^\circ(\mathbb{Q}_T)$. Thus, integrating the scalar product of the right-hand and left-hand sides of equality (3.43) with a function $g \in L_2^\circ(\mathbb{Q}_T)$ and passing to limit as $n = n_j \to \infty$, we deduce that function (3.47) satisfies the equation

$$\int_0^T \left(\frac{\partial u}{\partial t} - \nu \widetilde{\Delta} u + P(u, \nabla)u - Pf(x,t), g(x,t) \right) dt = 0, \qquad \text{for all } g \in L_2^\circ(\mathbb{Q}_T) \tag{3.48}$$

on the intervals $[0, T]$, $T < T_g$. Evidently, the function $u(x,t)$ possesses all properties of the GSNS solution. By Lemma 1.2 the GSNS solution is unique.

1d) Now, we prove that all the sequence $\{(u^n, \nabla)u^n\}$ converge in $L_2(\mathbb{Q}_T)$. Observe that $\{u^{n_j}\}$ converges in the norm $\|\cdot\|_{W(T)}$ to a unique GSNS $u(x,t)$, and

$$\int_0^T \|(u^{n_j}, \nabla)u^{n_j} - (u, \nabla)u\|^2 \, dt \to 0 \;\text{ as }\; j \to \infty.$$

So, for any subsequence $\{n_q\}$, such that the sequence $\{(u^{n_j}, \nabla)u^{n_j}\}$ converges in $L_2(\mathbb{Q}_T)$, this sequence converges to the same limit $(u, \nabla)u$, where the function u is the GSNS function obtained above.

Now, suppose the opposite, i.e., that all the sequence $\{(u^n, \nabla)u^n\}$ do not converge in $L_2(\mathbb{Q}_T)$. Then there exist $\varepsilon_0 > 0$ and such a subsequence $\{\widetilde{n}_q\}$ that

$$\int_0^T \left\| (u^{\widetilde{n}_q}, \nabla)u^{\widetilde{n}_q} - (u, \nabla)u \right\|^2 dt \geq \varepsilon_0 \;\text{ for all }\; \{\widetilde{n}_q\}.$$

By the above-mentioned considerations we can find a subsequence $\{\widehat{n}_i\} \subset \{\widetilde{n}_q\}$ such that

$$\int_0^T \left\| (u^{\widehat{n}_i}, \nabla)u^{\widehat{n}_i} - (u, \nabla)u \right\|^2 dt \to 0 \;\text{ as }\; j \to \infty.$$

The obtained contradiction proves that all the sequence $\{(u^n, \nabla)u^n\}$ converges in $L_2(\mathbb{Q}_T)$ to the function $(u, \nabla)u$.

2a) Now we prove inequality (3.25). The Galerkin approximations u^n satisfy inequality (3.28). Recall that $\|u^n - u\|_{W(T)} \to 0$ as $n \to \infty$. So we can pass to limit in inequality (3.28) and obtain inequality (3.25).

2b) In order to prove inequality (3.27) we consider inequality (2.14) at the Galerkin approximation u^n and apply inequality (3.17) to the functions u^n. Therefore, we obtain

$$-(\|u^n(\cdot, t)\|_{1,2}^2 + b)^{-2} + (\|u^n(\cdot, \tau)\|_{1,2}^2 + b)^{-2} \leq \gamma(t - \tau), \tag{3.49}$$

where $b := c_6 \nu^{2/3} \sup_{t \in [0, T_g]} \|f(\cdot, t)\|^{2/3}$. By definition of the value T_g, there exist sequences $n_m \to \infty$ and $t_{n_m} \to T_g$ such that $\|u^{n_m}(\cdot, t_{n_m})\|_{1,2}^2 \to \infty$. Let us set $t = t_{n_m}$ and $n = n_m$ in inequality (3.49). Passing in inequality (3.49) to limit as $n_m \to \infty$ we obtain inequality (3.27).

2c) The statement 3) of the theorem was proved in part 1c). □

References

[1] R.A. Adams, *Sobolev Spaces.* Academic Press, New York 1973.

[2] C. Fefferman, *Navier-Stokes Equation.* Clay Mathematical Institute, Official Problem Description, http://www.claymath.org/millennium/Navier-Stokes_Equations, 2011.

[3] E. Gagliardo, *Ulteriori proprietà di alcune classi di funzioni in piu variablili.* Ricerche Mat. **8** (1959), 24–52.

[4] E. Hopf, *Über die Anfangswertaufgaben der Hydrodynamik zäher Flüssigkeiten.* Math. Nachrichten **4** (1950–1951), 213–231.

[5] O.A. Ladizhenskaya, *The Mathematical Theory of Viscous Incompressible Flow.* Gordon and Breach Science Publishers, New York, London, Paris, 1969.

[6] O.A. Ladizhenskaya, V.A. Solonnikov, and N.N. Ural'ceva, *Linear and Quasi-linear Equations of Parabolic Type.* American Mathematical Society, Providence, RI, 1968.

[7] J.L. Lions and E. Magenes, *Problèmes aux Limites non Homogènes et Applications.* Dunod · Paris, 1968.

[8] A.J. Majda and A.L. Bertozzi, *Vorticity and Incompressible Flow.* Cambridge University Press, 2002.

[9] L. Nirenberg, *On elliptic partial differential equations.* Ann. Scuola Norm. Sup. Pisa **3** (13) (1959), 115–162.

[10] V.A. Solonnikov, *On general boundary-value problem for elliptic systems in the sense of Douglis-Nirenberg, I.* Izv. Akad. Nauk SSSR, ser. mat. **28** (1964), 665–706.

[11] V.A. Solonnikov, *On general boundary-value problem for elliptic systems in the sense of Douglis-Nirenberg, II.* Trudy Mat. Inst. Steklov **92** (1966), 233–297.

Valeri V. Kucherenko
Instituto Politecnico Nacional – ESFM
Av. IPN S/N, U.P.A.L.M
Edif. 9, C.P. 07738
Mexico, D.F., Mexico
e-mail: valeri@esfm.ipn.mx

Operator Theory:
Advances and Applications, Vol. 228, 261–274

Some New Scales of Weight Characterizations of Hardy-type Inequalities

Alois Kufner, Lars-Erik Persson and Natasha Samko

Dedicated to the 70th anniversary of Professor Vladimir Rabinovich

Abstract. In this paper we present, discuss and illustrate some new scales of conditions to characterize modern forms of Hardy's inequalities which can not be found in the newest books in this area. Moreover, some results of importance as motivation for these scales are presented and discussed in a historical perspective.

Mathematics Subject Classification (2010). Primary 26D10; Secondary 26D15.

Keywords. Inequalities, Hardy's inequality, weights, Hardy-type inequalities, Muckenhoupt type conditions, scales of conditions.

1. Introduction

We consider the general one-dimensional Hardy inequality

$$\left(\int_0^b\left(\int_0^x f(t)dt\right)^q u(x)dx\right)^{1/q}\le C\left(\int_0^b f^p(x)v(x)dx\right)^{1/p} \tag{1.1}$$

with a fixed b, $0<b\le\infty$, for measurable functions $f\ge 0$, weights u and v and for the parameters p,q satisfying $0<q<\infty$ and $p\ge 1$. The validity of this inequality can be characterized by some single conditions which are different for the case $p\le q$ (then we call it the Muckenhoupt-Bradley condition and denote it $A_{MB}<\infty$) and for the case $q<p$ (then we call it the Maz'ya-Rozin condition and denote it $A_{MR}<\infty$).

The first author was supported by RVO: 67985840. The third author gratefully acknowledges financial support by *Fundação para a Ciência e a Tecnologia* (FCT), Portugal, (Grant SFRH/BPD/34258/2006) and Luleå University of Technology for financial support for the research visits in 2011.

The main goal of the paper is to give a survey of recent results, not going into details, which could be found in the references.

We want to show that the necessary and sufficient conditions mentioned above can be extended (or more precisely replaced) by a whole SCALE of conditions depending on some additional parameters and, hence, provide the reader with infinitely many equivalent conditions.This is important for example due to the fact that the validity of the Hardy inequality is closely connected with the solvability of some boundary value problems, in particular with spectral problems for (nonlinear) ordinary differential equations, and with extremal problems. To be more concrete, the best constant C in (1.1) describes the (first) eigenvalue of a differential operator, and equivalent conditions from the scales mentioned provide us also with different estimates of the eigenvalues (for details see, e.g., [12], Sections 4 and 14, or [6], Chapters 7 and 8, and several papers mentioned in [6]). Moreover, among the estimates for the constant C resulting from our equivalent conditions, there are also estimates expressed in term of the gamma-function (see, e.g., the reference [202] in [6]) and the conditions of the validity of inequality (1.1) allow even to decide about the discreteness of the spectrum. Therefore, it is important to obtain estimates – as good as possible – for the best constant in (1.1).

The fact that scales of conditions can improve the estimate of the best constant was first illustrated in the Ph.D. thesis of A. Wedestig [18], Example 3.1, p. 29. In this thesis it was also proved that Hardy type inequalities described by a scale of conditions can be used to derive a characterization of the corresponding limiting inequalities (where the (arithmetic mean) Hardy operator is replaced by the geometric mean operator). Such a result can not be obtained in this way by using the standard Muckenhoupt-Bradley condition (see (2.3)). In fact, this was the crucial motivation already when L.E. Persson and V. Stepanov derived their alternative conditions (see (2.6) and (2.8)).

In the following sections, we divide our description into three cases: $p = q$, $p < q$ and $q < p$, and after some historical remarks, we present and illustrate the main results concerning the scales. Finally, in Section 4 we make a final discussion and present some examples and illustrations, which support some previous statements in the text.

2. Some historical results

2a) The case $p = q$. A classical result here reads:

Theorem 2.1. *Let $1 \le p < \infty$. Then the inequality* (1.1) *holds for all measurable functions $f \ge 0$ on $(0, b), 0 < b \le \infty$, if and only if*

$$A := \sup_{r\in(0,b)} \left(\int_r^b u(x)dx\right)^{1/p} \left(\int_0^r v^{1-p'}(x)dx\right)^{1/p'} < \infty, \tag{2.1}$$

where as usual $p' = p/(p-1)$ when $p > 1$ and $p' = \infty$ when $p = 1$ (so the second integral must be interpreted as a supremum).

Remark 2.2. The condition (2.1) is frequently called the Muckenhoupt condition since B. Muckenhoupt [11] presented in 1972 a nice and direct proof, which absolutely has influenced the further development in a crucial way. However, Muckenhoupt mentioned that G. Talenti [16] and G.A. Tomaselli [17] had already proved this result in 1969, but in these papers the result was not so explicitly stated as in [11]. More of this history, including some surprising details, can be found in the book [6] by A. Kufner, L. Maligranda and L.E. Persson.

Remark 2.3. In the paper [17] Tomaselli also derived two other conditions for characterizing the Hardy inequality, namely the following:

$$A^* := \sup_{r\in(0,b)} \left(\int_0^r u(x) \left(\int_0^x v^{1-p'}(t)dt\right)^p dx\right) \left(\int_0^r v^{1-p'}(x)dx\right)^{-1} < \infty$$

and

$$A^{**} := \inf_{f>0} \sup_{x\in(0,b)} \frac{1}{f(x)} \int_0^x u(t) \left[f(t) + \int_0^t v^{1-p'}(s)ds\right]^p dt < \infty.$$

Also this result has absolutely influenced the further development and in fact it was proved in [2] that these two conditions can be replaced by infinitely many conditions (indeed even by 14 different scales of conditions, of course even for the case $p \leq q$, see our Theorem 3.1 with $p = q$).

Moreover, for the best constant C in (1.1) it yields that

$$C \approx A \approx A^* \approx A^{**}.$$

2b) The case $1 < p \leq q$. Inequality (1.1) is usually characterized by the (so-called Muckenhoupt-Bradley) condition

$$A_{MB} := \sup_{0<x<b} A_{MB}(x) < \infty, \tag{2.2}$$

where

$$A_{MB}(x) := \left(\int_x^b u(t)dt\right)^{1/q} \left(\int_0^x v^{1-p'}(t)dt\right)^{1/p'}. \tag{2.3}$$

Here and in the sequel $p' = p/(p-1)$. Further, let us denote

$$U(x) := \int_x^b u(t)dt, \qquad V(x) := \int_0^x v^{1-p'}(t)dt, \tag{2.4}$$

and assume that $U(x) < \infty$, $V(x) < \infty$ for every $x \in (0,b)$.

The index MB in $A_{MB} := \sup_{0<x<b} U^{1/q}(x)V^{1/p'}(x)$ indicates the efforts of B. Muckenhoupt and J.S. Bradley. In 1972 B. Muckenhoupt [11] gave a nice proof of the fact that $A_{MB} < \infty$ is necessary and sufficient for (1.1) to hold for the case $p = q$ and in 1978 J.S. Bradley [1] extended the Muckenhoupt result to the case $p \leq q$ and gave a complete and simple proof of Muckenhoupt type of this result. However, this result was also independently derived in 1979 by V. Maz'ya and L. Rozin (see [9] and [10]) and by V. Kokilashvili (see [5]).

Besides the condition $A_{MB} < \infty$, some other equivalent conditions have been derived during the next decades, e.g., the conditions $A_G < \infty$ or $A_G^* < \infty$, where

$$\begin{aligned} A_G &:= \inf_{h>0} \sup_{0<x<b} \left(\frac{1}{h(x)} \int_0^x u(t)(h(t)+V(t))^{\frac{q}{p'}+1} dt \right)^{1/q}; \\ A_G^* &:= \inf_{h>0} \sup_{0<x<b} \left(\frac{1}{h(x)} \int_x^b v^{1-p'}(t)(h(t)+U(t))^{\frac{p'}{q}+1} dt \right)^{1/p'}. \end{aligned} \tag{2.5}$$

This result was proved by P. Gurka in 1984 in [4]; he extended to the case $p \leq q$ the result proved for $p = q$ in 1969 by G.A. Tomaselli [17].

Some other alternative conditions are that $A_{PS} < \infty$ or $A_{PS}^* < \infty$, where

$$\begin{aligned} A_{PS} &:= \sup_{0<x<b} \left(\int_0^x u(t) V^q(t) dt \right)^{1/q} V^{-1/p}(x); \\ A_{PS}^* &:= \sup_{0<x<b} \left(\int_x^b v^{1-p'}(t) U^{p'}(t) dt \right)^{1/p'} U^{-1/q'}(x). \end{aligned} \tag{2.6}$$

This result was proved in 2002 in [13] by L.E. Persson and V. Stepanov but as we have seen it was proved for the case $p = q$ already in 1969 in [17].

Moreover, for the best constant C in (1.1) it yields that

$$C \approx A_{MB} \approx A_G \approx A_G^* \approx A_{PS} \approx A_{PS}^*.$$

2c) The case $1 \leq q < p < \infty$. A necessary and sufficient condition for (1.1) to hold in this case was derived by V. Maz'ya and L. Rozin in the late seventies (see [9] and [10]) and it reads:

$$B_{MR} := \left(\int_0^\infty U^{r/p}(x) V^{r/p'}(x) u(x) dx \right)^{1/r} < \infty, \tag{2.7}$$

where $1/r := 1/q - 1/p$. An alternative condition was found by L.E. Persson and V. Stepanov in 2002 (see [13]) and it reads:

$$B_{PS} := \left(\int_0^\infty \left[\int_0^x u(t) V^q(t) dt \right]^{r/p} u(x) V^{q-r/p}(x)\, dx \right)^{1/r} < \infty. \tag{2.8}$$

Moreover, for the best constant C in (1.1) it yields that

$$C \approx B_{MR} \approx B_{PS}.$$

Some complementary history to this section can be found in the book [6]; see also [7] and [12].

3. Unification and extensions: Scales of conditions

3a) The case $1 < p \leq q$. The first scale of conditions was derived in 2004 in [8] by A. Kufner, L.E. Persson and A. Wedestig. It reads $A(r) < \infty$ (with $1 < r < p$) or $A^*(r) < \infty$ (with $1 < r < q'$), where

$$A(r) := \sup_{0<x<b} \left(\int_x^b u(t) V^{q(p-r)/p}(t)dt \right)^{1/q} V^{(r-1)/p}(x), \quad 1 < r < p; \tag{3.1}$$

$$A^*(r) := \sup_{0<x<b} \left(\int_0^x v^{1-p'}(t) U^{p'(q'-r)/q'}(t)dt \right)^{1/p'} U^{(r-1)/q'}(x), \ 1 < r < q'.$$

Note that the end point condition $A(p) < \infty$ is just the Muckenhoupt-Bradley condition $A_{MB} < \infty$ mentioned above.

In 2004 in [3] four new scales of equivalent integral conditions were derived by A. Gogatishvili et al. This result was used to characterize the inequality (1.1) by four scales of conditions, namely the scales including the Muckenhoupt-Bradley condition, the Persson-Stepanov condition and the dual of these scales.

Here, we will present, discuss and extend the existing list of (equivalent) scales with 10 new scales of conditions where also, e.g., the Gurka result mentioned above (see [7]) appears as a special case.

Theorem 3.1. *Let $1 < p \leq q < \infty$, $0 < s < \infty$, and define, for the weight functions u, v, the functions U and V by (2.4), and the functions $A_i(s)$, $i = 1, 2, \ldots, 14$, as follows:*

$$A_1(s) := \sup_{0<x<b} \left(\int_x^b u(t) V^{q(\frac{1}{p'}-s)}(t)dt \right)^{1/q} V^s(x); \tag{3.2}$$

$$A_2(s) := \sup_{0<x<b} \left(\int_0^x v^{1-p'}(t) U^{p'(\frac{1}{q}-s)}(t)dt \right)^{1/p'} U^s(x);$$

$$A_3(s) := \sup_{0<x<b} \left(\int_0^x u(t) V^{q(\frac{1}{p'}+s)}(t)dt \right)^{1/q} V^{-s}(x);$$

$$A_4(s) := \sup_{0<x<b} \left(\int_x^b v^{1-p'}(t) U^{p'(\frac{1}{q}+s)}(t)dt \right)^{1/p'} U^{-s}(x);$$

$$A_5(s) := \sup_{0<x<b} \left(\int_x^b u(t) V^{\frac{q}{p'(1+sq)}}(t)dt \right)^{\frac{1+sq}{q}} U^{-s}(x);$$

$$A_6(s) := \sup_{0<x<b} \left(\int_0^x v^{1-p'}(t) U^{\frac{p'}{q(1+sp')}}(t)dt \right)^{\frac{1+sp'}{p'}} V^{-s}(x);$$

$$A_7(s) := \sup_{0<x<b} \left(\int_0^x u(t) V^{\frac{q}{p'(1-sq)}}(t)dt \right)^{\frac{1-sq}{q}} U^s(x), \quad qs < 1;$$

$$A_8(s) := \sup_{0<x<b} \left(\int_x^b u(t) V^{\frac{q}{p'(1-sq)}}(t) dt \right)^{\frac{1-sq}{q}} U^s(x), \quad qs > 1;$$

$$A_9(s) := \sup_{0<x<b} \left(\int_x^b v^{1-p'}(t) U^{\frac{p'}{q(1-sp')}}(t) dt \right)^{\frac{1-sp'}{p'}} V^s(x), \quad p's < 1;$$

$$A_{10}(s) := \sup_{0<x<b} \left(\int_0^x v^{1-p'}(t) U^{\frac{p'}{q(1-sp')}}(t) dt \right)^{\frac{1-sp'}{p'}} V^s(x), \quad p's > 1;$$

$$A_{11}(s) := \inf_{h>0} \sup_{0<x<b} \left(\int_x^b u(t) h(t)^{q(\frac{1}{p'}-s)} dt \right)^{1/q} (h(x) + V(x))^s, \quad p's > 1;$$

$$A_{12}(s) := \inf_{h>0} \sup_{0<x<b} \left(\int_0^x v^{1-p'}(t) h(t)^{p'(\frac{1}{q}-s)} dt \right)^{1/p'} (h(x) + U(x))^s, \quad qs > 1;$$

$$A_{13}(s) := \inf_{h>0} \sup_{0<x<b} \left(\int_0^x u(t)(h(t) + V(t))^{q(\frac{1}{p'}+s)} dt \right)^{1/q} h^{-s}(x);$$

$$A_{14}(s) := \inf_{h>0} \sup_{0<x<b} \left(\int_x^b v^{1-p'}(t)(h(t) + U(t))^{p'(\frac{1}{q}+s)}(t) \right)^{1/p'} h^{-s}(x).$$

Then the Hardy inequality (1.1) *holds for all measurable functions* $f \geq 0$ *if and only if any of the quantities* $A_i(s)$, $i = 1, 2, 3, \ldots, 14$, *is finite for some* $0 < s < \infty$. *Moreover, for the best constant* C *in* (1.1) *we have* $C \approx A_i(s)$, $i = 1, 2, 3, \ldots, 14$. *The constants in the equivalence relations can depend on* s.

Remark 3.2. The constants in (2.2), (2.5) (2.6) and (3.1) can be described in the following way:

$$A_{MB} = A_1\left(\frac{1}{p'}\right), \quad A_{PS} = A_3\left(\frac{1}{p}\right), \quad A(r) = A_1\left(\frac{r-1}{p}\right) \text{ with } 1 < r < p,$$

$$A^*_{PS} = A_4\left(\frac{1}{q'}\right), \quad A^*(r) = A_2\left(\frac{r-1}{q'}\right) \text{ with } 1 < r < q',$$

$$A_G = A_{13}\left(\frac{1}{q}\right), \quad A^*_G = A_{14}\left(\frac{1}{p'}\right).$$

Hence, Theorem 3.1 generalizes the corresponding results in [3] and [8] and also all previous results of this type.

The first 4 scales were those proved in [3] and the scale in [8] is just the interval to the left of the Muckenhoupt-Bradley point (see Figure 1). The proof of Theorem 3.1 can be found in [2].

3b) The case $0 < q < p < \infty$, $p > 1$, $q \neq 1$. The main result here (Theorem 3.3) is taken from the paper [14] by L.E. Persson, V. Stepanov and P. Wall, where the complete proof and further information can be found. We remark that there is a substantial difference with the case $1 < p \leq q < \infty$, because no duality exists for $0 < q < 1 < p < \infty$. Moreover, also in the case $1 < q < p < \infty$ we have only found 4 different scales of conditions corresponding to the first four conditions in Theorem 2.1. For simplicity we here also only consider the case $b = \infty$.

For simplicity we again suppose (see [6]) the following concerning the involved weight functions:

$$0 < U(x) := \int_x^\infty u(t)dt < \infty, \ 0 < V(x) := \int_0^x v^{1-p'}(t)dt < \infty \text{ for all } x > 0. \tag{3.3}$$

Let $1/r := 1/q - 1/p$.

We now introduce the following scales of constants related to previous constants and their dual ones:

For $s > 0$ we define the following functionals:

$$B_{MR}^{(1)}(s) := \left(\int_0^\infty \left[\int_t^\infty uV^{q(1/p'-s)}\right]^{r/p} V^{q(1/p'-s)+rs}(t)u(t)\,dt\right)^{1/r},$$

$$B_{PS}^{(1)}(s) := \left(\int_0^\infty \left[\int_0^t uV^{q(1/p'+s)}\right]^{r/p} u(t)V^{q(1/p'+s)-sr}(t)\,dt\right)^{1/r},$$

$$B_{MR}^{(2)}(s) := \left(\int_0^\infty \left[\int_0^t U^{p'(1/q-s)}dV\right]^{r/p'} U^{rs-1}(t)u(t)\,dt\right)^{1/r},$$

$$B_{PS}^{(2)}(s) := \left(\int_0^\infty \left[\int_t^\infty U^{q(1/p'+s)}dV\right]^{r/p} U^{q(1/p'+s)-rs}(t)\,dV(t)\right)^{1/r}.$$

The main theorem in this case reads:

Theorem 3.3.

a) *Let $0 < q < p < \infty$, $1 < p < \infty$ and $q \neq 1$. Then the Hardy inequality* (1.1) *with $b = \infty$ holds for some finite constant $C \geq 0$ if and only if any of the constants $B_{MR}^{(1)}(s)$ or $B_{PS}^{(1)}(s)$ is finite for some $s > 0$. Moreover, for the best constant C in* (1.1) *we have*

$$C \approx B_{MR}^{(1)}(s) \approx B_{PS}^{(1)}(s). \tag{3.4}$$

b) *Let $1 < q < p < \infty$. Then the Hardy inequality* (1.1) *with $b = \infty$ holds for some finite constant $C > 0$ if and only if any of the constants $B_{MR}^{(2)}(s)$ or $B_{PS}^{(2)}(s)$ is finite for some $s > 0$. Moreover, for the best constant C in* (1.1) *we have*

$$C \approx B_{MR}^{(2)}(s) \approx B_{PS}^{(2)}(s).$$

Remark 3.4. Note that Theorem 3.3 is a generalization of the original results of Maz'ya-Rozin and Persson-Stepanov since $B_{MR}^{(1)}(\frac{1}{p'}) = B_{MR}$ and $B_{PS}^{(1)}(\frac{1}{p}) = B_{PS}$ (see Figure 2).

Remark 3.5. The statement in b) is just a corollary of a statement similar as in a) in a dual situation, namely when $\int_0^x$ in inequality (1.1) is replaced by $\int_x^\infty$ (see Theorem 2 in [13]).

Remark 3.6. It is known ([15], Remark on p. 93), that under the condition (3.3) the Maz'ya-Rozin constant has an equivalent form

$$\mathcal{B}_{MR} := \left(\int_0^\infty U^{r/q} V^{r/q'} dV\right)^{1/r}, \quad \text{and} \quad B_{MR}^r = \frac{q}{p'} \mathcal{B}_{MR}^r.$$

Similarly, a counterpart to the Persson-Stepanov constant is

$$\mathcal{B}_{PS} := \left(\int_0^\infty \left[\int_0^x u(t) V^q(t) dt\right]^{r/q} V^{-r/q} dV(x)\, dx\right)^{1/r}.$$

Moreover (see [13])

$$B_{PS}^r = \frac{q}{p} \mathcal{B}_{PS}^r \quad \text{if } V(\infty) = \infty,$$

and

$$B_{PS}^r = \frac{q}{r} \left(\int_0^\infty u V^q\right)^{r/q} V^{-r/p}(\infty) + \frac{q}{p} \mathcal{B}_{PS}^r \quad \text{if } 0 < V(\infty) < \infty.$$

These observations motivate us to introduce the following new alternative scales of constants:

$$\mathcal{B}_{MR}^{(1)}(s) := \left(\int_0^\infty \left[\int_t^\infty u V^{q(1/p'-s)}\right]^{r/q} V^{rs-1}(t)\, dV(t)\right)^{1/r},$$

$$\mathcal{B}_{PS}^{(1)}(s) := \left(\int_0^\infty \left[\int_0^t u V^{q(1/p'+s)}\right]^{r/q} V^{-rs-1}(t)\, dV(t)\right)^{1/r}.$$

It can be proved as above that

$$\left[B_{MR}^{(1)}(s)\right]^r = qs \left[\mathcal{B}_{MR}^{(1)}(s)\right]^r,$$

$$\left[B_{PS}^{(1)}(s)\right]^r = qs \left[\mathcal{B}_{PS}^{(1)}(s)\right]^r \quad \text{if } V(\infty) = \infty$$

and

$$\left[B_{PS}^{(1)}(s)\right]^r = \frac{q}{r} \left(\int_0^\infty u V^{q(1/p'+s)}\right)^{r/q} V^{-rs}(\infty) + qs \left[\mathcal{B}_{PS}^{(1)}(s)\right]^r,$$

if $0 < V(\infty) < \infty$. Similarly,

$$\left[B_{MR}^{(2)}(s)\right]^r = \frac{1}{sp'} \left[\mathcal{B}_{MR}^{(2)}(s)\right]^r,$$

where

$$\mathcal{B}_{MR}^{(2)}(s) := \left(\int_0^\infty \left[\int_0^t U^{p'(1/q-s)} dV\right]^{r/q'} U^{p'(1/q-s)+rs}(t)\, dV(t)\right)^{1/r}$$

and

$$\left[B_{PS}^{(2)}(s)\right]^r = qs\left[\mathcal{B}_{PS}^{(2)}(s)\right]^r, \qquad U(0)=\infty,$$

$$\left[B_{PS}^{(2)}(s)\right]^r = \frac{q}{r}\left(\int_0^\infty U^{q(1/p'+s)}\,dV\right)^{r/q} U^{-rs}(0) + qs\left[\mathcal{B}_{PS}^{(2)}(s)\right]^r,$$

where $0<U(0)<\infty$, and

$$\mathcal{B}_{PS}^{(2)}(s) := \left(\int_0^\infty \left[\int_t^\infty U^{q(1/p'+s)}dV\right]^{r/q} U^{-rs-1}(t)u(t)\,dt\right)^{1/r}.$$

Hence it is possible to complement Theorem 3.3 with some additional scales of conditions.

4. Final discussion, examples and some illustrations

First we illustrate the various conditions for characterizing the inequality

$$\left(\int_0^b \left(\int_0^x f(t)\,dt\right)^q u(x)\,dx\right)^{\frac{1}{q}} \le C\left(\int_0^b f^p(x)v(x)\,dx\right)^{\frac{1}{p}},$$

which are stated and discussed in the previous sections.

The case $1<p\le q<\infty$ (Figure 1)

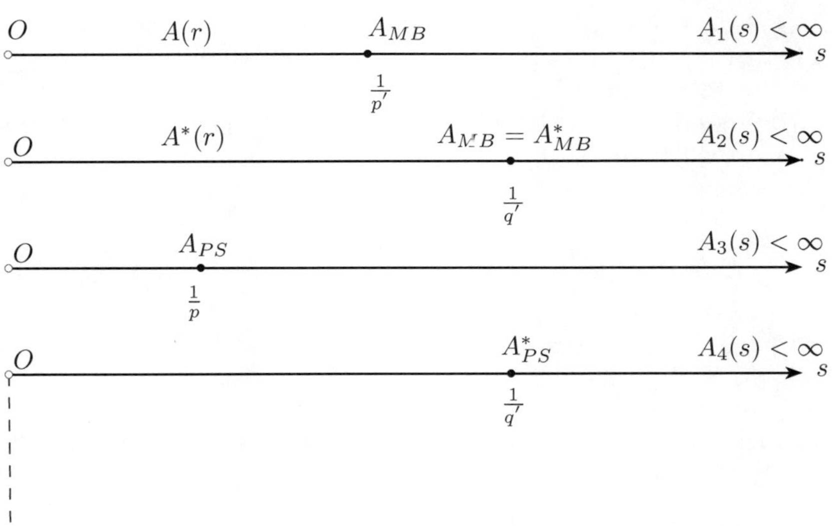

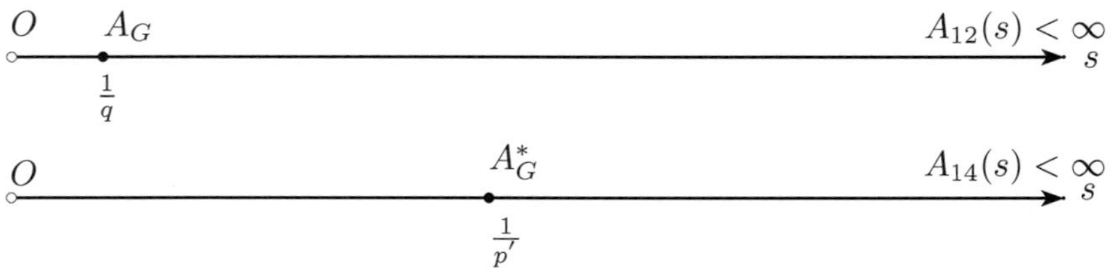

The case $0 < q < p < \infty$, $p > 1$, $q \neq 1$ (Figure 2)

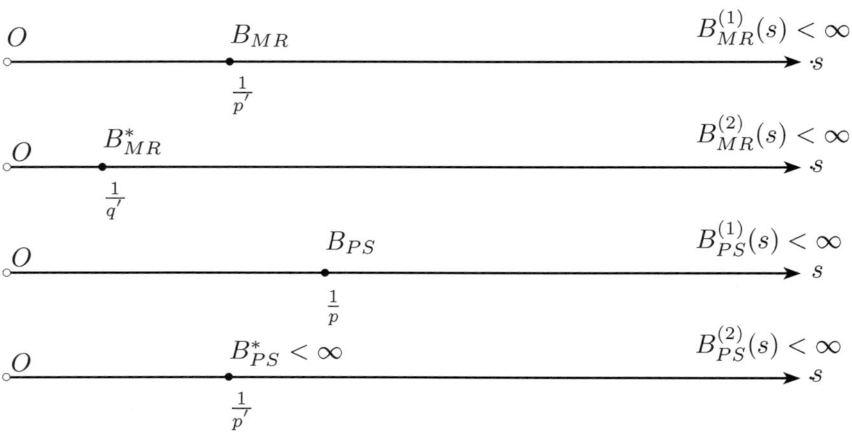

We have not given explicit estimates of the best constants C in (1.1). We shall continue by shortly discussing also this important aspect. First we note that by analyzing the proof of the theorems in this review paper we can state also some estimates of the best constants C in (1.1). By then just using these estimates we conclude that $\alpha \leq C \leq \beta$, where α is the maximum of all lower estimates of C and β is the minimum of all upper estimates of C. We refer to the paper [8] and the Ph.D. thesis [18], where this aspect was developed and illustrated via concrete estimates and examples.

For example, for the condition $A(r)$ on the first scale in Figure 1 it yields that

$$\sup_{1<s<p} \left(\frac{\left(\frac{p}{p-s}\right)^p}{\left(\frac{p}{p-s}\right)^p + \frac{1}{s-1}} \right)^{1/p} A(s) \leq C \leq \inf_{1<s<p} \left(\frac{p-1}{p-s} \right)^{1/p'} A(s), \tag{4.1}$$

see [8], Theorem 1 and cf. [18]. Here and in the sequel $p' = \frac{p}{p-1}$. Moreover, for the condition A_{MB} at the right endpoint we have the standard estimate (see the books [6, 7, 12])

$$A_{MB} \leq C \leq A_{MB} \cdot k(p,q), \tag{4.2}$$

where $k(p,q) = \left(1+\frac{q}{p'}\right)^{1/q}\left(1+\frac{p'}{q}\right)^{1/p'}$. Later on V.M. Manakov improved this constant to

$$k(p,q) = \left(\frac{\Gamma\left(\frac{pq}{q-p}\right)}{\Gamma\left(\frac{q}{q-p}\right)\Gamma\left(\frac{p(q-1)}{q-p}\right)}\right)^{\frac{q-p}{qp}}, \quad p < q.$$

Concerning the condition A_{PS} we have the following estimate (see [13, Theorem 1]):

$$A_{PS} \le C \le p' \, A_{PS}. \tag{4.3}$$

By using this information and making elementary estimates we can see that by using A_{PS} we get that

$$A_{MB} \cdot (p-1)^{1/q} \le C \le A_{MB} \cdot p'(p-1)^{1/q}, \tag{4.4}$$

see [18], p. 29. In particular, this always improves the lower bound in (4.2) for all $p > 2$. We only give the following elementary example how this information can be used:

Example 1 (cf. [18], Example 3.1). Let $p = 3, q = 4$ and choose $s = 1.15$ in the lower bound in (4.1). Then we obtain the following estimates:

a) $A_{MB} \le C \le A_{MB} \cdot 1.530348452$,

by using (4.2) with Manakov's constant.

b) $A_{MB} \cdot 1.189207115 \le C$,

by using (4.3).

c) $A_{MB} \cdot 1.396254480 \le C$,

by using the scale of conditions in (4.1).

Summing up, we find that

$$A_{MB} \cdot 1.396254480 \le C \le A_{MB} \cdot 1.530348452$$

which is a better estimate than those we can find in the standard books (see [6, 7, 12]).

We have also mentioned that alternative conditions for characterizing (1.1) are useful for deriving limit cases of Hardy-type inequalities. To illustrate this idea we first give the following elementary example.

Example 2. The classical form of Hardy's inequality reads:

$$\int_0^\infty \left(\frac{1}{x}\int_0^x f(y)dy\right)^p dx \le \left(\frac{p}{p-1}\right)^p \int_0^\infty f^p(x)dx, \ p > 1. \tag{4.5}$$

This inequality was stated in 1920 and proved in 1925 by G.H. Hardy. The constant $C = \left(\frac{p}{p-1}\right)^p$ is sharp.

Replacing $f(x)$ by $(f(x))^{1/p}$ in (4.5) we find that

$$\int_0^\infty \left(\frac{1}{x}\int_0^x (f(y))^{1/p}dy\right)^p dx \le \left(\frac{p}{p-1}\right)^p \int_0^\infty f(x)dx, \ p > 1.$$

By now letting $p \to \infty$ we find that

$$\left(\frac{1}{x}\int_0^x (f(y))^{1/p}dy\right)^p \to \exp\left(\frac{1}{x}\int_0^x \ln f(y)dy\right), \text{ and } \left(\frac{p}{p-1}\right)^p \to e$$

(the scale of Power means converges to the geometric mean when $p \to \infty$). Hence, we find that the inequality

$$\int_0^\infty \exp\left(\frac{1}{x}\int_0^x \ln f(y)dy\right)dx \le e\int_0^\infty f(x)dx \tag{4.6}$$

may be regarded as a limit case of (4.5).

In fact, (4.6) is known in the literature as the Pólya-Knopp inequality but the relation above is not always pointed out. The constant $C = e$ in (4.6) is sharp.

Guided by this example it is natural to find a characterization of the inequality (1.1) with the arithmetic mean operator $Hf(x) := \frac{1}{x}\int_0^x \ln f(y)dy$ replaced by the geometric mean operator $Gf(x) := \exp\left(\frac{1}{x}\int_0^x \ln f(y)dy\right)$ via some limit procedure like that in Example 2.

However, this is impossible by using the standard A_{MB} condition. This was the motivation when Persson and Stepanov derived the alternative condition A_{PS} in [13]. In fact, by using this condition the suggested limit algorithm above works perfectly and the following result can be proved (cf. [13], Theorem 2): Let $0 < p \le q < \infty$. Then the inequality

$$\left(\int_0^b \left(\exp\left(\frac{1}{x}\int_0^x \ln f(y)dy\right)\right)^q u(x)dx\right)^{1/q} \le C\left(\int_0^b f^p(x)v(x)dx\right)^{1/p} \tag{4.7}$$

holds if and only if

$$D := \sup_{0<x<b} x^{-1/p}\left(\int_0^x w(y)dy\right)^{1/q} < \infty, \tag{4.8}$$

where

$$w(x) := \left[G\left(\frac{1}{v(x)}\right)\right]^{q/p} u(x).$$

Moreover, $D \le C \le e^{1/p}D$.

Example 3. By using this result with $v(x) \equiv u(x) \equiv 1$, $b = \infty$ and $p = q = 1$ we see that $w(y) \equiv 1$ so that $D = 1$ and we obtain (4.6).
Moreover, by considering the Hardy inequality (1.1) with $b = \infty$, $u(x) \equiv x^{-p}$, $v(x) \equiv 1$ and $p = q > 1$ and using (2.6) we find that $A_{PS} = 1$ and via the corresponding estimate (4.3) we obtain (4.5). This fact does not follow by using the condition A_{MB} with the corresponding estimate (4.2).

In the Ph.D. thesis [18] it was proved that also by using the scale $A(s)$, $1 < s < p$, in Figure 1 it is possible to obtain other characterizations of the inequality (4.7) (see [18], Theorem 3.2).

Acknowledgment

We thank all three referees for some good suggestions and remarks, which have improved the final version of this paper.

References

[1] J.S. Bradley, *Hardy inequalities with mixed norms.* Canad. Math. Bull. **21** (1978), 405–408.

[2] A. Gogatishvili, A. Kufner, and L.E. Persson, *Some new scales of characterizations of Hardy's inequality.* Proc. Estonian Acad. Sci. **59** (2010), 7–18.

[3] A. Gogatishvili, A. Kufner, L.E. Persson, and A. Wedestig, *An equivalence theorem for integral conditions related to Hardy's inequality with applications.* Real Anal. Exchange 29 (2004), no. 2, 867–880.

[4] P. Gurka, *Generalized Hardy's inequality.* Casopis Pest. Mat. **109** (1984), 194–203.

[5] V. Kokilashvili, *On Hardy's inequality in weighted spaces.* Soobshch. Acad. Nauk Gruzin. SSR **96** (1979), no. 1, 37–40.

[6] A. Kufner, L. Maligranda, and L.E. Persson, *The Hardy Inequality. About Its History and Some Related Results.* Vydavatelsky Servis Publishing House, Pilsen, 2007.

[7] A. Kufner and L.E. Persson, *Weighted Inequalities of Hardy Type.* World Scientific Publishing Co. Inc., Singapore, 2003.

[8] A. Kufner, L.E. Persson, and A. Wedestig, *A study of some constants characterizing the weighted Hardy inequality.* Banach Center Publ., Orlicz Centenary volume. **64** (2004), 135–146.

[9] V.G. Maz'ya, *Einbettungssätze für Sobolewsche Räume. Teil* 1 (*Imbedding Theorems for Sobolev Spaces. Part* 1). Teubner-Texte zur Mathematik, Leipzig, 1979.

[10] V.G. Maz'ya, *Sobolev Spaces.* Springer Series in Soviet Mathematics. Springer-Verlag, Berlin, 1985.

[11] B. Muckenhoupt, *Hardy's inequality with weights.* Studia Math. **44** (1972), 31–38.

[12] B. Opic and A. Kufner, *Hardy-type Inequalities.* Pitman Research Notes in Mathematics, 219. Longman Scientific & Technical, Harlow, 1990.

[13] L.E. Persson and V.D. Stepanov, *Weighted integral inequalities with the geometric mean operator.* J. Inequal. Appl. **7** (2002), 727–746.

[14] L.E. Persson, V.D. Stepanov, and P. Wall, *Some scales of equivalent weight characterizations of Hardy's inequality: the case $q < p$.* Math. Inequal. Appl. **10** (2007), no. 2, 267–279.

[15] G. Sinnamon and V.D. Stepanov, *The weighted Hardy inequality: new proofs and the case $p = 1$.* J. London Math. Soc. **54** (1996), 89–101.

[16] G. Talenti, *Osservazioni sopra una classa di diseguaglianza.* Rend. Sem. Mat. Fis. Milano **39** (1969), 171–185.

[17] G.A. Tomaselli, *A class of inequalities.* Boll. Un. Mat. Ital. **2** (1969), 622–631.

[18] A. Wedestig, *Weighted Inequalities of Hardy-Type and their Limiting Inequalities.* Ph.D. thesis, Department of Mathematics, Luleå University of Technology, ISSN. 1402-1544, 2003.

Alois Kufner
Math. Inst. Czech. Acad. Sci.,
Prague, Czech Republic
e-mail: kufner@math.cas.cz

Lars-Erik Persson
Luleå University of Technology, Sweden
Narvik University College, Norway
e-mail: larserik@ltu.se

Natasha Samko
Centro CEAF, Instituto Superior Técnico,
Lisbon, Portugal
e-mail: nsamko@gmail.com

Operator Theory:
Advances and Applications, Vol. 228, 275–292

Almost Periodic Elliptic Equations: Sub- and Super-solutions

Gaston M. N'Guérékata and Alexander Pankov

Dedicated to Professor Vladimir Rabinovich on the occasion of his 70th Birthday

Abstract. The method of sub- and super-solutions is a classical tool in the theory of second-order differential equations. It is known that this method does not have a direct extension to almost periodic equations. We show that if an almost periodic second-order semi-linear elliptic equation possesses an ordered pair of almost periodic sub- and super-solutions, then very many equations in the envelope have either almost automorphic solutions, or Besicovitch almost periodic solutions. In addition, we provide an application to almost periodically forced pendulum equations.

Mathematics Subject Classification (2010). Primary 35B15; Secondary 35J15, 34C27, 34C15.

Keywords. Sub- and super-solutions, almost periodicity, almost automorphy.

1. Introduction

The method of sub- and super-solutions (alternative terms are upper and lower solutions) is a popular and powerful tool in the existence theory of boundary value problems for which a maximum principle holds. In the framework of ordinary differential equations basic ideas can be traced back to E. Picard and O. Perron. Since that time hundreds papers have been published in this direction. A detailed account of the existing results in the case of ordinary differential equations can be found in [3] (see also references therein). The method extends to elliptic and parabolic partial differential equations. Simplest results of such kind can be found, *e.g.*, in [2, Ch. IV, Appendix] and [21, Ch. 1]. For further development and applications to real world problems we refer to [11, 24] and references therein.

In particular, the sub- and super-solution method provides results on existence of solutions to periodic boundary value problems, both in one and many spatial dimensions. Therefore, it is quite natural to try to develop the method for existence of almost periodic solutions to almost periodic differential equations. First results in the case of nonlinear ordinary differential equations have been obtained by M. Krasnosel'skii, V. Burd and Yu. Kolesov [8, Sect. 10] by using the theory of monotone operators. Those authors found existence results for almost periodic solutions under certain assumptions that, in particular, guarantee the uniqueness of the solution.

In 1983 A. Pankov [19] (see also [20, Sect. 5.1.2]) has considered almost periodic semi-linear elliptic equations of second order. Assuming the existence of a properly ordered pair of sub- and super-solutions, he was able to prove the existence of a very weak almost periodic solution in the sense of Besicovitch-Sobolev spaces. This solution is almost periodic in classical sense under certain additional assumptions that ensure the uniqueness of solution.

In the case of ordinary differential equations, the authors of [22] made an attempt to prove that an ordered pair of Bohr almost periodic sub- and super-solutions gives rise to the existence of a Bohr almost periodic solution without any uniqueness assumption. Unfortunately, that result is wrong. R. Ortega and M. Tarallo [18] have constructed an example of almost periodic equation of the form

$$-u'' + cu = g(t, u)$$

that possesses an ordered pair of *constant* sub- and super-solutions, but has no almost periodic solution between them. Actually, in their example the function g is quasi-periodic in t with two independent frequencies.

Our main aim in this paper is to understand what happens if there are ordered sub- and super-solutions, but the uniqueness does not hold. Basically, we prove that in the envelop of the equation under consideration there exist a residual set Ω_{aa} and a set of full measure Ω_b such that for every equation in Ω_{aa} there is an almost automorphic solution, while equations in Ω_b possess bounded almost periodic in the sense of Besicovitch solutions. We do not know whether at least one equation in the envelop has a solution that is, at the same time, almost automorphic and almost periodic in Besicovitch sense, *i.e.*, $\Omega_{aa} \bigcap \Omega_b \neq \emptyset$. In addition, we weaken regularity assumptions made in [19] and show that the frequency modulus of solution obtained is contained in the frequency modulus of the equation. Let us point out that the main ingredients of the paper are monotone iteration techniques and the metrizability of appropriate Bohr compactifications.

The organization of the paper is as follows. In Section 2 we sketch basic facts on Bohr almost periodic functions. Our approach is based on the notion of Bohr compactification and follows [20, 23]. A standard presentation of the theory can be found in [12]. In Section 3 we remind the notion of almost automorphy (see, *e.g.*, [15, 16] for more details) and prove a technical result needed later on. Section 4 is devoted to a brief account of Besicovitch almost periodicity. In Section 5 we study almost periodic second-order elliptic equations. The main result of the section, Theorem 5.1, provides sufficient conditions for the existence of almost periodic solutions, including the modulus containment property. The central part of the paper is Section 6 in which the main result, Theorem 6.1, is proven. In Theorem 6.2 we give certain sufficient conditions for the existence of constant sub- and super-solutions. In addition, we show that if each equation in the envelop has at most one bounded solution, the solution constructed in Theorem 6.1 is almost periodic in the sense of Bohr. In Section 7 we give an application of Theorem 6.1 to the pendulum equation with almost periodic forcing term.

Note that to simplify the notation we often denote by C a *generic* positive constant.

2. Bohr almost periodic functions and Bohr compactification

Let $C_b(\mathbb{R}^n)$ denote the space of all bounded continuous functions on $\mathbb{R}^n$. Endowed with the norm

$$\|f\| = \sup_{x\in\mathbb{R}^n} |f(x)|,$$

this is a Banach space. According to the Bochner definition, a function $f \in C_b(\mathbb{R}^n)$ is *almost periodic* (shortly, a.p.) in the sense of Bohr if the family of shifts $\{f(\cdot + y)\}_{y\in\mathbb{R}^n}$ is precompact in $C_b(\mathbb{R}^n)$. The set of all a.p. functions is a closed linear subspace of $C_b(\mathbb{R}^n)$, hence, a Banach space. We denote by $CAP(\mathbb{R}^n)$ the space of all almost periodic functions on $\mathbb{R}^n$.

An important property of a.p. functions is the existence of mean value. Let

$$K_T = \{x \in \mathbb{R}^n : |x_k| \le T, k = 1, 2, \dots, k\}.$$

The *mean value* of an almost periodic function f is defined by

$$\langle f\rangle = \lim_{T\to\infty} \frac{1}{(2T)^n} \int_{a+K_T} f(x)\, dx. \tag{2.1}$$

The limit in (2.1) exists uniformly with respect to $a \in \mathbb{R}^n$ and is independent of a.

The following statement is often considered as the main result on almost periodic functions. Let $\mathrm{Trig}(\mathbb{R}^n)$ be the space of all trigonometric polynomials, *i.e.*, finite sums of the form

$$\sum a_j \exp(\mathrm{i}\xi_j \cdot x),$$

where $a_j \in \mathbb{C}$, $\xi_j \in \mathbb{R}^n$ and

$$x \cdot y = \sum_{k=1}^{n} x_k y_k$$

is the standard dot product in $\mathbb{R}^n$.

Proposition 2.1 (Approximation Theorem). *The space* $\mathrm{Trig}(\mathbb{R}^n)$ *is a dense subspace of the Banach space* $CAP(\mathbb{R}^n)$.

See [20, Proposition 1.3 of Ch. 1].

The *Fourier-Bohr transform* of an almost periodic function f is defined by

$$\hat{f}(\xi) = \langle f(x) \exp(-\mathrm{i}\xi \cdot x)\rangle . \tag{2.2}$$

The set

$$\sigma(f) = \{\xi \in \mathbb{R}^n : \hat{f}(\xi) \neq 0\} \tag{2.3}$$

is called the *spectrum* of an almost periodic function f. It follows immediately from Proposition 2.1 that, for any a.p. function f, the set $\sigma(f)$ is at most countable. The additive subgroup $\mathrm{Mod}(f)$ of $\mathbb{R}^n$ generated by $\sigma(f)$ is called the *modulus* of the function f.

Now we give a brief description of Bohr compactifications of the (additive group of) space $\mathbb{R}^n$. The standard approach uses Pontryagin's duality theory (see, *e.g.*, [7, Ch. 6] for a detailed presentation of the theory of locally compact abelian groups including Pontryagin's duality). Consider $\mathbb{R}^n$ as a locally compact abelian group. Its dual group, $(\mathbb{R}^n)'$, consists of all characters which are, in this case, functions of the form $\exp(\mathrm{i}\xi \cdot x)$. The correspondence $\exp(\mathrm{i}\xi \cdot x) \mapsto \xi$ is an isomorphisms $(\mathbb{R}^n)' \simeq \mathbb{R}^n$. Denote by $(\mathbb{R}^n)'_d$ the group $(\mathbb{R}^n)'$ endowed with the discrete topology. We set $\mathbb{R}^n_B = ((\mathbb{R}^n)'_d)'$. This is a compact abelian group called the *Bohr compactification* of $\mathbb{R}^n$. Also we introduce the dual homomorphism

$$i_B : \mathbb{R}^n = (\mathbb{R}^n)'' \to \mathbb{R}^n_B = ((\mathbb{R}^n)'_d)'$$

to the identity homomorphism $(\mathbb{R}^n)'_d \to (\mathbb{R}^n)'$. The homomorphism i_B is injective and its image $i_B(\mathbb{R}^n)$ is a dense subgroup in $\mathbb{R}^n_B$.

In what follows, we need a more general notion of relative Bohr compactification. Let $\Gamma \subseteq (\mathbb{R}^n)'$ be an nonzero additive subgroup considered as a discrete group (later on we always suppose that $\Gamma \neq \{0\}$). The *Bohr compactification* of $\mathbb{R}^n$ *relative to* Γ is defined as $\mathbb{R}^n_{B,\Gamma} = \Gamma'$. The homomorphism

$$i_{B,\Gamma} : \mathbb{R}^n \to \mathbb{R}^n_{B,\Gamma}$$

is defined as the dual to the identity map $\Gamma \to \mathbb{R}^n$. Its image is still a dense subgroup of $\mathbb{R}^n_{B,\Gamma}$. The kernel $\ker i_{B,\Gamma}$ is a linear subspace of $\mathbb{R}^n$ orthogonal to the linear subspace of $(\mathbb{R}^n)'$ generated by Γ. If $\Gamma = (\mathbb{R}^n)'$, we return to the original Bohr compactification.

The main result on Bohr compactifications is the following

Proposition 2.2. *A function f on $\mathbb{R}^n$ is almost periodic, with* $\mathrm{Mod}(f) \subseteq \Gamma$, *if and only if f is of the form*

$$f(x) = \tilde{f}(i_{B,\Gamma}x) ,$$

where $\tilde{f}$ is a (unique) continuous function on $\mathbb{R}^n_{B,\Gamma}$.

See [20, Proposition 3.5 of Ch. 1]

Let f be an a.p. function and $\Gamma \supseteq \mathrm{Mod}(f)$. We use $\tilde{f}$ as a standard notation for the function on the Bohr compactification given in Proposition 2.2. For any $s \in \mathbb{R}^n_{B,\Gamma}$, we set

$$f^{(s)}(x) = \tilde{f}(s + i_{B,\Gamma}x) , \quad x \in \mathbb{R}^n .$$

Notice, that the map $s \mapsto f^{(s)}$ is a continuous map from $\mathbb{R}^n_{B,\Gamma}$ into $CAP(\mathbb{R}^n)$. The set of a.p. functions $H(f) = \{f^{(s)}\}_{s\in\mathbb{R}^n_{B,\Gamma}}$ is called the *envelope* of f. The envelope $H(f)$ is independent of the choice of $\Gamma \supseteq \mathrm{Mod}(f)$. Actually, $H(f)$ is the closure of the set of shifts $\{f(\cdot+y)\}_{y\in\mathbb{R}^n}$. In classical literature the later property is accepted as the definition of the envelope.

Also let us point out that, for any additive subgroup $\Gamma \subseteq \mathbb{R}^n$,

$$CAP_\Gamma(\mathbb{R}^n) = \{f \in CAP(\mathbb{R}^n) : \mathrm{Mod}(f) \subseteq \Gamma\}$$

is a closed linear subspace of $CAP(\mathbb{R}^n)$. By Proposition 2.2, the operator $J_\Gamma : f \mapsto \tilde{f}$ is an isometric isomorphism from the Banach space $CAP_\Gamma(\mathbb{R}^n)$ onto the Banach space $C(\mathbb{R}^n_{B,\Gamma})$.

Now we complement Proposition 2.2 with the following result that expresses the mean value of an almost periodic function in terms of Bohr compactification (see [20, 23]). Let $\mu = \mu_\Gamma$ be the Haar measure on $\mathbb{R}^n_{B,\Gamma}$, *i.e.*, a unique positive translation invariant measure such that $\mu(\mathbb{R}^n_{B,\Gamma}) = 1$ (see, *e.g.*, [7, Ch. 4]).

Proposition 2.3. *For every $f \in CAP_\Gamma(\mathbb{R}^n)$ we have*

$$\langle f\rangle = \int_{\mathbb{R}^n_{B,\Gamma}} \tilde{f}(s)\, d\mu(s)\,.$$

Also we need a refined version of Proposition 2.1.

Proposition 2.4. *Given a countable subgroup $\Gamma \subseteq \mathbb{R}^n$, there exists a sequence of trigonometric polynomials $P_m(x)$ with the following properties*

(a) $P_m(x) \geq 0$ *for all $x \in \mathbb{R}^n$;*
(b) $\langle P_m\rangle = 1$;
(c) *For any $f \in CAP_\Gamma(\mathbb{R}^n)$, the sequence of trigonometric polynomials*

$$f_m(x) = \langle f(y)P_m(x-y)\rangle_y = \langle f(x-y)P_m(y)\rangle_y$$

belongs to $CAP_\Gamma(\mathbb{R}^n)$ and converges to f in that space.

For a proof we refer to [20, 23]. The trigonometric polynomials P_m are called the *Bochner-Fejer kernels*, while f_m are the *Bochner-Fejer approximations* of f.

It is known (see, *e.g.*, [7, Section 24]) that a compact abelian group is metrizable if and only if its dual group is countable. Hence, the Bohr compactification $\mathbb{R}^n_{B,\Gamma}$ is metrizable whenever the subgroup $\Gamma \subseteq \mathbb{R}^n$ is countable. In the rest of the paper we accept the following convention: the symbol Γ denotes a *countable* subgroup of $\mathbb{R}^n$ so that $\mathbb{R}^n_{B,\Gamma}$ is metrizable.

3. Almost automorphic functions

Let us remind the notion of almost automorphic function due to S. Bochner. For details we refer to [15, 16] and references therein. A function $f \in C_b(\mathbb{R}^n)$ is *almost automorphic* if for every sequence $y_k \in \mathbb{R}^n$ there exists a subsequence $y_{k'}$ such that the pointwise limit

$$\lim f(x + y_{k'}) = g(x) \tag{3.1}$$

exists and

$$\lim g(x - y_{k'}) = f(x) \tag{3.2}$$

pointwise. Note that the function g in (3.1) is measurable, but not necessarily continuous.

An almost automorphic function f is *uniformly almost automorphic* if the limits in (3.1) and (3.2) are uniform on compact subsets of $\mathbb{R}^n$, *i.e.*, in the space $C(\mathbb{R}^n)$ which is a Fréchet space. Equivalently, f is uniformly almost automorphic if all functions g that appear in (3.1) are continuous (see [17]).

We denote by $AA(\mathbb{R}^n)$ (respectively, $AA_u(\mathbb{R}^n)$) the sets of all almost automorphic (respectively, uniformly almost automorphic) functions on $\mathbb{R}^n$. These are closed linear subspaces in $C_b(\mathbb{R}^n)$. Notice that

$$CAP(\mathbb{R}^n) \subset AA_u(\mathbb{R}^n) \subset AA(\mathbb{R}^n)$$

and all the inclusions are strict.

Proposition 3.1. *Let $\tilde{u}$ be a function on $\mathbb{R}^n_{B,\Gamma}$, where $\Gamma \subset \mathbb{R}^n$ is a countable subgroup. Suppose that for all $s \in \mathbb{R}^n_{B,\Gamma}$ the function $u^{(s)}(x) = \tilde{u}(s + i_{B,\Gamma}x)$ belongs to $C_b(\mathbb{R}^n)$ and is uniformly continuous. If the map*

$$U : \mathbb{R}^n_{B,\Gamma} \ni s \mapsto u^{(s)} \in C(\mathbb{R}^n)$$

is continuous at the point $s_0 \in \mathbb{R}^n_{B,\Gamma}$, then U is continuous at each point of the orbit $s_0 + i_{B,\Gamma}\mathbb{R}^n$ and $u^{(s_0)} \in AA_u(\mathbb{R}^n)$.

Proof. Suppose that $s'_m \to s'_0 = s_0 + t_0$ in $\mathbb{R}^n_{B,\Gamma}$, where $t_0 = i_{B,\Gamma}x_0$. Then $s_m = s'_m - t_0 \to s_0$ and, by continuity, $u^{(s_m)} \to u^{(s_0)}$ in the space $C(\mathbb{R}^n)$. Hence,

$$u^{(s'_m)}(\cdot) = u^{(s_m)}(\cdot + x_0) \to u^{(s_0)}(\cdot + x_0) = u^{(s'_0)}(\cdot)$$

in $C(\mathbb{R}^n)$, and the first statement of the proposition follows.

Denote by T the closure of the set

$$\{(s_0 + i_{B,\Gamma}y, u^{(s_0)}(\cdot + y)) : y \in \mathbb{R}^n\}$$

in the space $\mathbb{R}^n_{B,\Gamma} \times C(\mathbb{R}^n)$. Since the function $u^{(s_0)}$ is bounded and uniformly continuous, the Arzelà-Ascoli theorem implies that T is a compact subset of $\mathbb{R}^n_{B,\Gamma} \times C(\mathbb{R}^n)$. The projection of T on the first factor is a surjective map, while the image H of the other projection is the closure of the set $\{u^{(s_0)}(\cdot + y) : y \in \mathbb{R}^n\}$ in $C(\mathbb{R}^n)$, the so-called *hull* of $u^{(s_0)}$. We set

$$H_s = \{f \in C(\mathbb{R}^n) : (s, f) \in T\}$$

for any $s \in \mathbb{R}^n_{B,\Gamma}$. This is a non-empty closed subset of T.

First we show that $H_{s_0} = \{u^{(s_0)}\}$. Indeed, suppose that $(s_0, f) \in T$. Then there exists $s_m = i_{B,\Gamma} x_m$ such that $s_m \to 0$ in $\mathbb{R}^n_{B,\Gamma}$ and

$$u^{(s_0)}(\cdot + x_m) = u^{(s_0+s_m)} \to f .$$

The continuity of U at s_0 implies that $f = u^{(s_0)}$.

Now let us prove that the function $u^{(s_0)}$ is uniformly almost automorphic. Since T is compact, for any sequence $t_m = i_{B,\Gamma} x_m$, $x_m \in \mathbb{R}^n$, there exists a subsequence $t_{m'}$ such that $s_0 + t_{m'} \to t_0$ and $t_0 - t_{m'} \to s_0$ in $\mathbb{R}^n_{B,\Gamma}$, and $u^{(s_0)}(\cdot + x_{m'}) \to f$ and $f(\cdot - x_{m'}) \to h$ in $C(\mathbb{R}^n)$. Since T is closed, $(t_0, f) \in T$ and $(s_0, h) \in T$). Hence, $h \in H_{s_0}$. Since $H_{s_0} = \{u^{(s_0)}\}$, we conclude that $h = u^{(s_0)}$ and, therefore, $u^{(s_0)} \in AA_u(\mathbb{R}^n)$. This completes the proof. □

4. Besicovitch almost periodic functions

Let $L^p_{\text{loc}}(\mathbb{R}^n)$, $1 \le p \le \infty$, stand for the local Lebesgue space with the exponent p. For any $f \in L^p_{\text{loc}}(\mathbb{R}^n)$, $p < \infty$, we introduce the quantity

$$\|f\|_{(p)} = \limsup_{T\to\infty} \frac{1}{2T} \left[\int_{K_T} |f(x)|^p \, dx \right]^{1/p} . \tag{4.1}$$

Functions with finite semi-norm $\|f\|_{(p)}$ form the so-called *Marcinkiewicz space* $M^p(\mathbb{R}^n)$. It is easily seen that $M^p(\mathbb{R}^n) \subseteq M^q(\mathbb{R}^n)$ whenever $q \le p$.

A function $f \in M^p(\mathbb{R}^n)$, $p < \infty$, is *Besicovitch almost periodic*, with the exponent p, if there is a sequence $f_n \in CAP(\mathbb{R}^n)$ such that

$$\lim_{n\to\infty} \|f - f_n\|_{(p)} = 0 .$$

The set of all such functions is denoted by $B^p(\mathbb{R}^n)$. Obviously, $B^p(\mathbb{R}^n) \subseteq B^q(\mathbb{R}^n)$ if $q \le p$. It is not difficult to verify that, for any $f \in B^p(\mathbb{R}^n)$, 'lim sup' in (4.1) can be replaced by 'lim'.

The spaces $M^p(\mathbb{R}^n)$ and $B^p(\mathbb{R}^n)$ are complete semi-normed spaces, but *not* Banach spaces because the semi-norm $\|\cdot\|_{(p)}$ has a nontrivial kernel.

For Besicovitch a.p. functions the definition of mean value given in (2.1) makes sense. The only difference is that, in general, the limit is not uniform with respect to a. Therefore, the notions of Fourier-Bohr transform and spectrum extend immediately to Besicovitch a.p. functions. Moreover, the spectrum is at most countable. The modulus, $\operatorname{Mod}(f)$, of a Besicovitch a.p. function f is well defined as well. Moreover, for any subgroup $\Gamma \subseteq \mathbb{R}^n$, we set

$$B^p_\Gamma(\mathbb{R}^n) = \{f \in B^p(\mathbb{R}^n) : \operatorname{Mod}(f) \subseteq \Gamma\} .$$

Obviously, this is a linear subspace of $B^p(\mathbb{R}^n)$ closed in the sense that if $f_n \in B^p_\Gamma(\mathbb{R}^n)$ and $\|f_n - f\|_{(p)} \to 0$, then $f \in B^p_\Gamma(\mathbb{R}^n)$.

By Proposition 2.3, the operator J_Γ initially defined on $CAP_\Gamma(\mathbb{R}^n)$ extends uniquely to an isometric epimorphism

$$J_\Gamma : B^p_\Gamma(\mathbb{R}^n) \to L^p(\mathbb{R}^n_{B,\Gamma}) , \quad p \in [1, \infty) .$$

Its kernel consists of all functions $f \in B^p_\Gamma(\mathbb{R}^n)$ such that $\|f\|_{(p)} = 0$. However, the relation between Besicovitch a.p. functions and functions on Bohr compactifications is less straightforward than in the case of Bohr a.p. functions (Proposition 2.2). The following statement is a direct consequence of the Birkhoff ergodic theorem (see, *e.g.*, [4, Section VIII.7]).

Proposition 4.1. *Suppose that $\tilde{f} \in L^p(\mathbb{R}^n_{B,\Gamma})$. Then there exists a measurable subset $\Omega \subseteq \mathbb{R}^n_{B,\Gamma}$ such that $\mu(\Omega) = 1$, and for all $s \in \Omega$ the function*

$$f^{(s)}(x) = \tilde{f}(s + i_{B,\Gamma}x), \quad x \in \mathbb{R}^n,$$

belongs to $B^p_\Gamma(\mathbb{R}^n)$ and

$$\langle f^{(s)} \rangle = \int_{\mathbb{R}^n_{B,\Gamma}} \tilde{f}(z)\, d\mu(z).$$

Now we notice that the Bochner-Fejer approximations introduced in Proposition 2.4, (c), make sense for Besicovitch a.p. functions. Moreover, the following statement holds.

Proposition 4.2. *If $f \in B^p_\Gamma(\mathbb{R}^n)$, $p \in [1, \infty)$, and f_k is the sequence of Bochner-Fejer approximations for f, then $\|f - f_k\|_{(p)} \to 0$ as $k \to \infty$.*

See [20, Theorem 2.4 of Ch. 1].

Surprisingly enough, we did not find the following simple proposition in the existing literature.

Proposition 4.3. *Suppose that $f \in B^1(\mathbb{R}^n) \cap L^\infty(\mathbb{R}^n)$. Then $f \in B^p(\mathbb{R}^n)$ for all $p \in [1, \infty)$.*

Proof. Let $\Gamma = \mathrm{Mod}(f)$ and f_k be the sequence of Bochner-Fejer approximations for f. Making use of the properties of Bochner-Fejer kernels listed in Proposition 2.4, we deduce easily that

$$\|f_k\|_{L^\infty} \leq \|f\|_{L^\infty}.$$

Hence,

$$\|f - f_k\|^p_{(p)} = \langle |f - f_k|^p \rangle = \langle |f - f_k| |f - f_k|^{p-1} \rangle \leq (2\|f\|_{L^\infty})^{p-1} \|f - f_k\|_{(1)}.$$

This, together with Proposition 4.2, implies the required. □

5. Linear almost periodic problem

First we introduce certain functional spaces. For a detailed account of Hölder spaces on an arbitrary, not necessarily bounded, domain we refer to [9, Section 3.1]. In this and subsequent sections we consider real-valued functions only. Let $\alpha \in (0, 1)$. The space $C^\alpha_b(\mathbb{R}^n)$ consists of all functions $f \in C_b(\mathbb{R}^n)$ that satisfy the uniform Hölder condition with the exponent α:

$$[f]_\alpha = \sup_{x,y \in \mathbb{R}^n, x \neq y} \frac{|f(x) - f(y)|}{|x - y|^\alpha} < \infty.$$

This is a Banach space with respect to the norm

$$\|f\|_{C_b^\alpha} = \|f\| + [f]_\alpha .$$

The space

$$CAP^\alpha(\mathbb{R}^n) = CAP(\mathbb{R}^n) \cap C_b^\alpha(\mathbb{R}^n)$$

is a closed subspace of $C_b^\alpha(\mathbb{R}^n)$.

For any positive integer m, we denote by $C_b^m(\mathbb{R}^n)$ the space of all functions $f \in C_b(\mathbb{R}^n)$ such that all derivatives of f up to order m belong to $C_b(\mathbb{R}^n)$. This is a Banach space with respect to the norm

$$\|f\|_{C_b^m} = \sum_{k=0}^{m} \|D^k f\| ,$$

where $D^k f$ is a vector that consists of all kth derivatives of f, $D^0 f = f$. Similarly, we denote by $CAP^m(\mathbb{R}^n)$ the space of all a.p. functions having a.p. derivatives up to order m. This is a closed subspace of $C_b^m(\mathbb{R}^n)$.

Finally, the space $C_b^{m+\alpha}(\mathbb{R}^n)$ is the space of all functions $f \in C_b^m(\mathbb{R}^n)$ such that $D^m f \in C_b^\alpha(\mathbb{R}^m)$. Endowed with the norm

$$\|f\|_{C_b^{m+\alpha}} = \|f\|_{C_b^m} + [D^m f]_\alpha ,$$

this is a Banach space. We set

$$CAP^{m+\alpha}(\mathbb{R}^n) = CAP^m(\mathbb{R}^n) \cap C_b^{m+\alpha}(\mathbb{R}^n) .$$

This is a closed subspace of $C_b^{m+\alpha}(\mathbb{R}^n)$. Actually, it is easily seen that

$$CAP^{m+\alpha}(\mathbb{R}^n) = CAP(\mathbb{R}^n) \cap C_b^{m+\alpha}(\mathbb{R}^n) .$$

We use the following convention:

$$C_b^0(\mathbb{R}^n) = C_b(\mathbb{R}^n) \quad \text{and} \quad CAP^0(\mathbb{R}^n) = CAP(\mathbb{R}^n) .$$

Now we consider second-order elliptic operators of the form

$$Au(x) = -\sum_{i,j=1}^{n} a_{ij}(x)\frac{\partial^2 u(x)}{\partial x_i \partial x_j} + \sum_{i=1}^{n} b_i(x)\frac{\partial u(x)}{\partial x_i} + c(x)u(x) . \tag{5.1}$$

More precisely, we assume that

(i) *The matrix (a_{ij}) of leading coefficients is symmetric and there exists a constant $\lambda_0 > 0$ such that*

$$\sum_{i,j=1}^{n} a_{ij}(x)\xi_i\xi_j \geq \lambda_0|\xi|^2$$

for all $\xi = (\xi_1, \ldots, \xi_n) \in \mathbb{R}^n$ and $x \in \mathbb{R}^n$.

(ii) *There exists a constant $c_0 > 0$ such that $c(x) \geq c_0$ for all $x \in \mathbb{R}^n$.*

We do not exclude the case when $n = 1$. In this case, without loss of generality, we may assume that the leading coefficient is equal to 1 and the operator becomes

$$Au(x) = -u''(x) + b(x)u'(x) + c(x)u(x) . \tag{5.2}$$

We start with the following

Proposition 5.1. *Suppose that the operator A satisfies* (i) *and* (ii), *and its coefficients belong to $C_b^\alpha(\mathbb{R}^n)$, with $\alpha \in (0,1)$ if $n>1$ and $\alpha \in [0,1)$ if $n=1$. Then for any $f \in C_b^\alpha(\mathbb{R}^n)$ there exists a unique solution $u \in C_b^{2+\alpha}(\mathbb{R}^n)$ of the equation*

$$Au = f\,. \tag{5.3}$$

Moreover,

$$\|u\| \le c_0^{-1}\|f\|\,, \tag{5.4}$$

$$\|u\|_{C_b^1} \le C\|f\| \tag{5.5}$$

and

$$\|u\|_{C_b^{2+\alpha}} \le C\|f\|_{C_b^\alpha}\,, \tag{5.6}$$

where the constant $C>0$ depends only on λ_0, c_0 and norms of the coefficients in $C_b^\alpha(\mathbb{R}^n)$. In addition, if $f \ge 0$, then $u \ge 0$.

In the case when $\alpha \in (0,1)$, this is a well-known result (see, *e.g.*, [9, Theorems 4.3.1 and 4.3.2]) based on the so-called Schauder's *a priori* estimates. We mention that estimate (5.4) and the positivity result follow from the maximum principle (see [9, Theorem 2.9.2]). Less known estimate (5.5) follows from interior L^p estimates for elliptic equations [6, Section 9.5] and the Sobolev embedding theorem. In the case when $n=1$ and $\alpha=0$ the statement of the proposition is also well known and can be found, *e.g.*, in [8].

In the rest of the paper, in addition to (i) we impose the following almost periodicity assumption.

(iii) *The coefficients a_{ij}, b_j and c, $i,j=1,\dots,n$, belong to $CAP^\alpha(\mathbb{R}^n)$, where $\alpha \in (0,1)$ if $n>1$ and $\alpha \in [0,1)$ if $n=1$.*

We denote by $\mathrm{Mod}(A)$ the smallest additive subgroup of $\mathbb{R}^n$ that contains the spectra of all coefficients of A. Now let $\Gamma \subseteq \mathbb{R}^n$ be any countable subgroup that contains $\mathrm{Mod}(A)$. We introduce the *envelope* $H(A)=\{A^{(s)}\}_{s\in\mathbb{R}^n_{B,\Gamma}}$ of A by

$$A^{(s)}u(x) = -\sum_{i,j=1}^n a_{ij}^{(s)}(x)\frac{\partial^2 u(x)}{\partial x_i \partial x_j} + \sum_{i=1}^n b_i^{(s)}(x)\frac{\partial u(x)}{\partial x_i} + c^{(s)}(x)u(x)\,.$$

As in the case of functions, the *set* $H(A)$ is independent of the choice of $\Gamma \supseteq \mathrm{Mod}(A)$, while the parametrization of the envelope does depend on Γ. It is easily seen that if A satisfies (i), (ii) and (iii), then all operators $A^{(s)}$ in the envelope satisfy the same assumptions.

The key result of the section is

Theorem 5.1. *Assume* (i)–(iii). *If $f \in CAP^\alpha(\mathbb{R}^n)$, then equation* (5.3) *has a unique solution $u \in CAP^{2+\alpha}(\mathbb{R}^n)$. In addition,*

$$\mathrm{Mod}(u) \subseteq \mathrm{Mod}(A) + \mathrm{Mod}(f)\,. \tag{5.7}$$

Proof. Let $\Gamma \supseteq \mathrm{Mod}(A)$ be any countable subgroup of $\mathbb{R}^n$ such that $f \in CAP_\Gamma(\mathbb{R}^n)$. By Proposition 5.1, equation (5.3) has a unique solution $u \in C_b^{2+\alpha}(\mathbb{R}^n)$.

We have to show that $u \in CAP_\Gamma(\mathbb{R}^n)$ which implies that $u \in CAP_\Gamma^{2+\alpha}(\mathbb{R}^n)$. With this aim we consider the family of equations

$$A^{(s)}u_s = f^{(s)}, \quad s \in \mathbb{R}^n_{B,\Gamma}.$$

By Proposition 5.1, each of these equations has a unique solution $u_s \in C_b^{2+\alpha}(\mathbb{R}^n)$, with $u_0 = u$.

Notice that the map $s \mapsto u_s$ is a continuous mapping from $\mathbb{R}^n_{B,\Gamma} \to C_b(\mathbb{R}^n)$. Indeed, by inequality (5.4), for any $s_0, s \in \mathbb{R}^n_{B,\Gamma}$

$$\|u_s - u_{s_0}\| \le c_0^{-1}\|A^{(s)}(u_s - u_{s_0})\| \le c_0^{-1}(\|f^{(s)} - f^{(s_0)}\| + \|(A^{(s_0)} - A^{(s)})u_{s_0}\|).$$

Since $f \in CAP_\Gamma(\mathbb{R}^n)$ and the coefficients of A satisfy almost periodicity assumption (ii), the right-hand side of the last inequality tends to zero as $s \to s_0$ in $\mathbb{R}^n_{B,\Gamma}$ and the conclusion follows.

As consequence, $\tilde{u}(s) = u_s(0)$ is a well-defined continuous function on $\mathbb{R}^n_{B,\Gamma}$. Furthermore, due to the uniqueness of bounded solution (see Proposition 5.1),

$$u_{s+i_{B,\Gamma}x}(y) = u_s(y+x), \quad \forall s \in \mathbb{R}^n_{B,\Gamma}, x \in \mathbb{R}^n \text{ and } y \in \mathbb{R}^n.$$

Hence,

$$u(x) = u_0(x) = \tilde{u}(i_{B,\Gamma}x),$$

and, by Proposition 2.2, $u \in CAP_\Gamma(\mathbb{R}^n)$. □

6. Semi-linear problem

In this section we consider semi-linear equations

$$Au(x) = g(x, u(x)), \quad x \in \mathbb{R}^n, \tag{6.1}$$

where A is a second-order elliptic operator of the form (5.1). We always suppose that A satisfies assumptions (i) and (iii). Assumption (ii) is not needed in general. In addition, we impose the following assumption on the nonlinearity g.

(iv) *For any $R > 0$ the function $g(x,u)$ is almost periodic in $x \in \mathbb{R}^n$ uniformly with respect to $u \in \mathbb{R}$, with $|u| \le R$, and there exists a constant $C_R > 0$ such that*

$$|g(x,u) - g(x,v)| \le C_R|u - v|$$

for all $x \in \mathbb{R}^n$ and $u, v \in \mathbb{R}$, with $|u| \le R$ and $|v| \le R$, and

$$|g(x,u) - g(y,v)| \le C_R|x-y|^\alpha$$

for all $x, y \in \mathbb{R}^n$ and $u \in \mathbb{R}$, with $|u| \le R$, provided $\alpha \ne 0$ in assumption (iii).

In particular, assumption (iv) implies that the function $g(x,u)$ can be considered as an a.p. function of $x \in \mathbb{R}^n$ with values in the Fréchet space $C(\mathbb{R})$ of continuous functions on $\mathbb{R}$ endowed with the topology of uniform convergence on compact intervals. Hence, $\bigcup_{u \in \mathbb{R}} \sigma(g(\cdot,u))$ generates a countable subgroup $\mathrm{Mod}(g) \subseteq \mathbb{R}^n$.

Moreover, for any $\Gamma \supseteq \mathrm{Mod}(g)$ there exists a unique continuous function $\tilde{g}$ on $\mathbb{R}^n_{B,\Gamma}$ such that

$$g(x,u) = \tilde{g}(i_{B,\Gamma}x, u), \quad (x,u) \in \mathbb{R}^n \times \mathbb{R}.$$

If $\Gamma \supseteq \mathrm{Mod}(g)$ is a countable subgroup of $\mathbb{R}^n$, we set

$$g^{(s)}(x,u) = \tilde{g}(s + i_{B,\Gamma}x, u), \quad (x,u) \in \mathbb{R}^n \times \mathbb{R}, \quad s \in \mathbb{R}^n_{B,\Gamma}.$$

Notice that all the functions $g^{(s)}$ satisfy assumption (iv).

Together with equation (6.1) we consider the following family of equations

$$A^{(s)}u(x) = g^{(s)}(x, u(x)), \quad x \in \mathbb{R}^n, \tag{6.2}$$

where $s \in \mathbb{R}^n_{B,\Gamma}$ and $\Gamma \supseteq \mathrm{Mod}(A) + \mathrm{Mod}(g)$ will be fixed later.

A function $\overline{u} \in CAP^{2+\alpha}(\mathbb{R}^n)$ (respectively, $\underline{u} \in CAP^{2+\alpha}(\mathbb{R}^n)$) is called a *super-solution* (respectively, a *sub-solution*) of equation (6.1) if

$$A\overline{u}(x) \geq g(x, \overline{u}(x))$$

(respectively,

$$A\underline{u}(x) \leq g(x, \underline{u}(x)) \;)$$

for all $x \in \mathbb{R}^n$. Given super- and sub-solutions, $\overline{u}$ and $\underline{u}$, we set

$$\Gamma = \mathrm{Mod}(A) + \mathrm{Mod}(g) + \mathrm{Mod}(\overline{u}) + \mathrm{Mod}(\underline{u}).$$

Notice that the functions $\overline{u}^{(s)}$ and $\underline{u}^{(s)}$ are super- and sub-solutions for equation (6.2) for all $s \in \mathbb{R}^n_{B,\Gamma}$.

Theorem 6.1. *Under assumptions* (i), (iii) *and* (iv) *suppose that there exist sub- and super-solutions for equation* (6.1) *such that* $\underline{u} \leq \overline{u}$. *Then for every* $s \in \mathbb{R}^n_{B,\Gamma}$ *there exists a solution* $u_s \in C_b^{2+\alpha}(\mathbb{R}^n)$ *of equation* (6.2) *such that* $\underline{u}^{(s)} \leq u_s \leq \overline{u}^{(s)}$. *Furthermore, there exist a residual set* $\Omega_{aa} \subseteq \mathbb{R}^n_{B,\Gamma}$ *and a set* $\Omega_b \subseteq \mathbb{R}^n_{B,\Gamma}$ *of measure* 1 *both translation invariant and such that* u_s *is uniformly almost automorphic is* $s \in \Omega_{aa}$ *and* $u_s \in B^p_\Gamma(\mathbb{R}^n)$ *for all* $p \in [1,\infty)$ *if* $s \in \Omega_b$.

Proof. Replacing $c(x)$ by $c(x)+\theta$ and $g(x,u)$ by $g(x,u)+\theta u$, we may suppose, due to assumption (iv), that $c(x) \geq c_0 > 0$ and the nonlinearity $g(x,u)$ is increasing in $u \in [\inf \underline{u}, \sup \overline{u}]$.

Let us consider a sequence of functions u_k defined recurrently as follows. We set $u_0 = \overline{u}$. Next, $u_{k+1} \in CAP^{2+\alpha}_\Gamma(\mathbb{R}^n)$ is defined as a unique solution of the equation

$$Au_{k+1}(x) = g(x, u_k(x)). \tag{6.3}$$

The sequence u_k is well defined. Indeed, if $u_k \in CAP^{2+\alpha}_\Gamma(\mathbb{R}^n)$, then, by assumption (iv), $g(x, u_k(x)) \in CAP^\alpha_\Gamma(\mathbb{R}^n)$. By Theorem 5.1, equation (6.3) has a unique solution in $CAP^{2+\alpha}_\Gamma(\mathbb{R}^n)$. Moreover, since $\overline{u}$ is a super-solution, the positivity statement of Proposition 5.1 implies that the sequence u_k is monotone decreasing, *i.e.*, $u_{k+1} \leq u_k$. Since $u_k \in CAP^{2+\alpha}(\mathbb{R}^n)$, the function u_k extends to a unique function $\tilde{u}_k \in C(\mathbb{R}^n_{B,\Gamma})$, and the sequence of functions $\tilde{u}_k$ is monotone decreasing and bounded. Hence, the sequence $\tilde{u}_k$ converges to a measurable function $\tilde{u}$ pointwise on $\mathbb{R}^n_{B,\Gamma}$.

We are going to prove that the function

$$u^{(s)}(x) = \tilde{u}(s + i_{B,\Gamma}x)$$

is actually a $C_b^{2+\alpha}$-solution of equation (6.2). It is easily seen that, for any $s \in \mathbb{R}^n_{B,\Gamma}$, the functions

$$u_k^{(s)}(x) = \tilde{u}_k(s + i_{B,\Gamma}x)$$

satisfy

$$A^{(s)}u_{k+1}^{(s)}(x) = g^{(s)}(x, u_k^{(s)}(x)) . \tag{6.4}$$

As a consequence, $u_k^{(s)} \in CAP^{2+\alpha}(\mathbb{R}^n)$ for all $s \in \mathbb{R}^n_{B,\Gamma}$ and integer k.

We claim that there exists a constant $C > 0$ independent of s and k such that

$$\|u_k^{(s)}\|_{C_b^{2+\alpha}} \le C . \tag{6.5}$$

Indeed, denoting by C a generic positive constant independent of s and k, we have that $\|u_k^{(s)}\| \le C$. Assumption (iv) implies easily that

$$\|g^{(s)}(\cdot, u_k^{(s)})\| \le C .$$

Equation (6.4) and estimate (5.5) of Proposition 5.1 imply that

$$\|u_k^{(s)}\|_{C_b^\alpha} \le \|u_k^{(s)}\|_{C_b^1} \le C$$

(the first estimate is trivial). By assumption (iv),

$$\|g^{(s)}(\cdot, u_k^{(s)})\|_{C_b^\alpha} \le C ,$$

and estimate (6.5) follows from inequality (5.6) of Proposition 5.1.

Suppose that $\alpha > 0$. By (6.5), the functions $u_k^{(s)}$ and their derivatives up to second order are equicontinuous. Since $u_k^{(s)} \to u^{(s)}$ pointwise, by the Arzelà-Ascoli theorem, $u^{(s)} \in C_b^{2+\alpha}(\mathbb{R}^n)$ and $u_k^{(s)}$ converges to $u^{(s)}$ uniformly on compact sets together with derivatives up to second order. Passing to the limit in equation (6.4), we obtain that $u^{(s)}$ is a solution of equation (6.2).

Now suppose that $\alpha = 0$ and $n = 1$. By (6.5), the functions $u_k^{(s)}$ and their first derivatives are equicontinuous. As above, $u^{(s)} \in C_b^1(\mathbb{R}^n)$ and $u_k^{(s)}$ converges to $u^{(s)}$ uniformly on compact sets together with first derivatives. Equation (6.4) can be expressed as follows

$$-(u_{k+1}^{(s)})'' = -b^{(s)}(x)(u_{k+1}^{(s)})' - c^{(s)}(x)u_{k+1}^{(s)} + g^{(s)}(x, u_k^{(s)}) .$$

Hence, second derivatives converges uniformly on compact sets, and $u^{(s)} \in C_b^2(\mathbb{R}^n)$ satisfies equation (6.2).

Since $\tilde{u} \in \mathbb{R}^n_{B,\Gamma}$, the existence of the set Ω_b follows from Proposition 4.1. On the other hand, the map $U : s \mapsto u^{(s)}$ from $\mathbb{R}^n_{B,\Gamma}$ into $C(\mathbb{R}^n)$ is of the first Baire class as a pointwise limit of continuous maps. It is well known that any map of the first Baire class from a complete metric space into a separable metric space is continuous on a residual set (see, *e.g.*, [10, Section 31.X, Theorem 1]). Hence, the existence of the set Ω_{aa} follows from Proposition 3.1. □

Remark 6.1. Bounded solutions $u^{(s)}$ obtained in the proof of Theorem 6.1 are maximum solutions between $\underline{u}^{(s)}$ and $\overline{u}^{(s)}$. Starting the iteration process with $u_0 = \underline{u}$, we obtain the minimum solution.

Remark 6.2. The function $\tilde{u}$ constructed in the proof of Theorem 6.1 can be considered as a generalized solution of equation (6.1) in the sense of Sobolev-Besicovitch spaces [20].

Now we give general sufficient conditions for sub- and super-solutions to exist.

Theorem 6.2. *Assume* (i)–(iii). *Suppose that the nonlinearity g is of the form*

$$g(x,u) = g_1(x,u) + g_2(x,u)\,,$$

where both g_1 and g_2 satisfy assumption (iv), *$\frac{\partial g_2}{\partial u}$ is a continuous function on $\mathbb{R}^n \times \mathbb{R}$, and*

$$|g_1(x,u)| \le C \tag{6.6}$$

and

$$\frac{\partial g_2}{\partial u}(x,u) \le 0 \tag{6.7}$$

for all $(x,u) \in \mathbb{R}^n \times \mathbb{R}$. Then the conclusion of Theorem 6.1 *holds with*

$$\Gamma = \mathrm{Mod}(A) + \mathrm{Mod}(g)\,.$$

Proof. (a) *Reduction to the case when $g_2 = 0$.* If $u \in C_b^{2+\alpha}(\mathbb{R}^n)$ is a solution of equation (6.1), then

$$Au(x) + h(x)u = g_1(x,u(x))\,, \quad \text{where} \quad h(x) = \int_0^1 \frac{\partial g_2}{\partial u}(x, tu(x))\,dt\,.$$

It is easy to see that $h(x) \ge 0$ and, by estimate (5.4) of Proposition 5.1,

$$\|u\| \le C_1\,, \tag{6.8}$$

where $C_1 > 0$ is independent of the solution. Modifying g_2 outside the region $|u| \ge 2C_1$, we may assume that $g = g_1$ is a bounded function.

(b) *The case when $g = g_1$ is bounded.* Let

$$\theta = \sup\left\{\left|\frac{\partial g}{\partial u}(x,u)\right| \,:\, x \in \mathbb{R}^n, |u| \le C_1\right\}\,,$$

where C_1 is the constant in estimate (6.8). Obviously, we may assume that $C_1 \ge C$. Equation (6.1) is equivalent to the equation

$$Au + \theta u = g(x,u) + \theta u$$

and the function $g(x,u) + \theta u$ is increasing in the region $|u| \le C_1$. We define $\overline{u} \in CAP_\Gamma^{2+\alpha}(\mathbb{R}^n)$ as a unique nonnegative solution of equation $Au = C$ which exists by Proposition 5.1 and set $\underline{u} = -\overline{u}$. It is easy to verify that these are super- and sub-solutions, respectively, and we conclude by Theorem 6.1. □

Remark 6.3. If equation (6.1) possesses at most one solution between $\underline{u}$ and $\overline{u}$, then the map $U : \mathbb{R}^n_{B,\Gamma} \mapsto C(\mathbb{R}^n)$ considered in the proof of Theorem 6.1 is continuous at the point $s = 0$ and, hence, the solution $u = u^{(0)}$ of equation (6.1) is uniformly almost automorphic. If the uniqueness of solution between $\inf \underline{u}$ and $\sup \overline{u}$ holds for all equations (6.2), $s \in \mathbb{R}^n_{B,\Gamma}$, then all solutions $u^{(s)}$ are almost periodic because the function $\tilde{u}(s) = u(s)(0)$ is continuous on $\mathbb{R}^n_{B,\Gamma}$.

Corollary 6.1. *Under the assumptions of Theorem* 6.1, *suppose in addition that* $c(x) \geq 0$ *on* $\mathbb{R}^n$, *while*

$$\frac{\partial g}{\partial u} \leq -\kappa < 0$$

and is uniformly continuous on the strip $\mathbb{R}^n \times [\inf \underline{u}, \sup \overline{u}]$. *Then equation* (6.1) *has a unique solution* $u \in CAP^{2+\alpha}_{\Gamma}$ *between* $\underline{u}$ *and* $\overline{u}$.

Proof. Due to Remark 6.3, we have to verify the uniqueness needed there. Let us mention that $\partial g^{(s)}/\partial u$ is bounded and continuous on $\mathbb{R}^n \times [\inf \underline{u}, \sup \overline{u}]$ for all $s \in \mathbb{R}^n_{B,\Gamma}$. If u_1 and u_2 are two solutions of equation (6.2) between $\underline{u}$ and $\overline{u}$, then $v = u_1 - u_2$ satisfies

$$A^{(s)}v + h(x)v = 0$$

where

$$h(x) = -\int_0^1 \frac{\partial g^{(s)}}{\partial u}(x, tu_1(x) + (1-t)u_2(x))\, dt \geq \kappa > 0\,.$$

By Proposition 5.1, $v = 0$ and we conclude. □

Remark 6.4. The statement of Corollary 6.1 remains valid if we replace the assumptions $c(x) \geq 0$ and $\partial g/\partial u < -\kappa$ by $c(x) \geq c_0 > 0$ and $\partial g/\partial u \leq 0$, respectively.

Finally, we mention that Theorem 6.1 covers the case when the coefficients and sub- and super-solution are periodic. In that case it implies the existence of a periodic solution – a statement well known in the literature. Indeed, in the periodic case $\mathbb{R}^n_{B,\Gamma}$ is a torus. Hence, the sets $\Omega_a a$ and Ω_b, being translation invariant, coincide with the whole of the torus.

7. Almost periodically forced pendulum

As an application of our main result, consider the pendulum equation

$$u'' + cu' + a \sin u = h(t) \tag{7.1}$$

with an almost periodic forcing term $h(t)$. Here $a > 0$ and the damping coefficient $c \geq 0$ so that the undamped case is allowed. The envelope of equation (7.1) is

$$u'' + cu' + a \sin u = h^{(s)}(t)\,, \quad s \in \mathbb{R}_{B,\Gamma}\,, \tag{7.2}$$

where $\Gamma = \mathrm{Mod}(h)$.

Theorem 7.1. *Let $h \in CAP(\mathbb{R})$ and $\Gamma = \mathrm{Mod}(h)$.*

(a) *If $\|h\| = a$, then for every $s \in \mathbb{R}_{B,\Gamma}$ there exists a solution $u_s \in C_b^2(\mathbb{R})$ of equation (7.2) such that*

$$\frac{\pi}{2} \leq u_s(t) \leq \frac{3\pi}{2}, \quad \forall t \in \mathbb{R}.$$

Furthermore, there exist a residual subset $\Omega_{aa} \subseteq \mathbb{R}_{B,\Gamma}$ and a measurable subset $\Omega_b \subseteq \mathbb{R}_{B,\Gamma}$ of measure 1 both translation invariant and such that u_s is uniformly almost automorphic if $s \in \Omega_{aa}$ and $u_s \in B^p_\Gamma(\mathbb{R}^n)$ for all $p \in [1,\infty)$ if $s \in \Omega_b$.

(b) *If $\|h\| < a$, then equation (7.1) has a unique solution $u \in CAP^2(\mathbb{R})$ such that*

$$\frac{\pi}{2} < u(t) < \frac{3\pi}{2}, \quad \forall t \in \mathbb{R}.$$

Moreover, $\mathrm{Mod}(u) \subseteq \Gamma$.

Proof. (a) It is easy to verify that $\underline{u} = \pi/2$ and $\overline{u} = 3\pi/2$ are sub- and super-solutions, and the result follows from Theorem 6.1.

(b) If $\delta > 0$ is sufficiently small, then $\underline{u} = \pi/2 + \delta$ and $\overline{u} = 3\pi/2 - \delta$ are sub- and super-solutions. Since the derivative $(a \sin u)' = a \cos u$ is strictly negative on $[\pi/2 + \delta, 3\pi/2 - \delta]$, Corollary 6.1 applies and we conclude. □

Let us point out that if $h(t)$ is periodic and $\|h\| \leq a$, then there is a periodic solution between $\pi/2$ and $3\pi/2$. The uniqueness of such solution takes place whenever $\|h\| < a$.

The existence of a bounded solution under the assumptions of Theorem 7.1(a), as well as the almost periodicity of a unique bounded solution in case (b), is obtained in [5, 13] (see also [14]). In [5] the sub- and super-solution approach based on the Schauder fixed point theorem is used, while the proofs of [13] make use of an early result of Z. Opial which can be considered as a simple version of the sub- and super-solution method. In our approach we employ relative Bohr compactifications together with monotone iteration techniques. This permits us to obtain an extra information about the almost automorphy and Besicovitch almost periodicity of solutions to the equations in the envelope of (7.1) as well as the modules containment property. According to a remark of Mawhin [13], all these results can be considered as an improvement of a result of [1] which provides the existence of a weak Besicovitch almost periodic solution to equation (7.1) (*cf.* Remark 6.2) by means of certain variational techniques. It is interesting that in [1] the same interval $[\pi/2, 3\pi/2]$ appears.

Acknowledgment

We are grateful to the referee for his/her careful reading and valuable suggestions.

References

[1] J. Blot, *Almost periodically forced pendulum*, Funkcialaj Ekvacioj, **36** (1993), 235–250.

[2] R. Courant, D. Gilbert, *Methods of Mathematical Physics, V. II. Partial Differential Equations*, Wiley, New York, 1962.

[3] C. De Coster, P. Habets, *Two-Point Boundary Value Problems: Lower and Upper Solutions*, Elsevier, Amsterdam, 2006.

[4] N. Dunford, J. Schwartz, *Linear Operators; Part I: General Theory*, Wiley, New York, 1957.

[5] G. Fournier, A. Szulkin, M. Willem, *Semilinear elliptic equations in $\mathbb{R}^n$ with almost periodic or unbounded forcing term*, SIAM J. Math. Anal., **27** (1996), 1653–1660.

[6] D. Gilbarg, N. Trudinger, *Elliptic Partial Differential Equations of Second Order*, Springer, Berlin, 1998.

[7] E. Hewitt, K. Ross, *Abstract Harmonic Analysis*, V. 1, Springer, Berlin, 1979.

[8] M.A. Krasnosel'skii, V.Sh. Burd, Yu.S. Kolesov, *Nonlinear Almost Periodic Oscillations*, Wiley, New York, 1973 (English Edition).

[9] N.V. Krylov, *Lectures on Elliptic and Parabolic Equations in Hölder Spaces*, Amer. Math. Soc., Providence, R.I., 1996.

[10] K. Kuratowski, *Topology*, V. 1, Academic Press, New York, 1996 (English Edition).

[11] A. Leung, *Nonlinear Systems of Partial Differential Equations. Applications to Life and Physical Sciences*, World Sci. Publ., Nackensack, NJ, 2009.

[12] B.M. Levitan, V.V. Zhikov, *Almost Periodic Functions and Differential Equations*, Cambridge Univ. Press, Campbridge, 1982 (English Edition).

[13] L. Mawhin, Remarques sur les solutions bornées ou quasi-périodiques de l'équation du pendule forcé, *Ann. Sci. Math. Québec*, **22** (1998), 213–224.

[14] J. Mawhin, *Global results for the forced pendulum equation*, Handbook of Differential Equations, Ordinary Differential Equations, 1 (eds. A. Cañada, P. Drabek, A. Fonda), Elsevier-North-Holland, Amsterdam, 2004, 533–589.

[15] G.M. N'Guérékata, *Almost Automorphic and Almost Periodic Functions in Abstract Spaces*, Kluwer, New York, 2001.

[16] G.M. N'Guérékata, *Topics in Almost Automorphy*, Springer, New York, 2005.

[17] G.M. N'Guérékata, *Comments on almost automorphic and almost periodic functions in Banach spaces*, Far East J. Math. Sci. (FJMS), **17** no. 3 (2005), 337–344.

[18] D. Ortega, M. Tarallo, *Almost periodic upper and lower solutions*, J. Differential Equations, **193** (2003), 343–358.

[19] A. Pankov, *Nonlinear second order elliptic equations with almost periodic coefficients*, Ukr. Mat. Zh., **35** no 5 (1983), 649–652; English translation: Ukr. Math. J., **35** no 5 (1983), 564–567.

[20] A. Pankov, *Bounded and Almost Periodic Solutions of Nonlinear Operator Differential Equations*, Kluwer, Dordrecht, 1990 (English Edition).

[21] V. Radŭlescu, *Qualitative Analysis of Nonlinear Partial Differential Equations: Monotonicity, Analytic, and Variational Methods*, Hindawi Publ. Corp., New York, 2008.

[22] K. Schmitt, J.R. Ward, *Almost periodic solutions of nonlinear second order differential equations*, Results Math., **21** (1992), 190–199.

[23] M.A. Shubin, *Almost periodic functions and differential equations*, Uspekhi Mat. Nauk, **33** no 2 (1978), 3–47; English translation: Russian Math. Surv., **33** no 2 (1978), 1–52.

[24] J. Smoller, *Shock Waves and Reaction-Diffusion Equations*, Springer, New York, 1994.

Gaston M. N'Guérékata and Alexander Pankov
Department of Mathematics
Morgan State University
1700 E. Cold Spring Lane
Baltimore, MD 21251, USA
e-mail: Gaston.N'Guerekata@morgan.edu
Alexander.Pankov@morgan.edu

Operator Theory:
Advances and Applications, Vol. 228, 293–323

Morrey-Campanato Spaces: an Overview

Humberto Rafeiro, Natasha Samko and Stefan Samko

Dedicated to the 70th anniversary of Professor Vladimir Rabinovich

Abstract. In this paper we overview known and recently obtained results on Morrey-Campanato spaces with respect to the properties of the spaces themselves, that is, we do not touch the study of operators in these spaces. In particular, we overview equivalent definitions of various versions of the spaces, the so-called φ- and θ-generalizations, structure of the spaces, embeddings, dual spaces, etc.

Mathematics Subject Classification (2010). Primary 46E30; Secondary 46E35.

Keywords. Morrey spaces; Campanato spaces; embeddings; preduals; interpolation; variable exponents.

Contents

1. Introduction 294
2. Morrey spaces 295
2.1. Classical Morrey spaces 295
2.1.1. Embeddings in Morrey spaces 296
2.1.2. Hölder's inequality 297
2.1.3. Weak Morrey spaces 297
2.1.4. Interpolation 298
2.1.5. Preduals 298
2.1.6. Vanishing Morrey spaces $VL^{p,\lambda}$ 299
2.1.7. Different underlying spaces 299
2.1.8. Anisotropic Morrey spaces 300
2.1.9. Miscellaneous 301
2.2. Morrey spaces over $\mathbb{R}^n$ in case of a general measure 302

H.R. (Grant SFRH/BPD/63085/2009) and N.S. (Grant SFRH/BPD/34258/2006) gratefully acknowledge financial support by *Fundação para a Ciência e a Tecnologia* (FCT), Portugal.

3. Generalized Morrey spaces 302
3.1. φ-generalizations 303
3.2. θ-generalizations 305
4. Campanato spaces 306
4.1. Definitions and basic facts 307
4.2. DeVore-Sharpley-Christ versions of Campanato-type spaces 309
4.3. φ-generalization 309
4.4. Interpolation results 312
4.5. Other characterizations of Campanato spaces 313
4.6. Miscellaneous 314
5. Variable exponent Morrey and Campanato spaces 315
References 317

1. Introduction

We started our studies of various operators in Morrey and Campanato-type spaces several years ago, mainly in the case of maximal, singular and potential operators in such spaces with variable exponents and Hardy operators in Morrey spaces with constant exponents. We discovered that there existed a vast bibliography on the subject counting many hundreds of publications, especially on applications to differential equations. They include in particular the books A. Kufner, O. John and S. Fučík [63] (1977) and M. Giaquinta [40] (1983). We refer also to Section 27 of the book O.V. Besov, V.P. Il′in and S.M. Nikol′skiĭ [13] (1996) (see also the English translation [14, 15] of the first Russian edition of [13]) where an important overview on anisotropic Morrey type spaces may be found.

The earliest overview on Morrey-Campanato spaces seems to be first given in the paper J. Peetre [86] (1969). Probably the next one was M.H. Taibleson and G. Weiss [104] (1979).

During the last several decades there was a kind of a boom in studies in Morrey-Campanato-type spaces and their usage in applications, both enriching each other. Many of them, as well as various old results, were not covered in the existing surveys or books, but were of interest.

In the study of this topic and search of references, also in the historical retrospective, we made many notes in our notebooks. Our personal overview of those notes led us to the idea to collect and edit them, and publish it as a survey which may be useful for others involved into research around the Morrey-Campanato-type spaces.

This however led us to a manuscript exceeding one hundred pages, which is not well suited for a paper. About 4/5 of that overview was naturally related to the study of various operators, mainly classical operators of harmonic analysis, in Morrey-Campanato-type spaces, and about 1/5 of it was connected with the spaces themselves, i.e., proper definitions of various versions of the spaces, study of the structure of the spaces, preduals, etc. We made a decision to restrict ourselves to this first portion. It is presented in this paper. We hope to submit the remaining part for publication elsewhere. Note that in this paper we do not touch Sobolev-Morrey and Besov-Morrey type spaces as well as other generalizations of such a kind and refer a potential reader to Section 27 of the above cited books [13–15] and the recent book [114] (2010) titled "*Morrey and Campanato Meet Besov, Lizorkin and Triebel*".

The subjects we touch in this overview may be seen from Contents. Inside every Subsection we mainly follow the chronological order which more or less corresponds to a natural way of generalization from the simple to more advanced.

We could have lost some references. Anyway, we tried to do our best through a vast search in MathSciNet, MathNetRu and other sources. In the case the overview occasionally proves to be not complete in this or other item, we will be grateful to the readers for the indication of possible omissions. To be clear, we emphasize once again that in this survey we do not touch mapping properties of operators, so that many important papers on the behaviour of the classical operators of harmonic analysis in Morrey and Campanato spaces remained beyond this overview. We are aware of the fact that sometimes such a separation is rather relative because any property of an operator in a space may be considered as a property of the space. Nevertheless we had to follow the choice we made. Otherwise we would exceed any reasonable limit for this paper.

2. Morrey spaces

2.1. Classical Morrey spaces

The spaces which bear the name of Morrey spaces were introduced in 1938 by C. Morrey [71] in relation to regularity problems of solutions to partial differential equations.

We start from the definition of these spaces. Let $\Omega \subseteq \mathbb{R}^n$ be an open set. We denote $\widetilde{B}(x,r) = B(x,r) \cap \Omega, x \in \Omega, r > 0$, and $|A|$ will stand for the Lebesgue measure of a measurable subset in $\mathbb{R}^n$.

Definition 2.1 (Morrey spaces). Let $1 \leq p < \infty$ and $\lambda \geq 0$. The *Morrey space* $L^{p,\lambda}(\Omega)$ is defined as

$$L^{p,\lambda}(\Omega) = \left\{ f \in L^p(\Omega) : \sup_{x\in\Omega; r>0} \frac{1}{r^\lambda} \int_{\widetilde{B}(x,r)} |f(y)|^p \,\mathrm{d}y < \infty \right\}. \tag{1}$$

This is a Banach space with respect to the norm

$$\|f\|_{L^{p,\lambda}(\Omega)} := \sup_{x\in\Omega; r>0} \left(\frac{1}{r^\lambda} \int_{\widetilde{B}(x,r)} |f(y)|^p \,\mathrm{d}y \right)^{1/p}. \tag{2}$$

The space $L^{p,\lambda}(\Omega)$ is trivial when $\lambda > n$ ($L^{p,\lambda}(\Omega) = \{0\}$) and $L^{p,0}(\Omega) \cong L^p(\Omega)$ and $L^{p,n}(\Omega) \cong L^\infty(\Omega)$. In the case $\lambda \in (0, n]$, the space $L^{p,\lambda}(\Omega)$ is non-separable.

Note that for these spaces sometimes another notation, $M^{p,q}$, is used. Apart from the choice of a different letter M, the second parameter is also introduced into the norm in a way different from (2), namely

$$\|f\|_{M^{p,q}(\Omega)} := \sup_{x\in\Omega; r>0} r^{\frac{n}{q}-\frac{n}{p}} \|f\|_{L^p(\widetilde{B}(x,r))}.$$

In this survey we mainly follow the notation in (1)–(2).

The local version of such spaces, with only one point $x = 0$ taken into account, has a connection with studies of N. Wiener [111] (1930), [112] (1932), who considered functions f for which

$$\frac{1}{T^{1-\alpha}} \int_0^T |f(x)|^p \, \mathrm{d}x, \quad \alpha \in (0,1), \quad p = 1 \text{ or } p = 2$$

is limited in $T > 0$ or tends to zero as $T \to \infty$. In the multidimensional case such local spaces defined by the norm

$$\|f\|_{B^p} = \sup_{r>0} \left(\frac{1}{|B(0,r)|} \int_{B(0,r)} |f(x)|^p \, \mathrm{d}x \right)^{1/p} \tag{3}$$

appeared in A. Beurling [16] (1964) as the dual of the so-called *Beurling algebra*. He also considered similar spaces with $\sup_{r>1}$ instead of $\sup_{r>0}$. Similar local Morrey type spaces with the norm of type (3) where $\frac{1}{|B(0,r)|}$ is replaced by $\frac{1}{|B(0,r)|^\lambda}$ appeared in V.S. Guliev [47] (1994), see also [50] (1996), and in J. García-Cuerva and M.J.L. Herrero [38] (1994). In [38] and J. Alvarez, M. Guzmán-Partida and J. Lakey [8] (2000) there were introduced the function space $B^{q,\lambda}(\mathbb{R}^n)$ characterized by the norm

$$\|f\|_{B^{p,\lambda}} = \sup_{r>1} \left(\frac{1}{|B(0,r)|^{1+\frac{\lambda}{p}}} \int_{B(0,r)} |f(x)|^p \, \mathrm{d}x \right)^{1/p} \tag{4}$$

(called inhomogeneous) and also its homogeneous version $\dot{B}^{q,\lambda}(\mathbb{R}^n)$ of type (4) with the supremum taken over $r > 0$.

Morrey spaces are a particular case of Campanato spaces considered in Section 4 and we present many results for Morrey spaces in that section in the context of Campanato spaces. Nevertheless, in this section we dwell on some results just for Morrey spaces.

2.1.1. Embeddings in Morrey spaces. By application of the Hölder inequality to integrals over $\widetilde{B}(x,r)$ the embedding for Morrey spaces follows:

Theorem 2.2. *Let $1 \le p \le q < \infty$ and let λ, ν be non-negative numbers. Then*

$$L^{q,\nu}(\Omega) \hookrightarrow L^{p,\lambda}(\Omega) \tag{5}$$

under the condition

$$\frac{\lambda - n}{p} \leq \frac{\nu - n}{q} \tag{6}$$

if $|\Omega|$ *is finite and the condition*

$$\frac{\lambda - n}{p} = \frac{\nu - n}{q} \tag{7}$$

if $|\Omega|$ *is infinite.*

Condition (6) is necessary and sufficient for embedding (5) in case of "nice" sets Ω, see L.C. Piccinini [89] (1969), where $\Omega = Q_0$ was a cube in $\mathbb{R}^n$, see also a similar result for a modification $L_r^{p,\lambda}, p, r \in [1, \infty)$ of Morrey spaces in Y. Furusho [36] (1980). This modification is introduced as follows: let $\overline{S}$ be the family of all systems $S = \{Q_j \colon \bigcup Q_j \subset Q_0\}$ consisting of a finite number of non-intersecting parallel subcubes Q_j, and let $\|u\|_{L^{(p,\lambda)}(Q_j)} = \sup_{Q \subset Q_j} |Q|^{\frac{\lambda - n}{np}} \|u\|_{L^p(Q)}$, and

$$\|u\|_{L_r^{p,\lambda}(Q_0)} = \sup_{S \in \overline{S}} \left\{ \sum_{Q_j \in S} \|u\|^r_{L^{(p,\lambda)} Q_j} \right\}^{1/r};$$

there is proved a necessary and sufficient condition for the validity of the embedding $L_r^{p,\lambda} \hookrightarrow L_s^{q,\mu}$ in the case of $n/r - \lambda/p \leq 1$ and $n/s - \mu/q \leq 1$.

See also embedding theorems for Campanato spaces in Subsection 4.1.

2.1.2. Hölder's inequality. For Morrey spaces the following Hölder type inequality holds (obtained by application of the usual Hölder inequality to integrals over $\widetilde{B}(x, r)$, see for instance Lemma 11 in P. Olsen [78] (1995)).

Theorem 2.3 (Hölder's inequality in Morrey spaces). *Let* $f \in L^{p,\lambda}(\Omega)$ *and* $g \in L^{q,\mu}(\Omega)$. *Then*

$$\|fg\|_{L^{r,\nu}(\Omega)} \leq \|f\|_{L^{p,\lambda}(\Omega)} \|g\|_{L^{q,\mu}(\Omega)}, \tag{8}$$

where $1 \leq p < \infty, 1 \leq q < \infty, \ \frac{1}{p} + \frac{1}{q} \geq 1$ *and*

$$\frac{1}{r} = \frac{1}{p} + \frac{1}{q}, \quad \frac{\nu}{r} = \frac{\lambda}{p} + \frac{\mu}{q}.$$

2.1.3. Weak Morrey spaces. Weak Morrey-Campanato spaces appeared already in the paper by S. Spanne [100] (1966), see also Subsection 4.3. Such weak-type Morrey spaces defined by the condition

$$\sup_{\substack{t>0 \\ x \in \Omega}} t^p |\{y \in \Omega : |f(y)| > t\} \cap B(x, r)| \leq C r^\lambda$$

where $\Omega \subset \mathbb{R}^n$, were used by M. Ragusa [91] (1995). In the paper C. Miao and B. Yuan [70] (2007) weak Morrey spaces $M^*_{p,\lambda}$ were defined in a more general setting in terms of Lorentz spaces of functions initially defined on non-atomic measurable spaces. For the spaces $M^*_{p,\lambda} = \{f : \|f\|^*_{p,\lambda} < \infty\}$ introduced via the norm

$$\|f\|^*_{p,\lambda} = \sup_{x,r} r^{-\frac{\lambda}{p}} \sup_{t>0} t\mu\{y : |f(y)| > t, \quad y \in B(x,r)\}$$

there were proved an embedding theorem and a convexity property.

2.1.4. Interpolation. G. Stampacchia [101] (1964), [102] (1965) and S. Campanato & M. Murthy [21] (1965) proved interpolation properties of Morrey spaces (in fact they obtained the result for the more general space, now called Campanato space, see its definition in Section 4). Loosely speaking, they proved (in the spirit of Riesz-Thorin interpolation theorem) that if T is a bounded linear operator from L^{q_i} to L^{p_i,λ_i}, $i = 1, 2$, then T is bounded from L^q to $L^{p,\lambda}$ for the corresponding intermediate values of p, q and λ, see the precise formulation in Theorem 4.5 in the setting of Campanato spaces. The conclusion in the other direction is false, see the comments after Theorem 4.5.

2.1.5. Preduals. Recall that for a given normed space X, a normed space Y is called predual of X, if X is dual of Y.

Preduals of Morrey spaces were studied by some authors, namely by C. Zorko [115] (1986), D.A. Adams [3] (1988), E.A. Kalita [57] (1998) and D.R. Adams and J. Xiao [4] (2004). Following D.R. Adams and J. Xiao, we denote the preduals obtained in [115], [57] and [4] by $Z^{q,\lambda}$, $K^{q,\lambda}$ and $H^{q,\lambda}$, respectively, $q = \frac{p}{p-1}$. The first two spaces are defined by the following norms

$$\|f\|_{Z^{q,\lambda}} = \inf \left\{ \|\{c_k\}\|_{\ell^1} \, : \, f = \sum_k c_k a_k \right\}$$

where a_k is a (q, λ)-atom and the infimum is taken with respect to all possible atomic decompositions of f (a function a on $\mathbb{R}^n$ is called a (q, λ)-atom, if it is supported on a ball $B \subset \mathbb{R}^n$ and $\|a\|_q \leq |B|^{-\frac{\lambda}{np}}$); note that in C. Zorko [115] the predual was introduced in a more general setting of generalized Morrey spaces;

$$\|f\|_{K^{q,\lambda}} = \inf_\sigma \left(\int_{\mathbb{R}^n} |f(x)|^q \omega_\sigma^{1-q}(x) \, \mathrm{d}x \right)^{1/q},$$

with

$$\omega_\sigma(x) = \int_{\mathbb{R}^{n+1}_+} r^{-\lambda} 1_{\mathbb{R}^1_+}(r - |x - y|) \, \mathrm{d}\sigma(y, r),$$

where the infimum is taken over all non-negative Radon measures $\sigma(y, r)$ on $\mathbb{R}^{n+1}_+$ with the normalization $\sigma(\mathbb{R}^{n+1}_+) = 1$;

$$\|f\|_{H^{q,\lambda}} = \inf_\omega \left(\int_{\mathbb{R}^n} |f(x)|^q \omega^{1-q}(x) \, \mathrm{d}x \right)^{1/q},$$

where the infimum is taken over all nonnegative functions on $\mathbb{R}^n$ satisfying the condition

$$\|\omega\|_{L^1(\Lambda_\lambda^{(\infty)})} \leq 1, \tag{9}$$

with the λ-dimensional Hausdorff capacity $\Lambda_\lambda^{(\infty)}$, the introduction of the latter norm in [4] being based on the previous studies in [3].

As shown in [4], for $1 < p < \infty$, $0 < \lambda < n$,

$$Z^{q,\lambda} = K^{q,\lambda} = H^{q,\lambda} \quad \text{with} \quad \|f\|_{Z^{q,\lambda}} \sim \|f\|_{K^{q,\lambda}} \sim \|f\|_{H^{q,\lambda}}$$

and the Morrey space may be characterized in terms of its predual by the following theorem.

Theorem 2.4. *Let* $1 < p < \infty$, $0 < \lambda < n$. *Then*

$$\|f\|_{L^{p,\lambda}} = \sup_{\omega} \left(\int_{\mathbb{R}^n} |f(x)|^p \omega(x)\,\mathrm{d}x \right)^{1/p}$$

where the supremum is taken with respect to all nonnegative functions on $\mathbb{R}^n$ *satisfying the condition* (9).

An interested reader may be also referred to Sections 5–7 of [4] with respect to Morrey type capacities.

In the case of Campanato spaces, M.H. Taibleson and G. Weiss [105] (1980) proved that they are dual to some Hardy spaces.

2.1.6. Vanishing Morrey spaces $VL^{p,\lambda}$. Morrey space $L^{p,\lambda}$, as noted, is not separable in the case $\lambda > 0$. A version of Morrey space where it is possible to approximate by "nice functions" is the so-called vanishing Morrey space $VL^{p,\lambda}(\Omega)$ introduced by C. Vitanza [110] (1990). This is a subspace of functions in $L^{p,\lambda}(\Omega)$, which satisfy the condition

$$\lim_{r\to 0} \sup_{\substack{x\in\mathbb{R}^n \\ 0<\varrho<r}} \frac{1}{\varrho^\lambda} \int_{\widetilde{B}(x,\varrho)} |f(y)|^p \,\mathrm{d}y = 0. \tag{10}$$

2.1.7. Different underlying spaces. The spaces $L^{p,\lambda}$ may be introduced on sets of different nature, for instance, an n-dimensional compact manifold via local charts (see M. Geisler [39] (1988)) where the spaces introduced in this way were characterized in terms of geodesic distances and other quantities on the manifold. In Subsection 2.2 we touch a more general setting when the underlying space is a quasimetric measure space. Morrey spaces and their generalizations in the case where the underlying spaces is the Heisenberg group were studied in V. Gulyiev [50] (1996).

2.1.8. Anisotropic Morrey spaces. Morrey spaces corresponding to anisotropic distances appeared first in G. Barozzi [12] (1965) defined in the following way. Let $\Omega \subset \mathbb{R}^n$ be a bounded open set, $p \geq 1$ and $0 \leq \lambda \leq n$. Let $\overline{m} = (m_1, \dots, m_n)$ be an n-tuple of non-negative numbers, $m_j \geq 1$ and $m = \max(m_1, \dots, m_n)$. Let $B_{\overline{m}}(x,r) = \{y \in \Omega : d_{\overline{m}}(x,y) < r\}$ be an anisotropic ball defined by the distance

$$d_{\overline{m}}(x,y) = \left(\sum_{j=1}^{n} |x_j - y_j|^{m_j} \right)^{1/m}.$$

Then the corresponding Morrey space is introduced by the condition

$$\sup_{x,r} \frac{1}{r^\lambda} \int_{B_{\overline{m}}(x,r)} |f(y)|^p \, dy < \infty.$$

The corresponding anisotropic Sobolev spaces were also introduced in [12].

In a more general setting such anisotropic Morrey spaces were later studied by V.P. Il′in in [52] (1959), [53](1971), see the presentation of the latter results also in Section 27 of the book [13].

Morrey spaces with integral means over one-parametrical ellipsoids were introduced in L. Softova in [98] (2007) with the aim to study anisotropic singular integrals. Let $\overline{\alpha} = (\alpha_1, \dots, \alpha_n)$ be a given vector with $\alpha_i \geq 1, i = 1, \dots, n$, and

$$\mathscr{E}_{\overline{\alpha}}(x,r) = \left\{ y \in \mathbb{R}^n : \sum_{k=1}^{n} \frac{(x_k - y_k)^2}{r^{2\alpha_k}} < 1 \right\} \tag{11}$$

be an ellipsoid centered at the point $x \in \mathbb{R}^n$. Then the anisotropic space $L^{p,\lambda}(\mathbb{R}^n)$ localized at the origin and corresponding to the given vector $\overline{\alpha}$, is defined by the norm

$$\|f\|_{p,\lambda} = \sup_{r>0} \left(\frac{1}{r^\lambda} \int_{\mathscr{E}_{\overline{\alpha}}(0,r)} |f(y)|^p \, dy \right)^{1/p} < \infty. \tag{12}$$

See also Subsection 3.1 for the generalized anisotropic Morrey spaces of such a kind introduced in L. Softova [97] (2006).

Anisotropic Morrey spaces $L^{p,\overline{\lambda}}(\Omega)$, $\overline{\lambda} = (\lambda_1, \dots, \lambda_n)$ may be also introduced, with means taken over rectangles centered at the point x with independent lengths of sides. Such spaces $\mathcal{L}^{p,\lambda_1,\lambda_2}(\mathbb{R}^2_+)$ were introduced in L.-E. Persson and N. Samko [88] (2010) for the case $\Omega = \mathbb{R}^2_+$ by the norm

$$\|f\|_{\mathcal{L}^{p,\lambda_1,\lambda_2}} = \sup_{\substack{x_1>0, x_2>0 \\ r_1>0, r_2>0}} \left(\frac{1}{r_1^{\lambda_1} r_2^{\lambda_2}} \int_{(x_1-r_1)_+}^{x_1+r_1} \int_{(x_2-r_2)_+}^{x_2+r_2} |f(y_1,y_2)|^p \, dy_1 \, dy_2 \right)^{1/p} \tag{13}$$

with the aim to study two-dimensional Hardy operators in such spaces.

2.1.9. Miscellaneous. As is well known, Morrey spaces have been generalized or modified in various ways in order to obtain existence and uniqueness of solutions to partial differential equations. One of such modifications, $L^{p,\lambda}(\Omega,t)$ introduced in M. Transirico et al. [108] (1995) (with $t=1$) and A. Canale et al. [22] (1998), is aimed to better reflect the local nature of solutions, first of all for unbounded domains, being defined by the norm

$$\|f\|_{L^{p,\lambda}(\Omega,t)} = \sup_{\substack{x\in\Omega\\ 0<r<t}} \left(\frac{1}{r^\lambda}\int_{\widetilde{B}(x,r)} |f(y)|^p \, \mathrm{d}y\right)^{1/p} ;$$

in [22] the corresponding Sobolev spaces were also dealt with.

In P. Cavaliere, G. Manzo and A. Vitolo [23](1996) Morrey spaces were intentionally studied on unbounded domains with the main emphasis on the connection between Morrey type and BMO spaces and embedding and density results involving the continuity of the translation operator.

Another modification of Morrey spaces is known under the name of Stummel class introduced in M.A. Ragusa and P. Zamboni [92] (2001) (with the goal to obtain a better version of the Sobolev type embedding). The Stummel class is defined, for $0<p<n$, as

$$S_p = \left\{ f\in L^1_{\mathrm{loc}}(\mathbb{R}^n) : \lim_{r\to 0}\eta(r)=0, \quad \eta(r) = \sup_{x\in\mathbb{R}^n}\int_{|x-y|<r}\frac{|f(y)|}{|x-y|^{n-p}}\,\mathrm{d}y \right\},$$

which is the Stummel-Kato class in the case $p=2$. Note that $\eta(r) \geq \sup\limits_{x\in\mathbb{R}^n} \frac{1}{r^{n-p}} \int\limits_{|x-y|<r} |f(y)|\,\mathrm{d}y$.

In general $L^{1,\lambda}$ is contained in S_p, if $\lambda > n-p$, and in the case $\eta(r)\sim r^\alpha$ the following equivalence holds:

$$f\in S_p \quad \Longleftrightarrow \quad f\in L^{1,n-p+\alpha},$$

see Lemma 1.1 in [92]. Some versions of Stummel classes with η different from powers are also studied there, which corresponds to the generalized Morrey spaces studied in Subsection 3.

S. Leonardi [64] (2002) introduced a similar version of such a space, defined by the norm

$$\|f\|_{N^{p,\lambda}(\Omega)} := \sup_{x\in\Omega}\left\{\int_\Omega \frac{|f(y)|^p}{|x-y|^\lambda}\,\mathrm{d}y\right\}^{1/p}$$

and proved a certain version of the Miranda-Talenti inequality in terms of Sobolev type spaces related to the norms $\|f\|_{N^{p,\lambda}(\Omega)}$.

A more general hybrid of Morrey and Stummel type spaces, the space denoted by $M^{p,\lambda}_\beta(X,\mu)$, was introduced in Eridani, V. Kokilashvili and A. Meskhi [34] on a quasi-metric measure space (X,ρ,μ), with the norm defined by

$$\|f\|_{M^{p,\lambda}_\beta} := \sup_{\substack{x\in X\\ r>0}} \left(\frac{1}{r^\lambda}\int_{\rho(x,y)<r} |f(y)|^p \rho^\beta(x,y)\,\mathrm{d}\mu(y)\right)^{1/p}.$$

2.2. Morrey spaces over $\mathbb{R}^n$ in case of a general measure

Y. Sawano and H. Tanaka [95] (2005) introduced Morrey spaces in $\mathbb{R}^n$, but with a Radon measure μ as follows

$$\mathscr{M}^p_q(k,\mu) = \left\{ f\colon \sup_Q |\mu(kQ)|^{\frac{1}{p}-\frac{1}{q}} \left(\int_Q |f|^q\,\mathrm{d}\mu\right)^{1/q} < \infty \right\}, \tag{14}$$

where Q is a closed cube whose edges are parallel to the coordinate axes and it is supposed that the measure μ is not necessarily a doubling measure but satisfies the growth condition

$$\mu(B(x,r)) \le c_0\, r^\ell$$

for some fixed constants $c_0 > 0$ and $\ell \in (0,n]$, and $\mu(Q) > 0$. It is shown that the definition of the space does not depend on the choice of the parameter $k > 1$, that is,

$$\mathscr{M}^p_q(k_1,\mu) = \mathscr{M}^p_q(k_2,\mu) \tag{15}$$

for all $k_1 > 1, k_2 > 1$, up to equivalence of norms. More precisely

$$\|f\|_{\mathscr{M}^p_q(k_1,\mu)} \le \|f\|_{\mathscr{M}^p_q(k_2,\mu)} \le C_n \left(\frac{k_1-1}{k_2-1}\right)^n \|f\|_{\mathscr{M}^p_q(k_1,\mu)} \tag{16}$$

for $1 < k_1 < k_2 < \infty$, see formula (3) in [96]. In [96] there was also made a comparison of the space $\mathscr{M}^p_q(2,\mu)$ with the space $\mathscr{M}^p_q(1,\mu)$, the latter being defined with the usage of cubes Q which only satisfy the condition $\mu(kQ) \le \beta\mu(Q)$ with $\beta > k^{\frac{npq}{p-q}}$ where $k > 1$ is fixed and the measure μ does not necessarily satisfies the growth condition or the doubling condition. This comparison includes also the case of vector-valued Morrey spaces $\mathscr{M}^p_q(\ell^r,\mu)$ defined by

$$\|f_j\|_{\mathscr{M}^p_q(\ell^r,\mu)} := \sup_{Q\in\mathcal{Q}(\mu;k;\beta)} \mu(Q)^{\frac{1}{p}-\frac{1}{q}} \left(\int_Q \|f_j\|^q_{\ell^r}\,\mathrm{d}\mu\right)^{1/q} < \infty.$$

For similar results on Campanato spaces, we refer to Section 4.

For Morrey spaces in a more general setting of abstract quasimetric measure spaces see Subsection 3.1.

3. Generalized Morrey spaces

Recall that the classical Morrey space is defined by the norm.

$$\|f\|_{L^{p,\lambda}(\Omega)} := \sup_{x\in\Omega} \left\| \frac{1}{r^{\frac{\lambda}{p}}} \|f\|_{L^p(\widetilde{B}(x,r))} \right\|_{L^\infty(0,d)}, \quad d = \operatorname{diam}\Omega. \tag{17}$$

There are known two types of generalizations of Morrey spaces. The first is to replace the power function r^λ by a function $\varphi(r)$ (or more generally $\varphi(x,r)$), usually with some quasi monotonicity type conditions with respect to r. Another way is to replace the $L^\infty(0,d)$-norm by $L^\theta(0,d)$-norm, $0<\theta<\infty$. For brevity, we will call these by φ-generalizations and θ-generalizations. Both ways may be naturally mixed.

3.1. φ-generalizations

Let X be a quasimetric space with a Borel measure μ. The generalized Morrey space is defined by the (quasi)norm

$$\|f\|_{p,\varphi} = \sup_{x,r} \left(\frac{1}{\varphi(x,r)} \int_{B(x,r)} |f(y)|^p \, d\mu(y) \right)^{1/p}, \quad 0<p<\infty, \tag{18}$$

where $B(x,r)$ is a ball in X and the non-negative function φ is subject to some restrictions, usually related to monotonicity-type conditions in r. Generalized Morrey spaces, $L^{p,\varphi,S}$, of such a type seem to first appear in the paper G.T. Dzhumakaeva and K. Zh. Nauryzbaev [31] (1982), where the norm is introduced by

$$\|f\|_{p,\varphi,S} = \sup_{E\in S} \frac{1}{\varphi(|E|)} \left(\int_{E\cap\Omega} |f(y)|^p \, dy \right)^{1/p} < \infty,$$

$1\le p<\infty$, Ω is a domain of finite measure in $\mathbb{R}^n$, S is the family of all measurable subsets of Ω and $\varphi(r)$ is a positive nondecreasing function on $\mathbb{R}^1_+$. Under the assumption that $\varphi(r)=1$ for $r\ge 1$ and that $\varphi^p(r)$ is concave in $(0,1)$, in [31] there was proved that $L^{p,\varphi,S}(\Omega)\subset L^q(\Omega)$, $p<q\le\infty$, if and only if $\int_0^1 r^{-q/p}\varphi^q(r)\,dr<\infty$, with the corresponding interpretation for $q=\infty$.

The generalized Morrey spaces $L^{p,\varphi}(\Omega)$ defined by the norm

$$\|f\|_{p,\varphi} = \sup_{x,r} \left(\frac{1}{\varphi(r)} \int_{\widetilde{B}(x,r)} |f(y)|^p \, d\mu(y) \right)^{1/p}, \quad 1\le p<\infty, \tag{19}$$

were studied in the paper C. Zorko [115] (1986) in a more general setting of Campanato spaces, see Section 4. We mention the result from [115, Prop. 2] stating that the zero continuation of a function $f\in L^{p,\varphi}(\Omega)$ belongs to $L^{p,\varphi}(\mathbb{R}^n)$ under the assumption that the function φ is nondecreasing. In [115, Prop. 3] there was also shown a possibility to approximate by nice functions in the subspace of $L^{p,\varphi}(\mathbb{R}^n)$ defined by the condition $\lim_{y\to 0}\|f(\cdot-y)-f(\cdot)\|_{L^{p,\varphi}}=0$ (recall that Morrey spaces are not separable).

Often the (quasi)norm in such a generalized Morrey space is taken in the form

$$\|f\|_{L^p_\psi} = \sup_{B(x,r)} \frac{1}{\psi(x,r)} \left(\frac{1}{\mu(B(x,r))} \int_{B(x,r)} |f(y)|^p \, d\mu(y) \right)^{1/p}, \quad 0<p<\infty, \tag{20}$$

in particular in the form

$$\|f\|_{L^p_\psi(w)} = \sup_B \frac{1}{\psi(|B|)} \left(\frac{1}{|B|} \int_B |f(y)|^p \, w(y) \, \mathrm{d}y \right)^{1/p}, \quad 0 < p < \infty, \tag{21}$$

in the case $X = \mathbb{R}^n$.

With the norm of form (18), such spaces appeared in E. Nakai [72] (1994) for $X = \mathbb{R}^n$, and the spaces $L^p_\psi(X)$ with the (quasi)norm (21) in J. Alvarez and C. Pérez [9] (1994) and with the norm (20) in E. Nakai [73] (1997).

In [73] there were studied the pointwise multipliers from such a space $L^p_\psi(X)$ to another one of similar type. Let $\mathrm{PWM}(E,F)$ denote the set of pointwise multipliers from E to F. Under some assumptions on ψ_1 and ψ_2, it was proved that

$$\mathrm{PWM}(L^{p_1}_{\psi_1}, L^{p_2}_{\psi_2}) = L^{p_3}_{\psi_3}, \tag{22}$$

where $1/p_1 + 1/p_3 = 1/p_2$, $0 < p_2 < p_1 < \infty$ and $\psi_3 = \psi_2/\psi_1$. In E. Nakai [74] (2000) there were obtained necessary conditions on p_i and ψ_i for (22) to be valid, and sufficient conditions for $\mathrm{PWM}(L^{p_1}_{\psi_1}, L^{p_2}_{\psi_2}) = \{0\}$.

In the paper H. Arai and T. Mizuhara [10] (1997) the generalized Morrey spaces with the norm of the type (18) were considered within the framework of homogeneous underlying space, normal in the sense of Macías and Segovia [67], under the assumption that $\varphi(x,r)$ is increasing in r and satisfies the doubling condition uniformly in x. There was proved a general theorem which allows to obtain estimates of the form

$$\|F\|_{L^{p,\varphi}} \le C\|G\|_{L^{q,\varphi}}$$

from estimates of the form $\int F^p w d\mu \le C \int G^q w d\mu$, where w ranges some subclasses of the Muckenhoupt class $A_1(\mu)$. This important result was used to obtain Morrey space estimates for various classical operators.

Relations between the generalized Morrey spaces with the norm (21) and the corresponding Stummel classes (see section 2.1.9) were studied in Eridani and H. Gunawan [33] (2005), the results adjoin to those for the case where ψ is a power function.

In E. Nakai [75] (2006) the generalized Morrey spaces, with the norm defined as in (20), appeared in the case where the underlying space X was a homogeneous metric measure space.

In L. Softova [97] (2006) and [98] (2007) there were introduced the generalized anisotropic Morrey spaces with the aim to study anisotropic singular integrals. Let $\overline{\alpha} = (\alpha_1, \dots, \alpha_n)$ be a given vector with $\alpha_i \ge 1, i = 1, \dots, n$, and $\mathscr{E}_{\overline{\alpha}}(x,r)$ the ellipsoid defined in (11). Then the anisotropic space $L^{p,\varphi,\overline{\alpha}}(\mathbb{R}^n)$ is defined by the norm

$$\|f\|_{p,\varphi,\overline{\alpha}} = \sup_{x,r} \left(\frac{1}{\varphi(x,r)} \int_{\mathscr{E}_{\overline{\alpha}}(x,r)} |f(y)|^p \, \mathrm{d}y \right)^{1/p} < \infty.$$

As a generalization of results from Y. Sawano and H. Tanaka [95] (see Subsection 2.2), Y. Sawano in [94] (2008) dealt with the generalized Morrey spaces defined by the condition

$$\sup_Q \left(\frac{1}{\varphi(\mu(kQ))} \int_Q |f|^p \, \mathrm{d}\mu \right)^{1/p} < \infty,$$

where $1 \leq p < \infty, k > 1$, φ is an increasing function, Q is a cube with edges parallel to the coordinate axes, and μ is a positive Radon measure, non necessarily satisfying the doubling condition. The independence of such spaces on the choice of $k > 1$, as in (15)–(16), is extended to this setting.

Y. Komori and S. Shirai [60] (2009) considered the generalized Morrey spaces $L^{p,\kappa}(w)$, defined by the norm

$$\|f\|_{L^{p,\kappa}(w)} = \sup_Q \left(\frac{1}{w(Q)^\kappa} \int_Q |f(x)|^p w(x) \, \mathrm{d}x \right)^{1/p}, \quad w(Q) = \int_Q w(x) \, \mathrm{d}x, \qquad (23)$$

where $0 < \kappa < 1$ and the supremum is taken over all cubes in $\mathbb{R}^n$, which is nothing else, but the usual Morrey space with respect to the measure $\mu(E) = \int_E w(x) \, \mathrm{d}x$; the authors called this space *weighted*. Note that if we interpret the space $L^{p,\kappa}(w)$ as a weighted generalized Morrey space, then given the function w, the function $\varphi = w^\kappa$ already defines the generalized Morrey space, this meaning that the space $L^{p,\kappa}(w)$, introduced in this way, is not a space with an arbitrary weight, but with a special weight equal to a power of the function φ.

3.2. θ-generalizations

A Morrey-type space with $\sup_{r>0}$ replaced by the $\|\cdot\|_{L^\theta}(0,\infty)$-norm first appeared in D.R. Adams [5], p. 44 (1981) with the norm defined by

$$\|f\|_{L^{p,\theta,\lambda}(\mathbb{R}^n)} := \sup_{x\in\mathbb{R}^n} \left(\int_0^\infty \left(\frac{1}{r^\lambda} \int_{B(x,r)} |f(y)|^p \, \mathrm{d}y \right)^{\theta/p} \frac{\mathrm{d}r}{r} \right)^{1/\theta} \qquad (24)$$

where the corresponding Sobolev type theorem for the Riesz potential operator was stated. Spaces with both θ- and φ-generalization, but "localized" to the point $x = 0$, with the norm

$$\|f\|_{L^{p,\theta,\varphi}_{\mathrm{loc},0}(\mathbb{R}^n)} := \left(\int_0^\infty \left(\frac{1}{\varphi(r)} \int_{B(0,r)} |f(y)|^p \, \mathrm{d}y \right)^{\theta/p} \frac{\mathrm{d}r}{r} \right)^{1/\theta} \qquad (25)$$

were introduced and intensively studied by V.S. Guliyev [47] (1994) together with the study of the classical operators in these spaces, see also the books V.S. Guliyev [50] (1996) and [51] (1999) where these results were presented for the case when the underlying space is the Heisenberg group or a homogeneous group, respectively. Note that these investigations appeared in fact independently of the development of the main trends in the theory of Morrey spaces and their applications. They had as a background the usage of the local characteristics

$$\Omega(f,r) = \int_{\mathbb{R}^n\setminus B(x,r)} |f(y)|^p \,dy \quad \text{and} \quad \Omega^*(f,r) = \int_{B(x,r)} |f(y)|^p \,dy$$

widely used in Baku mathematical school (A.A. Babaev and his students) for a characterization of weighted Hölder and other spaces, we refer for instance to the papers [11] and [1], [2].

In the case $\theta = p$ the spaces $L^{p,\theta,\varphi}_{\mathrm{loc},0}(\mathbb{R}^n)$ coincide with a certain weighted Lebesgue spaces:

$$L^{p,p,\varphi}_{\mathrm{loc},0}(\mathbb{R}^n) = L^p(\mathbb{R}^n, w), \quad w(x) = \int_{|x|}^{\infty} \frac{dr}{r\varphi(r)}.$$

In a series of papers by V. Burenkov, H. Guliyev and V. Guliyev related to such spaces, this "localized" version with the norm (25), where $p,\theta \in (0,\infty)$, was called "local Morrey-type space" and the version with the norm

$$\|f\|_{L^{p,\theta,\varphi}(\mathbb{R}^n)} := \sup_{x\in\mathbb{R}^n} \left(\int_0^{\infty} \left(\frac{1}{\varphi(r)} \int_{B(x,r)} |f(y)|^p \,dy \right)^{\theta/p} \frac{dr}{r} \right)^{1/\theta}, \tag{26}$$

the "global Morrey-type space", with $p,\theta \in (0,\infty)$. As shown in V.I. Burenkov and H. Guliyev [18] (2004), such space $L^{p,\theta,\varphi}(\mathbb{R}^n)$ is "reasonable" under the assumptions

$$\left\| \frac{1}{\varphi^{1/p}} \right\|_{L^\theta(t_1,\infty)} < \infty \quad \text{and} \quad \left\| \frac{r^{\frac{n}{p}}}{\varphi^{1/p}} \right\|_{L^\theta(0,t_2)} < \infty$$

for some $t_1, t_2 \in (0,\infty)$, being trivial ($L^{p,\theta,\varphi}(\mathbb{R}^n) = \emptyset$) if one of these conditions is violated; the space $L^{p,\theta,\varphi}_{\mathrm{loc},0}$ is also trivial if the second condition is violated, and the function in $L^{p,\theta,\varphi}_{\mathrm{loc},0}$ must vanish in a sense at the origin, if the first condition does not hold.

4. Campanato spaces

Campanato spaces, also referred to sometimes as Morrey-Campanato spaces, were introduced by S. Campanato [19] (1963) (in the case of bounded domains in $\mathbb{R}^n$); in 1964 they also appeared in the paper of G. Stampacchia [101]. They are a generalization of the *BMO* spaces of functions of bounded mean oscillation introduced by F. John and L. Nirenberg [56] (1961) and defined, for open sets $\Omega \subseteq \mathbb{R}^n$, by the seminorm

$$[f]_{\mathrm{BMO}} := \sup_{x,r} \frac{1}{|\widetilde{B}(x,r)|} \int_{\widetilde{B}(x,r)} \left| f(y) - f_{\widetilde{B}(x,r)} \right| dy.$$

4.1. Definitions and basic facts

Definition 4.1 (Campanato spaces). Let $\Omega \subseteq \mathbb{R}^n$ be an open set, $1 \leq p < \infty$ and $\lambda \geq 0$. The *Campanato space* $\mathscr{L}^{p,\lambda}(\Omega)$ is defined as

$$\mathscr{L}^{p,\lambda}(\Omega) := \left\{ f \in L^p(\Omega) : [f]_{\mathscr{L}^{p,\lambda}(\Omega)} < \infty \right\} \tag{27}$$

the *Campanato seminorm* being given by

$$[f]_{\mathscr{L}^{p,\lambda}(\Omega)} := \sup_{x\in\Omega; r>0} \left(\frac{1}{r^\lambda} \int_{\widetilde{B}(x,r)} |f(y) - f_{\widetilde{B}(x,r)}|^p \, \mathrm{d}y \right)^{1/p}$$

or equivalently

$$\sup_{x\in\Omega; r>0} \left(\frac{1}{r^\lambda} \inf_{c\in\mathbb{R}^1} \int_{\widetilde{B}(x,r)} |f(y) - c|^p \, \mathrm{d}y \right)^{1/p}. \tag{28}$$

The embedding theorem for Campanato spaces reads as follows (see [63, p. 217])

Theorem 4.2. *Let $1 \leq p \leq q < \infty$ and let λ, ν be non-negative numbers. If $|\Omega|$ is finite then*

$$\mathscr{L}^{q,\nu}(\Omega) \hookrightarrow \mathscr{L}^{p,\lambda}(\Omega) \tag{29}$$

under the condition

$$\frac{\lambda - n}{p} \leq \frac{\nu - n}{q}. \tag{30}$$

In G. Stampacchia [102] (1965) there was introduced Campanato-type space $\mathscr{L}_r^{(p,\lambda)}(Q_0)$ where Q_0 is a cube in $\mathbb{R}^n$ defined by the set of seminorms

$$K(Q_j) := \sup_{Q\subset Q_j} \left(\frac{1}{|Q|^{1-\lambda/n}} \int_Q |u(x) - u_Q|^p \, \mathrm{d}x \right)^{1/p} \tag{31}$$

where $\{Q_j : \cup Q_j \subset Q_0\}$ is a given family of cubes parallel to the cube Q_0, no two of which have common interior points, and the condition

$$\sup_{\{Q_j\}} \left(\sum_j |K(Q_j)|^r \right)^{\frac{1}{r}} < \infty \tag{32}$$

holds, where the supremum is taken with respect to all admissible families of cubes. In some papers such spaces were called strong Campanato spaces, see, e.g., [79, 84].

The importance of Campanato spaces stems from the fact that, for λ greater than n (and less than $n + p$), they coincide with the spaces of Hölder continuous functions, providing an integral characterization of such functions, while in the case $\lambda < n$ they coincide with Morrey spaces, as the theorem below states, proved in S. Campanato [19] (1963) (in [19] the domain was supposed to satisfy the condition (A) and have Lipschitz boundary; for the proof under the only condition (A) we refer to Section 4.3 of the book by A. Kufner et al. [63]), where the proof of the coincidence of the Campanato spaces with the BMO space in the case $\lambda = n$ may be also found.

We say that an open set $\Omega \subset \mathbb{R}^n$ is of type (A), if there exists a constant $A > 0$ such that

$$|\widetilde{B}(x,r)| \geq Ar^n, \tag{33}$$

and by $H^\alpha(\overline{\Omega})$ we denote the space of functions satisfying the Hölder condition in $\overline{\Omega}$.

Theorem 4.3. *Let $1 \leq p < \infty$ and Ω be a bounded domain of type* (A). *Then*

1. $\mathscr{L}^{p,\lambda}(\Omega) \cong L^{p,\lambda}(\Omega)$, *when* $\lambda \in [0, n)$,
2. $\mathscr{L}^{p,\lambda}(\Omega) \cong BMO(\Omega)$ *when* $\lambda = n$,
3. $\mathscr{L}^{p,\lambda}(\Omega) \cong H^\alpha(\overline{\Omega})$ *with* $\alpha = \frac{\lambda - n}{p}$, *when* $\lambda \in (n, n+p]$.

Note that the statement (3) of Theorem 4.3 for the case $p = 1$ was also proved in N. Meyers [69] (1964).

For strong Campanato spaces defined by (31) and (32), in A. Ono [79] (1970) there were obtained relations with Lipschitz spaces $\mathrm{Lip}(\alpha, p)$ of functions Hölder continuous in L^p-norm, and in A. Ono [83] (1978) in the final form as the statement

$$\mathscr{L}_r^{(p,\lambda)}(Q_0) \cong \mathrm{Lip}\left(\frac{n}{r} - \frac{n-\lambda}{p}, r\right),$$

with $1 \leq r < \infty$ and $0 < n/r - (n-\lambda)/p < 1$.

We refer also to A. Ono [80] (1972), A. Ono and Y. Furusho [84], A. Ono [82] (1977/1978), and A. Ono [81] (1977/1978) with regards to other results around the strong Campanato spaces.

In [20] (1964) S. Campanato introduced spaces $\mathscr{L}_k^{p,\lambda}(\Omega)$ of "higher order" defined by the seminorm

$$[f]_{\mathscr{L}_k^{p,\lambda}} := \sup_{x \in \Omega; r > 0} \left(\frac{1}{r^\lambda} \inf_{P \in \mathcal{P}_k} \int_{\widetilde{B}(x,r)} |f(y) - P(y)|^p \, \mathrm{d}y \right)^{1/p} \tag{34}$$

where $\mathcal{P}_k$ is the class of polynomials of degree at most k and proved the following generalization of Theorem 4.3, where $C^{m,\alpha}(\Omega), m \geq 0, 0 < \alpha \leq 1$, stands for the class of functions continuous in $\overline{\Omega}$ with all the derivatives up to the order m and with the derivatives of order m in $H^\alpha(\overline{\Omega})$.

Theorem 4.4. *Let $1 \leq p < \infty, k \geq 0$ and Ω be a bounded domain of type* (A). *Then*

1. $\mathscr{L}_k^{p,\lambda}(\Omega) \cong L^{p,\lambda}(\Omega)$, *when* $\lambda \in [0, n)$,
2. $\mathscr{L}_k^{p,\lambda}(\Omega) \cong C^{m,\alpha}(\overline{\Omega})$ *with* $m = \left[\frac{n-\lambda}{p}\right], \alpha = \frac{\lambda - n}{p} - m$, *when* $n + mp < \lambda < n + (m+1)p$, $m = 0, 1, 2, \ldots, k$.

We refer to S. Janson et al. [55] (1983) for the alternative proof of Theorem 4.4 in the case $\Omega = \mathbb{R}^n$, which includes also the case $p = \infty$.

Note that the condition (A) is not necessary for the validity of the embedding $\mathscr{L}_k^{p,\lambda}(\Omega) \hookrightarrow C^{m,\alpha}(\overline{\Omega})$ but the inverse embedding in equivalence (2) in Theorem 4.4

essentially uses this condition. We refer to D. Opěla [85] (2003) for the study of the influence of the geometry of Ω on the inverse embedding.

4.2. DeVore-Sharpley-Christ versions of Campanato-type spaces

In R.A. DeVore and R.C. Sharpley [28] and M. Christ [25] there was introduced a version of Campanato-type spaces in which the L^∞-norm in x is replaced by L^p-norm, namely they introduced the space C_p^α defined for $1 \le q \le p$, by the norm

$$\|f\|_{C_p^\alpha} := \left[\int_\Omega \sup_{Q\ni x} \inf_{P\in\mathcal{P}_{[\alpha]}} \frac{1}{|Q|^{\frac{\alpha p}{n}+\frac{p}{q}}} \left(\int_Q |f(y)-P(y)|^q\,\mathrm{d}y\right)^{\frac{p}{q}} \mathrm{d}x\right]^{1/p}, \qquad (35)$$

where $\mathcal{P}_k$ stands for the class of polynomials of degree at most k, $k \ge 0$. This norm does not depend on $q \in [1,p]$, see [28, p. 36]. We refer to [28] for the study of various properties of these spaces such as comparison with Besov spaces, interpolation, embeddings, extension theorem, etc. These spaces may be also found in H. Triebel [109, Subsection 1.7.2.]. They are also known as *local approximation Campanato spaces*. In the case $p = 2$ we refer also to a paper [32] (2006) on a characterization of such spaces when α may be negative $(\alpha > -\frac{n}{2})$.

Spaces of the type $C_p^\alpha(X)$ were studied in D. Yang [113] (2005) in the case where the underlying space was a homogeneous metric measure spaces. A comparison of such spaces and some other Campanato related spaces with Besov and Triebel-Lizorkin spaces may be also found in that paper. We also mention a characterization of the Hajłasz-Sobolev spaces in terms of the Calderón-Scott maximal function $f_\alpha^\sharp$, obtained in [113].

4.3. φ-generalization

Following the long-standing traditions in the study of Campanato spaces, we use two forms to define them. Namely

$$\mathscr{L}_k^{p,\varphi} := \left\{f \in L^p : \sup_{x,r} \frac{1}{\varphi(r)} \inf_{P\in\mathcal{P}_k} \int_{\widetilde{B}(x,r)} |f(y)-P(y)|^p\,\mathrm{d}y < \infty\right\} \qquad (36)$$

and

$$\mathsf{L}_k^{p,\psi} = \left\{f \in L^p : \sup_{x,r} \frac{1}{r^n\psi(r)} \inf_{P\in\mathcal{P}_k} \int_{\widetilde{B}(x,r)} |f(y)-P(y)|^p\,\mathrm{d}y < \infty\right\}. \qquad (37)$$

Such a generalized Campanato space $\mathsf{L}^{1,\psi}(Q) := \mathsf{L}_0^{1,\psi}(Q)$, over cubes $Q \subset \mathbb{R}^n$, defined by the seminorm

$$[f]_{\mathsf{L}^{1,\psi}(\Omega)} := \sup_{x,r} \frac{1}{r^n\psi(r)} \int_{I(x,r)\subset Q} |f(y)-f_{I(x,r)}|\,\mathrm{d}y,$$

with $I(x,r) = \{u : |y-x| < r/2\}$, appeared in S. Spanne [99] (1965), where $\mathsf{L}^{1,\psi}(Q)$ was characterized in terms of rearrangements of the function $|f - f_{I(x,r)}|$, restricted to $I(x,r)$. Under the assumption that the function ψ is increasing on $(0,\infty)$ and the integral $\int_0^\varepsilon \frac{\psi(t)}{t}\,dt$ converges, he proved the embedding

$$\mathsf{L}^{1,\psi}(Q) \hookrightarrow H^{\psi_1}(Q), \tag{38}$$

where H^{ψ_1} is the generalized Hölder space

$$H^{\psi_1} = \{f : |f(x+h) - f(x)| \le C\psi_1(h)\}, \quad \psi_1(h) = \int_0^h \frac{\psi(t)}{t}\,\mathrm{d}t. \tag{39}$$

The generalized Campanato space $\mathscr{L}_k^{p,\varphi}(\Omega)$ of higher order defined by the seminorm

$$[f]_{\mathscr{L}_k^{p,\varphi}} := \sup_{x,r}\left(\frac{1}{\varphi(r)}\inf_{P\in\mathcal{P}_k}\int_{\widetilde{B}(x,r)}|f(y)-P(y)|^p\,\mathrm{d}y\right)^{1/p} \tag{40}$$

where $\mathcal{P}_k$ is the class of polynomials of degree at most k, $k \ge 0$ and Ω is an open set in $\mathbb{R}^n$, was studied by S. Spanne [100] (1966) who gave its equivalent characterization in terms of the seminorm

$$\sup_{x,r}\left(\frac{1}{\varphi(r)}\|f - \mathsf{P}_k(f)\|^p_{L^p(\widetilde{B}(x,r))}\right)^{1/p} \tag{41}$$

where P_k is the orthogonal projection of $L^2(\widetilde{B}(x,r))$ onto the space of restrictions of polynomials of order k on $\widetilde{B}(x,r)$, under the assumption that Ω is of type (A). He also considered weak generalized Morrey-type spaces with the L^p-norm in (41) replaced by the weak L^p-norm.

As shown in J. Alvarez [7] (1981) the generalized Campanato spaces $\mathscr{L}_0^{p,\varphi}$ are not better than the L^p space if one admits the function φ such that $\varphi(t)\to\infty$ as $t\to 0$. More precisely, let φ be a nonnegative function such that $\varphi(t)$ is nonincreasing and $t\varphi^p(t)$ is nondecreasing near zero and $\varphi(0)=\infty$; suppose also that $g:(0,1)\to\mathbb{R}$ is a nonnegative, nonincreasing p-integrable function such that $g(t)\to\infty$ as $t\to 0$. Then there exist a cube Q_0, a function $f\in\mathscr{L}_0^{p,\varphi}(Q_0)$ and two constants $C, t_0 > 0$ such that

$$\lambda_f(t) \ge C\lambda_g(t_0)$$

where $\lambda_f(t) = |\{x : |f(x)| > t\}|$ is the distribution function, so that $\mathscr{L}_0^{p,\varphi}(Q_0)$ contains functions whose distribution functions exceed that of any given function in $L^p(Q_0)$.

In the case where $\Omega\subset\mathbb{R}^n$ is a bounded open set, generalized Campanato spaces $\mathscr{L}_k^{p,\varphi}(\Omega)$ defined by condition (40), appeared in C. Zorko [115] (1986). As a generalization of the statement 1. of Theorem 4.4, there was proved that

$$\mathscr{L}_k^{p,\varphi}(\Omega) \cong L^{p,\varphi}(\Omega)$$

under the condition (A), see (33), and the following assumptions: $\varphi(r)$ is nondecreasing, $\varphi(r)r^{-n}$ is nonincreasing and $\varphi(2r)\le c\varphi(r)$ with $0 < c < 2^{\frac{n}{p}}$, with the generalized Morrey space $L^{p,\varphi}(\Omega)$ defined by the norm (19).

We refer also to Proposition 5 of [115] where the reader can find a statement on preduals of type of Theorem 2.4 for Campanato spaces.

As a generalization of Spanne's result (38), J. Kovats [61] (1999) proved the embedding

$$\mathsf{L}_k^{p,\psi}(\Omega) \hookrightarrow C^{k,\psi_1}(\Omega), \quad \psi_1(t) = \int_0^t \frac{\psi(r)^{1/p}}{r^{1+k}}\, \mathrm{d}r \tag{42}$$

where Ω is a domain of type (A) and C^{k,ψ_1} is the space of functions differentiable up to order k with the last derivative satisfying the Hölder condition as in (39), under the assumption that the integral defining the function ψ_1 converges.

The generalized Campanato spaces, in the case where the underlying space X was a normal homogeneous metric measure space, defined for $1 \leq p < \infty$ by

$$\|f\|_{\mathcal{L}^{p,\phi}} := \sup_{x,r} \frac{1}{\phi(x,r)} \left(\frac{1}{\mu B(x,r)} \int_{B(x,r)} |f(y) - f_{B(x,r)}|^p \, \mathrm{d}\mu(y) \right)^{1/p}$$

were introduced in E. Nakai [75] (2006). Recall that a homogeneous metric measure space is called *normal* if

$$K_1 r \leq \mu B(x,r) \leq K_2 r. \tag{43}$$

There were given relations between such generalized Campanato spaces and Morrey and Hölder spaces, the latter defined by the norm

$$\|f\|_{\Lambda_\phi} := \sup_{\substack{x,y\in X\\ x\neq y}} \frac{2|f(x)-f(y)|}{\phi(x,d(x,y)) + \phi(y,d(y,x))},$$

including necessary and sufficient conditions on the function ϕ for the relations

$$\mathcal{L}^{p,\phi}(X)/\mathcal{C} \cong L^{p,\phi}(X), \quad \mathcal{L}^{p,\phi}(X) \cong L^{p,\phi}(X), \quad \mathcal{L}^{p,\phi}(X) \cong \Lambda_\phi(X).$$

A modified version of (vector-valued) Campanato spaces, with non-doubling measures, in the language of the *RBMO* spaces of X. Tolsa [107] (2001) was introduced and studied in Y. Sawano and H. Tanaka [96] (2006).

P. Górka [41, Theor. 3.1] (2009) gave a simple proof of a statement of type (3) of Theorem 4.3 in the general setting of homogeneous metric measure spaces (X, ρ, μ), for the Campanato spaces defined by the condition

$$\frac{1}{\mu B(x,r)} \int_{\widetilde{B}(x,r)} |f(y) - f_{B(x,r)}|^p \, \mathrm{d}\mu(y) \leq C^p r^{\alpha p},$$

not requiring the space (X, ρ, μ) to be normal. A local version of this theorem was used in [41, Theor. 3.3] to prove some embeddings of Hajłasz-Sobolev space $M^{1,p}(X)$, $1 < p < \infty$, into Hölder spaces.

4.4. Interpolation results

G. Stampacchia [101] (1964), [102] (1965) and S. Campanato and M. Murthy [21] (1965) proved a Riesz-Thorin-type interpolation theorem for operators acting from L^p into Campanato spaces $\mathscr{L}^{q,\lambda}$ (at that time, Morrey and Campanato spaces were simply called Morrey spaces). The result in a more complete form obtained in S. Campanato and M. Murthy [21] (1965) is the following, where $\mathscr{L}_k^{p,\lambda}(\Omega)$ is the space defined by (34) and Ω is a bounded open set in $\mathbb{R}^n$.

Theorem 4.5. *Let* $1 \le p_i \le \infty, 1 \le q_i \le \infty,\ 0 \le \lambda_i < n+p,\ i = 1,2,$ *and for* $0 < \theta < 1$ *define* p, q *and* λ *by*

$$\frac{1}{p} = \frac{1-\theta}{p_1} + \frac{\theta}{p_2}, \quad \frac{1}{q} = \frac{1-\theta}{q_1} + \frac{\theta}{q_2}, \quad \frac{1}{\lambda} = \frac{1-\theta}{\lambda_1} + \frac{\theta}{\lambda_2}. \tag{44}$$

If T *is a bounded linear operator from* $L^{q_i}(\Omega)$ *to* $\mathscr{L}_k^{p_i,\lambda_i}(\Omega)$, $i = 1,2$ *with the operator norm* K_i, *then* T *is bounded from* $L^q(\Omega)$ *to* $\mathscr{L}_k^{p,\lambda}(\Omega)$ *with the norm at most* $CK_1^{1-\theta}K_2^{\theta}$, *with* C *depending only on* θ, λ_i, p_i *and* q_i.

Interpolation in the other direction fails, as first shown by E. Stein and A. Zygmund [103] (1967) who constructed a bounded linear operator on H^α and L^2 but not on $L^q, q > 2$ and BMO. Further results on such a failure may be found in the papers by A. Ruiz and L. Vega [93] (1995) and O. Blasco et al. [17] (1999), where there were given examples of operators bounded from $L^{p_i,\lambda}$ to L^{q_i}, which are not bounded in the intermediate spaces.

Note that a version of Marcinkiewicz type theorem was obtained in G. Stampacchia [101] (1964) for spaces $\mathscr{L}^{p,\lambda}(Q_0)$, where Q_0 is a cube in $\mathbb{R}^n$. The linear operator T was defined to be of strong type (p,q,λ), if $\|Tf\|_{\mathscr{L}^{q,\lambda}} \le K\|f\|_{L^p}$ and of weak type (p,q,λ), if

$$\sup_Q r^{-\lambda} |\{x \in Q : |Tf - (Tf)_Q| > \sigma\}| \le \left(\frac{K}{\sigma}\|f\|_{L^p}\right)^q,$$

where Q is a cube with sides parallel to Q_0, and the following interpolation theorem was proved

Theorem 4.6. *If* T *is of weak types* (p_1,q_1,λ_1) *and* (p_2,q_2,λ_2), *where* $p_i \ge 1$, $p_i \le q_i$, $i = 1,2$, $q_1 \ne q_2$, $p_1 \ne p_2$, *then* T *is of strong type* (p,q,λ) *with* p,q,λ *defined in* (44).

For some related interpolation statements we also refer to the thesis of P. Grisvard [44] (1965), published in [45, 46] (1966) and the paper J. Peetre [87] (1966). S. Spanne [100] (1966) generalized and simplified the proofs of the interpolation theorem in the setting of generalized Campanato space. In fact, he reduced the validity of the interpolation to the L^p case. Namely, let

$$\frac{1}{p} = \frac{1-\theta}{p_0} + \frac{\theta}{p_1}, \quad \varphi(r) = \varphi_0(r)^{1-\theta}\varphi_1(r)^{\theta}, \quad 0 < \theta < 1$$

and let A_0, A_θ, A_1 be normed spaces such that the interpolation theorem is valid for the two triplets (A_0, A_θ, A_1) and (L^{p_0}, L^p, L^{p_1}). Then the interpolation theorem is valid also for (A_0, A_θ, A_1) and $(\mathscr{L}_k^{p_0,\varphi_0}, \mathscr{L}_k^{p,\varphi}, \mathscr{L}_k^{p_1,\varphi_1})$, with the same convexity constant. A similar result holds for the corresponding weak Campanato spaces.

4.5. Other characterizations of Campanato spaces

B. Grevholm [43] (1970) used the interpolation theorem for Campanato spaces to characterize the Campanato spaces as the Besov spaces, namely

$$\mathscr{L}_k^{p,\lambda}(\Omega) = B^\alpha(\Omega), \quad 0 < \alpha = \frac{\lambda - n}{p} < k,$$

where Ω is an open set in $\mathbb{R}^n$ satisfying some conditions and $B^\alpha(\Omega)$, in the case $\Omega = \mathbb{R}^n$, is defined by the seminorm

$$\sup_{t>0,|y|<1} \frac{\|\Delta_{ty}^k f\|_{L^\infty}}{t^\alpha}$$

while in the case $\Omega \neq \mathbb{R}^n$ the space $B^\alpha(\Omega)$ is defined as the interpolation space

$$B^\alpha(\Omega) = \big(C^0(\Omega), C^k(\Omega)\big)_{\frac{\alpha}{k},\infty}$$

under a certain interpolation method.

A result similar in a sense was obtained by different means in H.C. Greenwald [42] (1983) who proved the coincidence of the Campanato space $\mathscr{L}_k^{p,\lambda}(\mathbb{R}^n)$ with the Lipschitz-type space $\Lambda(\alpha, k)$ defined in terms of Gauss-Weierstrass integral:

$$\|f\|_{\alpha,k+1} = \sum_{|\nu|=k} \sup_{t\in\mathbb{R}^+} \sup_{x\in\mathbb{R}^n} t^{(k-\alpha)/2}|D^\nu f(x,t)| < \infty,$$

where $f(x,t)$ is the Gauss-Weierstrass integral of f and D stands for the differentiation with respect to x.

Consider also the space $L(\alpha, p, k-1)$ of equivalence classes modulo P_{k-1} of locally integrable functions f for which

$$\|f\|_{L(\alpha,p,k-1)} = \sup_{Q\subset\mathbb{R}^n} |Q|^{-\alpha/n} \left[\frac{1}{|Q|}\int_Q |f(x) - P_Q f(x)|^p \, dx\right]^{1/p} < \infty, \tag{45}$$

where Q is a ball and $P_Q f$ is the unique element of P_{k-1} such that

$$\int_Q [f(x) - P_Q f(x)] x^\nu \, dx = 0, \quad 0 \le |\nu| \le k-1. \tag{46}$$

Such spaces occur in the duality theory of Hardy spaces as discussed by M. Taibleson and G. Weiss [105] (1980); we refer also to a related paper M.H. Taibleson and G. Weiss [104] (1979). The main result of [42] asserts that the spaces $\Lambda_{\alpha,k}$ and $L(\alpha, p, k-1)$ coincide and that their norms are equivalent. An earlier result of similar nature was obtained by B. Grevholm [43] (1970) for p in the range $1 \le p < \infty$ using interpolation theory. The result in [42] is valid for $1 \le p \le \infty$ and is proved by elementary methods.

X.T. Duong and L.X. Yan [30] (2005) studied identity approximations adapted to Morrey-Campanato spaces on quasimetric measure spaces.

In D. Deng, X.T. Duong and L. Yan [27], the authors gave an equivalent characterization of the spaces $L(\alpha, p, k-1)$ by using the identity approximations instead of the minimizing polynomial in the definition of the norm (45) in the case $\alpha > 0$, $k > [n\alpha] + 1$ when these spaces do not depend on $p \in [1, \infty]$.

X.T. Duong, J. Xiao and L. Yan [29] (2006) studied the Morrey-Campanato spaces defined with the constant $c = f_B$ in the definition in (28) replaced by a semigroup of operators. They studied relations with the usually defined Morrey-Campanato spaces and showed that under appropriate choice of a semigroup, the new definition coincides with the old one.

L. Tang [106] (2007) used the ideas of [30] to define the Campanato spaces by the norm

$$\sup_B \frac{1}{\mu(B)^{\alpha+1}} \int_B |f(x) - A_B(f)| \, \mathrm{d}x,$$

where $A_B(f)$ is an identity approximation from [30]. There is shown that in some cases such different norms are equivalent but there were also given examples where they are not.

4.6. Miscellaneous

The *central mean oscillation space* CMO^q, introduced in Y.Z. Chen and K.S. Lau [24] (1989) and J. García-Cuerva [37] (1989), defined by

$$\|f\|_{\mathrm{CMO}^q} = \sup_{r\geq 1} \left(\frac{1}{|B(0,r)|} \int_{B(0,r)} |f(x) - f_{B(0,r)}|^q \, \mathrm{d}x \right)^{1/q}$$

was shown to be the dual space of an atomic space HA^q associated with the Beurling algebra. The *central bounded mean oscillation space* CBMO^q introduced in S. Lu and D. Yang [65] (1992) and S. Lu and D. Yang [66] (1995) is defined by

$$\|f\|_{\mathrm{CBMO}^q} = \sup_{r>0} \left(\frac{1}{|B(0,r)|} \int_{B(0,r)} |f(x) - f_{B(0,r)}|^q \, \mathrm{d}x \right)^{1/q}.$$

A generalization of CMO^q and CBMO^q, introduced in J. García-Cuerva and M.J.L. Herrero [38] (1994) and J. Alvarez, M. Guzmán-Partida and J. Lakey [8] (2000), are the so-called λ-*central mean oscillation spaces* $\mathrm{CMO}^{q,\lambda}$ and λ-*central bounded mean oscillation spaces* $\mathrm{CBMO}^{q,\lambda}$, defined by

$$\|f\|_{\mathrm{CMO}^{q,\lambda}} = \sup_{r\geq 1} \frac{1}{|B(0,r)|^\lambda} \left(\frac{1}{|B(0,r)|} \int_{B(0,r)} |f(x) - f_{B(0,r)}|^q \, \mathrm{d}x \right)^{1/q}$$

and

$$\|f\|_{\mathrm{CBMO}^{q,\lambda}} = \sup_{r>0} \frac{1}{|B(0,r)|^\lambda} \left(\frac{1}{|B(0,r)|} \int_{B(0,r)} |f(x) - f_{B(0,r)}|^q \, \mathrm{d}x \right)^{1/q}.$$

M. Kronz [62] (2001) introduced Morrey and Campanato spaces for elements which are mappings between metric measure spaces.

A classical Morrey inequality states that in the case $p > n$, the following embedding of a Sobolev space into Hölder space holds

$$W^{1,p}(\Omega) \hookrightarrow C^{0,\alpha}(\Omega).$$

In the paper A. Cianchi and L. Pick [26] (2003), in the case $p = 1$, there was given a detailed study of more general embeddings of Sobolev spaces into Morrey and Campanato spaces for the case Ω is a cube in $\mathbb{R}^n$. For a weakly differentiable function f on Q they gave optimal integrability conditions on the gradient of f, to belong to Morrey or Campanato space. More generally they gave a characterization of the rearrangement-invariant Banach function spaces such that the corresponding Sobolev space $W^1X(Q)$ is continuously embedded into Morrey or Campanato space. This enabled the authors to find the largest space $X(Q)$ for which such an embedding holds (the so-called optimal range partner). Such an optimal space is of Marcinkiewicz type in the case of Campanato spaces and have a different nature in the case of Morrey spaces. In particular, the following theorem was proved in [26], where $M_\psi(Q)$ is the Marcinkiewicz space defined by the norm

$$\|f\|_{M_\psi(Q)} = \sup_{0<t<1} \psi(t) f^{**}(t), \quad \psi(t) = \frac{t^{\frac{1}{n}+1}}{\varphi(t^{\frac{1}{n}})}.$$

Theorem 4.7. *Let φ be a strictly positive continuous function on $(0,\infty)$. Then the space $X(Q) = M_\psi(Q)$ is the largest rearrangement invariant space for which the embedding*

$$\|f\|_{\mathscr{L}^{1,\varphi}(Q)} \le C\|\nabla f\|_{X(Q)}$$

holds.

A version of grand Morrey spaces $L^{p),\lambda}(X)$ over homogeneous-type space X, which turns into the grand Lebesgue space $L^{p)}(X)$ introduced in T. Iwaniec and C. Sbordone [54] (1992) when $\lambda = 0$, was suggested in A. Meskhi [68] (2009). It is defined by the norm

$$\|f\|_{L^{p),\lambda}(X)} := \sup_{0<\varepsilon<p-1} \left(\sup_{x\in X, r>0} \frac{\varepsilon}{(\mu(B(x,r)))^\lambda} \int_{B(x,r)} |f(y)|^{p-\varepsilon}\, \mathrm{d}\mu(y) \right)^{1/(p-\varepsilon)}.$$

5. Variable exponent Morrey and Campanato spaces

The Morrey spaces $L^{p(\cdot),\lambda(\cdot)}(\Omega)$ with variable exponents $\lambda(\cdot)$ and $p(\cdot)$ over an open set $\Omega \subset \mathbb{R}^n$, were recently introduced almost simultaneously by different authors in A. Almeida, J. Hasanov and S. Samko [6] (2008), V. Kokilashvili and A. Meskhi [58] (2008), [59] (2010), T. Ohno [77] (2008), X. Fan [35] (2010).

In A. Almeida, J. Hasanov and S. Samko [6] (2008) the space $L^{p(\cdot),\lambda(\cdot)}(\Omega)$ was introduced as the space of functions with the finite norm

$$\|f\|_{L^{p(\cdot),\lambda(\cdot)}(\Omega)} = \inf\left\{\nu : I^{p(\cdot),\lambda(\cdot)}\left(\frac{f}{\nu}\right) \le 1\right\}$$

and the modular $I^{p(\cdot),\lambda(\cdot)}(f)$ defined by

$$I^{p(\cdot),\lambda(\cdot)}(f) := \sup_{x\in\Omega, r>0} \frac{1}{r^{\lambda(x)}} \int_{\widetilde{B}(x,r)} |f(y)|^{p(y)}\, \mathrm{d}y.$$

In the case of a bounded Ω they gave several equivalent norms and proved embedding theorems for such Morrey spaces under the assumption that $p(x)$ satisfies the log-condition well known in the variable exponent analysis. Similar embedding theorem for variable Campanato spaces may be found in [90] (2011) within the frameworks of the general setting of metric measure spaces.

V. Kokilashvili and A. Meskhi [58] (2008), see also [59] (2010), introduced Morrey-type spaces $M^{q(\cdot)}_{p(\cdot)}$ in the general setting when the underlying space is a homogeneous-type space (X, ρ, μ), with the norm defined by

$$\|f\|_{M^{q(\cdot)}_{p(\cdot)}} = \sup_{x\in X, r>0} (\mu(B(x,r)))^{1/p(x)-1/q(x)} \|f\|_{L^{q(\cdot)}(B(x,r))}$$

where $1 < \inf_X q \le q(\cdot) \le p(\cdot) \le \sup_X p < \infty$. In the case where X is bounded, some equivalence of norms and embedding theorems were obtained.

A φ-generalization $L^{p(\cdot),\nu,\varphi}(\mathbb{R}^n)$ of Morrey spaces with variable exponent $p(x)$ and constant $0 \le \nu \le n$, was given in T. Ohno [77] (2008) by the condition

$$\frac{\varphi(r)}{r^\nu} \int_{B(x,r)} \left|\frac{f(y)}{\lambda}\right|^{p(y)} \mathrm{d}y \le 1$$

for some $\lambda > 0$.

A more general version $\mathcal{M}^{p(\cdot),\omega}(\Omega)$, $\Omega \subseteq \mathbb{R}^n$ of such generalized variable exponent Morrey spaces was introduced in V. Guliev, J. Hasanov and S. Samko [49] (2010), defined by the norm

$$\|f\|_{\mathcal{M}^{p(\cdot),\omega}} = \sup_{x\in\Omega, r>0} \frac{r^{-\frac{n}{p(x)}}}{\omega(x,r)} \|f\|_{L^{p(\cdot)}(\widetilde{B}(x,r))}.$$

They recover the space $L^{p(\cdot),\lambda(\cdot)}(\Omega)$ under the choice $\omega(x,r) = r^{\frac{\lambda(x)-n}{p(x)}}$.

Both φ- and θ-generalizations of Morrey spaces of variable order were introduced in V. Guliev, J. Hasanov and S. Samko [48] (2010), as the space of functions with the finite norm

$$\sup_{x\in\Omega} \left\| \frac{\omega(x,r)}{r^{\frac{n}{p(x)}}} \|f\|_{L^{p(\cdot)}(\widetilde{B}(x,r))} \right\|_{L^{\theta(\cdot)}(0,\ell)},$$

where $\ell = \operatorname{diam} \Omega$.

The corresponding variable exponent Campanato spaces are interesting because they in general contain functions which are locally in $L^{p(\cdot),\lambda(\cdot)}$ on one subset,

BMO-functions locally on another subset and variable order Hölder continuous on the third one.

Such spaces appeared in X. Fan [35] (2010), where besides variable exponent Morrey spaces there were also introduced Campanato spaces $\mathscr{L}^{p(\cdot),\lambda(\cdot)}$ of variable order, in the Euclidean case, via the norm

$$\|f\|_{\mathscr{L}^{p(\cdot),\lambda(\cdot)}(\Omega)} := \|f\|_{L^{p(\cdot)}(\Omega)} + \sup_{x_0\in\Omega, r>0} \left\| r^{-\frac{\lambda(\cdot)}{p(\cdot)}}(f - f_{B(x_0,r)})\right\|_{L^{p(\cdot)}(B(x_0,r))},$$

where $f_B = |B|^{-1}\int_B f(x)\,\mathrm{d}x$. The equivalence of such Campanato spaces to variable exponent Hölder spaces is shown when $\inf_{x\in\Omega}\lambda(x) > n$ and to variable exponent Morrey spaces, when $\sup_{x\in\Omega}\lambda(x) < n$. In the latter result, the proof of the embedding of Morrey spaces into Campanato spaces was based on the notion of $p(\cdot)$-average of a function introduced in this paper.

Similar results for variable exponent Campanato spaces $\mathscr{L}^{p(\cdot),\lambda(\cdot)}(X)$ in a more general setting of metric measure spaces were obtained in H. Rafeiro and S. Samko [90] (2011). In [90], in the setting of an arbitrary quasimetric measure spaces, the log-Hölder condition for $p(x)$ is introduced with the distance $d(x,y)$ replaced by $\mu B(x, d(x,y))$, which provides a weaker restriction on $p(x)$ in the general setting. Some initial basic facts for variable exponent Lebesgue spaces hold without the assumption that X is homogeneous or even Ahlfors lower or upper regular, but the main results for Campanato spaces are proved in the case of homogeneous spaces X.

In E. Nakai [76] (2010) there were introduced φ-generalizations of such spaces on a space of homogeneous-type, normal in the sense of Macías and Segovia. In [76] φ was admited to be variable, but p constant and the norm defined by

$$\|f\|_{\mathscr{L}^{p,\varphi}} = \sup_{x,r>0} \frac{1}{\varphi(B(x,r))}\left(\frac{1}{\mu(B(x,r))}\int_{B(x,r)} |f(y) - f_{B(x,r)}|^p\,\mathrm{d}\mu(y)\right)^{1/p}.$$

We note also the embedding $L^{p(\cdot)}(X) \hookrightarrow L^{1,\varphi} \hookrightarrow \mathscr{L}^{1,\varphi}$ proved in [76], where $L^{1,\varphi}$ stands for the corresponding Morrey space and $\varphi(B(x,r)) = r^{-\frac{1}{p_*(x)}}$, where $p_*(x) = p(x)$ when $0 < r < 1/2$ and $p_*(x) = p_+$ when $1/2 \le r < \infty$.

References

[1] S.K. Abdullaev and A.A. Babaev, *Certain estimates for the singular integral with summable density* (Russian). Dokl. Acad. Nauk **188**(2) (1969), 263–265.

[2] S.K. Abdullaev and A.A. Babaev, *On a singular integral with summable density* (Russian). Funkts.nyi Anal. Primen., Baku **3** (1978), 3–32.

[3] D.R. Adams, *A note on Choquet integrals with respect to Hausdorff capacity.* In: Function Spaces and Applications (Lund, 1986), Lecture Notes in Math., vol. 1302, pp. 115–124, Springer, Berlin, 1988.

[4] D.R. Adams and J. Xiao, *Nonlinear potential analysis on Morrey spaces and their capacities.* Indiana Univ. Math. J. **53** (2004), no. 6, 1629–1663.

[5] D.R. Adams, *Lectures on L^p-potential theory.* Umeå University Reports, no. 2, 1981.

[6] A. Almeida, J. Hasanov and S. Samko, *Maximal and potential operators in variable exponent Morrey spaces.* Georgian Math. J. **15** (2008), no. 2, 195–208.

[7] J. Alvarez, *The distribution function in the Morrey space.* Proc. Amer. Math. Soc. **83** (1981), no. 4, 693–699.

[8] J. Alvarez, M. Guzmán-Partida and J. Lakey, *Spaces of bounded λ-central mean oscillation, Morrey spaces, and λ-central Carleson measures.* Collect. Math. **51** (2000), 1–47.

[9] J. Alvarez and C. Pérez, *Estimates with A_∞ weights for various singular integral operators.* Boll. Un. Mat. Ital. A (7) **8**, (1994), no. 1, 123–133.

[10] H. Arai and T. Mizuhara, *Morrey spaces on spaces of homogeneous type and estimates for $\Box_b$ and the Cauchy-Szegö projection.* Math. Nachr. **185** (1997), 5–20.

[11] A.A. Babaev, *Certain estimates for the singular integral* (Russian). Dokl. Akad. Nauk SSSR **170**, (1966), no. 5, 1003–1005.

[12] G.C. Barozzi, *Su una generalizzazione degli spazi $L^{(q,\lambda)}$ di Morrey.* Ann. Scuola Norm. Sup. Pisa (3) **19** (1965), 609–626.

[13] O.V. Besov, V.P. Il′in and S.M. Nikol′skiĭ, *Integral Representations of Functions and Embedding Theorems* (Russian). 2nd Edition, Fizmatlit "Nauka", Moscow, 1996.

[14] O.V. Besov, V.P. Il′in and S.M. Nikol′skiĭ, *Integral Representations of Functions and Imbedding Theorems. Vol. I.* V.H. Winston & Sons, Washington, D.C., 1978. Translated from the Russian, Scripta Series in Mathematics, edited by Mitchell H. Taibleson.

[15] O.V. Besov, V.P. Il′in and S.M. Nikol′skiĭ, *Integral Representations of Functions and Imbedding Theorems. Vol. II.* V.H. Winston & Sons, Washington, D.C., 1979. Scripta Series in Mathematics, edited by Mitchell H. Taibleson.

[16] A. Beurling, *Construction and analysis of some convolution algebra.* Ann. Inst. Fourier **14** (1964), 1–32.

[17] O. Blasco, A. Ruiz and L. Vega, *Non-interpolation in Morrey-Campanato and block spaces.* Ann. Scuola Norm. Sup. Pisa Cl. Sci. (4) **28** (1999), no. 1, 31–40.

[18] V.I. Burenkov and H.V. Guliyev, *Necessary and sufficient conditions for boundedness of the maximal operator in local Morrey-type spaces.* Studia Math. **163** (2004), no. 2, 157–176.

[19] S. Campanato, *Proprietà di hölderianità di alcune classi di funzioni.* Ann. Scuola Norm. Sup. Pisa (3) **17** (1963), 175–188.

[20] S. Campanato, *Proprietà di una famiglia di spazi funzionali.* Ann. Scuola Norm. Sup. Pisa (3) **18** (1964), 137–160.

[21] S. Campanato and M.K.V. Murthy, *Una generalizzazione del teorema di Riesz-Thorin.* Ann. Scuola Norm. Sup. Pisa (3) **19** (1965), 87–100.

[22] A. Canale, P. Di Gironimo and A. Vitolo, *Functions with derivatives in spaces of Morrey type and elliptic equations in unbounded domains.* Studia Math. **128** (1998), no. 3, 199–218.

[23] P. Cavaliere, G. Manzo and A. Vitolo, *Spaces of Morrey type and BMO spaces in unbounded domains of* $\mathbf{R}^n$. Rend. Accad. Naz. Sci. XL Mem. Mat. Appl. (5) **20** (1996), 123–140.

[24] Y.Z. Chen and K.S. Lau, *Some new classes of Hardy spaces.* J. Funct. Anal. **84** (1989), 255–278.

[25] M. Christ, *The extension problem for certain function spaces involving fractional orders of differentiability.* Ark. Mat. **22** (1984), no. 1, 63–81.

[26] A. Cianchi and L. Pick, *Sobolev embeddings into spaces of Campanato, Morrey and Hölder type.* J. Math. Anal. Appl. **282** (2003), no. 1, 128–150.

[27] D. Deng, X.T. Duong and L. Yan, *A characterization of the Morrey-Campanato spaces.* Math. Z. **250** (2005), no. 3, 641–655.

[28] R.A. DeVore and R.C. Sharpley, *Maximal functions measuring smoothness.* Mem. Amer. Math. Soc. 47 **293** :viii+115, 1984.

[29] X.T. Duong, J. Xiao and L. Yan, *Old and new Morrey spaces with heat kernel bounds.* J. Fourier Anal. Appl. **13** (2007), no. 1, 87–111.

[30] X.T. Duong and L. Yan, *New function spaces of BMO type, the John-Nirenberg inequality, interpolation and applications.* Comm. Pure Appl. Math. **58** (2005), no. 10, 1375–1420.

[31] G.T. Dzhumakaeva and K. Zh. Nauryzbaev, *Lebesgue-Morrey spaces.* Izv. Akad. Nauk Kazakh. SSR Ser. Fiz.-Mat. **79** (1982), no. 5, 7–12.

[32] A. El Baraka, *Littlewood-Paley characterization for Campanato spaces.* J. Funct. Spaces Appl. **4** (2006), no. 2, 193–220.

[33] A. Eridani and H. Gunawan, *Stummel class and Morrey spaces.* Southeast Asian Bull. Math. **29** (2005), no. 6, 1053–1056.

[34] A. Eridani, V. Kokilashvili and A. Meskhi, *Morrey spaces and fractional integral operators.* Expo. Math. **27** (2009), no. 3, 227–239.

[35] X. Fan, *Variable exponent Morrey and Campanato spaces.* Nonlinear. Anal. **72** (2010), no. 11, 4148–4161.

[36] Y. Furusho, *On inclusion property for certain* $\mathcal{L}^{(p,\lambda)}$ *spaces of strong type.* Funkcial. Ekvac. **23** (1980), no. 2, 197–205.

[37] J. García-Cuerva, *Hardy spaces and Beurling algebras.* J. London Math. Soc. **39** (1989), no. 2, 499–513.

[38] J. García-Cuerva and M.J.L. Herrero, *A theory of Hardy spaces associated to the Herz spaces.* Proc. London Math. Soc. **69** (1994), 605–628.

[39] M. Geisler, *Morrey-Campanato spaces on manifolds.* Comment. Math. Univ. Carolin. **29** (1988), no. 2, 309–318.

[40] M. Giaquinta, *Multiple Integrals in the Calculus of Variations and Nonlinear Elliptic Systems.* Annals of Mathematics Studies, vol. 105. Princeton University Press, Princeton, NJ, 1983.

[41] P. Górka, *Campanato theorem on metric measure spaces.* Ann. Acad. Sci. Fenn. Math. **34** (2009), no. 2, 523–528.

[42] H. Greenwald, *On the theory of homogeneous Lipschitz spaces and Campanato spaces.* Pacific J. Math. **106** (1983), no. 1, 87–93.

[43] B. Grevholm, *On the structure of the spaces* $\mathscr{L}_k^{p,\lambda}$. Math. Scand. **26** (1970), 241–254.

[44] P. Grisvard, *Commutativité de deux foncteurs d'interpolation et applications.* Thèse, Paris, 1965.

[45] P. Grisvard, *Commutativité de deux foncteurs d'interpolation et applications.* J. Math. Pures Appl. (9) **45** (1966), 143–206.

[46] P. Grisvard, *Commutativité de deux foncteurs d'interpolation et applications.* J. Math. Pures Appl. (9) **45** (1966), 207–290.

[47] V.S. Guliev, *Integral operators on function spaces defined on homogeneous groups and domains in* $\mathbf{R}^n$ (Russian). D.Sc. Dissertation, Steklov Math. Inst., 1994.

[48] V. Guliyev, J. Hasanov and S. Samko, *Boundedness of maximal, potential type and singular integral operators in the generalized variable exponent Morrey type spaces.* J. Math. Sci. (N.Y.) **170** (2010), no. 4, 423–443.

[49] V.S. Guliyev, J.J. Hasanov and S.G. Samko, *Boundedness of the maximal, potential and singular operators in the generalized variable exponent Morrey spaces.* Math. Scand. **107** (2010), no. 2, 285–304.

[50] V.S. Guliyev, *Integral Operators, Function Spaces and Questions of Approximation on the Heisenberg Group* (Russian). Elm, Baku, 1996.

[51] V.S. Guliyev, *Function Spaces, Integral Operators and two Weighted Inequalities on Homogeneous Groups. Some Applications* (Russian). Çaşioğlu, Baku, 1999.

[52] V.P. Il′in, *On theorems of "imbedding".* Trudy Mat. Inst. Steklov **53** (1959), 359–386.

[53] V.P. Il′in, *Certain properties of functions of the spaces* $W^l_{p,\,a,\,\kappa}(G)$. Zap. Naučn. Sem. Leningrad. Otdel. Mat. Inst. Steklov. (LOMI) **23** (1971), 33–40.

[54] T. Iwaniec and C. Sbordone, *On the integrability of the Jacobian under minimal hypotheses.* Arch. Rational Mech. Anal. **119** (1992), no. 2, 129–143.

[55] S. Janson, M. Taibleson and G. Weiss, *Elementary characterizations of the Morrey-Campanato spaces.* In: Harmonic Analysis (Cortona, 1982), Lecture Notes in Math., vol. 992, pp. 101–114. Springer, Berlin, 1983.

[56] F. John and L. Nirenberg, *On functions of bounded mean oscillation.* Comm. Pure Appl. Math. **14** (1961), 415–426.

[57] E.A. Kalita, *Dual Morrey spaces.* Dokl. Akad. Nauk **361** (1998), no. 4, 447–449.

[58] V. Kokilashvili and A. Meskhi, *Boundedness of maximal and singular operators in Morrey spaces with variable exponent.* Armen. J. Math. **1** (2008), no. 1, 18–28.

[59] V. Kokilashvili and A. Meskhi, *Maximal functions and potentials in variable exponent Morrey spaces with non-doubling measure.* Complex Var. Elliptic Equ. **55** (2010), no. 8–10, 923–936.

[60] Y. Komori and S. Shirai, *Weighted Morrey spaces and a singular integral operator.* Math. Nachr. **282** (2009), no. 2, 219–231.

[61] J. Kovats, *Dini-Campanato spaces and applications to nonlinear elliptic equations.* Electron. J. Differential Equations, **37** (1999), 20 pp. (electronic).

[62] M. Kronz, *Some function spaces on spaces of homogeneous type.* Manuscripta Math. **106** (2001), no. 2, 219–248.

[63] A. Kufner, O. John and S. Fučík, *Function Spaces.* Noordhoff International Publishing, Leyden, 1977. Monographs and Textbooks on Mechanics of Solids and Fluids; Mechanics: Analysis.

[64] S. Leonardi, *Weighted Miranda-Talenti inequality and applications to equations with discontinuous coefficients.* Comment. Math. Univ. Carolin. **43** (2002), no. 1, 43–59.

[65] S. Lu and D. Yang, *The Littlewood-Paley function and ϕ-transform characterizations of a new Hardy space HK_2 associated with the Herz space.* Studia Math. **101** (1992), no. 3, 285–298.

[66] S. Lu and D. Yang, *The central BMO spaces and Littlewood-Paley operators.* Approx. Theory Appl. **11** (1995), 72–94.

[67] R.A. Macías and C. Segovia, *Lipshitz functions on spaces of homogeneous type.* Adv. Math. **33** (1979), 257–270.

[68] A. Meskhi, *Maximal functions and singular integrals in Morrey spaces associated with grand Lebesgue spaces.* Proc. A. Razmadze Math. Inst. **151** (2009), 139–143.

[69] N.G. Meyers, *Mean oscillation over cubes and Hölder continuity.* Proc. Amer. Math. Soc. **15** (1964), no. 5, 717–721.

[70] C.-x. Miao and B.-q. Yuan, *Weak Morrey spaces and strong solutions to the Navier-Stokes equations.* Sci. China Ser. A **50** (2007), no. 10, 1401–1417.

[71] C.B. Morrey, *On the solutions of quasi-linear elliptic partial differential equations.* Trans. Amer. Math. Soc. **43** (1938), no. 1, 126–166.

[72] E. Nakai, *Hardy-Littlewood maximal operator, singular integral operators and the Riesz potentials on generalized Morrey spaces.* Math. Nachr. **166** (1994), 95–103.

[73] E. Nakai, *Pointwise multipliers on the Morrey spaces.* Mem. Osaka Kyoiku Univ. III Natur. Sci. Appl. Sci. **46** (1997), no. 1, 1–11.

[74] E. Nakai, *A characterization of pointwise multipliers on the Morrey spaces.* Sci. Math. **3** (2000), no. 3, 445–454.

[75] E. Nakai, *The Campanato, Morrey and Hölder spaces on spaces of homogeneous type.* Studia Math. **176** (2006), no. 1, 1–19.

[76] E. Nakai, *Singular and fractional integral operators on Campanato spaces with variable growth conditions.* Rev. Mat. Complut. **23** (2010), no. 2, 355–381.

[77] T. Ohno, *Continuity properties for logarithmic potentials of functions in Morrey spaces of variable exponent.* Hiroshima Math. J. **38** (2008), no. 3, 363–383.

[78] P.A. Olsen, *Fractional integration, Morrey spaces and a Schrödinger equation.* Comm. Partial Differential Equations **20** (1995), no. 11–12, 2005–2055.

[79] A. Ono, *On imbedding between strong Hölder spaces and Lipschitz spaces in L^p.* Mem. Fac. Sci. Kyushu Univ. Ser. A **24** (1970), 76–93.

[80] A. Ono, *On imbedding theorems in strong $\mathfrak{L}^{(q,\mu)}$ spaces.* Rend. Ist. Mat. Univ. Trieste **4** (1972), 53–65.

[81] A. Ono, *Note on the Morrey-Sobolev-type imbedding theorems in the strong $\mathcal{L}^{(p,\lambda)}$ spaces.* Math. Rep. Kyushu Univ. **11** (1977/78), no. 1, 31–37.

[82] A. Ono, *Remarks on the Morrey-Sobolev type imbedding theorems in the strong $\mathcal{L}^{(p,\lambda)}$ spaces.* Math. Rep. Kyushu Univ. **11** (1977/78), no. 2, 149–155.

[83] A. Ono, *On isomorphism between certain strong $\mathcal{L}^{(p,\lambda)}$ spaces and the Lipschitz spaces and its applications.* Funkcial. Ekvac. **21** (1978), no. 3, 261–270.

[84] A. Ono and Y. Furusho, *On isomorphism and inclusion property for certain $\mathfrak{L}^{(p,\lambda)}$ spaces of strong type.* Ann. Mat. Pura Appl. (4) **114** (1977), 289–304.

[85] D. Opěla, *Spaces of functions with bounded and vanishing mean oscillation.* In: Function Spaces, Differential Operators and Nonlinear Analysis (Teistungen, 2001), pp. 403–413. Birkhäuser, Basel, 2003.

[86] J. Peetre, *On the theory of $\mathscr{L}_{p,\lambda}$ spaces.* J. Funct. Anal. **4** (1969), 71–87.

[87] J. Peetre, *Espaces d'interpolation et théorème de Soboleff.* Ann. Inst. Fourier (Grenoble) **16** (1966), no. 1, 279–317.

[88] L.-E. Persson and N. Samko, *Some remarks and new developments concerning Hardy-type inequalities.* Rend. Circ. Mat. Palermo, serie II **82** (2010), 1–29.

[89] L.C. Piccinini, *Inclusioni tra spazi di Morrey.* Boll. Un. Mat. Ital. (4) **2** (1969), 95–99.

[90] H. Rafeiro and S. Samko, *Variable exponent Campanato spaces.* J. Math. Sci. (N.Y.) **172** (2011), no. 1, 143–164.

[91] M.A. Ragusa, *Regularity for weak solutions to the Dirichlet problem in Morrey space.* Riv. Mat. Univ. Parma (5) **3** (1995), no. 2, 355–369.

[92] M.A. Ragusa and P. Zamboni, *A potential theoretic inequality.* Czechoslovak Math. J. **51** (2001), no. 1, 55–65.

[93] A. Ruiz and L. Vega, *Corrigenda to: "Unique continuation for Schrödinger operators with potential in Morrey spaces" and a remark on interpolation of Morrey spaces.* Publ. Mat. **39** (1995), no. 2, 405–411.

[94] Y. Sawano, *Generalized Morrey spaces for non-doubling measures.* NoDEA Nonlinear Differential Equations Appl. **15** (2008), no. 4–5, 413–425.

[95] Y. Sawano and H. Tanaka, *Morrey spaces for non-doubling measures.* Acta Math. Sin. (Engl. Ser.) **21** (2005), no. 6, 1535–1544.

[96] Y. Sawano and H. Tanaka, *Equivalent norms for the Morrey spaces with non-doubling measures.* Far East J. Math. Sci. **22** (2006), no. 3, 387–404.

[97] L. Softova, *Singular integrals and commutators in generalized Morrey spaces.* Acta Math. Sin. (Engl. Ser.) **22** (2006), no. 3, 757–766.

[98] L. Softova, *Singular integral operators in functional spaces of Morrey type.* In: Advances in deterministic and stochastic analysis, pp. 33–42, World Sci. Publ., Hackensack, NJ, 2007.

[99] S. Spanne, *Some function spaces defined using the mean oscillation over cubes.* Ann. Scuola Norm. Sup. Pisa (3) **19** (1965), 593–608.

[100] S. Spanne, *Sur l'interpolation entre les espaces $\mathcal{L}_k^{p\Phi}$.* Ann. Scuola Norm. Sup. Pisa (3) **20** (1966), 625–648.

[101] G. Stampacchia, *$\mathcal{L}^{(p,\lambda)}$-spaces and interpolation.* Comm. Pure Appl. Math. **17** (1964), 293–306.

[102] G. Stampacchia, *The spaces $\mathcal{L}^{(p,\lambda)}$, $N^{(p,\lambda)}$ and interpolation.* Ann. Scuola Norm. Sup. Pisa (3) **19** (1965), 443–462.

[103] E.M. Stein and A. Zygmund, *Boundedness of translation invariant operators on Hölder spaces and L^p-spaces.* Ann. of Math. (2) **85** (1967), 337–349.

[104] M.H. Taibleson and G. Weiss, *The molecular characterization of Hardy spaces.* In: Harmonic analysis in Euclidean spaces (Proc. Sympos. Pure Math., Williams Coll., Williamstown, Mass., 1978), Part 1, Proc. Sympos. Pure Math., XXXV, Part, pp. 281–287. Amer. Math. Soc., Providence, R.I., 1979.

[105] M.H. Taibleson and G. Weiss, *The molecular characterization of certain Hardy spaces.* Astérisque **77** (1980), 67–151.

[106] L. Tang, *New function spaces of Morrey-Campanato type on spaces of homogeneous type.* Illinois J. Math. **51** (2007), no. 2, 625–644 (electronic).

[107] X. Tolsa, *BMO, H^1 and Calderón-Zygmund operators for non doubling measures.* Math. Ann. **319** (2001), no. 1, 89–149.

[108] M. Transirico, M. Troisi, and A. Vitolo, *Spaces of Morrey type and elliptic equations in divergence form on unbounded domains.* Boll. Un. Mat. Ital. B (7) **9** (1995), no. 1, 153–174.

[109] H. Triebel, *Theory of Functional Spaces, II*, Monographs in Mathematics, vol. 84. Birkhäuser Verlag, Basel, 1992.

[110] C. Vitanza, *Functions with vanishing Morrey norm and elliptic partial differential equations.* In: Proceedings of Methods of Real Analysis and Partial Differential Equations, Capri, pp. 147–150. Springer, 1990.

[111] N. Wiener, *Generalized Harmonic Analysis.* Acta Math. **55** (1930), 117–258.

[112] N. Wiener, *Tauberian theorems.* Ann. of Math. **33** (1932), 1–100.

[113] D. Yang, *Some function spaces relative to Morrey-Campanato spaces on metric spaces.* Nagoya Math. J. **177** (2005), 1–29.

[114] W. Yuan, W. Sickel, and D. Yang, *Morrey and Campanato Meet Besov, Lizorkin and Triebel.* Lecture Notes in Mathematics, 2005. Springer, 2010.

[115] C.T. Zorko, *Morrey space.* Proc. Amer. Math. Soc. **98** (1986), no. 4, 586–592.

Humberto Rafeiro
Instituto Superior Técnico, Centro CEAF
Av. Rovisco Pais
1049–001 Lisboa, Portugal
e-mail: hrafeiro@math.ist.utl.pt

Natasha Samko and Stefan Samko
Universidade do Algarve
Campus de Gambelas
8005–139 Faro, Portugal
e-mail: nsamko@ualg.pt
ssamko@ualg.pt

Operator Theory:
Advances and Applications, Vol. 228, 325–342

Arveson Dichotomy and Essential Fractality

Steffen Roch

Dedicated to Vladimir S. Rabinovich on the occasion of his 70th birthday

Abstract. The notions of fractal and essentially fractal algebras of approximation sequences and of the Arveson dichotomy have proved extremely useful for several spectral approximation problems. The purpose of this short note is threefold: to present a short new proof of the fractal restriction theorem, to relate essential fractality with Arveson dichotomy, and to derive a restriction theorem for essential fractality.

Mathematics Subject Classification (2010). Primary 65J10; Secondary 46L99, 47N40.

Keywords. Arveson dichotomy, essential spectral approximation, essential fractality, essential fractal restriction of approximation sequences.

1. Preliminaries

Let H be an infinite-dimensional separable Hilbert space. We denote by $L(H)$ the C^*-algebra of the bounded linear operators and by $K(H)$ the ideal of the compact operators on H.

A sequence $\mathcal{P} = (P_n)_{n\geq 1}$ of orthogonal projections of finite rank which converges strongly to the identity operator on H is called a *filtration* on H. Given a filtration $\mathcal{P}$, let $\mathcal{F}^{\mathcal{P}}$ stand for the set of all sequences $\mathbf{A} = (A_n)$ of operators $A_n : \operatorname{im} P_n \to \operatorname{im} P_n$ such that the sequence $(A_n P_n)$ converges strongly to an operator $W^{\mathcal{P}}(\mathbf{A}) \in L(H)$. Since every sequence in $\mathcal{F}^{\mathcal{P}}$ is bounded by the Banach-Steinhaus theorem, one can introduce pointwise defined operations

$$(A_n) + (B_n) := (A_n + B_n), \quad (A_n)(B_n) := (A_n B_n), \quad (A_n)^* := (A_n^*) \qquad (1.1)$$

and the supremum norm $\|(A_n)\|_{\mathcal{F}} := \sup_n \|A_n\|$, which make $\mathcal{F}^{\mathcal{P}}$ to a unital C^*-algebra and $W^{\mathcal{P}} : \mathcal{F}^{\mathcal{P}} \to L(H)$ to a unital *-homomorphism. This homomorphism is also known as the consistency map associated with the filtration $\mathcal{P}$.

Set $\delta(n) := \operatorname{rank} P_n := \dim \operatorname{im} P_n < \infty$ for every n and choose an orthonormal basis in each of the spaces $\operatorname{im} P_n$. Every operator $A_n \in L(\operatorname{im} P_n)$ can be identified with its matrix representation with respect to the chosen basis and, thus, with an element of the C^*-algebra $\mathbb{C}^{\delta(n)\times\delta(n)}$ of all $\delta(n)\times\delta(n)$ matrices with complex entries. The choice of a basis in each space $\operatorname{im} P_n$ makes $\mathcal{F}^{\mathcal{P}}$ to a special instance of an algebra of matrix sequences in the following sense. Given a sequence δ of positive integers, we let $\mathcal{F}^\delta$ stand for the set of all bounded sequences (A_n) of matrices $A_n \in \mathbb{C}^{\delta(n)\times\delta(n)}$. Introducing again pointwise operations and the supremum norm, we make $\mathcal{F}^\delta$ to a C^*-algebra with identity element $(I_{\delta(n)})$, the *algebra of matrix sequences with dimension function* δ. The set of all sequences in $\mathcal{F}^\delta$ which tend to zero in the norm forms a closed ideal of $\mathcal{F}^\delta$. We denote this ideal by $\mathcal{G}^\delta$ and refer to sequences in $\mathcal{G}^\delta$ as *zero sequences*. For example, the algebra of matrix sequences with constant dimension function $\delta = 1$ is $l^\infty(\mathbb{N})$, but in what follows we will be mainly interested in strictly increasing dimension functions, as they occur in the context of filtrations.

When passing from $\mathcal{F}^{\mathcal{P}}$ to $\mathcal{F}^\delta$ with $\delta(n) := \operatorname{rank} P_n$, one loses the embedding of the matrix algebras $L(\operatorname{im} P_n) \cong \mathbb{C}^{\delta(n)\times\delta(n)}$ into a common Hilbert space. It makes thus no sense to speak about strong convergence of a sequence in $\mathcal{F}^\delta$. But it will turn out that algebras of matrix sequences provide a suitable frame to formulate and study stability problems as well as a lot of other problems which do not depend upon an embedding into a Hilbert space. Moreover, some of the notions and assertions discussed in this paper remain meaningful in the much more general context, when $\mathcal{F}^{\mathcal{C}}$ is the direct product of a sequence $\mathcal{C} = (\mathcal{C}_n)_{n\ge 1}$ of unital C^*-algebras. The associated ideal of zero sequences in $\mathcal{F}^{\mathcal{C}}$, which can be identified with the direct sum of the family $\mathcal{C}$ in a natural way, will then be denoted by $\mathcal{G}^{\mathcal{C}}$. Sequences in $\mathcal{G}^{\mathcal{C}}$ will be called zero sequences again.

The following will serve as a running example in this paper. We consider the algebra of the finite sections discretization for Toeplitz operators with continuous generation function. For a continuous function a on the complex unit circle $\mathbb{T}$, the associated *Toeplitz operator* is the operator $T(a)$ on $l^2(\mathbb{Z}^+)$ which is given by the infinite matrix $(a_{i-j})_{i,j=0}^\infty$, with a_k denoting the kth Fourier coefficient of a. Note that $T(a)$ is a bounded operator and $\|T(a)\| = \|a\|_\infty$. For $n \in \mathbb{N}$, put

$$P_n : l^2(\mathbb{Z}^+) \to l^2(\mathbb{Z}^+), \quad (x_n)_{n\ge 0} \mapsto (x_0, x_1, \dots, x_{n-1}, 0, 0, \dots).$$

Then $\mathcal{P} = (P_n)$ is a filtration on $l^2(\mathbb{Z}^+)$. We let $\mathcal{S}(\mathsf{T}(C))$ stand for the smallest closed subalgebra of $\mathcal{F}^{\mathcal{P}}$ which contains all sequences $(P_n T(a)|_{\operatorname{im} P_n})$ of finite sections of Toeplitz operators $T(a)$ with $a \in C(\mathbb{T})$. Let $R_n : \operatorname{im} P_n \to \operatorname{im} P_n$ be the reflection operator

$$(x_0, x_1, \dots, x_{n-1}, 0, 0, \dots) \mapsto (x_{n-1}, \dots, x_1, x_0, 0, 0, \dots).$$

It is not hard to see that for each sequence $\mathbf{A} = (A_n) \in \mathcal{S}(\mathsf{T}(C))$, the strong limit $\widetilde{W}(\mathbf{A}) := \text{s-lim} R_n A_n R_n P_n$ exists and that $\widetilde{W}$ is a unital and fractal *-homomorphism from $\mathcal{S}(\mathsf{T}(C))$ to $L(l^2\mathbb{Z}^+)$. The following is a by now classical result by Böttcher and Silbermann [5], see also Chapter 2 in [6] and Sections 1.3, 1.4 and 1.6 in [7].

Theorem 1.1.

(a) *The algebra* $\mathcal{S}(\mathsf{T}(C))$ *consists of all sequences* $(P_nT(a)P_n+P_nKP_n+R_nLR_n+G_n)$ *where* $a \in C(\mathbb{T})$, $K, L \in K(l^2(\mathbb{Z}^+))$, *and* $(G_n) \in \mathcal{G}^{\mathcal{P}}$.

(b) *For every sequence* $\mathbf{A} \in \mathcal{S}(\mathsf{T}(C))$, *the coset* $\mathbf{A} + \mathcal{G}^{\mathcal{P}}$ *is invertible in the quotient algebra* $\mathcal{S}(\mathsf{T}(C))/\mathcal{G}^{\mathcal{P}}$ *if and only if the operators* $W^{\mathcal{P}}(\mathbf{A})$ *and* $\widetilde{W}(\mathbf{A})$ *are invertible.*

Due to its transparent structure, the algebra $\mathcal{S}(\mathsf{T}(C))$ served as a basic example for the development of algebraic methods in asymptotic numerical analysis. These methods have found fruitful applications in the stability analysis of different approximation methods for numerous classes of operators; see the monographs [7, 8, 17] for an overview. In particular, I would like to emphasize the finite sections method for band-dominated operators, a topic which was mainly influenced and shaped by Vladimir S. Rabinovich and the limit operator techniques developed by him, see [9–13] and [15] for an overview. In fact, the algebra of the finite sections method for band-dominated operators is the first real-life example of an essentially fractal, but not fractal, algebra (these notions will be introduced below).

2. Fractality

As it was observed in [14, 16], several natural approximation procedures lead to C^*-subalgebras $\mathcal{A}$ of the algebra $\mathcal{F}$ which are distinguished by the property of self-similarity: Given a subsequence of a sequence in $\mathcal{A}$, one can uniquely reconstruct the full sequence up to a sequence which tends to zero in the norm. These algebras were called fractal in [16]. The goals of this section is to recall the basic definitions and some consequences of fractality, and to give a short proof of the known fact that every separable subalgebra of $\mathcal{F}$ possesses a fractal restriction.

In this section, we let $\mathcal{F} := \mathcal{F}^{\mathcal{C}}$ be the product of a family $\mathcal{C} = (\mathcal{C}_n)_{n\in\mathbb{N}}$ of unital C^*-algebras and $\mathcal{G} := \mathcal{G}^{\mathcal{C}}$ the associated ideal of zero sequences.

2.1. Definition and first consequences

For each strictly increasing sequence $\eta : \mathbb{N} \to \mathbb{N}$, let $\mathcal{F}_\eta$ stand for the product of the family $(\mathcal{C}_{\eta(n)})_{n\in\mathbb{N}}$ of C^*-algebras, and write $\mathcal{G}_\eta$ for the associated ideal of zero sequences. The elements of $\mathcal{F}_\eta$ can be viewed of as subsequences of sequences in $\mathcal{F}$. The canonical restriction mapping $R_\eta : \mathcal{F} \to \mathcal{F}_\eta$, $(A_n) \mapsto (A_{\eta(n)})$ is a *-homomorphism from $\mathcal{F}$ onto $\mathcal{F}_\eta$ and maps $\mathcal{G}$ onto $\mathcal{G}_\eta$. More generally, for each C^*-subalgebra $\mathcal{A}$ of $\mathcal{F}$, we let $\mathcal{A}_\eta$ denote the image of $\mathcal{A}$ under R_η. Clearly, $\mathcal{A}_\eta$ is a C^*-subalgebra of $\mathcal{F}_\eta$. We call algebras obtained in this way *restrictions* of $\mathcal{A}$.

Definition 2.1.

(a) Let $\mathcal{A}$ be a C^*-subalgebra of $\mathcal{F}$. A *-homomorphism W from $\mathcal{A}$ into a C^*-algebra $\mathcal{B}$ is called *fractal* if it factors through $R_\eta|_{\mathcal{A}}$ for *every* strictly increasing sequence $\eta : \mathbb{N} \to \mathbb{N}$, i.e., if for each such η, there is a mapping $W_\eta : \mathcal{A}_\eta \to \mathcal{B}$ such that $W = W_\eta R_\eta|_{\mathcal{A}}$.

(b) A C^*-subalgebra $\mathcal{A}$ of $\mathcal{F}$ is *fractal* if the canonical homomorphism

$$\mathcal{A} \to \mathcal{A}/(\mathcal{A} \cap \mathcal{G}), \quad \mathbf{A} \mapsto \mathbf{A} + (\mathcal{A} \cap \mathcal{G})$$

is fractal.

(c) A sequence $\mathbf{A} \in \mathcal{F}$ is *fractal* if the smallest C^*-subalgebra of $\mathcal{F}$ which contains the sequence $\mathbf{A}$ and the identity sequence is fractal.

For example, if $\mathcal{P}$ is a filtration, then the associated consistency map $W^{\mathcal{P}}$ is fractal (since the strong limit of a sequence $(A_n) \in \mathcal{F}^{\mathcal{P}}$ can be determined from each subsequence of (A_n)). For the same reason, the homomorphism $\widetilde{W}$ appearing in Theorem 1.1 is fractal.

The fractal subalgebras of $\mathcal{F}$ are distinguished by their property that every sequence in the algebra can be rediscovered from each of its (infinite) subsequences up to a sequence tending to zero. Note that, by Definition 2.1, a fractal sequence always lies in a *unital* fractal algebra, whereas a fractal algebra needs not to be unital.

Assertion (a) of the following theorem provides an equivalent characterization of the fractality of an algebra. Proofs of Theorems 2.2 and 2.3 are given in [16] and in Section 1.6 of [7].

Theorem 2.2.

(a) *A C^*-subalgebra $\mathcal{A}$ of $\mathcal{F}$ is fractal if and only if the implication*

$$R_\eta(\mathbf{A}) \in \mathcal{G}_\eta \;\Rightarrow\; \mathbf{A} \in \mathcal{G} \tag{2.1}$$

holds for every sequence $\mathbf{A} \in \mathcal{A}$ and every strictly increasing sequence η.

(b) *If $\mathcal{A}$ is a fractal C^*-subalgebra of $\mathcal{F}$, then $\mathcal{A}_\eta \cap \mathcal{G}_\eta = (\mathcal{A} \cap \mathcal{G})_\eta$ for each strictly increasing sequence η.*

(c) *A unital C^*-subalgebra of $\mathcal{F}$ is fractal if and only if each of its elements is fractal.*

The following criterion will prove to be useful in order to verify the fractality of many specific algebras of approximation methods.

Theorem 2.3. *A unital C^*-subalgebra $\mathcal{A}$ of $\mathcal{F}$ is fractal if and only if there is a family $\{W_t\}_{t \in T}$ of unital and fractal *-homomorphisms W_t from $\mathcal{A}$ into unital C^*-algebras $\mathcal{B}_t$ such that the following equivalence holds for every sequence $\mathbf{A} \in \mathcal{A}$: The coset $\mathbf{A} + \mathcal{A} \cap \mathcal{G}$ is invertible in $\mathcal{A}/(\mathcal{A} \cap \mathcal{G})$ if and only if $W_t(\mathbf{A})$ is invertible in $\mathcal{B}_t$ for every $t \in T$.*

For example, since $W^{\mathcal{P}}$ and $\widetilde{W}$ are fractal homomorphisms, we conclude from Theorem 1.1 (b) and from the previous theorem that the algebra $\mathcal{S}(\mathsf{T}(C))$ is fractal. □

The property of fractality has striking consequences for asymptotic spectral properties of a sequence $\mathbf{A} = (A_n)$, see [14, 16] and Chapter 3 in [7]. Here we only mention a few of them which are relevant for what follows. For every element a of a unital C^*-algebra $\mathcal{A}$, we let $\sigma_2(a)$ denote the set of all non-negative square roots of points in the spectrum of a^*a. In case $\mathcal{A} = \mathbb{C}^{n\times n}$, the numbers in $\sigma_2(a)$ are known as the singular values of a.

Proposition 2.4. *Let $\mathcal{A}$ be a fractal C^*-subalgebra of $\mathcal{F}$ and $\mathbf{A} = (A_n)$ a sequence in $\mathcal{A}$. Then*

- (a) *the sequence $\mathbf{A}$ is stable if and only if it possesses a stable subsequence;*
- (b) *the limit $\lim_{n\to\infty} \|A_n\|$ exists and is equal to $\|\mathbf{A} + \mathcal{G}\|$;*
- (c) *the limit $\lim_{n\to\infty} \sigma_2(A_n)$ exists with respect to the Hausdorff distance on $\mathbb{R}$ and is equal to $\sigma_2(\mathbf{A} + \mathcal{G})$.*

2.2. The fractal restriction theorem

The preceding proposition and related results from [7] indicate that it is a question of vital importance in numerical analysis to single out fractal subsequences of a given sequence in $\mathcal{F}$. The following theorem states that such subsequences always exist.

Theorem 2.5. *Let $\mathcal{A}$ be a separable C^*-subalgebra of $\mathcal{F}$. Then there exists a strictly increasing sequence $\eta : \mathbb{N} \to \mathbb{N}$ such that the restricted algebra $\mathcal{A}_\eta = R_\eta \mathcal{A}$ is a fractal subalgebra of $\mathcal{F}_\eta$.*

Since finitely generated C^*-algebras are separable, this result immediately implies:

Corollary 2.6. *Every sequence in $\mathcal{F}$ possesses a fractal subsequence.*

Theorem 2.5 was first proved in [14]. We shall give a much shorter proof here, which is based on the following converse of assertion (b) of Proposition 2.4 (whereas the original proof used the converse of assertion (c) of this proposition).

Proposition 2.7. *Let $\mathcal{A}$ be a C^*-subalgebra of $\mathcal{F}$ and $\mathcal{L}$ a dense subset of $\mathcal{A}$. If the sequence of the norms $\|A_n\|$ converges for each sequence $(A_n) \in \mathcal{L}$, then the algebra $\mathcal{A}$ is fractal.*

Proof. First we show that if the sequence of the norms converges for each sequence in $\mathcal{L}$, then it converges for each sequence in $\mathcal{A}$. Let $(A_n) \in \mathcal{A}$ and $\varepsilon > 0$. Choose $(L_n) \in \mathcal{L}$ such that $\|(A_n - L_n)\| = \sup \|A_n - L_n\| < \varepsilon/3$, and let $n_0 \in \mathbb{N}$ be such that $|\|L_n\| - \|L_m\|| < \varepsilon$ for all $m, n \geq n_0$. Then, for $m, n \geq n_0$,

$$\begin{aligned} |\|A_n\| - \|A_m\|| &\leq |\|A_n\| - \|L_n\|| + |\|L_n\| - \|L_m\|| + |\|L_m\| - \|A_m\|| \\ &\leq \|A_n - L_n\| + |\|L_n\| - \|L_m\|| + \|L_m - A_m\| \leq \varepsilon. \end{aligned}$$

Thus, $(\|A_n\|)$ is a Cauchy sequence, hence convergent. But the convergence of the norms for each sequence in $\mathcal{A}$ implies the fractality of $\mathcal{A}$ by Theorem 2.2. Indeed, if a subsequence of a sequence $(A_n) \in \mathcal{A}$ tends to zero, then $0 = \liminf \|A_n\| = \lim \|A_n\|$, whence $(A_n) \in \mathcal{G}$. □

Proof of Theorem 2.5. Let $\{\mathbf{A}^m\}_{m\in\mathbb{N}}$ be a dense countable subset of $\mathcal{A}$ which consists of sequences $\mathbf{A}^m = (A_n^m)_{n\in\mathbb{N}}$. Let $\eta_1 : \mathbb{N} \to \mathbb{N}$ be a strictly increasing sequence such that the sequence of the norms $\|A^1_{\eta_1(n)}\|$ converges. Next let η_2

be a strictly increasing subsequence of η_1 such that the sequence $(\|A^2_{\eta_2(n)}\|)_{n\in\mathbb{N}}$ converges. We proceed in this way and find, for each $k \geq 2$, a strictly increasing subsequence η_k of η_{k-1} such that the sequence $(\|A^k_{\eta_k(n)}\|)_{n\in\mathbb{N}}$ converges. Define the sequence η by $\eta(n) := \eta_n(n)$. Then η is strictly increasing, and the sequence $(\|A^k_{\eta(n)}\|)_{n\in\mathbb{N}}$ converges for every $k \in \mathbb{N}$.

Since the sequences $R_\eta(\mathbf{A}^m)$ with $k \in \mathbb{N}$ form a dense subset of the restricted algebra $\mathcal{A}_\eta$, and since each sequence $R_\eta(\mathbf{A}^m) = (A^k_{\eta(n)})_{n\in\mathbb{N}}$ has the property that the sequence of the norms $\|A^k_{\eta(n)}\|$ converges, the assertion follows from Proposition 2.7. □

3. Essential fractality

Recall that a C^*-subalgebra $\mathcal{A}$ of $\mathcal{F}$ is fractal if each sequence $(A_n) \in \mathcal{A}$ can be rediscovered from each of its (infinite) subsequences modulo a sequence in the ideal $\mathcal{G}$. There are plenty of subalgebras of $\mathcal{F}$ which arise from concrete discretization methods and which are fractal (the finite sections algebra $\mathcal{S}(\mathsf{T}(C))$ for Toeplitz operators is one example). On the other hand, the algebra of the finite sections method for band-dominated operators is an example of an algebra which fails to be fractal. But the latter algebra enjoys a weaker form of fractality which we called essential fractality in [15]. Basically, a C^*-subalgebra $\mathcal{A}$ of $\mathcal{F}$ is essentially fractal if each sequence $(A_n) \in \mathcal{A}$ can be rediscovered from each of its (infinite) subsequences modulo a sequence in the ideal $\mathcal{K}$ of the compact sequences. The role of this ideal in numerical analysis can be compared with the role of the ideal of the compact operators in operator theory.

In this section, we first recall the definition of a compact sequence and state some useful characterizations of compactness and the definitions of $\mathcal{J}$-fractality and essential fractality from [15]. The main goal of this section is to derive an analogue of the fractal restriction theorem for essential fractality.

Unless otherwise stated, we let $\mathcal{F} = \mathcal{F}^\delta$ be an algebra of matrix sequences with dimension function δ and $\mathcal{G} := \mathcal{G}^\delta$ the associated ideal of zero sequences in this section.

3.1. Compact sequences

Slightly abusing the notation, we call a sequence $(K_n) \in \mathcal{F}$ a *sequence of rank one matrices* if the rank of every matrix K_n is less than or equal to one. The product of a sequence of rank one matrices with a sequence in $\mathcal{F}$ is a sequence of rank one matrices again. Hence, the set of all finite sums of sequences of rank one matrices forms an (in general, non-closed) ideal of $\mathcal{F}$. We let $\mathcal{K}$ denote the closure of this ideal and refer to the elements of $\mathcal{K}$ as *compact sequences*. Thus, $\mathcal{K}$ is the smallest closed ideal of $\mathcal{F}$ which contains all sequences of rank one matrices, and a sequence $(A_n) \in \mathcal{F}$ is compact if, and only if, for every $\varepsilon > 0$, there is a sequence $(K_n) \in \mathcal{F}$ such that

$$\sup_n \|A_n - K_n\| < \varepsilon \quad \text{and} \quad \sup_n \operatorname{rank} K_n < \infty. \tag{3.1}$$

Note that $\mathcal{K}$ contains the ideal $\mathcal{G}$, and that the restriction of a compact sequence is compact. More precisely, if $\mathbf{K}$ is a compact sequence in the algebra $\mathcal{F}^\delta$ of matrix sequences with dimension function δ and if η is a strictly increasing sequence, then the restriction $R_\eta \mathbf{K}$ is a compact sequence in the algebra $R_\eta \mathcal{F}^\delta \cong \mathcal{F}^{\delta \circ \eta}$ of matrix sequences with dimension function $\delta \circ \eta$.

An appropriate notion of the rank of a sequence in $\mathcal{F}$ can be introduced as follows. A sequence $\mathbf{A} \in \mathcal{F}$ has *finite essential rank* if it is the sum of a sequence in $\mathcal{G}$ and a sequence (K_n) with $\sup_n \operatorname{rank} K_n < \infty$. If $\mathbf{A}$ is of finite essential rank, then there is a smallest integer $r \geq 0$ such that $\mathbf{A}$ can be written as $(G_n) + (K_n)$ with $(G_n) \in \mathcal{G}$ and $\sup_n \operatorname{rank} K_n \leq r$. We call this integer the *essential rank* of $\mathbf{A}$ and write $\operatorname{ess\,rank} \mathbf{A} = r$. Thus, the sequences of essential rank 0 are just the sequences in $\mathcal{G}$. If $\mathbf{A}$ is not of finite essential rank, we set $\operatorname{ess\,rank} \mathbf{A} = \infty$. Clearly, the sequences of finite essential rank form an ideal of $\mathcal{F}$ which is dense in $\mathcal{K}$, and for arbitrary sequences $\mathbf{A}, \mathbf{B} \in \mathcal{F}$ one has

$$\operatorname{ess\,rank} (\mathbf{A} + \mathbf{B}) \leq \operatorname{ess\,rank} \mathbf{A} + \operatorname{ess\,rank} \mathbf{B},$$
$$\operatorname{ess\,rank} (\mathbf{AB}) \leq \min \{\operatorname{ess\,rank} \mathbf{A}, \operatorname{ess\,rank} \mathbf{B}\}.$$

Given a filtration $\mathcal{P} = (P_n)$ on a Hilbert space H, we identify the algebra $\mathcal{F}^\mathcal{P}$ with the algebra $\mathcal{F}$ of matrix sequences with dimension function $\delta(n) := \operatorname{rank} P_n$. Note that this identification requires the choice of an orthogonal basis in each space $\operatorname{im} P_n$. We define the *ideal $\mathcal{K}^\mathcal{P}$ of the compact sequences in $\mathcal{F}^\mathcal{P}$* in the same way as before. It is clear that then the ideal $\mathcal{K}^\mathcal{P}$ can be identified with $\mathcal{K}$, independently of the choice of the bases.

For example, using the explicit description of the finite sections algebra of Toeplitz operators in Theorem 1.1 (a), it is not hard to show that the intersection $\mathcal{S}(\mathsf{T}(C)) \cap \mathcal{K}$ consists of all sequences

$$(P_n K P_n + R_n L R_n + G_n) \quad \text{with } K,\, L \text{ compact and } (G_n) \in \mathcal{G} \tag{3.2}$$

and that the essential rank of the sequence (3.2) is equal to $\operatorname{rank} K + \operatorname{rank} L$. □

There are several equivalent characterizations of compact sequences, see [15]. In what follows we shall need a characterization of a compact sequence (K_n) in terms of the asymptotic behavior of the singular values of the entries K_n. To state this criterion, we denote the decreasingly ordered singular values of an $n \times n$ matrix A by

$$\|A\| = \Sigma_1(A) \geq \Sigma_2(A) \geq \cdots \geq \Sigma_n(A) \geq 0 \tag{3.3}$$

and recall from Linear Algebra that A^*A and AA^* are unitarily equivalent, whence $\Sigma_k(A) = \Sigma_k(A^*)$, and that every matrix A has a singular value decomposition (SVD)

$$A = E^* \operatorname{diag} (\Sigma_1(A), \ldots, \Sigma_n(A)) F \tag{3.4}$$

with unitary matrices E and F.

The announced characterization of compact sequences in terms of singular values reads as follows. See Sections 4.2 and 5.1 in [15] for the proof of this and the following theorem.

Theorem 3.1. *The following assertions are equivalent for a sequence* $(K_n) \in \mathcal{F}$*:*

(a) $\lim_{k\to\infty} \sup_{n\geq k} \Sigma_k(K_n) = 0$;

(b) $\lim_{k\to\infty} \limsup_{n\to\infty} \Sigma_k(K_n) = 0$;

(c) *the sequence* (K_n) *is compact.*

A sequence in $\mathcal{F}$ is called a *Fredholm sequence* if it is invertible modulo $\mathcal{K}$. As the compact sequences, Fredholm sequences can be characterized in terms of singular values. Let $\sigma_1(A) \leq \cdots \leq \sigma_n(A)$ denote the increasingly ordered singular values of an $n \times n$-matrix A.

Theorem 3.2. *The following assertions are equivalent for a sequence* $(A_n) \in \mathcal{F}$*:*

(a) (A_n) *is a Fredholm sequence.*

(b) *There are sequences* $(B_n) \in \mathcal{F}$ *and* $(J_n) \in \mathcal{K}$ *with* $\sup_n \operatorname{rank} J_n < \infty$ *such that* $B_n A_n = I_n + J_n$ *for all* $n \in \mathbb{N}$.

(c) *There is a* $k \in \mathbb{Z}^+$ *such that* $\liminf_{n\to\infty} \sigma_{k+1}(A_n) > 0$.

3.2. $\mathcal{J}$-fractal algebras

Our next goal is to introduce fractality of an algebra $\mathcal{A}$ with respect to an arbitrary ideal $\mathcal{J}$ in place of $\mathcal{G}$. The results presented in this subsection hold in the general case, when $\mathcal{F}$ is the product of a family $(\mathcal{C}_n)_{n\in\mathbb{N}}$ of unital C^*-algebras. We start with a criterion for the fractality of the canonical quotient map $\mathcal{A} \to \mathcal{A}/\mathcal{J}$.

Theorem 3.3. *Let* $\mathcal{A}$ *be a* C^**-subalgebra of* $\mathcal{F}$ *and* $\mathcal{J}$ *a closed ideal of* $\mathcal{A}$*. The canonical homomorphism* $\pi^{\mathcal{J}} : \mathcal{A} \to \mathcal{A}/\mathcal{J}$ *is fractal if and only if the following implication holds for every sequence* $\mathbf{A} \in \mathcal{A}$ *and every strictly increasing sequence* $\eta : \mathbb{N} \to \mathbb{N}$

$$R_\eta(\mathbf{A}) \in \mathcal{J}_\eta \quad \Longrightarrow \quad \mathbf{A} \in \mathcal{J}. \tag{3.5}$$

Proof. Let $\pi^{\mathcal{J}}$ be fractal, i.e., for each η, there is a mapping $\pi^{\mathcal{J}}_\eta$ such that $\pi^{\mathcal{J}} = \pi^{\mathcal{J}}_\eta R_\eta|_{\mathcal{A}}$. Let $R_\eta(\mathbf{A}) \in \mathcal{J}_\eta$ for a sequence $\mathbf{A} \in \mathcal{A}$. We choose a sequence $\mathbf{J} \in \mathcal{J}$ such that $R_\eta(\mathbf{A}) = R_\eta(\mathbf{J})$. Applying the homomorphism $\pi^{\mathcal{J}}_\eta$ to both sides of this equality we obtain $\pi^{\mathcal{J}}(\mathbf{A}) = \pi^{\mathcal{J}}(\mathbf{J}) = 0$, whence $\mathbf{A} \in \mathcal{J}$.

For the reverse implication, let $\mathbf{A}$ and $\mathbf{B}$ be sequences in $\mathcal{A}$ with $R_\eta(\mathbf{A}) = R_\eta(\mathbf{B})$. Then $R_\eta(\mathbf{A} - \mathbf{B}) = 0 \in \mathcal{J}_\eta$, and (3.5) implies that $\mathbf{A} - \mathbf{B} \in \mathcal{J}$. Thus, the mapping

$$\pi^{\mathcal{J}}_\eta : \mathcal{A}_\eta \to \mathcal{A}/\mathcal{J}, \quad R_\eta(\mathbf{A}) \mapsto \mathbf{A} + \mathcal{J}$$

is correctly defined, and it satisfies $\pi^{\mathcal{J}}_\eta R_\eta|_{\mathcal{A}} = \pi^{\mathcal{J}}$. □

Let now $\mathcal{J}$ be a closed ideal of $\mathcal{F}$. Then $\mathcal{A} \cap \mathcal{J}$ is a closed ideal of $\mathcal{A}$, and the preceding theorem states that the canonical mapping $\pi^{\mathcal{A}\cap\mathcal{J}} : \mathcal{A} \to \mathcal{A}/(\mathcal{A} \cap \mathcal{J})$ is fractal if and only if the implication

$$R_\eta(\mathbf{A}) \in (\mathcal{A} \cap \mathcal{J})_\eta \quad \Longrightarrow \quad \mathbf{A} \in \mathcal{J} \tag{3.6}$$

holds for every sequence $\mathbf{A} \in \mathcal{A}$ and every strictly increasing sequence η. It would be much easier to check this implication if one would have

$$(\mathcal{A} \cap \mathcal{J})_\eta = \mathcal{A}_\eta \cap \mathcal{J}_\eta \tag{3.7}$$

for every η, in which case the implication (3.6) reduces to $R_\eta(\mathbf{A}) \in \mathcal{J}_\eta \Rightarrow \mathbf{A} \in \mathcal{J}$. Recall from Theorem 2.2 (b) that (3.7) indeed holds if $\mathcal{J} = \mathcal{G}$ and if the canonical homomorphism $\pi^{\mathcal{A}\cap\mathcal{G}} : \mathcal{A} \to \mathcal{A}/(\mathcal{A}\cap\mathcal{G})$ is fractal. One cannot expect an analogous result for arbitrary closed ideals $\mathcal{J}$ of $\mathcal{F}$, as the following example shows.

Example. Let $\mathcal{A} := \mathcal{S}(\mathsf{T}(C))$ the algebra of the finite sections method for Toeplitz operators and $\mathcal{K}$ the ideal of the compact sequences in the corresponding algebra $\mathcal{F}$. Then

$$\mathcal{J} := \{(K_n) \in \mathcal{K} : \lim_{n\to\infty} \|K_{2n}\| = 0\}$$

is a closed ideal of $\mathcal{F}$. Employing again the explicit description of $\mathcal{S}(\mathsf{T}(C))$ in Theorem 1.1 (a), it is not hard to see that $\mathcal{S}(\mathsf{T}(C)) \cap \mathcal{J} = \mathcal{G}$. Consequently, the canonical homomorphism $\pi^{\mathcal{S}(\mathsf{T}(C))\cap\mathcal{J}}$ coincides with $\pi^{\mathcal{G}}$ and is, thus, fractal. But $\mathcal{G}_\eta = (\mathcal{S}(\mathsf{T}(C)) \cap \mathcal{J})_\eta$ is a proper subset of $\mathcal{S}(\mathsf{T}(C))_\eta \cap \mathcal{J}_\eta$ for the sequence $\eta(n) := 2n + 1$. Indeed, the sequence $(P_{2n+1}KP_{2n+1})$ belongs to $\mathcal{S}(\mathsf{T}(C))_\eta \cap \mathcal{J}_\eta$ for each compact operator K. □

The previous considerations suggest the following definitions. Note that both definitions coincide if $\mathcal{J}$ is a closed ideal of $\mathcal{A}$ *and* $\mathcal{F}$.

Definition 3.4. Let $\mathcal{A}$ be a C^*-subalgebra of $\mathcal{F}$.

(a) If $\mathcal{J}$ is a closed ideal of $\mathcal{A}$ then $\mathcal{A}$ is called $\mathcal{J}$*-fractal* if the canonical homomorphism $\pi^{\mathcal{J}} : \mathcal{A} \to \mathcal{A}/\mathcal{J}$ is fractal.
(b) If $\mathcal{J}$ is a closed ideal of $\mathcal{F}$ then $\mathcal{A}$ is called $\mathcal{J}$*-fractal* if $\mathcal{A}$ is $(\mathcal{A} \cap \mathcal{J})$-fractal and if $(\mathcal{A} \cap \mathcal{J})_\eta = \mathcal{A}_\eta \cap \mathcal{J}_\eta$ for every strictly increasing sequence $\eta : \mathbb{N} \to \mathbb{N}$.

The following results show that $\mathcal{J}$-fractality implies what one expects: A sequence in a $\mathcal{J}$-fractal algebra belongs to $\mathcal{J}$ or is invertible modulo $\mathcal{J}$ if and only if at least one of its subsequences has this property.

Theorem 3.5. *Let $\mathcal{J}$ be a closed ideal of $\mathcal{F}$. A C^*-subalgebra $\mathcal{A}$ of $\mathcal{F}$ is $\mathcal{J}$-fractal if and only if the following implication holds for every sequence $\mathbf{A} \in \mathcal{A}$ and every strictly increasing sequence η:*

$$R_\eta(\mathbf{A}) \in \mathcal{J}_\eta \quad \Longrightarrow \quad \mathbf{A} \in \mathcal{J}. \tag{3.8}$$

Proof. Let $\mathcal{A}$ be $\mathcal{J}$-fractal and $\mathbf{A} \in \mathcal{A}$ a sequence with $R_\eta(\mathbf{A}) \in \mathcal{J}_\eta$. Then $R_\eta(\mathbf{A}) \in \mathcal{A}_\eta \cap \mathcal{J}_\eta = (\mathcal{A}\cap\mathcal{J})_\eta$, and the $(\mathcal{A}\cap\mathcal{J})$-fractality of $\mathcal{A}$ implies $\mathbf{A} \in \mathcal{J}$ via Theorem 3.3.

Conversely, let (3.8) hold for every strictly increasing sequence η. From Theorem 3.3 we conclude that $\mathcal{A}$ is $(\mathcal{A} \cap \mathcal{J})$-fractal. Further, the inclusion $\subseteq$ in $(\mathcal{A} \cap \mathcal{J})_\eta = \mathcal{A}_\eta \cap \mathcal{J}_\eta$ is obvious. For the reverse inclusion, let $\mathbf{A}$ be a sequence in $\mathcal{F}$ with $R_\eta(\mathbf{A}) \in \mathcal{A}_\eta \cap \mathcal{J}_\eta$. Then there are sequences $\mathbf{B} \in \mathcal{A}$ and $\mathbf{J} \in \mathcal{J}$ such that $R_\eta(\mathbf{A}) = R_\eta(\mathbf{B}) = R_\eta(\mathbf{J})$. Since $R_\eta(\mathbf{B}) \in \mathcal{J}_\eta$, the implication (3.8) gives $\mathbf{B} \in \mathcal{J}$. Hence, $R_\eta(\mathbf{B}) \in (\mathcal{A} \cap \mathcal{J})_\eta$, and since $R_\eta(\mathbf{B}) = R_\eta(\mathbf{A})$, one also has $R_\eta(\mathbf{A}) \in (\mathcal{A} \cap \mathcal{J})_\eta$. □

Theorem 3.6. *Let $\mathcal{J}$ be a closed ideal of $\mathcal{F}$ and $\mathcal{A}$ a $\mathcal{J}$-fractal and unital C^*-subalgebra of $\mathcal{F}$. Then the following implication holds for every sequence $\mathbf{A} \in \mathcal{A}$ and every strictly increasing sequence η:*

$$R_\eta(\mathbf{A}) + \mathcal{J}_\eta \text{ is invertible in } \mathcal{F}_\eta/\mathcal{J}_\eta \implies \mathbf{A} + \mathcal{J} \text{ is invertible in } \mathcal{F}/\mathcal{J}. \tag{3.9}$$

Proof. Let $\mathbf{A} \in \mathcal{A}$ be such that $R_\eta(\mathbf{A}) + \mathcal{J}_\eta$ is invertible in $\mathcal{F}_\eta/\mathcal{J}_\eta$. Since C^*-algebras are inverse closed, this coset is also invertible in $(\mathcal{A}_\eta + \mathcal{J}_\eta)/\mathcal{J}_\eta$. The latter algebra is canonically *-isomorphic to $\mathcal{A}_\eta/(\mathcal{A}_\eta \cap \mathcal{J}_\eta)$, hence, to $\mathcal{A}_\eta/(\mathcal{A} \cap \mathcal{J})_\eta$ by $\mathcal{J}$-fractality of $\mathcal{A}$. Thus, the coset $R_\eta(\mathbf{A}) + (\mathcal{A} \cap \mathcal{J})_\eta$ is invertible in $\mathcal{A}_\eta/(\mathcal{A} \cap \mathcal{J})_\eta$. Choose sequences $\mathbf{B} \in \mathcal{A}$ and $\mathbf{J} \in \mathcal{A} \cap \mathcal{J}$ such that

$$R_\eta(\mathbf{A})\, R_\eta(\mathbf{B}) = R_\eta(\mathbf{I}) + R_\eta(\mathbf{J})$$

where $\mathbf{I}$ denotes the identity element of $\mathcal{F}$. Applying the homomorphism $\pi_\eta^{\mathcal{A} \cap \mathcal{J}}$ to both sides of this equality one gets

$$\pi^{\mathcal{A} \cap \mathcal{J}}(\mathbf{A})\, \pi^{\mathcal{A} \cap \mathcal{J}}(\mathbf{B}) = \pi^{\mathcal{A} \cap \mathcal{J}}(\mathbf{I}) + \pi^{\mathcal{A} \cap \mathcal{J}}(\mathbf{J})$$

which shows that $\mathbf{AB} - \mathbf{I} \in \mathcal{J}$. Hence, $\mathbf{A}$ is invertible modulo $\mathcal{J}$ from the right-hand side. The invertibility from the left-hand side follows analogously. □

Corollary 3.7. *Let $\mathcal{J}$ be a closed ideal of $\mathcal{F}$ and $\mathcal{A}$ a $\mathcal{J}$-fractal and unital C^*-subalgebra of $\mathcal{F}$. Then a sequence $\mathbf{A} \in \mathcal{A}$*

(a) *belongs to $\mathcal{J}$ if and only if there is a strictly increasing sequence η such that $\mathbf{A}_\eta$ belongs to $\mathcal{J}_\eta$.*
(b) *is invertible modulo $\mathcal{J}$ if and only if there is a strictly increasing sequence η such that $\mathbf{A}_\eta$ is invertible modulo $\mathcal{J}_\eta$.*

We still mention the following simple facts for later reference.

Proposition 3.8. *Let $\mathcal{J}$ be a closed ideal of $\mathcal{F}$ and $\mathcal{A}$ a $\mathcal{J}$-fractal C^*-subalgebra of $\mathcal{F}$. Then*

(a) *every C^*-subalgebra of $\mathcal{A}$ is $\mathcal{J}$-fractal.*
(b) *if $\mathcal{I}$ is an ideal of $\mathcal{F}$ with $\mathcal{J} \subseteq \mathcal{I}$ and if $(\mathcal{A} \cap \mathcal{I})_\eta = \mathcal{A}_\eta \cap \mathcal{I}_\eta$ for each strictly increasing sequence $\eta : \mathbb{N} \to \mathbb{N}$, then $\mathcal{A}$ is $\mathcal{I}$-fractal.*

Proof. (a) Let $\mathcal{B}$ be a C^*-subalgebra of $\mathcal{A}$, and let $\mathbf{B}$ be a sequence in $\mathcal{B}$ with $R_\eta(\mathbf{B}) \in \mathcal{J}_\eta$ for a certain strictly increasing sequence η. Then $R_\eta(\mathbf{B}) \in \mathcal{A}_\eta \cap \mathcal{J}_\eta$. Since $\mathcal{A}$ is $\mathcal{J}$-fractal, Theorem 3.5 implies that $\mathbf{B} \in \mathcal{J}$. Hence $\mathcal{B}$ is $\mathcal{J}$-fractal, again by Theorem 3.5.

(b) Let $R_\eta(\mathbf{A}) \in \mathcal{I}_\eta$ for a sequence $\mathbf{A} \in \mathcal{A}$ and a strictly increasing sequence η. By hypothesis, $R_\eta(\mathbf{A}) \in (\mathcal{A} \cap \mathcal{I})_\eta$. Choose a sequence $\mathbf{J} \in \mathcal{A} \cap \mathcal{I}$ with $R_\eta(\mathbf{A}) = R_\eta(\mathbf{J})$. The $\mathcal{J}$-fractality of $\mathcal{A}$ implies that $\mathbf{A} - \mathbf{J} \in \mathcal{J}$, whence $\mathbf{A} \in \mathbf{J} + \mathcal{J} \subseteq \mathcal{I}$. By Theorem 3.5, $\mathcal{A}$ is $\mathcal{I}$-fractal. □

3.3. Essential fractality and Fredholm property

Let again $\mathcal{F}$ be the algebra of matrix sequences with dimension function δ and $\mathcal{K}$ the associated ideal of compact sequences. We call the $\mathcal{K}$-fractal C^*-subalgebras of $\mathcal{F}$ *essentially fractal.*

Note that each restriction $\mathcal{F}_\eta$ of $\mathcal{F}$ is again an algebra of matrix sequences (with dimension function $\delta \circ \eta$); hence, the restriction $\mathcal{K}_\eta$ of $\mathcal{K}$ is just the ideal of the compact sequences related with $\mathcal{F}_\eta$. If we speak on *compact subsequences* and *Fredholm subsequences* in what follows, we thus mean sequences $R_\eta \mathbf{A} \in \mathcal{K}_\eta$ and sequences $R_\eta \mathbf{A}$ which are invertible modulo $\mathcal{K}_\eta$, respectively. In these terms, Corollary 3.7 reads as follows.

Corollary 3.9. *Let $\mathcal{A}$ be an essentially fractal and unital C^*-subalgebra of $\mathcal{F}$. Then a sequence $\mathbf{A} \in \mathcal{A}$ is compact (resp. Fredholm) if and only if one of the subsequences of $\mathbf{A}$ is compact (resp. Fredholm).*

The following is a consequence of Proposition 3 8.

Corollary 3.10. *Let $\mathcal{A}$ be a fractal C^*-subalgebra of $\mathcal{F}$. If $(\mathcal{A} \cap \mathcal{K})_\eta = \mathcal{A}_\eta \cap \mathcal{K}_\eta$ for each strictly increasing sequence $\eta : \mathbb{N} \to \mathbb{N}$, then $\mathcal{A}$ is essentially fractal.*

Essential fractality has striking consequences for the behavior of the smallest singular values.

Theorem 3.11. *Let $\mathcal{A}$ be an essentially fractal and unital C^*-subalgebra of $\mathcal{F}$. A sequence $(A_n) \in \mathcal{A}$ is Fredholm if and only if there is a $k \in \mathbb{N}$ such that*

$$\limsup_{n\to\infty} \sigma_k(A_n) > 0. \tag{3.10}$$

Proof. If (A_n) is Fredholm then $\liminf_{n\to\infty} \sigma_k(A_n) > 0$ for some $k \in \mathbb{N}$ by Theorem 3.2 (c), whence (3.10). Conversely, let (3.10) hold for some k. We choose a strictly increasing sequence η such that $\lim_{n\to\infty} \sigma_k(A_{\eta(n)}) > 0$. Thus, the restricted sequence $(A_{\eta(n)})_{n\geq 1}$ is Fredholm by Theorem 3.2. Since $\mathcal{A}$ is essentially fractal, Corollary 3.9 (b) implies the Fredholm property of the sequence (A_n) itself. □

Consequently, if a sequence (A_n) in an essentially fractal and unital C^*-subalgebra of $\mathcal{F}$ is *not* Fredholm, then

$$\lim_{n\to\infty} \sigma_k(A_n) = 0 \qquad \text{for each } k \in \mathbb{N}. \tag{3.11}$$

In analogy with operator theory, we call a sequence (A_n) with property (3.11) *not normally solvable.*

Corollary 3.12. *Let $\mathcal{A}$ be an essentially fractal and unital C^*-subalgebra of $\mathcal{F}$. Then a sequence in $\mathcal{A}$ is either Fredholm or not normally solvable.*

Example. Consider the finite sections algebra $\mathcal{S}(\mathsf{T}(C))$ for Toeplitz operators. It is a simple consequence of Theorem 1.1 (a) that $(\mathcal{S}(\mathsf{T}(C)) \cap \mathcal{K})_\eta = \mathcal{S}(\mathsf{T}(C))_\eta \cap \mathcal{K}_\eta$ for each strictly increasing sequence η. Since $\mathcal{S}(\mathsf{T}(C))$ is fractal and $\mathcal{G} \subset \mathcal{K}$, the algebra $\mathcal{S}(\mathsf{T}(C))$ is essentially fractal by Corollary 3.10. □

3.4. Essential fractal restriction

Our final goal is an analogue of Theorem 2.5 for essential fractality. Recall that we based the proof of Theorem 2.5 on the fact that there is a sequence η such that the norms $\|A_{\eta(n)}\|$ converge for each sequence (A_n). We start with showing that η can be even chosen such that not only the sequences $(\|A_{\eta(n)}\|) = (\Sigma_1(A_{\eta(n)}))$ converge, but *every* sequence $(\Sigma_k(A_{\eta(n)}))$ with $k \in \mathbb{N}$. Here, $\Sigma_1(A) \geq \cdots \geq \Sigma_n(A)$ denote the decreasingly ordered singular values of the $n \times n$-matrix A.

Proposition 3.13. *Let $\mathcal{A}$ be a separable C^*-subalgebra of $\mathcal{F}$. Then there is a strictly increasing sequence $\eta : \mathbb{N} \to \mathbb{N}$ such that the sequence $(\Sigma_k(A_{\eta(n)}))_{n\geq 1}$ converges for every sequence $(A_n)_{n\geq 1} \in \mathcal{A}$ and every $k \in \mathbb{N}$.*

Proof. First consider a single sequence $(A_n) \in \mathcal{A}$. We choose a strictly increasing sequence $\eta_1 : \mathbb{N} \to \mathbb{N}$ such that the sequence $(\Sigma_1(A_{\eta_1(n)}))_{n\geq 1}$ converges, then a subsequence η_2 of η_1 such that the sequence $(\Sigma_2(A_{\eta_2(n)}))_{n\geq 1}$ converges, and so on. The sequence $\eta(n) := \eta_n(n)$ has the property that the sequence $(\Sigma_k(A_{\eta(n)}))_{n\geq 1}$ converges for every $k \in \mathbb{N}$.

Now let $(\mathbf{A}^m)_{m\geq 1}$ be a countable dense subset of $\mathcal{A}$, consisting of sequences $\mathbf{A}^m = (A_n^m)_{n\geq 1}$. We use the result of the previous step to find a strictly increasing sequence η_1 such that the sequences $(\Sigma_k(A^1_{\eta_1(n)}))_{n\geq 1}$ converge for every $k \in \mathbb{N}$, then a subsequence η_2 of η_1 such that the sequences $(\Sigma_k(A^2_{\eta_2(n)}))_{n\geq 1}$ converge for every k, and so on. Then the sequence $\eta(n) := \eta_n(n)$ has the property that the sequences $(\Sigma_k(A^m_{\eta(n)}))_{n\geq 1}$ converge for every pair $k, m \in \mathbb{N}$.

Let η be as in the previous step, i.e., the sequences $(\Sigma_k(A^m_{\eta(n)}))_{n\geq 1}$ converge for every $k \in \mathbb{N}$ and for every sequence $\mathbf{A}^m = (A_n^m)_{n\geq 1}$ in a countable dense subset of $\mathcal{A}$. We show that then the sequences $(\Sigma_k(A_{\eta(n)}))_{n\geq 1}$ converge for every $k \in \mathbb{N}$ and every sequence $\mathbf{A} = (A_n)$ in $\mathcal{A}$. Fix $k \in \mathbb{N}$ and let $\varepsilon > 0$. Using the well-known inequality $|\Sigma_k(A) - \Sigma_k(B)| \leq \|A - B\|$ we obtain

$$\begin{aligned}
&|\Sigma_k(A_{\eta(n)}) - \Sigma_k(A_{\eta(l)})| \\
&\quad \leq |\Sigma_k(A_{\eta(n)}) - \Sigma_k(A^m_{\eta(n)})| + |\Sigma_k(A^m_{\eta(n)}) - \Sigma_k(A^m_{\eta(l)})| \\
&\qquad + |\Sigma_k(A^m_{\eta(l)}) - \Sigma_k(A_{\eta(l)})| \\
&\quad \leq \|A_{\eta(n)} - A^m_{\eta(n)}\| + |\Sigma_k(A^m_{\eta(n)}) - \Sigma_k(A^m_{\eta(l)})| + \|A^m_{\eta(l)} - A_{\eta(l)}\| \\
&\quad \leq 2\,\|\mathbf{A} - \mathbf{A}^m\|_{\mathcal{F}} + |\Sigma_k(A^m_{\eta(n)}) - \Sigma_k(A^m_{\eta(l)})|.
\end{aligned}$$

Now choose $m \in \mathbb{N}$ such that $\|\mathbf{A} - \mathbf{A}^m\|_{\mathcal{F}} < \varepsilon/3$ and then $N \in \mathbb{N}$ such that $|\Sigma_k(A^m_{\eta(n)}) - \Sigma_k(A^m_{\eta(l)})| < \varepsilon/3$ for all $n, l \geq N$. Then $|\Sigma_k(A_{\eta(n)}) - \Sigma_k(A_{\eta(l)})| < \varepsilon$ for all $n, l \geq N$. Thus, $(\Sigma_k(A_{\eta(n)}))_{n\geq 1}$ is a Cauchy sequence, hence convergent. □

Proposition 3.14. *Let $\mathcal{A}$ be a C^*-subalgebra of $\mathcal{F}$ with the property that the sequences $(\Sigma_k(A_n))_{n\geq 1}$ converge for every sequence $(A_n) \in \mathcal{A}$ and every $k \in \mathbb{N}$. Then $\mathcal{A}$ is essentially fractal.*

Proof. Let $\mathbf{K} = (K_n) \in \mathcal{A}$ and let $\eta : \mathbb{N} \to \mathbb{N}$ be a strictly increasing sequence such that $\mathbf{K}_\eta \in \mathcal{K}_\eta$. Then, by Theorem 3.1 (b),

$$\lim_{k\to\infty} \limsup_{n\to\infty} \Sigma_k(K_{\eta(n)}) = 0. \tag{3.12}$$

By hypothesis, $\limsup_{n\to\infty} \Sigma_k(K_{\eta(n)}) = \lim_{n\to\infty} \Sigma_k(K_n)$. Hence, (3.12) implies $\lim_{k\to\infty} \lim_{n\to\infty} \Sigma_k(K_n) = 0$, whence $\mathbf{K} \in \mathcal{K}$ by assertion (a) of Theorem 3.1. Thus, every sequence in $\mathcal{A}$ which has a compact subsequence is compact itself. Thus $\mathcal{A}$ is essentially fractal by Theorem 3.5. □

Theorem 3.15. *Let $\mathcal{A}$ be a separable C^*-subalgebra of $\mathcal{F}$. Then there is a strictly increasing sequence $\eta : \mathbb{N} \to \mathbb{N}$ such that the restricted algebra $\mathcal{A}_\eta = R_\eta \mathcal{A}$ is essentially fractal.*

Indeed, if η is as in Proposition 3.13, then the restriction $\mathcal{A}_\eta$ is essentially fractal by Proposition 3.14. □

We know from Theorems 2.5 and 3.15 that every separable C^*-subalgebra of $\mathcal{F}$ has both a fractal and an essentially fractal restriction. It is an open question whether this fact holds for arbitrary closed ideals $\mathcal{J}$ of $\mathcal{F}$ in place of $\mathcal{G}$ or $\mathcal{K}$, i.e., whether one can always force $\mathcal{J}$-fractality by a suitable restriction.

4. Essential spectral approximation

In a series of papers [1–3], Arveson studied the question of whether one can discover the essential spectrum of a self-adjoint operator A from the behavior of the eigenvalues of the finite sections P_nAP_n of A. More generally, one might ask whether one can discover the *essential spectrum* of a self-adjoint sequence $\mathbf{A} = (A_n) \in \mathcal{F}$ (i.e., the spectrum of the coset $\mathbf{A} + \mathcal{K}$, considered as an element of the quotient algebra $\mathcal{F}/\mathcal{K}$) from the behavior of the eigenvalues of the matrices A_n? To answer this question, Arveson introduced the notions of essential and transient points, and he discovered (under an additional condition) a certain dichotomy: if A is a self-adjoint band-dominated operator, then every point in $\mathbb{R}$ is either transient or essential; see Subsection 4.2. The goal of this section is to relate the essential spectral approximation with the property of essential fractality. In particular, we will see that a subalgebra $\mathcal{A}$ of $\mathcal{F}$ is essentially fractal if and only if every self-adjoint sequence in $\mathcal{A}$ has Arveson dichotomy.

4.1. Essential spectra of self-adjoint sequences

Given a self-adjoint matrix A and a subset M of $\mathbb{R}$, let $N(A, M)$ denote the number of eigenvalues of A which lie in M, counted with respect to their multiplicity. If $M = \{\lambda\}$ is a singleton, we write $N(A, \lambda)$ in place of $N(A, \{\lambda\})$. Thus, if λ is an eigenvalue of A, then $N(A, \lambda)$ is its multiplicity.

Let $\mathbf{A} = (A_n) \in \mathcal{F}$ be a self-adjoint sequence. Following Arveson [1–3], a point $\lambda \in \mathbb{R}$ is called *essential* for this sequence if, for every open interval U

containing λ,

$$\lim_{n\to\infty} N(A_n, U) = \infty,$$

and $\lambda \in \mathbb{R}$ is called *transient* for $\mathbf{A}$ if there is an open interval U which contains λ such that

$$\sup_{n\in\mathbb{N}} N(A_n, U) < \infty.$$

Thus, $\lambda \in \mathbb{R}$ is *not* essential for $\mathbf{A}$ if and only if λ is transient for a subsequence of $\mathbf{A}$, and λ is *not* transient for $\mathbf{A}$ if and only if λ is essential for a subsequence of $\mathbf{A}$. Moreover, if a point λ is transient (resp. essential) for $\mathbf{A}$, then is also transient (resp. essential) for every subsequence of $\mathbf{A}$.

Theorem 4.1. *Let $\mathbf{A} \in \mathcal{F}$ be a self-adjoint sequence. A point $\lambda \in \mathbb{R}$ belongs to the essential spectrum of $\mathbf{A}$ if and only if it is not transient for the sequence $\mathbf{A}$.*

Proof. Let $\mathbf{A} = (A_n)$ be a bounded sequence of self-adjoint matrices. First let $\lambda \in \mathbb{R} \setminus \sigma(\mathbf{A} + \mathcal{K})$. We set $B_n := A_n - \lambda I_n$ and have to show that 0 is transient for the sequence (B_n). Since $\lambda \in \mathbb{R} \setminus \sigma(\mathbf{A} + \mathcal{K})$, the sequence (B_n) is Fredholm. By Theorem 3.2 (c), there is a $k \in \mathbb{Z}^+$ such that

$$\liminf_{n\to\infty} \sigma_{k+1}(B_n) =: C > 0 \quad \text{and} \quad \liminf_{n\to\infty} \sigma_k(B_n) = 0.$$

Let $U := (-C/2, C/2)$. Since the singular values of a self-adjoint matrix are just the absolute values of the eigenvalues of that matrix, we conclude that $N(B_n, U) \leq k$ for all sufficiently large n. Thus, 0 is transient.

Conversely, let $\lambda \in \mathbb{R}$ be transient for (A_n). We claim that $(A_n - \lambda I_n)$ is a Fredholm sequence. By transiency, there is an interval $U = (\lambda - \varepsilon, \lambda + \varepsilon)$ with $\varepsilon > 0$ such that $\sup_{n\in\mathbb{N}} N(A_n, U) =: k < \infty$. Let T_n denote the orthogonal projection from $\mathbb{C}^{\delta(n)}$ onto the U-spectral subspace of A_n. Then $\operatorname{rank} T_n$ is not greater than k. It is moreover obvious that the matrices $C_n := (A_n - \lambda I_n)(I - T_n) + T_n$ are invertible for all $n \in \mathbb{N}$ and that their inverses are uniformly bounded by the maximum of $1/\varepsilon$ and 1. Hence, $(C_n^{-1}) \in \mathcal{F}$ and

$$(A_n - \lambda I_n)(I - T_n)C_n^{-1} = I - T_n C_n^{-1}.$$

Since (T_n) is a compact sequence (of essential rank not greater than k), this identity shows that the coset $(A_n - \lambda I_n) + \mathcal{K}$ is invertible from the right-hand side. Since this coset is self-adjoint, it is then invertible from both sides. Thus, $(A_n - \lambda I_n)$ is a Fredholm sequence. □

Proposition 4.2. *The set of the non-transient points and the set of the essential points of a self-adjoint sequence $\mathbf{A} \in \mathcal{F}$ are compact.*

Proof. The first assertion is an immediate consequence of Theorem 4.1. The second assertion will follow once we have shown that the set of the essential points of $\mathbf{A}$ is closed.

Let (λ_k) be a sequence of essential points for $\mathbf{A} = (A_n)$ with limit λ. Assume that λ is not essential for $\mathbf{A}$. Then there is a strictly increasing sequence $\eta : \mathbb{N} \to \mathbb{N}$ such that λ is transient for $\mathbf{A}_\eta$. Let U be an open neighborhood of λ

with $\sup_{n\in\mathbb{N}} N(A_{\eta(n)}, U) =: c < \infty$. Since $\lambda_k \to \lambda$ and U is open, there are a $k \in \mathbb{N}$ and an open neighborhood U_k of λ_k with $U_k \subseteq U$. Clearly, $N(A_{\eta(n)}, U_k) \leq N(A_{\eta(n)}, U) \leq c$. On the other hand, since λ_k is also essential for the restricted sequence $\mathbf{A}_\eta$, one has $N(A_{\eta(n)}, U_k) \to \infty$ as $n \to \infty$, a contradiction. □

Note that the set of the non-transient points of a self-adjoint sequence is non-empty by Theorem 4.1, whereas it is easy to construct self-adjoint sequences without any essential point: take a sequence which alternates between the zero and the identity matrix. In contrast to this observation, the following result shows that sequences which arise by discretization of a self-adjoint operator, always possess essential points. Let H be an infinite-dimensional separable Hilbert space with filtration $\mathcal{P} := (P_n)$, and define the algebra $\mathcal{F}^\mathcal{P}$ as in Section 1. One can think of $\mathcal{F}^\mathcal{P}$ as a C^*-subalgebra of the algebra $\mathcal{F}^\delta$ with dimension function $\delta(n) := \operatorname{rank} P_n$.

Theorem 4.3. *Let $\mathbf{A} := (A_n) \in \mathcal{F}^\mathcal{P}$ be a self-adjoint sequence with strong limit A. Then every point in the essential spectrum of A is an essential point for $\mathbf{A}$.*

Proof. We show that $A - \lambda I$ is a Fredholm operator if $\lambda \in \mathbb{R}$ is not essential for $\mathbf{A}$. Then λ is transient for a subsequence of $\mathbf{A}$, i.e., there are an infinite subset $\mathbb{M}$ of $\mathbb{N}$ and an interval $U = (\lambda - \varepsilon, \lambda + \varepsilon)$ with $\varepsilon > 0$ such that

$$\sup_{n\in\mathbb{M}} N(A_n, U) =: k < \infty. \tag{4.1}$$

Let T_n denote the orthogonal projection from H onto the U-spectral subspace of A_nP_n. By (4.1), the rank of the projection T_n is not greater than k if $n \in \mathbb{M}$. So we conclude that the operators $C_n := (A_n - \lambda I_n)(I - T_n) + T_n$ are invertible for all $n \in \mathbb{M}$ and that their inverses are uniformly bounded by the maximum of $1/\varepsilon$ and 1. Hence,

$$(A_n - \lambda I_n)(I - T_n)C_n^{-1} = I - T_nC_n^{-1} \tag{4.2}$$

for all $n \in \mathbb{M}$. By the weak sequential compactness of the unit ball of $L(H)$, one finds weakly convergent subsequences $((I - T_{n_r})C_{n_r}^{-1})_{r\geq 1}$ of $((I - T_n)C_n^{-1})_{n\in\mathbb{M}}$ and $(T_{n_r}C_{n_r}^{-1})_{r\geq 1}$ of $(T_nB_n^{-1})_{n\in\mathbb{M}}$ with limits C and T, respectively. The product of a weakly convergent sequence with limit X and a *-strongly convergent sequence with limit Y is weakly convergent with limit XY. Thus, passing to subsequences and taking the weak limit in (4.2) yields $(A - \lambda I)C = I - T$. Further, the rank of T is not greater than k by Lemma 5.7 in [4]. Thus, $(A - \lambda I)C - I$ is a compact operator. The compactness of $C(A - \lambda I) - I$ follows similarly. Hence, A is a Fredholm operator. □

Arveson gave a first example where the inclusion in Theorem 4.3 is proper. He constructed a self-adjoint unitary operator $A \in L(l^2(\mathbb{N}))$ with

$$\sigma(A) = \sigma_{ess}(A) = \{-1, 1\} \tag{4.3}$$

such that 0 is an essential point of the sequence (P_nAP_n).

4.2. Arveson dichotomy and essential fractality

We say that a self-adjoint sequence $\mathbf{A} \in \mathcal{F}$ enjoys *Arveson's dichotomy* if every real number is either essential or transient for this sequence. Note that Arveson dichotomy is preserved when passing to subsequences. Arveson introduced and studied this property in [1–3]. In particular, he proved the dichotomy of the finite sections sequence $(P_n A P_n)$ when A is a self-adjoint band-dominated operator which satisfies a Wiener and a Besov space condition. A generalization to arbitrary band-dominated operators was obtained in [15].

Theorem 4.4. *The set of all self-adjoint sequences in $\mathcal{F}$ with Arveson dichotomy is closed in $\mathcal{F}$.*

Proof. Let $(\mathbf{A}_n)_{n\in\mathbb{N}}$ be a sequence of self-adjoint sequences in $\mathcal{F}$ with Arveson dichotomy which converges to a (necessarily self-adjoint) sequence $\mathbf{A}$ in the norm of $\mathcal{F}$. Then $\mathbf{A}_n + \mathcal{K} \to \mathbf{A} + \mathcal{K}$ in the norm of $\mathcal{F}/\mathcal{K}$. Since $\mathbf{A}_n + \mathcal{K}$ and $\mathbf{A} + \mathcal{K}$ are self-adjoint elements of $\mathcal{F}/\mathcal{K}$, this implies that the spectra of $\mathbf{A}_n + \mathcal{K}$ converge to the spectrum of $\mathbf{A} + \mathcal{K}$ in the Hausdorff metric. Thus, by Theorem 4.1, the sets of the non-transient points of $\mathbf{A}_n$ converge to the set of the non-transient points of $\mathbf{A}$. Since the $\mathbf{A}_n$ have Arveson dichotomy by hypothesis, this finally implies that the sets of the essential points of $\mathbf{A}_n$ converge to the set of the non-transient points of $\mathbf{A}$ in the Hausdorff metric.

Let now λ be a non-transient point for $\mathbf{A}$ and assume that λ is not essential for $\mathbf{A}$. Then there is a strictly increasing sequence $\eta : \mathbb{N} \to \mathbb{N}$ such that λ is transient for the restricted sequence $\mathbf{A}_\eta$. As we have seen above, there is a sequence (λ_n), where λ_n is an essential point for $\mathbf{A}_n$, with $\lambda_n \to \lambda$. Since the property of being an essential is preserved under passage to a subsequence, λ_n is also essential for the restricted sequence $(\mathbf{A}_n)_\eta$.

Since the sequences $(\mathbf{A}_n)_\eta$ also have Arveson dichotomy and since $(\mathbf{A}_n)_\eta$ tends to $\mathbf{A}_\eta$ in the norm of $\mathcal{F}_\eta$, we can repeat the above arguments to conclude that the sets M_n of the essential points for $(\mathbf{A}_n)_\eta$ converge to the set M of the non-transient points for $\mathbf{A}_\eta$ in the Hausdorff metric. Since $\lambda_n \in M_n$ by construction, this implies that $\lambda \in M$. This means that λ in not transient for $\mathbf{A}_\eta$, a contradiction. □

Here is the announced result which relates Arveson dichotomy with essential fractality.

Theorem 4.5. *Let $\mathcal{A}$ be a unital C^*-subalgebra of $\mathcal{F}$. Then $\mathcal{A}$ is essentially fractal if and only if every self-adjoint sequence in $\mathcal{A}$ has Arveson dichotomy.*

Proof. First let $\mathcal{A}$ be essentially fractal. Let $\mathbf{A}$ be a self-adjoint sequence in $\mathcal{A}$ and $\lambda \in \mathbb{R}$ a point which is not essential for $\mathbf{A}$. The λ is transient for a subsequence of $\mathbf{A}$, thus, 0 is transient for a subsequence of $\mathbf{A} - \lambda\mathbf{I}$. From Theorem 4.1 we conclude that this subsequence has the Fredholm property. Then, by Corollary 3.9 (b) and since $\mathcal{A}$ is essentially fractal, the sequence $\mathbf{A} - \lambda\mathbf{I}$ itself is a Fredholm sequence.

Thus, 0 is transient for $\mathbf{A} - \lambda\mathbf{I}$ by Theorem 4.1 again, whence finally follows that λ is transient for $\mathbf{A}$. Hence, $\mathbf{A}$ has Arveson dichotomy.

Now assume that $\mathcal{A}$ is not essentially fractal. Then, by Theorem 3.5, there are a sequence $\mathbf{A} = (A_n) \in \mathcal{A}$ and a strictly increasing sequence η $\mathbb{N} \to \mathbb{N}$ such that the restricted sequence $\mathbf{A}_\eta$ belongs to $\mathcal{K}_\eta$ but $\mathbf{A} \notin \mathcal{K}$. The self-adjoint sequence $\mathbf{A}^*\mathbf{A}$ has the same properties, i.e., $(\mathbf{A}^*\mathbf{A})_\eta = \mathbf{A}_\eta^*\mathbf{A}_\eta \in \mathcal{K}_\eta$, but $\mathbf{A}^*\mathbf{A} \notin \mathcal{K}$.

Since $\mathbf{A}_\eta^*\mathbf{A}_\eta \in \mathcal{K}_\eta$, the essential spectrum of $\mathbf{A}_\eta^*\mathbf{A}_\eta$ (i.e., the spectrum of the coset $\mathbf{A}_\eta^*\mathbf{A}_\eta + \mathcal{K}_\eta$ in $\mathcal{F}_\eta/\mathcal{K}_\eta$) consists of the point 0 only. Thus, by Theorem 4.1, 0 is the only non-transient point for the restricted sequence $\mathbf{A}_\eta^*\mathbf{A}_\eta$.

Since $\mathbf{A}^*\mathbf{A} \notin \mathcal{K}$, there is a strictly increasing sequence $\mu : \mathbb{N} \to \mathbb{N}$ such that $\mu(\mathbb{N}) \cap \eta(\mathbb{N}) = \emptyset$ and $\mathbf{A}_\mu^*\mathbf{A}_\mu \notin \mathcal{K}_\eta$. Hence, the essential spectrum of $\mathbf{A}_\mu^*\mathbf{A}_\mu$ contains at least one point $\lambda \neq 0$, and this point is non-transient for $\mathbf{A}_\mu^*\mathbf{A}_\mu$ by Theorem 4.1 again. Hence, there is a subsequence ν of μ such that λ is essential for $\mathbf{A}_\nu^*\mathbf{A}_\nu$, but $\lambda \neq 0$ is transient for $\mathbf{A}_\eta^*\mathbf{A}_\eta$ as we have seen above. Thus, λ is neither transient nor essential for $\mathbf{A}^*\mathbf{A}$. Hence, the sequence $\mathbf{A}^*\mathbf{A}$ does not have Arveson dichotomy. □

Corollary 4.6. *Every self-adjoint sequence in $\mathcal{F}$ possesses a subsequence with Arveson dichotomy.*

Proof. Let $\mathbf{A}$ be a self-adjoint sequence in $\mathcal{F}$. The smallest closed subalgebra $\mathcal{A}$ of $\mathcal{F}$ which contains $\mathbf{A}$ is separable. By Theorem 3.15, there is an essentially fractal restriction $\mathcal{A}_\eta$ of $\mathcal{A}$. Then $\mathbf{A}_\eta$ is a subsequence of $\mathbf{A}$ with Arveson dichotomy by the previous theorem. □

References

[1] W. Arveson, *Improper filtrations for C^*-algebras: Spectra of unilateral tridiagonal operators.* Acta Sci. Math. (Szeged) **57** (1993), 11–24.

[2] W. Arveson, *C^*-algebras and numerical linear algebra.* J. Funct. Anal. **122** (1994), 333–360.

[3] W. Arveson, *The role of C^*-algebras in infinite dimensional numerical linear algebra.* Contemp. Math. **167** (1994), 115–129.

[4] A. Böttcher, S.M. Grudsky, *Toeplitz Matrices, Asymptotic Linear Algebra and Functional Analysis.* Hindustan Book Agency, New Delhi 2000.

[5] A. Böttcher, B. Silbermann, *The finite section method for Toeplitz operators on the quarter-plane with piecewise continuous symbols.* Math. Nachr. **110** (1983), 279–291.

[6] A. Böttcher, B. Silbermann, *Introduction to Large Truncated Toeplitz Matrices.* Springer-Verlag, Berlin, Heidelberg 1999.

[7] R. Hagen, S. Roch, B. Silbermann, *C^*-Algebras and Numerical Analysis.* Marcel Dekker, Inc., New York, Basel 2001.

[8] S. Prössdorf, B. Silbermann, *Numerical Analysis for Integral and Related Operator Equations.* Akademie-Verlag, Berlin 1991, and Birkhäuser Verlag, Basel, Boston, Stuttgart 1991.

[9] V.S. Rabinovich, S. Roch, *Fredholm properties of band-dominated operators on periodic discrete structures.* Complex Anal. Oper. Theory **2** (2008), 4, 637–681.

[10] V.S. Rabinovich, S. Roch, B. Silbermann, *Fredholm theory and finite section method for band-dominated operators.* Integral Equations Oper. Theory **30** (1998), 452–495.

[11] V.S. Rabinovich, S. Roch, B. Silbermann, *Band-dominated operators with operator-valued coefficients, their Fredholm properties and finite sections.* Integral Eq. Oper. Theory **40** (2001), 3, 342–381.

[12] V.S. Rabinovich, S. Roch, B. Silbermann, *Limit Operators and Their Applications in Operator Theory.* Oper. Theory: Adv. Appl. **150**, Birkhäuser, Basel 2004.

[13] V.S. Rabinovich, S. Roch, B. Silbermann, *On finite sections of band-dominated operators.* In: Oper. Theory: Adv. Appl. **181**, Birkhäuser 2008, 385–391.

[14] S. Roch, *Algebras of approximation sequences: Fractality.* In: Oper. Theory: Adv. Appl. **121**, Birkhäuser, Basel 2001, 471–497.

[15] S. Roch, *Finite sections of band-dominated operators.* Memoirs AMS Vol. **191**, 895, Providence, R.I., 2008.

[16] S. Roch, B. Silbermann, *C^*-algebra techniques in numerical analysis.* J. Oper. Theory **35** (1996), 2, 241–280.

[17] S. Roch, P.A. Santos, B. Silbermann, *Non-commutative Gelfand Theories. A Toolkit for Operator Theorists and Numerical Analysts.* Universitext, Springer, London 2011.

Steffen Roch
Technische Universität Darmstadt
Schlossgartenstrasse 7
D-64289 Darmstadt, Germany
e-mail: roch@mathematik.tu-darmstadt.de

Operator Theory:
Advances and Applications, Vol. 228, 343–354

Lower Bounds for the Counting Function of Integral Operators

Y. Safarov

To Vladimir Rabinovich on his 70th birthday

Abstract. The paper presents a lower bound for the number of eigenvalues of an integral operator K with continuous kernel $\mathcal{K}$ lying in the interval $(-\infty, t)$ with $t \leqslant 0$. The estimate is given in terms of some integrals of $\mathcal{K}$.

Mathematics Subject Classification (2010). 45C05, 35P15.

Keywords. Eigenvalues, counting function, integral operators.

Introduction

Consider a self-adjoint integral operator K in the space $L_2(M,\nu)$ on a domain or a manifold M provided with a finite measure ν. If its integral kernel $\mathcal{K}$ is continuous then the operator is compact and its spectrum consists of eigenvalues accumulating to zero. Such operators have been considered by many authors, most of whom studied the rate of convergence of eigenvalues and obtained various quantitative versions of the following general statement: *the smoother the kernel is, the faster the eigenvalues tend to zero* (see, for instance, [2] or [8]).

The paper deals with a different, seemingly simple question: *how many negative eigenvalues are there?* More precisely, we are interested in obtaining explicit lower bounds for the number of negative eigenvalues in terms of the integral kernel $\mathcal{K}$.

One can argue that in the generic case the dimensions of positive and negative eigenspaces must be the same, so that both of them are infinite dimensional and there are infinitely many negative eigenvalues. However, this argument is of little use when we need to study a particular integral operator.

The research was partly carried out during my visit to Academia Sinica, Taipei. I am very grateful to my hosts, especially to Dr. Jin-Cheng Jiang, for their financial and scientific support.

It is not immediately clear what properties of $\mathcal{K}$ guarantee that there are many negative eigenvalues. The fact that $\mathcal{K}$ is real and negative on a large set is clearly insufficient (for instance, the operator with constant integral kernel $\mathcal{K} \equiv -1$ has only one negative eigenvalue). On the positive side, if $\mathcal{K}$ takes large negative values on the diagonal and the Hilbert–Schmidt norm of K is relatively small, one can estimate the number of its negative eigenvalues as follows.

Example. Let $M_- = \{\xi \in M : \mathcal{K}(\xi,\xi) < 0\}$. If $M_- \neq \varnothing$ then the number of negative eigenvalues of the operator K is not smaller than

$$C_- := \left(\int_{M_-} \mathcal{K}(\xi,\xi)\,\mathrm{d}\nu(\xi)\right)^2 \left(\int_{M_-}\int_{M_-} |\mathcal{K}(\xi,\eta)|^2\,\mathrm{d}\nu(\xi)\,\mathrm{d}\nu(\eta)\right)^{-1}.$$

Indeed, if K_- is the truncation of K to the subspace $L_2(M_-,\nu)$ then $C_- = (\operatorname{Tr} K_-)^2\,\|K_-\|_2^{-2}$ where Tr and $\|\cdot\|_2$ stand for the trace and the Hilbert–Schmidt norm. Since $\operatorname{Tr} K_- < 0$, we have

$$(\operatorname{Tr} K_-)^2\,\|K_-\|_2^{-2} \leqslant \left(\sum_j \lambda_j\right)^2 \left(\sum_j \lambda_j^2\right)^{-2} \leqslant \#\{\lambda_j\},$$

where λ_j are the negative eigenvalues of K_-. Thus C_- estimates the number of negative eigenvalues of K_- and, consequently, of K from below.

The main result of the paper is Theorem 1.2 which provides a similar estimate involving some integrals of $\mathcal{K}$. Unlike in the previous example, it does not rely only on the behaviour of $\mathcal{K}$ on the diagonal and takes into account the contribution of its off-diagonal part.

Theorem 1.2 is stated and proved in Section 1. It is formulated in a very general setting but even in the simplest situation (say, for integral operators on a line segment) the result is not obvious. Section 2 contains some comments and examples. In particular, in Subsection 2.3 we discuss the link between the problems of estimating the number of negative eigenvalues and the difference between the Dirichlet and Neumann counting functions of the Laplace operator on a domain.

1. The main theorem

Throughout the paper $\mathcal{N}(A;t)$ denotes the dimension of the eigenspace of a self-adjoint operator (or a Hermitian matrix) A corresponding to the interval $(-\infty,t)$.

Let M be a Hausdorff topological space equipped with a locally finite Borel measure ν. We shall always be assuming that M and ν satisfy the following condition,

(C$_1$) every open set $U \subset M$ contains infinitely many elements and has non-zero measure.

Let us consider the symmetric integral operator K_0 in the space $L_2(M,\nu)$ given by a continuous kernel $\mathcal{K}(\eta,\xi)=\overline{\mathcal{K}(\xi,\eta)}$,

$$K_0 \;:\; u(\eta) \;\mapsto\; K_0u(\xi) \;:=\; \int_M \mathcal{K}(\xi,\eta)\,u(\eta)\,\mathrm{d}\nu(\eta)\,. \tag{1.1}$$

We assume that the domain of K_0 consists of L_2-functions u such that the integral on the right-hand side of (1.1) is absolutely convergent for almost all $\xi\in M$ and the function K_0u, defined by this integral, belongs to the space $L_2(M,\nu)$. Let K be an arbitrary self-adjoint extension of K_0.

Let $\varkappa(\xi,\eta)$ be the smaller eigenvalue of the Hermitian 2×2-matrix

$$\mathcal{K}^{(2)}(\xi,\eta) \;:=\; \begin{pmatrix}\mathcal{K}(\xi,\xi) & \mathcal{K}(\xi,\eta)\\ \mathcal{K}(\eta,\xi) & \mathcal{K}(\eta,\eta)\end{pmatrix}, \tag{1.2}$$

that is,

$$\varkappa(\xi,\eta) \;=\; \frac{\mathcal{K}(\xi,\xi)+\mathcal{K}(\eta,\eta)}{2}-\frac{1}{2}\sqrt{\left(\mathcal{K}(\xi,\xi)-\mathcal{K}(\eta,\eta)\right)^2+4\,|\mathcal{K}(\xi,\eta)|^2}\,. \tag{1.3}$$

Obviously, $\varkappa(\xi,\eta)$ is a continuous real-valued function on $M\times M$ such that $\varkappa(\xi,\eta)=\varkappa(\eta,\xi)$.

Remark 1.1. By (1.3), if $\mathcal{K}(\xi,\xi)$ is identically equal to a constant C then $\varkappa(\xi,\eta)=C-|\mathcal{K}(\xi,\eta)|$.

We shall say that a measure μ on $M\times M$ is *symmetric* if it is invariant with respect to the transformation $(\xi,\eta)\mapsto(\eta,\xi)$. If μ is a symmetric measure on $M\times M$, we shall denote by μ' its marginal, that is, the measure on M such that $\mu'(S)=\mu\,(S\times M)$ for all measurable $S\subset M$. Finally, assuming that

(C$_2$) $\quad 0<\int_{M\times M}(t-\varkappa(\xi,\eta))_+\,\mathrm{d}\mu(\xi,\eta)<\infty$

where

$$(t-\varkappa(\xi,\eta))_+ \;:=\; \begin{cases} t-\varkappa(\xi,\eta) & \text{if } t-\varkappa(\xi,\eta)\geqslant 0\,,\\ 0 & \text{if } t-\varkappa(\xi,\eta)<0\,,\end{cases}$$

let us denote

$$C_t(\mu) \;:=\; \frac{\left(\int_{M\times M}(t-\varkappa(\xi,\eta))_+\,\mathrm{d}\mu(\xi,\eta)\right)^2}{\int_M\int_M|\mathcal{K}(\xi,\eta)|^2\,\mathrm{d}\mu'(\xi)\,\mathrm{d}\mu'(\eta)}\,. \tag{1.4}$$

Theorem 1.2. *Let the condition* **(C$_1$)** *be fulfilled. If* $\inf\varkappa<t\leqslant 0$ *then*

$$\mathcal{N}(K,t) \;\geqslant\; \frac{1}{2}+\frac{C_t(\mu)}{16} \tag{1.5}$$

for all symmetric Borel measures μ *satisfying the condition* **(C$_2$)**.

Proof. Consider the open set

$$\Sigma_t \;:=\; \{(\xi,\eta)\in M\times M \;:\; \varkappa(\xi,\eta)<t\}\,,$$

and let M_t be its projection onto M,

$$M_t \ := \ \left\{\xi\in M\,:\,(\xi,M)\bigcap\Sigma_t\neq\varnothing\right\}.$$

Further on, without loss of generality, we shall be assuming that μ is supported on Σ_t (if not, we replace μ with its restriction to Σ_t).

Given a collection

$$\theta_n \ = \ ((\xi_1,\eta_1),\ldots,(\xi_n,\eta_n))\in\Sigma_t^n \ := \ \underbrace{\Sigma_t\times\cdots\times\Sigma_t}_{n\ \text{times}}$$

of points $(\xi_j,\eta_j)\in\Sigma_t$, let us consider the Hermitian $2n\times 2n$-matrix

$$\mathcal{K}^{(2n)}(\theta_n) \ := \ \begin{pmatrix}\mathcal{K}_{1,1} & \mathcal{K}_{1,2} & \cdots & \mathcal{K}_{1,n}\\ \mathcal{K}_{2,1} & \mathcal{K}_{2,2} & \cdots & \mathcal{K}_{2,n}\\ \vdots & \vdots & \ddots & \vdots\\ \mathcal{K}_{n,1} & \mathcal{K}_{n,2} & \cdots & \mathcal{K}_{n,n}\end{pmatrix}$$

where

$$\mathcal{K}_{i,j} \ := \ \begin{pmatrix}\mathcal{K}(\xi_i,\xi_j) & \mathcal{K}(\xi_i,\eta_j)\\ \mathcal{K}(\eta_i,\xi_j) & \mathcal{K}(\eta_i,\eta_j)\end{pmatrix} \ = \ (\mathcal{K}_{j,i})^*\,.$$

Let

$$\tilde{\mathcal{K}}^{(2n)}(\theta_n) \ := \ \Lambda\left(\mathcal{K}^{(2n)}(\theta_n)-tI\right)\Lambda\,,$$

where I is the identity $2n\times 2n$-matrix and $\Lambda=\operatorname{diag}\{\Lambda_1,\Lambda_2,\ldots,\Lambda_n\}$ is the block diagonal matrix formed by the 2×2-matrices

$$\Lambda_j \ = \ (t-\varkappa(\xi_j,\eta_j))^{-1/2}\begin{pmatrix}1 & 0\\ 0 & 1\end{pmatrix}.$$

Since $\Lambda>0$, we have

(i) $\mathcal{N}\left(\mathcal{K}^{(2n)}(\theta_n)\,,\,t\right)=\mathcal{N}\left(\tilde{\mathcal{K}}^{(2n)}(\theta_n)\,,\,0\right)$ for all $\theta_n\in\Sigma_t^n$.

By direct calculation,

$$\tilde{\mathcal{K}}^{(2n)}(\theta_n) \ := \ \begin{pmatrix}\tilde{\mathcal{K}}_{1,1}-t\Lambda_1^2 & \tilde{\mathcal{K}}_{1,2} & \cdots & \tilde{\mathcal{K}}_{1,n}\\ \tilde{\mathcal{K}}_{2,1} & \tilde{\mathcal{K}}_{2,2}-t\Lambda_2^2 & \cdots & \tilde{\mathcal{K}}_{2,n}\\ \vdots & \vdots & \ddots & \vdots\\ \tilde{\mathcal{K}}_{n,1} & \tilde{\mathcal{K}}_{n,2} & \cdots & \tilde{\mathcal{K}}_{n,n}-t\Lambda_n^2\end{pmatrix},$$

where

$$\tilde{\mathcal{K}}_{i,j} \ := \ (t-\varkappa(\xi_i,\eta_i))^{-1/2}\,(t-\varkappa(\xi_j,\eta_j))^{-1/2}\,\mathcal{K}_{i,j} \ = \ (\tilde{\mathcal{K}}_{j,i})^*\,.$$

Let us split $\tilde{\mathcal{K}}^{(2n)}(\theta_n)$ into the sum of the block diagonal matrix

$$\tilde{\mathcal{K}}^{(2n)}_{\text{diag}}(\theta_n) \ := \ \operatorname{diag}\left\{\tilde{\mathcal{K}}_{1,1}-t\Lambda_1^2\,,\,\tilde{\mathcal{K}}_{2,2}-t\Lambda_2^2\,,\,\ldots\,,\,\tilde{\mathcal{K}}_{n,n}-t\Lambda_n^2\right\}$$

and the matrix $\tilde{\mathcal{K}}^{(2n)}_{\text{off}}(\theta_n):=\tilde{\mathcal{K}}^{(2n)}(\theta_n)-\tilde{\mathcal{K}}^{(2n)}_{\text{diag}}(\theta_n)$. The equalities

$$\tilde{\mathcal{K}}_{j,j}-t\Lambda_j^2 \ = \ (t-\varkappa(\xi_j,\eta_j))^{-1}\left(\mathcal{K}^{(2)}(\xi_j,\eta_j)-tI\right), \qquad j=1,\ldots,n\,,$$

imply that

(ii) -1 is an eigenvalue of $\tilde{\mathcal{K}}^{(2n)}_{\mathrm{diag}}(\theta_n)$ of multiplicity n or higher for each $\theta_n \in \Sigma_t^n$.

On the other hand,

$$\|\tilde{\mathcal{K}}^{(2n)}_{\mathrm{off}}(\theta_n)\|_2^2 = \sum_{i\neq j} \frac{|\mathcal{K}(\xi_i,\xi_j)|^2 + |\mathcal{K}(\xi_i,\eta_j)|^2 + |\mathcal{K}(\eta_i,\xi_j)|^2 + |\mathcal{K}(\eta_i,\eta_j)|^2}{(t-\varkappa(\xi_i,\eta_i))\,(t-\varkappa(\xi_j,\eta_j))}\,, \quad (1.6)$$

where $\|\cdot\|_2$ is the Hilbert–Schmidt norm. Let us consider the absolutely continuous with respect to μ measure $\tilde{\mu}$ with the density $(t-\varkappa(\xi,\eta))$, so that $\mathrm{d}\tilde{\mu}(\xi,\eta) = (t-\varkappa(\xi,\eta))\,\mathrm{d}\mu(\xi,\eta)$. Then

$$\int_{\Sigma_t^n} (t-\varkappa(\xi_i,\eta_i))^{-1}\,(t-\varkappa(\xi_j,\eta_j))^{-1}\,|\mathcal{K}(\xi_i,\xi_j)|^2\,\mathrm{d}\tilde{\mu}(\xi_1,\eta_1)\ldots\mathrm{d}\tilde{\mu}(\xi_n,\eta_n)$$

$$= (\tilde{\mu}(\Sigma_t))^{n-2}\int_{\Sigma_t}\int_{\Sigma_t} |\mathcal{K}(\xi_i,\xi_j)|^2\,\mathrm{d}\mu(\xi_i,\eta_i)\,\mathrm{d}\mu(\xi_j,\eta_j) = (C_t(\mu))^{-1}\,(\tilde{\mu}(\Sigma_t))^n$$

for all $i\neq j$, where $C_t(\mu)$ is defined by (1.4) and $\tilde{\mu}(\Sigma_t)$ is finite in view of **(C$_2$)**. Similar calculations show that the integrals over Σ_t^n with respect to $\mathrm{d}\tilde{\mu}(\xi_1,\eta_1)\ldots\mathrm{d}\tilde{\mu}(\xi_n,\eta_n)$ of all other term in the right-hand side of (1.6) are also equal to $(C_t(\mu))^{-1}\,(\tilde{\mu}(\Sigma_t))^n$. Therefore

$$(\tilde{\mu}(\Sigma_t))^{-n}\int_{\Sigma_t^n} \|\tilde{\mathcal{K}}^{(2n)}_{\mathrm{off}}(\theta_n)\|_2^2\,\mathrm{d}\tilde{\mu}(\xi_1,\eta_1)\ldots\mathrm{d}\tilde{\mu}(\xi_n,\eta_n) \;=\; 4\,n(n-1)\,(C_t(\mu))^{-1}$$

and, consequently, there exists a point $\theta_{n,0}\in\Sigma_t^n$ such that

$$\|\tilde{\mathcal{K}}^{(2n)}_{\mathrm{off}}(\theta_{n,0})\|_2^2 \;\leqslant\; 4\,n(n-1)\,(C_t(\mu))^{-1}.$$

Since $\|\tilde{\mathcal{K}}^{(2n)}_{\mathrm{off}}(\theta_n)\|_2^2$ continuously depends on θ_n and every open set contains infinitely many elements, for each $\varepsilon>0$ there exists a point

$$\theta_{n,\varepsilon} \;=\; ((\xi_{1,\varepsilon},\eta_{1,\varepsilon}),\ldots,(\xi_{n,\varepsilon},\eta_{n,\varepsilon})) \;\in\; \Sigma_t^n$$

such that

$$\|\tilde{\mathcal{K}}^{(2n)}_{\mathrm{off}}(\theta_{n,\varepsilon})\|_2^2 \;\leqslant\; 4\,n(n-1)\,(C_t(\mu))^{-1}+\varepsilon \qquad (1.7)$$

and all the entries $\xi_{i,\varepsilon}$ and $\eta_{j,\varepsilon}$ are distinct. The estimate (1.7) implies that the number of eigenvalues of the matrix $\tilde{\mathcal{K}}^{(2n)}_{\mathrm{off}}(\theta_{n,\varepsilon})$ lying in the interval $[1,\infty)$ does not exceed $4\,n(n-1)\,(C_t(\mu))^{-1}+\varepsilon$. Therefore, in view of **(i)** and **(ii)**,

$$\mathcal{N}\left(\mathcal{K}^{(2n)}(\theta_{n,\varepsilon})\,,\,t\right) = \mathcal{N}\left(\tilde{\mathcal{K}}^{(2n)}(\theta_{n,\varepsilon})\,,\,0\right) \geqslant n - 4\,n(n-1)\,(C_t(\mu))^{-1} - \varepsilon\,.$$

Since the measure ν is locally finite and the function $\mathcal{K}$ is continuous, for every $\delta>0$ and $(\xi,\eta)\in M\times M$ there exist open neighbourhoods $U_{\xi,\delta}\subset M$ and $U_{\eta,\delta}\subset M$ of the points ξ and η such that $\nu(U_{\xi,\delta})<\infty$, $\nu(U_{\eta,\delta})<\infty$ and

$$|\mathcal{K}(\xi,\eta)-\mathcal{K}(\xi',\eta')| \;<\; \delta\,, \qquad \forall\xi'\in U_{\xi,\delta}\,,\quad \forall\eta'\in U_{\eta,\delta}\,.$$

In view of $(\mathbf{C}_1)$, $\nu(U_{\xi,\delta}) > 0$ and $\nu(U_{\eta,\delta}) > 0$. Let $u_{\xi,\delta} := (\nu(U_{\xi,\delta}))^{-1}\chi_{\xi,\delta}$ and $u_{\eta,\delta} := (\nu(U_{\eta,\delta}))^{-1}\chi_{\eta,\delta}$, where $\chi_{\xi,\delta}$ and $\chi_{\eta,\delta}$ are the characteristic functions of the sets $U_{\xi,\delta}$ and $U_{\eta,\delta}$. Then $u_{\xi,\delta}$, $u_{\eta,\delta} \in L_2(M,\nu)$ and

$$\left|\mathcal{K}(\xi,\eta) - (Ku_{\xi,\delta}, u_{\eta,\delta})_{L_2(M,\nu)}\right| < \delta. \tag{1.8}$$

Let us choose the neighbourhoods $U_{\xi,\delta}$ and $U_{\eta,\delta}$ so small that all the functions $u_{\xi_{i,\varepsilon},\delta}$ and $u_{\eta_{j,\varepsilon},\delta}$ have disjoint supports, and let $K_{\varepsilon,\delta}$ be the contraction of K to the $2n$-dimensional subspace spanned by these functions. In view of (1.8), in the basis

$$\{u_{\xi_{1,\varepsilon},\delta}, \ldots, u_{\xi_{n,\varepsilon},\delta}, u_{\eta_{1,\varepsilon},\delta}, \ldots, u_{\eta_{n,\varepsilon},\delta}\}$$

the operator $K_{\varepsilon,\delta}$ is represented by a $2n\times 2n$-matrix that converges to $\mathcal{K}^{(2n)}(\theta_{n,\varepsilon})$ as $\delta \to 0$. Consequently,

$$\mathcal{N}(K_{\varepsilon,\delta}, t) \geqslant n - 4\,n(n-1)\,(C_t(\mu))^{-1} - \varepsilon$$

for all sufficiently small δ. By the variational principle, the same estimate holds $\mathcal{N}(K,t)$. Letting $\varepsilon \to 0$, we see that

$$\begin{aligned}\mathcal{N}(K,t) &\geqslant n - 4\,n(n-1)\,(C_t(\mu))^{-1}\\ &= \frac{1}{C_t(\mu)}\left(\left(\frac{C_t(\mu)+4}{4}\right)^2 - \left(2n - \frac{C_t(\mu)+4}{4}\right)^2\right)\end{aligned}$$

for all positive integers n. Choosing $n_\mu \geqslant 1$ such that $\left|2n_\mu - \frac{C_t(\mu)+4}{4}\right| \leqslant 1$ and substituting $n = n_\mu$ in the above inequality, we obtain (1.5). □

2. Comments and examples

2.1. General comments

The estimate (1.5) implies that K has at least one eigenvalue below a negative t whenever $\inf \varkappa < t$. Indeed, in this case (1.5) with any symmetric measure μ satisfying $(\mathbf{C}_2)$ shows that $\mathcal{N}(K,t) \geqslant \frac{1}{2}$. Since the function $\mathcal{N}(K,t)$ is integer valued, it follows that $\mathcal{N}(K,t) \geqslant 1$.

If $C_t(\mu) \leqslant 8$ then (1.5) implies only the obvious estimate $\mathcal{N}(K,t) \geqslant 1$. In order to obtain a better result, one has to increase the constant $C_t(\mu)$ by choosing an appropriate measure μ. In particular, Theorem 1.2 gives a good estimate when the function $\varkappa(\xi,\eta)$ is takes large negative values on a "thin" subset $\Sigma' \subset M \times M$, the measure μ is supported on Σ' and $|\mathcal{K}|$ is relatively small outside a neighbourhood of Σ'. On the contrary, if $\mathcal{K}$ is almost constant on $M \times M$ then $\varkappa \approx \mathcal{K} - |\mathcal{K}|$ and $C_t(\mu) \approx (t - \mathcal{K} + |\mathcal{K}|)_+^2\,|\mathcal{K}|^{-2} \leqslant 4$.

A possible strategy of optimizing the choice of μ is to fix the marginal μ' and to maximize $\int (t - \varkappa(\xi,\eta))_+\,\mathrm{d}\mu(\xi,\eta)$ over the set of symmetric measures μ with the fixed marginal. The minimization (or maximization) of an integral of the form $\int f(\xi,\eta)\,\mathrm{d}\mu(\xi,\eta)$ over the set of measures with fixed marginals is known as

Kantorovich's problem. It has been solved for some special functions $f(\xi,\eta)$ (see, for instance, [6] and references therein).

I won't elaborate further on this problem, as it requires different techniques (and a different author). Instead, in the rest of the paper we shall consider a couple of examples demonstrating possible applications of Theorem 1.2.

2.2. Operators with difference kernels in $\mathbb{R}^n$

Let ν be a Borel measure on $\mathbb{R}^n$, and let h be a continuous function on $\mathbb{R}^n$ such that $h(-\theta) = \overline{h(\theta)}$. Consider the symmetric operator

$$u(\eta) \mapsto K_0 u(\xi) := \int h(\xi-\eta)\, u(\eta)\, \mathrm{d}\nu(\eta) \tag{2.1}$$

in the space $L_2(\mathbb{R}^n,\nu)$. In the notation of Section 1, $M=\mathbb{R}^n$, $\mathcal{K}(\xi,\eta)=h(\xi-\eta)$ and $\varkappa(\xi,\eta) = h(0) - |h(\xi-\eta)|$ (see Remark 1.1).

Let us fix $t \leqslant 0$ and $\theta \in \mathbb{R}^n$ such that $|h(\theta)| > h(0) - t$, and define a measure μ_θ on $\mathbb{R}^{2n}$ by the identity

$$\int f(\xi,\eta)\, \mathrm{d}\mu_\theta(\xi,\eta) = \int (f(\eta,\eta+\theta) + f(\eta+\theta,\eta))\, \mathrm{d}\tilde{\mu}(\eta),$$

where $\tilde{\mu}$ is a probability measure on $\mathbb{R}^n$. Clearly, the measure μ_θ is symmetric, and its marginal coincides with the measure μ'_θ on $\mathbb{R}^n$ given by the equality

$$\int v(\eta)\, \mathrm{d}\mu'_\theta(\eta) = \int (v(\eta) + v(\eta+\theta))\, \mathrm{d}\tilde{\mu}(\eta).$$

We have

$$\int (t - \varkappa(\xi,\eta))_+\, \mathrm{d}\mu_\theta(\xi,\eta) = 2\,(|h(\theta)| - h(0) + t) \tag{2.2}$$

and

$$\begin{aligned} &\iint |\mathcal{K}(\xi,\eta)|^2\, \mathrm{d}\mu'_\theta(\xi)\, \mathrm{d}\mu'_\theta(\eta) \\ &= \iint \left(2|h(\xi-\eta)|^2 + |h(\xi-\eta+\theta)|^2 + |h(\xi-\eta-\theta)|^2\right) \mathrm{d}\tilde{\mu}(\xi)\, \mathrm{d}\tilde{\mu}(\eta). \end{aligned} \tag{2.3}$$

Since $|h(\xi-\eta-\theta)| = |h(\eta-\xi+\theta)|$ and

$$\iint |h(\eta-\xi+\theta)|^2\, \mathrm{d}\tilde{\mu}(\xi)\, \mathrm{d}\tilde{\mu}(\eta) = \iint |h(\xi-\eta+\theta)|^2\, \mathrm{d}\tilde{\mu}(\xi)\, \mathrm{d}\tilde{\mu}(\eta),$$

the equality (2.3) can be rewritten in the form

$$\iint |\mathcal{K}(\xi,\eta)|^2\, \mathrm{d}\mu'(\xi)\, \mathrm{d}\mu'(\eta) = 2\iint \left(|h(\xi-\eta)|^2 + |h(\xi-\eta+\theta)|^2\right) \mathrm{d}\tilde{\mu}(\xi)\, \mathrm{d}\tilde{\mu}(\eta). \tag{2.4}$$

In view of (2.2) and (2.4), Theorem 1.2 implies that

$$\mathcal{N}(K,t) \geqslant \frac{1}{2} + \frac{(|h(\theta)| - h(0) + t)^2}{8 \iint (|h(\xi-\eta)|^2 + |h(\xi-\eta+\theta)|^2)\, \mathrm{d}\tilde{\mu}(\xi)\, \mathrm{d}\tilde{\mu}(\eta)} \tag{2.5}$$

for all probability measures $\tilde{\mu}$ on $\mathbb{R}^n$.

Let $\mathrm{d}\tilde{\mu}(\xi) = \varepsilon^n \chi(\varepsilon\xi)\,\mathrm{d}\xi$ where χ is the characteristic function of the unit ball. One can easily see that

$$\limsup_{\varepsilon\to 0} \iint \left(|h(\xi-\eta)|^2 + |h(\xi-\eta+\theta)|^2\right) \varepsilon^{2n}\chi(\varepsilon\xi)\,\chi(\varepsilon\eta)\,\mathrm{d}\xi\,\mathrm{d}\eta \leqslant 2\limsup_{\theta\to\infty}|h(\theta)|^2 .$$

Passing to the limit in (2.5) and optimizing the choice of θ, we obtain

Corollary 2.1. *Let K be a self-adjoint extension of the operator* (2.1). *If the measure ν satisfies the condition* $(\mathbf{C}_1)$ *and $h(0) - \sup_{\theta\in\mathbb{R}^n}|h(\theta)| < t \leqslant 0$ then*

$$\mathcal{N}(K,t) \;\geqslant\; \frac{1}{2} + \left(\frac{\sup_{\theta\in\mathbb{R}^n}|h(\theta)| - h(0) + t}{4\limsup_{\theta\to\infty}|h(\theta)|}\right)^2 . \tag{2.6}$$

In particular, (2.6) implies that $\mathcal{N}(K,0) = \infty$ whenever

$$h \not\equiv 0 \quad \text{and} \quad \lim_{\theta\to\infty}|h(\theta)| = 0.$$

2.3. Dirichlet and Neumann counting functions

Consider the Laplace operator Δ on an open domain $\Omega \subset \mathbb{R}^d$, and denote by $N_{\mathrm{D}}(\lambda)$ and $N_{\mathrm{N}}(\lambda)$ the numbers of its Dirichlet and Neumann eigenvalues lying in the interval $[0,\lambda^2)$.

Let $G_\lambda := \left\{f\in L_2(\Omega)\ :\ -\Delta f = \lambda^2 f\right\}$, where the equality $-\Delta u = \lambda^2 f$ is understood in the sense of distributions, and let $\mathcal{B}_\lambda$ be the self-adjoint operator in G_λ generated by the truncation of the quadratic form $\|\nabla f\|^2_{L_2(\Omega)} - \lambda^2\,\|f\|^2_{L_2(\Omega)}$ to the subspace G_λ.

Lemma 2.2. *For any open bounded set Ω,*

(1) *the kernel of $\mathcal{B}_\lambda$ is spanned by the Dirichlet and Neumann eigenfunctions corresponding to λ^2;*
(2) $N_{\mathrm{N}}(\lambda) - N_{\mathrm{D}}(\lambda) = n_{\mathrm{D}}(\lambda) + g^-(\lambda)$ *, where $n_{\mathrm{D}}(\lambda)$ is the number of linearly independent Dirichlet eigenfunctions corresponding to the eigenvalue λ^2, and $g^-(\lambda)$ is the dimension of the negative eigenspace of $\mathcal{B}_\lambda$.*

Proof. This is a particular case of [9, Lemma 1.2] and [9, Theorem 1.7]. □

Remark 2.3. For domains smooth boundaries, Lemma 2.2(2) was proved in [5]. In this section, we shall only need the estimate $N_{\mathrm{N}}(\lambda) - N_{\mathrm{D}}(\lambda) \leqslant n_{\mathrm{D}}(\lambda) + g_\lambda$ which can easily be deduced from the variational principle, using integration by parts [3].

In order to obtain effective estimates with the use of Lemma 2.2, we need some information about the space G_λ. It is not easy to describe, as it depends on Ω. However, the subspace G_λ always contains restrictions to Ω of the functions f satisfying the equation $-\Delta f = \lambda^2 f$ on the whole space $\mathbb{R}^d$. In particular, G_λ contains restrictions to Ω of the functions

$$f_u(x) \;=\; \int_{\mathbb{S}_\lambda^{d-1}} e^{-ix\cdot\xi}\,u(\xi)\,\mathrm{d}\nu(\xi)\,, \tag{2.7}$$

where $\nu(\xi)$ is a finite Borel measure on the sphere $\mathbb{S}_\lambda^{d-1} := \{\xi \in \mathbb{R}^d : |\xi| = \lambda\}$ and u is a function from $L_2(\mathbb{S}_\lambda^{d-1},\nu)$. Note that the integral in the right-hand side of (2.7) defines a real analytic function on $\mathbb{R}^n$, so that $f_u|_\Omega \not\equiv 0$ for all nonzero $u \in L_2(\mathbb{S}_\lambda^{d-1},\nu)$.

Let $K_{\lambda,\nu}$ be the operator in the space $L_2(\mathbb{S}_\lambda^{d-1},\nu)$ given by the integral kernel

$$\mathcal{K}(\xi,\eta) \ := \ -|\xi-\eta|^2\,\hat\chi_\Omega(\xi-\eta)\,,$$

where $\hat\chi_\Omega$ is the Fourier transform of the characteristic function χ_Ω of the set Ω. One can easily see that

$$\|\nabla f_u\|^2_{L_2(\Omega)} - \lambda^2\,\|f_u\|^2_{L_2(\Omega)} \ = \ \frac{1}{2}\,(K_{\lambda,\nu}u,u)_{L_2(\mathbb{S}_\lambda^{d-1},\nu)} \tag{2.8}$$

for all $u \in L_2(\mathbb{S}_\lambda^{d-1},\nu)$

Corollary 2.4. *For all open sets Ω, all $\lambda>0$ and all Borel measures ν on $\mathbb{S}_\lambda^{d-1}$ we have*

$$N_{\mathrm{N}}(\lambda) - N_{\mathrm{D}}(\lambda) \ \geqslant\ \mathcal{N}(K_{\lambda,\nu},0) + n_{\mathrm{D}}(\lambda)\,. \tag{2.9}$$

Proof. Denote by $\mathcal{L}_\lambda^-$ the negative eigensubspace of the operator $K_{\lambda,\nu}$, and let $L_\lambda^- = \{f_u\ :\ u\in\mathcal{L}_\lambda^-\}$. In view of (2.8), $(\mathcal{B}_\lambda f,f)_{L_2(\Omega)}<0$ for all nonzero $f\in L_\lambda^-$. By the variational principle, $g_-(\lambda)\geqslant \dim L_\lambda^- = \mathcal{N}(K_{\lambda,\nu},0)$. This inequality and Lemma 2.2(2) imply (2.9). □

One can slightly improve the estimate (2.9) assuming that

(C$_3$) the subspace L_λ does not contain a Dirchlet or Neumann eigenfunction of the form f_u with $u\in L_2(\mathbb{S}_\lambda^{d-1},\nu)$.

Corollary 2.5. *If the condition* **(C$_3$)** *is fulfilled then*

$$N_{\mathrm{N}}(\lambda) - N_{\mathrm{D}}(\lambda) \ \geqslant\ \mathcal{N}(K_{\lambda,\nu},0) + \dim\ker K_{\lambda,\nu} + n_{\mathrm{D}}(\lambda)\,. \tag{2.10}$$

Proof. Let $L_\lambda^0 = \{f_u\ :\ u\in\ker K_{\lambda,\nu}\}$. By (2.8), $(\mathcal{B}_\lambda f,f)_{L_2(\Omega)}\leqslant 0$ for all functions $f\in L_\lambda^- + L_\lambda^0$. Also, Lemma 2.2(1) and the condition **(C$_3$)** imply that $\ker\mathcal{B}_\lambda \bigcap \left(L_\lambda^- + L_\lambda^0\right) = \{0\}$. Now the standard variational arguments show that

$$g_-(\lambda)\geqslant \dim\left(L_\lambda^- + L_\lambda^0\right) = \mathcal{N}(K_{\lambda,\nu},0) + \dim\ker K_{\lambda,\nu}\,,$$

and (2.10) follows from Lemma 2.2(2). □

Remark 2.6. Since $\mathcal{K}_\lambda(\xi,\xi)\equiv 0$, we have $\varkappa(\xi,\eta) = -|\mathcal{K}_\lambda(\xi,\eta)|$ (see Remark 1.1). Thus $\inf\varkappa<0$ and, consequently, $\mathcal{N}(K_{\lambda,\nu},0)\geqslant 1$. Therefore (2.9) implies the estimate $N_{\mathrm{N}}(\lambda)-N_{\mathrm{D}}(\lambda)\geqslant 1+n_{\mathrm{D}}(\lambda)$, which was obtained in [5] and [3].

Remark 2.7. If $\hat\chi_\Omega(\theta) = 0$ for some $\theta\in\mathbb{R}^d$ and ν is the sum of δ-measures at any two points $\xi,\eta\in\mathbb{S}_\lambda^{d-1}$ such that $\xi-\eta=\theta$, then $K_{\lambda,\nu}=0$. Applying Corollary 2.5, we see that $N_{\mathrm{N}}(\lambda)-N_{\mathrm{D}}(\lambda)\geqslant 2+n_{\mathrm{D}}(\lambda)$ for all $\lambda\geqslant|\theta|/2$. This estimate was discussed in [1].

Since the function $\mathcal{K}$ is continuous, it is almost constant for small ξ and η. Therefore Theorem 1.2 is not well suited for estimating $\mathcal{N}(K_{\lambda,\nu},0)$ with small λ (see the remark in Subsection 2.1). However, it is useful for studying the behaviour of $N_{\mathrm{N}}(\lambda)-N_{\mathrm{D}}(\lambda)$ for large values of λ.

Lemma 2.8. *Denote*

$$C_\Omega(\lambda,r) \;=\; \frac{c_{d-1}\,r^4}{18}\left(\inf_{|\theta|=r}|\hat\chi_\Omega(\theta)|^2\right)\lambda^{d-4}\,|\Omega_{\lambda^{-1}}|^{-1},$$

where c_{d-1} is the volume of the unit $(d-1)$-dimensional sphere in $\mathbb{R}^d$ and $|\Omega_{\lambda^{-1}}|$ is the volume of the set $\{x\in\Omega\,:\,\operatorname{dist}(x,\partial\Omega)<\lambda^{-1}\}$. If ν is the Euclidean measure on $\mathbb{S}^{d-1}_\lambda$ then $\mathcal{N}(K_{\lambda,\nu},0)\geqslant\frac12+\frac{C_\Omega(\lambda,r)}{16}$ for all $r\in(0,2\lambda)$.

Proof. Let m_n be the normalized Euclidean measure on an n-dimensional sphere $\mathbb{S}^n_t:=\{\xi\in\mathbb{R}^{n+1}\,:\,|\xi|=t\}$, such that $m_n(\mathbb{S}^n_t)=1$. Consider the symmetric probability measure μ_r on $\mathbb{S}^{d-1}_\lambda\times\mathbb{S}^{d-1}_\lambda$ defined by the equality

$$\begin{aligned}&\int_{\mathbb{S}^{d-1}_\lambda\times\mathbb{S}^{d-1}_\lambda} f(\xi,\eta)\,\mathrm{d}\mu_r(\xi,\eta)\\ &\quad=\frac12\int_{\mathbb{S}^{d-1}_\lambda}\int_{\eta\in\mathbb{S}^{d-1}_\lambda:|\xi-\eta|=r}(f(\xi,\eta)+f(\eta,\xi))\,\mathrm{d}m_{d-2}(\eta)\,\mathrm{d}m_{d-1}(\xi)\,.\end{aligned}$$

For all functions g on $\mathbb{S}^{d-1}_\lambda$, we obviously have

$$\int_{\mathbb{S}^{d-1}_\lambda}\int_{\eta\in\mathbb{S}^{d-1}_\lambda:|\xi-\eta|=r} g(\xi)\,\mathrm{d}m_{d-2}(\eta)\,\mathrm{d}m_{d-1}(\xi)=\int_{\mathbb{S}^{d-1}_\lambda} g(\xi)\,\mathrm{d}m_{d-1}(\xi)\,. \tag{2.11}$$

On the other hand,

$$\begin{aligned}&\int_{\mathbb{S}^{d-1}_\lambda}\int_{\eta\in\mathbb{S}^{d-1}_\lambda:|\xi-\eta|\leqslant r} g(\eta)\,\mathrm{d}m_{d-1}(\eta)\,\mathrm{d}m_{d-1}(\xi)\\ &\quad=\iint_{\mathbb{S}^{d-1}_\lambda\times\mathbb{S}^{d-1}_\lambda}\psi_r(\xi,\eta)\,g(\eta)\,\mathrm{d}m_{d-1}(\eta)\,\mathrm{d}m_{d-1}(\xi)\\ &\quad=C_\lambda(r)\int_{\mathbb{S}^{d-1}_\lambda} g(\eta)\,\mathrm{d}m_{d-1}(\eta)\,,\end{aligned}\tag{2.12}$$

where ψ_r is the characteristic function of the set

$$\{(\xi,\eta)\in\mathbb{S}^{d-1}_\lambda\times\mathbb{S}^{d-1}_\lambda\,:\,|\xi-\eta|\leqslant r\}$$

and $C_\lambda(r)=\int_{\xi\in\mathbb{S}^{d-1}_\lambda:|\xi-\eta|\leqslant r}\mathrm{d}m_{d-1}(\xi)$. Since

$$\frac{\mathrm{d}}{\mathrm{d}r}\left(\int_{\eta\in\mathbb{S}^{d-1}_\lambda:|\xi-\eta|\leqslant r} g(\eta)\,\mathrm{d}m_{d-1}(\eta)\right)=\frac{c_{d-2}\,r^{d-2}}{c_{d-1}\,\lambda^{d-1}}\int_{\eta\in\mathbb{S}^{d-1}_\lambda:|\xi-\eta|=r} g(\eta)\,\mathrm{d}m_{d-2}(\eta),$$

differentiating the right- and left-hand sides of the identity (2.12), we obtain

$$\int_{\mathbb{S}^{d-1}_\lambda}\int_{\eta\in\mathbb{S}^{d-1}_\lambda:|\xi-\eta|=r} g(\eta)\,\mathrm{d}m_{d-2}(\eta)\,\mathrm{d}m_{d-1}(\xi)=\int_{\mathbb{S}^{d-1}_\lambda} g(\eta)\,\mathrm{d}m_{d-1}(\eta)\,. \tag{2.13}$$

The equalities (2.11) and (2.13) imply that the marginal μ_r' of the measure μ_r coincides with m_{d-1}.

Using Remark 1.1, we obtain

$$\begin{aligned}
&\int_{\mathbb{S}_\lambda^{d-1}\times\mathbb{S}_\lambda^{d-1}} (-\varkappa(\xi,\eta))_+ \,\mathrm{d}\mu_r(\xi,\eta)\\
&\quad= \int_{\mathbb{S}_\lambda^{d-1}\times\mathbb{S}_\lambda^{d-1}} |\xi-\eta|^2\,|\hat\chi_\Omega(\xi-\eta)|\,\mathrm{d}\mu_r(\xi,\eta)\\
&\quad= r^2\int_{\mathbb{S}_\lambda^{d-1}}\int_{\eta\in\mathbb{S}_\lambda^{d-1}:|\xi-\eta|=r} |\hat\chi_\Omega(\xi-\eta)|\,\mathrm{d}m_{d-2}(\eta)\,\mathrm{d}m_{d-1}(\xi)\\
&\quad\geqslant r^2\inf_{|\theta|=r}|\hat\chi_\Omega(\theta)|.
\end{aligned}$$

As was shown in [4],

$$\begin{aligned}
&\int_{\mathbb{S}_\lambda^{d-1}}\int_{\mathbb{S}_\lambda^{d-1}} |\mathcal{K}(\xi,\eta)|^2\,\mu_r'(\xi)\,\mu_r'(\eta)\\
&\quad= \int_{\mathbb{S}_\lambda^{d-1}}\int_{\mathbb{S}_\lambda^{d-1}} |\xi-\eta|^4\,|\hat\chi_\Omega(\xi-\eta)|^2\,\mathrm{d}m_{d-1}(\xi)\,\mathrm{d}m_{d-1}(\eta)\\
&\quad\leqslant 18\,c_{d-1}^{-1}\,\lambda^{4-d}\,|\Omega_{\lambda^{-1}}|\,.
\end{aligned}$$

Now the required estimate follows from Theorem 1.2. □

Corollary 2.4 and Lemma 2.8 imply that

$$N_{\mathrm{N}}(\lambda)-N_{\mathrm{D}}(\lambda)\ \geqslant\ \mathrm{const}\,\lambda^{d-4}\,|\Omega_{\lambda^{-1}}|^{-1} \tag{2.14}$$

for all sufficiently large λ. This estimate was obtained by a different method in [4]. So far it is unknown whether one can get a better result in terms of growth as $\lambda\to\infty$ for a general domain Ω.

For domains with smooth boundaries, the two-term Weyl asymptotic formula (see, for instance, [7] or [10]) implies that $N_{\mathrm{N}}(\lambda)-N_{\mathrm{D}}(\lambda)\geqslant O(\lambda^{d-1})$. There are reasons to believe that the same is true for all domains but the standard techniques, which work for domains with irregular boundaries, fail to produce such results. It is possible that (2.14) can be improved by applying Theorem 1.2 with some other measures ν and μ to the operator $K_{\lambda,\nu}$ or/and more careful analysis of the asymptotic behaviour of the integrals in (1.4) as $\lambda\to\infty$.

References

[1] R. Benguria, M. Levitin and L. Parnovski, *Fourier transform, null variety, and Laplacian's eigenvalues.* J. Func. Anal., **257:7** (2009), 2088–2123.

[2] M. Birman and M. Solomyak, *Estimates of singular numbers of integral operators.* Uspekhi Mat. Nauk **32:1** (1977), 15–89 (Russian). English translation in Russian Math. Surveys **32:1** (1977), 15–89.

[3] N. Filonov, *On an inequality between Dirichlet and Neumann eigenvalues for the Laplace operator.* Algebra Anal. **16:2**, 172–176 (Russian). English translation in St. Petersbg. Math. J. **16:2**, (2005), 413–416.

[4] N. Filonov and Y. Safarov, *Asymptotic estimates for the difference between Dirichlet and Neumann counting functions.* Funktsional. Anal. i Prilozhen. **44:4** (2010), 54–64 (Russian). English translation in Funct. Anal. Appl. **44:4** (2010), 286–294.

[5] L. Friedlander, *Some inequalities between Dirichlet and Neumann eigenvalues.* Arch. Ration. Mech. Anal. **116** (1991), 153–160.

[6] W. Gangbo and R. McCann, *The geometry of optimal transportation.* Acta Math. **177** (1996), 113–161.

[7] V. Ivrii, *On the second term of the spectral asymptotics for the Laplace-Beltrami operator on manifolds with boundary.* Funktsional. Anal. i Prilozhen. **14:2** (1980), 25–34 (Russian). English transl. in Functional Anal. Appl. **14** (1980).

[8] H. König, *Eigenvalue distribution of compact operators.* Birkhäuser Verlag, 1986.

[9] Y. Safarov, *On the comparison of the Dirichlet and Neumann counting functions.* AMS Translations (2), Advances in Mathematical Sciences, vol. 225 (2008), 191–204.

[10] Y. Safarov and D. Vassiliev, *The asymptotic distribution of eigenvalues of partial differential operators.* American Mathematical Society, 1996.

Y. Safarov
Department of Mathematics
King's College London
Strand, London WC2R 2LS
United Kingdom
e-mail: yuri.safarov@kcl.ac.uk

Operator Theory:
Advances and Applications, Vol. 228, 355–374

Toeplitz Operators on the Bergman Spaces with Pseudodifferential Defining Symbols

Armando Sánchez-Nungaray and Nikolai Vasilevski

To Vladimir Rabinovich in occasion of his 70th anniversary

Abstract. We study the C^*-algebra generated by Toeplitz operators acting on the Bergman or poly-Bergman space over the unit disk $\mathbb{D}$ on the complex plane, whose pseudodifferential defining symbols belong to the algebra $\mathcal{R} = \mathcal{R}(C(\overline{\mathbb{D}}); S_{\mathbb{D}}, S_{\mathbb{D}}^*)$. The algebra $\mathcal{R}$ is generated by the multiplication operators aI, where $a \in C(\overline{\mathbb{D}})$, and the following two operators

$$(S_{\mathbb{D}}\varphi)(z) = -\frac{1}{\pi}\int_{\mathbb{D}} \frac{\varphi(\zeta)}{(\zeta - z)^2} d\nu(\zeta) \quad \text{and} \quad (S_{\mathbb{D}}^*\varphi)(z) = -\frac{1}{\pi}\int_{\mathbb{D}} \frac{\varphi(\zeta)}{(\overline{\zeta} - \overline{z})^2} d\nu(\zeta).$$

In the Bergman space case, both algebras $\mathcal{T}(C(\overline{\mathbb{D}}))$, generated by Toeplitz operators T_a with defining symbols $a \in C(\overline{\mathbb{D}})$, and $\mathcal{T}(\mathcal{R}(C(\overline{\mathbb{D}}); S_{\mathbb{D}}, S_{\mathbb{D}}^*))$, generated by Toeplitz operators T_A with defining symbols $A \in \mathcal{R}(C(\overline{\mathbb{D}}); S_{\mathbb{D}}, S_{\mathbb{D}}^*)$, consist of the same operators, and the Fredholm symbol algebra for both of them is isomorphic and isometric to $C(\partial\mathbb{D})$. At the same time, their generating Toeplitz operators possess quite different properties.

Mathematics Subject Classification (2010). Primary 47B35; Secondary 47L80, 47G10.

Keywords. Toeplitz operator, poly-Bergman space, pseudodifferential operator.

1. Introduction

The idea of considering Toeplitz operators with pseudodifferential symbols is not new, see for example [2, 8, 12, 15], where operators related to the Hardy spaces were studied. While for the Bergman space case, this is probably a first attempt to treat such a question. To be more precise we study the C^*-algebra generated by Toeplitz operators acting on the Bergman or poly-Bergman space over the unit

The second named author has been partially supported by CONACYT Project 102800, México.

disk $\mathbb{D}$ on the complex plane, whose pseudodifferential defining symbols belong to the algebra $\mathcal{R} = \mathcal{R}(C(\overline{\mathbb{D}}); S_{\mathbb{D}}, S_{\mathbb{D}}^*)$. The last algebra is generated by the multiplication operators aI, where $a \in C(\overline{\mathbb{D}})$, and the following two singular integral (pseudodifferential) operators

$$(S_{\mathbb{D}}\varphi)(z) = -\frac{1}{\pi}\int_{\mathbb{D}} \frac{\varphi(\zeta)}{(\zeta - z)^2} d\nu(\zeta) \quad \text{and} \quad (S_{\mathbb{D}}^*\varphi)(z) = -\frac{1}{\pi}\int_{\mathbb{D}} \frac{\varphi(\zeta)}{(\overline{\zeta} - \overline{z})^2} d\nu(\zeta), \tag{1.1}$$

where $d\nu$ is the standard Lebesgue plane measure.

The choice of the algebra $\mathcal{R}$ is not accidental, moreover it is quite natural due to the deep internal connection between the action of the operators (1.1), considered in the upper half-plane Π, and the decomposition of $L_2(\Pi)$ onto the direct sum of poly-Bergman type spaces (see, for example, [17] and Section 3 of this paper).

The main qualitative results of the paper are as follows. The algebra $C(\overline{\mathbb{D}})$ is obviously commutative, while the algebra $\mathcal{R}(C(\overline{\mathbb{D}}); S_{\mathbb{D}}, S_{\mathbb{D}}^*)$ is even not essentially commutative, its Fredholm symbol algebra has infinite-dimensional irreducible representations. Nevertheless, for the Bergman space case, both algebras $\mathcal{T}(C(\overline{\mathbb{D}}))$, which is generated by Toeplitz operators T_a with defining symbols $a \in C(\overline{\mathbb{D}})$, and $\mathcal{T}(\mathcal{R}(C(\overline{\mathbb{D}}); S_{\mathbb{D}}, S_{\mathbb{D}}^*))$, which is generated by Toeplitz operators T_A with defining symbols $A \in \mathcal{R}(C(\overline{\mathbb{D}}); S_{\mathbb{D}}, S_{\mathbb{D}}^*)$, are, in fact, the same; and the Fredholm symbol algebra for both of them is isomorphic and isometric to $C(S^1)$, where S^1 is the boundary of the unit disk $\mathbb{D}$.

At the same time, although both algebras $\mathcal{T}(\mathcal{R}(C(\overline{\mathbb{D}}); S_{\mathbb{D}}, S_{\mathbb{D}}^*))$ and $\mathcal{T}(C(\overline{\mathbb{D}}))$ consist of the same operators, being thought as generated by Toeplitz operators with defining symbols from $\mathcal{R}(C(\overline{\mathbb{D}}); S_{\mathbb{D}}, S_{\mathbb{D}}^*)$ and $C(\overline{\mathbb{D}})$ respectively, they possess quite different properties.

Contrary to the case of $\mathcal{T}(C(\overline{\mathbb{D}}))$, the first algebra $\mathcal{T}(\mathcal{R}(C(\overline{\mathbb{D}}); S_{\mathbb{D}}, S_{\mathbb{D}}^*))$ does not obey compact semi-commutator property; the Toeplitz operators T_A, with $A \in \mathcal{R}(C(\overline{\mathbb{D}}); S_{\mathbb{D}}, S_{\mathbb{D}}^*)$ can be zero for non zero A; there are symbols $A \in \mathcal{R}(C(\overline{\mathbb{D}}); S_{\mathbb{D}}, S_{\mathbb{D}}^*)$ such that the Toeplitz operator T_A has finite rank; etc.

At the same time, for the n-poly-Bergman space case, the corresponding algebras $\mathcal{T}_n(C(\overline{\mathbb{D}}))$, which is generated by Toeplitz operators T_a with defining symbols $a \in C(\overline{\mathbb{D}})$, and $\mathcal{T}_n(\mathcal{R}(C(\overline{\mathbb{D}}); S_{\mathbb{D}}, S_{\mathbb{D}}^*))$, which is generated by Toeplitz operators T_A with defining symbols $A \in \mathcal{R}(C(\overline{\mathbb{D}}); S_{\mathbb{D}}, S_{\mathbb{D}}^*)$, are quite different. The first one is essentially quite similar to $\mathcal{T}(C(\overline{\mathbb{D}}))$, while the second one is unitarily equivalent to the matrix algebra $\mathcal{T}(C(\overline{\mathbb{D}})) \otimes M_{n\times n}(\mathbb{C})$. The Fredholm symbol algebra of $\mathcal{T}_n(C(\overline{\mathbb{D}}))$ is isomorphic and isometric to $C(S^1)$, while the Fredholm symbol algebra of $\mathcal{T}_n(\mathcal{R}(C(\overline{\mathbb{D}}); S_{\mathbb{D}}, S_{\mathbb{D}}^*))$ is isomorphic and isometric to the matrix algebra $M_{n\times n}(C(S^1)) = C(S^1) \otimes M_{n\times n}(\mathbb{C})$.

2. Algebra $\mathcal{R}(C(\overline{\mathbb{D}}); S_{\mathbb{D}}, S_{\mathbb{D}}^*)$

Let $\mathbb{D}$ be the unit disk on the complex plane. We consider the Hilbert space $L_2(\mathbb{D})$ with the standard Lebesgue plane measure $d\nu(z) = dxdy$, where $z = x + iy$. The following singular integral operators

$$(S_{\mathbb{D}}\varphi)(z) = -\frac{1}{\pi}\int_{\mathbb{D}} \frac{\varphi(\zeta)}{(\zeta - z)^2} d\nu(\zeta) \quad \text{and} \quad (S_{\mathbb{D}}^*\varphi)(z) = -\frac{1}{\pi}\int_{\mathbb{D}} \frac{\varphi(\zeta)}{(\overline{\zeta} - \overline{z})^2} d\nu(\zeta)$$

are known to be bounded on $L_2(\mathbb{D})$ and mutually adjoint. It is known as well that, for each $a \in C(\overline{\mathbb{D}})$, both commutators $[S_{\mathbb{D}}, aI]$ and $[S_{\mathbb{D}}^*, aI]$ are compact, and being considered in the whole complex plane, these operators obey the relation $S_{\mathbb{C}}^* = S_{\mathbb{C}}^{-1}$.

We denote by $\mathcal{R} = \mathcal{R}(C(\overline{\mathbb{D}}); S_{\mathbb{D}}, S_{\mathbb{D}}^*)$ the C^*-algebra generated by all operators of the form aI, where $a \in C(\overline{\mathbb{D}})$, $S_{\mathbb{D}}$ and $S_{\mathbb{D}}^*$. This algebra is irreducible and contains the ideal $\mathcal{K}$ of all operators compact on $L_2(\mathbb{D})$. We will denote by π the natural projection

$$\pi\,:\, \mathcal{R} \longrightarrow \widehat{\mathcal{R}} = \mathcal{R}/\mathcal{K}.$$

The structure of the Fredholm symbol (or Calkin) algebra $\operatorname{Sym}\mathcal{R} = \widehat{\mathcal{R}}$ is known for a long time and well understood. We give briefly its description based on the simple functional model for the operators S_{Π} and S_{Π}^*, considered in the upper half-plane Π in $\mathbb{C}$. According to [17], *both operators S_{Π} and S_{Π}^* are unitarily equivalent to the direct sum of two unilateral shifts, forward and backward, both taken with the infinite multiplicity.*

To describe the algebra $\widehat{\mathcal{R}}$ we will use the standard Douglas-Varela local principle [7, 14, 18]. The algebra $\pi(C(\overline{\mathbb{D}})) \cong C(\overline{\mathbb{D}})$ is obviously a central commutative C^*-subalgebra of the algebra $\widehat{\mathcal{R}}$, thus we will localize by the points $z_0 \in \overline{\mathbb{D}}$. Denote by J_{z_0} the maximal ideal of the algebra $\pi(C(\overline{\mathbb{D}}))$ which corresponds to the point $z_0 \in \overline{\mathbb{D}}$, and by $J(z_0)$ the closed two-sided ideal of the algebra $\widehat{\mathcal{R}}$ generated by J_{z_0}. Then the local algebra $\widehat{\mathcal{R}}(z_0)$ is defined as $\widehat{\mathcal{R}}/J(z_0)$, and let $\pi(z_0)$ be the natural straight-through projection

$$\pi(z_0) : \mathcal{R} \longrightarrow \widehat{\mathcal{R}} \longrightarrow \widehat{\mathcal{R}}(z_0).$$

For each $z_0 \in \mathbb{D}$, the local algebra $\widehat{\mathcal{R}}(z_0)$ has quite a simple description. Indeed, we have the following local equivalences: $\pi(z_0)(aI) \overset{z_0}{\sim} a(z_0) \in \mathbb{C}$ and $\pi(z_0)(S_{\mathbb{D}}^*) \overset{z_0}{\sim} [\pi(z_0)(S_{\mathbb{D}})]^{-1}$, moreover the spectrum of $\pi(z_0)(S_{\mathbb{D}})$ coincides with the unit circle: $\operatorname{spec}\pi(z_0)(S_{\mathbb{D}}) = S^1 = \partial\mathbb{D}$. Thus the algebra $\widehat{\mathcal{R}}(z_0)$, being a C^*-algebra with identity generated by a single normal element, is isomorphic and isometric to $C(\operatorname{spec}\pi(z_0)(S_{\mathbb{D}})) = C(S^1)$, and under their identification the homomorphism $\pi(z_0)$ is defined on generators of the algebra $\mathcal{R} = \mathcal{R}(C(\overline{\mathbb{D}}); S_{\mathbb{D}}, S_{\mathbb{D}}^*)$ as follows

$$\begin{aligned} \pi(z_0) \;&:\; aI \longmapsto a(z_0) \in \mathbb{C}, \\ \pi(z_0) \;&:\; S_{\mathbb{D}} \longmapsto t \in C(S^1), \\ \pi(z_0) \;&:\; S_{\mathbb{D}}^* \longmapsto \overline{t} \in C(S^1), \end{aligned} \tag{2.1}$$

here t and $\bar{t}$ stand for the continuous functions defined on S^1 by $t \mapsto t$ and $t \mapsto \bar{t}$, respectively.

The case $z_0 \in S^1 = \partial\mathbb{D}$ is more delicate. Given $z_0 \in S^1$, introduce the Möbius transformation

$$w = i\frac{z_0 - z}{z_0 + z},$$

which maps the unit disk $\mathbb{D}$ onto the upper half-plane Π and sends the point $z_0 \in \partial\mathbb{D}$ to $0 \in \Pi$. The inverse mapping has obviously the form

$$z = z_0\frac{i - w}{i + w}.$$

Further, the operator $V_{z_0} : L_2(\Pi) \longrightarrow L_2(\mathbb{D})$, which is given by

$$(V_{z_0}\varphi)(z) = \frac{2iz_0}{(z_0 + z)^2}\,\varphi\left(i\frac{z_0 - z}{z_0 + z}\right),$$

is unitary, and

$$(V_{z_0}^{-1}\varphi)(w) = (V_{z_0}^{*}\varphi)(w) = \frac{2iz_0}{(i + w)^2}\,\varphi\left(z_0\frac{i - w}{i + w}\right).$$

It is straightforward to check that

$$V_{z_0}^{-1}S_{\mathbb{D}}V_{z_0} = (i\overline{z_0})^2 S_{\Pi}h(w)I \quad \text{and} \quad V_{z_0}^{-1}S_{\mathbb{D}}^{*}V_{z_0} = (iz_0)^2\overline{h(w)}S_{\Pi}^{*},$$

where $h(w) = (i + w)^4/|i + w|^4$. We note that the function $h(w)$ is continuous on $\Pi \cup \partial\Pi$ (except the point ∞), $h(0) = 1$, and $\overline{h(w)} = 1/h(w)$. As $h(0) = 1$ we have, for each $z_0 \in S^1$, the following local equivalences at the point $0 \in \Pi$:

$$V_{z_0}^{-1}S_{\mathbb{D}}V_{z_0} = (i\overline{z_0})^2 S_{\Pi}h(w)I \overset{0}{\sim} (i\overline{z_0})^2 S_{\Pi},$$
$$V_{z_0}^{-1}S_{\mathbb{D}}^{*}V_{z_0} = (iz_0)^2\overline{h(w)}S_{\Pi}^{*} \overset{0}{\sim} (iz_0)^2 S_{\Pi}^{*},$$

which implies that the local algebra $\widehat{\mathcal{R}}(z_0)$ is isomorphic and isometric to the C^*-algebra with identity $\mathcal{R}(\mathbb{C}; S_{\Pi}, S_{\Pi}^{*})$, which is generated by the operators S_{Π} and S_{Π}^{*} (or by $(i\overline{z_0})^2 S_{\Pi}$ and $(iz_0)^2 S_{\Pi}^{*}$), acting on $L_2(\Pi)$. Identifying the last algebras, the homomorphism

$$\pi(z_0) : \mathcal{R} \longrightarrow \widehat{\mathcal{R}}(z_0) = \mathcal{R}(\mathbb{C}; S_{\Pi}, S_{\Pi}^{*})$$

is defined on generators of the algebra $\mathcal{R} = \mathcal{R}(C(\overline{\mathbb{D}}); S_{\mathbb{D}}, S_{\mathbb{D}}^{*})$ as follows

$$\begin{aligned}
\pi(z_0) &: aI \longmapsto a(z_0) \in \mathbb{C},\\
\pi(z_0) &: S_{\mathbb{D}} \longmapsto (i\overline{z_0})^2 S_{\Pi} \in \mathcal{R}(\mathbb{C}; S_{\Pi}, S_{\Pi}^{*}),\\
\pi(z_0) &: S_{\mathbb{D}}^{*} \longmapsto (iz_0)^2 S_{\Pi}^{*} \in \mathcal{R}(\mathbb{C}; S_{\Pi}, S_{\Pi}^{*}).
\end{aligned}$$

Recall now necessary ingredients from [17]. We define the unitary operator

$$U_1 = F \otimes I : L_2(\Pi) = L_2(\mathbb{R}) \otimes L_2(\mathbb{R}_+) \longrightarrow L_2(\mathbb{R}) \otimes L_2(\mathbb{R}_+),$$

where the Fourier transform $F : L_2(\mathbb{R}) \to L_2(\mathbb{R})$ is given by

$$(Ff)(x) = \frac{1}{\sqrt{2\pi}}\int_{\mathbb{R}} e^{-ix\xi} f(\xi)\,d\xi.$$

The next unitary operator

$$U_2 : L_2(\Pi) = L_2(\mathbb{R}) \otimes L_2(\mathbb{R}_+) \longrightarrow L_2(\mathbb{R}) \otimes L_2(\mathbb{R}_+)$$

is given by

$$(U_2\varphi)(x,y) = \frac{1}{\sqrt{2|x|}}\,\varphi(x, \frac{y}{2|x|}).$$

The inverse operator $U_2^{-1} = U_2^* : L_2(\mathbb{R}) \otimes L_2(\mathbb{R}_+) \longrightarrow L_2(\mathbb{R}) \otimes L_2(\mathbb{R}_+)$ acts as follows,

$$(U_2^{-1}\varphi)(x,y) = \sqrt{2|x|}\,\varphi(x, 2|x| \cdot y).$$

We introduce then the following integral operators

$$(S_+f)(y) = -f(y) + e^{-\frac{y}{2}} \int_0^y e^{\frac{t}{2}}\, f(t)\, dt,$$

$$(S_-f)(y) = -f(y) + e^{\frac{y}{2}} \int_y^\infty e^{-\frac{t}{2}}\, f(t)\, dt,$$

which are bounded on $L_2(\mathbb{R}_+)$ and are mutually adjoint.

Theorem 2.1 ([17]). *The unitary operator $U = U_2U_1$ gives an isometric isomorphism of the space $L_2(\Pi) = [L_2(\mathbb{R}_+) \otimes L_2(\mathbb{R}_+)] \oplus [L_2(\mathbb{R}_-) \otimes L_2(\mathbb{R}_+)]$ under which the two-dimensional singular integral operators S_Π and S_Π^* are unitary equivalent to the following operators*

$$U\, S_\Pi\, U^{-1} = (I \otimes S_+) \oplus (I \otimes S_-),$$
$$U\, S_\Pi^*\, U^{-1} = (I \otimes S_-) \oplus (I \otimes S_+).$$

Recall (see, for example, [1]), that the Laguerre polynomial $L_n(y)$ of degree n, $n = 0, 1, 2, \dots$, and type 0 is defined by

$$L_n(y) = L_n^0(y) = \frac{e^y}{n!}\frac{d^n}{dy^n}(e^{-y}\, y^n)$$
$$= \sum_{k=0}^n \frac{n!}{k!(n-k)!}\,\frac{(-y)^k}{k!}, \qquad y \in \mathbb{R}_+,$$

and that the system of functions

$$\ell_n(y) = e^{-y/2} L_n(y), \qquad n \in \mathbb{Z}_+ = \mathbb{N} \cup \{0\},$$

forms an orthonormal basis in the space $L_2(\mathbb{R}_+)$.

Theorem 2.2 ([17]). *For each admissible n, the following equalities hold:*

$$(S_+\ell_n)(y) = -\ell_{n+1}(y), \qquad (S_-\ell_n)(y) = -\ell_{n-1}(y), \qquad \text{and} \qquad (S_-\ell_0)(y) = 0.$$

Corollary 2.3. *The unitary operator $U = U_2U_1$ gives an isometric isomorphism of the space $L_2(\Pi) = [L_2(\mathbb{R}_+) \otimes L_2(\mathbb{R}_+)] \oplus [L_2(\mathbb{R}_-) \otimes L_2(\mathbb{R}_+)]$ under which*

$$U\,(i\overline{z_0})^2 S_\Pi\, U^{-1} = (I \otimes (i\overline{z_0})^2 S_+) \oplus (I \otimes (i\overline{z_0})^2 S_-),$$
$$U\,(iz_0)^2 S_\Pi^*\, U^{-1} = (I \otimes (iz_0)^2 S_-) \oplus (I \otimes (iz_0)^2 S_+).$$

The operators $(i\overline{z_0})^2 S_+$ *and* $(iz_0)^2 S_-$ *are the forward and backward unilateral shift operators on* $L_2(\mathbb{R}_+)$ *with respect to the orthonormal basis formed by the functions* $\ell'_n(y) = (-1)^n (i\overline{z_0})^{2n} \ell_n(y)$, $n = 0, 1, 2, \ldots$.

The operators $(iz_0)^2 S_+$ *and* $(i\overline{z_0})^2 S_-$ *are the forward and backward unilateral shift operators on* $L_2(\mathbb{R}_+)$ *with respect to the orthonormal basis formed by the functions* $\ell''_n(y) = (-1)^n (iz_0)^{2n} \ell_n(y)$, $n = 0, 1, 2, \ldots$.

Finally we denote by $\mathcal{T}(C(S^1))$ the C^*-algebra generated by all Toeplitz operators T_a with continuous defining symbols $a(t) \in C(S^1)$ and acting on the Hardy space $H^2(S^1)$ over the unit circle S^1. Note that we use the standard normalized measure on S^1 so that the system of functions $\{t^k\}_{k\in\mathbb{Z}_+}$ forms an orthonormal base in $H^2(S^1)$.

It is well known (see, for example, [6]) that the algebra $\mathcal{T}(C(S^1))$ coincides with the set of all operators of the form $T_a + K$, where $a(t) \in C(S^1)$ and K is a compact operator on $H^2(S^1)$. Then by the obtained description of the local algebra $\widehat{\mathcal{R}}(z_0)$ and by [4, 5] we have

Corollary 2.4. *For any* $z_0 \in S^1$, *the local algebra* $\widehat{\mathcal{R}}(z_0)$ *is isomorphic and isometric to a subalgebra of the* C^**-algebra* $\mathcal{T}(C(S^1)) \oplus \mathcal{T}(C(S^1))$. *The homomorphism*

$$\pi(z_0) \; : \; \mathcal{R} \longrightarrow \widehat{\mathcal{R}}(z_0) \subset \mathcal{T}(C(S^1)) \oplus \mathcal{T}(C(S^1))$$

is defined on generators of the algebra $\mathcal{R} = \mathcal{R}(C(\overline{\mathbb{D}}); S_{\mathbb{D}}, S^*_{\mathbb{D}})$ *as follows*

$$\begin{aligned}
\pi(z_0) \; &: \; aI \longmapsto a(z_0) \in \mathbb{C}, \\
\pi(z_0) \; &: \; S_{\mathbb{D}} \longmapsto (T_t,\, T_{\overline{t}}) \in \mathcal{T}(C(S^1)) \oplus \mathcal{T}(C(S^1)), \\
\pi(z_0) \; &: \; S^*_{\mathbb{D}} \longmapsto (T_{\overline{t}},\, T_t) \in \mathcal{T}(C(S^1)) \oplus \mathcal{T}(C(S^1)).
\end{aligned}$$

We note as well that the local algebra $\widehat{\mathcal{R}}(z_0)$ possesses many one-dimensional irreducible representations. In particular, for each $t_0 \in S^1$, we denote by $\iota(t_0) : \widehat{\mathcal{R}}(z_0) \to \mathbb{C}$ the representation which is defined on the generators by

$$\iota(t_0) : \; (T_t,\, T_{\overline{t}}) \longmapsto t_0 \quad \text{and} \quad \iota(t_0) : \; (T_{\overline{t}},\, T_t) \longmapsto \overline{t_0}.$$

Thus, for each $(z_0, t_0) \in \partial\mathbb{D} \times S^1$, the homomorphism

$$\iota(t_0) \circ \pi(z_0) \; : \; \mathcal{R}(C(\overline{\mathbb{D}}); S_{\mathbb{D}}, S^*_{\mathbb{D}}) \longrightarrow \mathbb{C}$$

is a one-dimensional representation of the algebra $\mathcal{R}$ which is defined on generators of the algebra $\mathcal{R} = \mathcal{R}(C(\overline{\mathbb{D}}); S_{\mathbb{D}}, S^*_{\mathbb{D}})$ as follows

$$\begin{aligned}
\iota(t_0) \circ \pi(z_0) \; &: \; aI \longmapsto a(z_0), \\
\iota(t_0) \circ \pi(z_0) \; &: \; S_{\mathbb{D}} \longmapsto t_0, \\
\iota(t_0) \circ \pi(z_0) \; &: \; S^*_{\mathbb{D}} \longmapsto \overline{t_0}.
\end{aligned}$$

These representations define the homomorphism (2.1) for all boundary points $z_0 \in \partial\mathbb{D}$. Now, gluing together the obtained characterizations of the local algebras we come to the following description of the Fredholm symbol algebra $\operatorname{Sym}\mathcal{R} = \widehat{\mathcal{R}}$ of the algebra $\mathcal{R} = \mathcal{R}(C(\overline{\mathbb{D}}); S_{\mathbb{D}}, S^*_{\mathbb{D}})$.

Let $\mathfrak{M} = \overline{\mathbb{D}} \times S^1 \sqcup S^1 \times \{0, \infty\}$. We denote by $\mathfrak{S}$ the set of all vector-functions σ, continuous on $\mathfrak{M}$ and having the form

$$\sigma = \begin{cases} c(z,t) \in \mathbb{C}, & (z,t) \in \overline{\mathbb{D}} \times S^1 \\ T_{c(z,t)} + K_0(z) \in \mathcal{T}(C(S^1)), & (z,0) \in S^1 \times \{0,\infty\} \\ T_{c(z,\overline{t})} + K_\infty(z) \in \mathcal{T}(C(S^1)), & (z,\infty) \in S^1 \times \{0,\infty\} \end{cases} .$$

The set $\mathfrak{S}$ is a C^*-algebra with respect to the component-wise operations and the norm $\|\sigma\| = \sup_{\mathfrak{M}} \|\sigma(\cdot,\cdot)\|$.

Theorem 2.5. *The Fredholm symbol algebra* $\operatorname{Sym}\mathcal{R} = \widehat{\mathcal{R}}$ *of the algebra* $\mathcal{R} = \mathcal{R}(C(\overline{\mathbb{D}}); S_{\mathbb{D}}, S_{\mathbb{D}}^*)$ *is isomorphic and isometric to the algebra* $\mathfrak{S}$*. Under their identification, the symbol homomorphism*

$$\operatorname{sym} : \mathcal{R} \longrightarrow \operatorname{Sym}\mathcal{R} = \mathfrak{S}$$

is generated by the following mapping: if $A = a_1(z)I + a_2(z)S_{\mathbb{D}} + a_3(z)S_{\mathbb{D}}^*$ *and* $c(z,t) = a_1(z) + a_2(z)t + a_3(z)\overline{t}$*, then*

$$\operatorname{sym} A = \begin{cases} c(z,t) \in \mathbb{C}, & (z,t) \in \overline{\mathbb{D}} \times S^1 \\ T_{c(z,t)} \in \mathcal{T}(C(S^1)), & (z,0) \in S^1 \times \{0,\infty\} \\ T_{c(z,\overline{t})} \in \mathcal{T}(C(S^1)), & (z,\infty) \in S^1 \times \{0,\infty\} \end{cases} .$$

3. Poly-Bergman type spaces and action of the operators S_Π and S_Π^*

Recall that the space $\mathcal{A}_n^2(\Pi)$ of n-analytic functions as the subspace of $L_2(\Pi)$ consisting of all functions $\varphi = \varphi(z,\overline{z}) = \varphi(x,y)$, which satisfy the equation

$$\left(\frac{\partial}{\partial \overline{z}}\right)^n \varphi = \frac{1}{2^n}\left(\frac{\partial}{\partial x} + i\frac{\partial}{\partial y}\right)^n \varphi = 0.$$

Similarly, the space $\widetilde{\mathcal{A}}_n^2(\Pi)$ of n-anti-analytic functions as the subspace of $L_2(\Pi)$ consisting of all functions $\varphi = \varphi(z,\overline{z}) = \varphi(x,y)$, which satisfy the equation

$$\left(\frac{\partial}{\partial z}\right)^n \varphi = \frac{1}{2^n}\left(\frac{\partial}{\partial x} - i\frac{\partial}{\partial y}\right)^n \varphi = 0.$$

Of course, we have $\mathcal{A}_1^2(\Pi) = \mathcal{A}^2(\Pi)$ and $\widetilde{\mathcal{A}}_1^2(\Pi) = \widetilde{\mathcal{A}}^2(\Pi)$, for $n = 1$, as well as $\mathcal{A}_n^2(\Pi) \subset \mathcal{A}_{n+1}^2(\Pi)$ and $\widetilde{\mathcal{A}}_n^2(\Pi) \subset \widetilde{\mathcal{A}}_{n+1}^2(\Pi)$, for each $n \in \mathbb{N}$.

We introduce as well the space $\mathcal{A}_{(n)}^2(\Pi)$ of true-n-analytic functions by

$$\mathcal{A}_{(n)}^2(\Pi) = \mathcal{A}_n^2(\Pi) \ominus \mathcal{A}_{n-1}^2(\Pi),$$

for $n > 1$, and by $\mathcal{A}_{(1)}^2(\Pi) = \mathcal{A}_1^2(\Pi)$; and, symmetrically, introduce the space $\widetilde{\mathcal{A}}_{(n)}^2(\Pi)$ of true-n-anti-analytic functions by

$$\widetilde{\mathcal{A}}_{(n)}^2(\Pi) = \widetilde{\mathcal{A}}_n^2(\Pi) \ominus \widetilde{\mathcal{A}}_{n-1}^2(\Pi),$$

for $n > 1$, and by $\widetilde{\mathcal{A}}_{(1)}^2(\Pi) = \widetilde{\mathcal{A}}_1^2(\Pi)$, for $n = 1$.

We have obviously

$$\mathcal{A}^2_n(\Pi) = \bigoplus_{k=1}^{n} \mathcal{A}^2_{(k)}(\Pi) \qquad \text{and} \qquad \widetilde{\mathcal{A}}^2_n(\Pi) = \bigoplus_{k=1}^{n} \widetilde{\mathcal{A}}^2_{(k)}(\Pi).$$

It is known as well (see, for example [16]) that

$$L_2(\Pi) = \bigoplus_{k=1}^{\infty} \mathcal{A}^2_{(k)}(\Pi) \oplus \bigoplus_{k=1}^{\infty} \widetilde{\mathcal{A}}^2_{(k)}(\Pi).$$

Theorem 3.1 ([9, 17]). *For all admissible indices, we have*

$$\begin{aligned}
(S_\Pi)^k|_{\mathcal{A}^2_{(n)}(\Pi)} &: \mathcal{A}^2_{(n)}(\Pi) \to \mathcal{A}^2_{(n+k)}(\Pi),\\
(S^*_\Pi)^k|_{\mathcal{A}^2_{(n)}(\Pi)} &: \mathcal{A}^2_{(n)}(\Pi) \to \mathcal{A}^2_{(n-k)}(\Pi),\\
(S_\Pi)^k|_{\widetilde{\mathcal{A}}^2_{(n)}(\Pi)} &: \widetilde{\mathcal{A}}^2_{(n)}(\Pi) \to \widetilde{\mathcal{A}}^2_{(n-k)}(\Pi),\\
(S^*_\Pi)^k|_{\widetilde{\mathcal{A}}^2_{(n)}(\Pi)} &: \widetilde{\mathcal{A}}^2_{(n)}(\Pi) \to \widetilde{\mathcal{A}}^2_{(n+k)}(\Pi).
\end{aligned}$$

Corollary 3.2 ([17]). *For all $n \in \mathbb{N}$ we have*

$$(S_\Pi)^n(S^*_\Pi)^n(S_\Pi)^n = (S_\Pi)^n \ \text{and}\ (S^*_\Pi)^n(S_\Pi)^n(S^*_\Pi)^n = (S^*_\Pi)^n.$$

Corollary 3.3. *For $n, m \in \mathbb{Z}_+$*

$$B_\Pi(S^*_\Pi)^m(S_\Pi)^n B_\Pi = \begin{cases} B_\Pi, & m = n \\ 0, & m \neq n \end{cases},$$

while for $n, m \in \mathbb{N}$

$$B_\Pi(S_\Pi)^n(S^*_\Pi)^m B_\Pi = 0,$$

where B_Π is the Bergman projection of $L_2(\Pi)$ onto the Bergman space $\mathcal{A}^2(\Pi)$.

4. Toeplitz operators on the Bergman space with defining symbol in $\mathcal{R}(\mathbb{C}; S_\Pi, S^*_\Pi)$

We describe here the algebra $\mathcal{T}_0 = \mathcal{T}_0(\mathcal{R}(\mathbb{C}; S_\Pi, S^*_\Pi)) = B_\Pi\mathcal{R}(\mathbb{C}; S_\Pi, S^*_\Pi)B_\Pi$, which is generated by all Toeplitz operators $T_A = B_\Pi A B_\Pi$ acting on the Bergman space $\mathcal{A}^2(\Pi)$ and with $A \in \mathcal{R}(\mathbb{C}; S_\Pi, S^*_\Pi)$.

Lemma 4.1. *Given $k_1, \dots, k_N, n_1, \dots, n_N \in \mathbb{Z}_+$ such that*

$$\sum_{i=1}^{m} n_i \leq \sum_{i=1}^{m} k_i \quad \text{for} \ \ m = 1, \dots, N-1 \tag{4.1}$$

and

$$\sum_{i=1}^{N} n_i = \sum_{i=1}^{N} k_i, \tag{4.2}$$

there exists $s \in \mathbb{Z}_+$ such that

$$(S^*_\Pi)^{n_N}(S_\Pi)^{k_N} \cdots (S^*_\Pi)^{n_1}(S_\Pi)^{k_1} = (S^*_\Pi)^s(S_\Pi)^s.$$

Proof. Given $k_1, \dots, k_N, n_1, \dots, n_N \in \mathbb{Z}_+$, which satisfy (4.1) and (4.2), it is easy to see that there is $j \in \{1, \dots, N-1\}$ such that

$$(k_j \geq n_j \ \ and \ \ k_{j+1} \geq n_j) \quad \text{or} \quad (n_j \geq k_{j+1} \ \ and \ \ n_{j+1} \geq k_{j+1}). \tag{4.3}$$

If $k_j \geq n_j$ and $k_{j+1} \geq n_j$, then by Corollary 3.2 we have that

$$\begin{aligned}
&(S^*_\Pi)^{n_N}(S_\Pi)^{k_N} \dots (S^*_\Pi)^{n_{j+1}}(S_\Pi)^{k_{j+1}}(S^*_\Pi)^{n_j}(S_\Pi)^{k_j} \dots (S^*_\Pi)^{n_1}(S_\Pi)^{k_1} \\
&\quad = (S^*_\Pi)^{n_N}(S_\Pi)^{k_N} \dots (S^*_\Pi)^{n_{j+1}}(S_\Pi)^{k_{j+1}-n_j}(S_\Pi)^{n_j}(S^*_\Pi)^{n_j}(S_\Pi)^{n_j}(S_\Pi)^{k_j-n_j} \\
&\qquad \times \dots \times (S^*_\Pi)^{n_1}(S_\Pi)^{k_1} \\
&\quad = (S^*_\Pi)^{n_N}(S_\Pi)^{k_N} \dots (S^*_\Pi)^{n_{j+1}}(S_\Pi)^{k_{j+1}+k_j-n_j}(S^*_\Pi)^{n_{j-1}}(S_\Pi)^{k_{j-1}} \dots (S^*_\Pi)^{n_1}(S_\Pi)^{k_1}.
\end{aligned}$$

If $n_j \geq k_{j+1}$ and $n_{j+1} \geq k_{j+1}$, then similarly by Corollary 3.2, we have that

$$\begin{aligned}
&(S^*_\Pi)^{n_N}(S_\Pi)^{k_N} \dots (S^*_\Pi)^{n_{j+1}}(S_\Pi)^{k_{j+1}}(S^*_\Pi)^{n_j}(S_\Pi)^{k_j} \dots (S^*_\Pi)^{n_1}(S_\Pi)^{k_1} \\
&\quad = (S^*_\Pi)^{n_N}(S_\Pi)^{k_N} \dots (S^*_\Pi)^{n_{j+1}-k_{j+1}}(S^*_\Pi)^{k_{j+1}}(S_\Pi)^{k_{j+1}}(S^*_\Pi)^{k_{j+1}}(S^*_\Pi)^{n_j-k_{j+1}}(S_\Pi)^{k_j} \\
&\qquad \times \dots \times (S^*_\Pi)^{n_1}(S_\Pi)^{k_1} \\
&\quad = (S^*_\Pi)^{n_N}(S_\Pi)^{k_N} \dots (S^*_\Pi)^{n_{j+1}+n_j-k_{j+1}}(S_\Pi)^{k_j}(S^*_\Pi)^{n_{j-1}}(S_\Pi)^{k_{j-1}} \dots (S^*_\Pi)^{n_1}(S_\Pi)^{k_1}.
\end{aligned}$$

Applying the above arguments inductively $(N-1)$-times we obtain the result. □

Given a multi-index $J = (n_1, k_1, \dots, n_N, k_N)$, where $n_i, k_i \in \mathbb{Z}_+$, we define the non-commutative monomial $m_J(x,y)$ by

$$m_J(x,y) = y^{n_N} x^{k_N} \cdots y^{n_1} x^{k_1}$$

and set its degree by

$$\deg m_J = |J| = n_N + k_N + \dots + n_1 + k_1.$$

The following corollary is a consequence of the above lemma and Corollary 3.3.

Corollary 4.2. *Let $m_J(x,y)$ be a non-commutative monomial, then*

$$B_\Pi m_J(S_\Pi, S^*_\Pi) B_\Pi = \begin{cases} B_\Pi, & \text{if } J \text{ satisfies to (4.1) and (4.2)} \\ 0, & \text{otherwise} \end{cases}.$$

Lemma 4.3. *Let $P(x,y)$ be a non-commutative polynomial of degree k*

$$P(x,y) = \sum_{|J| \leq k} a_J m_J(x,y),$$

where $a_J \in \mathbb{C}$. Then

$$B_\Pi P(S_\Pi, S^*_\Pi) B_\Pi = b_P B_\Pi,$$

where

$$b_P = \sum_{|J| \leq k, \ J \in I_{0,0}} a_J$$

and $I_{0,0}$ is the set of multi-indexes that satisfy (4.1) and (4.2).

Proof. We split the polynomial P as follows

$$P(x,y) = \sum_{|J|\le k,\ J\in I_{0,0}} a_J m_J(x,y) + \sum_{|J|\le k,\ J\notin I_{0,0}} a_J m_J(x,y)$$

and evaluate it on S_Π and S_Π^*

$$P(S_\Pi, S_\Pi^*) \quad = \quad \sum_{|J|\le k, J\in I_{0,0}} a_J m_J(S_\Pi, S_\Pi^*) + \sum_{|J|\le k, J\notin I_{0,0}} a_J m_J(S_\Pi, S_\Pi^*). \qquad (4.4)$$

Then by Corollary 3.2 we have

$$\begin{aligned} B_\Pi P(S_\Pi, S_\Pi^*) B_\Pi &= \sum_{\substack{|J|\le k,\\ J\in I_{0,0}}} a_J B_\Pi m_J(S_\Pi, S_\Pi^*) B_\Pi + \sum_{\substack{|J|\le k,\\ J\notin I_{0,0}}} a_J B_\Pi m_J(S_\Pi, S_\Pi^*) B_\Pi \\ &= b_P B_\Pi + 0 = b_P B_\Pi. \end{aligned}$$

□

Theorem 4.4. *Let A be an element of $\mathcal{R}(\mathbb{C}; S_\Pi, S_\Pi^*)$. Then the Toeplitz operator T_A acting on $\mathcal{A}^2(\Pi)$ is equal to $b_A B_\Pi$, where b_A is given by*

$$b_A = \langle A f_0, f_0 \rangle,$$

where f_0 is any function from $\mathcal{A}^2(\Pi)$ having norm 1.

Proof. The set of non-commutative polynomials $P(S_\Pi, S_\Pi^*)$ is dense in the algebra $\mathcal{R}(\mathbb{C}; S_\Pi, S_\Pi^*)$. By Lemma 4.3 we have

$$T_{P(S_\mathbb{D}, S_\mathbb{D}^*)} = B_\Pi P(S_\mathbb{D}, S_\mathbb{D}^*) B_\Pi = b_P B_\Pi$$

with $b_P \in \mathbb{C}$. On the other hand,

$$\begin{aligned} b_P &= \langle b_P B_\Pi f_0, f_0 \rangle = \langle T_{P(S_\mathbb{D}, S_\mathbb{D}^*)} f_0, f_0 \rangle = \langle B_\Pi P(S_\mathbb{D}, S_\mathbb{D}^*) B_\Pi f_0, f_0 \rangle \\ &= \langle P(S_\mathbb{D}, S_\mathbb{D}^*) f_0, f_0 \rangle. \end{aligned}$$

The functional $A \longmapsto \langle A f_0, f_0 \rangle$ is continuous on $\mathcal{R}(\mathbb{C}; S_\Pi, S_\Pi^*)$, thus the result follows. □

To get an alternative formula for b_A we proceed as follows. Analogously to Corollary 3.3 we have

Lemma 4.5. *For Toeplitz operators T_t and $T_{\bar{t}}$ from the algebra $\mathcal{T}(C(S^1))$ and for $n, m \in \mathbb{Z}_+$, we have*

$$(I - T_t T_{\bar{t}})\, T_{\bar{t}}^m\, T_t^n\, (I - T_t T_{\bar{t}}) = \begin{cases} I - T_t T_{\bar{t}}, & m = n \\ 0, & m \ne n \end{cases},$$

while for $n, m \in \mathbb{N}$

$$(I - T_t T_{\bar{t}})\, T_t^m\, T_{\bar{t}}^n\, (I - T_t T_{\bar{t}}) = 0.$$

We note that the operator $K_0 = I - T_t T_{\bar{t}}$ is the one-dimensional projection onto the subspace of $H^2(S^1)$ generated by 1, and that $I - T_{\bar{t}} T_t = 0$, which implies $(I - T_{\bar{t}} T_t)\, T\, (I - T_{\bar{t}} T_t) = 0$, for all $T \in \mathcal{T}(C(S^1))$.

The Bergman projection $B_\Pi = I - S_\Pi S_\Pi^*$ obviously belongs to the algebra $\mathcal{R}(\mathbb{C}; S_\Pi, S_\Pi^*)$, and by Corollary 2.4, under the isomorphic inclusion

$$\sigma : \mathcal{R}(\mathbb{C}; S_\Pi, S_\Pi^*) \longrightarrow \mathcal{T}(C(S^1)) \oplus \mathcal{T}(C(S^1)),$$

the image of the Bergman projection B_Π has the form $(K_0, 0)$.

Now, given $A \in \mathcal{R}(\mathbb{C}; S_\Pi, S_\Pi^*)$, consider the Toeplitz operator

$$T_A = (I - S_\Pi S_\Pi^*)A(I - S_\Pi S_\Pi^*) \in \mathcal{R}(\mathbb{C}; S_\Pi, S_\Pi^*).$$

Let $\sigma(A) = (\sigma_0(A), \sigma_\infty(A)) \in \mathcal{T}(C(S^1)) \oplus \mathcal{T}(C(S^1))$, then we have that $\sigma(T_A) = (K_0\, \sigma_0(A) K_0, 0)$, and thus, being considered as acting on the Bergman space $\mathcal{A}^2(\Pi)$, the Toeplitz operator T_A is scalar, $T_A = b_A I$, with the following alternative formula for b_A:

$$\begin{aligned} b_A &= \langle K_0\, \sigma_0(A) K_0 1, 1\rangle_{H^2(S^1)} = \langle \sigma_0(A) K_0 1, K_0\, 1\rangle_{H^2(S^1)} \\ &= \langle \sigma_0(A) 1, 1\rangle_{H^2(S^1)}. \end{aligned} \tag{4.5}$$

5. Toeplitz operators on the Bergman space with defining symbol in $\mathcal{R}(C(\overline{\mathbb{D}}); S_\mathbb{D}, S_\mathbb{D}^*)$

We consider here the C^*-algebra $\mathcal{T}(\mathcal{R}(C(\overline{\mathbb{D}}); S_\mathbb{D}, S_\mathbb{D}^*))$ which is generated by all Toeplitz operators of the form T_A with symbols $A \in \mathcal{R}(C(\overline{\mathbb{D}}); S_\mathbb{D}, S_\mathbb{D}^*)$ and acting on the Bergman space $\mathcal{A}^2(\mathbb{D})$ over the unit disk $\mathbb{D}$.

Given $A \in \mathcal{R}(C(\overline{\mathbb{D}}); S_\mathbb{D}, S_\mathbb{D}^*)$, consider its Fredholm symbol (see Theorem 2.5)

$$\operatorname{sym} A = \begin{cases} c(z,t) \in \mathbb{C}, & (z,t) \in \overline{\mathbb{D}} \times S^1 \\ \sigma_0(A,z) = T_{c(z,t)} + K_0(z) \in \mathcal{T}(C(S^1)), & (z,0) \in S^1 \times \{0,\infty\} \\ \sigma_\infty(A,z) = T_{c(z,\bar{t})} + K_\infty(z) \in \mathcal{T}(C(S^1)), & (z,\infty) \in S^1 \times \{0,\infty\} \end{cases}.$$

Combining the local description of Corollary 2.4 with formula (4.5) and the global description provided by Theorem 2.5, we arrive to the following result.

Theorem 5.1. *The Fredholm symbols algebra*

$$\operatorname{Sym} \mathcal{T}(\mathcal{R}(C(\overline{\mathbb{D}}); S_\mathbb{D}, S_\mathbb{D}^*)) = \mathcal{T}(\mathcal{R}(C(\overline{\mathbb{D}}); S_\mathbb{D}, S_\mathbb{D}^*))/\mathcal{K}$$

of the algebra $\mathcal{T}(\mathcal{R}(C(\overline{\mathbb{D}}); S_\mathbb{D}, S_\mathbb{D}^*))$ *is isomorphic and isometric to* $C(S^1)$*. Under their identification the symbol homomorphism*

$$\operatorname{sym} : \mathcal{T}(\mathcal{R}(C(\overline{\mathbb{D}}); S_\mathbb{D}, S_\mathbb{D}^*)) \longrightarrow C(S^1)$$

is generated by the following mapping of generators of $\mathcal{T}(\mathcal{R}(C(\overline{\mathbb{D}}); S_\mathbb{D}, S_\mathbb{D}^*))$*: for any* $A \in \mathcal{R}(C(\overline{\mathbb{D}}); S_\mathbb{D}, S_\mathbb{D}^*)$,

$$\operatorname{sym} T_A = b_A(z) = \langle \sigma_0(A,z)1, 1\rangle_{H^2(S^1)} \in C(S^1).$$

Remark 5.2. Given $A \in \mathcal{R}(C(\overline{\mathbb{D}}); S_\mathbb{D}, S_\mathbb{D}^*)$, denote by $\widetilde{b}_A(z)$ an extension of the function $b_A(z)$ continuous on S^1 to a function continuous on the closed unit disk $\overline{\mathbb{D}}$. Then the Toeplitz operator T_A with pseudodifferential symbol $A \in \mathcal{R}(C(\overline{\mathbb{D}}); S_\mathbb{D}, S_\mathbb{D}^*)$

is nothing but a compact perturbation of the Toeplitz operator $T_{\widetilde{b}_A(z)}$ whose symbol is a function continuous on $\overline{\mathbb{D}}$:

$$T_A = T_{\widetilde{b}_A(z)} + K,$$

where K is a compact operator.

That is, if fact, both classes of symbols $\mathcal{R}(C(\overline{\mathbb{D}}); S_{\mathbb{D}}, S_{\mathbb{D}}^*)$ and $C(\overline{\mathbb{D}})$ generate the same Toeplitz operator algebra:

$$\mathcal{T}(\mathcal{R}(C(\overline{\mathbb{D}}); S_{\mathbb{D}}, S_{\mathbb{D}}^*)) = \mathcal{T}(C(\overline{\mathbb{D}})).$$

Example. Consider the operator

$$A = \sum_{i,j=0}^{n} a_{ij}(z)(S_{\mathbb{D}}^*)^i(S_{\mathbb{D}})^j + \sum_{i,j=1}^{n} b_{ij}(z)(S_{\mathbb{D}})^i(S_{\mathbb{D}}^*)^j,$$

where $a_{ij}, b_{ij} \in C(\overline{\mathbb{D}})$. Its Fredholm symbol is given by $\operatorname{sym} T_A = \widetilde{a}(z) = \sum_{i=0}^{n} a_{ii}(z)$. Thus we have

$$T_A = T_{\widetilde{a}} + K$$

where K is a compact operator.

Example. Consider now

$$A = \sum_{i,j,m,k=0}^{n} a_{i,j,m,k}(z)(S_{\mathbb{D}}^*)^i(S_{\mathbb{D}})^j(S_{\mathbb{D}}^*)^m(S_{\mathbb{D}})^k,$$

where $a_{i,j,m,k} \in C(\overline{\mathbb{D}})$. The Fredholm symbol of T_A is given by

$$\operatorname{sym} T_A = \widetilde{a}(z) = \sum_{j,k=0}^{n} \left(\sum_{m=0}^{k} a_{j+k-m,j,m,k}(z) \right).$$

Thus we have

$$T_A = T_{\widetilde{a}} + K,$$

where K is a compact operator.

We mention that both algebras $\mathcal{T}(\mathcal{R}(C(\overline{\mathbb{D}}); S_{\mathbb{D}}, S_{\mathbb{D}}^*))$ and $\mathcal{T}(C(\overline{\mathbb{D}}))$ consist of the same operators, although their generators are different, being the Toeplitz operators with defining symbols from $\mathcal{R}(C(\overline{\mathbb{D}}); S_{\mathbb{D}}, S_{\mathbb{D}}^*)$ and from $C(\overline{\mathbb{D}})$, respectively. Moreover these generating Toeplitz operators possess quite different properties.

For example, the last algebra $\mathcal{T}(C(\overline{\mathbb{D}}))$ possesses the compact semi-commutator property:

$$[T_a, T_b) = T_a T_b - T_{ab} \in \mathcal{K}, \qquad \text{for all} \qquad a, b \in C(\overline{\mathbb{D}}),$$

while the first algebra $\mathcal{T}(\mathcal{R}(C(\overline{\mathbb{D}}); S_{\mathbb{D}}, S_{\mathbb{D}}^*))$ does not, i.e., the semi-commutator $[T_A, T_B) = T_A T_B - T_{AB}$ is not necessarily compact for each A and B from $\mathcal{R}(C(\overline{\mathbb{D}}); S_{\mathbb{D}}, S_{\mathbb{D}}^*)$.

Indeed, consider the semi-commutator $[T_{S_{\mathbb{D}}^*}, T_{S_{\mathbb{D}}}) = T_{S_{\mathbb{D}}^*} T_{S_{\mathbb{D}}} - T_{S_{\mathbb{D}}^* S_{\mathbb{D}}}$. Its compactness is equivalent to $\operatorname{sym} [T_{S_{\mathbb{D}}^*}, T_{S_{\mathbb{D}}}) = 0$. At the same time, we have

$$\sigma_0(S_{\mathbb{D}}^*, z) = T_{\overline{t}}, \qquad \sigma_0(S_{\mathbb{D}}, z) = T_t, \qquad \sigma_0(S_{\mathbb{D}}^* S_{\mathbb{D}}, z) = T_{\overline{t}} T_t = I,$$

and thus

$$\operatorname{sym}[T_{S_{\mathbb{D}}^*}, T_{S_{\mathbb{D}}}) = \langle T_{\bar{t}}1, 1\rangle_{H^2(S^1)} \cdot \langle T_t 1, 1\rangle_{H^2(S^1)} - \langle 1, 1\rangle_{H^2(S^1)} = 0\cdot 0 - 1 = -1.$$

That is, $[T_{S_{\mathbb{D}}^*}, T_{S_{\mathbb{D}}}) = -I + K$, for some compact operator K.

The exact form of K can be easily figured out. By [11, Lemma 2] we have that $T_{S_{\mathbb{D}}^*} = 0$ and $T_{S_{\mathbb{D}}} = 0$. It is well known (see, for example, [11, Lemma 1]) that the orthogonal projection $\widetilde{B}_{\mathbb{D}}$ of $L_2(\mathbb{D})$ onto the anti-analytic Bergman space $\widetilde{\mathcal{A}}^2(\mathbb{D})$ has the form $\widetilde{B}_{\mathbb{D}} = I - S_{\mathbb{D}}^* S_{\mathbb{D}}$. Then by [3], we have

$$B_{\mathbb{D}}\widetilde{B}_{\mathbb{D}}|_{\mathcal{A}^2(\mathbb{D})} = T_{I - S_{\mathbb{D}}^* S_{\mathbb{D}}} = I - T_{S_{\mathbb{D}}^* S_{\mathbb{D}}} = K_{\ell_1},$$

where

$$(K_{\ell_1}\varphi)(z) = \frac{1}{\pi}\int_{\mathbb{D}} \varphi(\zeta)dv(\zeta) = \langle \varphi, \ell_1\rangle \ell_1$$

is the one-dimensional projection of $\mathcal{A}^2(\mathbb{D})$ onto the one-dimensional space L_1 generated by the first element of the standard orthonormal monomial basis $\ell_k(z) = \sqrt{\frac{k}{\pi}}z^{k-1}$, $k \in \mathbb{N}$, of $\mathcal{A}^2(\mathbb{D})$. Thus finally,

$$[T_{S_{\mathbb{D}}^*}, T_{S_{\mathbb{D}}}) = T_{S_{\mathbb{D}}^*}T_{S_{\mathbb{D}}} - T_{S_{\mathbb{D}}^* S_{\mathbb{D}}} = -I + K_{\ell_1}.$$

The above suggests two observations. First, contrary to the case of Toeplitz operators with defining symbols from $C(\overline{\mathbb{D}})$, the Toeplitz operator T_A, with $A \in \mathcal{R}(C(\overline{\mathbb{D}}); S_{\mathbb{D}}, S_{\mathbb{D}}^*)$ can be the zero-operator for non zero A. Two examples, $T_{S_{\mathbb{D}}^*} = 0$ and $T_{S_{\mathbb{D}}} = 0$, have been just considered. We note that, in particular, such symbols A are those whose kernel contains $\mathcal{A}^2(\mathbb{D})$ or those for which the image of their restriction on $\mathcal{A}^2(\mathbb{D})$ is orthogonal to $\mathcal{A}^2(\mathbb{D})$.

The second observation is as follows. By a result of D. Luecking [13], there are no symbols $a \in C(\overline{\mathbb{D}})$ such that the Toeplitz operator T_a has a finite rank. At the same time, such symbols $A \in \mathcal{R}(C(\overline{\mathbb{D}}); S_{\mathbb{D}}, S_{\mathbb{D}}^*)$ do exist. As we have just shown,

$$T_{I - S_{\mathbb{D}}^* S_{\mathbb{D}}} = K_{\ell_1} \in \mathcal{T}(C(\overline{\mathbb{D}})) = \mathcal{T}(\mathcal{R}(C(\overline{\mathbb{D}}); S_{\mathbb{D}}, S_{\mathbb{D}}^*))$$

is a rank one operator. Similarly, by [10, Theorem 2.3, Lemma 3.3], we have

$$T_{(S_{\mathbb{D}}^*)^{k-1}(S_{\mathbb{D}})^{k-1} - (S_{\mathbb{D}}^*)^k (S_{\mathbb{D}})^k} = K_{\ell_k}, \qquad \text{for all} \quad k \in \mathbb{N},$$

where $K_{\ell_k}\varphi = \langle \varphi, \ell_k\rangle \ell_k$ is the one-dimensional projection onto the one-dimensional subspace generated by $\ell_k(z) = \sqrt{\frac{k}{\pi}}z^{k-1}$.

Further, to cover a set of finite rank operators, which is dense in the set $\mathcal{K}$ of all compact operators in $\mathcal{A}^2(\mathbb{D})$, it is sufficient to add the rank one operators of the form $K_{\ell_k,\ell_l}\varphi = \langle \varphi, \ell_k\rangle \ell_l$, where $k, l \in \mathbb{N}$, and then consider all their linear combinations. These last operators (for $k \neq l$) are just the products of two Toeplitz operators:

$$K_{\ell_k,\ell_l} = \begin{cases} K_{\ell_l} T_{\sqrt{\frac{k}{l}}\bar{z}^{k-l}} = T_{(S_{\mathbb{D}}^*)^{l-1}(S_{\mathbb{D}})^{l-1} - (S_{\mathbb{D}}^*)^l (S_{\mathbb{D}})^l} T_{\sqrt{\frac{k}{l}}\bar{z}^{k-l}}, & k > l, \\ K_{\ell_k} = T_{(S_{\mathbb{D}}^*)^{k-1}(S_{\mathbb{D}})^{k-1} - (S_{\mathbb{D}}^*)^k (S_{\mathbb{D}})^k}, & k = l, \\ T_{\sqrt{\frac{l}{k}}z^{l-1}} K_{\ell_k} = T_{\sqrt{\frac{l}{k}}z^{l-1}} T_{(S_{\mathbb{D}}^*)^{k-1}(S_{\mathbb{D}})^{k-1} - (S_{\mathbb{D}}^*)^k (S_{\mathbb{D}})^k}, & k < l. \end{cases}$$

6. Toeplitz operators on the poly-Bergman space with defining symbol in $\mathcal{R}(\mathbb{C}; S_\Pi, S_\Pi^*)$

Now we describe the algebra $B_{\Pi,n}\mathcal{R}(\mathbb{C}; S_\Pi, S_\Pi^*)B_{\Pi,n}$. This algebra is generated by Toeplitz operators acting on the poly-Bergman space over Π and with defining symbols from $\mathcal{R}(\mathbb{C}; S_\Pi, S_\Pi^*)$.

Recall that the poly-Bergman space admits the representation

$$\mathcal{A}_n^2(\Pi) = \bigoplus_{m=1}^{n} \mathcal{A}_{(m)}^2(\Pi),$$

where $\mathcal{A}_{(m)}^2(\Pi)$ is the subspace of the true-m-analytic functions.

By Theorem 3.1, the operator $(S_\Pi)^k : \mathcal{A}_{(n)}^2(\Pi) \to \mathcal{A}_{(n+k)}^2(\Pi)$ is unitary, thus we define

$$U \; : \; \bigoplus_{m=1}^{n} \mathcal{A}^2(\Pi) \longrightarrow \mathcal{A}_n^2(\Pi) \tag{6.1}$$

as follows

$$U(\phi_1, \cdots, \phi_n) = \phi_1 + (S_\Pi)(\phi_2) + \ldots + (S_\Pi)^{n-1}(\phi_n),$$

the adjoint operator

$$U^* \; : \; \mathcal{A}_n^2(\Pi) \longrightarrow \bigoplus_{m=1}^{n} \mathcal{A}^2(\Pi) \tag{6.2}$$

is given by

$$U^*\psi = (B_\Pi\psi, \ldots, B_\Pi(S_\Pi^*)^{n-1}\psi).$$

Theorem 6.1. *Let A be an element of $\mathcal{R}(\mathbb{C}; S_\Pi, S_\Pi^*)$. Then the Toeplitz operator $T_{A,n} = B_{\Pi,n}AB_{\Pi,n}$ acting on $\mathcal{A}_n^2(\Pi)$ is unitarily equivalent to the matrix operator $M_{A,n} = U^*T_{A,n}U$ acting on $\bigoplus_{m=1}^{n} \mathcal{A}^2(\Pi)$, where U and U^* are given by* (6.1) *and* (6.2) *respectively. The entries of the matrix-operator $M_{A,n}$ are given by*

$$M_{A,n}(i,j) = T_{(S_\Pi^*)^{i-1}A(S_\Pi)^{j-1}}, \qquad i,j = 1,2,\ldots,n,$$

where $T_{(S_\Pi^*)^{i-1}A(S_\Pi)^{j-1}}$ is a Toeplitz operator acting on $\mathcal{A}^2(\Pi)$ and with defining symbol $(S_\Pi^*)^{i-1}A(S_\Pi)^{j-1}$.

Proof. We start with (see, for example, [17])

$$B_{\Pi,n} = \sum_{m=1}^{n} B_{\Pi,(m)} \quad \text{and} \quad B_{\Pi,(m)} = (S_\Pi)^{m-1}B_\Pi(S_\Pi^*)^{m-1},$$

where $B_{\Pi,(m)}$ is the orthogonal projection onto the space $\mathcal{A}_{(m)}^2(\Pi)$ of true-m-analytic functions.

Let us calculate $U^*T_{A,n}U$. For $\phi = (\phi_1, \dots, \phi_n)$, we have

$$\begin{aligned}
M_{A,n}\phi &= U^*T_{A,n}U(\phi_1, \dots, \phi_n)\\
&= U^*B_{\Pi,n}(A(\phi_1) + A((S_\Pi)(\phi_2)) + \dots + A((S_\Pi)^{n-1}(\phi_n)))\\
&= U^*\left(\sum_{m=1}^{n} B_{\Pi,(m)}\right)\left(\sum_{j=1}^{n} A((S_\Pi)^{j-1}(\phi_j))\right)\\
&= U^*\left(\sum_{m=1}^{n} (S_\Pi)^{m-1}B_\Pi(S_\Pi^*)^{m-1}\right)\left(\sum_{j=1}^{n} A((S_\Pi)^{j-1}(\phi_j))\right)\\
&= U^*\left(\sum_{m=1}^{n}\sum_{j=1}^{n}((S_\Pi)^{m-1}B_\Pi(S_\Pi^*)^{m-1}A(S_\Pi)^{j-1})(\phi_j)\right)\\
&= \left(B_\Pi(S_\Pi^*)^{i-1}\left(\sum_{m=1}^{n}\sum_{j=1}^{n}((S_\Pi)^{m-1}B_\Pi(S_\Pi^*)^{m-1}A(S_\Pi)^{j-1}B_\Pi)(\phi_j)\right)\right)_{i=1}^{n}\\
&= \left(\sum_{m=1}^{n}\sum_{j=1}^{n}(B_\Pi(S_\Pi^*)^{i-1}(S_\Pi)^{m-1}B_\Pi(S_\Pi^*)^{m-1}A(S_\Pi)^{j-1}B_\Pi)(\phi_j)\right)_{i=1}^{n}.
\end{aligned}$$

By Corollary 3.3 we obtain

$$\begin{aligned}
U^*T_{A,n}U(\phi) &= \left(\sum_{j=1}^{n}(B_\Pi(S_\Pi^*)^{i-1}A(S_\Pi)^{j-1}B_\Pi)(\phi_j)\right)_{i=1}^{n}\\
&= \left(\sum_{j=1}^{n} T_{(S_\Pi^*)^{i-1}A(S_\Pi)^{j-1}}(\phi_j)\right)_{i=1}^{n}\\
&= \left(T_{(S_\Pi^*)^{i-1}A(S_\Pi)^{j-1}}\right)_{i,j=1}^{n}\phi. \qquad \square
\end{aligned}$$

Since $T_{(S_\Pi^*)^{i-1}A(S_\Pi)^{j-1}} \in B_\Pi\mathcal{R}(\mathbb{C}; S_\Pi, S_\Pi^*)B_\Pi$, by Theorem 4.4 we have that $T_{(S_\Pi^*)^{i-1}A(S_\Pi)^{j-1}} = m_{i,j}(A)B_\Pi$, where $m_{i,j}(A) \in \mathbb{C}$. Hence we have the following corollary.

Corollary 6.2. *Let A be an element of $\mathcal{R}(\mathbb{C}; S_\Pi, S_\Pi^*)$. Then the Toeplitz operator $T_{A,n}$ acting on $\mathcal{A}_n^2(\Pi)$ is unitarily equivalent to the matrix $M_{A,n} = U^*T_{A,n}U$ acting on $\bigoplus_{m=1}^{n}\mathcal{A}^2(\Pi)$, where U and U^* are given by (6.1) and (6.2) respectively. The entries of the matrix $M_{A,n}$ are given by*

$$M_{A,n}(i,j) = m_{i,j}(A)B_\Pi, \qquad i,j = 1,2,\dots,n,$$

where

$$m_{i,j}(A) = \langle AS_\Pi^{j-1}f_0, S_\Pi^{i-1}f_0\rangle,$$

and f_0 is any function from $\mathcal{A}^2(\Pi)$ having norm 1.

Moreover, the operator $T_{A,n} = UM_{A,n}U^*$ *has the following form*

$$T_{A,n} = \sum_{i,j=1}^{n} m_{i,j}(A)(S_\pi)^{i-1} B_\Pi (S_\Pi^*)^{j-1}.$$

To get an alternative formula for $m_{i,j}(A)$ we proceed as follows. Analogously to Corollary 3.3 and Lemma 4.5 we have

Lemma 6.3. *For Toeplitz operators* T_t *and* $T_{\bar{t}}$ *from the algebra* $\mathcal{T}(C(S^1))$ *and for* $i, j, n, m \in \mathbb{Z}_+$, *we have*

$$T_t^i\,(I - T_tT_{\bar{t}})\,T_{\bar{t}}^i\; T_{\bar{t}}^m\,T_t^n\,T_t^j\,(I - T_tT_{\bar{t}})T_{\bar{t}}^j = \begin{cases} T_t^i(I - T_tT_{\bar{t}})T_{\bar{t}}^j, & i+m = n+j \\ 0, & i+m \neq n+j \end{cases},$$

while

$$\begin{aligned} &T_t^i\,(I - T_tT_{\bar{t}})\,T_{\bar{t}}^i\; T_t^nT_{\bar{t}}^m\; T_t^j\,(I - T_tT_{\bar{t}})T_{\bar{t}}^j \\ &\quad = \begin{cases} T_t^i(I - T_tT_{\bar{t}})T_{\bar{t}}^j, & i+m = n+j \text{ and } m \le j \\ 0, & i+m \neq n+j \text{ or } m > j \end{cases}. \end{aligned}$$

We note that the operator $K_n = I - T_t^nT_{\bar{t}}^n$ is the n-dimensional projection onto the subspace of $H^2(S^1)$ generated by $1, t, \dots, t^{n-1}$, and that $I - T_{\bar{t}}^nT_t^n = 0$, which implies $(I - T_{\bar{t}}^nT_t^n)\,T\,(I - T_{\bar{t}}^nT_t^n) = 0$, for all $T \in \mathcal{T}(C(S^1))$.

The Bergman projection $B_{\Pi,n} = I - (S_\Pi)^n(S_\Pi^*)^n$ belongs to the algebra $\mathcal{R}(\mathbb{C}; S_\Pi, S_\Pi^*)$, and by Corollary 2.4, under the isomorphic inclusion

$$\sigma:\; \mathcal{R}(\mathbb{C}; S_\Pi, S_\Pi^*) \longrightarrow \mathcal{T}(C(S^1)) \oplus \mathcal{T}(C(S^1)),$$

the image of the Bergman projection $B_{\Pi,n}$ has the form $(K_n, 0)$.

Now, given $A \in \mathcal{R}(\mathbb{C}; S_\Pi, S_\Pi^*)$, consider the Toeplitz operator

$$T_A = (I - (S_\Pi)^n(S_\Pi^*)^n)A(I - (S_\Pi)^n(S_\Pi^*)^n) \in \mathcal{R}(\mathbb{C}; S_\Pi, S_\Pi^*).$$

Let $\sigma(A) = (\sigma_0(A), \sigma_\infty(A)) \in \mathcal{T}(C(S^1)) \oplus \mathcal{T}(C(S^1))$, then we have $\sigma(T_{A,n}) = (K_n\,\sigma_0(A)K_n, 0)$, and thus, being considered as acting on the poly-Bergman space $\mathcal{A}_n^2(\Pi)$, the Toeplitz operator $T_{A,n}$ has the form

$$T_{A,n} = \sum_{i,j=1}^{n} m_{i,j}(A)(S_\Pi)^{i-1} B_\Pi (S_\Pi^*)^{j-1},$$

with the following alternative formula for $m_{i,j}(A)$:

$$\begin{aligned} m_{i,j}(A) &= \langle K_n\,\sigma_0(A)K_nt^{j-1}, t^{i-1}\rangle_{H^2(S^1)} = \langle \sigma_0(A)K_nt^{j-1}, K_nt^{i-1}\rangle_{H^2(S^1)} \\ &= \langle \sigma_0(A)t^{j-1}, t^{i-1}\rangle_{H^2(S^1)}. \end{aligned} \tag{6.3}$$

7. Toeplitz operators on the poly-Bergman space with defining symbols in $\mathcal{R}(C(\overline{\mathbb{D}}); S_{\mathbb{D}}, S_{\mathbb{D}}^*)$

Given $A \in \mathcal{R}(C(\overline{\mathbb{D}}); S_{\mathbb{D}}, S_{\mathbb{D}}^*)$, we consider the Toeplitz operator on the poly-Bergman space $\mathcal{A}_n^2(\mathbb{D})$ with pseudodifferential defining symbols A:

$$T_{A,n} : \mathcal{A}_n^2(\mathbb{D}) \longrightarrow \mathcal{A}_n^2(\mathbb{D}),$$

$$\varphi \longmapsto B_{\mathbb{D},n}(A\varphi),$$

where $B_{\mathbb{D},n}$ is the orthogonal projection of $L_2(\mathbb{D})$ onto the poly-Bergman space $\mathcal{A}_n^2(\mathbb{D})$.

Introduce now the C^*-algebra $\mathcal{T}_n = \mathcal{T}_n(\mathcal{R}(C(\overline{\mathbb{D}}); S_{\mathbb{D}}, S_{\mathbb{D}}^*))$ which is generated by all Toeplitz operators $T_{A,n}$ with symbols $A \in \mathcal{R}(C(\overline{\mathbb{D}}); S_{\mathbb{D}}, S_{\mathbb{D}}^*)$. Given $A \in \mathcal{R}(C(\overline{\mathbb{D}}); S_{\mathbb{D}}, S_{\mathbb{D}}^*)$, consider its Fredholm symbol (see Theorem 2.5):

$$\operatorname{sym} A = \begin{cases} c(z,t) \in \mathbb{C}, & (z,t) \in \overline{\mathbb{D}} \times S^1 \\ \sigma_0(A,z) = T_{c(z,t)} + K_0(z) \in \mathcal{T}(C(S^1)), & (z,0) \in S^1 \times \{0,\infty\} \\ \sigma_\infty(A,z) = T_{c(z,\bar{t})} + K_\infty(z) \in \mathcal{T}(C(S^1)), & (z,\infty) \in S^1 \times \{0,\infty\}. \end{cases}$$

Combining the local description of Corollary 2.4 with formula (6.3) and the global description provided by Theorem 2.5, we arrive to the following result.

Theorem 7.1. *The Fredholm symbols algebra*

$$\operatorname{Sym} \mathcal{T}_n(\mathcal{R}(C(\overline{\mathbb{D}}); S_{\mathbb{D}}, S_{\mathbb{D}}^*)) = \mathcal{T}_n(\mathcal{R}(C(\overline{\mathbb{D}}); S_{\mathbb{D}}, S_{\mathbb{D}}^*))/\mathcal{K}$$

of the algebra $\mathcal{T}_n(\mathcal{R}(C(\overline{\mathbb{D}}); S_{\mathbb{D}}, S_{\mathbb{D}}^*))$ is isomorphic and isometric to the algebra $M_{n\times n}(C(S^1))$ of all $n \times n$ matrix-functions continuous on S^1. Under their identification the symbol homomorphism

$$\operatorname{sym} : \mathcal{T}_n(\mathcal{R}(C(\overline{\mathbb{D}}); S_{\mathbb{D}}, S_{\mathbb{D}}^*)) \longrightarrow M_{n\times n}(C(S^1))$$

is generated by the following mapping of generators of $\mathcal{T}_n(\mathcal{R}(C(\overline{\mathbb{D}}); S_{\mathbb{D}}, S_{\mathbb{D}}^*))$: for any $A \in \mathcal{R}(C(\overline{\mathbb{D}}); S_{\mathbb{D}}, S_{\mathbb{D}}^*)$,

$$\begin{aligned} \operatorname{sym} T_{A,n} &= \{m_{i,j}[A](z)\}_{i,j=1}^n \\ &= \{\langle \sigma_0(A,z)t^{j-1}, t^{i-1}\rangle_{H^2(S^1)}\}_{i,j=1}^n \in M_{n\times n}(C(S^1)). \end{aligned}$$

Remark 7.2. Given $A \in \mathcal{R}(C(\overline{\mathbb{D}}); S_{\mathbb{D}}, S_{\mathbb{D}}^*)$, denote by $\widetilde{m}_{i,j}[A](z)$ an extension of the function $m_{i,j}[A](z)$ continuous on S^1 to a function continuous on the closed unit disk $\overline{\mathbb{D}}$. Then the Toeplitz operator $T_{A,n}$ with pseudodifferential symbol $A \in \mathcal{R}(C(\overline{\mathbb{D}}); S_{\mathbb{D}}, S_{\mathbb{D}}^*)$ is nothing but a compact perturbation of the Toeplitz operator $T_{A_1,n}$ whose defining symbol has the form

$$A_1 = \sum_{i,j=1}^n \widetilde{m}_{i,j}[A](z)(S_{\mathbb{D}})^{i-1} B_{\mathbb{D}} (S_{\mathbb{D}}^*)^{j-1},$$

where $\widetilde{m}_{i,j}[A](z)$ are the above functions continuous on $\overline{\mathbb{D}}$. We calculate now the Fredholm symbol $\operatorname{sym} T_{A_1,n}$.

$$\begin{aligned}
\operatorname{sym} T_{A_1,n} &= \left(\left\langle \sum_{l,s=1}^{n} \widetilde{m}_{l,s}[A](z)(T_t)^{l-1}(1-T_tT_{\bar t})(T_{\bar t})^{s-1}t^{j-1}, t^{i-1} \right\rangle_{H^2(S^1)} \right)_{i,j=1}^{n} \\
&= \left(\sum_{l,s=1}^{n} \widetilde{m}_{l,s}[A](z) \left\langle (1-T_tT_{\bar t})(T_{\bar t})^{s-1}t^{j-1}, (T_{\bar t})^{l-1}t^{i-1} \right\rangle_{H^2(S^1)} \right)_{i,j=1}^{n} \\
&= \left(\sum_{l=1}^{n} \widetilde{m}_{l,j}[A](z) \left\langle 1, t^{i-l} \right\rangle_{H^2(S^1)} \right)_{i,j=1}^{n} \\
&= (\widetilde{m}_{i,j}[A](z))_{i,j=1}^{n} = \operatorname{sym} T_{A,n}.
\end{aligned}$$

Therefore, from the above equation we have that

$$T_{A,n} = T_{A_1,n} + K,$$

where K is a compact operator.

At the same time the Toeplitz operator $T_{A,n}$ is unitarily equivalent (via U of the form (6.1)) to the following matrix-operator

$$T_{\widetilde{M}[A](z)} + K \in \mathcal{T}(C(\overline{\mathbb{D}})) \otimes M_{n\times n}(\mathbb{C}),$$

where K is compact, and $\widetilde{M}[A](z) = \{\widetilde{m}_{i,j}[A](z)\}_{i,j=1}^{n} \in M_{n\times n}(C(\overline{\mathbb{D}}))$. Moreover, contrary to the Bergman space case, for $n \geq 2$ the algebra $\mathcal{T}_n(\mathcal{R}(C(\overline{\mathbb{D}}); S_{\mathbb{D}}, S_{\mathbb{D}}^*))$ is unitarily equivalent to the matrix algebra $\mathcal{T}(C(\overline{\mathbb{D}})) \otimes M_{n\times n}(\mathbb{C})$.

Remark 7.3. Let us consider the operator

$$A_0 = \sum_{i,j=1}^{n} a_{i,j}(z)(S_{\mathbb{D}})^{i-1}(S_{\mathbb{D}}^*)^{j-1},$$

and calculate the Fredholm symbol $\operatorname{sym} T_{A_0,n}$,

$$\begin{aligned}
\operatorname{sym} T_{A_0,n} &= \left(\left\langle \sum_{l,s=1}^{n} a_{l,s}(z)(T_t)^{l-1}(T_{\bar t})^{s-1}t^{j-1}, t^{i-1} \right\rangle_{H^2(S^1)} \right)_{i,j=1}^{n} \\
&= \left(\sum_{l=1}^{i} \sum_{s=1}^{j} a_{l,s}(z) \langle t^{j-s}, t^{i-l} \rangle_{H^2(S^1)} \right)_{i,j=1}^{n} = \left(\sum_{l=1}^{i} \sum_{s=1}^{j} a_{l,s}(z)\delta_{j-s,i-l} \right)_{i,j=1}^{n} \\
&= \left(\sum_{u=0}^{i-1} \sum_{k=0}^{j-1} a_{i-u,j-k}(z)\delta_{k,u} \right)_{i,j=1}^{n} = \left(\sum_{k=0}^{min\{i,j\}-1} a_{i-k,j-k}(z) \right)_{i,j=1}^{n},
\end{aligned}$$

here $\delta_{k,u}$ is the Kronecker symbol.

Consider now the functions $\widetilde{m}_{i,j}[A](z)$ as in the above remark. If we take

$$a_{i,j}(z) = \widetilde{m}_{i,j}[A](z) - \widetilde{m}_{i-1,j-1}[A](z),$$

where $\widetilde{m}_{i,j}[A](z) = 0$ if $i \leq 0$ or $j \leq 0$, then

$$\operatorname{sym} T_{A_0,n} = \{\widetilde{m}_{i,j}[A](z)\}_{i,j=1}^{n} = \operatorname{sym} T_{A,n}.$$

Thus we have that

$$T_{A,n} = T_{A_0,n} + K,$$

where K is compact operator.

References

[1] Harry Bateman and Arthur Erdélyi, *Higher Transcendental Functions.* Vol. 2, McGraw-Hill, 1954.

[2] Louis Boutet de Monvel and Victor Guillemin, *The Spectral Theory of Toeplitz Operators.* Priceton University Press, 1985.

[3] Boo Rim Choe and Young Joo Lee, *Commuting Toeplitz operators on the harmonic Bergman space.* Michigan Math. J. **46** (1999), no. 1, 163–174.

[4] L.A. Coburn, *The C^*-algebra generated by an isometry.* Bull. Amer. Math. Soc. **73** (1967), 722–726.

[5] L.A. Coburn, *The C^*-algebra generated by an isometry. II.* Trans. Amer. Math. Soc. **137** (1969), 211–217.

[6] L.A. Coburn, *Singular integral operators and Toeplitz operators on odd spheres.* Indiana Univ. Math. J. **23** (1973), no. 5, 433–439.

[7] Ronald G. Douglas, *Banach Algebra Techniques in Operator Theory.* Academic Press, 1972.

[8] Victor Guillemin, *Toeplitz operators in n dimensions.* Integr. Equat. Oper. Th. **7** (1984), no. 2, 145–205.

[9] Yu.I. Karlovich and Luís Pessoa, *C^*-algebras of Bergman type operators with piecewise continuous coefficients.* Integr. Equat. Oper. Th. **57** (2007), no. 4, 521–565.

[10] Yuri I. Karlovich and Luís V. Pessoa, *Poly-Bergman projections and orthogonal decompositions of L^2-spaces over bounded domains.* In: Operator algebras, operator theory and applications, Oper. Theory Adv. Appl., vol. 181, pp. 263–282. Birkhäuser, 2008.

[11] I.I. Komyak, *Conditions for Noethericity and a formula for the index of a class of singular integral equations over a circular domain.* Diff. Uravneniya **16** (1980), no. 2, 328–343 (Russian). English translation: Differential Equations **16** (1980), no. 2, 215–225.

[12] Bernard Lascar and Richard Lascar, *Un lien explicite entre les opérateurs pseudo-différentiels et les opérateurs de Toeplitz.* C. R. Acad. Sci. Paris Sér. I Math. **300** (1985), no. 11, 339–341.

[13] Daniel H. Luecking, *Finite rank Toeplitz operators on the Bergman space.* Proc. Amer. Math. Soc. **136** (2008), no. 5, 717–723.

[14] J. Varela, *Duality of C^*-algebras.* Memories Amer. Math. Soc. **148** (1974), 97–108.

[15] N.L. Vasilevski, *An algebra generated by Toeplitz operators with zero-order pseudodifferential presymbols.* Dokl. Akad. Nauk SSSR **289** (1986), no. 1, 14–18.

[16] N.L. Vasilevski, *On the structure of Bergman and poly-Bergman spaces.* Integr. Equat. Oper. Th. **33** (1999), 471–488.

[17] Nikolai Vasilevski, *Poly-Bergman spaces and two-dimensional singular integral operators.* In: *The extended field of operator theory*, Oper. Theory Adv. Appl., vol. 171, pp. 349–359. Birkhäuser, 2007.

[18] Nikolai Vasilevski, *Commutative Algebras of Toeplitz Operators on the Bergman Space.* Birkhäuser Verlag, 2008.

Armando Sánchez-Nungaray
Facultad de Matemáticas
Universidad Veracruzana
Lomas del estadio s/n, Zona Universitaria
Xalapa, Veracruz, México
e-mail: armsanchez@uv.mx

Nikolai Vasilevski
Departamento de Matemáticas
CINVESTAV del I.P.N.
Av. IPN 2508, Col. San Pedro Zacatenco
México D.F. 07360, México
e-mail: nvasilev@math.cinvestav.mx

Operator Theory:
Advances and Applications, Vol. 228, 375–390

Finite Sections of Band-dominated Operators – Norms, Condition Numbers and Pseudospectra

Markus Seidel and Bernd Silbermann

Dedicated to our friend and colleague Vladimir Rabinovich on the occasion of his seventieth birthday

Abstract. This paper is devoted to the asymptotic behavior of the norms, condition numbers and pseudospectra of the finite sections of band-dominated operators acting on the spaces $l^p(\mathbb{Z}, X)$ with $1 \le p \le \infty$ and X being a Banach space.

Mathematics Subject Classification (2010). 65J10; 47A58.

Keywords. Operator sequence, condition numbers, pseudospectrum, band-dominated operator, finite section method.

Introduction

Let X be a Banach space and $l^p = l^p(\mathbb{Z}, X)$ denote the set of all p-summable sequences $(x_i)_{i\in\mathbb{Z}} \subset X$, that is all functions $x : \mathbb{Z} \to X, i \mapsto x_i$ with $\sum_{i\in\mathbb{Z}} \|x_i\|^p < \infty$. Provided with pointwise defined operations and the norm $\|(x_i)\| := \left(\sum_{i\in\mathbb{Z}} \|x_i\|^p\right)^{\frac{1}{p}}$, l^p becomes a Banach space for every $1 \le p < \infty$. Analogously, one introduces the space $l^\infty = l^\infty(\mathbb{Z}, X)$ of all bounded sequences $(x_i) \subset X$ with the norm $\|(x_i)\| := \sup_{i\in\mathbb{Z}} \|x_i\|$.

Define the linear and bounded shift operators V_k and the operators aI of multiplication with a function $a \in l^\infty(\mathbb{Z}, \mathcal{L}(X))$ by the rules

$$V_k(x_i) := (x_{i-k}) \quad \text{and} \quad (aI)(x_i) := (a_i x_i), \quad \text{respectively.}$$

Here $\mathcal{L}(X)$ denotes the Banach algebra of all bounded linear operators on X. Every finite sum of the form $\sum a_k V_k$ is said to be banded and all operators which are limits of sequences of band operators in the operator norm are referred to as band-dominated operators. Notice that the set $\mathcal{A}_{l^p}$ of all band-dominated operators depends on p but always forms a closed and inverse closed subalgebra of $\mathcal{L}(l^p)$ (cf. [12, Section 2.1]).

Let $\mathcal{P} = (L_n)_{n\in\mathbb{N}}$ be the sequence of the projections $L_n := \chi_{\{-n,\dots,n\}}I$, where χ_U stands for the characteristic function of the set U. In [11, 12] and [18] the Fredholm properties and the finite sections L_nAL_n of band-dominated operators A have been of particular interest. For this, in the latter paper the authors introduce certain sequence algebras which contain the finite section sequences, as well as homomorphisms on these algebras which provide snapshots of a given sequence carrying crucial information on its asymptotic behavior. More precisely, this setting is as follows:

Firstly, let $\mathcal{F}$ denote the Banach algebra of all bounded sequences $\mathbb{A} = \{A_n\}$ of bounded linear operators $A_n \in \mathcal{L}(\operatorname{im} L_n)$ equipped with entry-wise defined operations and the norm $\|\mathbb{A}\| := \sup_n \|A_n\|$. The sequence $\mathbb{A}$ is said to be stable if $\limsup_n \|A_n^{-1}\| < \infty$, where we set $\|A_n^{-1}\| := \infty$ if A_n is not invertible. It is well known that stability is equivalent to invertibility of the coset $\mathbb{A}+\mathcal{G}$ in the quotient $\mathcal{F}/\mathcal{G}$, where $\mathcal{G}$ stands for the closed ideal in $\mathcal{F}$ of all sequences $\{G_n\}$ with $\|G_n\| \to 0$ as $n \to \infty$. If $h = (h_n)_{n\in\mathbb{N}}$ is a strictly increasing sequence of positive integers then we analogously define $\mathcal{F}_h$ and $\mathcal{G}_h$ as algebras of (sub)sequences $\mathbb{A}_h = \{A_{h_n}\}$.

Of course, the finite section sequence $\{L_nAL_n\}$ of a band-dominated operator always belongs to $\mathcal{F}$ and we let $\mathcal{F}_{\mathcal{A}_{l^p}}$ denote the smallest closed subalgebra of $\mathcal{F}$ containing all of these $\{L_nAL_n\}$ with so-called rich band-dominated operators (see Definition 1.3). Due to the special structure of the elements $\mathbb{A} = \{A_n\} \in \mathcal{F}_{\mathcal{A}_{l^p}}$ one easily finds that the sequence (A_nL_n) converges in $\mathcal{L}(l^p)$ in the sense of $\mathcal{P}$-strong convergence (a precise definition will also be given later on). Denote the limit by $W^0(\mathbb{A})$. Further, at least for certain subsequences $\mathbb{A}_h$ of $\mathbb{A}$, also the shifted copies $(V_{\mp h_n}A_{h_n}L_{h_n}V_{\pm h_n})$ of $\mathbb{A}_h$ converge. Their limits can be considered as operators acting on $l^p(\mathbb{Z}_\mp, X)$ and will be denoted by $W^{\pm1}(\mathbb{A}_h)$. They somehow capture the asymptotic behavior of $\mathbb{A}$ at the edges of the truncation process.

One of the main results of [11, 12] and [18] states that the stability of a sequence $\mathbb{A} \in \mathcal{F}_{\mathcal{A}_{l^p}}$ is equivalent to the invertibility of all snapshots $W^t(\mathbb{A}_h)$ with $t \in \{-1, 0, 1\}$ and suitable h.

After repeating some important notions and results from [18] in Section 1 we state this theorem exactly.

In Section 2 we turn our attention to the main goal of the present paper, the convergence of the norms $\|A_n\|$ and condition numbers $\operatorname{cond}(A_n)$ for a class of sequences $\mathbb{A} = \{A_n\} \in \mathcal{F}$. Besides that we even determine the convergence of the ϵ-pseudospectra $\operatorname{sp}_\epsilon A_n$. Such results on pseudospectral approximation have been proved in [12], Sections 6.3 and 6.4 in the case l^2, based on an abstract C^*-algebra approach. The case l^p, which is subject of the present text, is much more involved and requires advanced techniques.

The final section is devoted to the application of our results to $\mathcal{F}_{\mathcal{A}_{l^p}}$.

1. $\mathcal{P}$-notions and the algebraic framework

For the treatment of the given approximation method it turned out to be valuable to replace compactness, the usual Fredholm property, and strong convergence by the following similar concepts of $\mathcal{P}$-compactness, $\mathcal{P}$-Fredholmness and $\mathcal{P}$-strong convergence. For details and proofs we refer to [18] and, under additional restrictions, its predecessors [12, 16] and [11]. The major benefit of these substitutes is the unified treatment of all cases $p \in [1, \infty]$.

1.1. $\mathcal{P}$-compact and $\mathcal{P}$-Fredholm operators

We say that an operator $A \in \mathcal{L}(l^p)$ is $\mathcal{P}$-**compact** if the norms $\|(I - L_n)A\|$ and $\|A(I - L_n)\|$ tend to zero as $n \to \infty$. In what follows let $\mathcal{K}(l^p, \mathcal{P})$ denote the set of all $\mathcal{P}$-compact operators and $\mathcal{L}(l^p, \mathcal{P})$ the set of all operators $A \in \mathcal{L}(l^p)$ for which KA and AK are $\mathcal{P}$-compact whenever K is so. Then $\mathcal{L}(l^p, \mathcal{P})$ is a Banach algebra and $\mathcal{K}(l^p, \mathcal{P})$ forms a closed ideal in $\mathcal{L}(l^p, \mathcal{P})$.

An operator $A \in \mathcal{L}(l^p, \mathcal{P})$ is said to be $\mathcal{P}$-**Fredholm** if the respective coset $A + \mathcal{K}(l^p, \mathcal{P})$ is invertible in the quotient algebra $\mathcal{L}(l^p, \mathcal{P})/\mathcal{K}(l^p, \mathcal{P})$. The elements in $(A + \mathcal{K}(l^p, \mathcal{P}))^{-1}$ are called $\mathcal{P}$-**regularizers** for A.

Let us mention that the picture which one obtains with these modifications is still surprisingly similar to what we know from the classical setting, as the following theorem reveals.

Theorem 1.1 ([18, Theorem 1.28]).

- *$\mathcal{L}(l^p, \mathcal{P})$ is an inverse closed subalgebra of $\mathcal{L}(l^p)$ and $\mathcal{K}(l^p, \mathcal{P})$ forms a closed ideal in $\mathcal{L}(l^p, \mathcal{P})$.*
- *Let $A \in \mathcal{L}(l^p, \mathcal{P})$. Then the following are equivalent.*
  - *A is $\mathcal{P}$-Fredholm.*
  - *There is an operator $B \in \mathcal{L}(l^p)$ with $I - AB, I - BA \in \mathcal{K}(l^p, \mathcal{P})$.*
  - *There is an operator $B \in \mathcal{L}(l^p, \mathcal{P})$ with $I - AB, I - BA \in \mathcal{K}(l^p, \mathcal{P})$.*
- *If $A \in \mathcal{L}(l^p, \mathcal{P})$ is Fredholm then A is $\mathcal{P}$-Fredholm. In case $\dim X < \infty$ $\mathcal{P}$-Fredholmness also implies Fredholmness.*

Also the band-dominated operators perfectly fit into that framework.

Theorem 1.2 ([18, Theorem 1.30]). *The set $\mathcal{A}_{l^p}$ of band-dominated operators forms a closed and inverse closed subalgebra of $\mathcal{L}(l^p, \mathcal{P})$ containing $\mathcal{K}(l^p, \mathcal{P})$ as a closed ideal. Furthermore $\mathcal{A}_{l^p}/\mathcal{K}(l^p, \mathcal{P})$ is inverse closed in $\mathcal{L}(l^p, \mathcal{P})/\mathcal{K}(l^p, \mathcal{P})$.*

1.2. $\mathcal{P}$-strong convergence

A sequence $(A_n) \subset \mathcal{L}(l^p)$ is said to converge $\mathcal{P}$-**strongly** to $A \in \mathcal{L}(l^p)$ if, for all $K \in \mathcal{K}(l^p, \mathcal{P})$, both $\|K(A_n - A)\|$ and $\|(A_n - A)K\|$ tend to zero as $n \to \infty$. We write $A = \mathcal{P}\text{-lim}_n A_n$ in this case. From [12, Proposition 1.1.17] we learn that the $\mathcal{P}$-strong limit A of a sequence $(A_n) \subset \mathcal{L}(l^p, \mathcal{P})$ is uniquely determined, belongs to $\mathcal{L}(l^p, \mathcal{P})$ and fulfills

$$\|A\| \leq \liminf_{n\to\infty} \|A_n\|.$$

Definition 1.3. Let $\mathcal{H}_\pm$ denote the set of all strictly increasing (decreasing) sequences $h = (h_n)$ of positive (negative) integers h_n and set $\mathcal{H} := H_+ \cup \mathcal{H}_-$. For $A \in \mathcal{L}(l^p, \mathcal{P})$ and $h \in \mathcal{H}$ the operator A_h is called **limit operator** of A with respect to h if $A_h = \mathcal{P}\text{-}\lim_n V_{-h_n} A V_{h_n}$. Further we say that an operator $A \in \mathcal{L}(l^p, \mathcal{P})$ is **rich** if every sequence $h \in \mathcal{H}$ has a subsequence $g \subset h$ such that the limit operator A_g exists.

Notice that in case $\dim X < \infty$ all band-dominated operators are rich by [12, Corollary 2.1.17].

1.3. Sequences and snapshots

Set $T := \{-1, 0, 1\}$, $I^0 := I$, $I^{\pm 1} := \chi_{\mathbb{Z}_\mp} I$, and for $t \in T$ and $n \in \mathbb{N}$ introduce $L_n^t := V_{-nt} L_n V_{nt}$ together with $E_n^t : \mathcal{L}(\operatorname{im} L_n^t) \to \mathcal{L}(\operatorname{im} L_n)$ being the isometric isomorphism given by $E_n^t(B_n) := V_{nt} B_n V_{-nt}$.

Let $\mathcal{F}^T$ denote the set of all sequences $\mathbb{A} = \{A_n\} \in \mathcal{F}$ for which the $\mathcal{P}$-strong limits

$$W^t(\mathbb{A}) := \mathcal{P}\text{-}\lim_{\mathrm{n}\to\infty} E_n^{-t}(A_n) L_n^t, \quad t \in T,$$

exist. Notice that $W^{+1}(\mathbb{A}) \chi_{\mathbb{Z}_+} I = \chi_{\mathbb{Z}_+} W^{+1}(\mathbb{A}) = 0$, hence $W^{+1}(\mathbb{A})$ can be considered as operator acting on $\mathbf{E}^{+1} := l^p(\mathbb{Z}_-, X)$. Similarly, $W^{-1}(\mathbb{A})$ can be regarded as operator on $\mathbf{E}^{-1} := l^p(\mathbb{Z}_+, X)$, whereas $W^0(\mathbb{A})$ acts on $\mathbf{E}^0 := l^p(\mathbb{Z}, X)$. With the definition $\mathcal{P}^t = (L_n^t)_{n\in\mathbb{N}}$ we analogously get the notions of $\mathcal{P}^t$-compact and $\mathcal{P}^t$-Fredholm operators on $\mathbf{E}^t$, $t \in T$.

Moreover we let $\mathcal{J}^T$ stand for the set

$$\left\{ \sum_{t\in T} \{E_n^t(L_n^t K^t L_n^t)\} + \mathbb{G} : \, K^t \in \mathcal{K}(\mathbf{E}^t, \mathcal{P}^t), \, \mathbb{G} \in \mathcal{G} \right\}.$$

It can be easily derived that $\mathcal{F}^T$ is a closed subalgebra of $\mathcal{F}$ containing $\mathcal{G}$ and $\mathcal{J}^T$ as closed ideals. If $h \in \mathcal{H}_+$ we write $\mathbb{A}_h$ for the subsequence $\{A_{h_n}\}$ and we analogously introduce the respective algebras $\mathcal{F}_h$, $\mathcal{G}_h$, $\mathcal{F}_h^T$ and $\mathcal{J}_h^T$. Following [18] we call a sequence $\mathbb{A}_h \in \mathcal{F}_h^T$ $\mathcal{J}_h^T$-Fredholm if $\mathbb{A}_h + \mathcal{J}_h^T$ is invertible in the quotient $\mathcal{F}_h^T / \mathcal{J}_h^T$.

In the sequel, for given $\mathbb{A} \in \mathcal{F}$ and $h \in \mathcal{H}_+$, we let $\mathcal{H}_{\mathbb{A}_h}$ stand for the set of all subsequences $g \subset h$ such that $\mathbb{A}_g \in \mathcal{F}_g^T$, and we call the operators $W^t(\mathbb{A}_g)$, $g \in \mathcal{H}_\mathbb{A}$ **snapshots** of $\mathbb{A}$. Also notice that the mappings which send a sequence $\mathbb{A}_g \in \mathcal{F}_g^T$ to a snapshot are unital algebra homomorphisms on $\mathcal{F}_g^T$.

1.4. An example

Consider the space $l^p(\mathbb{Z}, \mathbb{C})$ and the bounded linear operator $A = I + aV_{-1}$ with a sequence $a = (a_i) \in l^\infty$. Its matrix representation (with respect to the standard basis) is

$$A = \left(\begin{array}{cccc|cccc} \ddots & \ddots & & & & & & \\ & 1 & a_{-2} & & & & & \\ & & 1 & a_{-1} & & & & \\ & & & 1 & a_0 & & & \\ \hline & & & & 1 & a_1 & & \\ & & & & & 1 & a_2 & \\ & & & & & & 1 & \ddots \\ & & & & & & & \ddots \end{array}\right).$$

Further suppose that a_n tends to zero as n goes to ∞, whereas, for $n \leq 0$, $a_{2n} = 1$ and $a_{2n+1} = 0$ hold. Clearly, the finite sections $L_n A L_n$ can be regarded as finite matrices in the above sense. With g (and h) being the increasing sequence of all positive odd (even) numbers, we get the following snapshots of $\mathbb{A} = \{L_n A L_n\}$: $W^0(\mathbb{A}) = A$, $W^1(\mathbb{A}_g) = W^1(\mathbb{A}_h) = W^1(\mathbb{A}) = I^1$,

$$W^{-1}(\mathbb{A}_g) = \begin{pmatrix} 1 & 0 & & & \\ & 1 & 1 & & \\ & & 1 & 0 & \\ & & & 1 & \ddots \\ & & & & \ddots \end{pmatrix}, \quad W^{-1}(\mathbb{A}_h) = \begin{pmatrix} 1 & 1 & & & \\ & 1 & 0 & & \\ & & 1 & 1 & \\ & & & 1 & \ddots \\ & & & & \ddots \end{pmatrix}$$

and there are no further ones.

1.5. The snapshots of $\mathbb{A} \in \mathcal{F}_{\mathcal{A}_{l^p}}$

Recall that $\mathcal{F}_{\mathcal{A}_{l^p}}$ stands for the smallest closed subalgebra of $\mathcal{F}$ containing the finite section sequences $\{L_n A L_n\}$ of rich band-dominated operators A. While for $\mathbb{A} = \{A_n\} \in \mathcal{F}_{\mathcal{A}_{l^p}}$ the $\mathcal{P}$-strong limit $W^0(\mathbb{A}) = \mathcal{P}\text{-lim}\, A_n L_n$ always exists, this is in general not guaranteed for $W^{\pm 1}(\mathbb{A})$. However, the following holds.

Proposition 1.4 ([18, Proposition 3.1]). *Let $\mathbb{A} \in \mathcal{F}_{\mathcal{A}_{l^p}}$ and $h \in \mathcal{H}_+$. Then there exists a subsequence g of h such that $\mathbb{A}_g \in \mathcal{F}_g^T$, that is $\mathcal{H}_{\mathbb{A}_h}$ is not empty.*

Here comes the announced criterion for the stability.

Theorem 1.5 ([18, Corollary 3.4]). *A sequence $\mathbb{A} \in \mathcal{F}_{\mathcal{A}_{l^p}}$ is stable if and only if all of its snapshots $W^t(\mathbb{A}_h)$, $t \in T$, $h \in \mathcal{H}_{\mathbb{A}}$ are invertible in $\mathcal{L}(\mathbf{E}^t)$, respectively.*

2. Localizable sequences and their properties

The aim of this section is to identify a class of sequences $\mathbb{A} = \{A_n\} \in \mathcal{F}^T$ which offer stronger connections between the operators A_n and the snapshots $W^t(\mathbb{A})$. The main tools for this are the local principle of Allan and Douglas and the concept of KMS-algebras which is due to Böttcher, Krupnik and one of the authors [3].

2.1. Localization and the KMS property

Let $\mathcal{A}$ be a Banach algebra with identity and $\mathcal{C}$ be a closed C^*-subalgebra of the center of $\mathcal{A}$ which contains the identity. By the Gelfand-Naimark Theorem, $\mathcal{C}$ is isomorphic to the algebra of continuous functions on the maximal ideal space $\mathcal{M}_\mathcal{C}$ of $\mathcal{C}$. Therefore the elements of $\mathcal{C}$ will be called functions. For each maximal ideal $x \in \mathcal{M}_\mathcal{C}$ we introduce $\mathcal{J}_x$, the smallest closed ideal in $\mathcal{A}$ containing x, and let ϕ_x denote the canonical mapping from $\mathcal{A}$ to $\mathcal{A}/\mathcal{J}_x$. In the case $\mathcal{J}_x = \mathcal{A}$ we define that $\phi_x(A)$ is invertible in $\mathcal{A}/\mathcal{J}_x$ for each $A \in \mathcal{A}$. The local principle of Allan and Douglas [12, Theorem 2.3.16] states that $A \in \mathcal{A}$ is invertible if and only if $\phi_x(A)$ is invertible in $\mathcal{A}/\mathcal{J}_x$ for every $x \in \mathcal{M}_\mathcal{C}$.

Definition 2.1. The algebra $\mathcal{A}$ is a **KMS-algebra** with respect to $\mathcal{C}$ [1] if for every $A \in \mathcal{A}$ and $\varphi, \psi \in \mathcal{C}$ with disjoint supports

$$\|(\varphi - \psi)A\| \leq \max(\|\varphi A\|, \|\psi A\|).$$

From [3, Theorem 5.3] we know that $\mathcal{A}$ is a KMS-algebra w.r.t. $\mathcal{C}$ if and only if

$$\|A\| = \max_{x \in \mathcal{M}_\mathcal{C}} \|\phi_x(A)\| \quad \text{for every} \quad A \in \mathcal{A}. \tag{2.1}$$

Also notice that, by Proposition 5.1 in [3],

$$\|\phi_x(A)\| = \inf\{\|\varphi A\| : \varphi \in \mathcal{C}, 0 \leq \varphi \leq 1, \varphi \equiv 1 \text{ in a neighborhood of } x\}. \tag{2.2}$$

2.2. A family of central (sub)algebras of sequence algebras

For our concrete setting of l^p-spaces and within the algebra $\mathcal{F}^T$ we introduce central subalgebras $\mathcal{C}_\gamma$ as follows. Let $\gamma : \mathbb{N} \to \mathbb{R}_+$ be a non-decreasing sequence of positive numbers $\gamma_n \leq \frac{1}{2}n$ with $\gamma_n \to \infty$ as $n \to \infty$ and let $b_n^\gamma : \mathbb{R} \to [-1, 1]$ denote the continuous piecewise linear spline which is given by $b_n^\gamma(\pm n) = \pm 1$ and $b_n^\gamma(\pm(n - \gamma_n)) = \pm\frac{1}{2}$ and constant outside the interval $[-n, n]$, for every n, respectively. For every continuous function $\varphi \in C[-1, 1]$ we let φ_n^γ stand for the restriction of the inflated copy $\varphi \circ b_n^\gamma$ of φ to $\mathbb{Z}$. It straightforwardly follows that the set

$$\mathcal{C}^\gamma := \{\{\varphi_n^\gamma L_n\} : \varphi \in C[-1, 1]\}$$

forms a Banach subalgebra of $\mathcal{F}^T$ with $W^t(\{\varphi_n^\gamma L_n\}) = \varphi(t)I^t$, $t \in T$. At this point we mention that the sequence γ provides a certain flexibility in the inflation process which will permit to adapt it to the sequence $\mathbb{A}$ under consideration. This will play a crucial role in the application to the finite section sequences of band-dominated operators.

Introduce the Banach algebra $\mathcal{B}^\gamma$ of all sequences $\mathbb{A} \in \mathcal{F}$ for which the commutator $[\mathbb{A}, \mathbb{C}] := \mathbb{AC} - \mathbb{CA}$ belongs to $\mathcal{G}$ for every $\mathbb{C} \in \mathcal{C}^\gamma$. Let $\mathcal{B}^{\gamma,T}$ denote the Banach algebra $\mathcal{B}^\gamma \cap \mathcal{F}^T$ and notice that $\mathcal{B}^{\gamma,T}$ includes $\mathcal{C}^\gamma$ as well as $\mathcal{J}^T$.

Proposition 2.2. *The set $\mathcal{C}^\gamma/\mathcal{G} := \{\mathbb{C} + \mathcal{G} : \mathbb{C} \in \mathcal{C}^\gamma\}$ forms a closed central C^*-subalgebra of both $\mathcal{B}^\gamma/\mathcal{G}$ and $\mathcal{B}^{\gamma,T}/\mathcal{G}$, and it is isometrically isomorphic to $C[-1, 1]$.*

---

[1] In recent literature (e.g., [15]) such pairs $(\mathcal{A}, \mathcal{C})$ are also called "faithful localizing pairs".

Proof. Let $\varphi \in C[-1,1]$. Then we obviously have

$$\|\varphi\|_\infty = \|\{\varphi_n^\gamma L_n\}\|_{\mathcal{F}} \geq \|\{\varphi_n^\gamma L_n\} + \mathcal{G}\|_{\mathcal{F}/\mathcal{G}}.$$

On the other hand $\|\varphi\|_\infty = \sup_{x\in[-1,1]} |\varphi(x)| \leq \|\{\varphi_n^\gamma L_n\} + \mathcal{G}\|_{\mathcal{F}/\mathcal{G}}$ follows from

$$\begin{aligned}|\varphi(x)| = \|\varphi(x)I\| &= \|\mathcal{P}\text{-}\lim_{n\to\infty} V_{-\lfloor (b_n^\gamma)^{-1}(x)\rfloor}\varphi_n^\gamma L_n V_{\lfloor (b_n^\gamma)^{-1}(x)\rfloor}\| \\ &\leq \liminf_{n\to\infty} \|V_{-\lfloor (b_n^\gamma)^{-1}(x)\rfloor}\varphi_n^\gamma L_n V_{\lfloor (b_n^\gamma)^{-1}(x)\rfloor}\| \\ &\leq \limsup_{n\to\infty} \|\varphi_n^\gamma L_n\| = \|\{\varphi_n^\gamma L_n\} + \mathcal{G}\|_{\mathcal{F}/\mathcal{G}},\end{aligned}$$

with $(b_n^\gamma)^{-1}$ being the inverse of the bijective function $b_n^\gamma : [-n,n] \to [-1,1]$, and $\lfloor\cdot\rfloor$ the floor function. Thus, $\mathcal{C}^\gamma/\mathcal{G} \cong C[-1,1]$ and the rest easily follows. □

Corollary 2.3. *The set $\mathcal{C}^\gamma/\mathcal{J}^T := \{\mathbb{C} + \mathcal{J}^T : \mathbb{C} \in \mathcal{C}^\gamma\}$ is a closed central C^*-subalgebra of $\mathcal{B}^{\gamma,T}/\mathcal{J}^T$, and it is isometrically isomorphic to $C[-1,1]$.*

Proof. Clearly, $\mathcal{C}^\gamma/\mathcal{J}^T$ is commutative and inherits the involution from $\mathcal{C}^\gamma/\mathcal{G}$. We only need to show that $\|\{\varphi_n^\gamma L_n\} + \mathcal{G}\| = \|\{\varphi_n^\gamma L_n\} + \mathcal{J}^T\|$ holds for every $\varphi \in C[-1,1]$. The estimate "$\geq$" is obvious. Assume that it is even proper, which means that there is an $\epsilon > 0$ and a sequence $\mathbb{J} \in \mathcal{J}^T$ such that $\|\{\varphi_n^\gamma L_n\} + \mathcal{G}\| > \|\{\varphi_n^\gamma L_n\} + \mathbb{J}\| + \epsilon \geq \|\{\varphi_n^\gamma L_n\} + \mathbb{J} + \mathcal{G}\| + \epsilon$. WLOG we can also assume that $\|\{\varphi_n^\gamma L_n\} + \mathcal{G}\| = 1$. Fix $x_0 \in (-1,1) \setminus \{0\}$ such that $|\varphi(x_0)| \geq 1-\epsilon$ and choose a function $\psi \in C[-1,1]$ equal to 1 in a neighborhood of x_0, equal to zero on T and of norm one. Then $\mathbb{B} + \mathcal{G} := \varphi(x_0)\mathbb{I} - \{\psi_n^\gamma L_n\}(\{\varphi_n^\gamma L_n\} + \mathbb{J}) + \mathcal{G}$ is invertible in $\mathcal{F}^T$ where its inverse is given by a Neumann series. Further, $\{\psi_n^\gamma L_n\}\mathbb{J} \in \mathcal{G}$, hence $\phi_{x_0}(\mathbb{B} + \mathcal{G})$ is zero, a contradiction. □

Proposition 2.4. *The set $\mathcal{B}^\gamma/\mathcal{G}$ is a KMS-algebra with respect to $\mathcal{C}^\gamma/\mathcal{G}$, hence for every $\mathbb{A} = \{A_n\} \in \mathcal{B}^\gamma$*

$$\limsup_{n\to\infty} \|A_n\| = \max_{x\in[-1,1]} \|\phi_x(\mathbb{A} + \mathcal{G})\|.$$

Proof.[2] Choose $\varphi, \psi \in C[-1,1]$ with disjoint supports, define $Y_n := A_n\varphi_n^\gamma L_n$ and $Z_n := A_n\psi_n^\gamma L_n$, as well as $N_n := \operatorname{supp}\varphi_n^\gamma$, $M_n := \operatorname{supp}\psi_n^\gamma$, and prove that

$$\limsup \|Y_n\chi_{N_n}L_n + Z_n\chi_{M_n}L_n\| \leq \max(\limsup \|Y_n\|, \limsup \|Z_n\|).$$

For this and in case $p \in [1,\infty)$ let $x \in l^p$ and observe

$$\begin{aligned}&\|(\chi_{N_n}Y_n\chi_{N_n}L_n + \chi_{M_n}Z_n\chi_{M_n}L_n)x\|^p \\ &\quad= \|\chi_{N_n}Y_n\chi_{N_n}L_nx\|^p + \|\chi_{M_n}Z_n\chi_{M_n}L_nx\|^p \\ &\quad\leq \max(\|Y_n\|, \|Z_n\|)^p[\|\chi_{N_n}L_nx\|^p + \|\chi_{M_n}L_nx\|^p] \\ &\quad\leq \max(\|Y_n\|, \|Z_n\|)^p\|x\|^p.\end{aligned} \tag{2.3}$$

Thus, for every $\epsilon > 0$ there is an integer N such that

$$\|\chi_{N_k}Y_k\chi_{N_k}L_k + \chi_{M_k}Z_k\chi_{M_k}L_k\| \leq \max(\limsup \|Y_n\|, \limsup \|Z_n\|) + \epsilon$$

[2]This idea already appeared in [3].

for every $k > N$. Consequently, $\limsup \|\chi_{N_n} Y_n \chi_{N_n} L_n + \chi_{M_n} Z_n \chi_{M_n} L_n\|$ is not greater than $\max(\limsup \|Y_n\|, \limsup \|Z_n\|)$. From $\mathbb{A} \in \mathcal{B}^\gamma$ we deduce that

$$\chi_{N_n} Y_n \chi_{N_n} L_n - Y_n \chi_{N_n} L_n \quad \text{and} \quad \chi_{M_n} Z_n \chi_{M_n} L_n - Z_n \chi_{M_n} L_n$$

belong to $\mathcal{G}$ and thus the assertion follows. For the case $p = \infty$ simply replace (2.3) by

$$\begin{aligned}
&\|(\chi_{N_n} Y_n \chi_{N_n} L_n + \chi_{M_n} Z_n \chi_{M_n} L_n)x\| \\
&\quad = \max(\|\chi_{N_n} Y_n \chi_{N_n} L_n x\|, \|\chi_{M_n} Z_n \chi_{M_n} L_n x\|) \\
&\quad \leq \max(\|Y_n\|, \|Z_n\|) \max(\|\chi_{N_n} L_n x\|, \|\chi_{M_n} L_n x\|) \\
&\quad \leq \max(\|Y_n\|, \|Z_n\|)\|x\|.
\end{aligned}$$

□

2.3. Localizable sequences

Roughly speaking, we call a sequence $\mathbb{A} \in \mathcal{B}^{\gamma,T}$ localizable, if its snapshots describe $\mathbb{A}$ locally sufficiently well and our aim is to replace the local cosets in Proposition 2.4 by the snapshots. For the precise definition recall the local homomorphisms ϕ_x which were defined in the beginning of this section.

Definition 2.5. Let $\mathcal{L}^{\gamma,T}$ denote the set of all sequences $\mathbb{A} \in \mathcal{B}^{\gamma,T}$ such that for every $x \in [-1,1]$ and $\varphi \in C[-1,1]$

- $\|[W^t(\mathbb{A}), V_{-tn}\varphi_n^\gamma V_{tn} I^t]\| \to 0$ as $n \to \infty$
- $\phi_x(\mathbb{A} + \mathcal{G}) = \phi_x(\{E_n^t(L_n^t W^t(\mathbb{A}) L_n^t\} + \mathcal{G})$,

where $t := x$ if $x \in \{\pm 1\}$, and $t := 0$ otherwise. In what follows, the sequences in $\mathcal{L}^{\gamma,T}$ are said to be **localizable** (with respect to $\mathcal{C}^\gamma$).

Proposition 2.6. *We have*

1. *$\mathcal{L}^{\gamma,T}$ is a Banach algebra containing $\mathcal{G}$ and $\mathcal{J}^T$ as closed ideals.*
2. *$\mathcal{L}^{\gamma,T}/\mathcal{G}$ is inverse closed in $\mathcal{F}/\mathcal{G}$, and $\mathcal{L}^{\gamma,T}/\mathcal{J}^T$ is inverse closed in $\mathcal{F}^T/\mathcal{J}^T$.*
3. *A sequence $\mathbb{A} \in \mathcal{L}^{\gamma,T}$ is stable if and only if its snapshots are invertible. It is $\mathcal{J}^T$-Fredholm iff its snapshots are $\mathcal{P}^t$-Fredholm, respectively.*

Proof. The first assertion is quite obvious and from [18] we know that all snapshots of a $\mathcal{J}^T$-Fredholm sequence are $\mathcal{P}^t$-Fredholm and further that a sequence in $\mathcal{F}^T$ is stable if and only if it is $\mathcal{J}^T$-Fredholm and all snapshots are invertible.

Let $\varphi \in C[-1,1]$, $x \in [-1,1]$ with $t \in T$ as in Definition 2.5, further set $W_n := V_{-tn}\varphi_n^\gamma V_{tn} I^t$ and suppose that the operator $A := W^t(\mathbb{A})$ is $\mathcal{P}^t$-Fredholm with B one of its $\mathcal{P}^t$-regularizers. Then $T_1 := BA - I^t$, $T_2 := AB - I^t$ and $K := B - BAB$ are $\mathcal{P}^t$-compact and

$$\begin{aligned}
BW_n &= (B(T_2 + I^t) + K)W_n =_{\mathcal{G}} BW_n(T_2 + I^t) + KW_n \\
&= BW_n AB + KW_n =_{\mathcal{G}} BAW_n B + KW_n \\
&=_{\mathcal{G}} W_n BAB + W_n K = W_n B,
\end{aligned}$$

where $=_{\mathcal{G}}$ means equality up to a sequence of operators tending to zero in the norm. Fix a function $\varphi^x \in C[-1,1]$ with $\|\varphi^x\|_\infty = 1$, which equals 1 in a neighborhood of x and which is equal to zero in a neighborhood of every $\tau \in T$, $\tau \neq t$. Then, both sequences

$$\mathbb{A}^x := \{\varphi_n^{x,\gamma} E_n^t (L_n^t A L_n^t) \varphi_n^{x,\gamma} L_n\} \quad \text{and} \quad \mathbb{B}^x := \{\varphi_n^{x,\gamma} E_n^t (L_n^t B L_n^t) \varphi_n^{x,\gamma} L_n\}$$

belong to $\mathcal{B}^{\gamma,T}$ and

$$\begin{aligned}
&\phi_x(\mathbb{A}^x \mathbb{B}^x - \mathbb{I} + \mathcal{G}) \\
&\quad = \phi_x(\{\varphi_n^{x,\gamma} E_n^t (L_n^t A V_{-tn} \varphi_n^{x,\gamma} \varphi_n^{x,\gamma} V_{tn} I^t B L_n^t) \varphi_n^{x,\gamma}\} - \mathbb{I} + \mathcal{G}) \\
&\quad = \phi_x(\{(\varphi_n^{x,\gamma})^3 E_n^t (L_n^t A B L_n^t) \varphi_n^{x,\gamma}\} - \mathbb{I} + \mathcal{G}) \\
&\quad = \phi_x(\{E_n^t (L_n^t (T_2 + I^t) L_n^t\} - \mathbb{I} + \mathcal{G}) \\
&\quad = \phi_x(\{E_n^t (L_n^t T_2 L_n^t\} + \mathcal{G}),
\end{aligned}$$

as well as $\phi_x(\mathbb{B}^x \mathbb{A}^x - \mathbb{I} + \mathcal{G}) = \phi_x(\{E_n^t (L_n^t T_1 L_n^t\} + \mathcal{G})$.

From Corollary 2.3 we know that we can apply the local principle of Allan/Douglas to the elements in the algebra $\mathcal{B}^{\gamma,T}/\mathcal{J}^T$. Let Φ_x, $x \in [-1,1]$ denote the respective local homomorphisms. Since for the localizable sequence $\mathbb{A}$ the cosets $\Phi_x(\mathbb{A} + \mathcal{J}^T)$, $\Phi_x(\mathbb{A}^x + \mathcal{J}^T)$ coincide and since $\Phi_x(\mathbb{A}^x\mathbb{B}^x + \mathcal{J}^T)$, $\Phi_x(\mathbb{B}^x\mathbb{A}^x + \mathcal{J}^T)$ both equal $\Phi_x(\mathbb{I} + \mathcal{J}^T)$ we find that $\mathbb{A} + \mathcal{J}^T$ is invertible in $\mathcal{B}^{\gamma,T}/\mathcal{J}^T$ if all snapshots are $\mathcal{P}^t$-Fredholm. This particularly yields that $\mathbb{A}$ is $\mathcal{J}^T$-Fredholm in this case.

So, let $\mathbb{A} \in \mathcal{L}^{\gamma,T}$ be $\mathcal{J}^T$-Fredholm and $\mathbb{B} \in \mathcal{F}^T$ be a regularizer. We show that $\mathbb{B} \in \mathcal{L}^{\gamma,T}$. The operator $A := W^t(\mathbb{A})$ is $\mathcal{P}^t$-Fredholm and $B := W^t(\mathbb{B})$ is one of its $\mathcal{P}^t$-regularizers [18, Theorem 2.4]. Check that

$$\phi_x(\mathcal{J}^T) = \phi_x(\mathcal{J}^t) := \phi_x(\{\{E_n^t(L_n^t K L_n^t)\} + \mathcal{G} : K \in \mathcal{K}(\mathbf{E}^t, \mathcal{P}^t)\}).$$

With the notions as above we get $\phi_x(\mathbb{A}^x\mathbb{B} - \mathbb{I} + \mathcal{G}), \phi_x(\mathbb{A}^x\mathbb{B}^x - \mathbb{I} + \mathcal{G}) \in \phi_x(\mathcal{J}^t)$ and then we successively deduce that $\phi_x(\mathcal{J}^t)$ further contains the following elements: $\phi_x(\mathbb{A}^x(\mathbb{B} - \mathbb{B}^x) + \mathcal{G})$, $\phi_x(\mathbb{B}^x\mathbb{A}^x(\mathbb{B} - \mathbb{B}^x) + \mathcal{G})$, $\phi_x(\mathbb{B} - \mathbb{B}^x + \mathcal{G})$. Consequently, there is a $\mathcal{P}^t$-compact operator M on the space $\mathbf{E}^t$ such that $\phi_x(\mathbb{B} - \mathbb{B}^x - \{E_n^t(L_n^t M L_n^t)\} + \mathcal{G})$ equals zero, and hence the respective snapshot $W^t(\mathbb{B} - \mathbb{B}^x - \{E_n^t(L_n^t M L_n^t)\})$ is zero as well. Since $W^t(\mathbb{B}^x) = W^t(\mathbb{B})$ and $W^t(\{E_n^t(L_n^t M L_n^t)\}) = M$, this operator M equals zero and we see that $\phi_x(\mathbb{B} - \mathbb{B}^x + \mathcal{G}) = 0$. This proves that $\mathbb{B} \in \mathcal{L}^{\gamma,T}$. □

Following many standard references such as [4, 5, 8] or [12] we now slightly change the perspective, consider the Cartesian product

$$\mathcal{S} := \mathcal{L}(\mathbf{E}_{-1}) \times \mathcal{L}(\mathbf{E}_0) \times \mathcal{L}(\mathbf{E}_{+1})$$

and equip it with componentwise defined algebraic operations as well as the norm

$$\|(A, B, C)\| := \max\{\|A\|, \|B\|, \|C\|\}$$

to obtain a Banach algebra. For a sequence $\mathbb{A} \in \mathcal{L}^{\gamma,T}$ we denote by $\operatorname{smb} \mathbb{A}$ the triple

$$\operatorname{smb} \mathbb{A} := (W^{-1}(\mathbb{A}), W^0(\mathbb{A}), W^{+1}(\mathbb{A})) \in \mathcal{S}$$

and call it the **symbol** of $\mathbb{A}$.

Proposition 2.7. *The mapping*

$$\mathrm{Smb} : \mathcal{L}^{\gamma,T}/\mathcal{G} \to \mathcal{S}, \quad \mathbb{A} + \mathcal{G} \mapsto \mathrm{smb}\,\mathbb{A}$$

is an isometric (and hence injective) algebra homomorphism. Moreover,

$$\|\mathbb{A} + \mathcal{G}\| = \lim_{n\to\infty} \|A_n\|$$

holds for every $\mathbb{A} = \{A_n\} \in \mathcal{L}^{\gamma,T}$.

Proof. Obviously, the mapping $\mathcal{L}^{\gamma,T} \to \mathcal{S}$, $\mathbb{A} \mapsto \mathrm{smb}\,\mathbb{A}$ is an algebra homomorphism and $\mathcal{G}$ belongs to its kernel, hence Smb proves to be a homomorphism as well. For every $x \in [-1,1]$, the respective local homomorphisms ϕ_x and t chosen as in Definition 2.5 we have

$$\begin{aligned}\|\phi_x(\mathbb{A} + \mathcal{G})\| &= \|\phi_x(\{E_n^t(L_n^t W^t(\mathbb{A}) L_n^t)\} + \mathcal{G})\| \\ &\leq \|\{E_n^t(L_n^t W^t(\mathbb{A}) L_n^t)\} + \mathcal{G}\| \\ &= \limsup_{n\to\infty} \|E_n^t(L_n^t W^t(\mathbb{A}) L_n^t)\| \leq \|W^t(\mathbb{A})\|.\end{aligned}$$

On the other hand, Theorem 1.13. in [18] and Equation (2.1) together with Proposition 2.4 yield

$$\begin{aligned}\|W^t(\mathbb{A})\| &\leq \liminf_{n\to\infty} \|E_{g_n}^{-t}(A_{g_n}) L_{g_n}^t\| \\ &\leq \limsup_{n\to\infty} \|A_{g_n}\| \leq \limsup_{n\to\infty} \|A_n\| \\ &= \|\mathbb{A} + \mathcal{G}\| = \max_{x\in[-1,1]} \|\phi_x(\mathbb{A} + \mathcal{G})\|\end{aligned}$$

for every $t \in T$ and every subsequence (A_{g_n}) of (A_n). Thus, we have proved that $\limsup \|A_{g_n}\|$ equals $\max_{t\in T} \|W^t(\mathbb{A})\|$ for every subsequence (A_{g_n}), and therefore the limit $\lim \|A_n\|$ exists and has the same value. □

Now we can prove the announced asymptotic behavior for localizable sequences. For this we introduce the notation $\|B^{-1}\| := \infty$ if B is not invertible.

Corollary 2.8. *Let* $\mathbb{A} = \{A_n\} \in \mathcal{L}^{\gamma,T}$. *Then the norms* $\|A_n\|$ *and* $\|A_n^{-1}\|$ *converge and*

$$\lim_{n\to\infty} \|A_n\| = \max_{t\in T} \|W^t(\mathbb{A})\|, \quad \lim_{n\to\infty} \|A_n^{-1}\| = \max_{t\in T} \|(W^t(\mathbb{A}))^{-1}\|.$$

Proof. The limit $\lim \|A_n\|$ is already supplied by the previous proposition, and we now consider the "inverses". Suppose that one snapshot of $\mathbb{A} = \{A_n\}$ is not invertible. Then Theorem 3.2. in [18] yields that $s_1^r(A_n)$ or $s_1^l(A_n)$ tend to zero as $n \to \infty$ and by Corollary 2.11. in [18] it follows that $\lim \|A_n^{-1}\| = \infty$. Conversely, if all snapshots are invertible then the sequence is stable. Set $B_n := A_n^{-1}$ if A_n is invertible and $B_n := 0$ otherwise. Then $\{B_n\}$ proves to be a $\mathcal{G}$-regularizer for $\mathbb{A}$, hence belongs to $\mathcal{L}^{\gamma,T}$ by Proposition 2.6, and its snapshots are $(W^t(\mathbb{A}))^{-1}$. This provides the second asserted limit as well. □

Corollary 2.9. *Let $\mathbb{A} = \{A_n\} \in \mathcal{L}^{\gamma,T}$ be stable. Then the condition numbers of the operators A_n which are defined by* $\operatorname{cond}(A_n) := \|A_n\|\|A_n^{-1}\|$ *converge and their limit is*

$$\lim_{n\to\infty} \operatorname{cond}(A_n) = \max_{t\in T} \|W^t(\mathbb{A})\| \cdot \max_{t\in T} \|(W^t(\mathbb{A}))^{-1}\|.$$

Remark 2.10. We want to point out that the idea to study an extension of the natural finite section algebra (arising from a certain class of operators) by additional sequences of "inflated functions" in order to obtain an algebra with a well-suited center which permits to employ the KMS-techniques of [3] goes back to Roch [14].

2.4. Approximation of pseudospectra

Within this paragraph we suppose X to be a (complex) space of the type $L^p(S, \Sigma, \mu)$, with (S, Σ, μ) being a measure space such that $\mu(S) < \infty$. This particularly includes $X = \mathbb{C}^N$, $N \in \mathbb{N}$, equipped with the p-norm and the usual counting measure, as well as $X = L^p([0, 1))$. The latter is important for the treatment of band-dominated operators over the real axis (see [18, Section 3.3]).

Then $\mathbf{X} := l^p(\mathbb{Z}, X)$ can be identified with $L^p(\hat{S}, \hat{\Sigma}, \hat{\mu})$ where the measure space $(\hat{S}, \hat{\Sigma}, \hat{\mu})$ is given by $\hat{S} := \mathbb{Z} \times S$, $\hat{\Sigma} := \{\bigcup_{k\in\mathbb{Z}}\{k\} \times A_k : A_k \in \Sigma\}$ and $\hat{\mu}(\bigcup_{k\in\mathbb{Z}}\{k\} \times A_k) := \sum_{k\in\mathbb{Z}} \mu(A_k)$. For the following result see the discussion in [20], preliminary to Theorem 2.5 and Theorem 2.6.

Theorem 2.11. *Let Ω be a connected open subset of $\mathbb{C}$ and $A : \Omega \to \mathcal{L}(\mathbf{X})$ an analytic operator-valued function. Suppose that there exists $\lambda_0 \in \Omega$ such that the derivative $A'(\lambda_0)$ of A in λ_0 is invertible.*
If $\|A(\lambda)\| \le M$ for all $\lambda \in \Omega$ then $\|A(\lambda)\| < M$ for all $\lambda \in \Omega$.

This theorem stands in the end of a series of results dealing with the question "Can the resolvent norm of an operator be constant on an open set?" As some milestones in this development we further refer to Globevnik [7], Böttcher [2] and Daniluk [6], and Shargorodsky [20].

Definition 2.12. For $N \in \mathbb{N}_0$ and $\epsilon > 0$ the (N, ϵ)-**pseudospectrum** of a bounded linear operator A is defined as the set (we again use the convention $\|B^{-1}\| = \infty$ if B is not invertible)

$$\operatorname{sp}_{N,\epsilon} A := \{z \in \mathbb{C} : \|(A - zI)^{-2^N}\|^{2^{-N}} \ge 1/\epsilon\}.$$

Let $M_1, M_2, \ldots$ be a sequence of nonempty subsets of $\mathbb{C}$. The **uniform (partial) limiting set**

$$\text{u-}\lim_{n\to\infty} M_n \quad \left(\text{p-}\lim_{n\to\infty} M_n\right)$$

of this sequence is the set of all $\lambda \in \mathbb{C}$ that are (partial) limits of a sequence (λ_n) with $\lambda_n \in M_n$.

Remark 2.13. Notice that (for $N = 0$) this definition of the (N, ϵ)-pseudospectrum includes the definition of the (classical) ϵ-pseudospectrum

$$\operatorname{sp}_\epsilon A := \{z \in \mathbb{C} : \|(A - zI)^{-1}\| \ge 1/\epsilon\}.$$

This has gained attention after it was discovered in [13] and [2] that, on the one hand the ϵ-pseudospectra approximate the spectrum but are less sensitive to perturbations, and on the other hand the pseudospectra of discrete convolution operators mimic exactly the pseudospectra of an appropriate limiting operator, which is in general not true for the "usual" spectrum. See also [1, 5, 8] and the references cited there.

Later on, Hansen [9, 10] introduced the (N, ϵ)-pseudospectra for linear operators on separable Hilbert spaces and pointed out that they share the nice properties with case $N = 0$, but offer a better approximation of the spectrum. Furthermore, it was shown how the spectrum can be approximated numerically, based on the consideration of singular values of certain finite matrices. Recently, one of the authors extended this to the Banach space case [17].

Here, we restrict our considerations to the asymptotic connection between the (N, ϵ)-pseudospectra of the operators A_n and the respective snapshots. In analogy to [5, Theorem 3.17] we prove

Theorem 2.14. *Let $X = L^p(S, \Sigma, \mu)$ with (S, Σ, μ) being a measure space such that $\mu(S) < \infty$. For $\mathbb{A} = \{A_n\} \in \mathcal{L}^{\gamma,T}$ and every $N \in \mathbb{N}_0$, $\epsilon > 0$*

$$u\text{-}\lim_{n\to\infty} \operatorname{sp}_{\mathrm{N},\epsilon} A_n = p\text{-}\lim_{n\to\infty} \operatorname{sp}_{\mathrm{N},\epsilon} A_n = \bigcup_{t\in T} \operatorname{sp}_{\mathrm{N},\epsilon} W^t(\mathbb{A}).$$

Proof. Suppose $z \in \operatorname{sp} W^t(\mathbb{A})$, that is $W^t(\mathbb{A} - z\mathbb{I})$ and hence $W^t((\mathbb{A} - z\mathbb{I})^{2^N})$ are not invertible. Then by Proposition 2.6 $\mathbb{A} - z\mathbb{I}$ is not stable and Proposition 2.7 yields that $\lim \|(A_n - zL_n)^{-2^N}\| = \infty$ which implies $z \in \operatorname{sp}_{\mathrm{N},\epsilon} A_n$ for sufficiently large n. Thus $z \in u\text{-}\lim_n \operatorname{sp}_{\mathrm{N},\epsilon} A_n$.

So, now suppose that $\mathbb{A} - z\mathbb{I}$ is stable, but $z \in \operatorname{sp}_{\mathrm{N},\epsilon} W^t(\mathbb{A})$, which means that $\|(W^t(\mathbb{A}) - zI)^{-2^N}\| \geq \epsilon^{-2^N}$. Let U be an arbitrary open ball around z such that $W^t(\mathbb{A}) - yI$ is invertible for all $y \in U$. If $\|(W^t(\mathbb{A}) - yI)^{-2^N}\|$ would be less than or equal to ϵ^{-2^N} for every $y \in U$ then Theorem 2.11 would imply that $\|(W^t(\mathbb{A}) - zI)^{-2^N}\| < \epsilon^{-2^N}$, a contradiction. For this notice that the function $y \mapsto (W^t(\mathbb{A}) - yI)^{-2^N}$ is analytic on U and its first derivative in z is invertible. Hence there is a $y \in U$ such that $\|(W^t(\mathbb{A}) - yI)^{-2^N}\| > \epsilon^{-2^N}$, that is we can find a k_0 such that

$$\|(W^t(\mathbb{A}) - yI)^{-2^N}\| > \left(\epsilon - \frac{1}{k}\right)^{-2^N} \quad \text{for all} \quad k \geq k_0.$$

Because U was arbitrary we can choose a sequence $(z_m)_{m\in\mathbb{N}}$ of complex numbers $z_m \in \operatorname{sp}_{N,\epsilon-1/m} W^t(\mathbb{A})$ such that $z_m \to z$. Since $\lim_n \|(A_n - z_m L_n)^{-2^N}\|$ exists and equals $\max_{t\in T} \|(W^t(\mathbb{A}) - z_m I)^{-2^N}\|$, due to Proposition 2.7, it is greater than or equal to $(\epsilon - 1/m)^{-2^N}$. Consequently, for sufficiently large n, $\|(A_n - z_m I)^{-2^N}\| \geq \epsilon^{-2^N}$ and thus $z_m \in \operatorname{sp}_{\mathrm{N},\epsilon} A_n$. This shows that $z = \lim_m z_m$ belongs to the closed set $u\text{-}\lim_n \operatorname{sp}_{\mathrm{N},\epsilon} A_n$.

Finally consider the case that $\|(W^t(\mathbb{A}) - zI)^{-2^N}\| < \epsilon^{-2^N}$ for all $t \in T$. Then

$$\lim_{n\to\infty} \|(A_n - zL_n)^{-2^N}\| = \max_{t\in T} \|(W^t(\mathbb{A}) - zI)^{-2^N}\| < \epsilon^{-2^N},$$

hence there are a $\delta > 0$ and an $n_0 \in \mathbb{N}$ such that $\|(A_n - zL_n)^{-2^N}\| \leq \epsilon^{-2^N} - \delta$ for all $n \geq n_0$. If $|y - z|$ is sufficiently small and $n \geq n_0$ we then get

$$\begin{aligned}
\|(A_n - yL_n)^{-2^N}\| &= \|((A_n - zL_n)(I + (z-y)(A_n - zL_n)^{-1}))^{-2^N}\| \\
&= \|(A_n - zL_n)^{-2^N}(I + (z-y)(A_n - zL_n)^{-1})^{-2^N}\| \\
&\leq \frac{\|(A_n - zL_n)^{-2^N}\|}{(1 - |z-y|\|(A_n - zL_n)^{-1}\|)^{2^N}} \\
&\leq \frac{\epsilon^{-2^N} - \delta}{(1 - |z-y|\|(A_n - zL_n)^{-1}\|)^{2^N}} \\
&< \epsilon^{-2^N}.
\end{aligned}$$

Thus, $z \notin p\text{-}\lim_n \mathrm{sp}_{N,\epsilon} A_n$. Since $u\text{-}\lim_n \mathrm{sp}_{N,\epsilon} A_n \subset p\text{-}\lim_n \mathrm{sp}_{N,\epsilon} A_n$ this completes the proof. □

Proposition 3.6 in [8] states that for compact sets M_n the limits $u\text{-}\lim M_n$ and $p\text{-}\lim M_n$ coincide if and only if M_n converge w.r.t. the Hausdorff distance (to the same limiting set). Thus, we can reformulate the preceding theorem as follows.

Corollary 2.15. *For a sequence $\mathbb{A} = \{A_n\} \in \mathcal{L}^{\gamma,T}$ the (N,ϵ)-pseudospectra of the elements A_n converge with respect to the Hausdorff distance to the union of the (N,ϵ)-pseudospectra of all snapshots $W^t(\mathbb{A})$.*

Remark 2.16. It is not hard to check that the previous results remain true, if one considers subsequences $\mathbb{A}_g$ of $\mathbb{A}$ and the respective algebras $\mathcal{L}_g^{\gamma,T}$.

3. Finite sections of band-dominated operators

Now, we reap the fruit of our labor and we recover and extend the results of [12, Section 6.3] on the finite sections of band-dominated operators on l^2.

Proposition 3.1. *Let $\mathbb{A} = \{A_n\} \in \mathcal{F}_{\mathcal{A}_{l^p}}$. Then*

$$\begin{aligned}
\limsup_{n\to\infty} \|A_n\| &= \max_{g\in\mathcal{H}_{\mathbb{A}}} \max_{t\in T} \|W^t(\mathbb{A}_g)\| \\
\liminf_{n\to\infty} \|A_n\| &= \min_{g\in\mathcal{H}_{\mathbb{A}}} \max_{t\in T} \|W^t(\mathbb{A}_g)\|.
\end{aligned}$$

Proof. The idea is simple and straightforward: Given $\mathbb{A}$ and $g \in \mathcal{H}_{\mathbb{A}}$ we are going to construct a sequence γ such that $\mathbb{A}_g$ is localizable with respect to $\mathcal{C}^\gamma$, and apply the abstract results of the previous section. Choose $g \in \mathcal{H}_{\mathbb{A}}$, set $i_1 := 1$ and construct a subsequence $i = (i_k)$ of g such that

$$\max\Big\{\|(E_{g_n}^{-t}(A_{g_n})L_{g_n}^t - W^t(\mathbb{A}_g))L_k\|, \\ \|L_k(E_{g_n}^{-t}(A_{g_n})L_{g_n}^t - W^t(\mathbb{A}_g))\| : t \in T, g_n \geq i_k\Big\} < \frac{1}{k}$$

for every $k \geq 2$. This is possible, since $E_{g_n}^{-t}(A_{g_n})L_{g_n}^t$ tend $\mathcal{P}^t$-strongly to $W^t(\mathbb{A}_g)$. Now, for $k \in \mathbb{N}$, define $\gamma_n := k/2$ for all $n \in \{i_k, \ldots, i_{k+1} - 1\}$ and consider $\mathcal{C}^\gamma$, $\mathcal{B}^\gamma$ with respect to $\gamma = (\gamma_k)_k$.

Firstly, let $x \in (-1, 1)$ and $t = 0$, and check that for every $\varphi \in C[-1, 1]$ and every operator $B \in \mathcal{A}_{l^p}$ we have that $\|[B, \varphi_n^\gamma I]\| \to 0$ as $n \to \infty$. Indeed, this is easily seen if B is a shift or an operator of multiplication, hence it is clear for band operators and follows for band-dominated ones by a simple approximation argument. Consequently, $[\{L_n B L_n\}, \{\varphi_n^\gamma L_n\}] \in \mathcal{G}$ for all $B \in \mathcal{A}_{l^p}$ and all $\varphi \in C[-1, 1]$. Since $\mathbb{A}$ can be approximated (in the norm) by sequences $\mathbb{A}^m$ which consist of finite sums and products of pure finite section sequences $\{L_n B_\alpha^m L_n\}$, and since $\mathcal{G}$ is a closed ideal, we even find that $[\mathbb{A}, \{\varphi_n^\gamma L_n\}] \in \mathcal{G}$. Thus, $\mathbb{A}_i \in \mathcal{B}_i^{\gamma,T}$. Moreover, for every continuous function φ being equal to one in a neighborhood of x and vanishing in the endpoints ± 1, we conclude that $(I - L_n)B\varphi_n^\gamma I$ and $\varphi_n^\gamma B(I - L_n)$ tend to zero in the norm as $n \to \infty$ and therefore $\{\varphi_n^\gamma L_n\}(\{L_n B L_n\}\{L_n C L_n\} - \{L_n BC L_n\}) \in \mathcal{G}$ for every $B, C \in \mathcal{A}_{l^p}$. Applying this observation to the sequences $\mathbb{A}^m$ and utilizing the approximation $\|\mathbb{A}^m - \mathbb{A}\| \to 0$ as $m \to \infty$, we easily get that the sequence $\{\varphi_n^\gamma L_n\}(\mathbb{A} - \{L_n W^0(\mathbb{A}) L_n\})$ belongs to the ideal $\mathcal{G}$, hence $\phi_x(\mathbb{A}_i + \mathcal{G}_i)$ equals $\phi_x(\{E_{i_n}^0(L_{i_n}^0 W^0(\mathbb{A}_i) L_{i_n}^0)\} + \mathcal{G}_i)$. Here ϕ_x is the local homomorphism on $\mathcal{B}_i^{\gamma,T}/\mathcal{G}_i$ in the point x.

For $x = t = 1$ (and similarly for $x = t = -1$) we note that the snapshot $W^1(\mathbb{A}_i)$ is always (the compression to the space $\mathbf{E}^1$ of) a band-dominated operator, and therefore it follows that $\|[W^1(\mathbb{A}_i), V_{-n}\varphi_n^\gamma V_n I^1]\| \to 0$ as $n \to \infty$, by the same arguments as above. In order to verify the relation

$$\phi_x(\mathbb{A}_i + \mathcal{G}_i) = \phi_x(\{E_{i_n}^1(L_{i_n}^1 W^1(\mathbb{A}_i) L_{i_n}^1)\} + \mathcal{G}_i)$$

we simply fix a continuous function φ being equal to one in $x = 1$, having its support in $[1/2, 1]$ and derive from the choice of $\mathbb{A}_i$ and the construction of the blowing-up process for φ that

$$\|\varphi_{i_k}^\gamma(A_{i_k} - E_{i_k}^1(L_{i_k}^1 W^1(\mathbb{A}_i) L_{i_k}^1)\| = \|\varphi_{i_k}^\gamma L_k(A_{i_k} - E_{i_k}^1(L_{i_k}^1 W^1(\mathbb{A}_i) L_{i_k}^1)\| \to 0$$

as $k \to \infty$. Thus $\mathbb{A}_i$ belongs to $\mathcal{L}_i^{\gamma,T}$ and, by Proposition 2.7, the limit of $\|A_{i_n}\|$ exists and

$$\limsup_{n\to\infty} \|A_n\| \geq \lim_{n\to\infty} \|A_{i_n}\| = \max_{t\in T} \|W^t(\mathbb{A}_i)\| = \max_{t\in T} \|W^t(\mathbb{A}_g)\|.$$

Since g is chosen arbitrarily, we see that this estimate holds for every $g \in \mathcal{H}_{\mathbb{A}}$. On the other hand, choose h such that the sequence $(\|A_{h_n}\|)$ converges and realizes the lim sup, that is $\lim_n \|A_{h_n}\| = \limsup_n \|A_n\|$. Pass to a subsequence $g \in \mathcal{H}_{\mathbb{A}_h}$ to deduce the asserted equality. The lim inf can be tackled analogously. □

In the same way we get

Corollary 3.2. *Let* $\mathbb{A} = \{A_n\} \in \mathcal{F}_{\mathcal{A}_{l^p}}$ *be stable. Then*

$$\limsup_{n\to\infty} \operatorname{cond}(A_n) = \max_{g\in\mathcal{H}_{\mathbb{A}}} \left(\max_{t\in T} \|W^t(\mathbb{A}_g)\| \cdot \max_{t\in T} \|(W^t(\mathbb{A}_g))^{-1}\| \right)$$

$$\liminf_{n\to\infty} \operatorname{cond}(A_n) = \min_{g\in\mathcal{H}_{\mathbb{A}}} \left(\max_{t\in T} \|W^t(\mathbb{A}_g)\| \cdot \max_{t\in T} \|(W^t(\mathbb{A}_g))^{-1}\| \right).$$

Finally, again by considering suitable subsequences, one obtains

Corollary 3.3. *Let* $X = L^p(S, \Sigma, \mu)$ *with* (S, Σ, μ) *being a measure space such that* $\mu(S) < \infty$. *For* $\mathbb{A} = \{A_n\} \in \mathcal{F}_{\mathcal{A}_{l^p}}$ *and every* $N \in \mathbb{N}_0$, $\epsilon > 0$

$$p\text{-}\lim_{n\to\infty} \operatorname{sp}_{N,\epsilon} A_n = \bigcup_{g\in\mathcal{H}_{\mathbb{A}},\, t\in T} \operatorname{sp}_{N,\epsilon} W^t(\mathbb{A}_g).$$

Remark 3.4. If for the whole sequence $\mathbb{A} = \{A_n\} \in \mathcal{F}_{\mathcal{A}_{l^p}}$ the snapshots $W^{\pm 1}(\mathbb{A})$ already exist then these relations simplify to

$$\lim_{n\to\infty} \|A_n\| = \max_{t\in T} \|W^t(\mathbb{A})\| \quad \text{and}$$

$$\lim_{n\to\infty} \operatorname{cond}(A_n) = \max_{t\in T} \|W^t(\mathbb{A})\| \cdot \max_{t\in T} \|(W^t(\mathbb{A}))^{-1}\|$$

and we again get that p-$\lim \operatorname{sp}_{N,\epsilon} A_n$ and u-$\lim \operatorname{sp}_{N,\epsilon} A_n$ coincide and equal to the union of the (N, ϵ)-pseudospectra of all snapshots. This result particularly includes the sequences in the algebra generated by the finite sections of Toeplitz or block Toeplitz operators with continuous symbol by the construction as in Section 2.4.5 of [19]. These have been considered in lots of papers. A comprehensive survey is given in the book [5]. See, in particular, its Sections 3.2 and 7.3.

If we consider $X = L^p[0, 1)$ the above results can be applied to convolution type operators on $L^p(\mathbb{R})$. For more details see [18] and the references cited there.

References

[1] A. Böttcher, *Norms of inverses, spectra, and pseudospectra of large truncated Wiener-Hopf operators and Toeplitz matrices.* New York J. Math. **3** (1997), 1–31.

[2] A. Böttcher, *Pseudospectra and singular values of large convolution operators.* J. Integral Equations Appl. **6** (1994), No. 3, 267–301.

[3] A. Böttcher, N. Krupnik, B. Silbermann, *A general look at local principles with special emphasis on the norm computation aspect.* Integral Equations Oper. Theory **11** (1988), No. 4, 455–479.

[4] A. Böttcher, B. Silbermann, *Analysis of Toeplitz Operators.* 2nd Edition, Springer-Verlag, Berlin, 2006.

[5] A. Böttcher, B. Silbermann, *Introduction to Large Truncated Toeplitz Matrices.* Springer-Verlag, New York, 1999.

[6] A. Daniluk, *The maximum principle for holomorphic operator functions.* Integral Equations Oper. Theory **69** (2011), No. 3, 365–372.

[7] J. Globevnik, *Norm-constant analytic functions and equivalent norms.* Illinois J. Math. **20** (1976), 503–506.

[8] R. Hagen, S. Roch, B. Silbermann, *C^*-Algebras and Numerical Analysis.* Marcel Dekker, Inc., New York, Basel, 2001.

[9] A.C. Hansen, *On the approximation of spectra of linear operators on Hilbert spaces.* J. Funct. Anal. **254** (2008), No. 8, 2092–2126.

[10] A.C. Hansen, *On the solvability complexity index, the n-pseudospectrum and approximations of spectra of operators.* J. Amer. Math. Soc. **24** (2011), 81–124.

[11] M. Lindner, *Infinite Matrices and their Finite Sections.* Birkhäuser Verlag, Basel, Boston, Berlin, 2006.

[12] V.S. Rabinovich, S. Roch, B. Silbermann, *Limit Operators and Their Applications in Operator Theory.* Birkhäuser Verlag, Basel, Boston, Berlin, 2004.

[13] L. Reichel, L.N. Trefethen, *Eigenvalues and pseudo-eigenvalues of Toeplitz matrices.* Lin. Alg. Appl. **162** (1992), 153–185.

[14] S. Roch, *Spectral approximation of Wiener-Hopf operators with almost periodic generating function.* Proceedings of the International Conference on Fourier Analysis and Applications (Kuwait, 1998), Numer. Funct. Anal. Optim. **21** (2000), No. 1-2, 241–253.

[15] S. Roch, P. Santos, B. Silbermann, *Non-commutative Gelfand Theories.* Universitext, Springer-Verlag, London, 2011.

[16] S. Roch, B. Silbermann, *Non-strongly converging approximation methods.* Demonstr. Math. **22** (1989), 3, 651–676.

[17] M. Seidel, *On (N,ϵ)-pseudospectra of operators on Banach spaces.* J. Funct. Anal. **262** (2012), 4916–4927.

[18] M. Seidel, B. Silbermann, *Banach algebras of operator sequences.* Oper. Matrices **6** (2012), No. 3, 385–432.

[19] M. Seidel, B. Silbermann, *Banach algebras of structured matrix sequences.* Lin. Alg. Appl. **430** (2009), 1243–1281.

[20] E. Shargorodsky, *On the level sets of the resolvent norm of a linear operator.* Bull. London Math. Soc. **40** (2008), 493–504.

Markus Seidel and Bernd Silbermann
Chemnitz University of Technology
Department of Mathematics
D-09107 Chemnitz, Germany
e-mail: markus.seidel@mathematik.tu-chemnitz.de
bernd.silbermann@mathematik.tu-chemnitz.de

Operator Theory:
Advances and Applications, Vol. 228, 391–406

On the Reduction of Linear Systems Related to Boundary Value Problems

Frank-Olme Speck

To Vladimir Rabinovich on the occasion of his 70th birthday

Abstract. The main topic of this work is the investigation of operator relations which appear during the reduction of linear systems, particularly in the study of boundary value problems. The first objective is to improve formulations like "equivalent reduction" by the help of operator relations. Then we describe how some of these operator relations can be employed to determine the regularity class and effective solution of boundary value problems. Furthermore operator relations are used to put boundary value problems into a correct space setting, e.g., by operator normalization.

Mathematics Subject Classification (2010). Primary 47Axx; Secondary 34B05.

Keywords. Reduction, equivalence, linear system, boundary value problem.

1. Introduction

In many applications problems are "reduced" to simpler problems. So it happens, e.g., in linear boundary value problems for partial differential equations of mathematical physics by potential methods leading to boundary integral equations or to semi-homogeneous boundary value problems where either the differential equation or the boundary condition is homogeneous. It is quite common to speak about "equivalent reduction", see [41, p. 174], for instance. Sometimes it is mentioned that there exists a one-to-one correspondence (substitution) between the solution spaces and another one between the given data spaces, see [16, Theorem 5.6.7] and connected remarks. Clearly, if these mappings are linear homeomorphisms, then well-posed problems are transformed into well-posed problems and ill-posed into ill-posed problems.

In the present paper we would like to illuminate the situation a little more. Typically an elliptic linear boundary value problem is written in the form

$$Au = f \quad \text{in } \Omega \qquad \text{(pde in nice domain)} \tag{1.1}$$

$$Bu = g \quad \text{on } \Gamma = \partial\Omega \qquad \text{(boundary condition).} \tag{1.2}$$

More precisely the problem is: Determine the (general*) solution of the system (1.1)–(1.2) (in a certain form**) where the following are given: Ω is a Lipschitz domain in $\mathbb{R}^n$ (e.g.), $A \in \mathcal{L}(\mathcal{X}, Y_1), B \in \mathcal{L}(\mathcal{X}, Y_2)$ are bounded linear operators in Banach spaces of function(al)s living on Ω or $\Gamma = \partial\Omega$. The data (f, g) are arbitrarily given in the (known) space $Y = Y_1 \oplus Y_2$ (denoting the direct sum considered as a Banach space). It is even more precise to mention that: *We are looking for *all* solutions for any given data in those indicated spaces and in a *specific form* that is **explicit, closed analytic, of series expansion, numerical (plenty possible choices), with error estimate etc., or just in any form. As a rule it is expected that the more detailed formulation is guessed by the reader.

The situation becomes a bit more transparent if we consider the *operator associated with the boundary value problem*

$$L = \begin{pmatrix} A \\ B \end{pmatrix} : \mathcal{X} \to Y = Y_1 \oplus Y_2 \tag{1.3}$$

where the data space Y and the solution space $\mathcal{X}$ are usually assumed to be known (eventually modified later for practical reasons and in contrast to free boundary problems or certain inverse problems). As a standard situation we shall work only with Banach spaces; other interesting frameworks could be topological vector spaces or Hilbert spaces. It is clear that a linear boundary value problem in the abstract setting (1.1)–(1.2) is well posed if and only if the operator L is boundedly invertible (a linear homeomorphism). Thus the main problem is: Find (in a certain form) the inverse (resolvent) of the *associated operator* L (or a generalized inverse etc.). Associated operators were systematically used, e.g., in the work of [5, 9, 10, 14–16, 27, 29, 34, 36, 41].

Who is not interested in the determination of the resolvent operator but only in the (unique) solution u for a single data set f, g may become more interested by the question if the solution depends continuously on the data, i.e., in the proof of the problem to be well posed and therefore the existence of a continuous resolvent operator (for continuity one needs to know the topologies).

This paper aims at discovering relations between associated operators, describing their properties in view of so-called "equivalent reduction" to simpler situations. Precise operator theoretical formulations allow the discovery of odd situations (like ill-posedness) and of convenient strategies for normalization (like inclusion of compatibility conditions or the identification of transmission properties). We prove that the reduction of linear systems to semi-homogeneous linear systems can be seen as an operator relation (Section 3). This kind of operator relation has strong transfer properties in what concerns (a) common regularity properties of operators (like invertibility, the Fredholm property etc.) and (b) the mutual computation of generalized inverses (Section 4). Finally three well-known classes of examples are discussed (Section 5) in order to underline the usefulness of the employment of operator relations.

If the reader is interested to see the explicit and efficient solution of concrete boundary value problems by the help of operator relations, we refer to further recent publications of the author and his collaborators [6, 8–10, 30, 40].

2. Some operator relations appearing in potential methods

The classical idea to present the possible solution $u \in \mathcal{X}$ by surface and/or volume potentials can be seen as an operator factorization:

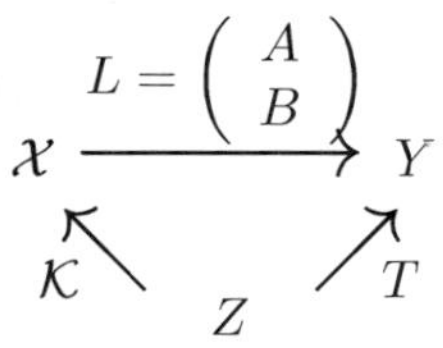

If $\mathcal{K}$ is a linear homeomorphism, then L is "equivalently reduced" to $T = \mathcal{K}L$ in the sense that the two operators are (*algebraically and topologically*) *equivalent*, i.e., by definition that T is representable as

$$T \quad = \quad E \quad L \quad F \tag{2.1}$$

where E, F are linear homeomorphisms. This defines an equivalence relation in the genuine mathematical sense (reflexive, symmetric and transitive) and practically it includes the idea of a substitution in the solution *and* in the data space. For the existence of a relation (2.1) we write $T \sim L$.

Obviously T has all the good properties that L has and vice versa. More precisely: the relation (2.1) implies the *transfer property* **TP1: Both operators belong to the same regularity class of bounded linear operators in Banach spaces** in the sense of the following classification, which was stimulated by [28, 32] and introduced in [38]:

	$\alpha(T) = 0$	$\alpha(T) < \infty$	$\ker T$ complemented	$\ker T$ closed
$\beta(T) = 0$	boundedly invertible	right invertible Fredholm	right invertible	surjective
$\beta(T) < \infty$	left invertible Fredholm	Fredholm	right regularizable	semi-Fredholm $\mathcal{F}_-$
$\operatorname{im} T$ complem.	left invertible	left regularizable	generalized invertible	no name
$\operatorname{im} T$ closed	injective	semi-Fredholm $\mathcal{F}_+$	no name	normally solvable

Herein $\alpha(T) = \dim \ker T$ and $\beta(T) = \operatorname{codim} \operatorname{im} T = \dim Y / \operatorname{im} T$. An operator T is said to be generalized invertible if there exists another bounded linear operator T^- such that $TT^-T = T$. This is equivalent to the fact that $\ker T$ and $\operatorname{im} T$ are complemented. For more details see [8, 28, 38]. The reason for TP1 to be valid is simply that equivalent operators have isomorphic kernels, cokernels and isomorphic related quotient spaces, as well. But there is another important *transfer property* **TP2: Pseudo-inverses can be computed from each other**, provided E, F or E^{-1}, F^{-1} are known. The word *pseudo-inverses* stands here for the collection of inverses, one-sided inverses, generalized inverses and Fredholm regularizers (yielding one-sided inversion up to compact or finite rank operators). TP2 is doubtless of particular interest in applications. Surely one finds plenty of further transfer properties such as possible normalization methods or asymptotic expansion of L and T.

Now let us think, instead of (2.1) about operator relations in more generality. Beside of the common definition of a relation between elements $S \in \mathcal{L}_1$, $T \in \mathcal{L}_2$ in two classes of operators $\mathcal{L}_1$ and $\mathcal{L}_2$ as a subset of $\mathcal{L}_1 \times \mathcal{L}_2$ one concretely meets relations defined

- by common properties (such as shown in the diagram),
- by isomorphic subspaces (like kernels etc.),
- by operator matrix identities.

For instance it makes sense in a certain context to consider two operators to be "equivalent" if and only if they are both Fredholm operators and have the same defect numbers [4] (which is quite different from the present notation). Also "local equivalence" [37] is an operator relation but does not directly fall into one of these categories.

Some important operator matrix identities are the following: Two operators acting in Banach spaces are called *equivalent after extension*, in brief $S \overset{*}{\sim} T$ [1], if there are Banach spaces Z_1, Z_2 and linear homeomorphisms E, F such that

$$\begin{pmatrix} S & 0 \\ 0 & I_{Z_1} \end{pmatrix} = E \begin{pmatrix} T & 0 \\ 0 & I_{Z_2} \end{pmatrix} F. \tag{2.2}$$

The relation (2.1) can be seen as a special case. Further the two operators are called *Δ-related operators*, in brief $S \ \Delta \ T$ [6, 8], if there is a *companion operator* S_Δ and linear homeomorphisms E, F such that

$$\begin{pmatrix} S & 0 \\ 0 & S_\Delta \end{pmatrix} = E\, T\, F. \tag{2.3}$$

If E or F are only linear bijections (not necessarily bi-continuous), then S and T are called *algebraically equivalent*, etc., writing

$$S \underset{alg}{\sim} T \quad , \quad S \underset{alg}{\overset{*}{\sim}} T \quad , \quad S \underset{alg}{\Delta} T \quad . \tag{2.4}$$

Properties of these relations are described in [3, 6, 8]. In the present context the most important relation is (2.2) as we shall see. A remarkable known result is the following

Theorem 2.1 (of Bart and Tsekanovskii [3]). *Let $T \in \mathcal{L}(X_1, X_2)$ and $S \in \mathcal{L}(Y_1, Y_2)$ be bounded linear operators in Banach spaces and assume $T \overset{*}{\sim} S$. Then* $\ker T \simeq$

$\ker S$. *Also* $\operatorname{im} T$ *is closed if and only if* $\operatorname{im} S$ *is closed, and in that case* $X_2/\operatorname{im} T \simeq Y_2/\operatorname{im} S$.

Assume moreover that T *and* S *are generalized invertible*. Then* $T \overset{*}{\sim} S$ *if and only if* $\ker T \simeq \ker S$ *and* $X_2/\operatorname{im} T \simeq Y_2/\operatorname{im} S$.

This assumption* is essential. There are cases where sufficiency fails otherwise, see Example 6 in [3]. The following result is known from [6, 8], the last conclusion was already observed in [3], Theorem 3.

Corollary 2.2. *If* $T \overset{*}{\sim} S$, *then the two operators have the two above-mentioned transfer properties TP1 and TP2.*

In case of Fredholm or semi-Fredholm operators the corresponding finite-dimensional defect spaces have the same dimension.

3. Reduction to semi-homogeneous systems

Consider the semi-homogeneous (abstract) boundary value problem

$$L^0 u = \begin{pmatrix} A \\ B \end{pmatrix} u = \begin{pmatrix} 0 \\ g \end{pmatrix} \in \{0\} \oplus Y_2 \simeq Y_2 \tag{3.1}$$

with associated operator

$$B|_{\ker A} : X_0 = \ker A \longrightarrow Y_2. \tag{3.2}$$

How is this operator related to the full thing

$$L = \begin{pmatrix} A \\ B \end{pmatrix} : \mathcal{X} \longrightarrow Y = Y_1 \oplus Y_2 \qquad ?$$

In general, they will not be equivalent operators, since Y and Y_2 may not be isomorphic. But, if A is surjective and $\ker A$ is complemented, i.e., $A : \mathcal{X} \longrightarrow Y_1$ is right invertible, then we have the following relation:

Lemma 3.1. *Let* $L = \begin{pmatrix} A \\ B \end{pmatrix} \in \mathcal{L}(\mathcal{X}, Y_1 \oplus Y_2)$ *be a bounded linear operator acting in Banach spaces. Further let* R *be a right inverse of* A, *i.e.,*

$$R \in \mathcal{L}(Y_1, \mathcal{X}) \quad , \quad AR = I|_{Y_1} \quad . \tag{3.3}$$

Then the following operator factorization holds

$$L = ETF = \begin{pmatrix} 0 & A|_{X_1} \\ I|_{Y_2} & B|_{X_1} \end{pmatrix} \begin{pmatrix} B|_{X_0} & 0 \\ 0 & I|_{X_1} \end{pmatrix} \begin{pmatrix} P \\ Q \end{pmatrix} \tag{3.4}$$

where $P = I - RA$, $Q = RA$ *are continuous projectors in* $\mathcal{X}$, $X_0 = \ker A = \operatorname{im} P = \ker Q$, $X_1 = \operatorname{im} Q = \ker P$. *The first and third factor in* (3.4) *are (boundedly) invertible as*

$$E = Y_2 \oplus X_1 \longrightarrow Y_1 \oplus Y_2$$
$$F = \mathcal{X} \longrightarrow X_0 \oplus X_1 .$$

Proof. This lemma is proved by verification. □

Obviously there is an analogous result for the semi-homogeneous (abstract) boundary value problem

$$L_0 u = \begin{pmatrix} A \\ B \end{pmatrix} u = \begin{pmatrix} f \\ 0 \end{pmatrix} \in Y_1 \oplus \{0\} \simeq Y_1 .$$

The following conclusion is known in special form from applications, see [16, 41] for instance, however never seen in this way, namely as an operator relation.

Theorem 3.2. *Let* $L = \begin{pmatrix} A \\ B \end{pmatrix} \in \mathcal{L}(\mathcal{X}, Y_1 \oplus Y_2)$ *be a bounded linear operator in Banach spaces. Then*

$$\begin{aligned} \exists_{R\in\mathcal{L}(Y_1,\mathcal{X})}\, AR = I \quad &\Rightarrow \quad L \overset{*}{\sim} B|_{\ker A}, \\ \exists_{R\in\mathcal{L}(Y_2,\mathcal{X})}\, BR = I \quad &\Rightarrow \quad L \overset{*}{\sim} A|_{\ker B}. \end{aligned}$$

Proof. The first statement follows from the lemma before, since the first and the third factor of (3.4) are invertible. Obviously, in the formulation of the lemma, A and B are interchangeable through a composition with permutation matrices. Hence an analogous relation holds also in the second case. For convenience we mention the corresponding formula: If $BR = I|_{Y_2}$, then

$$\begin{pmatrix} A \\ B \end{pmatrix} = \begin{pmatrix} I|_{Y_1} & A|_{X_1} \\ 0 & B|_{X_1} \end{pmatrix} \begin{pmatrix} A|_{X_0} & 0 \\ 0 & I|_{X_1} \end{pmatrix} \begin{pmatrix} P \\ Q \end{pmatrix} \tag{3.5}$$

where we put now $P = I - RB$, $Q = RB$ which are continuous projectors in $\mathcal{X}$, $X_0 = \ker B = \operatorname{im} P = \ker Q$, $X_1 = \operatorname{im} Q = \ker P$. □

Remark 3.3. Formula (3.4) holds also, if R is a generalized inverse of A, i.e., A is not necessarily right invertible. However, in this case, the first factor in (3.4) is not invertible, since $A|_{X_1}$ is not surjective. Therefore (3.4) is not anymore an equivalent after extension relation. Anyway, the two operators L and L^0 can be equivalent after extension. For instance, Fredholm operators A, B, L, where $B|_{\ker A}$ is invertible and both $A|_{\ker B}$ and L have the same defect numbers, satisfy a relation like (2.2), as a consequence of Theorem 2.1. It is not quite clear under which (interesting) conditions the inverse conclusions in Theorem 3.2 are valid.

Theorem 3.4. *Let L be defined as before. Then*

I. *L is boundedly invertible (and the abstract boundary value problem is well posed) if and only if*
   1. *the two semi-homogeneous problems are well posed,*
   2. *the solution of the abstract boundary value problem splits uniquely as $u = u_0 + u^0$ where*
   $$L_0 u_0 = \begin{pmatrix} f \\ 0 \end{pmatrix}, \qquad L^0 u^0 = \begin{pmatrix} 0 \\ g \end{pmatrix},$$
   3. *A and B are right invertible;*

II. *Each of the three conditions for its own is not sufficient for the boundary value problem to be well posed;*

III. *The first two or the last two conditions imply that L is invertible.*

Proof. I. If L is invertible, then it is surjective, the two semi-homogeneous problems are solvable for any data and the solutions are unique. Thus L_0 and L^0 are also bijective bounded linear mappings and boundedly invertible because of the inverse mapping theorem. This implies the properties 1 and 2. It further implies that A and B are surjective, since L_0 and L^0 are invertible, and that their kernels are complemented because of

$$\ker A + \ker B = \mathcal{X}\,, \qquad \ker A \ominus \ker B \simeq \mathcal{X}.$$

The latter norm equivalence follows from the fact that $L_0 u_1 = f$, $L^0 u_0 = g$ yields

$$\begin{aligned} |u_1| + |u_0| = \left|L_0^{-1} f\right| + \left|(L^0)^{-1} g\right| &\le 2\left|L^{-1}(f,g)\right| \\ \le 2\left|L^{-1}\right| |Lu| &\le 2\left|L^{-1}\right| |L|\,|u| \end{aligned}$$

beside of the triangular inequality $|u| \le |u_1| + |u_0|$ in $\mathcal{X}$. Therefore A and B are surjective and both have complemented kernels, i.e., they are right invertible.

The reverse implication is evident.

II. Condition 1 and 3 are both not sufficient, see Example 6.2 later on. Condition 2 is not sufficient, since there exist non-complemented, closed subspaces of Banach spaces, if they are not Hilbert spaces [20]. Thus, if $\mathcal{X} = X_1 + X_0$ is an algebraic and not topologic decomposition, then

$$L = \begin{pmatrix} I_{X_1} \\ I_{X_0} \end{pmatrix} : \mathcal{X} \to X_1 \oplus X_0$$

is not boundedly invertible, because the norms in $\mathcal{X}$ and $X_1 + X_0$ are not equivalent.

III. Both cases (if 1 and 2 or 2 and 3 are satisfied) imply the surjectivity of L and $L^0 \overset{*}{\gtrsim} L \overset{*}{\gtrsim} L_0$. Further L_0 and L^0 are injective. Thus L is bijective and equivalent after extension to a boundedly invertible or right invertible, injective operator, i.e., also boundedly invertible. □

Remark 3.5. What happens if A or B is not right invertible?

1. If A is not right invertible, then either
   (a) A is not surjective, the boundary value problem is not solvable for all data $f \in Y_1$, i.e., Y_1 is chosen too large for a well-posed problem; or
   (b) A is surjective but $\ker A$ is not complemented, in which case it helps to change the topology of Y_1 or of $\mathcal{X}$.
2. The right inverses R of A or B in applications are often a volume or surface potential or an extension operator, left invertible to a trace operator, see [16, 41].
3. Each of the formulations (corresponding with L, L^0 and L_0, respectively) has advantages in certain situations, see the examples in Section 5. A motivation for the consideration of the full problem (1.1)–(1.2) can be found in the theory of boundary-domain integro-differential equations [27].

4. Transfer properties

First we look at a consequence motivated by examples given in [3] which seems to be important if we think about formulations like "equivalent reduction" of linear boundary value problems or other linear systems.

Proposition 4.1. *Let T and S be two bounded linear operators in Banach spaces. If $T \underset{alg}{\overset{*}{\sim}} S$ but not $T \overset{*}{\sim} S$, then T, S do not necessarily belong to the same regularity class of operators.*

Proof. This follows from techniques with non-complemented subspaces which allow to construct operators with closed image, isomorphic kernels and co-kernels but only one of them being generalized invertible, see [3], Section 4. Further examples were given in [18]: convolution operators in Sobolev spaces on finite intervals. □

Hilbert spaces are of particular interest in applications, because of the energy norm, for instance. Here we have:

Proposition 4.2. *Let T and S be two normally solvable operators in Hilbert spaces. Then $T \overset{*}{\sim} S$ if and only if*

$$\ker T \simeq \ker S\,, \qquad \operatorname{coker} T \simeq \operatorname{coker} S.$$

If this is fulfilled, T, S have the transfer properties TP1 and TP2: they do belong to the same regularity class and generalized inverses can be computed from each other provided the mappings E, F or E^{-1}, F^{-1} in (2.2) are known.

Proof. Normally solvable operators are linear and bounded by definition and their images are closed according to a Lemma of Hausdorff [28]. In Hilbert spaces all closed subspaces are complemented [20]. Therefore T and S are generalized invertible (see the diagram), hence the second part of Theorem 2.1 applies. □

Proposition 4.3. *Let T and S be two bounded linear operators with closed image in separable Hilbert spaces. Then $T \overset{*}{\sim} S$ if and only if their defect numbers coincide:*

$$\alpha(T) = \alpha(S)\,, \qquad \beta(T) = \beta(S).$$

Proof. This is a consequence of the previous proposition since closed subspaces of separable Hilbert spaces are isomorphic if and only if they have the same dimension, finite or infinite. □

Remark 4.4. The stronger relation $T \sim S$ yields moreover

$$\begin{aligned} \ker T \simeq \ker S \quad &, \quad \operatorname{coker} T \simeq \operatorname{coker} S \\ X_1/\ker T \simeq Y_1/\ker S\,, &\quad \operatorname{im} T \simeq \operatorname{im} S \end{aligned}$$

(also in the Banach space case) and these conditions are obviously characteristic for the relation $T \sim S$ provided T and S are generalized invertible, i.e., if both kernels and both images are complemented, cf. Theorem 2.1.

According to the second transfer property (see Corollary 2.2) we have a kind of *reverse order law* [31] which reads in its simplest form: If $T = EF$ and E, F are invertible, then $T^{-1} = F^{-1}E^{-1}$. Now we have:

Theorem 4.5 (Reverse order law). *If the inverses of the operators E, F in the relation $S \overset{*}{\sim} T$ (see (2.2)) are known and if S is generalized invertible, a generalized inverse T^- of T can be computed from a generalized inverse S^- of S, by the formula*

$$T^- = R_{11}\, F^{-1} \begin{pmatrix} S^- & 0 \\ 0 & I_{Z_1} \end{pmatrix} E^{-1} \tag{4.1}$$

where R_{11} denotes the restriction to the first block of the operator matrix.

Proof. See [8] or verify directly that $TT^-T = T$. □

Similarly one obtains the following:

Corollary 4.6. *Assume again $S \overset{*}{\sim} T$.*

1. *If the operators T, S belong to the smaller class of invertible, left invertible or right invertible operators, a generalized inverse is automatically the inverse, a left or right inverse, respectively.*
2. *Regularizers of Fredholm operators or one-sided regularizers of semi-Fredholm operators (up to compact or finite rank operators) are obtained by the reverse order law (4.1) in the same way.*

Remark 4.7. In Theorem 4.5 it suffices even to assume only that E is left invertible and F is right invertible. However this case is less relevant for applications. If the order of the two factors is inverse: $T = FE$ where $E^-E = I$, $FF^- = I$, the operator T is not necessarily generalized invertible.

5. Normalization

If an operator L (associated to a boundary value problem) is not normally solvable, the question is: How to change the space setting $(\mathcal{X}, Y)$ such that the modified operator $\tilde{L}$ (defined by restriction and/or extension) is normally solvable or even of higher regularity in the sense of the diagram? We speak then about *normalization*. Somehow one likes to do this in a "natural way" by a "minimal change of spaces".

The idea is to normalize another, related operator $T \overset{*}{\sim} L$ (for instance) which belongs to a class of operators where the question can be answered more easily, and to transfer the normalization of T to a normalization of L.

Certainly there exist plenty of different normalization methods in various formulations, see [12, 17, 33] for instance. Here we shall describe only one representative concept called *minimal normalization*. It was realized in Sommerfeld diffraction problems and their reduction to Wiener-Hopf equations in Sobolev spaces [26, 39] with Fourier symbols in the class of invertible Hölder continuous functions with a possible jump at infinity $\mathcal{G}C^{\mu}(\ddot{\mathbb{R}})$ [30]. Actually the special form of the operator $S = L$ is not relevant.

Proposition 5.1. *Let $T : X_1 \to Y_1$ and $S : X_2 \to Y_2$ be bounded linear operators in Banach spaces and $S \overset{*}{\sim} T$ (see (2.2)). Further assume that*

(i) *T is not normally solvable,*

(ii) *T admits a* minimal image normalization, *i.e., there exists a linear subspace $Y_1^{\prec} \subset Y_1$ which is dense in Y_1 and a Banach space with respect to a different norm such that the (image) restricted operator*

$$T^{\prec} \;=\; \mathrm{Rst}\, T \;:\; X_1 \to Y_1^{\prec}$$

is normally solvable, and moreover $T^{\prec}$ is Fredholm with

$$\dim(Y_1^{\prec} / \operatorname{im} T^{\prec}) \;=\; \dim(Y_1 / \overline{\operatorname{im} T}).$$

Then

(j) *S is not normally solvable,*

(jj) *admits a minimal image normalization*

$$S^{\prec} \;=\; R_{11} E \begin{pmatrix} T^{\prec} & 0 \\ 0 & I \end{pmatrix} F \;:\; X_2 \to Y_2^{\prec}$$

where $Y_2^{\prec} = R_1 E(Y_1^{\prec} \oplus Z_2)$ is equipped with the norm topology induced by $Y_1^{\prec}$ and R_1 denotes restriction to the first component, further (γ) $S^{\prec}$ is Fredholm with

$$\alpha(S^{\prec}) = \alpha(T^{\prec}) \;=\; \alpha(T) \;=\; \alpha(S)$$
$$\beta(S^{\prec}) = \beta(T^{\prec}) \;=\; \dim(Y_1 / \overline{\operatorname{im} T}).$$

Proof. All statements are direct consequences of the relation $S \overset{*}{\sim} T$ and results of the previous section. Note that the last equality is not a definition but a statement. □

Corollary 5.2.

I. *A generalized inverse or Fredholm regularizer of $S^{\prec}$ can be computed from a corresponding one of $T^{\prec}$ by the reverse order law.*

II. *Roughly speaking: If $T^{\prec}$ is invertible, the problem $Su = g$ is well posed in the modified setting $(\mathcal{X}, Y_2^{\prec})$.*

Remark 5.3. 1. The formulation of Proposition 5.1 is a little long-winded but hits exactly the situation in Sommerfeld type and wedge diffraction problems [9, 10, 14, 26, 38] where the image of related Wiener-Hopf operators is made smaller by the postulation of so-called compatibility conditions between two given data.

2. There is a dual method called "minimal domain normalization" where the domain of T is enlarged to a space $X_1^{\succ} \supset X_1 = \operatorname{dom} T$ such that X_1 is dense in $X_1^{\succ}$, $T^{\succ} = \mathrm{Ext}\, T : X_1^{\succ} \to Y_1$ a continuous extension of T etc., see [30].

Example. The Sommerfeld type diffraction problem with two (possibly different) impedance conditions on the two banks of the boundary (which is a half-line in $\mathbb{R}^2$) leads to a Wiener-Hopf operator in the standard setting of [15]

$$W \;=\; r_+ A_\phi \;:\; H_+^{1/2} \to H^{1/2}(\mathbb{R}_+)$$

where $\phi(\xi) = 1 - ip(\xi^2 - k^2)^{-1/2} \neq 0$, $\xi \in \mathbb{R}$, and p is a suitable constant, see [22, 25, 29]. Here W maps the space $H_+^{1/2}$ (of $H^{1/2}$ functions supported on $\overline{\mathbb{R}_+}$) onto the space $\tilde{H}^{1/2}(\mathbb{R}_+)$ (of $H^{1/2}(\mathbb{R}_+)$ functions extensible by zero to a function in $H^{1/2}(\mathbb{R})$). As $\tilde{H}^{1/2}(\mathbb{R}_+)$ is a proper dense subspace of $H^{1/2}(\mathbb{R}_+)$, and the operator W restricted on $\tilde{H}^{1/2}(\mathbb{R}_+)$ is also bounded (with respect to a new norm) and injective, as well, the problem becomes well posed by minimal image normalization.

6. Examples of boundary value problems

Let us consider three well-known sceneries of elliptic boundary value problems with different functional analytic structure, however fitting the present framework, i.e., working with the operators associated to the boundary value problems, in contrast to more classical formulations, e.g., in [19].

6.1. Semi-classical formulation of an elliptic boundary value problem

The first class of boundary value problems is taken from the book of Wloka [41]. We call it a semi-classical formulation because the orders of the differential operators are not greater than the differentiability order of the solution space. Here the domain $\Omega \subset \mathbb{R}^n$ is bounded with $(2m + k, \kappa)$-smooth boundary where $m, k \in \mathbb{N}$, $k + \kappa \geq 1$. The spaces and operators are given by

$$\mathcal{X} = W_2^{2m+l}(\Omega)\ ,\ Y_1 = W_2^l(\Omega)\ ,\ Y_2 = \prod_{j=1}^{m} W_2^{2m+l-m_j-1/2}(\partial\Omega)$$

$$A = \sum_{|s|\leq 2m} a_s(x) D^s$$

$$B_j = \sum_{|s|\leq m_j} b_{j,s}(x) T_0 D^s$$

where A is uniformly elliptic and $B = (B_1, \dots, B_{2m-1})$ has $2m$-smooth coefficients, $\operatorname{ord} B_j \leq 2m - 1$ and the Lopatinskii-Shapiro condition is fulfilled, see [41, Section 11.1].

The main theorem is about the equivalence of (a) the boundary value problem is elliptic, (b) L is smoothable, (c) L is Fredholm, (d) an a priori estimate holds, see [41], Theorem 13.1. Surely, in general a (constructive) reduction to a semi-homogeneous problem is not possible. However, if the coefficients of A are constant and if an extension operator is known as

$$E_\Omega^\ell \quad : \quad W_2^l(\Omega) \quad \to \quad W_2^l(\mathbb{R}^n)$$

which is right invertible by the corresponding restriction operator, then a right inverse of A is given by

$$R \quad = \quad r_\Omega\, \mathcal{F}^{-1} \Phi^{-1} \cdot \mathcal{F}\, E_\Omega^\ell \qquad \text{where} \qquad \Phi(\xi) \;=\; \sum_{|s|\leq 2m} a_s (i\xi)^s\ ,\ \xi \in \mathbb{R},$$

and Theorem 3.2 is applicable. We find a relation $L \overset{*}{\sim} L^0 \;=\; B|_{\ker A}$.

This indeed is the strategy to construct resolvent operators in special situations, particularly for certain geometrical configurations [23, 26].

Conversely, for special boundary conditions such as Dirichlet, Neumann and others with constant coefficients it is often possible to find an extension operator that is left invertible by B. In this case we obtain an equivalent after extension relation between L and $L_0 = A|_{\ker B}$ as explained in Theorem 3.2.

6.2. Weak formulation of an elliptic boundary value problem

This class can be found in the book of Hsiao and Wendland [16], Chapter 5. The boundary value problems are put in a so-called variational or weak formulation. Here $\Omega \subset \mathbb{R}^n$ is a strong Lipschitz domain,

$$Au = -\sum_{j,k=1}^{n} \frac{\partial}{\partial x_j}(a_{jk}(x)\frac{\partial u}{\partial x_k}) + \sum_{j=1}^{n} b_j(x)\frac{\partial u}{\partial x_j} + c(x)u = f \quad \text{in} \quad \Omega$$

is an elliptic differential equation with f given in $\tilde{H}^{-1}(\Omega)$ and solution u wanted in $\mathcal{X} = H^1(\Omega)$. We consider the Dirichlet problem in the following sense. Defining the sesquilinear form:

$$a_\Omega(u,v) = \int_\Omega \Big\{ \sum_{j,k=1}^{n} \left(a_{jk}(x)\frac{\partial u}{\partial x_k}\right)^\top \frac{\partial \bar{v}}{\partial x_j} + \sum_{j=1}^{n} \left(b_j(x)\frac{\partial u}{\partial x_j}\right)^\top \bar{v} + (c(x)u)^\top \bar{v} \Big\} dx$$

we look (in the sense of the formulation in Section 2) for $u \in \mathcal{X}$ such that

$$\begin{aligned} a_\Omega(u,v) &= \langle f, \bar{v}\rangle_\Omega \qquad \text{for all} \quad v \in H_0^1(\Omega) \\ T_{0,\Gamma} u &= g \in H^{1/2}(\Gamma). \end{aligned}$$

Note that $\tilde{H}^{-1}(\Omega)$ is the space of $H^{-1}(\Omega)$ functionals u extensible by zero to a functional $\ell_0 f \in H^{-1}(\mathbb{R}^n)$ and $g \in H^{1/2}(\Gamma)$ is arbitrarily given.

To make the solution unique, one has to exclude functionals $f \in \tilde{H}_\Gamma^{-1}(\Omega)$ supported on $\Gamma = \partial\Omega$, which is possible by considering a smaller data space instead of Y_1: the orthogonal complement of $\tilde{H}_\Gamma^{-1}(\Omega)$ in $\tilde{H}^{-1}(\Omega)$ written as

$$\tilde{H}_0^{-1}(\Omega) = \tilde{H}^{-1}(\Omega) \ominus \tilde{H}_\Gamma^{-1}(\Omega).$$

It turns out that in this setting the problem is Fredholm or even well-posed (see [16], Chapter 5) and the previous results are applicable, if the corresponding right inverses can be constructed (for special configurations).

6.3. A class of canonical diffraction problems

The third class of problems is devoted to applications in the theory of wave propagation, see Meister et al. [9, 23]. There are plenty of so-called canonical problems which can be solved explicitly by Wiener-Hopf and related methods. One of the subclasses that admitted complete explicit solution consists of boundary value problems for the Helmholtz equation with a complex wave number k where

$\Im m\, k > 0$ in a quadrant $\Omega = \{x = (x_1, x_2) \in \mathbb{R}^2 \,:\, x_j > 0\}$

$$(\Delta + k^2)u = 0 \quad \text{in } \Omega$$
$$T_0(\alpha\, u + \beta\, \partial u/\partial x + \gamma\, \partial u/\partial y) = g \quad \text{on } \Gamma = \partial\Omega.$$

We are interested in weak solutions $u \in H^1(\Omega)$ for given $g \in H^{-1/2}(\mathbb{R}_+)^2$ identifying $\Gamma \setminus \{(0,0)\}$ with $\mathbb{R}_+ \times \mathbb{R}_+$ and admitting different sets (α, β, γ) of constant coefficients on the two half-lines.

The resulting boundary pseudo-differential operators (denoted by T in the beginning of Section 2) have the form [10]

$$T \quad = \quad \begin{pmatrix} T_1 & K_1 \\ K_2 & T_2 \end{pmatrix} \tag{6.1}$$

where T_j are convolution type operators with symmetry (alias Wiener-Hopf plus/minus Hankel operators acting in Sobolev spaces) and K_j are very special Fourier integral operators (appearing as compositions of certain extension and trace operators), provided the ansatz potentials satisfy some minimal assumption (kind of non-vanishing Fourier symbols called normal type). Precisely they have the form

$$T_j = r_+ A_{\phi_j} \ell^o \;:\; H^{-1/2}(\mathbb{R}_+) \to H^{-1/2}(\mathbb{R}_+)$$
$$K_j = C_0 A_{\psi_j} \ell^o \;:\; H^{-1/2}(\mathbb{R}_+) \to H^{-1/2}(\mathbb{R}_-).$$

Here ℓ^o denotes odd extension from $\mathbb{R}_+$ to $\mathbb{R}$, A_ϕ is the convolution operator with Fourier symbol ϕ as before, and C_0 is given by

$$C_0 f(x) = (2\pi)^{-1} \int_{\mathbb{R}} \exp[-t(\xi)x] \widehat{f}(\xi)\, d\xi \;,\; x > 0$$

where $\widehat{f}$ denotes the Fourier transform of f and $t(\xi) = (\xi^2 - k^2)^{1/2}\,,\, \xi \in \mathbb{R}$, with $t(\xi) \approx \xi$ at $+\infty$.

In brief, it is always possible to obtain generalized inverses of the (scalar) operators T_j by factorization methods provided the Fourier symbols do not vanish in $\mathbb{R}$ [9, 10]. Thus, if the matrix (6.1) is triangular with one of the $K_j = 0$, there is a chance to invert (in the generalized sense) the operator matrix (6.1). So it happens in all those boundary value problems (of normal type), but depending on a tricky choice of the ansatz, i.e., of the potential $\mathcal{K}$. For certain boundary value problems, e.g., the impedance problem with two different impedances on the two half-lines, it is only possible to obtain a triangular operator matrix T with a choice of $\mathcal{K}$ that is not a linear homeomorphism, but a left invertible Fredholm operator with index $\alpha(T) - \beta(T) = -1$. Thus the operator relations in Section 2 have to be modified by including a rank one operator, see [10, Section 5].

Beside of this, there appear compatibility conditions in most of the boundary value problems, up to the case where the two boundary conditions have different order: one is of order one (Neumann, Robin, oblique derivatives, etc.), the other of order zero (Dirichlet type). The compatibility conditions are automatically discovered via operator relations and sometimes they are of subtle nature, particularly in multimedia and interface problems (where more than two domains touch each other in a singular point), see [24, 30] for further study.

There are plenty of other sceneries where operator relations play a fundamental role, for instance in system theory [1], the theory of Wiener-Hopf plus Hankel operators [13], convolution equations on finite intervals [4, 7] and other singular equations [6, 8]. In this sense the following bibliography is kept short. Several other books and papers are relevant in the context of the present work. For instance, from the area of operator theory we should emphasize also the work of H. Bart, I. Gohberg, M. Kaashoek and collaborators where the notion of matricial coupling and Schur coupling is considered and its interaction with the notion of equivalence after extension, see [2].

The area of boundary value problems is so large that we can only refer indirectly to the bibliographies of the encyclopaedic work such as the books of O.A. Ladyzhenskaya [19] or G. Hsiao and W. Wendland [16]. It would be also interesting to know how the present idea can be applied in cases of general boundary value problems in the sense of [5, 35] and boundary value problems for pseudo-differential equations in non-Lipschitz domains [11, 36], for instance.

Conclusion

Operator relations in general and the equivalent after extension relation in particular represent a powerful tool for investigations in the theory of linear boundary value problems and other linear systems. The transfer property TP1 joins plenty of statements about common properties of two related operators such as to be Fredholm, semi-Fredholm etc. which were often listed separately in former publications, see [21] for instance.

TP2 enables results about explicit solution, simultaneously for different kinds of pseudo-inverses, i.e., in quite different functional analytic situations. Suitable normalization methods are discovered from the reduced systems, sometimes in a "natural way" like compatibility conditions in the image space of the related operators.

In the authors opinion, a remarkable value of the employment of operator relations consists in the possibility of a compact and clear formulation of results concerning the solution of linear systems. With the words of Albert Einstein: Make things as simple as possible, but not simpler.

Acknowledgment

The work was supported by FCT – Portuguese Science Foundation through the Center for Functional Analysis and Applications at Instituto Superior Técnico, Universidade Técnica de Lisboa.

References

[1] H. Bart, I. Gohberg and M. Kaashoek, *The coupling method for solving integral equations.* Oper. Theory Adv. Appl. **2** (1984), 39–73.

[2] H. Bart, I. Gohberg, M. Kaashoek and A.C.M. Ran, *Schur complements and state space realizations.* Lin. Alg. Appl. **399** (2005), 203–224.

[3] H. Bart and V.E. Tsekanovskii, *Matricial coupling and equivalence after extension.* Oper. Theory Adv. Appl. **59** (1991), 143–160.

[4] M.A. Bastos, A.F. dos Santos and R. Duduchava, *Finite interval convolution operators on the Bessel potential spaces* H_p^s. Math. Nachr. **173** (1995), 49–63.

[5] L. Boutet de Monvel, *Boundary problems for pseudo-differential operators.* Acta Math. **126** (1971), 11–51.

[6] L.P. Castro, *Relations between Singular Operators and Applications.* Ph.D. thesis, Universidade Técnica de Lisboa, 1998.

[7] L.P. Castro, R. Duduchava and F.-O. Speck, *Finite interval convolution operators with transmission property.* Integr. Equ. Oper. Theory **52** (2005), 165–179.

[8] L.P. Castro and F.-O. Speck, *Regularity properties and generalized inverses of delta-related operators.* Z. Anal. Anwend. **17** (1998), 577–598.

[9] L.P. Castro, F.-O. Speck and F.S. Teixeira, *On a class of wedge diffraction problems posted by Erhard Meister.* Oper. Theory Adv. Appl. **147** (2004), 211–238.

[10] L.P. Castro, F.-O. Speck and F.S. Teixeira, *Mixed boundary value problems for the Helmholtz equation in a quadrant.* Integr. Equ. Oper. Theory **56** (2006), 1–44.

[11] R. Duduchava and W.L. Wendland, *The Wiener-Hopf method for systems of pseudodifferential equations with an application to crack problems.* Integr. Equ. Oper. Theory **23** (1995), 294–335.

[12] V.B. Dybin, *Normalization of the Wiener-Hopf operator.* Dokl. Akad. Nauk. **191** (1970), 437–441.

[13] T. Ehrhardt, *Invertibility theory for Toeplitz plus Hankel operators and singular integral operators with flip.* J. Funct. Anal. **208** (2004), 64–106.

[14] T. Ehrhardt, A.P. Nolasco and F.-O. Speck, *Boundary integral methods for wedge diffraction problems: the angle* $2\pi/n$, *Dirichlet and Neumann conditions.* Operators and Matrices **5** (2011), 1–40.

[15] G.I. Eskin, *Boundary Value Problems for Elliptic Pseudodifferential Equations.* American Mathematical Society, Providence, Rhode Island, 1981.

[16] G.C. Hsiao and W.L. Wendland, *Boundary Integral Equations.* Springer, Berlin, 2008.

[17] V.G. Kravchenko, *On normalization of singular integral operators.* Sov. Math. Dokl. **32** (1985), 880–883.

[18] A.B. Kuijper, *A note on first kind convolution equations on a finite interval.* Integr. Equ. Oper. Theory **14** (1991), 146–152.

[19] O.A. Ladyzhenskaya, *The Boundary Value Problems of Mathematical Physics.* Applied Mathematical Sciences **32**. Springer, Berlin, 1985.

[20] J. Lindenstrauss and L. Tzafriri, *On the complemented subspaces problem.* Isr. J. Math. **9** (1971), 263–269.

[21] G.S. Litvinchuk, *Solvability Theory of Boundary Value Problems and Singular Integral Equations with Shift.* Kluwer Academic Publishers, Dordrecht, 2000.

[22] E. Lüneburg and R.A. Hurd, *On the diffraction problem on a half-plane with different face impedances.* Can. J. Phys. **62** (1984), 853–860.
[23] E. Meister, *Some solved and unsolved canonical problems in diffraction theory.* Lect. Notes Math. **1285** (Springer, Berlin, 1987), 320–336.
[24] E. Meister, F. Penzel, F.-O. Speck and F.S. Teixeira, *Two-media scattering problems in a half-space.* Research Notes in Mathematics **263** (Pitman, London, 1992), 122–146.
[25] E. Meister and F.-O. Speck, *Diffraction problems with impedance conditions.* Appl. Anal. **22** (1986), 193–211.
[26] E. Meister and F.-O. Speck, *Modern Wiener-Hopf methods in diffraction theory.* Ordinary and Partial Differential Equations **2** (Longman, London, 1989), 130–171.
[27] S.E. Mikhailov, *Analysis of united boundary-domain integro-differential and integral equations for a mixed BVP with variable coefficients.* Math. Meth. Appl. Sciences **29** (2006), 715–739.
[28] S.G. Mikhlin and S. Prössdorf, *Singular Integral Operators.* Springer, Berlin, 1986.
[29] A. Moura Santos, F.-O. Speck and F.S. Teixeira, *Compatibility conditions in some diffraction problems.* Pitman Research Notes in Mathematics Series **361** (Longman, London, 1996), 25–38.
[30] A. Moura Santos, F.-O. Speck, F.S. Teixeira, *Minimal normalization of Wiener-Hopf operators in spaces of Bessel potentials.* J. Math. Anal. Appl. **225** (1998), 501–531.
[31] M.Z. Nashed and L.B. Rall, *Annotated bibliography on generalized inverses and applications.* Generalized Inverses and Applications (Academic Press, New York, 1976), 771–1041.
[32] J. von Neumann, *On regular rings.* Proc. Natl. Acad. Sci. USA **22** (1936), 707–713.
[33] S. Prössdorf, *Some Classes of Singular Equations.* North-Holland, Amsterdam, 1978.
[34] V.S. Rabinovich, *Pseudodifferential operators on a class of noncompact manifolds.* Math. USSR, Sb. **18** (1972), 45–59.
[35] V.S. Rabinovich, *The Fredholm property of general boundary value problems on noncompact manifolds, and limit operators.* Russ. Acad. Sci. Dokl. Math. **46** (1993), 53–58.
[36] V.S. Rabinovich, B.-W. Schulze and N. Tarkhanov, *A calculus of boundary value problems in domains with non-Lipschitz singular points.* Math. Nachr. **215** (2000), 115–160.
[37] I.B. Simonenko, *A new general method of investigating linear operator equations of the type of singular integral equations.* Soviet Math. Dokl. **5** (1964), 1323–1326.
[38] F.-O. Speck, *General Wiener-Hopf Factorization Methods.* Pitman, London, 1985.
[39] F.-O. Speck, *Mixed boundary value problems of the type of Sommerfeld half-plane problem.* Proc. Royal Soc. Edinburgh **104 A** (1986), 261–277.
[40] F.-O. Speck, *Diffraction from a three-quarter-plane using an abstract Babinet principle.* Z. Ang. Math. Mech., to appear.
[41] J. Wloka, *Partial Differential Equations.* University Press, Cambridge, 1987.

Frank-Olme Speck
Departamento de Matemática, Instituto Superior Técnico Avenida Rovisco Pais
P-1049-001 Lisboa, Portugal
e-mail: fspeck@math.ist.utl.pt

Printed by Printforce, the Netherlands